FROM LAB MATERIALS TO PRINT AND AUDIO TOOLS... RESOURCES THAT HELP YOU SUCCEED

This online tool includes 14 modules with 130 activities that give you virtual experience gathering data and performing experiments through engaging simulations. Change parameters to see what happens in each simulation, generate your own data, and write up results. Each experiment includes a general introduction followed by a series of interactive laboratory activities, each with its own set of questions. With an easy-to-use design and unparalleled flexibility, **Virtual Biology Laboratory 3.0** will make you feel like you're in a real lab!

The following experimental modules are available for purchase at academic. cengage.com/biology:

Choose from the following lab modules:

- Biochemistry
- Cell Chemistry
- Cell Division
- Cell Membranes
- Cell Respiration
- Cell Structure
- Ecology
- Evolution
- Genetics
- Microscopy
- Molecular Biology
- Pedigree Analysis
- Photosynthesis
- Population Biology

ALSO AVAILABLE

Study Guide

This interactive workbook pairs text-specific concepts with questions, illustrations, and exercises that promote active learning, as well as topic maps, study strategies, and case studies to help you study more efficiently.

BIOLOGY

Exploring the Diversity of Life

First Canadian Edition

Volume Three

Peter J. Russell

Stephen L. Wolfe

Paul E. Hertz

Cecie Starr

M. Brock Fenton
University of Western Ontario

Heather Addy
University of Calgary

Denis Maxwell
University of Western Ontario

Tom Haffie
University of Western Ontario

Ken Davey
York University (Emeritus)

NELSON / EDUCATION

NELSON / EDUCATION

Biology: Exploring the Diversity of Life, First Canadian Edition, Volume Three

by Peter J. Russell, Stephen L. Wolfe, Paul E. Hertz, Cecie Starr, M. Brock Fenton, Heather Addy, Denis Maxwell, Tom Haffie, Ken Davey

Vice President, Editorial Director:
Evelyn Veitch

Editor-in-Chief, Higher Education:
Anne Williams

Executive Editor:
Paul Fam

Senior Marketing Manager:
Sean Chamberland

Managing Editor, Development:
Alwynn Pinard

Photo Researcher:
Indu Arora

Permissions Coordinator:
Indu Arora

Content Production Manager:
Christine Gilbert

Production Service:
PrePress PMG

Copy Editor:
Holly Dickinson

Proofreader:
Martha Ghent

Indexer:
Cindy Coan

Production Coordinator:
Ferial Suleman

Design Director:
Ken Phipps

Managing Designer:
Franca Amore

Interior Design:
Dianna Little

Cover Design:
Johanna Liburd, Cover Concept;
Jennifer Leung, Cover Design

Cover Image:
Main image (bat): Photo courtesy of
M. Brock Fenton

Background image (DNA double helix):
Grant Faint/Stone/Getty Images

Compositor:
PrePress PMG

Printer:
Courier

About the Cover: A flying little brown bat frozen in mid-wing stroke moves through the blackness of an underground passage. Echolocation allows the bat to collect information about its surroundings or to locate flying insects. In the background is the elegantly sinuous double helix of DNA, a widely recognized vernacular icon for life itself. The blurred DNA connotes the generative activity inherent in the molecule that carries the genetic code of all life into the future.

For, and because of, our generations of students.

About the Canadian Authors

M.B. (BROCK) FENTON received his Ph.D. from the University of Toronto in 1969. Since then, he has been a faculty member in biology at Carleton University, then at York University, and then at The University of Western Ontario. In addition to teaching parts of first-year biology, he has also taught vertebrate biology, animal biology, and conservation biology, as well as field courses in the biology and behaviour of bats. He has received awards for his teaching (Carleton University Faculty of Science Teaching Award, Ontario Confederation of University Faculty Associations Teaching Award, and a 3M Teaching Fellowship, Society for Teaching and Learning in Higher Education) in addition to recognition of his work on public awareness of science (Gordin Kaplan Award from the Canadian Federation of Biological Societies; Honourary Life Membership, Science North, Sudbury, Ontario; Canadian Council of University Biology Chairs Distinguished Canadian Biologist Award; The McNeil Medal for the Public Awareness of Science of the Royal Society of Canada; and the Sir Sanford Fleming Medal for public awareness of Science, the Royal Canadian Institute). He also received the C. Hart Merriam Award from the American Society of Mammalogists for excellence in scientific research. Bats and their biology, behaviour, evolution, and echolocation are the topic of his research, which has been funded by the Natural Sciences and Engineering Research Council of Canada (NSERC).

HEATHER ADDY is a graduate of the University of Alberta and received her Ph.D. in plant–soil relationships from the University of Guelph in 1995. During this training and in a subsequent postdoctoral fellowship focusing on mycorrhizas and other plant–fungus symbioses at the University of Alberta, she discovered a love of teaching. In 1998, she joined the Department of Biological Sciences at the University of Calgary in a faculty position that places emphasis on teaching and teaching-related scholarship. In addition to teaching introductory biology classes and an upper-level mycology class, she has led the development of investigative labs for introductory biology courses and the introduction of peer-assisted learning groups in large biology and chemistry classes. She received the Faculty of Science Award for Excellence in Teaching in 2005 and an Honourable Mention for the Student's Union Teaching Excellence Award in 2008.

DENIS MAXWELL received his Ph.D. from the University of Western Ontario in 1995. His thesis under the supervision of Norm Hüner focused on the role of the redox state of photosynthetic electron transport in photoacclimation in green algae. Following his doctorate, he was awarded an NSERC postdoctoral fellowship. He undertook postdoctoral training at the Department of Energy Plant Research Laboratory at Michigan State University, where he studied the function of the mitochondrial alternative oxidase. After taking up a faculty position at the University of New Brunswick in 2000, he moved in 2003 to the Department of Biology at The University of Western Ontario. His research program, which is supported by NSERC, is focused on understanding the role of the mitochondrion in intracellular stress sensing and signalling. In addition to research, he is passionate about teaching biology and science to first-year university students.

TOM HAFFIE is a graduate of the University of Guelph and the University of Saskatchewan in the area of microbial genetics. Currently the learning development coordinator for the Faculty of Science at the University of Western Ontario, Tom has devoted his 20-year career to teaching large biology classes in lecture, laboratory, and tutorial settings. He led the development of the innovative core laboratory course in the biology program, was an early adopter of computer animation in lectures and, most recently, has coordinated the implementation of personal response technology across campus. He holds a UWO Pleva Award for Excellence in Teaching, a UWO Fellowship in Teaching Innovation, a Province of Ontario Award for Leadership in Faculty Teaching (LIFT), and a national 3M Fellowship for Excellence in Teaching.

KEN DAVEY is a graduate of the University of Western Ontario and received his Ph.D. from Cambridge University. He is an emeritus professor of biology at York University and has by preference taught elementary courses in zoology at McGill and York and more advanced courses in invertebrate physiology, parasitology, and endocrinology. He has held a number of academic administrative positions at York. His research interests include invertebrate physiology and the endocrinology of insects and parasitic worms, supported by NSERC. Ken has accumulated a number of academic awards, including the Canadian Council of University Biology Chairs Distinguished Canadian Biologist Award and the Wigglesworth Award for Service to Entomology of the Royal Entomological Society. He is a Fellow of the Royal Society of Canada and an Officer of the Order of Canada.

About the U.S. Authors

Peter J. Russell received a B.Sc. in Biology from the University of Sussex, England, in 1968 and a Ph.D. in Genetics from Cornell University in 1972. He has been a member of the Biology faculty of Reed College since 1972; he is currently a Professor of Biology. He teaches a section of the introductory biology course, a genetics course, an advanced molecular genetics course, and a research literature course on molecular virology. In 1987 he received the Burlington Northern Faculty Achievement Award from Reed College in recognition of his excellence in teaching. Since 1986, he has been the author of a successful genetics textbook; current editions are *iGenetics: A Mendelian Approach, iGenetics: A Molecular Approach,* and *Essential iGenetics.* He wrote nine of the BioCoach Activities for The Biology Place. Peter Russell's research is in the area of molecular genetics, with a specific interest in characterizing the role of host genes in pathogenic RNA plant virus gene expression; yeast is used as the model host. His research has been funded by agencies including the National Institutes of Health, the National Science Foundation, and the American Cancer Society. He has published his research results in a variety of journals, including *Genetics, Journal of Bacteriology, Molecular and General Genetics, Nucleic Acids Research, Plasmid,* and *Molecular and Cellular Biology.* He has a long history of encouraging faculty research involving undergraduates, including cofounding the biology division of the Council on Undergraduate Research (CUR) in 1985. He was Principal Investigator/Program Director of an NSF Award for the Integration of Research and Education (AIRE) to Reed College, 1998–2002.

Stephen L. Wolfe received his Ph.D. from Johns Hopkins University and taught general biology and cell biology for many years at the University of California, Davis. He has a remarkable list of successful textbooks, including multiple editions of *Biology of the Cell, Biology: The Foundations, Cell Ultrastructure, Molecular and Cellular Biology,* and *Introduction to Cell and Molecular Biology.*

Paul E. Hertz was born and raised in New York City. He received a bachelor's degree in Biology at Stanford University in 1972, a master's degree in Biology at Harvard University in 1973, and a doctorate in Biology at Harvard University in 1977. While completing field research for the doctorate, he served on the Biology faculty of the University of Puerto Rico at Rio Piedras. After spending 2 years as an Isaac Walton Killam Postdoctoral Fellow at Dalhousie University, Hertz accepted a teaching position at Barnard College, where he has taught since 1979. He was named Ann Whitney Olin Professor of Biology in 2000, and he received The Barnard Award for Excellence in Teaching in 2007. In addition to his service on numerous college committees, Professor Hertz was Chair of Barnard's Biology Department for 8 years. He has also been the Program Director of the Hughes Science Pipeline Project at Barnard, an undergraduate curriculum and research program funded by the Howard Hughes Medical Institute, since its inception in 1992. The Pipeline Project includes the Intercollegiate Partnership, a program for local community college students that facilitates their transfer to 4-year colleges and universities. He teaches one semester of the introductory sequence for Biology majors and preprofessional students as well as lecture and laboratory courses in vertebrate zoology and ecology. Professor Hertz is an animal physiological ecologist with a specific research interest in the thermal biology of lizards. He has conducted fieldwork in the West Indies since the mid-1970s, most recently focusing on the lizards of Cuba. His work has been funded by the National Science Foundation, and he has published his research in such prestigious journals as *The American Naturalist, Ecology, Nature,* and *Oecologia.*

Cecie Starr is the author of best-selling biology textbooks. Her books include multiple editions of *Unity and Diversity of Life, Biology: Concepts and Applications,* and *Biology Today and Tomorrow.* Her original dream was to be an architect. She may not be building houses, but with the same care and attention to detail, she builds incredible books: *"I invite students into a chapter through an intriguing story. Once inside, they get the great windows that biologists construct on the world of life. Biology is not just another house. It is a conceptual mansion. I hope to do it justice."*

Beverly McMillan has been a science writer for more than 20 years and is coauthor of a college text in human biology, now in its seventh edition. She has worked extensively in educational and commercial publishing, including 8 years in editorial management positions in the college divisions of Random House and McGraw-Hill. In a multifaceted freelance career, Bev also has written or coauthored six trade books and numerous magazine and newspaper articles, as well as story panels for exhibitions at the Science Museum of Virginia and the San Francisco Exploratorium. She has worked as a radio producer and speechwriter for the University of California system and as a media relations advisor for the College of William and Mary. She holds undergraduate and graduate degrees from the University of California, Berkeley.

Preface

Welcome to an exploration of the diversity of life. The main goal of this text is to guide you on a journey of discovery about life's diversity across levels ranging from molecules to genes, cells to organs, and species to ecosystems. Along the way, we will explore many questions about the mechanisms underlying diversity as well as the consequences of diversity for our own species and for others.

At first glance, the riot of life that animates the biosphere overwhelms the minds of many who try to understand it. One way to begin to make sense of this diversity is to divide it into manageable sections on the basis of differences. In this book, we highlight the divisions between plants and animals, prokaryotes and eukaryotes, protostomes and deuterostomes, but we also consider features found in all life forms. We examine how different organisms solve the common problems of finding nutrients, energy, and mates on the third rock from our Sun. What basic evolutionary principles inform the relationships among life forms regardless of their different body plans, habitats, or life histories? Unlike many other first-year biology texts, this book has chapters integrating basic concepts such as genetic recombination, the effects of light, nutrition, and domestication across the breadth of life from microbes to mistletoe to moose. As you read this book, you will be referred frequently to other chapters for linked information that expands the ideas further.

Evolution provides a powerful conceptual lens for viewing and understanding the roots and history of diversity. We will demonstrate how knowledge of evolution helps us appreciate the changes we observe in organisms. Whether the focus is the conversion of free-living prokaryotes into mitochondria and chloroplasts or the steps involved in the domestication of rice, selection for particular traits over time can explain the current condition.

We hope that Canadian students will find the subject of biology as it is presented here accessible and engaging because it is presented in familiar contexts. We have highlighted the work of Canadian scientists, used examples of Canadian species, and referred to Canadian regulations and institutions, as well as discoveries made by Canadians.

Although many textbooks use the first few chapters to introduce and/or review background information, we have used the first chapters to convey the excitement and interest of biology itself. Within the centre of the book, we have placed important background information about biology and chemistry in the reference section entitled *The Chemical and Physical Foundations of Biology*. These pages are distinct and easy to find with their purple edges and have become affectionately known as the "Purple Pages." These pages enable information to be readily identifiable and accessible to students as they move through the textbook rather than information that is tied to a particular chapter. The purple background makes the pages easy to find when you need to check a topic. This section keeps background information out of the mainstream of the text, allowing you to focus on bigger pictures.

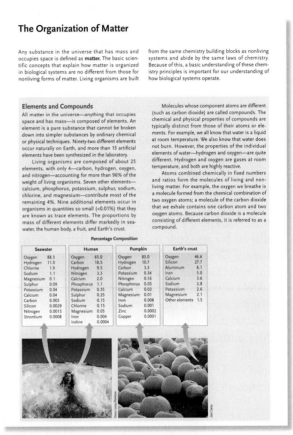

In addition to presenting material about biology, this book also makes a point of highlighting particular people, important molecules, interesting contexts, and examples of life in extreme conditions. Science that appears in textbooks is the product of people who have made careful and systematic observations, which led them to formulate hypotheses about these observations and, where appropriate, design and execute experiments to test these hypotheses. We illustrate this in each chapter with boxed stories about how particular people have used their ingenuity and creativity to expand our knowledge of biology. We have endeavoured to show not just the science itself but also the process behind the science.

Although biology is not simply chemistry, specific chemicals and their interactions can have dramatic effects on biological systems. From water to progesterone, amanitin, and DDT, each chapter features the activity of a relevant chemical.

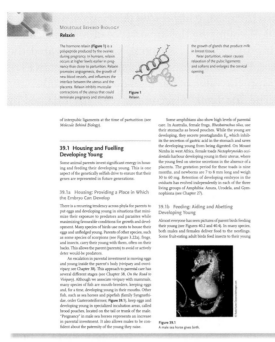

Figure 1
Relaxin.

of interpubic ligaments at the time of parturition (see *Molecule Behind Biology*).

39.1 Housing and Fuelling Developing Young

Some animal parents invest significant energy in housing and feeding their developing young. This is one aspect of the genetically selfish drive to ensure that their genes are represented in future generations.

39.1a Housing: Providing a Place in Which the Embryo Can Develop

There is a recurring tendency across phyla for parents to put eggs and developing young in situations that minimize their exposure to predators and parasites while maximizing favourable conditions for growth and development. Many species of birds use nests to house their eggs and unfledged young. Parents of other species, such as some species of scorpions (see Figure 3.22a), frogs, and insects, carry their young with them, often on their backs. This allows the parent (parents) to avoid or actively deter would-be predators.

An escalation in parental investment is moving eggs and young inside the parent's body (viviparity and ovoviviparity; see Chapter 38). This approach to parental care has several different stages (see Chapter 38, *On the Road to Viviparity*). Although we associate viviparity with mammals, many species of fish are mouth-breeders, keeping eggs and, for a time, developing young in their mouths. Other fish, such as sea horses and pipefish (family Syngnathidae, order Gasterosteiformes; Figure 39.1), keep eggs and developing young in specialized incubation areas, called brood pouches, located on the tail or trunk of the male. "Pregnancy" in sea horses represents an increase in parental investment. It also allows males to be confident about the paternity of the young they raise.

Some amphibians also show high levels of parental care. In Australia, female frogs, *Rheobatrachus silus*, use their stomachs as brood pouches. While the young are developing, they secrete prostaglandin E_2, which inhibits the secretion of gastric acid in the stomach and saves the developing young from being digested. On Mount Nimba in west Africa, female toads *Nectophrynoides occidentalis* harbour developing young in their uterus, where the young feed on uterine secretions in the absence of a placenta. The gestation period for these toads is nine months, and newborns are 7 to 8 mm long and weigh 30 to 60 mg. Retention of developing embryos in the oviducts has evolved independently in each of the three living groups of Amphibia: Anura, Urodela, and Gymnophiona (see Chapter 27).

39.1b Feeding: Aiding and Abetting Developing Young

Almost everyone has seen pictures of parent birds feeding their young (see Figures 40.2 and 40.4). In many species, both males and females deliver food to the nestlings. Some fruit-eating adult birds feed insects to their young

Figure 39.1
A male sea horse gives birth.

To help frame the material with an engaging context, we begin each chapter with a section called "Why It Matters." In addition, several chapters include boxed accounts of organisms thriving "on the edge" at unusual temperatures, pressures, radiation dosages, salt concentrations, etc. These brief articles explain how our understanding of "normal" can be increased through study of the "extreme."

Examining how biological systems work is another theme pervading this text and underlying the idea of diversity. We have intentionally tried to include examples that will tax your imagination, from sea slugs that steal chloroplasts for use as solar panels, to hummingbirds fuelling their hovering flight, to adaptive radiation of viruses. In each situation, we examine how biologists have explored and assessed the inner workings of organisms from gene regulation to the challenges of digesting cellulose.

Figure 1
Volvox carteri under heat stress. ROS indicated by green fluorescence.

Solving problems is another theme that runs through the book. Whether the topic is gene therapy to treat a disease in people, increasing crop production, or conserving endangered species, both the problem and the solution lie in biology. We will explore large problems facing planet Earth and the social implications that arise from them.

Science is by its nature a progressive enterprise in which answers to questions open new questions for consideration. Each chapter presents unanswered questions as well as questions for discussion to emphasize that biologists still have a lot to learn—topics for you to tackle should you decide to pursue a career in research.

"Study Breaks" occur after each section in the chapters. They contain questions written by students to identify some of the important features of the section. The answers are embedded in the "Review" section at the end of each chapter. Also included at the end of each chapter is a group of multiple-choice self-test questions, the answers to which can be found at the end of the book. "Questions for Discussion" at the end of each chapter challenge you to think more broadly about biology. You are encouraged to use these in discussions with other students and to explore potential answers by using the resources of the electronic library.

To maximize the chances of producing a useful text that draws in students (and instructors), we sought the advice of colleagues who teach biology (members of the Editorial Advisory Board). We also asked students (members of the Student Advisory Boards) for their advice and comments. Both groups read draft chapters and provided valuable feedback, but any mistakes are ours. The members of the Student Advisory Boards also wrote the Study Break questions found throughout the text.

We hope that you are as captivated by the biological world as we are and are drawn from one chapter to another. But don't stop there—use electronic resources to broaden your search for understanding.

18.8 Geography of Speciation

Geography has a huge impact on whether gene pools have the opportunity to mix. Biologists define three modes of speciation based on the geographic relationship of populations as they become reproductively isolated: allopatric speciation (*allo* = different; *patria* = homeland), parapatric speciation (*para* = beside), and sympatric speciation (*sym* = together).

18.8a Allopatric Speciation: New Species Develop from Isolated Populations

Allopatric speciation can occur when a physical barrier subdivides a large population or when a small population becomes separated from a species' main geographic distribution. Allopatric speciation, probably the most common mode of speciation in large animals, occurs in two stages. First, two populations become geographically isolated, preventing gene flow between them. Then, as the populations experience distinct mutations as well as different patterns of natural selection and genetic drift, they must accumulate genetic differences that isolate them reproductively.

Geographic separation sometimes occurs when a barrier divides a large population into two or more

STUDY BREAK

1. What is a reproductive isolating mechanism? Distinguish between two major types of reproductive isolating mechanisms.
2. Define five types of prezygotic isolating mechanisms that can prevent interspecific mating.
3. Prezygotic isolating mechanisms prevent gene pools of two species from mixing, so why do postzygotic isolating mechanisms exist? Distinguish between three types of postzygotic isolating mechanisms.

Supplementary Materials

An extensive array of supplemental materials is available to accompany this text. These supplements are designed to make teaching and learning more effective. For more information on any of these resources, please contact your local Nelson Education sales representative or call Nelson Education Limited Customer Support at 1-800-268-2222.

Instructor Resources

These resources are available to qualified adopters. Please consult your local Nelson Education sales representative for details.

Instructor's Resource DVD

The *Instructor's Resource DVD* contains the following resources:

Instructor's Resource Manual

The *Instructor's Resource Manual* for this First Canadian Edition has been dramatically revised by Tanya Noel, Tamara Kelly, and Julie Clark from York University to include tips on teaching using cases as well as suggestions on how to present material and use technology and other resources effectively, integrating the other supplements available to both students and instructors. This manual doesn't simply reinvent what's currently in the text; it helps the instructor make the material relevant and engaging to students.

ExamView® Computerized Test Bank

Create, deliver, and customize tests (both print and online) in minutes with this easy-to-use assessment and tutorial system. ExamView® offers both a Quick Test Wizard and an Online Test Wizard that guide you step-by-step through the process of creating tests, while its "what you see is what you get" capability allows you to see the test you are creating on the screen exactly as it will print or display online. You can build tests of up to 250 questions using up to 12 question types. Using *ExamView's* complete word-processing capabilities, you can enter an unlimited number of new questions or edit existing questions.

Nelson Education Testing Advantage

Nelson Education
Testing Advantage

In most postsecondary courses, a large percentage of student assessment is based on multiple-choice testing. Many instructors use multiple-choice testing reluctantly, believing that it is a methodology best used for testing what a student *remembers* rather than what she or he has *learned*.

Nelson Education Ltd. understands that a good-quality multiple-choice test bank can provide the means to measure *higher level thinking* skills as well as recall. Recognizing the importance of multiple-choice testing in today's classroom, we have created the Nelson Education Testing Advantage program (NETA) to ensure the value of our high-quality test banks.

The *Test Bank* to accompany *Biology*, adapted by Ivona Mladenovic of Simon Fraser University and Ian Dawe of Selkirk College, offers the Premium Nelson Education Testing Advantage. NETA was created in partnership with David DiBattista, a 3M National Teaching Fellow, professor of psychology at Brock University, and researcher in the area of multiple-choice testing. NETA ensures that subject-matter experts who author test banks have had training in two areas: avoiding common errors in test construction and developing multiple-choice test questions that "get beyond remembering" to assess higher level thinking. In addition, Professor DiBattista confirms the subject-matter expert's understanding of and adherence to the NETA principles through a review.

All Premium NETA test banks include David DiBattista's guide for instructors, "Multiple Choice Tests: Getting Beyond Remembering." This guide has been designed to assist you in using Nelson test banks to achieve your desired outcomes in your course.

Customers who adopt a Premium Nelson Education Testing Advantage title may also qualify for additional faculty training opportunities in multiple-choice testing and assessment. Please contact your local Nelson Education sales and editorial representative for more details about our Premium NETA.

Microsoft PowerPoint® Slides This one-stop lecture tool makes it easy to assemble, edit, publish, and present custom lectures. Adapted by Jane Young of the University of Northern British Columbia, this resource brings together text-specific lecture outlines, art, video, and animations, culminating in a powerful, personalized, PowerPoint® presentation.

Also included on the *Instructor's Resource DVD* are Word files of the *Test Bank,* as well as a full *Image Bank* of the art and photos from the text book. ISBN: 978-0-17-647529-1.

Student Resources

Study Guide

The *Study Guide* for the First Canadian Edition has been adapted by Colin Montpetit of the University of Ottawa, Julie Smit of the University of Windsor, and Wendy J. Keenleyside of the University of Guelph. The *Study Guide* contains unique case studies to integrate the concepts within the text, study strategies, interactive exercises, self-test questions, and more. ISBN: 978-0-17-647474-4.

CengageNOW™ CENGAGENOW

CengageNOW Personalized Study is a diagnostic tool (featuring a chapter-specific Pretest, Study Plan, and

Post-test) that empowers students to master concepts, prepare for exams, and be more involved in class. Results to *Personalized Study* provide immediate and ongoing feedback regarding what students are mastering and why they're not to both the instructor and the student. *CengageNOW Personalized Study* links to an integrated eBook so that students can easily review topics and also contains animations, links to websites, videos, and more as part of their Study Plan. *CengageNOW* has been adapted by Dora Cavallo-Medved, University of Windsor; Todd Nickle, Mount Royal College; and Edward Andrews, Sir Wilfred Grenfell College.

CengageNOW with Premium eBook

Want to take your biology experience to the next level? Our *Premium eBook* allows students access to an integrated, interactive learning environment with advanced learning tools and a user interface that gives students control over their learning experience. *CengageNOW Personalized Study* is included with the Premium eBook for the ultimate online study experience.

JoinIn™ on TurningPoint®

Transform your lecture into an interactive student experience with JoinIn™. Combined with your choice of keypad systems, JoinIn turns your Microsoft® PowerPoint® application into audience response software. With a click on a handheld device, students can respond to multiple-choice questions, short polls, interactive exercises, and peer-review questions. You can also take attendance, check student comprehension of concepts, collect student demographics to better assess student needs, and even administer quizzes. In addition, there are interactive text-specific slide sets that you can modify and merge with any of your own PowerPoint® lecture slides. These have been adapted by Jane Young of the University of Northern British Columbia and contain poll slides and pre- and post-test slides for each chapter in the text. This tool is available to qualified adopters at **http://www.turningtechnologies.com/**.

Students and Instructors

Visit the website to accompany *Biology: Exploring the Diversity of Life*, First Canadian Edition, at **http://biologyedl.nelson.com.** This website contains quizzes, flashcards, weblinks, and more.

Prospering in Biology

Using This Book

The following are things you will need to know in order to use this text and prosper in Biology.

Names

What's in a name? People are very attached to names—their own names, the names of other people, the names of flowers and food and cars, and so on. It is not surprising that biologists would also be concerned about names. Take, for example, our use of scientific names. Scientific names are always italicized and Latinized.

Castor canadensis Kuhl is the scientific name of the Canadian beaver. *Castor* is the genus name, *canadensis* is the species name, and Kuhl is the name of the person who described the species. "Beaver" by itself is not enough because there is a European beaver, *Castor fiber*, and an extinct giant beaver, *Castoides ohioensis*. Furthermore, common names can vary from place to place (*Myotis lucifugus* is sometimes known as the "little brown bat" or the "little brown myotis").

Biologists prefer scientific names because the name (Latinized) tells you about the organism. There are strict rules about the derivation and use of scientific names. Common names are not so restricted, so they are not precise. For example, in *Myotis lucifugus*, *Myotis* means mouse-eared and *lucifugus* means flees the light; hence, this species is a mouse-eared bat that flees the light.

Birds can be an exception. There are accepted "standard" common names for birds. The American robin is *Turdus migratorius*. The common names for birds are usually capitalized because of the standardization. However, the common names of mammals are not capitalized, except for geographic names or patronyms (*geographic* = named after a country; *patronym* = named after someone; e.g., Canadian beaver or Ord's kangaroo rat, respectively).

Although a few plants that have very broad distributions may have accepted standard common names (e.g., white spruce, *Picea glauca*), most plants have many common names. Furthermore, the same common name is often used for more than one species. Several species in the genus *Taraxacum* are referred to as "dandelion." It is important to use the scientific names of plants to be sure that it is clear exactly which plant we mean. The scientific names of plants also tell us something about the plant. The scientific name for the weed quack grass, *Elymus repens*, tells us that this is a type of wild rye (*Elymus*) and that this particular species spreads or creeps (*repens* = creeping). Anyone who has tried to eliminate this plant from their garden or yard knows how it creeps! Unlike for animals, plant-naming rules forbid the use of the same word for both genus and species names for a plant; thus, although *Bison bison* is an acceptable scientific name for buffalo, such a name would never be accepted for a plant.

In this book, we present the scientific names of organisms when we mention them. We follow standard abbreviations; for example, although the full name of an organism is used the first time it is mentioned (e.g., *Castor canadensis*), subsequent references to that same organism abbreviate the genus name and provide the full species name (e.g., *C. canadensis*).

In some areas of biology, the standard representation is of the genus, for example, *Chlamydomonas*. In other cases, names are so commonly used that only the abbreviation may be used (e.g., *E. coli* for *Escherichia coli*).

Units

The units of measure used by biologists are standardized (metric or SI) units, used throughout the world in science.

Definitions

The science of biology is replete with specialized terms (sometimes referred to as "jargon") used to communicate specific information. It follows that, as with scientific names, specialized terms increase the precision with which biologists communicate among themselves and with others. Be cautious about the use of terms because jargon can be a veneer of precision. When we encounter a "slippery" term (such as species or gene), we explain why one definition for all situations is not feasible.

Time

In this book, we use C.E. (Common Era) to refer to the years since year 1 and B.C.E. (Before the Common Era) to refer to years before that.

Geologists think of time over very long periods. A geologic time scale (**see Table 1.1**) shows that the age of Earth could be measured in years, but it's challenging to think of billions of years expressed in days (or hours, etc.). With the advent of using the decay rates of radioisotopes to measure the age of rocks, geologists adopted 1950 as the baseline, the "Present," and the past is referred to as B.P. ("Before Present"). A notation of 30 000 years B.P. (^{14}C) indicates 30 000 years before 1950 using the ^{14}C method of dating.

Other dating systems are also used. Some archaeologists use PPNA (PrePottery Neolithic A, where A is the horizon or stratum). In deposits along the Euphrates River, 11 000 PPNA appears to be the same as 11 000

B.P. In this book, we use B.C.E. or B.P. as the time units, except when referring to events or species from more than 100 000 years ago. For those dates, we refer you to the geologic time scale (see Table 1.1 on page xiv).

Sources

Where does the information presented in a text or in class come from? What is the difference between what you read in a textbook or an encyclopedia and the material you see in a newspaper or tabloid? When the topic relates to science, the information should be based on material that has been published in a scholarly journal. In this context, "scholarly" refers to the process of review. Scholars submit their manuscripts reporting their research findings to the editor (or editorial board) of a journal. The editor, in turn, sends the manuscript out for comment and review by recognized authorities in the field. The process is designed to ensure that what is published is as accurate and appropriate as possible. The review process sets the scholarly journal apart from the tabloid.

There are literally thousands of scholarly journals, which, together, publish millions of articles each year. Some journals are more influential than others, for example, *Science* and *Nature*. These two journals are published weekly and invariably contain new information of interest to biologists.

To collect information for this text, we have drawn on published works that have gone through the process of scholarly review. Specific references (citations) are provided, usually in the electronic resources designed to complement the book.

A citation is intended to make the information accessible. Although there are many different formats for citations, the important elements include (in some order) the name(s) of the author(s), the date of publication, the title, and the publisher. When the source is published in a scholarly journal, the journal name, its volume number, and the pages are also provided. With the citation information, you can visit a library and locate the original source. This is true for both electronic (virtual) and real libraries.

Students of biology benefit by making it a habit to look at the most recent issues of their favourite scholarly journals and use them to keep abreast of new developments.

M. Brock Fenton
Heather Addy
Denis Maxwell
Tom Haffie
Ken Davey

London, Calgary and Toronto
February 2009

Table 1.1 The Geological Time Scale and Major Evolutionary Events

Eon	Era	Period	Epoch	Millions of Years Ago	Major Evolutionary Events
Phanerozoic	Cenozoic	Quaternary	Holocene	0.01	
Phanerozoic	Cenozoic	Quaternary	Pleistocene	1.7	Origin of humans; major glaciations
Phanerozoic	Cenozoic	Tertiary	Pliocene	5.2	Origin of ape-like human ancestors
Phanerozoic	Cenozoic	Tertiary	Miocene	23	Angiosperms and mammals further diversify and dominate terrestrial habitats
Phanerozoic	Cenozoic	Tertiary	Oligocene	33.4	Divergence of primates; origin of apes
Phanerozoic	Cenozoic	Tertiary	Eocene	55	Angiosperms and insects diversify; modern orders of mammals differentiate
Phanerozoic	Cenozoic	Tertiary	Paleocene	65	Grasslands and deciduous woodlands spread; modern birds and mammals diversify; continents approach current positions
Phanerozoic	Mesozoic	Cretaceous		144	Many lineages diversify: angiosperms, insects, marine invertebrates, fishes, dinosaurs; asteroid impact causes mass extinction at end of period, eliminating dinosaurs and many other groups
Phanerozoic	Mesozoic	Jurassic		206	Gymnosperms abundant in terrestrial habitats; first angiosperms; modern fishes diversify; dinosaurs diversify and dominate terrestrial habitats; frogs, salamanders, lizards, and birds appear; continents continue to separate
Phanerozoic	Mesozoic	Triassic		251	Predatory fishes and reptiles dominate oceans; gymnosperms dominate terrestrial habitats; radiation of dinosaurs; origin of mammals; Pangaea starts to break up; mass extinction at end of period

Eons (Duration drawn to scale)

Cenozoic
Mesozoic
Paleozoic
Phanerozoic
Proterozoic

Eon / Era	Period	Millions of years ago	Events
Phanerozoic (continued) — Paleozoic	Permian	290	Insects, amphibians, and reptiles abundant and diverse in swamp forests; some reptiles colonize oceans; fishes colonize freshwater habitats; continents coalesce into Pangaea, causing glaciation and decline in sea level; mass extinction at end of period eliminates 85% of species
	Carboniferous	354	Vascular plants form large swamp forests; first seed plants and flying insects; amphibians diversify; first reptiles appear
	Devonian	417	Terrestrial vascular plants diversify; fungi and invertebrates colonize land; first insects appear; first amphibians colonize land; major glaciation at end of period causes mass extinction, mostly of marine life
	Silurian	443	Jawless fishes diversify; first jawed fishes; first vascular plants on land
	Ordovician	490	Major radiations of marine invertebrates and fishes; major glaciation at end of period causes mass extinction of marine life
	Cambrian	543	Diverse radiation of modern animal phyla (Cambrian explosion); simple marine communities
Proterozoic		2500	High concentration of oxygen in atmosphere; origin of aerobic metabolism; origin of eukaryotic cells; evolution and diversification of protists, fungi, soft-bodied animals
Archaean		3800	Evolution of prokaryotes, including anaerobic bacteria and photosynthetic bacteria; oxygen starts to accumulate in atmosphere
		4600	Formation of Earth at start of era; Earth's crust, atmosphere, and oceans form; origin of life at end of era

Archaean

Acknowledgements

We thank the many people who have worked with us on the production of this text, particularly Paul Fam, Executive Editor, whose foresight brought the idea to us and whose persistence saw the project through. Thanks go to those who reviewed the U.S. text to provide us with feedback for the Canadian edition including Logan Donaldson, York University; Robert Holmberg, Athabasca University; and Thomas H. MacRae, Dalhousie University. We also are grateful to the members of the Editorial Advisory Board and the Student Advisory Board, who provided us with valuable feedback and alternate perspectives (special acknowledgements to these individuals are listed below). We also thank Richard Walker at the University of Calgary, who began this journey with us but who was unable to continue. We thank Carl Lowenberger for contributing Chapter 44 (on defences). We are especially grateful to Alwynn Pinard, Managing Developmental Editor, and James Polley, who kept us moving through the chapters at an efficient pace, along with Tracy Duff, Project Manager, and Christine Gilbert, Content Production Manager. We thank Rosemary Tanner, who provided a thoughtful substantive edit of the entire manuscript, Holly Dickinson for her careful copy editing, and Sandra Peters, who did a cold read as a further check on our presentation. Finally, we thank Sean Chamberland, Senior Marketing Manager, for making us look good.

Brock Fenton would like to thank Allan Noon, who offered much advice about taking pictures; Laura Barclay, Jeremy McNeil, Tony Percival-Smith, C.S. (Rufus) Churcher, and David and Meg Cumming for the use of their images; and Karen Campbell for providing a critical read on the domestication chapter.

It is never easy to be in the family of an academic scientist. We are especially grateful to our families for their sustained support over the course of our careers, particularly during those times when our attentions were fully captivated by bacteria, algae, fungi, parasites, or bats. Saying "yes" to a textbook project means saying "no" to a variety of other pursuits. We appreciate the patience and understanding of those closest to us that enabled the temporary reallocation of considerable time from other endeavours and relationships.

Many of our colleagues have contributed to our development as teachers and scholars by acting as mentors, collaborators, and, on occasion, "worthy opponents." Like all teachers, we owe particular gratitude to our students. They have gathered with us around the discipline of biology, sharing their potent blend of enthusiasm and curiosity that leaves us energized and optimistic for the future.

Editorial and Student Advisory Boards

We were very fortunate to have the assistance of some extraordinary students and instructors of biology across Canada who provided us with feedback that helped shape this textbook into what you see before you. As such, we would like to say a very special thank you to the following people:

Editorial Advisory Board

Mark Brigham, University of Regina
Dion Durnford, University of New Brunswick
Wendy Keenleyside, University of Guelph
Marty Leonard, Dalhousie University
Cindy Paszkowski, University of Alberta
Carol Pollock, University of British Columbia
Kevin Scott, University of Manitoba
Paula Wilson, York University

Student Advisory Boards
University of Western Ontario (pictured above)
Rachael Danielson
Dalal Dharouj
Yvonne Dzal
Liam McGuire
Aimee McMillan
Errin Pfeiffer
Max Rachinsky
Nina Veselka
Ivana Vilimonovic
Marisol Wilcox

University of Calgary (pictured above)
Kristina Birkholz
Jobran Chebib
Liam Cummings

Aravind Ganesh
Shaista Hashem
Colleen Michael
Simon Sun
Camilla Tapp
Anita Tieu
Sahar Zaidi

University of New Brunswick
Maria Correia
Kelvin Gilliland
Jonathon Neilson
Allison Ritcey
Faith Shannon
Brittany Timberlake
Coleman Ward
Corey Willis

Thanks go as well to the high school students who participated, Meghan Harris and Lindsay Patton. Anne Duguay, a teacher from Queen Elizabeth High School in Calgary, and her student, Saskia, also participated. They provided a unique perspective on what entering students would expect from a text for an introductory course in biology. Finally, we wish to thank the student review boards from the University of Victoria, University of Toronto, Erindale Campus, Ryerson University, Sir Wilfred Grenfell College, and the University of Windsor. High school students, university students, and university instructors together provided us with an amazingly diverse array of feedback that allowed us to understand our audience and create a resource best suited to their needs.

Brief Contents

Contents

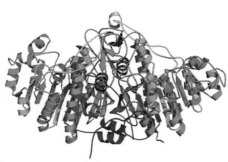
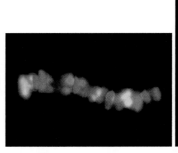
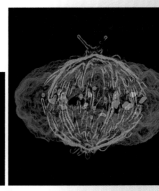

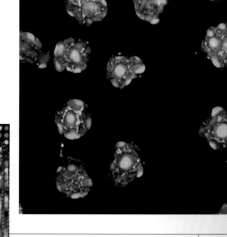

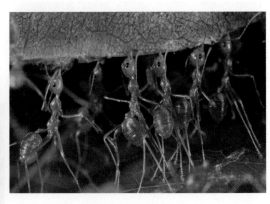

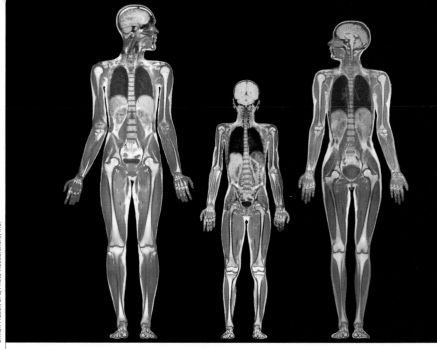

Magnetic resonance imaging (MRI) whole body scans of a man (left), a 9-year-old boy (middle), and a woman. Various organs can be seen in the scans: the whitish skeleton throughout the bodies, the brains within the skulls, lungs (dark) in the chests, lobes of the liver (green and brown ovals) in the abdomens, and bladders (dark ovals) in the lower abdomens.

Simon Fraser/SPL/Photo Researchers, Inc.

32 Introduction to Animal Organization and Physiology

WHY IT MATTERS

After a cold night in Africa's Kalahari Desert, gray meerkats (*Suricata suricatta*), a type of mongoose, awaken in their burrows. Although, like all mammals, meerkats regulate their body temperature, their internal temperature falls during cold nights. If the sun is shining in the morning and warms their burrows, the meerkats emerge and stand on their hind legs facing east, warming their bodies in the rays of the sun **(Figure 32.1, p. 766)**. This sunning behaviour helps raise their body temperature.

Once the meerkats warm up, they fan out from their burrows looking for food, mainly insects and an occasional lizard. Their highly integrated body systems allow them to move about, sense the presence of prey, react with speed and precision to capture those prey, and consume them. Within their bodies, the food is broken down into glucose and other nutrient molecules, which are transported throughout the body to provide energy for living. At the same time, mechanisms are constantly at work to maintain the animals' internal environment at a level that keeps body cells functioning. The maintenance of the internal environment in a stable state is called **homeostasis** (*homeo* = the same; *stasis* = standing or stopping). The processes and activities

Figure 32.1
Meerkats lining up to warm themselves in sunlight.

responsible for homeostasis are called **homeostatic mechanisms.** These mechanisms compensate both for the external environmental changes that the meerkats encounter as they explore places with differences in temperature, humidity, and other physical conditions, as well as for changes in their own body systems.

All animals have body systems for acquiring and digesting nutrients to provide energy for life, growth, reproduction, and movement. Biologists are interested in the structures and functions of these systems. **Anatomy** is the study of the structures of organisms, and **physiology** is the study of their functions. An understanding of structure is essential to an understanding of physiology, and an understanding of normal physiology is essential to the diagnosis and treatment of many diseases, such as Parkinson disease.

In this chapter, we begin with the organization of individual cells into tissues, organs, and organ systems, the major body structures that carry out animal activities. Our discussion continues with the coordination of the processes and activities of organ systems that accomplish homeostasis. The other chapters in this unit discuss the individual organ systems that carry out major body functions such as digestion, movement, and reproduction. Although we emphasize vertebrates throughout the unit, with particular reference to human physiology, we also make comparisons with invertebrates, to keep the structural and functional diversity of the animal kingdom in perspective and to understand the evolution of the structures and processes involved.

32.1 Organization of the Animal Body

32.1a In Animals, Specialized Cells Are Organized into Tissues, Tissues into Organs, and Organs into Organ Systems

The individual cells of animals have the same requirements as cells of any kind. They must be surrounded by an aqueous solution that contains ions and molecules required by the cells, including complex organic molecules that can be used as energy sources. The concentrations of these molecules and ions must be balanced to keep cells from shrinking or swelling excessively due to osmotic water movement. Most animal cells also require oxygen to serve as the final acceptor for electrons removed in oxidative reactions. Animal cells must be able to release waste molecules and other by-products of their activities, such as carbon dioxide, to their environment. The physical conditions of the cellular environment, such as temperature, must also remain within tolerable limits.

The evolution of multicellularity (see Section 2.5) made it possible for organisms to create an *internal fluid environment* that supplies all the needs of individual cells, including nutrient supply, waste removal, and osmotic balance. This internal environment allows multicellular organisms to occupy diverse habitats, including dry terrestrial habitats that would be lethal to single cells. Multicellular organisms can also become relatively large because their individual cells remain small enough to exchange ions and molecules with the internal fluid. The fluid occupying the spaces between cells in multicellular animals is called **interstitial fluid** or **extracellular fluid.**

The evolution of multicellularity also allowed major life functions to be subdivided among specialized groups of cells, with each group concentrating on a single activity. In animals, some groups of cells became specialized for movement, others for food capture, digestion, internal circulation of nutrients, excretion of wastes, reproduction, and other functions. Specialization greatly increases the efficiency by which animals carry out these functions.

In most animals, these specialized groups of cells are organized into tissues, the tissues into organs, and the organs into organ systems **(Figure 32.2).** A **tissue** is a group of cells with the same structure and function, working together as a unit to carry out one or more activities. The tissue lining the inner surface of the intestine, for example, is specialized to absorb nutrients released by digestion of food in the intestinal cavity.

An **organ** integrates two or more different tissues into a structure that carries out a specific function. The eye, liver, and stomach are examples of organs. Thus, the stomach integrates several different tissues into an organ specialized for processing food.

An **organ system** coordinates the activities of two or more organs to carry out a major body function such as movement, digestion, or reproduction. The organ system carrying out digestion, for example, coordinates the activities of organs, including the mouth, stomach, pancreas, liver, and small and large intestines. Some organs contribute functions to more than one organ system. For instance, the pancreas forms part of the endocrine system as well as the digestive system.

STUDY BREAK

1. What are some advantages for an organism being multicellular?
2. What is the difference between a tissue, an organ, and an organ system?

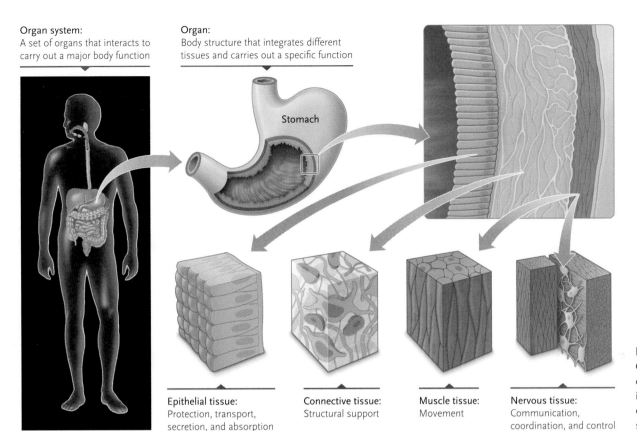

Organ system:
A set of organs that interacts to
carry out a major body function

Organ:
Body structure that integrates different
tissues and carries out a specific function

Stomach

Epithelial tissue:
Protection, transport,
secretion, and absorption

Connective tissue:
Structural support

Muscle tissue:
Movement

Nervous tissue:
Communication,
coordination, and control

Figure 32.2
Organization
of animal cells
into tissues,
organs, and organ
systems.

32.2 Animal Tissues

Although the most complex animals may contain hundreds of distinct cell types, all can be classified into one of four basic tissue groups: *epithelial, connective, muscle,* and *nervous* (see Figure 32.2). Each tissue type is assembled from individual cells. The properties of those cells determine the structure and, therefore, the function of the tissue. More specifically, the structure and integrity of a tissue depend on the structure and organization of the cytoskeleton within the cell, the type and organization of the extracellular matrix surrounding the cell, and the junctions holding cells together. The **extracellular matrix** (ECM) is nonliving material secreted by cells consisting of a variety of proteins and glycoproteins. The ECM provides support and shape for tissues and organs. The cell walls of plants and the cuticle of arthropods are examples of specialized ECM.

Junctions of various kinds link cells into tissues (see **Figure 32.3, p. 768**). *Anchoring junctions* form buttonlike spots or belts that weld cells together. They are most abundant in tissues subject to stretching, such as skin and heart muscle. *Tight junctions* seal the spaces between cells, keeping molecules and even ions from leaking between cells. For example, tight junctions in the tissue lining the urinary bladder prevent waste molecules and ions from leaking out of the bladder into other body tissues.

Gap junctions are open channels between cells in the same tissue, allowing ions and small molecules to flow freely from one to another. For example, gap junctions between muscle cells help muscle tissue to function as a unit.

Let us now consider the structural and functional features that distinguish the four types of tissues, with primary emphasis on the forms they take in vertebrates.

32.2a Epithelial Tissue Forms Protective, Secretory, and Absorptive Coverings and Linings of Body Structures

Epithelial tissue (*epi* = over; *thele* = covering) consists of sheetlike layers of cells that are usually joined tightly together, with little ECM material between them **(Figure 32.4, p. 769)**. Also called *epithelia* (singular, *epithelium*), these tissues cover body surfaces and the surfaces of internal organs, as well as line cavities and ducts within the body. They protect body surfaces from invasion by bacteria and viruses and secrete or absorb substances. For example, the epithelium covering a fish's gill structures serves as a barrier to bacteria and viruses and exchanges oxygen, carbon dioxide, and ions with the aqueous environment. The epithelium of the external surface of arthropods secretes the tough cuticle that, in addition to acting as a barrier to the environment, functions

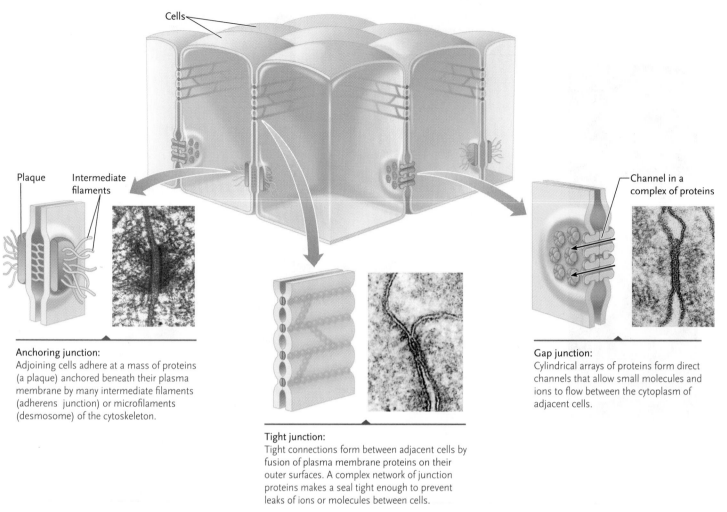

Cells

Plaque — **Intermediate filaments**

Anchoring junction:
Adjoining cells adhere at a mass of proteins (a plaque) anchored beneath their plasma membrane by many intermediate filaments (adherens junction) or microfilaments (desmosome) of the cytoskeleton.

Tight junction:
Tight connections form between adjacent cells by fusion of plasma membrane proteins on their outer surfaces. A complex network of junction proteins makes a seal tight enough to prevent leaks of ions or molecules between cells.

Channel in a complex of proteins

Gap junction:
Cylindrical arrays of proteins form direct channels that allow small molecules and ions to flow between the cytoplasm of adjacent cells.

Figure 32.3

Anchoring junctions, tight junctions, and gap junctions, which connect cells in animal tissues. Anchoring junctions reinforce the cell-to-cell connections made by cell adhesion molecules, tight junctions seal the spaces between cells, and gap junctions create direct channels of communication between animal cells.

as their skeleton. Nematodes also have a cuticle, but their skeleton is hydrostatic (see Chapter 26).

Some epithelia, such as those lining the capillaries of the circulatory system, act as filters, allowing ions and small molecules to leak from the blood into surrounding tissues while barring the passage of blood cells and large molecules such as proteins.

Because epithelia form coverings and linings, they have one free (or outer) surface, which may be exposed to water, air, or fluids within the body. In internal cavities and ducts, the free surface is often covered with *cilia,* which beat like oars to move fluids through the cavity or duct. The epithelium lining the oviducts in mammals, for example, is covered with cilia that generate fluid currents to move eggs from the ovaries to the uterus. In free-living flatworms, the ventral epithelium of the animal is frequently ciliated, allowing the worm to glide over surfaces. In some epithelia, including the lining of the small intestine, the free surface is crowded with *microvilli,* fingerlike extensions of the plasma membrane that increase the area available for secretion or absorption.

The inner surface of an epithelium adheres to a layer of glycoproteins secreted by the epithelial cells called the **basal lamina**, which is secreted by the epithelial cells. In many cases, such as the intestinal epithelium of vertebrates, there is a further layer of fibres secreted by underlying connective tissue, but this is lacking in most invertebrate epithelia. The entire assemblage is the **basement membrane.** The basal lamina is an example of an ECM.

Epithelial Cell Structure. Epithelia are classified as *simple*—formed by a single layer of cells—or *stratified*—formed by multiple cell layers (see **Figure 32.4a**). The shapes of cells within an epithelium may be *squamous* (mosaic, flattened, and spread out), *cuboidal* (shaped roughly like dice or cubes), or *columnar* (elongated, with the long axis perpendicular to the epithelial layer; see **Figure 32.4b**). For example, the outer epithelium of mammalian skin is stratified and contains columnar, cuboidal, and squamous cells; the epithelium lining blood vessels is

simple and squamous; and the intestinal epithelium is simple and columnar.

The cells of some epithelia, such as those forming the skin and the lining of the intestine, divide constantly to replace worn and dying cells. New cells are produced through division of stem cells in the basal (lowest) layer of the skin. *Stem cells* are undifferentiated (unspecialized) cells in the tissue that divide to produce more stem cells as well as cells that differentiate (that is, become specialized into one of the many cell types of the body). Stem cells are found in both adult organisms and embryos. Besides the skin, adult stem cells are found in tissues of the brain, bone marrow, blood vessels, skeletal muscle, and liver. Stem cells are an important aspect of development in many invertebrates. In some cases, these may be already programmed for a specific cell type as in the eye or wing disks of insect pupae, whereas in others, the stem cells may be totipotent, as in flatworms (Chapter 26).

Glands Formed by Epithelia. Epithelia typically contain or give rise to cells that are specialized for secretion. Some of these secretory cells are scattered among nonsecretory cells within the epithelium. Others form structures called **glands**, which are derived from pockets of epithelium during embryonic development.

Some glands, called **exocrine glands** (*exo* = external; *crine* = secretion), remain connected to the epithelium by a duct, which empties their secretion at the epithelial surface. Exocrine secretions include mucus, saliva, digestive enzymes, sweat, earwax, oils,

a. Patterns by which cells are arranged in epithelia

Simple epithelium

Stratified epithelium

Free surface

Epithelium

b. The three common shapes of epithelial cells

Squamous epithelium

Cuboidal epithelium

Columnar epithelium

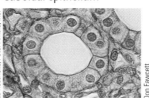

Ray Simmons/Photo Researchers, Inc.

Ed Reschke/Peter Arnold, Inc.

Don Fawcett

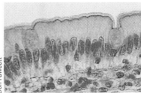

Description: Layer of flattened cells

Common locations: Walls of blood vessels; air sacs of lungs

Function: Diffusion

Description: Layer of cubelike cells; free surface may have microvilli

Common locations: Glands and tubular parts of nephrons in kidneys

Function: Secretion, absorption

Description: Layer of tall, slender cells; free surface may have microvilli

Common locations: Lining of gut and respiratory tract

Function: Secretion, absorption

Figure 32.4
Structure of epithelial tissues.

milk, and venom (**Figure 32.5a, p. 770** shows an exocrine gland in the skin of a poisonous tree frog). Other glands, called **endocrine** glands, may not be composed of epithelial cells. They have no ducts but secrete their products, hormones, into the interstitial fluid to be picked up by the blood for circulation to the organs

Figure 32.5
Exocrine and endocrine glands. The poison secreted by the blue poison frog (*Dendrobates azureus*) is one of the most lethal glandular secretions known.

Gregory Dimijian/Photo Researchers, Inc.

Thyroid

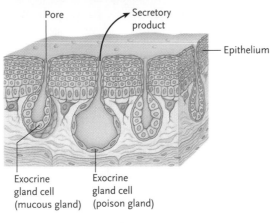

Pore

Secretory product

Epithelium

Exocrine gland cell (mucous gland)

Exocrine gland cell (poison gland)

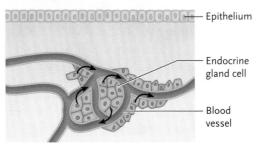

Epithelium

Endocrine gland cell

Blood vessel

a. Examples of exocrine glands: The mucus- and poison-secreting glands in the skin of a blue poison frog

b. Example of an endocrine gland: The thyroid gland, which secretes hormones that regulate the rate of metabolism and other body functions

and tissues of the body **(Figure 32.5b)**. The endocrine glands are considered in detail in Chapter 35.

Some glands contain both exocrine and endocrine elements. The pancreas, for example, has an exocrine function of secreting pancreatic juice through a duct into the small intestine, where it plays an important role in food digestion (see Chapter 41), and different cells provide an endocrine function by secreting the hormones insulin and glucagon into the bloodstream to help regulate glucose levels in the blood (see Chapter 35).

Some epithelial cells, particularly in the epidermis of vertebrates, contain a network of fibres of keratin, a family of tough proteins. Keratin forms the scales of fish and reptiles (including the shells of turtles), the feathers of birds, and the hair, claws, hooves, horns, and fingernails of mammals.

32.2b Connective Tissue Supports Other Body Tissues

Most animal body structures contain one or more types of **connective tissue.** Connective tissues support other body tissues, transmit mechanical and other forces, and in some cases act as filters. They consist of cells that form networks or layers in and around body structures and that are separated by nonliving material, specifically the ECM secreted by the cells of the tissue. Many forms of connective tissue have more ECM material (both by weight and by volume) than cellular material.

The mechanical properties of a connective tissue depend on the type and quantity of its ECM. The consistency of the ECM ranges from fluid (as in blood and lymph), through soft and firm gels (as in tendons), to the hard and crystalline (as in bone). In most connective tissues, the ECM consists primarily of the fibrous glycoprotein **collagen** embedded in a network of proteoglycans—glycoproteins that are very rich in carbohydrates. Collagen is the most abundant protein in animals. More than 25 different forms have been described, and some form of collagen occurs in all Metazoa, including Porifera (sponges). The collagen molecule is thus an ancient one that has been modified during evolution. In bone, the glycoprotein network surrounding the collagen is impregnated with mineral deposits that produce a hard, yet still somewhat elastic, structure. Another class of glycoproteins, **fibronectin,** aids in the attachment of cells to the ECM and helps hold the cells in position.

In some connective tissues, another rubbery protein, **elastin,** adds elasticity to the ECM. It is able to return to its original shape after being stretched, bent, or compressed. Elastin fibres, for example, help the skin return to its original shape when pulled or stretched and give the lungs the elasticity required for their alternating inflation and deflation. Resilin is a protein related to elastin that occurs only in insects and some Crustacea. It is the most elastic material known and is the basis for the jumping of fleas and locusts (see *Molecule Behind Biology*).

MOLECULE BEHIND BIOLOGY
Resilin: Insect Rubber

While Torkel Weis-Fogh, a young Danish comparative physiologist, was conducting his ground-breaking studies on insect flight in the Zoological Laboratory in Cambridge, England, in 1960, he discovered patches of highly elastic cuticle in the wing joints of locusts. Subsequently, the same elastic protein proved to be important to other insect movements. For example, it is involved in the movements of the membrane that produces the song of cicadas. Fleas are able to jump very large distances because muscle contractions store energy in pads of resilin in the hind legs **(Figure 1)**. The sudden release of this energy, in less than a millisecond, propels the jump with an instantaneous acceleration greater than that of the space shuttle.

Resilin belongs to the same family of proteins as elastin, an elastic protein in many animals, but it is restricted to insects and a few crustacea. Resilin's elastic properties have proved to be astonishing. It can be stretched to four times its length without breaking (elastin only manages two). It is 97% efficient, so that when energy stored in it by stretching is released, only 3% is released as heat. It survives huge numbers of cycles of stretch and relaxation. The membrane producing the song of cicadas vibrates several thousand times per second. Like other elastic proteins, resilin is composed of coiled protein molecules cross-linked to one another, and stretching involves uncoiling.

In 2005, a gene coding for *Drosophila* resilin was cloned in *E. coli* by a team led by Christopher Elvin of the Commonwealth Scientific and Industrial Research Organisation in Australia. That lab is now able to produce significant quantities of resilin. Ultimately, this insect rubber should find application, for example, as replacements for spinal disks in humans or as artificial blood vessels.

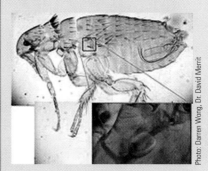

Figure 1 The pad of resilin (blue) in the coxa (part of the thigh) of the hind leg of a flea.

Photo: Darren Wong, Dr. David Merrit

Vertebrates have six major types of connective tissue: *loose connective tissue, fibrous connective tissue, cartilage, bone, adipose tissue,* and *blood.* Each type has a characteristic function correlated with its structure **(Figure 32.6, p. 772)**.

Loose Connective Tissue. **Loose connective tissue** consists of sparsely distributed cells surrounded by a more or less open network of collagen and other glycoprotein fibres (see Figure 32.6a). The cells, called **fibroblasts**, secrete most of the collagen and other proteins in this connective tissue.

In vertebrates, loose connective tissues support epithelia and form a corsetlike band around blood vessels, nerves, and some internal organs; they also reinforce deeper layers of the skin. Sheets of loose connective tissue, covered on both surfaces with epithelial cells, form the **mesenteries**, which hold the abdominal organs in place and provide lubricated, smooth surfaces that prevent chafing or abrasion between adjacent structures as the body moves. In insects, and perhaps some other invertebrates, the loose connective tissues suspending organs and providing support for epithelia are the products of specialized cells circulating in the blood.

Fibrous Connective Tissue. In **fibrous connective tissue**, fibroblasts are sparsely distributed among dense masses of collagen and elastin fibres that are lined up in highly ordered, parallel bundles (see Figure 32.5b). The parallel arrangement produces maximum tensile strength and elasticity. Examples include **tendons**, which attach muscles to bones, and **ligaments**, which connect bones to each other at a joint. The cornea of the eye is a transparent fibrous connective tissue formed from highly ordered collagen molecules.

In some invertebrates, fibrous connective tissue provides shape to the animal, as in many sponges (see Chapter 26) and echinoderms. In sea cucumbers (see Chapter 27), the rigidity of the connective tissue can be changed quickly by the animal, resulting in a loss or change of shape. This acts as an escape response.

Cartilage. **Cartilage** consists of sparsely distributed cells called **chondrocytes**, surrounded by networks of collagen fibres embedded in a tough but elastic matrix of the glycoprotein *chondroitin sulphate* (see Figure 32.6c). Elastin is also present in some forms of cartilage.

The elasticity of cartilage allows it to resist compression and stay resilient, like a piece of rubber. Bending your ear or pushing the tip of your nose, which are supported by cores of cartilage, gives a good idea of the flexible nature of this tissue. Cartilage also supports the larynx, trachea, and smaller air passages in the lungs. It forms the disks cushioning the vertebrae in the spinal column and the smooth, slippery capsules around the ends of bones in joints such as the hip and knee. Cartilage also serves as a precursor to bone during embryonic development;

Figure 32.6
The six major types of connective tissues in vertebrates.

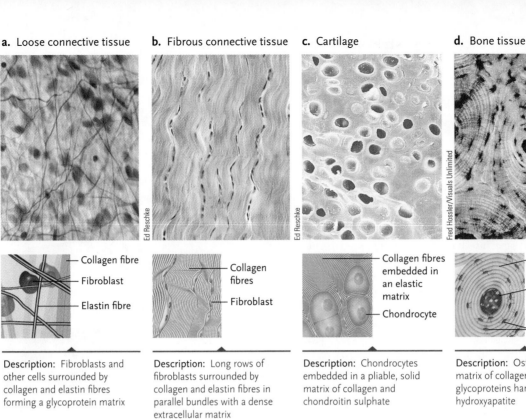

a. Loose connective tissue

- Collagen fibre
- Fibroblast
- Elastin fibre

Description: Fibroblasts and other cells surrounded by collagen and elastin fibres forming a glycoprotein matrix

Common locations: Under the skin and most epithelia

Function: Support, elasticity, diffusion

b. Fibrous connective tissue

- Collagen fibres
- Fibroblast

Description: Long rows of fibroblasts surrounded by collagen and elastin fibres in parallel bundles with a dense extracellular matrix

Common locations: Tendons, ligaments

Function: Strength, elasticity

c. Cartilage

- Collagen fibres embedded in an elastic matrix
- Chondrocyte

Description: Chondrocytes embedded in a pliable, solid matrix of collagen and chondroitin sulphate

Common locations: Ends of long bones, nose, parts of airways, skeleton of vertebrate embryos

Function: Support, flexibility, low-friction surface for joint movement

d. Bone tissue

- Fine canals
- Central canal containing blood vessel
- Osteocytes

Description: Osteocytes in a matrix of collagen and glycoproteins hardened with hydroxyapatite

Common locations: Bones of vertebrate skeleton

Function: Movement, support, protection

e. Adipose tissue

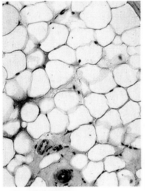

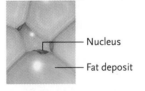

- Nucleus
- Fat deposit

Description: Large, tightly packed adipocytes with little extracellular matrix

Common locations: Under skin; around heart, kidneys

Function: Energy reserves, insulation, padding

f. Blood

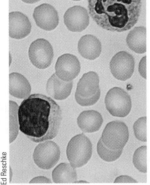

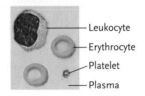

- Leukocyte
- Erythrocyte
- Platelet
- Plasma

Description: Leukocytes, erythrocytes, and platelets suspended in a plasma matrix

Common locations: Circulatory system

Function: Transport of substances

in sharks and rays and their relatives, almost the entire skeleton remains as cartilage in adults.

Bone. The densest form of connective tissue, **bone,** forms the skeleton, which supports the body, protects softer body structures such as the brain, and contributes to body movements.

Mature bone consists primarily of cells called **osteocytes** (*osteon* = bone) embedded in an ECM containing collagen fibres and glycoproteins impregnated with *hydroxyapatite,* a calcium–phosphate mineral (see Figure 32.6d). The collagen fibres give bone tensile strength and elasticity; the hydroxyapatite resists compression and allows bones to support body weight. Cells called **osteoblasts** (*blast* = bud or sprout) produce the collagen and mineral of bone—as much as 85% of the weight of bone is mineral deposits. Osteocytes, in fact, are osteoblasts that have become trapped and surrounded by the bone materials they themselves produce. **Osteoclasts** (*clast* = break) remove the minerals and recycle them through the bloodstream. Bone is not a stable tissue; it is reshaped continuously by the bone-building osteoblasts and the bone-degrading osteoclasts.

Although bones appear superficially to be solid, they are actually porous structures consisting of a system of microscopic spaces and canals. The structural unit of bone is the **osteon.** It consists of a minute central canal surrounded by osteocytes embedded in concentric layers of mineral matter (see Figure 32.6d). A blood vessel and extensions of nerve cells run through the central canal, which is connected to the spaces containing cells by very fine, radiating canals filled with interstitial fluid. The blood vessels supply nutrients to the cells with which the bone is built, and the nerve cells connect the bone and its cells to the body's nervous system.

Adipose Tissue. The connective tissue called **adipose tissue** mostly contains large, densely clustered cells called *adipocytes* that are specialized for fat storage (see Figure 32.6e). It has little ECM. Adipose tissue also cushions the body and, in mammals, forms an especially important insulating layer under the skin.

The animal body stores limited amounts of carbohydrates, primarily in muscle and liver cells. Excess carbohydrates are converted into the fats stored in adipocytes. The storage of chemical energy as fats offers animals a weight advantage. For example, the average human would weigh about 45 kg (100 pounds) more if the same amount of chemical energy were stored as carbohydrates instead of fats. Adipose tissue is richly supplied with blood vessels, which move fats or their components to and from adipose cells.

In invertebrates, fat or glycogen storage may occur in a variety of tissues. Insects have a fat body, an organ that functions both for storage and as an important structure for metabolism, much like the vertebrate liver. In nematodes and flatworms, the muscle cells take on this function.

Blood. Blood is considered to be a connective tissue because the fluid portion is essentially a fluid form of ECM. Blood functions as the principal transport vehicle to carry nutrients, oxygen (in most animals), and hormones to the tissues and to remove metabolic wastes for transport to the organs specialized for waste removal. It is also frequently involved in defence against disease (Chapter 44) and may be important in wound healing.

Vertebrates have two basic types of cells suspended in a straw-coloured fluid, the plasma (see Figure 32.6f). Erythrocytes (erythros = red), or red blood cells, contain haemoglobin, a protein to which O_2 binds; these are specialized for O_2 transport. Several types of **leukocytes** (leukos = white) protect the body against foreign elements such as viruses and bacteria. These are considered in Chapter 37. Vertebrate blood also contains platelets (often called thrombocytes), which are membrane-bound fragments of specialized leukocytes. They play an essential role in the formation of blood clots to heal wounds.

In invertebrates, the functions of vertebrate blood are carried out in several different ways. Some, such as annelids and cephalopods, have blood enclosed in blood vessels; others, such as the arthropods and many molluscs, have a more open system. Haemoglobin or other oxygen-carrying pigments may be present either within special cells, as in some annelids, or as part of the plasma. Blood cells in insects are known to take part in wound healing and protection against foreign bodies.

32.2c Muscle Tissue Produces the Force for Body Movements

Muscle tissue consists of cells that have the ability to contract (shorten). The contractions, which depend on the interaction of two proteins—*actin* and *myosin*—move body limbs and other structures, pump the blood, and produce a squeezing pressure in organs such as the intestine and uterus. Three types of muscle tissue, *skeletal, cardiac,* and *smooth,* produce body movements in vertebrates **(Figure 32.7, p. 774).**

Skeletal Muscle. **Skeletal muscle** is so called because most muscles of this type are attached by tendons to the skeleton. Skeletal muscle cells are also called **muscle fibres** because each is an elongated cylinder (see Figure 32.7a). These cells contain many nuclei and are packed with actin and myosin molecules arranged in highly ordered, parallel units that give the tissue a banded or striated appearance when viewed under a microscope. Muscle fibres packed side by side into parallel bundles surrounded by sheaths of connective tissue form many body muscles, such as the biceps.

Skeletal muscle contracts in response to signals carried by the nervous system. The contractions of skeletal muscles, which are characteristically rapid and powerful, move body parts and maintain posture. The contractions also release heat as a by-product of cellular metabolism. This heat helps mammals, birds, and some other vertebrates maintain their body temperatures when environmental temperatures fall. (Skeletal muscle is discussed further in Chapter 36.)

Cardiac Muscle. **Cardiac muscle** is the contractile tissue of the heart (see Figure 32.7b). Cardiac muscle has a striated appearance because it contains actin and myosin molecules arranged like those in skeletal muscle. However, cardiac muscle cells are short and branched, with each cell connecting to several neighbouring cells; the joining point between two such cells is called an *intercalated disk.* Cardiac muscle cells thus form an interlinked network, which is stabilized by **anchoring junctions** and gap junctions. This network makes heart muscle contract in all directions, producing a squeezing or pumping action rather than the lengthwise, unidirectional contraction characteristic of skeletal muscle.

Figure 32.7
Structure of skeletal, cardiac, and smooth muscle.

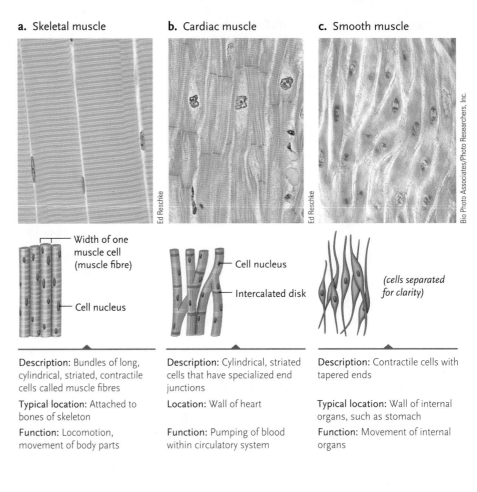

a. Skeletal muscle

b. Cardiac muscle

c. Smooth muscle

Width of one muscle cell (muscle fibre)

Cell nucleus

Cell nucleus

Intercalated disk

(cells separated for clarity)

Description: Bundles of long, cylindrical, striated, contractile cells called muscle fibres

Typical location: Attached to bones of skeleton

Function: Locomotion, movement of body parts

Description: Cylindrical, striated cells that have specialized end junctions

Location: Wall of heart

Function: Pumping of blood within circulatory system

Description: Contractile cells with tapered ends

Typical location: Wall of internal organs, such as stomach

Function: Movement of internal organs

Smooth Muscle. **Smooth muscle** is found in the walls of tubes and cavities in the body, including blood vessels, the stomach and intestine, the bladder, and the uterus. Smooth muscle cells are relatively small and spindle-shaped (pointed at both ends), and their actin and myosin molecules are arranged in a loose network rather than in bundles (see Figure 32.7c). This loose network makes the cells appear smooth rather than striated when viewed under a microscope. Smooth muscle cells are connected by gap junctions and enclosed in a mesh of connective tissue. The gap junctions transmit ions that make smooth muscles contract as a unit, typically producing a squeezing motion. Although smooth muscle contracts more slowly than skeletal and cardiac muscles do, its contractions can be maintained at steady levels for a much longer time. These contractions move and mix the stomach and intestinal contents, constrict blood vessels, and push the infant out of the uterus during childbirth.

Invertebrate Muscle. In general, most invertebrates have striated muscles throughout, even muscles involved with structures such as the intestine or reproductive ducts. There are, however, some differences in the way that these muscles are organized. In nematodes, the muscle cell includes both a contractile area and a large expansion of the cell containing the nucleus, which acts as a glycogen store **(Figure 32.8)**. The cell does not receive nerves but makes a connection with the nervous

system by an extension of the muscle to the nerve cords. The muscles of cestodes and possibly all flatworms have a similar form, but they are unstriated. In insects, the striated muscles that control the movements of some of the viscera, such as the ovaries and parts of the digestive system, are frequently branched and interconnected to form a lattice.

32.2d Nervous Tissue Receives, Integrates, and Transmits Information

Nervous tissue contains cells called **neurons** (also called *nerve cells*) that serve as lines of communication and control between body parts. Billions of neurons are packed into the human brain; others extend throughout the body. Nervous tissue also contains **glial cells** (*glia* = glue), which physically support and provide nutrients to neurons, provide electrical insulation between them, and scavenge cellular debris and foreign matter. Some neurons are specialized to form endocrine glands, as in the pituitary of vertebrates and the corpus cardiacum of insects (see Chapter 35).

A neuron consists of a *cell body,* which houses the nucleus and organelles, and two types of cell extensions, dendrites and axons **(Figure 32.9)**. *Dendrites* receive chemical signals from other neurons or from body cells of other types and convert them into an electrical signal that is transmitted to the cell body of the receiving neuron. Dendrites are usually highly

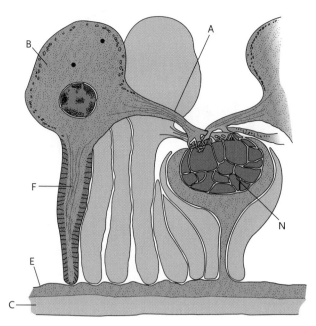

Figure 32.8

Muscle cell in *Ascaris lumbricoides*. The foot, F, of the muscle contains the contractile elements; the body, B, has the nucleus and is packed with glycogen; the arm, A, is an extension that makes contact with the nerve cord, N. The muscle cell is attached by fine fibres to the epidermis, E, beneath the cuticle, C.

branched. *Axons* conduct electrical signals away from the cell body to the axon terminals, or endings. At their terminals, axons convert the electrical signal to a chemical signal that stimulates a response in nearby muscle cells, gland cells, or other neurons. Axons are usually unbranched except at their terminals. Depending on the type of neuron and its location in the body, its axon may extend from a few micrometres or millimetres to more than a metre. (Neurons and their organization in body structures are discussed further in Chapter 33.)

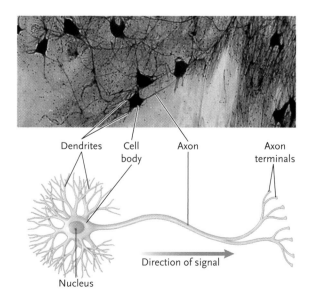

Figure 32.9

Neurons and their structure. The micrograph shows a network of motor neurons, which relay signals from the brain or spinal cord to muscles and glands. (Micrograph: Lennart Nilsson from Behold Man, © 1974 Albert Bonniers Forlag and Little, Brown and Company, Boston.)

All four major tissue types—epithelial, connective, muscle, and nervous—combine to form the organs and organ systems of animals. The next section depicts the major organs and organ systems of vertebrates and outlines their main tasks.

STUDY BREAK

1. Distinguish between exocrine and endocrine glands. What is the tissue type of each of these glands?
2. What are the six major types of connective tissue in vertebrates?

32.3 Coordination of Tissues in Organs and Organ Systems

32.3a Organs and Organ Systems Function Together to Enable an Animal to Survive

In the tissues, organs, and organ systems of an animal, each cell engages in the basic metabolic activities that ensure its own survival and performs one or more functions of the system to which it belongs. All vertebrates have 11 major organ systems, which are summarized in **Figure 32.10, p. 776–777**. Most invertebrates have the same systems but do not have a separate system of lymphatic ducts.

The functions of all these organ systems are coordinated and integrated to accomplish collectively a series of tasks that are vital to all animals, whether a flatworm, a salmon, a meerkat, or a human. These functions include

1. acquiring nutrients and other required substances, such as oxygen, coordinating their processing, distributing them throughout the body, and disposing of wastes.
2. synthesizing the protein, carbohydrate, lipid, and nucleic acid molecules required for body structure and function.
3. sensing and responding to changes in the environment, such as temperature, pH, and ion concentrations.
4. protecting the body against injury or attack from other animals and from viruses, bacteria, and other disease-causing agents.
5. reproducing and, in many instances, nourishing and protecting offspring through their early growth and development.

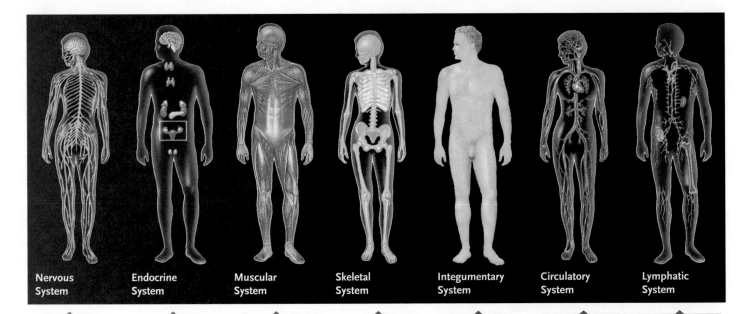

Nervous System	Endocrine System	Muscular System	Skeletal System	Integumentary System	Circulatory System	Lymphatic System
Main organs: Brain, spinal cord, peripheral nerves, sensory organs	**Main organs:** Pituitary, thyroid, adrenal, pancreas, and other hormone-secreting glands	**Main organs:** Skeletal, cardiac, and smooth muscle	**Main organs:** Bones, tendons, ligaments, cartilage	**Main organs:** Skin, sweat glands, hair, nails	**Main organs:** Heart, blood vessels, blood	**Main organs:** Lymph nodes, lymph ducts, spleen, thymus
Main functions: Principal regulatory system; monitors changes in internal and external environments and formulates compensatory responses; coordinates body activities. Nervous systems are present in all metazoans except sponges.	**Main functions:** Regulates and coordinates body activities through secretion of hormones. Endocrine systems are also present in most metazoans.	**Main functions:** Moves body parts; helps run bodily functions; generates heat. Specialized muscle cells do not appear in evolution until triploblastic animals.	**Main functions:** Supports and protects body parts; provides leverage for body movements. An internal skeleton composed of bone and/or cartilage occurs only in the vertebrates. Similar functions in invertebrates are carried out by an external skeleton or by internal hydrostatic pressure.	**Main functions:** Covers external body surfaces and protects against injury and infection; helps regulate water content and body temperature. All Metazoa except sponges have an integument of some sort.	**Main functions:** Distributes water, nutrients, oxygen, hormones, and other substances throughout body and carries away carbon dioxide and other metabolic wastes; helps stabilize internal temperature and pH. Specialized circulatory systems occur in all vertebrates and in the annelids, molluscs, and arthropods.	**Main functions:** Returns excess fluid to the blood; defends body against invading viruses, bacteria, fungi, and other pathogens as part of immune system. Invertebrates do not have a specialized lymphatic system.

Together these tasks maintain homeostasis, preserving the internal environment required for survival of the body. Homeostasis is the topic of the next section.

STUDY BREAK

1. What are the major functions of each of the 11 organ systems?
2. What are the major organ systems in a duck? in a shark? in an insect? in an earthworm?

32.4 Homeostasis

Homeostasis is the process by which animals maintain their internal environment in a steady state (constant level) or between narrow limits. Homeostasis depends on a number of the body's organ systems, with the nervous system and endocrine system being the most important. For example, blood pH is controlled by both the nervous and endocrine systems, blood glucose by the endocrine system, internal temperature by the nervous and endocrine systems, and oxygen and carbon dioxide concentrations by the nervous system.

Although the *stasis* part of homeostasis might suggest a static, unchanging process, homeostasis is actually a dynamic process, in which internal adjustments are made continuously to compensate for changes in the internal or external environment. For example, internal adjustments are needed to maintain homeostasis during exercise or hibernation. The factors controlled by homeostatic mechanisms all require energy.

32.4a Homeostasis Is Accomplished by Negative Feedback Mechanisms

The primary mechanism of homeostasis is **negative feedback**, in which a stimulus resulting from a

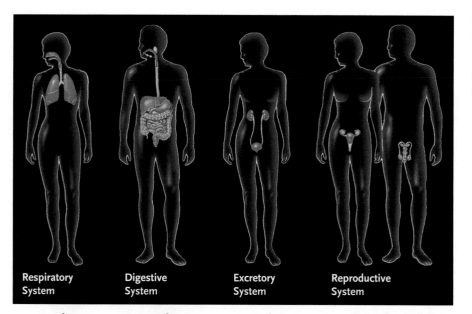

Figure 32.10
Organ systems of the human body. The immune system, which is primarily a cellular system, is not shown. The functions performed by these organ systems are also performed by all other animals, although different organs and systems may be involved.

Respiratory System

Main organs:
Lungs, diaphragm, trachea, and other airways

Main functions:
Exchanges gases with the environment, including uptake of oxygen and release of carbon dioxide. Fish have a respiratory system that involves gills. Some form of specialized respiratory system occurs in most invertebrates.

Digestive System

Main organs:
Pharynx, esophagus, stomach, intestines, liver, pancreas, rectum, anus

Main functions:
Converts ingested matter into molecules and ions that can be absorbed into body; eliminates undigested matter; helps regulate water content. Most metazoans, with the exception of some parasitic forms, have a digestive system.

Excretory System

Main organs:
Kidneys, bladder, ureter, urethra

Main functions:
Removes and eliminates excess water, ions, and metabolic wastes from body; helps regulate internal osmotic balance and pH. All animals perform these functions. All vertebrates have kidneys, and most invertebrates have specialized excretory organs and systems.

Reproductive System

Main organs:
Female: ovaries, oviducts, uterus, vagina, mammary glands
Male: testes, sperm ducts, accessory glands, penis

Main functions:
Maintains the sexual characteristics and passes on genes to the next generation. Most triploblastic animals have specialized reproductive organs and systems.

change in the external or internal environment triggers a response that compensates for the environmental change **(Figure 32.11)**. Homeostatic mechanisms typically include three elements: a sensor, an integrator, and an effector. The **sensor** consists of tissues or organs that detect a change in external or internal factors such as pH, temperature, or the concentration of a molecule such as glucose. The **integrator** is a control centre that compares the detected environmental change with a **set point**, the level at which the condition controlled by the pathway is to be maintained. The **effector** is a system, activated by the integrator, that returns the condition to the set point if it has strayed away. In most animals, the integrator is part of the central nervous system or endocrine system, whereas effectors may include parts of any body tissue or organ.

The Thermostat as a Negative Feedback Mechanism.
The concept of negative feedback may be most familiar in systems designed by human engineers. The thermostat maintaining temperature at a chosen level in a house provides an example. A sensor within the thermostat measures the temperature. If the room temperature changes more than a degree or so from the set point—the temperature that you set in the thermostat—an integrator circuit in the thermostat activates an effector that returns the room temperature

Figure 32.11
Components of a negative feedback mechanism maintaining homeostasis. The integrator coordinates a response by comparing the level of an environmental condition with a set point that indicates where the level should be.

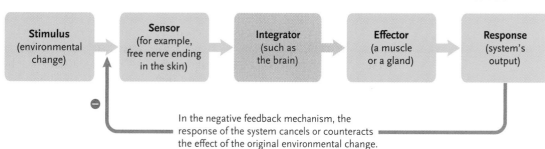

Stimulus (environmental change)	Sensor (for example, free nerve ending in the skin)	Integrator (such as the brain)	Effector (a muscle or a gland)	Response (system's output)

In the negative feedback mechanism, the response of the system cancels or counteracts the effect of the original environmental change.

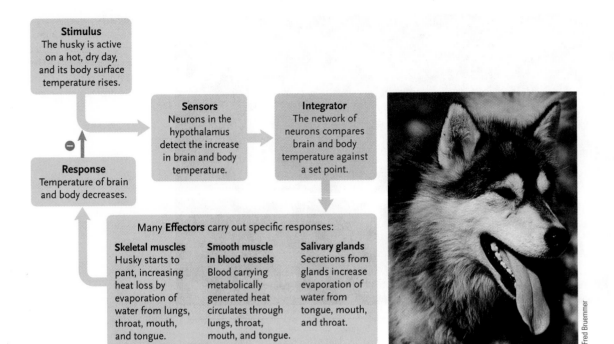

Figure 32.12
Homeostatic mechanisms maintaining the body temperature of a husky when environmental temperatures are high.

Stimulus
The husky is active on a hot, dry day, and its body surface temperature rises.

Sensors
Neurons in the hypothalamus detect the increase in brain and body temperature.

Integrator
The network of neurons compares brain and body temperature against a set point.

Response
Temperature of brain and body decreases.

Many **Effectors** carry out specific responses:

Skeletal muscles
Husky starts to pant, increasing heat loss by evaporation of water from lungs, throat, mouth, and tongue.

Smooth muscle in blood vessels
Blood carrying metabolically generated heat circulates through lungs, throat, mouth, and tongue.

Salivary glands
Secretions from glands increase evaporation of water from tongue, mouth, and throat.

Fred Bruemmer

to the set point. If the temperature has fallen below the set point, the effector is the furnace, which adds heat to the house until the temperature rises to the set point. If the temperature has risen above the set point, the effector is the air conditioner, which removes heat from the room until the temperature falls to the set point.

Negative Feedback Mechanisms in Animals. Mammals and birds also have a homeostatic mechanism that maintains body temperature within a relatively narrow range around a set point. The integrator (thermostat) for this mechanism is located in a brain centre called the *hypothalamus*. A group of neurons in the hypothalamus detects changes in the temperature of the brain and the rest of the body and compares it with a set point. For humans, the set point has a relatively narrow range centred at about 37°C.

One or more effectors are activated in humans if the temperature varies beyond the limits of the set point. If the temperature falls below the lower limit, the hypothalamus activates effectors that constrict the blood vessels in the skin. The reduction in blood flow means that less heat is conducted from the blood through the skin to the environment; in short, heat loss from the skin is reduced. Other effectors may induce shivering, a physical mechanism to generate body heat. Also, integrating neurons in the brain, stimulated by signals from the hypothalamus, make us consciously sense a chill, which we may counteract behaviourally by putting on more clothes or moving to a warmer area.

Conversely, if blood temperature rises above the set point, the hypothalamus triggers effectors that dilate the blood vessels in the skin, increasing blood flow to the skin and heat loss from it. Other effectors induce sweating, which cools the skin and the blood flowing through it as the sweat evaporates. And again,

through integrating neurons in the brain, we may consciously sense being overheated, which we may counteract by shedding clothes, moving to a cooler location, or taking a dip in a pool.

Sometimes the temperature set point changes, and the negative feedback mechanisms then operate to maintain body temperature at the new set point. For example, if you become infected by certain viruses and bacteria, the temperature set point increases to a higher level, producing a fever to help overcome the infection. Once the infection is combated, the set point is readjusted down again to its normal level.

All other mammals have similar homeostatic mechanisms that maintain or adjust body temperature. Dogs and birds pant to release heat from their bodies **(Figure 32.12)** and shiver to increase internal heat production. Many terrestrial vertebrates enter or splash water over their bodies to cool off. Also, recall from the beginning of the chapter how meerkats use behavioural mechanisms to regulate their body temperature.

Whereas mammals and birds regulate their internal body temperature within a narrow range around a set point, certain other vertebrates regulate over a broader range. These vertebrates use other negative feedback mechanisms for their temperature regulation. Snakes and lizards, for example, respond behaviourally to compensate for variations in environmental temperatures and use other, less precise negative feedback mechanisms for their temperature regulation. They may absorb heat by basking on sunny rocks in the cool early morning and move to cooler, shaded spots in the heat of the afternoon. Some fishes, such as the tuna, generate enough heat by contraction of the swimming muscles to maintain body temperature well above the temperature of the surrounding water.

Many insects employ similar mechanisms to raise their body temperature. Some caterpillars group together, increasing their body temperatures by a degree or two and shortening the time of development by as much as three days. Flight requires energy, and the flight muscles operate best at higher temperatures. Some insects bask in the sun to warm the muscles. Many, such as dragonflies, bumblebees, butterflies, and moths, contract the flight muscles rapidly in a process similar to shivering in order to warm them. This is particularly important in moths that fly at night, when the environmental temperature is lower. Honeybees form masses in the winter and maintain the temperature by contracting the wing muscles.

Once insects are in flight, however, the energy production is so high that they must dissipate the heat produced. In bees, the most important method is evaporative cooling by regurgitation of some of the intestinal contents onto the mouthparts, a process equivalent to panting in vertebrates.

32.4b Animals Also Have Positive Feedback Mechanisms That Do Not Result in Homeostasis

Under certain circumstances, animals respond to a change in internal or external environmental condition by a **positive feedback** mechanism that intensifies or adds to the change. Such mechanisms, with some exceptions, do not result in homeostasis. They operate when the animal is responding to life-threatening conditions (an attack, for instance) or as part of reproductive processes.

The birth process in mammals is a prime example. During human childbirth, initial contractions of the uterus push the head of the fetus against the cervix, the opening of the uterus into the vagina. The pushing causes the cervix to stretch. Sensors that detect the stretching signal the hypothalamus to release a hormone, oxytocin, from the pituitary gland. Oxytocin increases the uterine contractions, intensifying the squeezing pressure on the fetus and further stretching the cervix. The stretching results in more oxytocin release and stronger uterine contractions, repeating the positive feedback circuit and increasing the squeezing pressure until the fetus is pushed entirely out of the uterus.

Because positive feedback mechanisms such as the one triggering childbirth do not result in homeostasis, they occur less commonly than negative feedback in animals. They also operate as part of larger, more inclusive negative feedback mechanisms that ultimately shut off the positive feedback pathway and return conditions to normal limits.

We learned in this chapter about the various tissues and organ systems of the body and provided an example of the involvement of organ systems in homeostasis. Next, we begin a series of chapters describing the organ systems in detail, starting with the nervous system.

STUDY BREAK

What is the difference between a positive and a negative feedback loop?

UNANSWERED QUESTIONS

What Determines the Fate of Stem Cells?

The discovery in the early 1960s by Ernest McCullough and James Till of the Ontario Cancer Institute in Toronto that bone marrow contained cells that can be grown outside the body and that have the potential to differentiate into a number of different cell types stimulated a frenzy of research, much of it directed toward the possible use of such cells, **stem cells,** in medicine. It is known that there are two basic types of stem cells. **Embryonic stem cells** are from the early embryo, before the cells have begun to differentiate into the cell layers and tissues. These can be grown in culture indefinitely and have the potential to develop into any adult cell type. They're referred to as totipotent. **Adult stem cells** can also be grown in culture and are morphologically undifferentiated, but their potential for further differentiation is normally limited to the tissue in which they are found: for example, nerve stem cells can differentiate only into the cell types normally found in the nervous system. In some cases, adult stem cells can be induced in culture to form a variety of cell types: these are referred to as pluripotent. Embryonic stem cells have the greatest potential in medicine, but there are ethical debates about establishing cultures of them from humans

since the cultures require the destruction of an embryo, albeit at a very early stage. A good deal of research is focused on how adult stem cells might be deprogrammed so that they have the potential to make other types of tissues.

But there are other, more fundamental, biological questions. What are the signals that set embryonic stem cells on different developmental paths leading to different adult stem cell types? In some cases, scientists have found conditions that can reverse this process, but an understanding of the natural factors that lead to differentiation is fundamental to an understanding of development. What are the conditions in a particular tissue that activate the existing stem cells in that tissue to divide and differentiate into the functional cells? Given the potential importance of stem cells in the treatment of disease, there has been an understandable concentration on the mouse as a model system. But stem cells are important in a wide variety of animals. In the trematode flatworms, for example (see Chapter 26), stem cells are reserved at each larval stage, and these stem cells give rise to the next larval stage, which may be morphologically very different. Are systems such as this, far removed from mammals, able to provide useful clues?

Review

Go to CENGAGENOW™ at http://hed.nelson.com/ to access quizzing, animations, exercises, articles, and personalized homework help.

32.1 Organization of the Animal Body

- Multicellularity permits organisms to maintain an internal environment, allowing them to exploit a greater variety of environments; allows organisms to become larger; and allows for differentiation of cells specialized to perform specific functions.

- In most animals, cells are specialized and organized into tissues, tissues into organs, and organs into organ systems. A tissue is a group of cells with the same structure and function, working as a unit to carry out one or more activities. An organ is an assembly of tissues integrated into a structure that carries out a specific function. An organ system is a group of organs that carry out related steps in a major physiological process.

32.2 Animal Tissues

- Animal tissues are classified as epithelial, connective, muscle, or nervous. The properties of the cells of these tissues determine the structures and functions of the tissues.

- Various kinds of junctions link cells in a tissue. Anchoring junctions "weld" cells together. Tight junctions seal the cells into a leak-proof layer. Gap junctions form direct avenues of communication between the cytoplasm of adjacent cells in the same tissue.

- Epithelial tissue consists of sheetlike layers of cells that cover body surfaces and the surfaces of internal organs, and line cavities and ducts within the body.

- Exocrine glands are secretory structures derived from epithelia. Exocrine glands are connected by a duct that empties on the epithelial surface. Endocrine glands are ductless. Not all endocrine glands are derived from epithelium.

- Connective tissue consists of cell networks or layers and a prominent extracellular matrix (ECM) that separates the cells. It supports other body tissues and transmits mechanical and other forces.

- Loose connective tissue consists of sparsely distributed fibroblasts surrounded by an open network of collagen and other glycoproteins. It supports epithelia and organs of the body and forms a covering around blood vessels, nerves, and some internal organs.

- Fibrous connective tissue contains sparsely distributed fibroblasts in a matrix of densely packed, parallel bundles of collagen and elastin fibres. It forms high tensile-strength structures such as tendons and ligaments.

- Cartilage consists of sparsely distributed chondrocytes surrounded by a network of collagen fibres embedded in a tough but highly elastic matrix of branched glycoproteins. Cartilage provides support, flexibility, and a low-friction surface for joint movement.

- In bone, osteocytes are embedded in a collagen matrix hardened by mineral deposits. Osteoblasts secrete collagen and minerals for the ECM; osteoclasts remove the minerals and recycle them into the bloodstream.

- Adipose tissue consists of cells specialized for fat storage. It also cushions the body and provides an insulating layer under the skin.

- Blood in most animals consists of a fluid matrix, the plasma, in which cells may be suspended. In vertebrates, the erythrocytes carry oxygen to body cells and the leukocytes produce antibodies and initiate the immune response against disease-causing agents.

- Muscle tissue contains cells that have the ability to contract forcibly. Skeletal muscle, containing long cells called muscle fibres, moves body parts and maintains posture.

- Cardiac muscle, which contains short contractile cells with a branched structure, forms the heart.

- Smooth muscle consists of spindle-shaped contractile cells that form layers surrounding body cavities and ducts.

- Nervous tissue contains neurons and glial cells. Neurons communicate information between body parts in the form of electrical and chemical signals. Glial cells support the neurons or provide electrical insulation between them.

32.3 Coordination of Tissues in Organs and Organ Systems

- Organs and organ systems are coordinated to carry out vital tasks, including maintenance of internal body conditions; nutrient acquisition, processing, and distribution; waste disposal; molecular synthesis; environmental sensing and response; protection against injury and disease; and reproduction.

- In all vertebrates, the major organ systems that accomplish these tasks are the nervous, endocrine, muscular, skeletal, integumentary, circulatory, lymphatic, immune, respiratory, digestive, excretory, and reproductive systems. Many invertebrates also have these organ systems, with the exception of a lymphatic system.

32.4 Homeostasis

- Homeostasis is the process by which animals maintain their internal environment at conditions their cells can tolerate. It is a dynamic state, in which internal adjustments are made continuously to compensate for environmental changes.

- Homeostasis is accomplished by negative feedback mechanisms that include a sensor, which detects a change in an external or internal condition; an integrator, which compares the detected change with a set point; and an effector, which returns the condition to the set point if it has varied.

- Animals also have positive feedback mechanisms, in which a change in an internal or external condition triggers a response that intensifies the change, and typically does not result in homeostasis.

Questions

Self-Test Questions

1. Which tissue is a constant source of adult stem cells in a mammal?
   - a. bone marrow
   - b. pancreas
   - c. basal lamina
   - d. heart muscle
   - e. kidneys

2. The bones of an elderly woman break more easily than those of a younger person. You would surmise that with aging, the cell type that diminishes in activity is the
   - a. osteocyte.
   - b. osteoblast.
   - c. osteoclast.
   - d. chondrocyte.
   - e. fibroblast.

3. Which of the following is *not* a homeostatic response?
   - a. In a contest, a student eats an entire chocolate cake in 10 minutes. Due to hormonal secretions, his blood glucose level does not change dramatically.
   - b. The basketball players are dripping sweat at half time.
   - c. The pupils in the eyes constrict when looking at a light.
   - d. Slower breathing in sleep changes carbon dioxide and oxygen blood levels, which affect blood pH.
   - e. The brain is damaged when a fever rises above 40.5°C.

4. A decrease in body temperature causes the pituitary to release a hormone that stimulates the release of thyroxine from the thyroid gland. Thyroxine increases metabolism, generating heat. As the body temperature increases, the release of the pituitary hormone decreases and less thyroxine is released. This is an example of:
   - a. osmolarity.
   - b. environmental sensing.
   - c. integration.
   - d. positive feedback.
   - e. negative feedback.

5. The system that coordinates other organ systems is the
   - a. skeletal system.
   - b. reproductive system.
   - c. muscular system.
   - d. nervous system.
   - e. integumentary system.

Questions for Discussion

1. Astronauts lose bone mass during space travel. Why do you think this happens? To test your hypothesis, can you devise an experiment that does not involve space travel?

2. There are at least 25 known collagens. What information would you need to have and how would you use that information to propose a hypothesis that explains the way that evolution has acted to produce so many versions of the same molecule?

3. Positive feedback mechanisms are rare in animals compared with negative feedback mechanisms. Why do you think this is so?

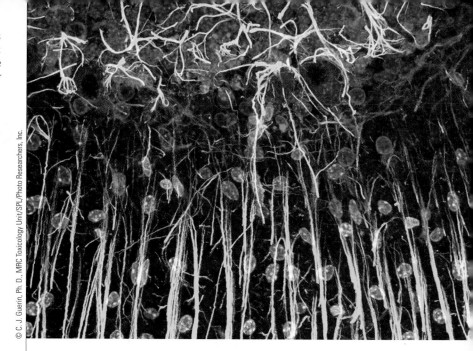

Section through the cerebellum, a part of the brain that integrates signals coming from particular regions of the body (confocal light micrograph). Neurons, the cells that send and receive signals, are red; glial cells, which provide structural and functional support for neurons, are yellow; and nuclei are purple.

© C. J. Guerin, Ph. D., MRC Toxicology Unit/SPL/Photo Researchers, Inc.

Continued on next page

33 Information Flow: Nerves, Ganglia, and Brains

WHY IT MATTERS

On a warm evening in early summer, the twilight in a garden in Montreal is punctuated by brief bursts of light from the abdomen of a flying male of the beetle *Photuris versicolor*, the "lightning bug." The flashes of light have a specific duration and come at specific intervals, constituting a code unique to that species. These visible mating calls are answered by a female perched on the vegetation below **(Figure 33.1, p. 784),** who emits flashes with the same code. The male orients himself toward the female and flies toward her. This photonic conversation continues until the male lands on the vegetation and mates with the female, who then ceases flashing. A day or two later, the mated female has begun to make eggs and again responds to flashes from males flying overhead. But now she responds to and mimics codes of flashing from males of other species of firefly. A male, lured to her by her mimicry of the flashing code for his species, lands and expects to mate but becomes prey and provides nutrition, enabling her to enhance egg production.

This "femme fatale" behaviour is a marvel of communication both within and between the beetle species. The brain of the male sends the appropriate rhythmic nervous signals to the

Figure 33.1
Photinus species female with abdomen flashing.

light-producing organ in his **abdomen**, whereas his lower nervous system controls the beating of his wings. The female's eyes detect the light flashes, and her nervous system processes the information, causing her brain to send the appropriate signals to her own light-producing organ so that she responds with the appropriate code. The male detects her signal, and his nervous system alters signals to the muscles controlling the wings so that he can fly toward the female. Mating is a complex behaviour involving coordinated movements not only of the genital apparatus but also of the other appendages. This act of mating turns off the flashing response of the female and signals the endocrine system of the female to release the hormones involved in egg production. A chemical transferred by the male in his semen acts on the brain of the female so that it no longer responds to the code for her species and causes her to mimic the codes of other species. The detection and flow of information from the environment and between the individuals, and the instantaneous analysis and processing of that information to produce specific behaviour, are astounding, even in relatively simple animals. Much of this information flow is mediated by the nervous system, but the endocrine system may also be involved in information flow.

In this chapter, we first examine the properties of the cells that make up the nervous system responsible for the reception, transmission, and analysis of the information. These functions result from the activities of only two major cell types: *neurons* and *glial cells*. In most animals, these cells are organized into complex networks called *nervous systems,* and we describe how these networks are organized into *ganglia* and brains.

33.1 Neurons and Their Organization in Nervous Systems: An Overview

An animal is constantly receiving stimuli from both internal and external sources. **Neural signalling**—communication by neurons—is the process by which an animal responds appropriately to a stimulus. In most animals, the four components of neural signalling are *reception, transmission, integration,* and *response.* **Reception**, the detection of a stimulus, is performed by **neurons**, the cellular components of nervous systems, and by specialized sensory receptors such as those in the eye and skin. **Transmission** is the sending of a message along a neuron and then to another neuron or to a muscle or gland. **Integration** is the sorting and interpretation of neural messages and determination of the appropriate response(s). **Response** is the "output" or action resulting from the integration of neural messages. For a *P. versicolor* male flying at dusk, for example, sensors in the eye (see Chapter 34) detect flashes of light, and this information is transmitted to the brain, where it is integrated with internal information about the positions of its wings, gravity, and other factors related to its orientation. This integration results in outputs along nerves controlling the flight apparatus to turn the male toward the source of the flashing.

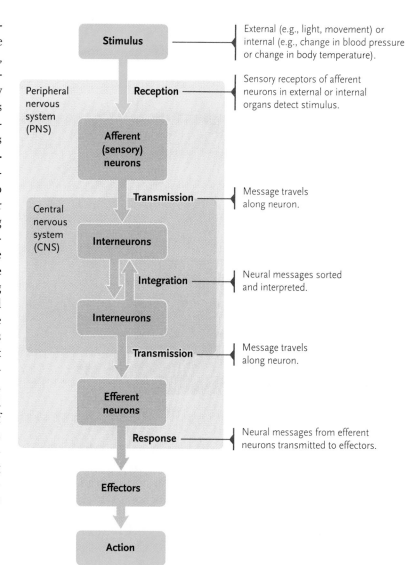

Figure 33.2
Neural signalling: information-processing steps in the nervous system

33.1a Neurons Are Cells Specialized for the Reception and Transmission of Informational Signals

Neural signalling involves three functional classes of neurons **(Figure 33.2)**. **Afferent neurons** (also called **sensory neurons**) transmit stimuli collected by sensory receptors on those neurons to **interneurons**, which integrate the information to formulate an appropriate response. In humans and some other primates, 99% of neurons are interneurons. **Efferent neurons** carry the signals indicating a response away from the interneuron networks to the **effectors**, the muscles and glands. Efferent neurons that carry signals to skeletal muscle are called **motor neurons**. The information-processing steps in the nervous system can be summarized as (1) reception by sensory receptors on afferent neurons; (2) transmission of messages by afferent neurons to interneurons; (3) integration of neural messages in interneurons; and (4) response by transmission of neural messages by efferent neurons to effectors where action appropriate to the stimulus occurs.

Neurons vary widely in shape and size. All have an enlarged cell body and two types of extensions or

processes, dendrites and axons **(Figure 33.3, p. 786)**. The **cell body**, which contains the nucleus and the majority of cell organelles, synthesizes most of the proteins, carbohydrates, and lipids of the neuron. Dendrites and axons conduct electrical signals that are produced by ions flowing down concentration gradients through channels in the plasma membrane of the neuron. **Dendrites** receive the signals and transmit them toward the cell body. They are generally highly branched, forming a treelike outgrowth at one end of the neuron (*dendros* = tree). **Axons** (*axon* = axis) conduct signals away from the cell body to another neuron or an effector. Neurons typically have a single axon that arises from a junction with the cell body called an **axon hillock**. The axon has branches at its tip that end as small, buttonlike swellings called **axon terminals**. The more terminals contacting a neuron, the greater its capacity to integrate incoming information.

Connections between the axon terminals of one neuron and the dendrites or cell body of a second neuron form the basic elements of a **neuronal circuit**. A typical neuronal circuit contains an afferent neuron,

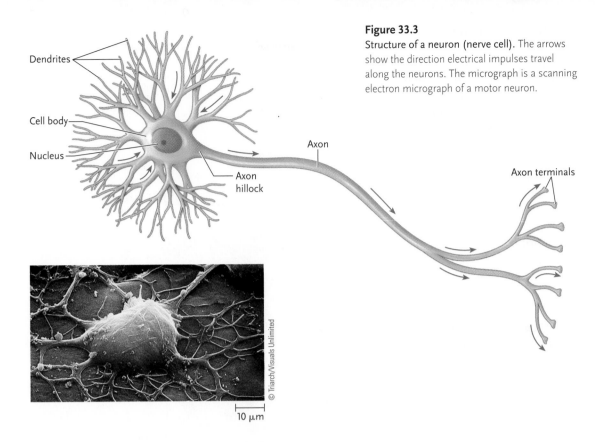

Figure 33.3
Structure of a neuron (nerve cell). The arrows show the direction electrical impulses travel along the neurons. The micrograph is a scanning electron micrograph of a motor neuron.

Dendrites

Cell body

Nucleus

Axon hillock

Axon

Axon terminals

© Triarch/Visuals Unlimited

10 μm

one or more interneurons, and an efferent neuron. Interneurons may receive input from several axons and may, in turn, connect to other interneurons and several efferent neurons. In this way, circuits combine into networks that interconnect the parts of the nervous system. In vertebrates, the afferent (sensory) neurons and efferent neurons collectively form the *peripheral nervous system (PNS)*. The interneurons form the brain and spinal cord, called the *central nervous system (CNS)*. As depicted in Figure 33.2, afferent (carrying toward) information is ultimately transmitted to the CNS, where efferent (carrying away) information is initiated. The nervous systems of most invertebrates are also divided into central and peripheral divisions.

33.1b Neurons Are Supported Structurally and Functionally by Glial Cells

Figure 33.4
An astrocyte from a rat brain.

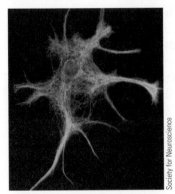

Society for Neuroscience

Glial cells are non-neuronal cells that provide nutrition and support to neurons. One type, called **astrocytes** because they are star-shaped **(Figure 33.4)**, were formerly thought to play only a supporting role in the CNS by maintaining ion concentrations in the interstitial fluid surrounding the neurons. More recently, however, scientists have realized that in vertebrates and some invertebrates, astrocytes communicate with neurons and may influence their activity.

Two other types of glial cells, **oligodendrocytes** in the CNS and **Schwann**

cells in the PNS, form tightly wrapped layers of plasma membrane, called myelin sheaths, around axons **(Figure 33.5)**. These myelin sheaths act as electrical insulators due to the membrane's high lipid content. The gaps between Schwann cells, called **nodes of Ranvier**, expose the axon membrane directly to extracellular fluids. This arrangement of insulated stretches of the axon punctuated by gaps speeds the rate at which electrical impulses move along the axons they protect.

Unlike most neurons, glial cells retain the capacity to divide throughout the life of the animal. This capacity allows glial tissues to replace damaged or dead cells but also makes them the source of almost all brain tumours, which are produced when regulation of glial cell division is lost.

33.1c Neurons Communicate via Synapses

A **synapse** (*synapsis* = juncture) is a site where a neuron makes a communicating connection with either another neuron or an effector such as a muscle fibre or gland. On one side of the synapse is the axon terminal of a **presynaptic cell**, the neuron that transmits the signal. On the other side is the dendrite or cell body of a **postsynaptic cell**, the neuron or the surface of an effector that receives the signal. Communication across a synapse may occur by the direct flow of an electrical signal or by means of a **neurotransmitter**, a chemical released by an axon terminal at a synapse.

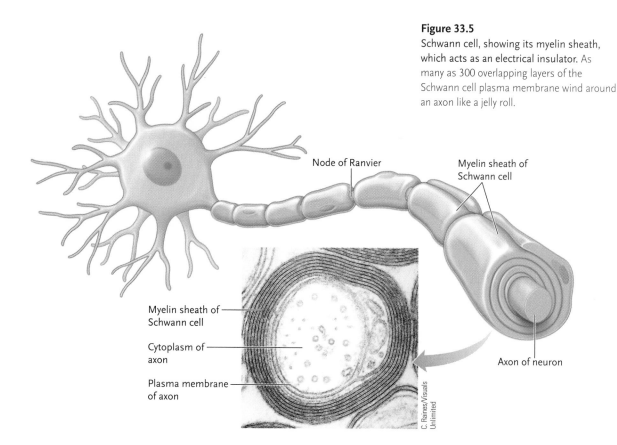

Figure 33.5
Schwann cell, showing its myelin sheath, which acts as an electrical insulator. As many as 300 overlapping layers of the Schwann cell plasma membrane wind around an axon like a jelly roll.

Node of Ranvier

Myelin sheath of Schwann cell

Myelin sheath of Schwann cell

Cytoplasm of axon

Plasma membrane of axon

C. Raines/Visuals Unlimited

Axon of neuron

In **electrical synapses**, the plasma membranes of the presynaptic and postsynaptic cells are in direct contact **(Figure 33.6a, p. 788).** When an electrical impulse arrives at the axon terminal, gap junctions (see Chapter 32) allow ions to flow directly between the two cells, leading to unbroken transmission of the electrical signal. Electrical synapses are useful for two types of functions.

- They allow for very rapid transmission. They were first discovered in the nervous system of crayfish, where they are involved in the rapid movements for escape from predators.
- They allow for synchronous activity in a group of neurons. For example, the neurons controlling the secretion of hormones from the hypothalamus of mammals are connected by electrical synapses, thus ensuring a coordinated burst of secretion of some hormones.

The vast majority of vertebrate neurons communicate by means of neurotransmitters **(Figure 33.6b, p. 788).** In these **chemical synapses**, the plasma membranes of the presynaptic and postsynaptic cells are separated by a narrow gap, about 25 nm wide, called the **synaptic cleft.** When an electrical impulse arrives at an axon terminal, it causes the release of a neurotransmitter into the synaptic cleft. The neurotransmitter diffuses across the synaptic cleft and binds to a receptor in the plasma membrane of the postsynaptic cell. If enough neurotransmitter molecules bind to these receptors, the postsynaptic cell generates a new

electrical impulse that travels along its axon to reach a synapse with the next neuron or effector in the circuit. A chemical synapse is more than a simple on–off switch because many factors can influence the generation of a new electrical impulse in the postsynaptic cell, including neurotransmitters that inhibit that cell rather than stimulate it. The balance of stimulatory and inhibitory effects in chemical synapses contributes to the integration of incoming information in a receiving neuron.

STUDY BREAK

1. Distinguish between the functions and locations of afferent neurons, efferent neurons, and interneurons.
2. What are the differences between an electrical synapse and a chemical synapse?

33.2 Signal Conduction by Neurons

All animal cells have a **membrane potential**, a separation of positive and negative charges across the plasma membrane. Outside the cell is positive, and inside the cell is negative. This charge separation in part produces voltage, an electrical potential difference, across the plasma membrane.

a. Electrical synapse

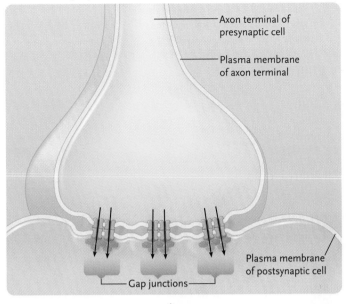

b. Chemical synapse

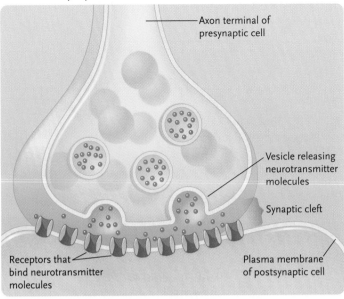

In an electrical synapse, the plasma membranes of the presynaptic and postsynaptic cells make direct contact. Ions flow through gap junctions that connect the two membranes, allowing impulses to pass directly to the postsynaptic cell.

Figure 33.6
The two types of synapses by which neurons communicate with other neurons or effectors.

In a chemical synapse, the plasma membranes of the presynaptic and postsynaptic cells are separated by a narrow synaptic cleft. Neurotransmitter molecules diffuse across the cleft and bind to receptors in the plasma membrane of the postsynaptic cell. The binding opens channels to ion flow that may generate an impulse in the postsynaptic cell.

The membrane potential is caused by the uneven distribution of Na^+ and K^+ inside and outside the cell. As you learned in Chapter 5, plasma membranes are *selectively* permeable in that they allow some ions but not others to move across the membrane through protein channels embedded in the phospholipid bilayer. Plasma membrane–embedded Na^+/K^+ active transport pumps use energy from ATP hydrolysis to pump simultaneously three Na^+ out of the cell for every two K^+ pumped in, generating a higher Na^+ concentration outside the cell than inside, and a higher K^+ concentration inside the cell than outside. This explains the positive charge outside the cell. The inside of the cell

is negatively charged because the cell also contains many **anions** (negatively charged molecules), such as proteins, amino acids, and nucleic acids.

For most cells, the membrane potential remains unchanged. However, neurons and muscle cells use the membrane potential in a specialized way. In response to electrical, chemical, mechanical, and certain other types of stimuli, their membrane potential changes rapidly and transiently. Cells with this property are said to be *excitable*. Excitability, produced by a sudden flow of ions across the plasma membrane, is the basis for nerve impulse generation.

33.2a Resting Potential Is the Unchanging Membrane Potential of an Unstimulated Neuron

The membrane of a neuron that is not conducting an impulse exhibits a steady negative membrane potential called the **resting potential** because the neuron is at rest. The resting potential has been measured at about −70 mV in isolated neurons **(Figure 33.7)**. A neuron exhibiting a resting potential is said to be *polarized*.

The distribution of ions inside and outside of an axon that produces the resting potential is shown in **Figure 33.8.** As described earlier, the Na^+/K^+ pump creates the imbalance of Na^+ and K^+ inside and outside the cell: the concentration of positively charged anions within the cell results in the inside being negatively charged and the outside being positively charged. As we will see in the following discussion of the changes in a neuron that occur when it is stimulated, the

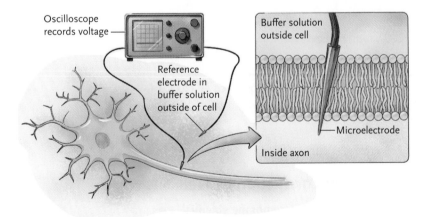

Figure 33.7
Membrane potential is measured by inserting a very fine electrode through the cell membrane of the axon. This electrode and another in the fluid bathing the neuron are connected to an oscilloscope, which measures the potential difference (volts) between the two electrodes and which can track the very rapid changes during an action potential.

Figure 33.8

The distribution of ions inside and outside an axon that produces the resting potential, –70 mV. The distribution of ions that do not directly affect the resting potential, such as Cl⁻, is not shown. The voltage-gated ion channels open and close when the membrane potential changes.

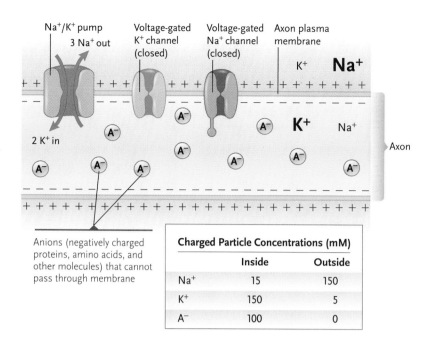

Anions (negatively charged proteins, amino acids, and other molecules) that cannot pass through membrane

Charged Particle Concentrations (mM)		
	Inside	Outside
Na⁺	15	150
K⁺	150	5
A⁻	100	0

voltage-gated ion channels for Na⁺ and K⁺ open and close when the membrane potential changes.

33.2b The Membrane Potential Changes from Negative to Positive during an Action Potential

When a neuron conducts an electrical impulse, there is an abrupt and transient change in membrane potential; this is called the **action potential**. An action potential begins as a stimulus that causes positive charges from outside the neuron to flow inward, making the cytoplasmic side of the membrane less negative **(Figure 33.9)**.

As the membrane potential becomes less negative, the membrane (which was polarized at rest) becomes **depolarized**. Depolarization proceeds relatively slowly until it reaches a level known as the **threshold potential**, about –50 to –55 mV in isolated neurons. Once the threshold is reached, the action potential fires, which causes the membrane potential to suddenly increase. In less than 1 ms (millisecond, one-thousandth of a second), it rises so high that the inside of the plasma membrane becomes positive because of an influx of positive ions across the cell membrane, momentarily reaching a value of +30 mV or more. The potential then falls, in many cases dropping to about –80 mV before rising again to the resting potential. When the potential is below the resting value, the membrane is said to be **hyperpolarized**. The entire change, from initiation of the action potential to the return to the resting potential, takes less than 5 ms in the fastest neurons. Action potentials take the same basic form in neurons of all types, with differences in the values of the resting potential and the peak of the action potential and in the time required to return to the resting potential.

All stimuli cause depolarization of a neuron, but an action potential is produced only if the stimulus is strong enough to cause the depolarization to reach the threshold. This is referred to as the **all-or-nothing principle**; once triggered, the changes in membrane potential take place independently of the strength of the stimulus.

Beginning at the peak of an action potential, the membrane enters a **refractory period** of a few milliseconds, during which the threshold required for generation of an action potential is much higher than normal. The refractory period lasts until the membrane has stabilized at the resting potential. As we will

see, the refractory period keeps impulses travelling in a one-way direction in neurons.

33.2c The Action Potential Is Produced by Ion Movements through the Plasma Membrane

The action potential is produced by movements of Na⁺ and K⁺ through the plasma membrane that are controlled by specific **voltage-gated ion channels**, membrane-embedded proteins that open and close as the membrane potential changes (see Figure 33.8).

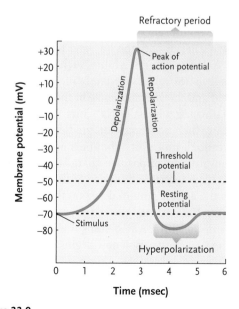

Figure 33.9

Changes in membrane potential during an action potential.

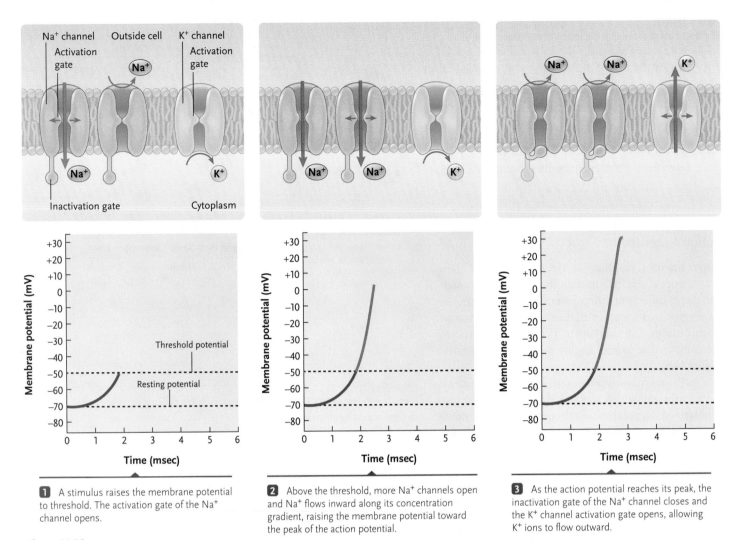

1 A stimulus raises the membrane potential to threshold. The activation gate of the Na⁺ channel opens.

2 Above the threshold, more Na⁺ channels open and Na⁺ flows inward along its concentration gradient, raising the membrane potential toward the peak of the action potential.

3 As the action potential reaches its peak, the inactivation gate of the Na⁺ channel closes and the K⁺ channel activation gate opens, allowing K⁺ ions to flow outward.

Figure 33.10

Changes in voltage-gated Na^+ and K^+ channels that produce the action potential.

Voltage-gated Na^+ channels have two gates, an *activation gate* and an *inactivation gate*, whereas voltage-gated K^+ channels have one gate, an *activation gate*.

Figure 33.10 shows how the two voltage-gated ion channels operate when generating an action potential. When the membrane is at the resting potential, the activation gates of both the Na^+ and K^+ channels are closed. A depolarizing stimulus, such as neurotransmitter substance, raises the membrane potential to the threshold and the activation gate of the Na^+ channels opens, allowing a burst of Na^+ ions to flow into the axon along their concentration gradient. Once above the threshold, more Na^+ channels open, causing a rapid inward flow of positive charges that raises the membrane potential to the peak of the action potential. As the action potential peaks, the inactivation gate of the Na^+ channel closes (resembling putting a stopper in the sink), which stops the inward flow of Na^+. The refractory period now begins.

At the same time, the activation gates of the K^+ channels begin to open, allowing K^+ ions to flow rapidly outward in response to their concentration gradient. The K^+ ions contribute to the refractory period and compensate for the inward movement of Na^+

ions, returning the membrane to the resting potential. As the resting potential is reestablished, the activation gates of K^+ channels close, as do those of Na^+ channels, and the inactivation gates of Na^+ channels open. These events end the refractory period and ready the membrane for another action potential.

In some neurons, closure of the gated K^+ channels lags, and K^+ continues to flow outward for a brief time after the membrane returns to the resting potential. This excess outward flow causes the hyperpolarization shown in Figure 33.9 and Figure 33.10 (step 6), in which the membrane potential dips briefly below the resting potential.

At the end of an action potential, the membrane potential has returned to its resting state, but the ion distribution has been changed slightly. That is, some Na^+ has entered the cell, and some K^+ has left the cell. Actually, relatively few of the total number of Na^+ and K^+ ions change locations during an action potential. Hence, additional action potentials can occur without the need to completely correct the altered ion distribution. In the long term, the Na^+/K^+ active transport pumps restore the Na^+ and K^+ to their original locations.

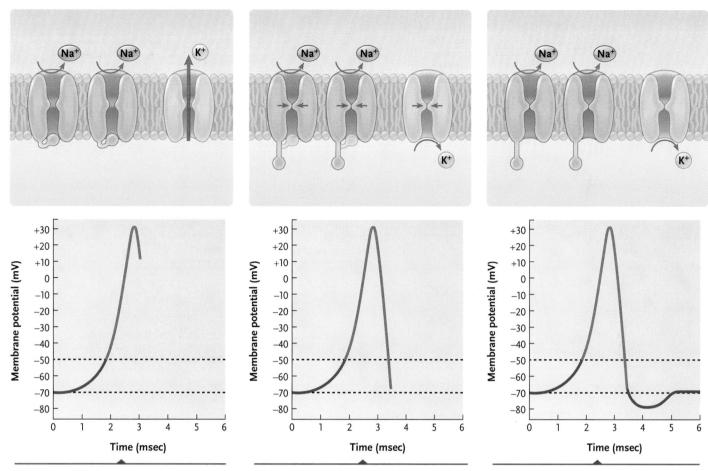

4 The outward flow of K$^+$ along its concentration gradient causes the membrane potential to begin to fall.

5 As the membrane potential reaches the resting value, the activation gate of the Na$^+$ channel closes and the inactivation gate opens. The K$^+$ activation gate also closes.

6 Closure of the K$^+$ activation gate stabilizes the membrane potential at the resting value.

33.2d Nerve Impulses Move by Propagation of Action Potentials

Once an action potential is initiated at the dendrite end of the neuron, it passes along the surface of a nerve or muscle cell as an automatic wave of depolarization. It travels away from the stimulation point without requiring further triggering events **(Figure 33.11, p. 792)**. This is called **propagation** of the action potential. In a segment of an axon generating an action potential, the outside of the membrane becomes temporarily negative and the inside positive. Because opposites attract, as the region outside becomes negative, local current flow occurs between the area undergoing an action potential and the adjacent downstream inactive area, both inside and outside the membrane (see arrows, Figure 33.11). This current flow makes nearby regions of the axon membrane less positive on the outside and more positive on the inside; in other words, the membrane of these adjacent regions depolarizes.

The depolarization is large enough to push the membrane potential past the threshold, opening the voltage-gated Na$^+$ and K$^+$ channels and starting an action potential in the downstream adjacent region.

In this way, each segment of the axon stimulates the next segment to fire, and the action potential moves rapidly along the axon as a nerve impulse.

The refractory period keeps an action potential from reversing direction at any point along an axon; only the region in front of the action potential can fire. The refractory period results from the properties of the voltage-gated ion channels. That is, once they are opened to their activated state, the upstream voltage-gated ion channels need time to reset to their original positions before they can open again. Therefore, only downstream voltage-gated ion channels are able to open, ensuring the one-way movement of the action potential along the axon toward the axon terminals. By the time the refractory period ends in a membrane segment that has just fired an action potential, the action potential has moved too far away to cause a second action potential to develop in the same segment.

The magnitude of an action potential stays the same as it travels along an axon, even where the axon branches at its tips. Thus, the propagation of an action potential resembles a burning fuse, which burns with the same intensity along its length and along

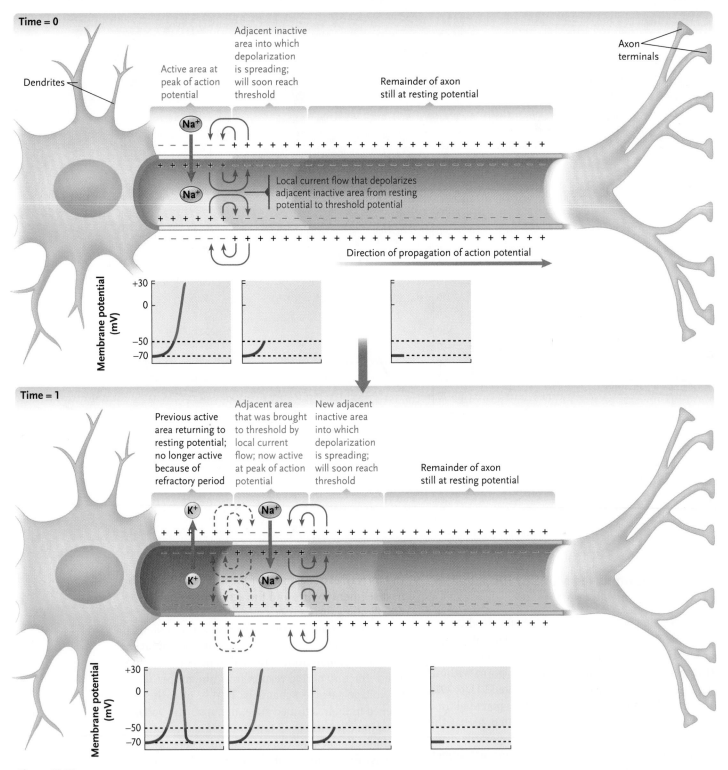

Figure 33.11

Propagation of an action potential along an unmyelinated axon by ion flows between a firing segment and an adjacent unfired region of the axon. Each firing segment induces the next to fire, causing the action potential to move along the axon.

any branches, once it is lit at one end. Unlike a fuse, however, an axon can fire another action potential of the same intensity within a few milliseconds after an action potential passes through.

The all-or-nothing principle of action potential generation means that the intensity of a stimulus is reflected in the *frequency* of action potentials rather than the size of the action potential. The greater the stimulus, the more action potentials per second, up to a limit depending on the axon type. For most neuron types, the limit lies between 10 and 100 action potentials per second.

33.2e Saltatory Conduction Increases Propagation Rate in Small-Diameter Axons

In the propagation pattern shown in Figure 33.11, an action potential spreads along every patch of the membrane along the length of the axon. The rate of conduction increases with the diameter of the axon. Some specialized axons with very large diameters occur in invertebrates such as lobsters, earthworms, and squids, as well as a few marine fishes. Giant axons typically carry signals that produce an escape or withdrawal response, such as the sudden flexing of the tail (abdomen) in lobsters that propels the animal backward. The largest known axons, 1.7 mm in diameter, occur in fanworms (Phylum Annelida, Class Polychaeta; see Figure 26.30a). The signals they carry contract a muscle that retracts the fanworm's body into a protective tube when the animal is threatened. The giant axons of the squid were used in the early experiments that led to the current conceptual model of the axon.

Although large-diameter axons can conduct impulses as rapidly as 25 m/s (over twice the speed of the world record 100-m race), they take up a great deal of space. In the jawed vertebrates, **saltatory conduction** (*saltere* = to leap) allows action potentials to "hop" rapidly along axons instead of burning smoothly like a fuse.

Saltatory conduction depends on the gaps in the insulating myelin sheath that surrounds many axons. These gaps, known as nodes of Ranvier, expose the axon membrane to extracellular fluids. Voltage-gated Na^+ and K^+ channels crowded into the nodes allow action potentials to develop at these positions **(Figure 33.12).** The inward movement of Na^+ ions produces depolarization, but the excess positive ions are unable to leave the axon through the membrane regions covered by the myelin sheath. Instead, they diffuse rapidly to the next node, where they cause depolarization, inducing an action potential at that node. As this mechanism repeats, the action potential jumps rapidly along the axon from

Figure 33.12
Saltatory conduction of the action potential by a myelinated axon. The action potential jumps from node to node, greatly increasing the speed at which it travels along the axon.

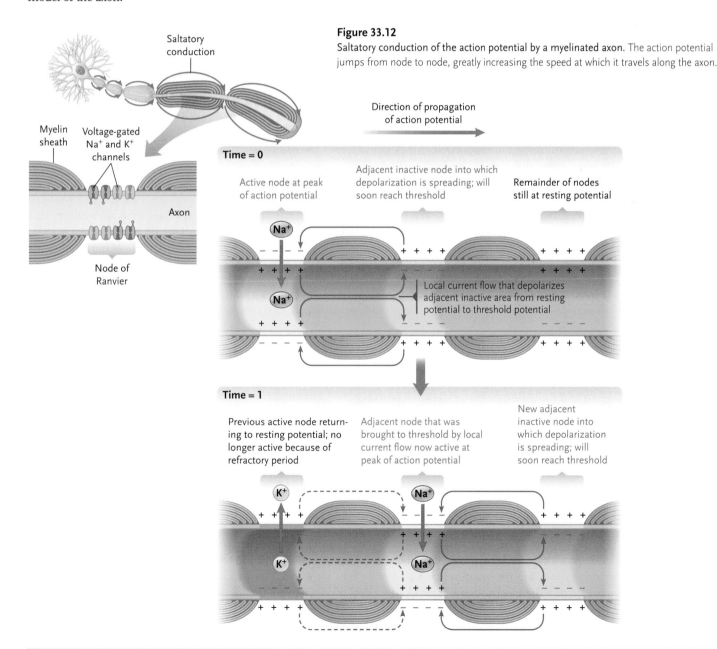

node to node. Saltatory conduction proceeds at rates up to 130 m/s, whereas an unmyelinated axon of the same diameter conducts action potentials at about 1 m/s.

Saltatory conduction allows thousands to millions of fast-transmitting axons to be packed into a relatively small diameter. For example, in humans, the 3-mm diameter optic nerve leading from the eye to the brain is packed with more than a million axons. If those axons were unmyelinated, each would have to be about 100 times thicker to conduct impulses at the same velocity, producing an optic nerve about 300 mm in diameter.

Among invertebrates, Schwann cells also form concentric coatings around many nerves, but the layers are not so compact, leaving some cytoplasm in the Schwann cells and some intercellular space between the layers. Nevertheless, fully myelinated fibres occur in oligochaete annelids and some crustaceans, complete with gaps to permit saltatory transmission. The occurrence of myelin in some protostome invertebrates and its absence from the lower vertebrates suggests that this important mechanism has evolved more than once, presenting another example of convergent evolution. The embryonic origin of Schwann cells is different in vertebrates and invertebrates, confirming the independent evolution of myelination in vertebrates and invertebrates.

The disease *multiple sclerosis* (*sclero* = hard) underscores the importance of myelin sheaths to the operation of the vertebrate nervous system. In this disease, myelin is attacked by the immune system and is progressively lost from axons and replaced by hardened scar tissue. The changes block or slow the transmission of action potentials, producing numbness, muscular weakness, faulty coordination of movements, and paralysis that worsens as the disease progresses. Although clear genetic factors are involved, the environment also plays a role: the incidence of the disease increases with the distance from the equator. The incidence in Canada, 2.4 people per 1000 population, is one of the highest in the world.

STUDY BREAK

1. What mechanism results in a membrane potential?
2. How is the directionality of an action potential achieved?
3. How is the intensity of action potential signals measured?

33.3 Conduction across Chemical Synapses

Action potentials are transmitted directly across electrical synapses, but they cannot jump across the synaptic cleft in a chemical synapse. Instead, the arrival of an action potential causes neurotransmitter molecules synthesized in the cell body of the neuron to be released by the plasma membrane of the axon terminal, called the **presynaptic membrane (Figure 33.13)**. The neurotransmitter diffuses across the cleft and alters ion conduction by activating *ligand-gated ion channels* in the **postsynaptic membrane**, the plasma membrane of the postsynaptic cell. **Ligand-gated ion channels** are channels that open or close when a specific chemical, the ligand, binds to the channel.

Neurotransmitters work in one of two ways. **Direct neurotransmitters** bind directly to a ligand-gated ion channel in the postsynaptic membrane, which opens or closes the channel gate and alters the flow of a specific ion or ions in the postsynaptic cell. The time between arrival of an action potential at an axon terminal and alteration of the membrane potential in the postsynaptic cell may be as little as 0.2 ms.

Indirect neurotransmitters work more slowly (on the order of hundreds of milliseconds). They act as *first messengers*, binding to G protein–coupled receptors (see Chapter 8) in the postsynaptic membrane, which activates the receptors and triggers the generation of a *second messenger* such as cyclic AMP or other processes. The cascade of second-messenger reactions opens or closes ion-conducting channels in the postsynaptic membrane. Indirect neurotransmitters typically have effects that may last for minutes or hours. Some substances can act as either direct or indirect neurotransmitters, depending on the types of receptors they bind to in the receiving cell. Not all of the chemicals released at nerve terminals directly stimulate the postsynaptic neuron to fire. Some may inhibit the neuron from firing, whereas others may enhance the action of other transmitters. Actions that modify the effects of other transmitters may not be confined to the single synaptic cleft but may act to coordinate groups of neurons. These transmitters are sometimes called neuromodulators.

The time required for the release, diffusion, and binding of neurotransmitters across chemical synapses delays transmission compared with the almost instantaneous transmission of impulses across electrical synapses. However, communication through chemical synapses allows neurons to receive inputs from hundreds to thousands of axon terminals at the same time. Some neurotransmitters have stimulatory effects, whereas others have inhibitory effects. All of the information received at a postsynaptic membrane is integrated to produce a response that consists of the receptor neuron firing with a particular frequency.

33.3a Neurotransmitters Are Released by Exocytosis

Neurotransmitters are stored in secretory vesicles, called **synaptic vesicles**, in the cytoplasm of an axon terminal. The arrival of an action potential at the terminal releases the neurotransmitters by *exocytosis*: the vesicles fuse with the presynaptic membrane and release the neurotransmitter molecules into the synaptic cleft.

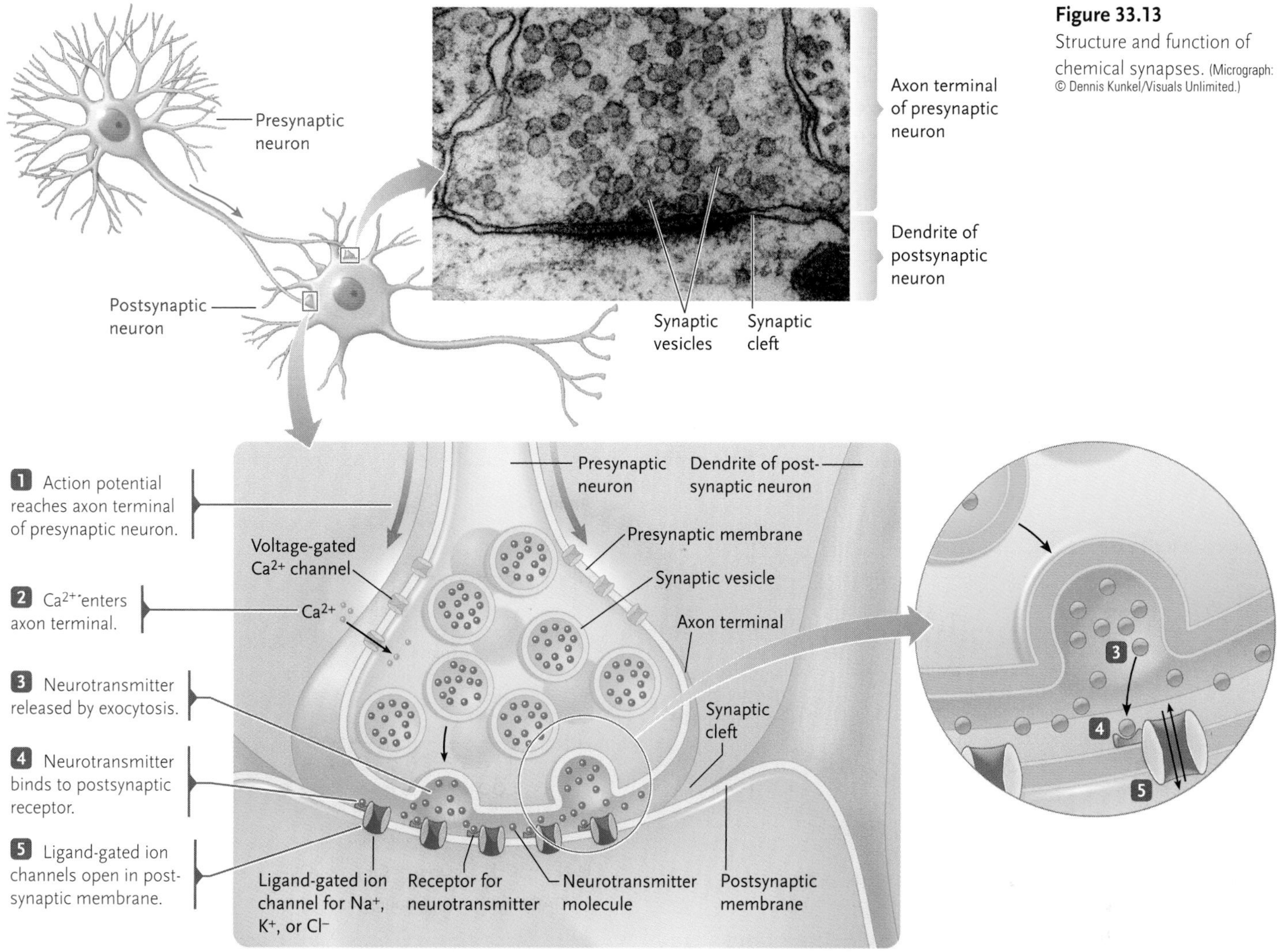

Axon terminal of presynaptic neuron

Dendrite of postsynaptic neuron

Synaptic vesicles

Synaptic cleft

Presynaptic neuron

Postsynaptic neuron

1 Action potential reaches axon terminal of presynaptic neuron.

2 Ca^{2+} enters axon terminal.

3 Neurotransmitter released by exocytosis.

4 Neurotransmitter binds to postsynaptic receptor.

5 Ligand-gated ion channels open in post-synaptic membrane.

Presynaptic neuron

Dendrite of post-synaptic neuron

Voltage-gated Ca^{2+} channel

Ca^{2+}

Presynaptic membrane

Synaptic vesicle

Axon terminal

Synaptic cleft

Ligand-gated ion channel for Na^+, K^+, or Cl^-

Receptor for neurotransmitter

Neurotransmitter molecule

Postsynaptic membrane

The release of synaptic vesicles depends on voltage-gated Ca^{2+} channels in the plasma membrane of an axon terminal (see Figure 33.13). Ca^{2+} ions are constantly pumped out of all animal cells by an active transport protein in the plasma membrane, keeping their concentration higher outside than inside. As an action potential arrives, the change in membrane potential opens the Ca^{2+} channel gates in the axon terminal, allowing Ca^{2+} to flow back into the cytoplasm. The rise in Ca^{2+} concentration triggers a protein in the membrane of the synaptic vesicle that allows the vesicle to fuse with the plasma membrane, releasing neurotransmitter molecules into the synaptic cleft.

Each action potential arriving at a synapse typically causes approximately the same number of synaptic vesicles to release their neurotransmitter molecules. For example, arrival of an action potential at one type of synapse causes about 300 synaptic vesicles to release a neurotransmitter called acetylcholine. Each vesicle contains about 10 000 molecules of the neurotransmitter, giving a total of some 3 million acetylcholine molecules released into the synaptic cleft by each arriving action potential.

When a stimulus is no longer present, action potentials are no longer generated. When action potentials stop arriving at the axon terminal, the voltage-gated Ca^{2+} channels in the axon terminal close, and the Ca^{2+} in the axon cytoplasm is quickly pumped to the outside. The drop in cytoplasmic Ca^{2+} stops vesicles from fusing with the presynaptic membrane, and no further neurotransmitter molecules are released. Any free neurotransmitter molecules remaining in the cleft are either broken down by enzymes in the cleft or reuptake occurs, meaning that they are pumped back into the axon terminals or into glial cells by active transport. Transmission of impulses across the synaptic cleft ceases within milliseconds after action potentials stop arriving at the axon terminal.

33.3b Most Neurotransmitters Alter Flow through Na^+ or K^+ Channels

Most neurotransmitters work by opening or closing membrane-embedded ligand-gated ion channels that conduct Na^+ or K^+ across the postsynaptic membrane, although some regulate chloride ions (Cl^-). The resulting ion flow may stimulate or inhibit the generation of action potentials by the postsynaptic cell. If Na^+ channels are opened, the inward Na^+ flow brings the membrane potential of the postsynaptic cell toward the threshold (the membrane becomes depolarized). If K^+ channels are opened, the outward flow of K^+ has the opposite effect (the membrane becomes hyperpolarized). The combined effects of the various stimulatory and inhibitory neurotransmitters at all the chemical synapses of a postsynaptic neuron or muscle cell determine whether the postsynaptic cell triggers an action potential.

33.3c Many Different Molecules Act as Neurotransmitters

Nearly 100 different substances are known or suspected to be neurotransmitters. Most of them are relatively small molecules that diffuse rapidly across the synaptic cleft. Some axon terminals release only one type of neurotransmitter, whereas others release several types. Depending on the type of receptor to which it binds, the same neurotransmitter may stimulate or inhibit the generation of action potentials in the postsynaptic cell. **Figure 33.14** depicts some examples of neurotransmitters.

Acetylcholine acts as a neurotransmitter in both invertebrates and vertebrates. In vertebrates, it acts as a direct neurotransmitter between neurons and muscle cells and as an indirect neurotransmitter between neurons carrying out higher brain functions such as memory, attention, perception, and learning. Acetylcholine-releasing neurons in the brain degenerate in people who develop Alzheimer disease, in which memory, speech, and perceptual abilities decline.

Figure 33.14

Chemical structures of the major neurotransmitter types.

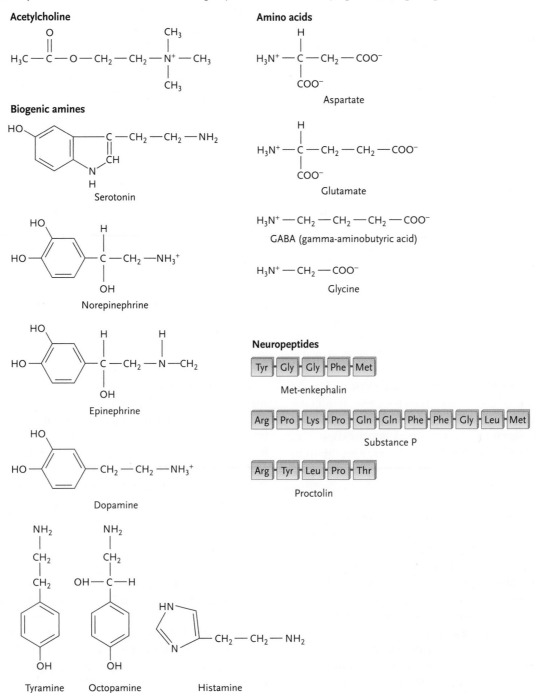

Botulinum Toxin: From Poison to Therapy

Botulinum toxin is the product of the bacterium *Clostridium botulinum*, a common organism that thrives in anaerobic environments, such as preserved foods. The toxin is a protein that is destroyed by heat, but if it is ingested, it is extraordinarily toxic— 1 000 000 times more toxic than strychnine. The protein can exist as up to seven serotypes, A to G. The protein is produced as a 150 kDa (1000 Daltons) molecule that is cleaved into a 100 kDa toxin and a 50 kDa protease. The heavy chain toxin acts at acetylcholine-mediated synapses (nerve–muscle junctions in vertebrates) by promoting the entry of the light chain protease into the neuron at the synapse, where the enzyme attacks one of the molecules essential for the release of acetycholine. The effect is long lasting: up to several months. Recovery from the effect has been hypothesized to involve the sprouting of new axon terminals.

These properties have been used since 1989 to treat a number of disorders that involve muscle spasms, such as strabismus ("crossed eyes"), by injecting minute quantities of the A serotype directly into the muscles. Ophthalmologists using the toxin in this way noted that "frown lines" around the eyes disappeared as a side effect, and the Botox cosmetic industry was born.

Acetylcholine is the target of many natural and artificial poisons. Curare, a plant extract used as an arrow poison by some indigenous peoples of South America, blocks muscle contraction and produces paralysis by competing directly with acetylcholine for binding sites in synapses that control muscle cells. Atropine, an ingredient in the drops an eye doctor uses to dilate your **pupils**, is also a plant extract; it relaxes the iris muscles by blocking their acetylcholine receptors. Nicotine also binds to acetylcholine receptors but acts as a stimulant by turning the receptors on rather than off.

Several amino acids operate as direct neurotransmitters in the CNS of vertebrates and in the nerve–muscle synapses of insects and crustaceans. *Glutamate* and *aspartate* stimulate action potentials in postsynaptic cells. They are directly involved in brain functions such as memory and learning, as well as some other functions. *Gamma-aminobutyric acid* (*GABA*), a derivative of glutamate, acts as an inhibitor by opening Cl^- channels in postsynaptic membranes. *Glycine* is also an inhibitor.

Other substances can block the operation of these neurotransmitters. For example, tetanus toxin, released by the bacterium *Clostridium tetani*, blocks GABA release in synapses that control muscle contraction. The body muscles contract so forcibly that the body arches painfully and the teeth become tightly clenched, giving the condition its common name of lockjaw. Once the effects extend to respiratory muscles, the victim quickly dies. The disease is entirely preventable, thanks to vaccination with inactivated tetanus toxin.

Biogenic amines are derived from amino acids. Norepinephrine, epinephrine, dopamine, tyramine, and octopamine are all derived from tyrosine. Serotonin is derived from tryptophan, and histamine is derived from histidine. Serotonin, histamine, and dopamine function in both vertebrates and invertebrates. Epinephrine and norepinephrine are characteristic of vertebrates, whereas octopamine and tyramine function in invertebrates.

These amines function primarily in the CNS, and in humans they have been associated with a diversity of brain activities such as consciousness, memory, mood, blood pressure, and sleep. For example, cocaine binds to the transporters for active reuptake of neurotransmitters such as norepinephrine, dopamine, and serotonin from the synaptic cleft. As a result, the concentrations of these neurotransmitters increase in the synapses, leading to amplification of their natural effects. That is, the affected neurons produce the symptoms characteristic of cocaine use: high energy from norepinephrine, euphoria from dopamine, and feelings of confidence from serotonin. Parkinson disease, in which there is progressive loss of muscle control, results from degeneration of dopamine-releasing neurons in regions of the brain coordinating movement.

Neuropeptides are short chains of two or more amino acids that act as indirect neurotransmitters in the central and peripheral nervous systems of both vertebrates and invertebrates. More than 50 neuropeptides are now known for vertebrates and as many as 2000 for invertebrates. Neuropeptides that act as neurotransmitters are also released into the general body circulation as peptide hormones. An example is the peptide proctolin, which occurs only in invertebrates.

Neuropeptides called *endorphins* ("endogenous morphines") are released during periods of pleasurable experience, such as eating or sexual intercourse, or physical stress, such as childbirth or extended physical exercise. These neurotransmitters have the opiate-like property of reducing pain and inducing euphoria, well known to exercise buffs as a pleasant by-product of their physical efforts. Most endorphins act on the PNS and effectors such as muscles, but *enkephalins*, a subclass of the endorphins, bind to particular receptors in the CNS. Morphine, a potent drug extracted from the opium poppy, blocks the sensation of pain and produces a sensation of well-being by binding to the same enkephalin receptors in the brain.

Another neuropeptide associated with pain response is *substance P*, which is released by special neurons in the spinal cord. Its effect is to increase messages associated with intense, persistent, or severe pain. If you put your hand on a hot barbecue grill, you snatch your hand away immediately by reflex action, but you don't feel the "ouch" of the pain until a little later. Why do events occur in this order? The reflex action is driven by rapid nerve impulse conduction along myelinated neurons. The neurons that release substance P are not myelinated, however, so their signal is conducted more slowly, and the feeling of pain is delayed. The action of endorphins is antagonistic to substance P, reducing the perception of pain.

In mammals and other animals, some neurons synthesize and release dissolved carbon monoxide (CO) and nitric oxide (NO), both gases, as neurotransmitters. For example, in the brain, CO regulates the release of hormones from the hypothalamus. NO contributes to many nervous system functions, such as learning, sensory responses, and muscle movements. By relaxing smooth muscles in the walls of blood vessels, NO causes the vessels to dilate, increasing the flow of blood. For example, when a male mammal is sexually aroused, neurons release NO into the erectile tissues in the penis. The relaxation of the muscles increases blood flow into the tissues, causing them to fill with blood and produce an erection. The drug sildenafil (Viagra) aids erection by inhibiting an enzyme that normally reduces NO concentration in the penis.

33.4 Integration of Incoming Signals by Neurons

Most neurons receive a multitude of stimulatory and inhibitory signals carried by both direct and indirect neurotransmitters. These signals are integrated by the postsynaptic neuron into a response that reflects their combined effects. The integration depends primarily on the patterns, number, types, and activity of the synapses that the postsynaptic neuron makes with presynaptic neurons. Inputs from other sources, such as indirect neurotransmitters and other signal molecules, can modify the integration. The response of the postsynaptic neuron is elucidated by the frequency of action potentials it generates.

33.4a Integration at Chemical Synapses Occurs by Summation

As mentioned earlier, depending on the type of receptor to which it binds, a neurotransmitter may stimulate or inhibit the generation of action potentials in the postsynaptic neuron. If a neurotransmitter opens a ligand-gated Na^+ channel, Na^+ enters the cell, causing a depolarization. This change in membrane potential pushes the neuron closer to threshold; that is, it is excitatory and is called an **excitatory postsynaptic potential**, or **EPSP**. On the other hand, if a neurotransmitter opens a ligand-gated ion channel that allows Cl^- to flow into the cell and K^+ to flow out, hyperpolarization occurs. This change in membrane potential pushes the neuron farther from threshold; that is, it is inhibitory and is called an **inhibitory postsynaptic potential**, or **IPSP**. In contrast to the all-or-nothing operation of an action potential, EPSPs and IPSPs are **graded potentials**, in which the membrane moves up or down in potential without necessarily triggering an action potential. There are no refractory periods for EPSPs and IPSPs.

A neuron typically has hundreds to thousands of chemical synapses formed by axon terminals of presynaptic neurons contacting its dendrites and cell body **(Figure 33.15)**. The events that occur at each synapse produce either an EPSP or an IPSP in that postsynaptic neuron. But how is an action potential produced if a single EPSP is not sufficient to push the postsynaptic neuron to threshold? The answer involves the summation of all the inputs received through all the

Cell body of postsynaptic neuron

Axon terminals

Figure 33.15
The multiple chemical synapses relaying signals to a neuron. The drying process used to prepare the neuron for electron microscopy has toppled the axon terminals and pulled them away from the neuron's surface.

E. R. Lewis, T. E. Everhart, Y. Y. Zevi/Visuals Unlimited

chemical synapses formed by presynaptic neurons. At any given time, some or many of the presynaptic neurons may be firing, producing EPSPs and/or IPSPs in the postsynaptic neuron. The sum of all the EPSPs and IPSPs at a given time determines the total potential in the postsynaptic neuron and, therefore, how that neuron responds. **Figure 33.16** shows, in a greatly simplified way, the effects of EPSPs and IPSPs on membrane potential and how the summation of inputs brings a postsynaptic neuron to threshold.

The postsynaptic neuron in Figure 33.16 has three neurons, N1 to N3, forming synapses with it. Suppose that the axon of N1 releases a neurotransmitter that produces an EPSP in the postsynaptic cell (see Figure 33.16a). The membrane depolarizes, but not enough to reach threshold. If N1 input causes a new EPSP after the first EPSP has died down, it will be of the same magnitude as the first EPSP, so no progression toward threshold happens because no summation has occurred. If, instead, N1 input causes a new EPSP before the first EPSP has died down, the second EPSP will sum with the first, leading to a greater depolarization (see Figure 33.16b). This summation of several EPSPs produced by successive firing of a single presynaptic neuron over a short period of time is called **temporal summation**. If the total depolarization achieved in this way reaches threshold, an action potential will be produced in the postsynaptic neuron.

The postsynaptic cell may also be brought to threshold by **spatial summation**, the summation of EPSPs produced by the firing of different presynaptic neurons, such as N1 and N2 (see Figure 33.16c). Lastly, EPSPs and IPSPs can cancel each other out. In the example shown in Figure 33.16d, firing of N1 alone produces an EPSP, firing of N3 alone produces an IPSP, whereas the simultaneous firing of N1 and N3 produces no change in the membrane potential.

The summation point for EPSPs and IPSPs is the axon hillock of the postsynaptic neuron. The greatest density of voltage-gated Na^+ channels occurs in that region, resulting in the lowest threshold potential in the neuron.

33.4b The Patterns of Synaptic Connections Contribute to Integration

The total number of connections made by a neuron may be very large. Some single interneurons in the human brain, for example, form as many as 100 000 synapses with other neurons. The synapses are not absolutely fixed; they can change through modification, addition, or removal of synaptic connections, or even entire neurons, as animals mature and experience changes in their environments. The combined activities of all the neurons in an animal provide the flow of information on which the integrated functioning of increasingly complex organisms depends. In the remainder of this chapter, we explore the ways that neurons are organized into nervous systems in the various major groups of animals.

STUDY BREAK

Differentiate between spatial and temporal summation.

33.5 Integration in Protostomes: Networks, Nerves, Ganglia, and Brains

In Chapter 26, we learned about some key innovations in animal evolution. Two of these, the appearance of bilateral symmetry and the independent evolution of

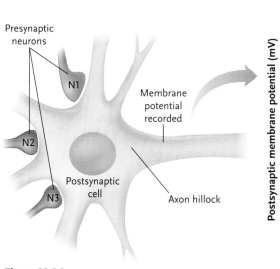

Figure 33.16
Summation of EPSPs and IPSPs by a postsynaptic neuron.

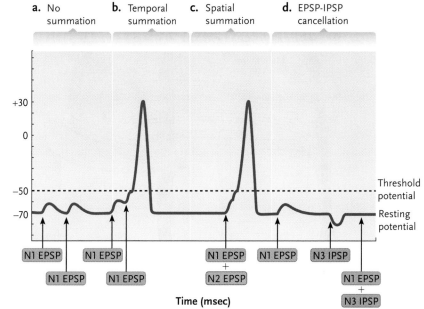

Figure 33.17
The nervous system of *Hydra*. **(a)** The entire nerve net. **(b)** The nerve ring around the hypostome. **(c)** The distribution of sensory neurons.

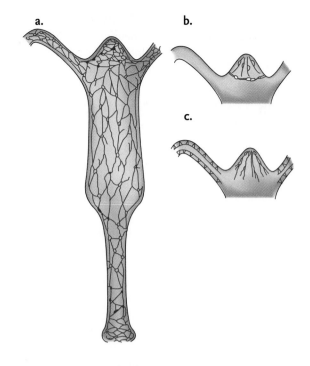
a.
b.
c.

neurons form a nerve ring around the hypostome. Some of the neurons are sensory. Neurons do not have dendrites and axons. Instead, there are synapses wherever the neurons cross ("en passant" synapses). Moreover, both neurons involved in a synapse may produce transmitters and have receptors for transmitters. This simple system permits the coordination of tentacles for feeding.

In medusoid cnidarians (jellyfish), some of which have coordinated swimming movements, the network is very extensive, forming a very fine meshwork of neurons in the entire animal. This network is connected to two nerve rings that circle the medusa **(Figure 33.18)**. Sense organs that detect light and gravity (see Chapter 34) are grouped together and have clusters of nerve cell bodies associated with them. These clusters of neurons ("pace makers") are responsible for generating the rhythmic action potentials that lead to the coordinated muscle contractions involved in swimming.

Even in these simple animals with radial symmetry, neurons are grouped into nerves, sensory structures are localized, and neurons are concentrated around the mouth. Neurons are imposing some degree of localized coordination or control.

segmentation in two phyla of the protostomes, were accompanied by evolutionary changes in the organization of the nervous systems. In radially symmetrical protostomes such as the cnidarian *Hydra*, a network of neurons extends over the entire organism just beneath the epithelium **(Figure 33.17)**. The neurons are more numerous toward the oral end, and several

33.5a Ganglia Enhance Integration in Invertebrates

Groups of nerve cell bodies with localized interconnections are called **ganglia** (singular, ganglion; **Figure 33.19)**. In protostomes, the cell bodies of a ganglion

Figure 33.18
Nervous system in a cnidarian medusa.

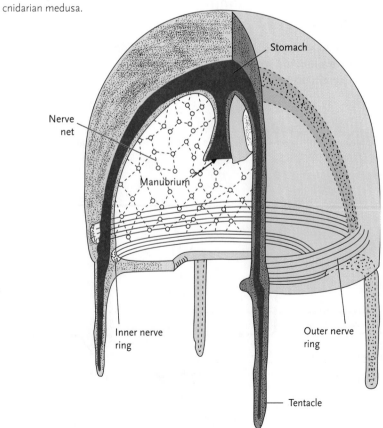

Stomach
Nerve net
Manubrium
Inner nerve ring
Outer nerve ring
Tentacle

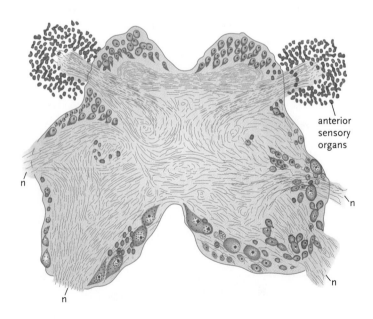

anterior sensory organs
n
n
n
n

Figure 33.19
A section of a typical invertebrate cerebral ganglion. The cell bodies (blue) are located on the periphery, with a mass of axons and dendrites forming the neuropile in the centre. Nerves (n) bring sensory information into the ganglion and carry motor information outward.

are located on its periphery, with the interconnections located in a tangled mass of axons and dendrites, the **neuropile**, in the centre. This anatomical localization of interconnections allows rapid integration of sensory information and more complex reactions to that information. Bilaterally symmetrical animals, which have an anterior and a posterior end, have a concentration of ganglia at the anterior end, forming a "brain" or cerebral ganglion. Many of the sense organs are also found at the anterior end.

The appearance of bilateral symmetry is accompanied by structures that ensure coordination between the right and left halves of the animal. In the most primitive bilateria, the flatworms **(Figure 33.20)**, the cerebral ganglion is frequently bilobed, with the two halves joined by a commissure. Flatworms always have a pair of prominent ventral nerves, and some have less prominent paired lateral and dorsal nerves leading to the cerebral ganglion. These are connected at intervals by transverse commissures. The cerebral ganglion, or brain, and its associated nerve cords represent the CNS. A nerve net connected to the nerve cords forms a complex PNS. This general pattern of anterior ganglia and nerve cords connected by commissures can be seen in many other invertebrates, although only the flatworms have a nerve net. In most of the protostomes, the anterior ganglia forming the brain surround the anterior digestive system. In cephalopod molluscs, cephalization is the most pronounced of any of the invertebrates. In the octopus **(Figure 33.21)**, for example, several ganglia surrounding the anterior digestive system fuse to form a brain with distinct motor and sensory areas and a series of paired nerves connect to sense organs and muscles. Octopuses are capable of rapid movement to hunt prey and have the capacity to learn complex behaviours.

33.5b Segmentation Includes the Nervous System

We saw in Chapter 26 that segmental development has occurred twice in the evolution of the protostomes, and the effects on the nervous system have been very similar. In both the annelids and the arthropods, each segment has a separate pair of ganglia, joined by a short commissure. In insects **(Figure 33.22, p. 802)**, for example, the ganglia of each segment are connected to those anterior and posterior to it by paired connectives, forming a chain of ganglia. Each ganglion gives rise to nerves that serve the segment and its appendages. Although there is some independence, the actions of one ganglion may be coordinated with those of adjacent ganglia and with the CNS through the paired intersegmental connectives. For example, a leg of a cockroach can make a stepping motion when the intersegmental connectives are cut, but walking requires an intact nervous system.

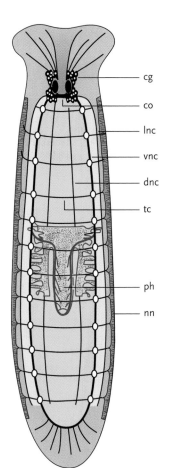

Figure 33.20
Nervous system in a free-living platyhelminth. The cerebral ganglion (cg) has two lobes connected by a commissure (co). It receives nerves from sense organs in the head and gives rise to three pairs of nerve cords: prominent ventral cords (vnc) and the less prominent lateral (lnc) and dorsal (dnc) cords, all linked to a nerve net (nn) and connected at intervals by transverse commissures (tc). The pharynx (ph) has prominent innervation.

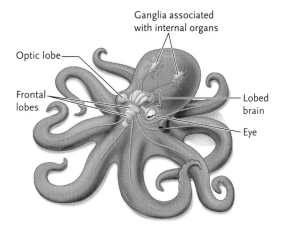

Figure 33.21
The nervous system of an octopus.

33.5c Brains in Segmental Animals Are Fused Segmental Ganglia

As cephalization proceeds during evolution, recognizable heads appear. Appendages have been modified to form sense organs or for feeding, and the paired ganglia associated with those structures are fused. In the insect head (see Figure 33.22), embryologists have recognized six ganglia. The protocerebrum is largely concerned with vision (the eyes are an extension of the brain); the deutocerebrum processes information

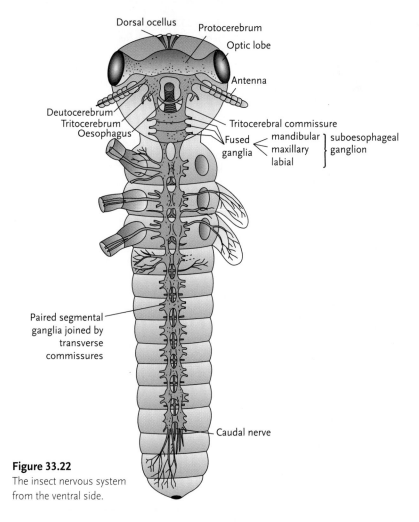

Dorsal ocellus
Protocerebrum
Optic lobe
Antenna
Deutocerebrum
Tritocerebrum
Oesophagus
Tritocerebral commissure
Fused ganglia — mandibular ⎫ suboesophageal
maxillary ⎬ ganglion
labial ⎭
Paired segmental ganglia joined by transverse commissures
Caudal nerve

Figure 33.22

The insect nervous system from the ventral side.

from the antennae, which are modified appendages of the second segment; and the tritocerebrum receives nerves from the labrum and anterior intestine. These three ganglia constitute the brain. Very few motor neurons are associated with the brain; most are involved with the information from the anterior sense organs or are associative neurons involved in processing information. As in all protostome ganglia, the cell bodies are at the periphery and the neuropile lies inside the ganglia.

The brain sends two connectives around the gut to connect with the suboesophageal ganglion. This is made up of three fused ganglia, one for each of the modified appendages that form the mouthparts. Thus, the insect head represents six primitive segments, three anterior to the mouth and three posterior. Similar analyses can be performed for other arthropods.

STUDY BREAK

1. Distinguish among nerve nets, nerves, and nerve cords.
2. What are segmental ganglia?

33.6 Vertebrates Have the Most Complex Nervous Systems

Vertebrates are thought to have evolved from the larval form of a primitive chordate, such as an ascidian (see Chapter 27). The ancestor probably resembled the lancelet *Branchiostoma*. The lancelet has a segmental organization, but the segmentation does not include the body surface. During development, blocks of mesodermal tissue arise that develop as segmental blocks of muscle, the myotomes. These develop before the nervous system and nerves grow out from the developing nerve cord to innervate them. Thus, the nervous system is not segmental in the same sense as arthropods or annelids. As in all chordates, the nerve cord is dorsal and contains a central cavity. This is a reflection of the different ways in which the nervous systems develop. In protostomes, differentiating nerve cells group together to form ganglia in the ventral part of the animal. In chordates, by contrast, the nervous system is formed dorsally as the hollow **neural tube**, the anterior end of which develops into the brain and the rest into the **spinal cord** (see Chapter 39). In the lancelet, the anterior end of the nerve cord is larger than the rest of it. Detailed anatomical studies by Thurston Lacalli at the University of Saskatchewan, supported by emerging molecular data, have identified three regions of the lancelet "brain" that are probably homologous with the forebrain, midbrain, and hindbrain regions of the vertebrate brain.

In vertebrates, the CNS consists of the brain and spinal cord, and the PNS consists of all the nerves that connect the brain and spinal cord to the rest of the body. The brain and nerve cord of vertebrates are hollow, fluid-filled structures located dorsally. The central cavity of the neural tube becomes the fluid-filled **ventricles** of the brain and the narrow **central canal** through the spinal cord.

Although the lancelet nervous system is simple and shows only minimal cephalization, all vertebrate nervous systems are highly cephalized, with major concentrations of neurons in a brain located in the head. During evolution, the complexity of the general structure of the brain has increased, and differences appear in the brains of the major groups.

The organization of the brain is exceedingly complex. One way to understand its evolution is to examine its embryological development from the embryonic neural tube. A generalized vertebrate brain approximately midway through its embryonic development **(Figure 33.23a)** shows the principal regions shared by all vertebrate brains. Early in embryonic development, the anterior part of the neural tube enlarges into three distinct regions. The **forebrain** was originally associated with olfaction or the sense of smell, the **midbrain** was primarily associated with vision, and the **hindbrain** was mainly associated with balance. Later, the embryonic hindbrain subdivided into the *metencephalon* (*met* = behind; *encephalon* = brain) and the *myelencephalon* (*myelo* = spinal cord), the midbrain developed into the *mesencephalon* (*mes* = middle), and the forebrain subdivided into the *telencephalon* (*tel* = distant) and the *diencephalon* (*di* = across).

Later still, the metencephalon, associated with the developing ear (when present) and balance organs, gave rise to the *cerebellum*, a major traffic centre that integrates sensory signals from the eyes, ears, and muscle spindles. The myelencephalon gave rise to the *medulla oblongata* (commonly shortened to medulla) that controls many vital involuntary tasks, such as respiration and blood circulation. The mesencephalon, or midbrain, received fibres from the optic nerves and from the ear and acts as a relay centre passing information forward for processing. The diencephalon is associated with the eyes, and the optic nerve is an outgrowth of it. The telencephalon, embryologically associated with olfaction, gave rise to the olfactory bulb and the cerebrum, the major processing centre of the brain. These events and the functions of the areas are summarized in **Figure 33.23b.**

The general pattern of brain development underwent major modification in the evolution of

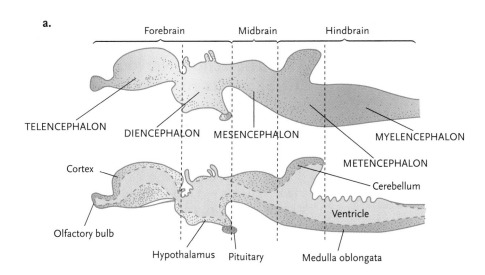

a.

Figure 33.23
(a) Diagram of a vertebrate brain midway in development. Upper, lateral view; lower, vertical section. At this stage, only a few of the adult structures are easily identified. During later development in birds and mammals, the elongate form of the brain becomes folded so that the very prominent forebrain lies above the other regions. During development to the adult, the size of the ventricle is reduced. **(b)** The regions of the brain of a bird or mammal during development and their functions in the adult.

b.

Regions in early embryo	Regions in mid-development	Neural tube	Regions in adult	Functions in adult
Forebrain	Telencephalon		Telencephalon (cerebrum)	Higher functions, such as thought, action, and communication
	Diencephalon		Thalamus	Coordinates sensory input and relays it to cerebellum
			Hypothalamus	Centre for homeostatic control of internal environment
Midbrain	Mesencephalon		Midbrain	Coordinates involuntary reactions and relays signals to telencephalon
Hindbrain	Metencephalon		Cerebellum	Integrates signals for muscle movement
			Pons	Centre for information flow between cerebellum and telencephalon
	Myelencephalon		Medulla oblongata	Controls many involuntary tasks

various groups of animals **(Figure 33.24).** In sharks, the cerebrum is relatively small, but the olfactory bulbs are prominent, testifying to the importance of olfaction in these very successful predators. Frogs are hunters that rely on vision, so the optic lobes of the mesencephalon are prominent, whereas the olfactory bulbs are less so. Birds also rely on vision for feeding and navigation, and their optic lobes reflect that.

One of the major trends in the evolution of the brain, however, is the increasing prominence of the cerebrum. Beginning with reptiles, it increased in size relative to the rest of the brain. In mammals, convolutions or folds appeared, increasing the amount of brain material in a particular volume. As well, the total mass of the brain relative to the size of the animal increased, permitting animals to undertake more complex tasks. The mass of bird and mammal brains is about 15 times greater than that of other taxa when corrected for the size of the animal. With their advanced locomotor and navigational skills, birds and mammals also exhibit an increase in the cerebellum, a major coordinating centre for automatic activities. Because we know most about the functioning of the human brain, the following sections examine the structure and function of the human nervous system, beginning with the CNS.

STUDY BREAK

1. From what part of the embryonic brain does the cerebellum come? What does it control?
2. How does the frog brain differ from that of sharks?

Figure 33.24

A comparison of brain structures in five different groups of vertebrates, illustrating the evolutionary trends described in the text.

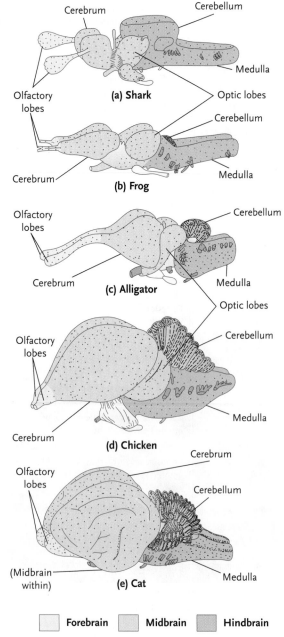

(a) Shark

(b) Frog

(c) Alligator

(d) Chicken

(e) Cat

Forebrain   Midbrain   Hindbrain

33.7 The Central Nervous System (CNS) and Its Functions

The brain and spinal cord are surrounded and protected by three layers of connective tissue, the **meninges** (*meninga* = membrane), and by the **cerebrospinal fluid**, which circulates through the central canal of the spinal cord, through the ventricles of the brain, and between two of the meninges. The fluid cushions the brain and spinal cord from jarring movements and impacts, nourishes the CNS, and protects it from toxic substances.

The CNS manages body activities by integrating incoming sensory information from the PNS into compensating responses. Our examination of the vertebrate CNS begins with the spinal cord and then considers the brain and its functions.

33.7a The Spinal Cord Relays Signals between the PNS and the Brain and Controls Reflexes

The spinal cord, which extends dorsally from the base of the brain, carries impulses between the brain and the PNS and contains the interneuron circuits that control motor reflexes. In cross section, the spinal cord has a butterfly-shaped core of **grey matter**, consisting of nerve cell bodies and dendrites. This is surrounded by **white matter**, consisting of axons, many of them surrounded by myelin sheaths **(Figure 33.25,** left side). Note that this arrangement is the reverse of that in invertebrate ganglia, where cell bodies are at the periphery (see Figure 33.19). Pairs of spinal nerves connect with the spinal cord at spaces between the vertebrae.

The afferent (incoming) axons entering the spinal cord make synapses with interneurons in the grey

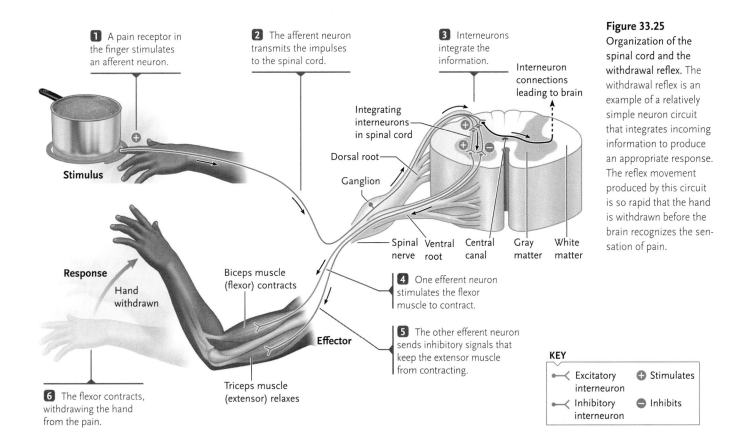

1 A pain receptor in the finger stimulates an afferent neuron.

2 The afferent neuron transmits the impulses to the spinal cord.

3 Interneurons integrate the information.

Interneuron connections leading to brain

Integrating interneurons in spinal cord

Dorsal root

Ganglion

Spinal nerve · Ventral root · Central canal · Gray matter · White matter

Stimulus

Response

Hand withdrawn

Biceps muscle (flexor) contracts

Effector

4 One efferent neuron stimulates the flexor muscle to contract.

5 The other efferent neuron sends inhibitory signals that keep the extensor muscle from contracting.

Triceps muscle (extensor) relaxes

6 The flexor contracts, withdrawing the hand from the pain.

KEY

Excitatory interneuron	⊕ Stimulates
Inhibitory interneuron	⊖ Inhibits

Figure 33.25
Organization of the spinal cord and the withdrawal reflex. The withdrawal reflex is an example of a relatively simple neuron circuit that integrates incoming information to produce an appropriate response. The reflex movement produced by this circuit is so rapid that the hand is withdrawn before the brain recognizes the sensation of pain.

matter, which send axons upward through the white matter of the spinal cord to the brain. Conversely, axons from interneurons of the brain pass downward through the white matter of the cord and make synapses with the dendrites and cell bodies of efferent neurons in the grey matter of the cord. The axons of these efferent (outgoing) neurons exit the spinal cord through the spinal nerves.

The grey matter of the spinal cord also contains interneurons of the pathways involved in **reflexes**, programmed movements that take place without conscious effort, such as the sudden withdrawal of a hand from a hot surface (shown in Figure 33.25). When your hand touches the hot surface, the heat stimulates an afferent neuron, which makes connections with at least two interneurons in the spinal cord. One of these interneurons stimulates an efferent neuron, causing the *flexor* muscle of the arm to contract. This bends the arm and withdraws the hand almost instantly from the hot surface. The other interneuron synapses with an efferent neuron connected to an *extensor* muscle, relaxing it so that the flexor can move more quickly. Interneurons connected to the reflex circuits also send signals to the brain, making you aware of the stimulus causing the reflex. You know from experience that when a reflex movement withdraws your hand from a hot surface or other damaging stimulus, you feel the pain shortly *after* the hand is withdrawn. This is the extra time required for impulses to travel from the neurons of the reflex to the brain.

33.7b The Brain Integrates Sensory Information and Formulates Compensating Responses

The brain is the major centre that receives, integrates, stores, and retrieves information. Its interneuron networks generate responses that provide the basis for our voluntary movements, consciousness, behaviour, emotions, learning, reasoning, language, and memory, among many other complex activities.

Major Brain Structures. We have noted that the three major divisions of the embryonic brain give rise to the structures of the adult brain. Like the spinal cord, each brain structure contains both grey matter and white matter and is surrounded by meninges and circulating cerebrospinal fluid **(Figure 33.26, p. 806)**.

The hindbrain of vertebrates develops into the *medulla oblongata* (the *medulla*) and the *cerebellum* (see Figure 33.23). In higher mammals, a mass of fibres connecting the cerebellum to higher centres in the brain is so prominent that it is identified as the *pons* (bridge). The medulla and pons, along with the midbrain, form a stalklike structure known as the **brain stem**, which connects the forebrain with the spinal cord. All but 2 of the 12 pairs of cranial nerves (see Section 33.8) also originate from the brain stem.

The forebrain, which makes up most of the mass of the brain in humans, forms the *telencephalon* (*cerebrum*). Its surface layer, the **cerebral cortex**, is a thin

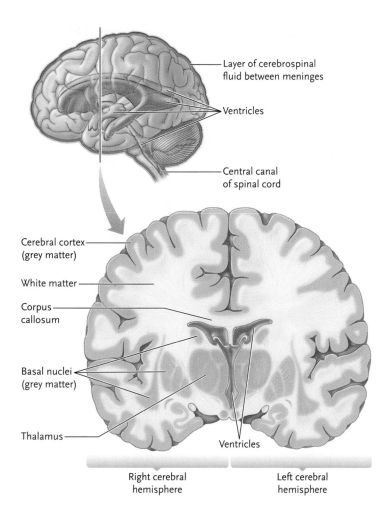

Layer of cerebrospinal fluid between meninges

Ventricles

Central canal of spinal cord

Cerebral cortex (grey matter)

White matter

Corpus callosum

Basal nuclei (grey matter)

Thalamus

Ventricles

Right cerebral hemisphere

Left cerebral hemisphere

Figure 33.26
The human brain, illustrating the distribution of grey matter, and the locations of the four ventricles (in blue) with their connection to the central canal of the spinal cord.

33.7c The Brain Stem Regulates Many Vital Housekeeping Functions of the Body

Physicians and scientists have learned much about the functions of various brain regions by studying patients with brain damage from stroke, infection, tumours, and mechanical disturbance. Techniques such as *functional magnetic resonance imaging* (*fMRI*) and *positron emission tomography* (*PET*) allow researchers to identify the normal functions of specific brain regions in noninvasive ways. The instruments record a subject's brain activity during various mental and physical tasks by detecting minute increases in blood flow or metabolic activity in specific regions **(Figure 33.27)**.

From such analyses, we know that grey-matter centres in the brain stem control many vital body functions without conscious involvement or control by the cerebrum. Among these functions are the heart and respiration rates, blood pressure, constriction and dilation of blood vessels, coughing, and reflex activities of the digestive system, such as vomiting.

33.7d The Cerebellum Integrates Sensory Inputs to Coordinate Body Movements

Although the cerebellum is connected to the pons, it is separate in structure and function from the brain stem. Through its extensive connections with other parts of the brain, the **cerebellum** receives sensory input originating from receptors in muscles and joints, from balance receptors in the inner ear, and from the receptors of touch, vision, and hearing. These signals

layer of grey matter in which numerous unmyelinated neurons are found. The telencephalon, which is divided into right and left *cerebral hemispheres*, is corrugated by fissures and folds that increase the surface area of the cerebral cortex (see Figure 33.26). This structure reflects two of the evolutionary tendencies in the brain of mammals: the corrugation of the hemispheres and the development of a layer of grey matter on the periphery.

The Blood–Brain Barrier. Unlike the epithelial cells that form capillary walls elsewhere in the body, which allow small molecules and ions to pass freely from the blood to surrounding fluids, those forming capillaries in the brain are sealed together by tight junctions (Chapter 32). The tight junctions set up a **blood–brain barrier** that prevents most substances dissolved in the blood from entering the cerebrospinal fluid, protecting the brain and spinal cord from viruses, bacteria, and toxic substances that may circulate in the blood. A few types of molecules and ions, such as oxygen, carbon dioxide, alcohol, and anaesthetics, can move directly across the lipid bilayer of the epithelial cell membranes by diffusion. A few other substances are moved across the plasma membrane by highly selective transport proteins. The most significant of these transported molecules is glucose, the important source of metabolic energy for the cells of the brain.

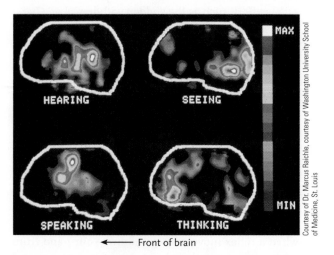

HEARING   SEEING

SPEAKING   THINKING

Front of brain

Courtesy of Dr. Marcus Raichle, courtesy of Washington University School of Medicine, St. Louis

Figure 33.27
PET scans showing regions of the brain active when a person performs specific mental tasks. The colours show the relative activity of the sections, with white being the most active.

convey information about how the body trunk and limbs are positioned, the degree to which different muscles are contracted or relaxed, and the direction in which the body or limbs are moving. The cerebellum integrates these sensory signals and compares them with signals from the cerebrum that control voluntary body movements. Outputs from the cerebellum to the cerebrum, brain stem, and spinal cord modify and fine-tune the movements to keep the body in balance and directed toward targeted positions in space. The cerebellum is particularly important in birds, and, like the mammalian cerebellum, it has a folded structure, increasing its relative size.

33.7e Basal Nuclei, Thalamus, and Hypothalamus Grey-Matter Centres Control a Variety of Functions

Grey-matter centres derived from the embryonic telencephalon include the thalamus, hypothalamus, basal nuclei, and limbic system **(Figure 33.28)**. They contribute to the control and integration of voluntary movements, body temperature, glandular secretions, osmotic balance of the blood and extracellular fluids, wakefulness, and the emotions, among other functions. Some of the grey-matter centres route information to and from the cerebral cortex and between the forebrain, brain stem, and cerebellum.

The **thalamus** (see Figure 33.28) forms a major switchboard that receives sensory information and relays it to the appropriate regions of the cerebral cortex. It also plays a role in alerting the cerebral cortex to full wakefulness or in inducing drowsiness or sleep.

The **hypothalamus** is a relatively small conical area that occurs in all vertebrates. It contains centres that regulate basic homeostatic functions of the body. Some centres set and maintain body temperature by triggering reactions such as shivering or sweating. Others constantly monitor the osmotic balance of the blood by testing its composition of ions and other substances. If departures from normal levels are detected, the hypothalamus triggers responses such as thirst or changes in urine output that restore the osmotic and fluid balance. The hypothalamus is an important part of the endocrine system (see Chapter 35). It produces some of the hormones released by the pituitary and governs the release of other pituitary hormones.

The centres of the hypothalamus that detect blood composition and temperature are directly exposed to the bloodstream: they are the only parts of the brain *not* protected by the blood–brain barrier. Parts of the hypothalamus also coordinate responses triggered by the autonomic system (see Section 33.8), making it an important link in such activities as control of the heartbeat, contraction of smooth muscle cells in the digestive system, and glandular secretion. Some regions of the hypothalamus establish a biological clock that sets up daily metabolic rhythms, such as the

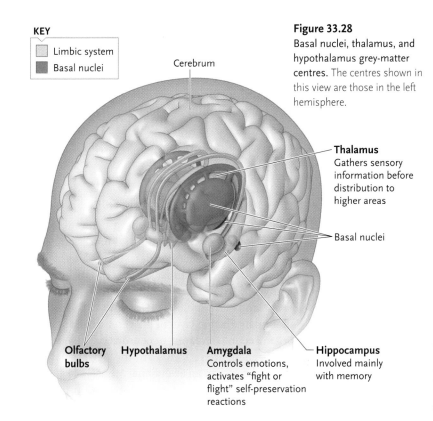

Figure 33.28
Basal nuclei, thalamus, and hypothalamus grey-matter centres. The centres shown in this view are those in the left hemisphere.

Cerebrum

Thalamus
Gathers sensory information before distribution to higher areas

Basal nuclei

Olfactory bulbs **Hypothalamus** **Amygdala**
Controls emotions, activates "fight or flight" self-preservation reactions

Hippocampus
Involved mainly with memory

KEY
▢ Limbic system
▪ Basal nuclei

regular changes in body temperature that occur on a daily cycle.

The **basal nuclei** are grey-matter centres that surround the thalamus on both sides of the brain (see Figure 33.28). They moderate voluntary movements directed by motor centres in the cerebrum and can be recognized in all amniotes. Damage to the basal nuclei can affect the planning and fine-tuning of movements, leading to stiff, rigid motions of the limbs and unwanted or misdirected motor activity, such as tremors of the hands and inability to start or stop intended movements at the intended place and time. Parkinson disease, in which affected individuals exhibit all of these symptoms, results from degeneration of centres in and near the basal nuclei.

Parts of the thalamus, hypothalamus, and basal nuclei, along with other nearby grey-matter centres—the amygdala, hippocampus, and olfactory bulbs—form a functional network called the **limbic system** (*limbus* = belt), sometimes called our "emotional brain" (see Figure 33.28). The **amygdala** works as a switchboard, routing information about experiences that have an emotional component through the limbic system. The **hippocampus** is involved in sending information to the frontal lobes, and the **olfactory bulbs** relay inputs from odour receptors to both the cerebral cortex and the limbic system. The olfactory connection to the limbic system may explain why certain odours can evoke particular, sometimes startlingly powerful, emotional responses.

The limbic system controls emotional behaviour and influences the basic body functions regulated by the hypothalamus and brain stem. Stimulation of different parts of the limbic system produces anger, anxiety, fear, satisfaction, pleasure, or sexual arousal. Connections between the limbic system and other brain regions bring about emotional responses such as smiling, blushing, or laughing.

33.7f The Cerebral Cortex Carries Out All Higher Brain Functions

Over the course of evolution, the surface area of the cerebral cortex increased by continuously folding in on itself, thereby expanding the structure into sophisticated information encoding and processing centres. Primates have cerebral cortices with the largest number of convolutions. In humans, each cerebral hemisphere is divided by surface folds into *frontal, parietal, temporal,* and *occipital* lobes (**Figure 33.29**). Uniquely in mammals, the top layer of the cerebral hemispheres is organized into six layers of neurons called the *neocortex* (*neo* = new; these layers are the newest part of the cerebral cortex in an evolutionary sense).

The two cerebral hemispheres can function separately, and each has its own communication lines internally and with the rest of the CNS and the body. The left cerebral hemisphere responds primarily to sensory signals from, and controls movements in, the right side of the body. The right hemisphere has the same relationships to the left side of the body. This opposite connection and control reflect the fact that the nerves carrying afferent and efferent signals cross from left to right within the spinal cord or brain

stem. Thick axon bundles, forming a structure called the **corpus callosum**, connect the two cerebral hemispheres and coordinate their functions.

Sensory Regions of the Cerebral Cortex. Areas that receive and integrate sensory information are distributed over the cerebral cortex. In each hemisphere, the **primary somatosensory area**, which registers information on touch, pain, temperature, and pressure, runs in a band across the parietal lobes of the brain (see Figure 33.29). Experimental stimulation of this band in one hemisphere causes prickling or tingling sensations in specific parts on the opposite side of the body, beginning with the toes at the top of each hemisphere and running through the legs, trunk, arms, and hands, to the head (**Figure 33.30** and *People Behind Biology*).

Other sensory regions of the cerebral cortex have been identified with hearing, vision, smell, and taste (see Figure 33.30). Regions of the temporal lobes on both sides of the brain receive auditory inputs from the ears, whereas inputs from the eyes are processed in the primary visual cortex in both occipital lobes. Olfactory input from the nose is processed in the olfactory lobes, located on the ventral side of the temporal lobes. Regions in the parietal lobes receive inputs from taste receptors on the tongue and other locations in the mouth.

Motor Regions of the Cerebral Cortex. The **primary motor area** of the cerebral cortex runs in a band just in front of the primary somatosensory area (see Figure 33.29). Experimental stimulation of points along this

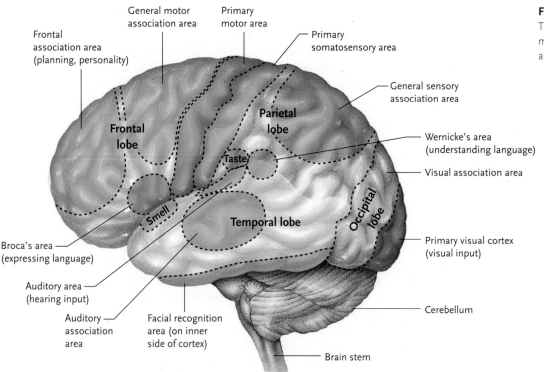

Figure 33.29

The lobes of the cerebrum, showing major regions and association areas of the cerebral cortex.

Frontal association area (planning, personality)

General motor association area

Primary motor area

Primary somatosensory area

General sensory association area

Frontal lobe

Parietal lobe

Wernicke's area (understanding language)

Visual association area

Taste

Broca's area (expressing language)

Smell

Temporal lobe

Occipital lobe

Primary visual cortex (visual input)

Auditory area (hearing input)

Auditory association area

Facial recognition area (on inner side of cortex)

Cerebellum

Brain stem

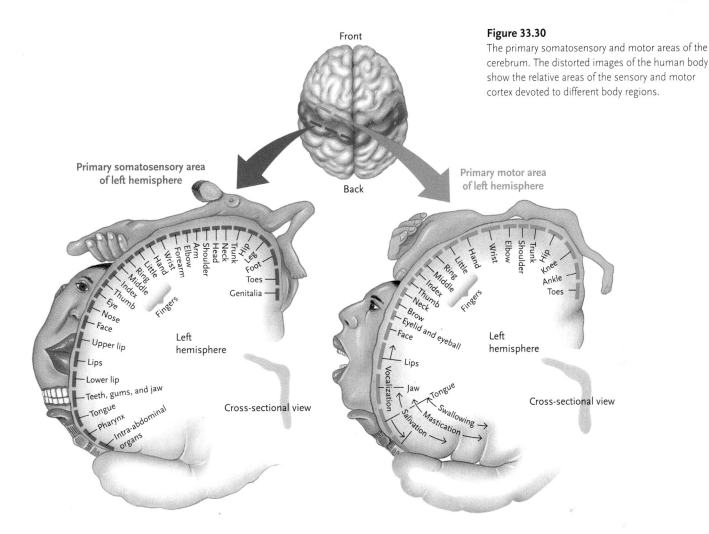

Figure 33.30
The primary somatosensory and motor areas of the cerebrum. The distorted images of the human body show the relative areas of the sensory and motor cortex devoted to different body regions.

Front

Back

Primary somatosensory area of left hemisphere

Primary motor area of left hemisphere

Left hemisphere

Cross-sectional view

Left hemisphere

Cross-sectional view

band in one hemisphere causes movement of specific body parts on the opposite side of the body, corresponding generally to the parts registering in the primary somatosensory area at the same level (see Figure 33.30). Other areas that integrate and refine motor control are located nearby.

In both the primary somatosensory and motor areas, some body parts, such as the lips and fingers, are represented by large regions, and others, such as the arms and legs, are represented by relatively small regions. As shown in Figure 33.30, the relative sizes produce a distorted image of the human body that is quite different from the actual body proportions. The differences are reflected in the precision of touch and movement in structures such as the lips, tongue, and fingers.

Association Areas. The sensory and motor areas of the cerebral cortex are surrounded by **association areas** (see Figure 33.29) that integrate information from the sensory areas, formulate responses, and pass them on to the primary motor area. Two of the most important association areas are *Wernicke's area* and *Broca's area*, which function in spoken and written language. They are usually present on only one side of the brain, in the left hemisphere in 97%

of the human population. Comprehension of spoken and written language depends on Wernicke's area, which coordinates inputs from the visual, auditory, and general sensory association areas. Interneuron connections lead from Wernicke's area to Broca's area, which puts together the motor program for coordination of the lips, tongue, jaws, and other structures producing the sounds of speech and passes the program to the primary motor area. The brain-scan images in Figure 33.27 dramatically illustrate how these brain regions participate as a person performs different linguistic tasks.

33.7g Some Higher Functions Are Distributed in Both Cerebral Hemispheres; Others Are Concentrated in One Hemisphere

Most of the other higher functions of the human brain, such as abstract thought and reasoning; spatial recognition; mathematical, musical, and artistic ability; and the associations forming the basis of personality, involve the coordinated participation of many regions of the cerebral cortex. Some of these regions are equally distributed in both cerebral hemispheres, and some are more concentrated in one hemisphere.

PEOPLE BEHIND BIOLOGY

Wilder Penfield, McGill University

Born in the United States at the end of the nineteenth century, Wilder Penfield graduated in literature from Princeton and won a Rhodes Scholarship to Oxford. He graduated as an M.D. from Johns Hopkins and worked in neurosurgery at Columbia. Attracted to McGill University in Montreal in 1928, he realized his dream by establishing the Montreal Neurological Institute, where scientists and clinicians could work together. It was here that he established the "Montreal Procedure" for the treatment of epilepsy. With patients under local anesthesia and fully conscious, he exposed the entire cerebrum and stimulated various parts of the brain while the patients described their sensations. In this way, he could identify the area of the brain responsible for the epileptic seizures and, if feasible, remove or destroy it. In the course of this work, he was able to identify those areas of the cerebral cortex that related to particular parts of the body and developed the well-known map shown in Figure 33.30. This was a remarkably courageous procedure at the time and required his patients to trust him absolutely.

Among the functions more or less equally distributed between the two hemispheres is the ability to recognize faces. Consciousness, the sense of time, and recognizing emotions also seem to be distributed in both hemispheres.

Typically, some brain functions are more localized in one of the two hemispheres, a phenomenon called **lateralization**. Studies of people with split hemispheres and surveys of brain activity by PET and fMRI have confirmed that, for the vast majority of people, the left hemisphere specializes in spoken and written language, abstract reasoning, and precise mathematical calculations. The right hemisphere specializes in nonverbal conceptualizing, intuitive thinking, musical and artistic abilities, and spatial recognition functions, such as fitting pieces into a puzzle. The right hemisphere also handles mathematical estimates and approximations that can be made by visual or spatial representations of numbers. Thus, the left hemisphere in most people is verbal and mathematical, and the right hemisphere is intuitive, spatial, artistic, and musical.

STUDY BREAK

1. What part of the brain is responsible for maintaining homeostasis? Where is it located? What is unique about this part of the brain?
2. Distinguish the functions of the cerebellum from those of the cortex.

33.8 The Peripheral Nervous System (PNS)

The PNS can be divided into two main systems. The afferent system of the PNS includes all the neurons that transmit sensory information from receptors to the CNS. The efferent system consists of the axons of neurons that carry signals to the muscles and glands acting as effectors. The efferent system is further divided into somatic and autonomic systems **(Figure 33.31)**.

33.8a The Somatic System Controls the Contraction of Skeletal Muscles

The **somatic nervous system** controls body movements that are primarily conscious and voluntary. In mammals, 31 pairs of **spinal nerves** carry signals between the spinal cord and the body trunk and limbs. These spinal nerves, emanating from between each of the vertebrae, reflect the segmental organization of the vertebrate body. Each spinal nerve is made up of a dorsal root and a ventral root (see Figure 33.25) that emerge from the nerve cord to form the nerve. In lower vertebrates, such as sharks, the ventral root contains motor axons for the somatic muscles (those associated with movement), and the dorsal root contains the afferent axons from the sense organs and the efferent axons to the visceral muscles. The cell bodies for the sensory nerves are located outside the spinal cord within the dorsal root, forming the dorsal root ganglion. In mammals, the visceral motor axons have become included in the ventral root. Its neurons, called motor neurons, carry efferent signals from the CNS to the skeletal muscles, exiting from the nerve cord in the ventral root. The dendrites and cell bodies of motor neurons are located in the spinal cord; their axons extend from the spinal cord to the skeletal muscle cells they control. As a result, the somatic portions of the spinal nerves consist only of axons.

Although the somatic system is primarily under conscious, voluntary control, some contractions of skeletal muscles are unconscious and involuntary. These include reflexes, shivering, and the constant muscle contractions that maintain body posture and balance.

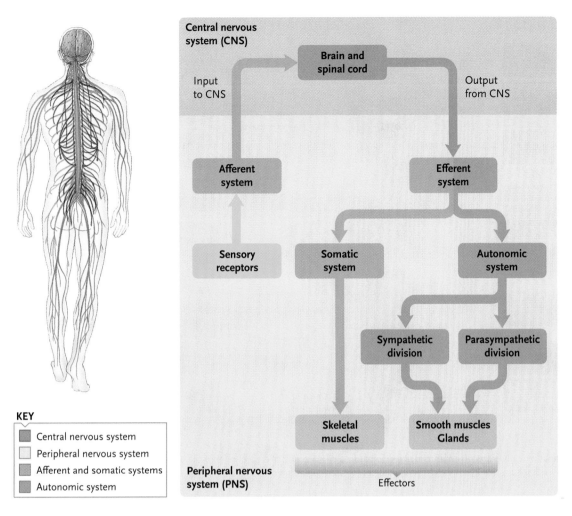

Figure 33.31
The central nervous system (CNS) and peripheral nervous system (PNS) and their subsystems.

Central nervous system (CNS)

Input to CNS

Brain and spinal cord

Output from CNS

Afferent system

Efferent system

Sensory receptors

Somatic system

Autonomic system

Sympathetic division

Parasympathetic division

Skeletal muscles

Smooth muscles Glands

Peripheral nervous system (PNS)

Effectors

KEY

- ▪ Central nervous system
- ▫ Peripheral nervous system
- ▪ Afferent and somatic systems
- ▪ Autonomic system

33.8b The Autonomic System Is Divided into Sympathetic and Parasympathetic Pathways

The **autonomic nervous system** controls largely involuntary processes such as digestion, secretion by sweat glands, circulation of the blood, many functions of the reproductive and excretory systems, and contraction of smooth muscles in all parts of the body. It is organized into *sympathetic* and *parasympathetic* divisions, which are always active and have opposing effects on the organs that they affect, thereby enabling precise control **(Figure 33.32, p. 812)**. For example, in the circulatory system, sympathetic neurons stimulate the force and rate of the heartbeat, and parasympathetic neurons inhibit these activities. In the digestive system, sympathetic neurons inhibit the smooth muscle contractions that move materials through the small intestine, whereas parasympathetic neurons stimulate the same activities. These opposing effects precisely control involuntary body functions.

The pathways of the autonomic nervous system include two neurons. The first neuron has its dendrites and cell body in the CNS, and its axon extends to a ganglion outside the CNS. There it synapses with the dendrites and cell body of the second neuron in the pathway. The axon of the second neuron extends from the ganglion to the effector carrying out the response.

The sympathetic division is associated primarily with the nerves of the **thorax** and abdomen, and the ganglia of the sympathetic division occur as a chain of segmental ganglia just ventral to the **vertebral column**.

The parasympathetic division has ganglia located within the brain, and the axons exit as cranial nerves or as the posterior or sacral parasympathetic nerves. The **cranial nerves** connect the brain directly to the head, neck, and body trunk. These are thought to represent the dorsal or ventral roots of the segmental nerves associated with the head and reflect the segmental origin of the head. The sacral parasympathetic nerves innervate the lower digestive tract and the external genitalia. The sacral nerves have additional ganglia in their target tissues.

The sympathetic division predominates in situations involving stress, danger, excitement, or strenuous

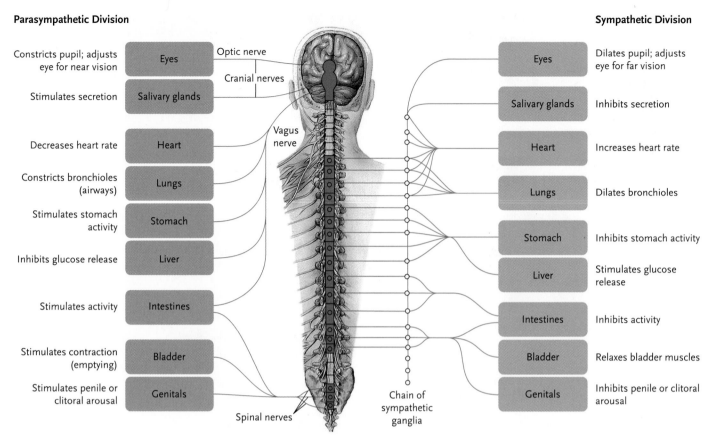

Parasympathetic Division

Constricts pupil; adjusts eye for near vision — Eyes
Stimulates secretion — Salivary glands
Decreases heart rate — Heart
Constricts bronchioles (airways) — Lungs
Stimulates stomach activity — Stomach
Inhibits glucose release — Liver
Stimulates activity — Intestines
Stimulates contraction (emptying) — Bladder
Stimulates penile or clitoral arousal — Genitals

Optic nerve
Cranial nerves
Vagus nerve
Spinal nerves

Sympathetic Division

Eyes — Dilates pupil; adjusts eye for far vision
Salivary glands — Inhibits secretion
Heart — Increases heart rate
Lungs — Dilates bronchioles
Stomach — Inhibits stomach activity
Liver — Stimulates glucose release
Intestines — Inhibits activity
Bladder — Relaxes bladder muscles
Genitals — Inhibits penile or clitoral arousal

Chain of sympathetic ganglia

Figure 33.32
Effects of sympathetic and parasympathetic divisions on organ and gland function. Only one side of each division is shown; both are duplicated on the left and right sides of the body.

physical activity. Signals from the sympathetic division increase the force and rate of the heartbeat, raise the blood pressure by constricting selected blood vessels, dilate air passages in the lungs, induce sweating, and open the pupils wide. Activities that are less important in an emergency, such as digestion, are suppressed by the sympathetic system. The parasympathetic division, in contrast, predominates during quiet, low-stress situations, such as relaxation. Under its influence, the effects of the sympathetic division, such as rapid heartbeat and elevated blood pressure, are reduced, and "housekeeping" (maintenance) activities such as digestion predominate.

STUDY BREAK

1. What two systems comprise the peripheral nervous system, and what do they generally control?
2. When an impala (*Aepyceros melampus,* a deerlike ungulate from Africa) attempts to evade an attacking pack of African wild dogs (*Lycaon pictus*), what division of its autonomic nervous system would dominate? What effects might result?

33.9 Memory, Learning, and Consciousness

We set memory, learning, and consciousness apart from the other CNS functions because they appear to involve coordination of structures from the brain stem to the cerebral cortex. **Memory** is the storage and retrieval of a thought or a sensory or motor experience. **Learning** involves a change in the response to a stimulus, based on information or experiences stored in memory. **Consciousness** is not easily defined. In a narrow sense, it involves awareness, a state of alertness to our surroundings. But there is a broader and deeper meaning that involves awareness of ourselves, our identity, and an understanding of the significance and likely consequences of events that we experience. In this section, we deal with sleep as a decrease in awareness.

33.9a Memory Takes Two Forms, Short Term and Long Term

Psychology research and our everyday experience indicate that humans have at least two types of memory. **Short-term memory** stores information for seconds, minutes, or at most an hour or so.

Long-term memory stores information from days to years or even for life. Short-term memory, but not long-term memory, is usually erased if a person experiences a disruption such as a sudden fright, a blow, a surprise, or an electrical shock. For example, a person knocked unconscious by an accident typically cannot recall the accident itself or the events just before it, but long-standing memories are not usually disturbed.

To explain these differences, investigators propose that short-term memories depend on transient changes in neurons that can be erased relatively easily, such as changes in the membrane potential of interneurons caused by EPSPs and IPSPs (excitatory and inhibitory postsynaptic potentials) and the action of indirect neurotransmitters that lead to reversible changes in ion transport. By contrast, storage of long-term memory is considered to involve more or less permanent molecular, biochemical, or structural changes in interneurons, which establish signal pathways that cannot be switched off easily.

All memories probably register initially in short-term form. They are then either erased and lost or committed to long-term form. The intensity or vividness of an experience, the attention focused on an event, emotional involvement, or the degree of repetition may all contribute to the conversion from short-term to long-term memory.

The storage pathway typically starts with an input at the somatosensory cortex that then flows to the amygdala, which relays information to the limbic system, and to the hippocampus, which sends information to the frontal lobes, a major site of long-term memory storage. People with injuries to the hippocampus cannot remember information for more than a few minutes; long-term memory is limited to information stored before the injury occurred. Squirrels hoard food for the winter in a number of caches and can locate these by remembering the location from landmarks rather than by tracking a smell. Each autumn, the hippocampus of a squirrel increases in size by about 15%.

How are neurons and neuron pathways permanently altered to create long-term memory? One change that has been much studied is **long-term potentiation**: a long-lasting increase in the strength of synaptic connections in activated neural pathways following brief periods of repeated stimulation. The synapses become increasingly sensitive over time, so that a constant level of presynaptic stimulation is converted into a larger postsynaptic output that can last hours, weeks, months, or years. Other changes consistently noted as part of long-term memory include more or less permanent alterations in the number and the area of synaptic connections between neurons, in the number and branches of dendrites, and in gene transcription and protein synthesis in interneurons. Experiments on both vertebrates and invertebrates

demonstrate that long-term memory depends on protein synthesis. For example, goldfish were trained to avoid an electrical shock by swimming to one end of an aquarium when a light was turned on. The fish could remember the training for about a month under normal conditions, but if they were exposed to a protein synthesis inhibitor while being trained, they forgot the training within a day.

33.9b Learning Involves Combining Past and Present Experiences to Modify Responses

As with memory, most animals appear to be capable of learning to some degree. Learning involves three sequential mechanisms: (1) storing memories, (2) scanning memories when a stimulus is encountered, and (3) modifying the response to the stimulus in accordance with the information stored as memory.

One of the simplest forms of memory is an increased responsiveness to mild stimuli after experiencing a strong stimulus, often called **sensitization**. The process was nicely illustrated by Eric Kandel of Columbia University and his associates in experiments with a shell-less marine snail, the Pacific sea hare, *Aplysia californica*. The first time the researchers administered a single sharp tap to the siphon (which admits water to the gills), the slug retracted its gills by a reflex movement. However, at the next touch, whether hard or gentle, the siphon retracted much more quickly and vigorously. Sensitization in *Aplysia* has been shown to involve changes in synapses, which become more reactive when more serotonin is released by action potentials. The cephalopod molluscs, such as the octopus, are capable of much more complex learning: they can distinguish and remember shapes and textures using only their tentacles.

33.9c Sleep Involves Different States of Awareness

Most animals that have been investigated, including some invertebrates, experience a daily rhythm of activity and inactivity. The inactive period, sleep, is essential to normal functioning. Sleep deprivation leads to disruption of a number of functions, including memory and learning, and, if prolonged, can be fatal. During sleep, there is some degree of awareness since external stimuli such as sound or internal stimuli such as a full bladder can interrupt sleep.

In humans and other mammals, sleep is accompanied by changes in the electrical activity of the cerebrum as detected by electrodes applied to the scalp during an *electroencephalogram*. The waking state is characterized by rapid, irregular *beta waves* **(Figure 33.33, p. 814).** As the eyes close and you become fully relaxed, these give way to slower, more regular *alpha waves*. As you become more drowsy, these are replaced by slower *theta waves*.

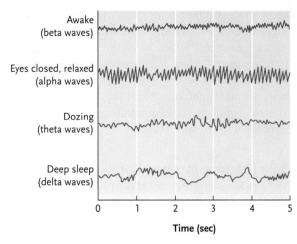

Figure 33.33
Brain waves characteristic of various states of consciousness.

Full sleep is characterized by even slower *delta waves*. The heart rate falls, and the muscles are relaxed.

During full sleep, the brain returns at intervals to periods of beta waves, during which the heart rate increases, the muscles may twitch, and the eyes move rapidly behind the closed lids, giving these periods the name **rapid eye movement**, or **REM, sleep**. This

brief period of about 10 to 15 minutes occurs every 90 minutes or so in healthy adults.

This pattern of alternating periods of greater or lesser cerebral activity is also characteristic of bird sleep, although birds may sleep on one side of the brain while the other remains fully alert as protection against predators. Reptiles, with their less developed cerebrum, experience alternating patterns of activity in the amygdala.

Although we know that sleep is essential, we do not understand the physiological basis for these effects. The fruit fly *Drosophila melanogaster* exhibits cycles of sleep. At night, it feeds and then seeks out an isolated place and becomes inactive for about eight hours. Interrupting the sleep interferes with memory and learning in the flies. Many labs are now using the flies as models to identify genes involved in the sleep process.

STUDY BREAK

In the sleep cycle, what are beta waves, and where do they originate? How do they differ from delta waves?

UNANSWERED QUESTIONS

Although great progress is being made in understanding many brain functions, it is in the area of "consciousness" that our understanding is very limited. If we restrict our definition of that word to alertness and awareness, then we have been able to describe some of the physiological characteristics of that state, if only by describing what happens when we are "unconscious" or asleep. But humans have an awareness of self and identity that is presumably a function of the brain. In some individuals, two or more "selfs" have been known to exist, or a new "self" suddenly appears to replace the original. The description of this broader consciousness, the self-awareness and its relationship to brain activity, is one of the major challenges of neuroscience.

Much of neuroscience is focused on humans. But we are a product of evolution, and we do not understand the evolution of many of the

functions that define humans. Our ability to learn and to pass the learning on to offspring is not unique—killer whales teach their young to hunt, and birds learn songs from their parents. Understanding how these functions arose requires novel approaches. Scientists are beginning to look at very simple models to understand memory, a prerequisite for learning. For example, the nematode *Caenorhabditis elegans* has been demonstrated to learn: offered a choice between pathogenic and nonpathogenic bacteria, they will at first not discriminate but will do so after experience. The simplicity of its nervous system and the extensive structural and genetic information available led Cathy Rankin at the University of British Columbia to explore the molecular basis of learning in *C. elegans*. She finds that learning can be associated with the level of expression of specific genes, depending on the response being studied.

Review

Go to CENGAGENOW™ at http://hed.nelson.com/ to access quizzing, animations, exercises, articles, and personalized homework help.

33.1 Neurons and Their Organization in Nervous Systems: An Overview

- The nervous system of an animal (1) receives information about conditions in the internal and external environment, (2) transmits the message along neurons, (3) integrates the information to formulate an appropriate response, and (4) sends out signals to effector organs.

- Neurons are cells specialized for the reception and transmission of signals. They have dendrites, which receive information and conduct signals toward the cell body, and axons, which conduct signals away from the cell body to another neuron or an effector.

- Afferent neurons conduct information from sensory receptors to interneurons, which integrate the information into a response. The response signals are passed to efferent neurons, which activate the effectors carrying out the response.

- Glial cells provide structural and functional support to neurons. They help maintain the balance of ions surrounding neurons and form insulating layers around the axons.
- Neurons make connections by two types of synapses, electrical and chemical. In an electrical synapse, impulses pass directly from the sending to the receiving cell. In a chemical synapse, neurotransmitter molecules released by the presynaptic neuron diffuse across a narrow synaptic cleft and bind to receptors in the plasma membrane of the postsynaptic cell. Binding of the neurotransmitters may generate an electrical impulse in the postsynaptic cell.

33.2 Signal Conduction by Neurons

- The membrane potential of a cell depends on the unequal distribution of positive and negative charges on either side of the membrane, which establishes a potential difference, the resting potential, across the membrane.
- The resting potential results from an active transport pump that sets up concentration gradients of Na^+ ions (higher outside) and K^+ ions (higher inside) and negatively charged proteins and other molecules inside the cell that cannot pass through the membrane.
- An action potential is generated when a stimulus pushes the resting potential to the threshold value at which voltage-gated Na^+ channels open in the plasma membrane. The inward flow of Na^+ changes membrane potential abruptly from a negative to a positive peak, which opens the voltage-gated K^+ channels. The potential falls to the resting value again as the gated K^+ channels allow this ion to flow out.
- Action potentials move along an axon as the ion flows generated in one location on the axon depolarize the potential in the adjacent location.
- Action potentials are prevented from reversing direction by a brief refractory period, during which a sector of membrane that has just generated an action potential cannot be stimulated to produce another for a few milliseconds and the action potential has moved too far away for its electrical disturbances to cause the preceding sector to depolarize again.
- In myelinated axons, ions can flow across the plasma membrane only at the nodes of Ranvier, where the insulating myelin sheath is interrupted.
- The intensity of a stimulus is reflected in the frequency of action potentials.

33.3 Conduction across Chemical Synapses

- Neurotransmitters released into the synaptic cleft bind to receptors in the plasma membrane of the postsynaptic cell, altering the flow of ions across the plasma membrane of the postsynaptic cell and pushing its membrane potential toward or away from the threshold potential.
- A direct neurotransmitter binds to a receptor associated with a ligand-gated ion channel in the postsynaptic membrane; the binding opens or closes the channel.
- An indirect neurotransmitter works as a first messenger, binding to a receptor in the postsynaptic membrane and triggering generation of a second messenger, which leads to the opening or closing of a gated channel.
- Neurotransmitters are released from synaptic vesicles into the synaptic cleft by exocytosis, which is triggered by entry of Ca^{2+} ions into the cytoplasm of the axon terminal through voltage-gated Ca^{2+} channels opened by the arrival of an action potential.
- Neurotransmitter release stops when action potentials cease arriving at the axon terminal. Neurotransmitters remaining in the synaptic cleft are broken down by enzymes or taken up by the axon terminal or glial cells.
- Types of neurotransmitters include acetylcholine, amino acids, biogenic amines, neuropeptides, and gases such as NO and CO.

33.4 Integration of Incoming Signals by Neurons

- Integration of incoming information by a neuron determines the frequency at which the receiving neuron sends out action potentials.
- Neurons carry out integration by summing excitatory postsynaptic potentials (EPSPs) and inhibitory postsynaptic potentials (IPSPs). The summation may occur over time (temporal) or from different neurons at the same time (spatial). This summation pushes the membrane potential of the receiving cell toward or away from the threshold for an action potential.

33.5 Integration in Protostomes: Networks, Ganglia, and Brains

- Identifiable nerves first appear in radially symmetrical animals as nerve nets composed of single neurons that are more concentrated around the mouth and that may have localized groupings of cell bodies that control particular functions.
- The development of bilateral symmetry resulted in the concentration of parts of the nerve net into several longitudinal nerve cords composed of several axons, with paired ventral cords becoming increasingly dominant.
- Ganglia, local concentrations of nerve cell bodies permitting enhanced coordination of sensory and motor functions, first appeared in the flatworms, with the anterior ganglia prominent and acting as a brain. Protostome ganglia have the cell bodies at the periphery and the neuropile of axons and dendrites in the interior.
- Molluscs have well-developed nervous systems with paired ventral nerve cords and a brain consisting of several fused ganglia that permits advanced behaviour in the cephalopods.
- In segmented protostomes, each segment has a pair of ganglia connected by a lateral commissure and connected to adjacent ganglia by paired ventral nerve cords. Segmental ganglia control the functions of the segment and its appendages and are subject to control from the brain.
- The brain in segmental animals consists of the fused ganglia of the segments that make up the head.

33.6 Vertebrates Have the Most Complex Nervous Systems

- The protovertebrate nervous system, represented by the lancelet, is a hollow dorsal tube with nerves leading to each segmental block of muscle. The anterior end of the tube is larger, and three regions equivalent to the forebrain, midbrain, and hindbrain of the vertebrate embryonic brain can be recognized.
- In the development of the vertebrate brain, the hindbrain subdivides into the myelencephalon, which becomes the medulla oblongata, or brain stem, responsible for many involuntary functions, and the metencephalon associated with hearing and balance. The cerebellum, a major processing center for balance and navigation, is an outgrowth of the metencephalon. The midbrain, or mesencephalon, coordinates hearing and vision. The forebrain subdivides into the diencephalon, which gives rise to the optic nerves, and the telencephalon, which is responsible for olfaction and gives rise to the cerebrum, the major processing centre of the brain.

- During evolution, some parts of the brain are more prominent, reflecting the lifestyle of the animal. Sharks have large olfactory centres, whereas frogs have larger optic centres.
- The evolution of the brain also involves an increase in its mass relative to the body weight of the animal and, in particular, an increase in the mass of the cerebrum relative to the rest of the brain. Birds and mammals exhibit an increased prominence of the cerebellum.
- In vertebrates, the CNS consists of a large brain located in the head and a hollow spinal cord, and the PNS consists of all the nerves and ganglia connecting the CNS to the rest of the body.

33.7 The Central Nervous System (CNS) and Its Functions

- The CNS consists of the brain and spinal cord. The spinal cord carries signals between the brain and the PNS. Its neuron circuits also control reflex muscular movements and some autonomic reflexes.
- The adult derivatives of the hindbrain—the pons, medulla oblongata, and cerebellum—together with the relatively reduced midbrain, form the brain stem, which connects the telencephalon with the spinal cord.
- The telencephalon (cerebrum) is divided into right and left cerebral hemispheres, which are connected by a thick band of nerve fibres, the corpus callosum. The cerebral cortex, the surface of the cerebrum, is formed by grey matter. Other collections of grey matter, such as the thalamus, hypothalamus, and basal nuclei, lie at deeper layers of the telencephalon.
- Cerebrospinal fluid provides nutrients to and cushions the CNS. A blood–brain barrier set up by tight junctions between the cells of the capillary walls in the CNS allows only selected substances to enter the cerebrospinal fluid.
- Grey-matter centres in the pons and medulla control involuntary functions such as heart rate, blood pressure, respiration rate, and digestion. Centres in the midbrain coordinate responses to visual and auditory sensory inputs.
- The cerebellum integrates sensory inputs on the positions of muscles and joints, along with visual and auditory information, to coordinate body movements.
- Certain grey-matter centres of the telencephalon control a number of functions. The thalamus receives, filters, and relays sensory and motor information to and from regions of the cerebral cortex. The hypothalamus, the only part of the brain not protected by the blood–brain barrier, regulates basic homeostatic functions of the body and contributes to the endocrine control of body functions. The basal nuclei affect the planning and fine-tuning of body movements.
- The limbic system includes parts of the thalamus, hypothalamus, and basal nuclei, as well as the amygdala and hippocampus. It controls emotional behaviour and influences the basic body functions controlled by the hypothalamus and brain stem.
- The primary somatosensory areas of the cerebral cortex register incoming information on touch, pain, temperature, and pressure from all parts of the body. The temporal lobes receive input from the ears, the primary visual cortex from the eyes, the olfactory lobes from the nose, and the parietal lobes from taste receptors

in the mouth. In general, the right cerebral hemisphere receives sensory information from the left side of the body, and vice versa.
- The primary motor areas of the cerebrum control voluntary movements of skeletal muscles in the body.
- The association areas integrate sensory information and formulate responses that are passed on to the primary motor areas. Importantly, Wernicke's area integrates visual, auditory, and other sensory information into the comprehension of language, whereas Broca's area coordinates movements of the lips, tongue, jaws, and other structures to produce the sounds of speech.
- Some functions, such as long-term memory and consciousness, are equally distributed between the two cerebral hemispheres. In contrast, the left hemisphere in most people specializes in spoken and written language, abstract reasoning, and precise mathematical calculations. The right hemisphere specializes in nonverbal conceptualizing, mathematical estimation, intuitive thinking, spatial recognition, and artistic and musical abilities.

33.8 The Peripheral Nervous System (PNS)

- Afferent neurons in the PNS conduct signals to the CNS, and signals from the CNS go via efferent neurons to the muscles and glands that carry out responses.
- The somatic system of the PNS controls the skeletal muscles that produce voluntary body movements, as well as involuntary muscle contractions that maintain balance, posture, and muscle tone.
- The autonomic system of the PNS controls involuntary functions such as heart rate and blood pressure, glandular secretion, and smooth muscle contraction.
- The autonomic system is organized into sympathetic and parasympathetic divisions that balance and fine-tune involuntary body functions. The sympathetic system predominates in situations involving stress, danger, or strenuous activities, whereas the parasympathetic system predominates during quiet, low-stress situations.

33.9 Memory, Learning, and Consciousness

- Memory is the storage and retrieval of a sensory or motor experience, or a thought. Short-term memory involves temporary storage of information, probably resulting from changes in the membrane potential of interneurons, whereas long-term memory is essentially permanent, involving molecular, biochemical, or structural changes in interneurons.
- Learning involves modification of a response through comparisons made with information or experiences that are stored in memory.
- Consciousness is the awareness of ourselves, our identity, and our surroundings. It varies through states from full alertness to sleep.
- Sleep is characterized by alternations in patterns of electrical activity in the cerebrum between slow, relatively regular delta waves, characteristic of deep sleep, and brief periods of rapid, irregular beta waves signalling REM sleep.

Questions

Self-Test Questions

1. Nerve signals travel in the following manner:
   a. A dendrite of a sensory neuron receives the signal; its cell body transmits the signal to a motor neuron's axon, and the signal is sent to the target.
   b. An axon of a motor neuron receives the signal; its cell body transmits the signal to a sensory neuron's dendrite, and the signal is sent to the target.
   c. Efferent neurons conduct nerve impulses toward the cell body of sensory neurons, which send them on to interneurons and, ultimately, to afferent motor neurons.
   d. A dendrite of a sensory neuron receives a signal; the cell's axon transmits the signal to an interneuron; the signal is then transmitted to dendrites of a motor neuron and sent via its axon to the target.
   e. The axons of oligodendrocytes transmit nerve impulses to the dendrites of astrocytes.

2. An example of a synapse could be the site where
   a. neurotransmitters released by an axon travel across a gap and are picked up by receptors on a muscle cell.
   b. an electrical impulse arrives at the end of a dendrite, causing ions to flow onto axons of presynaptic neurons.
   c. postsynaptic neurons transmit a signal across a cleft to a presynaptic neuron.
   d. the axons of a presynaptic neuron directly contact the dendrites of a postsynaptic neuron.
   e. an on–off switch stimulates an electrical impulse in a presynaptic cell to stimulate, not inhibit, other presynaptic cells.

3. Ganglia first became enlarged and fused into a lobed brain in the evolution of
   a. vertebrates.
   b. annelids.
   c. flatworms.
   d. cephalopods.
   e. mammals.

4. The metencephalon is the origin of the
   a. spinal cord.
   b. cerebellum.
   c. mesencephalon.
   d. medulla oblongata.
   e. cerebrum.

5. Persons who have had damage to the right side of the cerebellum
   a. will have difficulty reading.
   b. will be unable to distinguish colours.
   c. will have difficulty with balance.
   d. will be unable to speak.
   e. will be unable to hear.

Questions for Discussion

1. The mechanism for the propagation of the action potential along an axon was worked out using the giant axons of squids. How confident should we be that this model applies to vertebrates? Is there an evolutionary link between the nerves of vertebrates and those of molluscs, or did the mechanism arise twice?

2. Brain function in many animals changes with age, and not all of these changes are degenerative. How would you explore the hypothesis that changes in gene expression were involved, and what would be a useful model animal?

3. How did evolution of chemical synapses make higher brain function possible?

Male eastern red bat, *Lasiurus borealis*. This tree-roosting bat flies south for the winter, returning every summer to the same hunting grounds across much of North America.

M. B. Fenton

34 Sensory Systems

WHY IT MATTERS

Echolocation, also known as biosonar, allows some bats and other animals to operate in the dark and in situations where lighting is unpredictable. Many (but not all) bats, toothed whales, some birds, and some insectivores (insect-eating mammals) echolocate. Echolocation may have been a key development allowing the ancestors of bats to move into the niche of a nocturnal aerial insectivore perhaps about 60 million years ago. To echolocate, an animal produces a pulse of sound and listens for echoes of it. The differences between what the animal says and what it hears are the data used in **echolocation**.

A male pink moth (*Scoliopteryx libatrix*) flies silently through the night sky in search of a receptive female. Suddenly, it hears the faint echolocation calls of a hunting eastern red bat (as in the chapter opening photograph). The moth has a pair of ears, one on each side of its thorax. Vibrations of the moth's tympanic membranes (eardrums) generate stimuli in its auditory nerves. By comparing the left and right stimuli from the bat's calls, the moth can turn its back to the bat and fly away from it. Most of the time, moths never appear on the bat's echolocation screen. Had the bat been closer, its strong calls

Figure 34.1

A painted lichen moth, *Hypoprepia fucosa*. This colourful moth uses acoustic signals to warn would-be bat predators of its bad taste. It uses bright colours to provide the same warning to insectivorous birds.

Gord Temple

would have alerted the moth to the immediate danger and it would have dived for the ground.

The red bat flies on, oblivious to the insect it never detected. But picking up an echo from another insect, a painted lichen moth **(Figure 34.1)**, the bat turns and closes with its target. As it attacks, the bat adjusts its echolocation calls to ensure that its outgoing pulses do not deafen it to faint returning echoes. The bat increases the rate at which it produces echolocation calls, shortening each call and the intervals between them **(Figure 34.2)**. Just as the bat is about to make contact, the moth produces a sequence of clicks. The bat aborts its attack and the moth flies on, holding a steady course.

Caterpillars of painted lichen moths feed on lichens, sequestering toxic chemicals that remain in the adult moth. The moth's clicks warn the bat that it tastes bad. Experienced red bats get the message and abort their attacks. Inexperienced red bats (or those fooled by an experimenter into attacking a painted

lichen moth) finish the attack and grab the moth but quickly spit it out.

Many bats use echolocation to detect insect prey. Some insects have chemical defences and use sounds to warn bats of their bad taste. Painted lichen moths deal with two threats, using the clicking sounds for the bats and bright colours to warn birds that they taste bad.

The purpose of this chapter is to review information about sensory systems and describe how information from within and from without is acquired and used by animals.

34.1 Overview of Sensory Receptors and Pathways

Sensory systems begin with **sensory receptors (transducers)** that detect sensory information, convert it to neural activity, and pass the information along neurons to the central nervous system (CNS). Sensory receptors are formed by the dendrites of afferent neurons or by specialized receptor cells **(Figure 34.3)**. Receptors collect information about the internal and external environments of organisms. In organisms with a developed head region (cephalized), many receptors for external stimuli are located there so that the organism can collect information about where it is going. Receptors associated with eyes, ears, skin, and other surface organs detect stimuli from the external environment. Sensory receptors associated with internal organs detect stimuli arising in the body interior.

Sensory transduction occurs when stimuli cause changes in membrane potentials in the sensory receptors. Usually, this is achieved by changes in rates at which channels conduct positive ions (Na^+, K^+, or Ca^{2+}) across the plasma membrane. Stimuli may be in the form of light, heat, sound waves, mechanical stress, or chemicals. The change in membrane potential may generate one or more action potentials that travel along the axon of an afferent neuron to reach interneuron networks of the CNS. These interneurons integrate the action potentials, and the brain formulates a compensating response, that is, a response appropriate for the stimulus (see Chapter 33). In animals with complex nervous systems, interneuron networks may produce an awareness of a stimulus in the form of a conscious sensation or perception.

34.1a Basic Types of Receptors: What an Animal Needs to Know

Many sensory receptors are positioned individually in body tissues. Others are part of complex sensory organs, such as the eyes or ears, specialized for reception of physical or chemical stimuli. Receptors, particularly for external information, usually occur in pairs,

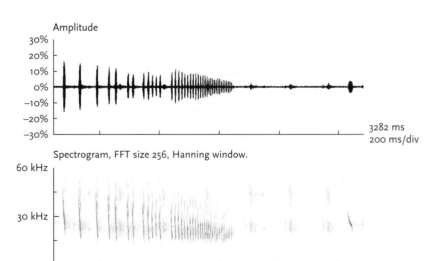

Amplitude

Spectrogram, FFT size 256, Hanning window.

Figure 34.2

Echolocation call sequence of a red bat attacking a flying prey.

a. Sensory receptor formed by dendrites of an afferent neuron

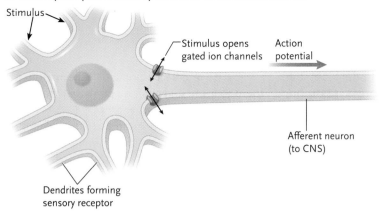

Stimulus

Stimulus opens gated ion channels

Action potential

Afferent neuron (to CNS)

Dendrites forming sensory receptor

In sensory receptors formed by the dendrites of afferent neurons, a stimulus causes a change in membrane potential that generates action potentials in the axon of the neuron. Temperature and pain receptors are among the receptors of this type.

Figure 34.3
Sensory receptors, formed **(a)** by the dendrites of an afferent neuron or **(b)** by a separate cell or structure that communicates with an afferent neuron via a neurotransmitter.

b. Sensory receptor formed by a cell that synapses with an afferent neuron

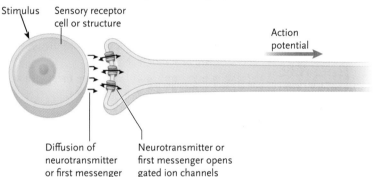

Stimulus

Sensory receptor cell or structure

Action potential

Diffusion of neurotransmitter or first messenger

Neurotransmitter or first messenger opens gated ion channels

In sensory receptors consisting of separate cells, a stimulus causes a change in membrane potential that releases a neurotransmitter from the cell. The neurotransmitter triggers an action potential in the axon of a nearby afferent neuron. Mechanoreceptors, photoreceptors, and chemoreceptors are examples of receptors of this type.

providing the opportunity for the animal to localize the stimulus. Eyes, ears, and antennae are examples of paired sensory organs. There are exceptions. *Opabinia* species from the Burgess Shales (see Chapter 26) had five eyes, and some species of preying mantis have only one ear. Some spiders have rows of simple eyes.

Sensory receptors are classified into five major types, based on the type of stimulus that each detects:

- **Mechanoreceptors** detect mechanical energy, such as changes in pressure, body position, or acceleration. The auditory receptors in the ears are examples of mechanoreceptors.
- **Photoreceptors** detect the energy of light. In vertebrates, photoreceptors are mostly located in the retina of the eye.
- **Chemoreceptors** detect specific molecules or chemical conditions such as acidity. Taste buds on the tongue are examples of chemoreceptors.
- **Thermoreceptors** detect the flow of heat energy. Receptors of this type are located in the skin, where they detect changes in the temperature of the body surface.
- **Nociceptors** detect tissue damage or noxious chemicals; their activity registers as pain. Pain receptors are located in the skin and in some internal organs.

Some animals also have receptors that detect electrical or magnetic fields. Traditionally, humans are

said to have five senses: vision, hearing, taste, smell, and touch. In reality, we can detect almost twice as many kinds of environmental stimuli as suggested by these labels. The traditional list should also include external heat, internal temperature, gravity, acceleration, the positions of muscles and joints, body balance, internal pH, and the internal concentrations of substances such as oxygen, carbon dioxide, salts, and glucose.

34.1b Afferent Links to the Central Nervous System (CNS)

Sensory pathways begin at a sensory receptor and proceed by afferent neurons to the CNS. Each type of receptor conveys information to a specific part of the CNS. Action potentials arising in the retina of the eye travel along the optic nerves to the visual cortex, where they are interpreted by the brain as differences in pattern, colour, and intensity of light. A blow to the eye is a stimulus that is interpreted in the visual cortex as differences in the colour and intensity of light detected by the eyes. Therefore, you "see stars" after receiving the blow even though the stimulus was mechanical.

The frequency of action potentials that the stimulus generates in the afferent neuron (number per unit time) can indicate the intensity and extent of the stimulus. Stronger stimuli cause more action potentials than weaker ones (see Chapter 33). A light touch to the hand, for example, causes action potentials to flow at low frequencies along the axons leading to the primary somatosensory area of the cerebral cortex. As the pressure increases, the number of action potentials per second rises in proportion. In the brain, the increase is interpreted as greater pressure on the hand. In the sensory cortex, maximum stimulus input is interpreted as pain.

The numbers of afferent neurons sending action potentials in response to a stimulus can also convey information about the intensity and extent of a stimulus. The more sensory receptors that are activated, the more axons carry information to the brain. A light touch activates a relatively small number of receptors in a small area near the surface of the finger. But as the pressure increases, the resulting indentation of the finger's surface increases in area and depth, activating more receptors. In the appropriate somatosensory area of the brain, the larger number of axons carrying action potentials is interpreted as an increase in pressure spread over a greater area of the finger.

34.1c Minimizing Sensory Overload: Reducing Background Noise

In many sensory systems, the effect of a stimulus is reduced if it continues at a constant level. This reduction is called **sensory adaptation** (do not confuse this with adaptation used in the context of evolution). Some receptors adapt quickly and broadly; other receptors adapt only slightly. In bed, you are initially aware of the touch and pressure of the covers on your skin. Within a few minutes, the sensations lessen or are lost, even though your position remains the same. The loss reflects adaptation of mechanoreceptors in your skin. If you move so that the stimulus changes, the mechanoreceptors again become active. In contrast, receptors detecting painful stimuli show little or no adaptation.

In some sensory receptors, biochemical changes in the receptor cell contribute to adaptation. When you move from a dark movie theatre into the bright sunshine, the photoreceptors of the eye adapt to the sudden bright light partly through breakdown of some of the pigments that absorb light.

Sensory adaptation is crucial to survival. Adaptation of photoreceptors in our eyes keeps us from being blinded indefinitely as we pass from the dark into bright sunlight. Sensory adaptation also increases the sensitivity of receptor systems to *changes* in environmental stimuli. These can be more important to survival than keeping track of constant environmental factors. Consider a cat sitting motionless, its attention focused on a stationary mouse. As long as the mouse stays still (environmental stimuli are constant), the cat also does not move. But if the mouse moves (change

in environmental stimuli), the cat responds rapidly, attempting to capture and kill it.

Many prey animals use adaptation by predators as a means for concealment or defence (see Chapters 40 and 46). These prey instinctively stop moving when they sense or detect a predator. Freezing often allows them to avoid detection by predators that depend on motion detectors to locate prey.

Pain detectors are examples of nonadapting receptors. They are also essential for survival. Pain signals a potential danger to some part of the body, and the signals are maintained until a response by the animal compensates for the stimulus causing the pain.

STUDY BREAK

1. What is the importance of sensory systems for an organism?
2. Name five sensory receptors and the type of stimulus each detects.
3. Why is it important that some receptors allow the effect of a stimulus to be reduced, whereas other receptors do not?

34.2 Mechanoreceptors and the Tactile and Spatial Senses

Mechanoreceptors detect mechanical stimuli such as touch and pressure. In this situation, mechanical forces of the stimulus distort proteins in the plasma membrane of receptors, altering the flow of ions through the membrane. Ion flows change the membrane potential of the receptors and generate action potentials in afferent neurons leading from the receptors to the CNS. Sensory information from these receptors informs the brain of the body's contact with objects in the environment, providing information on the movement, position, and balance of body parts and underlying the sense of hearing. The distinction between sound (vibrations in air) and seismic waves (vibrations in the substrate) may not always be clear (see *Good Vibrations*). The five basic types of mechanoreceptors are described below.

34.2a Touch and Pressure

In vertebrates, mechanoreceptors that detect touch and pressure are embedded in the skin and other surface tissues, in skeletal muscles, in the walls of blood vessels, and in internal organs. In humans, touch receptors in the skin are concentrated in greatest numbers in the fingertips, lips, and tip of the tongue, giving these regions the greatest sensitivity to mechanical stimuli. In other areas, such as the skin of the back, arms, and legs, the receptors are more widely spaced.

Good Vibrations

Seismic signals—vibrations in the ground—are used by a variety of animals for different purposes. Nocturnal scorpions (*Paruroctonus mesaensis*) use vibrations in sand to detect prey under the surface, as do some golden moles (Chrysochloridae family). Meanwhile, frogs, snakes, and various mammals use seismic vibrations as part of their communication repertoires.

The scorpion uses sense organs at the end of each of its six walking legs, a basitarsal compound slit sensillum and tarsal sensory hairs **(Figure 1).** Both receptors are stimulated by vibrations in sand: the tarsal hairs to compressional waves and the basitarsal compound slit sensillum to surface (Raleigh) waves. The tarsal hairs provide information about nearby (<15 cm) sources of vibration and the slit sensilla to more distant ones. Both receptors are sensitive to a very small amplitude of <10 Å (angstroms) mechanical stimuli, and receptor pairs on six legs give the scorpion information about the direction of the sources of detected vibrations.

Male white-lipped frogs (*Leptodactylus albilabris*) vocalize from within clumps of grass or from shallow depressions or burrows in the mud in their native Puerto Rico. These frogs are ground-dwelling, and males produce two distinct types of vocal signals: chirps that advertise the species' identity and chuckles that are aggressive signals to conspecifics. Males typically call with only the anterior half of their body above ground, and as the vocal sac expands explosively during the chirp call, it strikes the ground with enough force to generate a surface wave. The chirp and the associated surface waves indicate the position of calling males.

Vibration detectors in frogs **(Figure 2)** appear to be located in the sacculus of

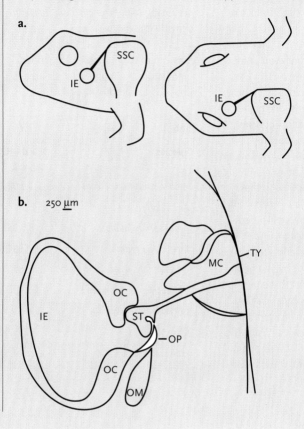

a.

b. 250 μm

Figure 2

Ranid frogs typically have this orientation of the opercularis muscle (OM), the m. opercularis connecting the suprascapular cartilage (SSC) of the pectoral girdle to the operculum in the middle ear. IE – inner ear; MC – middle ear; OC – otic capsule; OP – operculum; ST – stapes; TY – typanum.

Figure 1

Right fourth leg of *Paruroctonus mesaensis*, showing tarsal hairs (H) and basitarsal compound slit sensillum (BCSS). B – bristle hairs; BT – basitarsus; LC and MC – lateral and medial claws; PS – pedal spur; T – tarsus.

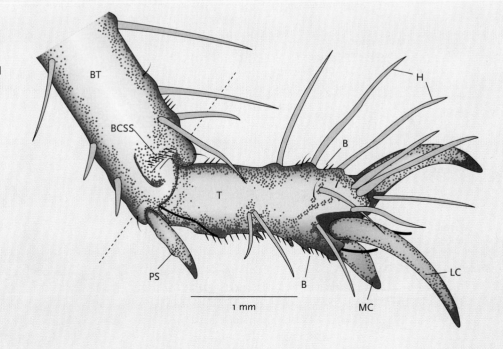

the inner ear. This otolithic organ contains a membranous sac containing a slurry of dense calcium carbonate crystals. Vibrations are conducted to the sacculus by a chain of connections that includes the opercularis muscle that connects the pectoral girdle with the operculum, which is a movable cartilaginous element in the oval window of the otic capsule.

Golden moles (family Chrysochloridae) are fossorial insectivorous mammals from southern Africa. They make and live in burrows, rarely venturing onto the surface. Species within this group show striking variation in the structure of the malleus, one of the three mammalian auditory ossicles **(Figure 3; Table 1).** The auditory ossicles of *Amblysomus hottentotus* are generally like those of other mammals, including humans. However, other species of golden moles show considerable enlargement of the malleus culminating in *Chrysospala trevelyani*, in which it is pea shaped and about twice as large as it is in a normal mammal. The differences suggest that some species of golden moles are specialized for detecting low-frequency ground vibrations (Rayleigh waves), using the combination of the horizontal orientation of the malleus and its expanded

size. The extended anterior process of the malleus also increases the overall size of the bone. Specialized golden moles, such as *C. stuhlmanni,* appear to use their sensitivity to surface waves to detect prey. Moles with these specializations may have much less acute hearing than less specialized species.

Do humans use seismic information? Why do people stamp their feet (see Chapter 40, *People Behind Biology*) in frustration and/or anger?

There are times when using vibrations can make you less conspicuous to predators. Although katydids are notorious for producing acoustic

signals, usually males trying to attract females, the cost of sound production can be predation. Many species of insect-eating bats listen for the courtship sounds of male katydids (see Figure 34.14) and use these signals to detect and home in on prey. Males of at least 13 species of katydids from the New World tropics use short songs supplemented by complex, species-specific vibrations (tremulations) to advertise themselves to females while avoiding marauding bats. The same general situation applies to frogs that are vulnerable to bat predation.

Table 1

Species	$M \bullet d \bullet IL^{-1}$, mg	dB re human
Amblysomus hottentotus	0.52	−28.91
Amblysomus gunningi	2.38	−15.70
Amblysomus julianae	0.60	−27.67
Eremitalpa granti	65.28	13.06
Chrysochloris asiatica	63.76	12.86
Chrysochloris stuhlmanni	49.78	10.71
Chrysospalax trevelyani	1145.35	37.95
Chrysospalax villosus	1545.90	40.55
Homo sapiens	14.51	0.00

Figure 3

The auditory ossicles (middle ear bones) of three species of golden moles, *Amblysomus hottentotus* **(a)**, *Chrysochloris stuhlmanni* **(b)**, and *Chrysospalax trevelyani* **(c)**. The scale bar is 5 mm with respect to *A. hottentotus* and *C. stuhlmanni* and 10 mm for *C. trevelyani*.

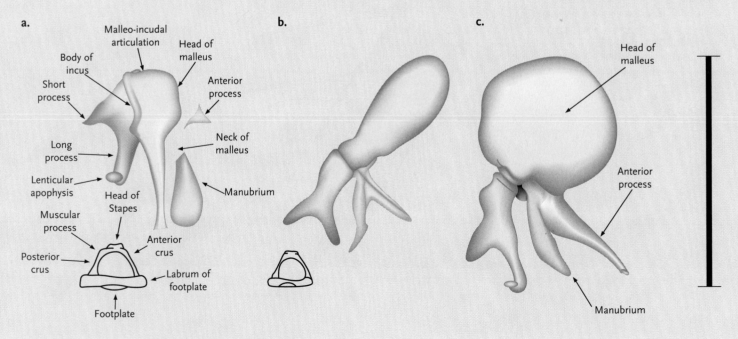

Forked Tongues

The forked tongues of serpents (see Figure 34.32) are deeply embedded in the world's religious iconography, often as a representation of deceit and malevolence. Aristotle thought that the forked tongue could double taste sensations. Hodierna proposed that the forked tongue allowed snakes to pick dirt out of both nostrils simultaneously.

Snake tongues are involved in chemoreception and serve as the delivery mechanism for paired sensors in Jacobson's organs (vomeronasal organs) on the roofs of snakes' mouths. Jacobson's organs connect with the oral cavity through two small openings in the palate (vomeronasal fenestrae).

However, forked tongues are not restricted to snakes, and lepidosaurian reptiles (see Chapter 27) show considerable variation in tongue structure **(Figure 1).** Forked tongues allow snakes (or lizards) to follow the pheromone trails of prey and conspecifics. The forked tongue specifically allows animals to use tropotaxis, which means simultaneously sampling chemical stimuli at two points. In some snakes, varanid lizards, and teiid lizards, the distance between the tongue tips exceeds the width of the head. Forked tongues have evolved at least twice and perhaps as many as four times in lepidosaurian reptiles **(Figure 2).**

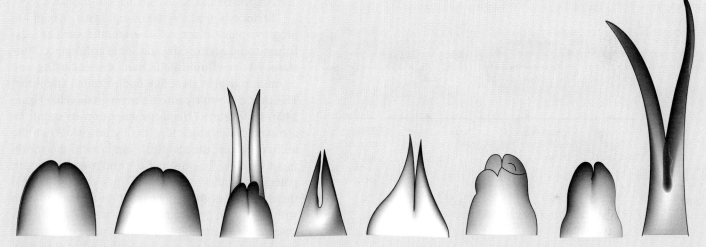

Figure 1
Tongue tips vary in squamate reptiles, from simple notches to deep forks. Shown here, from left to right, are the tongues of *Sceloporus* (Iguania), *Coleonyx* (Gekkonidae), *Cnemidophorus* (Teiidae), *Lacerta* (Lacertidae), *Bipes* (Amphisbaenia), *Scincella* (Scincidae), *Abronia* (Anguidae), and *Varanus* (Varanidae). Most snake tongues look like the *Varanus*.

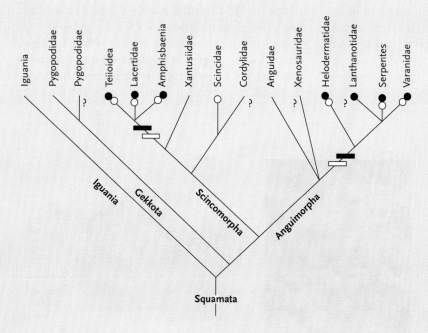

Figure 2
A cladistic phylogeny of Squamata. The black circles identify taxa with forked tongues.

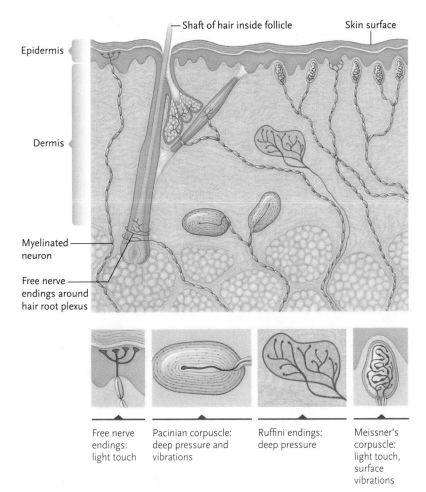

Epidermis

Dermis

Myelinated neuron

Free nerve endings around hair root plexus

Shaft of hair inside follicle

Skin surface

| Free nerve endings: light touch | Pacinian corpuscle: deep pressure and vibrations | Ruffini endings: deep pressure | Meissner's corpuscle: light touch, surface vibrations |

Figure 34.4

In human skin, four types of mechanoreceptors detect tactile stimulation.

You can compare the spacing of receptors by pressing two toothpicks lightly against a fingertip and then against the skin of your arm or leg. On your fingertip, two toothpicks separated by ~1 mm can be discerned as two separate points. On your arm or leg, the two toothpicks must be nearly 50 mm (almost 2 inches) apart to be distinguished as two separate points.

Human skin contains several types of touch and pressure receptors **(Figure 34.4)**. Free nerve endings

are dendrites of afferent neurons with no specialized structures surrounding them. In Pacinian corpuscles, structures surrounding the nerve endings contribute to reception of stimuli. Free nerve endings wrapped around hair follicles respond when the hair is bent, making you instantly aware of a spider exploring your arm or leg as it brushes against the hairs.

34.2b Proprioceptors: What's Happening Inside

Proprioceptors are mechanoreceptors (*proprius* = one's own) that detect stimuli that are used by the CNS to maintain body balance and equilibrium and to monitor changes in the position of the head and limbs. The activity of proprioceptors allows you to touch the tip of your nose with your eyes closed or to precisely reach and scratch an itch on your back.

Statocysts (*statos* = standing; *kystis* = bag) are proprioceptors in aquatic invertebrates such as jellyfishes, some gastropods, and some arthropods. Most statocysts are fluid-filled chambers enclosing one or more movable stonelike bodies called **statoliths**. The chamber walls contain **sensory hair cells (Figure 34.5)**. In lobsters (*Homarus americanus*), statoliths are sand grains stuck together by mucus. When the animal moves, the statoliths lag behind the movement, bending the sensory hairs and triggering action potentials in afferent neurons. Thus, statocysts signal the brain about the body's position and orientation with respect to gravity. If you replace the sand grain statoliths with iron filings, you can use a magnet and emulate the lobster's response to the pull of gravity. In plants, statoliths control the direction of growth (see Chapter 31).

Information about self-motion is particularly important for flying animals and must be quickly (almost instantaneously) available. Typical insects have two pairs of wings, but species in the order Diptera (flies) have one. The second pair are reduced and persist as *halteres* **(Figure 34.6)**. Halteres are club-shaped and oscillate at the wing beat frequency. They transduce information about pitch (oscillation around a

Figure 34.5

A statocyst, in invertebrates an organ of equilibrium, in this case located at the base of the antenna of a lobster. The statoliths inside are usually formed from fused grains of sand, as they are in the lobster, from calcium carbonate.

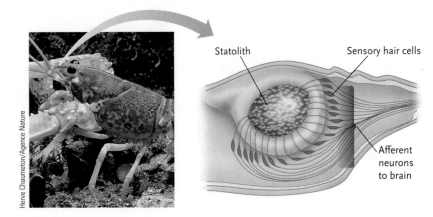

Statolith

Sensory hair cells

Afferent neurons to brain

Herve Chaumeton/Agence Nature

horizontal axis perpendicular to the direction of movement), roll (sway on the axis parallel to the direction of movement), and yaw (oscillation about a vertical axis) movements to the CNS. Coriolis (gyroscopic) forces cause the halteres to deviate in their plane of motion. Hawk moths, with two pairs of wings, use mechanosensors on their antennae to mediate flight control. Mechanical input to Johnston's organs **(Figure 34.7)** at the base of the antennae is essential for flight stability in moths. In bats, small hairs on the ventral surfaces of the wings are important for complex flight manoeuvres. Birds also must have mechanoreceptors associated with wings and flight, and, presumably, pterosaurs did as well.

Fishes and some aquatic amphibians use mechanoreceptors along the lateral line system to detect vibrations and currents in the water **(Figure 34.8, p. 828)**. Fishes have *neuromasts*, mechanoreceptors that provide information about the fish's orientation with respect to gravity, as well as its swimming velocity. In some fishes, neuromasts are exposed on the body surface; in others, they are recessed in water-filled canals with porelike openings to the outside **(Figure 34.9, p. 828; see also Figure 34.8)**. Sensory hairs are clustered at the base of each dome-shaped neuromast hair cell. One surface of the hair cell is covered with **stereocilia**, microvilli or cell processes reinforced by bundles of microfilaments. Stereocilia extend into a gelatinous structure, the **cupula** (*cupule* = little cup), which moves with pressure changes in the surrounding water. Movement of the cupula bends the stereocilia, causing depolarization of the hair cell's plasma membrane and release of neurotransmitter molecules that generate action potentials in associated afferent neurons.

Vibrations detected by the lateral line enable fishes to avoid obstacles, orient in a current, and monitor the presence of other moving objects in the water. The system is also responsible for the ability of schools of fish to move in unison, turning and diving in what appears to be a perfectly synchronized aquatic ballet. In actuality, the movement of each fish creates a pressure wave in the water that is detected by the lateral line systems of other fishes in the school. Schooling fishes can still swim in unison even if blinded, but if the nerves leading from the lateral line system to the brain are severed, the ability to school is lost.

Some fish neuromasts provide additional information. Blind cavefishes, such as *Typhlichthys subterraneus*

Figure 34.6
Halteres, vestigial hind wings of flies (enlarged image on the right at the end of the arrow), transduce information about pitch, roll, and yaw during flight. The fly shown here is a *Drosophila*.

A. Percival-Smith

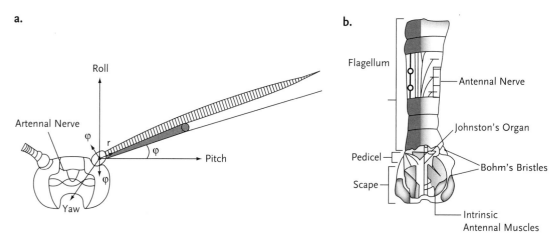

Figure 34.7
(a) Free flight kinetics of a hawk moth. Along the radial, elevation, and azimuthal directions, Φ, r, and q are unit vectors. Pitch, yaw, and roll are presented as a Cartesian coordinate system. **(b)** The head and brain of a hawk moth, including the anatomy of the antenna with the Johnston's organ.

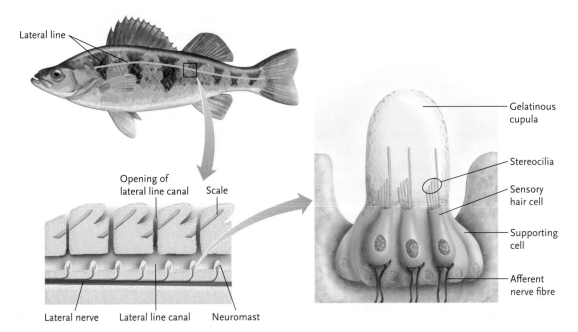

Figure 34.8

The lateral line system of fishes. Neuromasts are the sensory receptors in the lateral line system. Neuromasts have a gelatinous cupula that is pushed and pulled by vibrations and currents transmitted through the lateral line canal. As the cupula moves, the stereocilia of the sensory hair cells are bent, generating action potentials in afferent neurons that lead to the brain.

Figure 34.9
Pores along the lateral line of an arrowhana (*Sclero-pages* species).

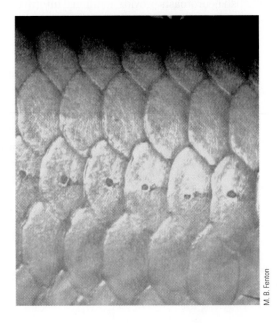

Figure 34.10
Cavefish. Whereas *Typlichthys subterraneus* uses specialized neuromast cells to detect the vibrations of swimming prey, *Astyanyx jordani* does not. **(a)** *T. subterraneus* lives in energy-poor caves, whereas **(b)** *A. jordani* lives in energy-rich caves where bat droppings are the basal energy source.

(Figure 34.10a), that live in energy-poor ecosystems use neuromast cells to detect the vibrations of swimming prey. The more commonly seen aquarium blind cavefishes (*Astyanax jordani*; **Figure 34.10b**) live in an energy-rich soup of bat droppings. They appear to find food by random searching and do not use neuromast cells to detect swimming prey.

34.2c Vestibular Apparatus of Vertebrates: Sense of Balance and Orientation

The inner ear of most terrestrial vertebrates has two specialized sensory structures, the *vestibular apparatus* and the *cochlea*. The **vestibular apparatus** is responsible for perceiving the position and motion of the head and is essential for maintaining equilibrium and for coordinating head and body movements. The cochlea is used in hearing (see Section 34.3).

The vestibular apparatus **(Figure 34.11)** consists of three **semicircular canals** and two chambers, the **utricle** and the **saccule**, filled with a fluid called *endolymph*. The semicircular canals are positioned at angles corresponding to the three planes of space. They detect rotational (spinning) motions. Each canal has an *ampulla*, a swelling at its base that is topped with sensory hair cells embedded in a cupula similar to that found in lateral line systems. Cupulas protrude into the endolymph of the canals. When the body or head rotates horizontally, vertically, or diagonally, endolymph in the semicircular canal corresponding to that direction lags behind, pulling the cupula with it. Displacement of the cupula bends the sensory hair cells and generates action potentials in afferent neurons that make synapses with the hair cells.

When the body is spinning at a constant rate and direction, fluid in the semicircular canal soon catches up with the movement, so the cupula is no longer displaced and the action potentials stop. When the spinning stops, the fluid in the canals continues to move for a time in the original direction, displacing the cupula and producing a new burst of signals to the brain.

The utricle and saccule provide information about the position of the head with respect to gravity (up versus down), as well as changes in the rate of linear movement of the body. The utricle and saccule are oriented approximately 30° to each other, and each contains sensory hair cells with stereocilia. The hair cells are covered with a gelatinous *otolithic membrane* (which is similar to a cupula) in which **otoliths**, small crystals of calcium carbonate (*oto* = ear; *lithos* = stone), are embedded (see Figure 34.4); the function of otoliths is analogous to that of statoliths of invertebrates.

When a tetrapod is standing in its normal posture, the sensory hairs in the utricle are oriented vertically and those in the saccule are oriented horizontally. When the head is tilted in any other direction or when there is a change in the linear motion of the body, the otolithic membrane of the utricle moves and bends the sensory hairs. Depending on the direction of movement, the hair cells release more or less neurotransmitter, and the brain integrates the signals it receives and generates a perception of the movement. In humans, the saccule responds to the tilting of the head away from the horizontal (such as in diving) and to a change in movement up and down (such as

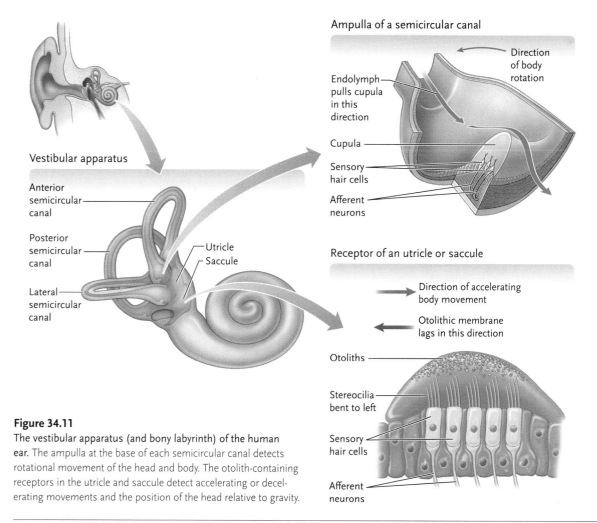

Figure 34.11

The vestibular apparatus (and bony labyrinth) of the human ear. The ampulla at the base of each semicircular canal detects rotational movement of the head and body. The otolith-containing receptors in the utricle and saccule detect accelerating or decelerating movements and the position of the head relative to gravity.

jumping up to dunk a basketball). The utricle and saccule adapt quickly to the body's motion, decreasing their response when there is no change in the rate and direction of movement.

Senses of up and down vary among animals, suggesting differences in how data from the utricle and saccule are interpreted. When it comes to posture and the definition of "dorsal," as bipeds, humans are aberrant compared with most other animals. The "normal" posture of upside-down catfish (*Synodontis nigriventris*), many bats (Chiroptera), and sloths (genera *Choloepus* and *Bradypus*) also differs from what is "normal" in other animals.

The bony labyrinths of mammals vary considerably **(Figure 34.12)**, reflecting lifestyle. Agile, arboreal mammals have semicircular canals with large radii. The large radii make the vestibular system extremely sensitive to changes in body position. Such sensitivity is not compatible with the lifestyles of cetacaeans because they frequently make fast body rotations. Cetaceans typi-

Figure 34.13
Muscle spindles, which detect the stretch and tension of muscles, and Golgi tendon organs, which detect the stretch of tendons.

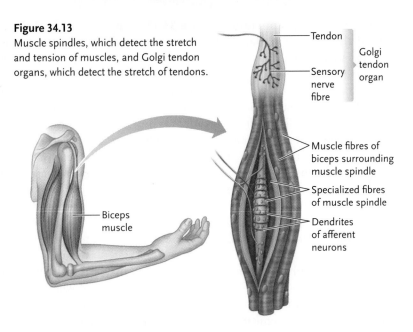

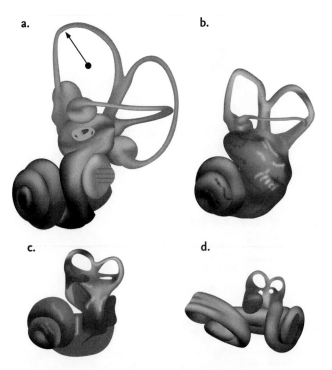

Figure 34.12
Bony labyrinths of four species of mammals. Lateral left view of the labyrinth organs of *Galago moholi* **(a)**, *Ichthyolestes pinfoldi* **(b)**, *Indocetus ramani* **(c)**, and *Tursiops truncatus* **(d)**. *I. pinfoldi* and *I. ramani* are fossil cetaceans (whales), *T. truncatus* is an extant species of dolphin, and *G. moholi* is an arboreal primate.

cally have semicircular canals with small radii (see Figure 34.12).

34.2d Stretch Receptors in Vertebrates: Keeping Track of Tension on Muscles

Stretch receptors are proprioceptors in the muscles and tendons of vertebrates that detect the position and movement, for example, of the limbs. Stretch receptors in muscles are **muscle spindles**, bundles of small, specialized muscle cells wrapped with the dendrites of afferent neurons and enclosed in connective tissue **(Figure 34.13)**. When the muscle stretches, the spindle stretches too, stimulating the dendrites and triggering the production of action potentials. The strength of the response of stretch receptors to stimulation depends on how much and how fast the muscle is stretched. Proprioceptors of tendons, called **Golgi tendon organs**, are dendrites that branch within the fibrous connective tissue of the tendon (shown in Figure 34.13). These nerve endings measure stretch and compression of the tendon as muscles contract and move limbs.

Proprioceptors allow the CNS to monitor the body's position and help keep the body in balance. They allow muscles to apply constant force under a constant load and to adjust almost instantly as the load changes. When you hold a cup while someone fills it with coffee, the muscle spindles in your biceps muscle detect the additional stretch as the cup becomes heavier. Signals from the spindles allow you to compensate for the additional weight by increasing the contraction of the muscle, keeping your arm level with no conscious effort on your part. Proprioceptors are typically slow to adapt, so the body's position and balance are constantly monitored.

1. If two fruit flies landed on you, one on your face and the other on your leg, which one would you be most likely to detect by touch receptors? Why?
2. What is the purpose of the vestibular apparatus in the inner ear of most vertebrates?
3. On what does the strength of the response of stretch receptors depend?

34.3 Mechanoreceptors and Hearing

34.3a Sound

Sounds are vibrations that travel as waves produced by the alternating compression and decompression of air or water. Although sound waves travel through air at about 340 m.s^{-1} at sea level, individual air molecules transmitting the waves move back and forth over only a short distance as the wave passes. Water is denser than air, so sounds move approximately three times faster under water.

We can measure several features of a sound. The **energy** in a sound is measured as intensity, expressed in decibels (dB) sound pressure level (SPL) at a specific distance from the sound source. Some echolocating eastern red bats produce signals of 130 dB SPL @ 10 cm. Compare this with the shriek of a typical smoke detector, 108 dB SPL @ 10 cm. Remember that decibels are a logarithmic scale.

The duration of a sound signal is the time it lasts, measured in seconds (s) or milliseconds (ms). The time between signals is also measured in seconds or milliseconds. Loudness refers to the way an animal perceives sound. To the eastern red bat, its echolocation calls are loud, but they are not to humans, who do not hear them because they are too high in pitch (frequency). The pitch of a sound is the perceived frequency that reflects the wavelength (λ) of the sound measured in Hertz (cycles per second). More cycles per second means higher pitch (1000 Hz versus 50 000 Hz). Humans are said to hear sounds between 40 Hz and 20 000 Hz (20 kHz), so we cannot hear the eastern red bat's echolocation calls, which have most of their energy around 35 kHz. Therefore, they are ultrasonic, above the range of human hearing. Infrasounds, frequencies below the range of human hearing (40 Hz), are used by African elephants to communicate.

34.3b Hearing in Invertebrates: Mechanoreceptors and Ears

Most invertebrates detect sound and other vibrations through mechanoreceptors in their skin or on other surface structures. An earthworm, for example, quickly retracts into its burrow at the smallest vibration of the surrounding earth, even though it has no specialized structures serving as ears. Cephalopods (squid and octopus) have a system of mechanoreceptors on their head and tentacles, similar to the lateral line of fishes. These mechanoreceptors detect vibrations in the surrounding water. In many insets and other arthropods, hairs or bristles are sensory receptors because they vibrate in response to sound waves, often at particular frequencies.

Insects such as grasshoppers and crickets have ears, complex auditory organs on each side of the abdomen or on the first pair of walking legs, whereas in moths, these "ears" have been found on the head (mouthpart), thorax, and abdomen **(Figure 34.14)**. These "ears" consist of a thinned region of the insect's exoskeleton forming a **tympanum** (*tympanum* = drum) over a hollow chamber. Sounds reaching the tympanum cause it to vibrate. Mechanoreceptors connected to the tympanum translate vibrations into nerve impulses. Some insect ears respond to sounds only at certain frequencies, such as to the pitch of a cricket's song.

34.3c Hearing in Vertebrates: Auditory Systems

The auditory structures of terrestrial vertebrates transduce vibrations in air (sound) to sensory hair cells that respond by triggering action potentials. The auditory system of humans is typical for mammals **(Figure 34.15, p. 832)**. The *pinna* (**outer ear**; *pinna* = wing or leaf) concentrates and focuses sound waves. Some

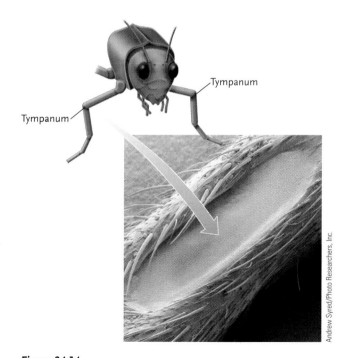

Figure 34.14
The tympanum or eardrum of a cricket, located on the front walking legs.

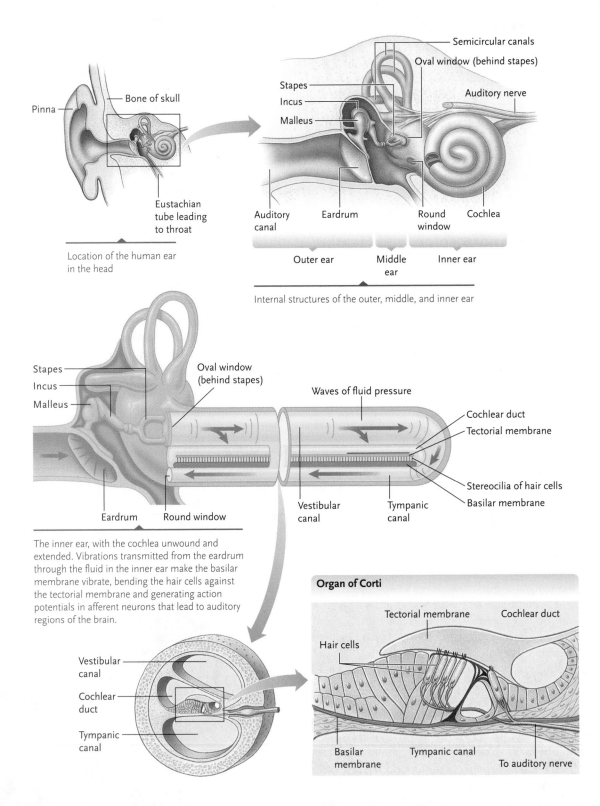

Figure 34.15
Structures of the human ear.

Location of the human ear in the head

- Pinna
- Bone of skull
- Eustachian tube leading to throat

Internal structures of the outer, middle, and inner ear

- Stapes
- Incus
- Malleus
- Semicircular canals
- Oval window (behind stapes)
- Auditory nerve
- Auditory canal
- Eardrum
- Round window
- Cochlea
- Outer ear
- Middle ear
- Inner ear

- Stapes
- Incus
- Malleus
- Oval window (behind stapes)
- Waves of fluid pressure
- Cochlear duct
- Tectorial membrane
- Stereocilia of hair cells
- Basilar membrane
- Eardrum
- Round window
- Vestibular canal
- Tympanic canal

The inner ear, with the cochlea unwound and extended. Vibrations transmitted from the eardrum through the fluid in the inner ear make the basilar membrane vibrate, bending the hair cells against the tectorial membrane and generating action potentials in afferent neurons that lead to auditory regions of the brain.

- Vestibular canal
- Cochlear duct
- Tympanic canal

Organ of Corti

- Tectorial membrane
- Cochlear duct
- Hair cells
- Basilar membrane
- Tympanic canal
- To auditory nerve

animals have pinnae; others lack them **(Figure 34.16)**. Sound waves enter the auditory canal and strike a thin sheet of tissue (tympanic membrane or eardrum) and start it vibrating.

Vibrations in the tympanic membrane generate vibrations in the auditory ossicles located in the middle ear, which is an air-filled cavity. Mammals have three auditory ossicles, the **malleus** (hammer), **incus** (anvil), and **stapes** (stirrup). The manubrium of the malleus sits immediately behind the eardrum,

and the eardrum's vibrations are conducted from the malleus to the incus and the stapes. The stapes abuts the inner ear at the **oval window**, a thin, elastic membrane where vibrations in bone are converted to vibrations in the fluid in the vestibular canal. Between the eardrum and the oval window, sounds are amplified at least 20 times.

The **inner ear** contains several fluid-filled compartments, the vestibular apparatus (see Section 34.2) and the **cochlea**, a spiral tube (*kochlias* = snail). In

d.

Figure 34.16

Pinnae (external ears) are lacking in mammals such as **(a)** the beluga (*Delphinapterus leucas*), **(b)** birds (*Struthio camelus*), and **(c)** reptiles (*Varanus komodoensis*), but large and conspicuous in **(d)** a bat (*Otonycteris hemprichii*).

humans, the cochlea twists through about 2.5 turns (if flattened, it would be about 3.5 cm long in an adult). The spiralling of the cochlea appears to make it more sensitive to lower frequency sounds. Thin membranes divide the cochlea into three longitudinal chambers, the *vestibular canal* at the top, the *cochlear duct* in the middle, and the *tympanic canal* at the bottom (see Figure 34.16). The vestibular canal and the tympanic canal join at the outer tip of the cochlea, so the fluid they contain is continuous. The **organ of Corti** lies within the cochlear duct. It contains sensory hair cells that detect vibrations transmitted to the inner ear (see Figure 34.16). Vibrations of the oval window pass through the fluid in the vestibular canal, make the turn at the end, and travel back through the fluid in the tympanic canal. At the end of the tympanic canal, they are transmitted to the **round window**, a thin membrane that faces the middle ear.

Vibrations in the fluid of the inner ear cause vibrations in the **basilar membrane**. The basilar membrane forms part of the floor of the cochlear duct and anchors the sensory hair cells in the organ of Corti. The stereocilia of these cells are embedded in the *tectorial membrane* extending the length of the cochlear canal. Vibrations of the basilar membrane cause the hair cells to bend, stimulating them to release a neurotransmitter that triggers action potentials in afferent neurons leading from the inner ear.

The basilar membrane is narrowest near the oval window and gradually widens toward the outer end of the cochlear duct. High-frequency vibrations produced by high-pitched sounds vibrate the basilar membrane most strongly near its narrow end, whereas vibrations of lower frequency vibrate the membrane nearer the outer

end. Thus, each frequency of sound waves causes hair cells in a different segment of the basilar membrane to initiate action potentials. More than 15 000 hair cells are distributed in small groups along the basilar membrane. Each group of hairs is connected by synapses to afferent neurons, the axons of which are bundled together in the *auditory nerve*, a cranial nerve leading to the thalamus. From there, the signals are routed to specific regions in the auditory centre of the temporal lobe.

The **eustachian tube**, a duct leading from the air-filled middle ear to the throat (see Figure 34.15), protects the eardrum from damage caused by changes in environmental atmospheric pressure. As we swallow or yawn, the tube opens, allowing air to flow into or out of the middle ear, equalizing pressure on both sides of the eardrum. When swelling or congestion prevents the tube from admitting air, we complain of having stopped-up ears because we sense a pressure difference between the outer and middle ear caused by the eardrum bulging inward or outward; this interferes with the transmission of sounds.

STUDY BREAK

1. How do most invertebrates detect sound? Give an example.
2. Explain in detail how a human hears.
3. Why are the echolocation calls of many bats inaudible to humans despite their high intensity?

34.4 Photoreceptors and Vision

Photoreceptors detect light at particular wavelengths, converting the stimuli to action potentials that move the information to visual centres in the CNS or the central ganglion, where the signals are integrated into a perception of light. Eyes are the organs that detect light (see Chapter 1). In their simplest forms, eyes distinguish only light from dark. The most complex eyes form images (have a lens), allowing the animal to distinguish shapes and to focus an accurate image of viewed objects on a layer of photoreceptors. Signals originating at the photoreceptors are integrated in the brain into an accurate, point-by-point perception of the viewed object. Image-forming eyes **(Figure 34.17)** have evolved in at least some species in five phyla: Cnidaria, Mollusca, Arthropoda, Onycophora, and Chordata. However, 99% of species with image-forming eyes belong to the phyla Arthropoda and Chordata. Pigments in some image-forming eyes allow animals to see in colour.

Annelids such as earthworms lack eyes but have photoreceptors in their skin. Input from the photoreceptors allows them to sense and respond to light. Earthworms respond negatively to light, as you can easily discover by shining a flashlight on an earthworm outside its burrow at night. The photoreceptors of nonchordates are depolarized when they absorb light and generate action potentials or increase their release of neurotransmitter molecules when stimulated.

34.4a Ocelli: Eyes that Detect Light

The **ocellus** (plural *ocelli*; also called an *eyespot* or *eyecup*) is the "simplest" eye, lacking a lens and not leading to image formation. Ocelli may each consist of <100 photoreceptor cells lining a cup or pit. Planarians (Platyhelminthes) are usually negatively phototropic. Their photoreceptor cells are located in two cuplike depressions below the epidermis. The photoreceptor cells are connected to the dendrites of afferent neurons, the axons of which are bundled into nerves that travel from the ocelli to the cerebral ganglion (see Figure 1.14). Each ocellus is covered on one side by a layer of pigment cells that blocks most light rays arriving from the opposite side of the animal. Therefore, a planarian

can identify which side is brightest, orient its body to equalize the stimuli, and then move away from the light source. Similar ocelli are found in a variety of animals, including a number of insects, other arthropods, and molluscs. There is evidence that a collection of ocelli can constitute an image-forming system, blurring the distinction between image-forming and non-image-forming eyes.

34.4b Image-Forming Eyes Get the Picture

There are two main types of image-forming eyes: compound eyes with multiple lenses and **single-lens eyes**. Compound eyes occur in arthropods (crustaceans, trilobites, insects). Each compound eye can contain hundreds to thousands of faceted visual units called **ommatidia** (*omma* = eye) fitted closely together **(Figure 34.18;** see also Figure 1.15). In insects, light entering an ommatidium is focused by a transparent **cornea** and a *crystalline cone* (just below the cornea) onto a bundle of photoreceptor cells. Microvilli in these cells interdigitate like the fingers of clasped hands, forming a central axis that contains **rhodopsin**, a photopigment also found in the rods of vertebrate eyes. Absorption of light by rhodopsin causes action potentials to be generated in afferent neurons connected to the base of the ommatidium. Each ommatidium samples a small part of the visual field. From these signals, the brain receives a mosaic image of the world. Because even the slightest motion is detected simultaneously by many ommatidia, compound eyes are sensitive movement detectors. Anyone who has tried to swat a fly knows this from personal experience.

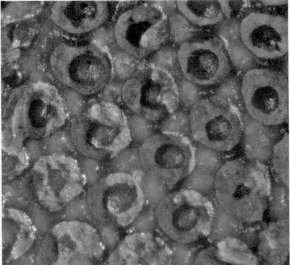

Figure 34.18
The compound eye of a trilobite, *Phacops iowensis*, consists of many ommatidia. Each of the 16 complete circles shown here is an ommatidium, each with a cornea that directs light into the crystalline cone. The cone focuses light on the photoreceptor cells. A light-blocking pigment layer at the sides of the ommatidium prevents light from scattering laterally in the compound eye.

Figure 34.17
This spider has a wonderful array of eyes.

Single-lens eyes of cephalopods (see Figure 1.16) and vertebrates (**Figure 34.19**) operate like a camera, but the eyes differ in developmental details and operation. They also differ in the molecular structure of the rhodopsin in the photoreceptors (**Figure 34.20**).

In both cephalopod and vertebrate eyes, light enters through the transparent cornea and passes through the pupil and then the lens, which focuses the image and projects it onto the retina, a layer of photoreceptors at the back of the eye. The iris is behind the cornea and surrounds the **pupil**. Muscles in the iris adjust the size of the pupil to vary the amount of light entering the eye. When the light is bright, circular muscles in the iris contract, shrinking the size of the pupil and reducing the amount of light that enters the eye. In dim light, radial muscles contract, enlarging the pupil and increasing the amount of light that enters the eye. Muscles move the lens forward and back with respect to the retina to focus the image. This process is called **accommodation**.

In most terrestrial vertebrates, accommodation is achieved through changes in the shape of the lens. The soft, flexible lens is held in place by the **ciliary body**, fine ligaments that anchor it to a surrounding layer of connective tissue and muscle. When the ciliary muscle is relaxed, the lens is put under tension and flattened, focusing light from distant objects onto the retina. When the ciliary muscles contract, tension on the lens is reduced, and the lens becomes more spherical in shape and focuses light from nearby objects onto the retina (**Figure 34.21, p. 836**).

In vertebrates and cephalopods, axons of afferent neurons originating in the retina converge to form the optic nerve leading from the eye to the brain. In cephalopods, the neural network lies under the retina, so light rays do not have to pass through the neurons to reach the photoreceptors. Some vertebrate eyes have the opposite arrangement, with retinal cells below the neural networks and blood vessels. This and other differences in structure and function indicate that mollusc and vertebrate eyes evolved independently.

In vertebrate eyes, the **aqueous humour** is a clear fluid that fills the space between the cornea and lens. This fluid carries nutrients to the lens and cornea, which do not contain any blood vessels. The main chamber of the eye, located between the lens and the retina, is filled with the jellylike **vitreous humour** (*vitrum* = glass). The outer wall of the eye contains a tough layer of connective tissue (the *sclera*). Inside the sclera is a darkly pigmented layer (the *choroid*) that prevents light from entering except through the pupil. It also contains the blood vessels that nourish the retina.

Rods and cones are two types of photoreceptors that occur in the retina along with layers of neurons that perform an initial integration of visual

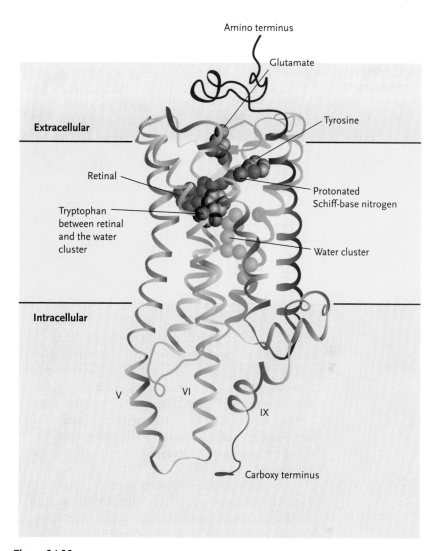

Figure 34.19

The rhodopsin of squid eyes is quite different from that of vertebrates. Like vertebrate rhodopsin, squid rhodopsin is a G protein–coupled receptor with seven transmembrane domains (helices that cross membranes). Helices V and VI protrude deeply into the cytoplasm of the photoreceptor cell. Retinol (red) is a light-sensitive chromatophore. The rhodopsin data demonstrate another difference between the eyes of squid and the eyes of vertebrates.

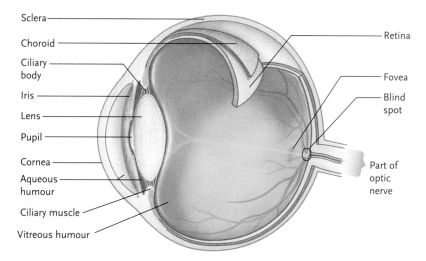

Figure 34.20
Structures of the human eye.

Figure 34.21

Accommodation in terrestrial vertebrates occurs when the lens changes shape to focus on distant **(a)** and near **(b)** objects.

a. Focusing on distant object

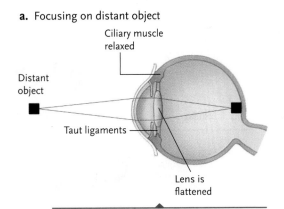

Ciliary muscle relaxed

Distant object

Taut ligaments

Lens is flattened

When the eye focuses on a distant object, the ciliary muscles relax, allowing the ligaments that support the lens to tighten. The tightened ligaments flatten the lens, bringing the distant object into focus on the retina.

b. Focusing on near object

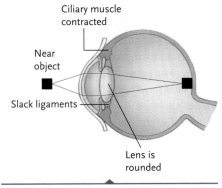

Ciliary muscle contracted

Near object

Slack ligaments

Lens is rounded

When the eye focuses on a near object, the ciliary muscles contract, loosening the ligaments and allowing the lens to become rounder. The rounded lens focuses a near object on the retina.

information before it is sent to the brain. Rods are specialized for the detection of light at low intensities and cones for the detection of different wavelengths (colours). The rods of mammals are much more sensitive than the cones to light of low intensity and in some species can respond to a single photon of light. The retina of a human eye contains about 120 million rods and 6 million cones, organized into a densely packed single layer.

In mammals and birds with eyes specialized for daytime vision, cones are concentrated in and around the **fovea**, a small, circular region of the retina that is <1 mm in diameter in humans (see Figure 34.19). Some birds have two foveas. Rods are spread over the remainder of the retina. The retinas of mammals and birds with eyes specialized for night vision contain mostly rods and no clearly defined fovea. In many nocturnal animals, the tapetum lucidum, a layer of tissue behind the retina, reflects light back through the retina. The tapetum lucidum accounts for conspicuous eyeshine in many nocturnal vertebrates. Some fishes and many reptiles have cones generally distributed throughout their retinas and very few rods.

34.4c Sensory Transduction by Rods and Cones: Converting Signals to Electrical Impulses

A photoreceptor cell has three parts:

- an outer segment consisting of stacked, flattened, membranous disks,
- an inner segment where the cell's metabolic activities occur, and
- the synaptic terminal where neurotransmitter molecules are stored and released **(Figure 34.22a).**

The photoreceptors of different animals contain different forms of *retinal*, a lipidlike pigment

synthesized from vitamin A as their light-absorbing pigment. Retinal is bonded covalently with an opsin protein to produce rhodopsin, by far the most common photopigment in the animal kingdom. Photopigments are embedded in the membranous disks of the photoreceptors' outer segments **(Figure 34.22b)**. Rhodopsin is the retinal–opsin photopigment in rods.

In the dark, the retinal segment of unstimulated rhodopsin is *cis*-retinal, an inactive form (see Figure 34.22b), and rods steadily release the neurotransmitter glutamate. When rhodopsin absorbs a photon of light, retinal converts to *trans*-retinal, the active form, and the rods *decrease* the amount of glutamate they release.

Rhodopsin is a membrane-embedded G protein–coupled receptor. Recall (see Chapter 8) that an extracellular signal received by a G protein–coupled receptor activates the receptor, triggering a signal transduction pathway within the cell and generating a cellular response. Here, activated rhodopsin triggers a signal transduction pathway that leads to the closure of Na$^+$ channels in the plasma membrane **(Figure 34.23, p. 838)**. Closure of the channels hyperpolarizes the photoreceptor's membrane, decreasing neurotransmitter release. The response is graded because as light absorption by photopigment molecules increases, the amount of neurotransmitter released is reduced proportionately. If light absorption decreases, neurotransmitter release by the photoreceptor increases proportionately. Transduction in rods works in the opposite way from most sensory receptors in which a stimulus increases neurotransmitter release.

34.4d Visual Processing in the Retina: Events at the Back of the Eye

In the vertebrate retina, the two types of photoreceptors are linked to a network of neurons that carry out initial integration and processing of visual information.

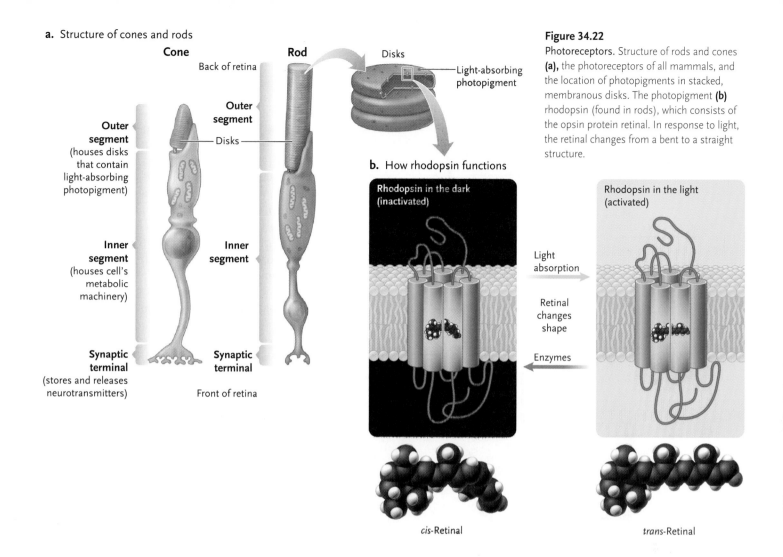

a. Structure of cones and rods

Cone

Rod

Back of retina

Outer segment
(houses disks that contain light-absorbing photopigment)

Outer segment

Disks

Inner segment
(houses cell's metabolic machinery)

Inner segment

Synaptic terminal
(stores and releases neurotransmitters)

Synaptic terminal

Front of retina

Disks

Light-absorbing photopigment

b. How rhodopsin functions

Rhodopsin in the dark (inactivated)

Rhodopsin in the light (activated)

Light absorption

Retinal changes shape

Enzymes

cis-Retinal

trans-Retinal

Figure 34.22

Photoreceptors. Structure of rods and cones **(a),** the photoreceptors of all mammals, and the location of photopigments in stacked, membranous disks. The photopigment **(b)** rhodopsin (found in rods), which consists of the opsin protein retinal. In response to light, the retinal changes from a bent to a straight structure.

The retina of mammals has four types of neurons **(Figure 34.24, p. 838).** There is a layer of bipolar cells just in front of the rods and cones. These neurons synapse with rods or cones at one end and with ganglion cells, a layer of neurons, at the other end. The axons of ganglion cells extend over the retina and collect at the back of the eyeball to form the optic nerve, which transmits action potentials to the brain. The point where the optic nerve exits the eye lacks photoreceptors. This *blind spot* can be several millimetres in diameter in humans. Horizontal cells connect photoreceptor cells, whereas **amacrine cells** connect bipolar and ganglion cells.

In the dark, the steady release of glutamate from rods and cones depolarizes some postsynaptic bipolar cells and hyperpolarizes others. In the light, the decrease in neurotransmitter release from rods and cones results in hyperpolarization of polarized bipolar cells. Changes in membrane potential in response to light are transmitted to the brain for processing.

Signals from the rods and cones may move vertically or laterally in the retina. Signals move vertically from the photoreceptors to bipolar cells and then to ganglion cells. Whereas the human retina has over 120 million photoreceptors, it has only about 1 million ganglion cells. This disparity is explained by the fact that each **ganglion cell** receives signals from a clearly defined set of photoreceptors constituting the *receptive field* for that cell. Therefore, stimulating numerous photoreceptors in a ganglion cell's receptive field results in only a single message to the brain from that cell. Receptive fields are typically circular and are of different sizes. Smaller receptive fields result in sharper images because they send more precise information to the brain about the location in the retina where the light was received.

Lateral movement of signals from a rod or cone proceeds to a horizontal cell and continues to **bipolar cells** with which the horizontal cell makes inhibitory connections. To understand this, consider a spot of light falling on the retina. Photoreceptors detect the light and send a signal to bipolar cells and horizontal cells. **Horizontal cells** inhibit more distant bipolar cells that are outside the spot of light, causing the light spot to appear lighter and its surrounding dark area to appear darker. This type of visual processing is called **lateral inhibition** and serves both to sharpen the edges of objects and enhance contrast in an image.

Figure 34.23

The signal transduction pathway that closes Na⁺ channels in photoreceptor plasma membranes when rhodopsin absorbs light.

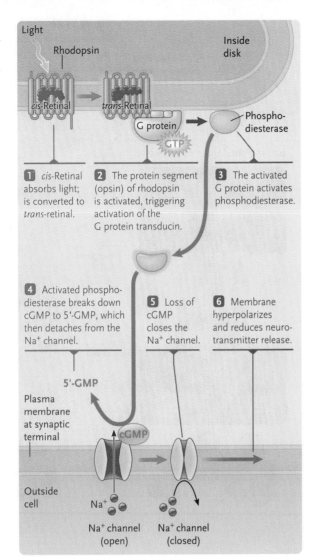

Light

Rhodopsin

cis-Retinal

trans-Retinal

Inside disk

G protein

GTP

Phospho-diesterase

1 *cis*-Retinal absorbs light; is converted to *trans*-retinal.

2 The protein segment (opsin) of rhodopsin is activated, triggering activation of the G protein transducin.

3 The activated G protein activates phosphodiesterase.

4 Activated phospho-diesterase breaks down cGMP to 5'-GMP, which then detaches from the Na⁺ channel.

5 Loss of cGMP closes the Na⁺ channel.

6 Membrane hyperpolarizes and reduces neuro-transmitter release.

5'-GMP

Plasma membrane at synaptic terminal

cGMP

Outside cell

Na⁺

Na⁺ channel (open)

Na⁺ channel (closed)

34.4e Opsins and Colour Vision

Many invertebrates and some species in each class of vertebrates have colour vision, which depends on cones in the retina. Most mammals have two types of cones, whereas humans and other primates have three types. Each human or primate cone cell contains one of three photopsins in which retinal is combined with different opsins. The three photopsins absorb light over different, but overlapping, wavelength ranges, with peak absorptions at 445 nm (blue light), 534 nm (green light), and 570 nm (red light). The farther a wavelength is from the peak colour absorbed, the less strongly the cone responds. Having more types of cones translates into better colour vision.

Overlapping wavelength ranges for the three photoreceptors means that light at any visible wavelength stimulates at least two of three types of cones. Maximal absorption by each type of cone at a different wavelength leads to differential stimulation of different types of cones. These differences are relayed to the visual centres of the brain, where they are integrated into the perception of a colour corresponding to the particular wavelength absorbed. Light stimulating all three receptor types equally is seen as white.

Colourblindness results from inherited defects in opsin proteins of one or more of the three types of cones. For example, people with a mutation preventing cones from making a functional form of red-absorbing opsin see orange, yellow, and red as the same grey or greenish colour.

Figure 34.24

Microscopic structure of the retina showing the network of neurons (bipolar cells, horizontal cells, amacrine cells, and ganglion cells) that carry out the initial integration of visual information.

Retina

Photoreceptors

Cone Rod

Retina

Front of retina

Back of retina

Optic nerve

Fibres of the optic nerve

Ganglion cell

Amacrine cell

Bipolar cell

Horizontal cell

Sclera

Choroid layer

Pigment layer

Direction of light

Direction of retinal visual processing

838 | UNIT EIGHT SYSTEMS AND PROCESS—ANIMALS

NEL

Donald R. Griffin, Rockefeller University

In 1794, the Italian scientist Lazzaro Spallananzi finished one stage of studying the nocturnal orientation behaviour of bats and owls. His experiments had progressively denied his study animals—barn owls (*Tyto alba*) and pipistrelle bats (*Pipistrellus pipistrellus*)—access to sensory information. Although the owls depended on vision to detect and avoid obstacles, the bats did not. Only when Spallanzani blocked one of the bats' ears did they become disoriented. Spallanzani concluded that "bats could see with their ears." In 1794, this suggestion was considered to be preposterous, and some scientists of note mocked him, asking if bats could see with their ears, could they hear with their eyes? But Spallanzani's data supported his conclusion, and his data had been collected by experimentation.

Fast-forward to the 1930s. Donald R. Griffin, then an undergraduate at Harvard University, knew of what had become known as "Spallanzani's bat problem." Using little brown bats (*Myotis lucifugus*), Griffin repeated many of Spallanzani's experiments.

Unlike Spallanzani, Griffin had the use of a "sonic detector," a microphone sensitive to the high-frequency acoustic signals (beyond the range of human hearing, or ultrasonic) of the bats. He and his colleague Robert Galambos determined that bats flying in the dark produced pulses of high-frequency sound. They coined the term "echolocation" to describe how bats used echoes of the sounds they produced to detect objects in their path.

By 1960, Griffin and his colleagues had demonstrated that little brown bats could use echolocation to detect insects as small as fruit flies (*Drosophila*) and mosquitoes. Both Lazzaro Spallanzani and Donald Griffin opened our eyes to animals' use of biosonar or echolocation. Each of them used the tools available at the time to design and conduct experiments whose results led to the conclusion that bats could "see with their ears."

The apparatus available to Spallanzani in 1794 was quite different from that available to Griffin in 1937. Griffin solved "Spallanzani's bat problem" because he could eavesdrop on sounds beyond the range of human hearing. But neither Spallanzani nor Griffin worked alone, and each benefitted from interactions with colleagues, Jurine for Spallanzani and Galambos and Pearce for Griffin.

Spallanzani is also known for other contributions, particularly relating to the question of spontaneous generation. In 1780, he demonstrated fertilization of frogs' eggs with semen collected from a male frog. In 1783, he used other experiments to disprove spontaneous generation of microscopic organisms.

Griffin, too, is well known for contributions across a wide range of topics. One central theme is the orientation behaviour of animals, a logical jump from echolocation. Another was the question of animal awareness—are animals more than "boxes of reflexes"?—a topic that remains controversial.

It is no wonder that curious and innovative scientists such as Spallanzani and Griffin led the way to the discovery of echolocation.

34.4f Visual Cortex: Images in the Brain

Just behind the eyes, the optic nerves converge before entering the base of the brain. A portion of each optic nerve crosses over to the opposite side, forming the **optic chiasm** (*chiasma* = crossing place). Most axons enter the **lateral geniculate nuclei** in the thalamus, where they synapse with interneurons leading to the visual cortex **(Figure 34.25, p. 840).**

Because of the optic chiasm, the left half of the image seen by both eyes is transmitted to the visual cortex in the right cerebral hemisphere, and the right half of the image is transmitted to the left cerebral hemisphere. The right hemisphere thus sees objects to the left of the centre of vision, and the left hemisphere sees objects to the right of the centre of vision. Communication between the right and left hemispheres integrates this information into a perception of the entire visual field seen by the two eyes.

If you look at a nearby object with one eye and then the other, you will notice that the point of view is slightly different. Integration of the visual field by the brain creates a single picture with a sense of distance and depth. The greater the difference between the images seen by the two eyes, the closer the object appears to the viewer.

When both eyes are used together, the animal has a wider field of view (in humans 180° versus 150°). Binocular vision also enhances the ability to see faint objects (binocular summation). Perhaps most importantly, binocular vision allows stereopsis because of the overlap in the angles of view between two eyes. Stereopsis allows the animal to make fine depth discriminations from parallax, the apparent difference in position of an object viewed from the different position of the eyes. In some cases, stereopsis involves neurons with binocular receptive fields. Humans have excellent stereopsis, but so do many other primates, as well as some birds and fishes.

Archerfish **(Figure 34.26, p. 840)** live in fresh water and knock flying or resting insects onto the water's surface with spit droplets. The fish then catches and eats the insects. During the spitting attacks, the fish's

Figure 34.25

Neural pathways for vision. Because half of the axons carried by the optic nerves cross over in the optic chiasma, the left half of the field seen by both eyes (blue segment) is transmitted to the visual cortex in the right cerebral hemisphere. The right half of the field seen by both eyes (red segment) is transmitted to the visual cortex in the left cerebral hemisphere. As a result, the right hemisphere of the brain sees objects to the left of the centre of vision, and the left hemisphere sees objects to the right of the centre of vision.

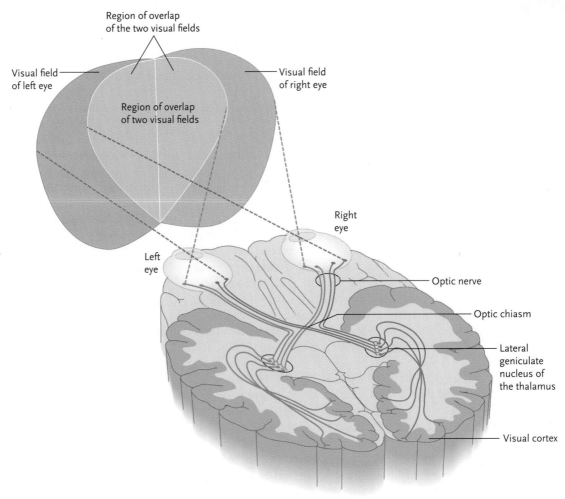

Region of overlap of the two visual fields

Visual field of left eye

Region of overlap of two visual fields

Visual field of right eye

Right eye

Left eye

Optic nerve

Optic chiasm

Lateral geniculate nucleus of the thalamus

Visual cortex

Figure 34.26
An archerfish, *Toxotes chatareus*.

© A & J Visage/Alamy

eyes are below the surface of the water, posing a potentially serious problem because of refraction, the deflection of rays of light at the air–water interface. Some evidence suggests that archerfish spit from directly under the prey, but further observations showed that

this is not always true. Archerfish correctly set their spitting angle to compensate for the refraction they experience at different positions. They also correct for curvature of the water droplet's trajectory. Other fish also spit at aerial prey, and some birds hunt fish from above the water's surface; both deal with the problems of refraction from a different standpoint.

The two optic nerves together contain more than a million axons, more than all other afferent neurons of the body put together. Almost one-third of the grey matter of the cerebral cortex is devoted to visual information. These numbers give some idea of the complexity of the information integrated into the visual image formed by the brain.

STUDY BREAK

1. What is the "simplest" eye? Why is it an eye, and how does it differ from image-forming eyes?
2. What causes colourblindness?
3. Why are compound eyes so adept at detecting motion?
4. What is accommodation?

34.5 Chemoreceptors

Chemoreceptors provide information about taste (gustation) and smell (olfaction), as well as measures of intrinsic levels of molecules such as oxygen, carbon dioxide, and hydrogen ions. All chemoreceptors probably work through membrane receptor proteins that are stimulated when they bind with specific molecules in the environment and generate action potentials in afferent nerves leading to the CNS.

34.5a Invertebrate Animals: A Rich World of Odours

In many invertebrates, the same receptors serve for sensing smell and taste. These receptors may be concentrated around the mouth or distributed over the body surface. The cnidarian *Hydra* has chemoreceptors around its mouth that respond to glutathione, a chemical released from prey organisms ensnared in the cnidarian's tentacles. Stimulation of chemoreceptors by glutathione causes the tentacles to retract, resulting in **ingestion** of the prey. In contrast, earthworms have taste and smell receptors distributed over the entire body surface.

Some terrestrial invertebrates have clearly differentiated receptors for taste and smell. In insects, taste receptors occur inside hollow sensory bristles called *sensilla* (singular, *sensillum*), usually located on the antennae, mouthparts, or feet **(Figure 34.27)**. Pores in the sensilla admit molecules from potential food to the chemoreceptors, which are specialized to detect sugars, salts, amino acids, or other chemicals. Many female insects have chemoreceptors on their ovipositors, allowing them to lay their eggs on food appropriate for the larvae when they hatch.

Pheromones are chemicals used in communication by both animals and plants (see Chapter 40). Insects are excellent examples of animals that make extensive use of pheromones. Female insects use pheromones to attract males, or vice versa. Olfactory

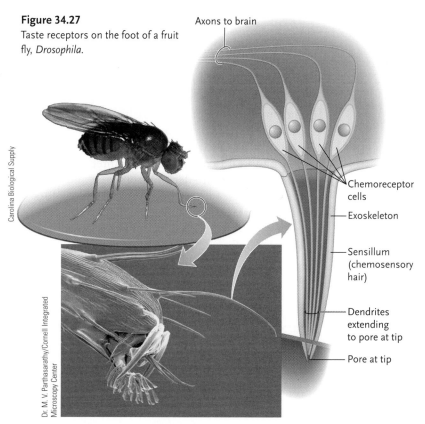

Figure 34.27
Taste receptors on the foot of a fruit fly, *Drosophila*.

Axons to brain

Chemoreceptor cells

Exoskeleton

Sensillum (chemosensory hair)

Dendrites extending to pore at tip

Pore at tip

Carolina Biological Supply

Dr. M. V. Parthasarathy/Cornell Integrated Microscopy Center

receptors in the bristles on the antennae of male silkworm moths (*Bombyx mori*) **(Figure 34.28)** bind a pheromone released by female conspecifics. If an antenna from a silkworm moth is connected to electrodes at its base and tip, action potentials from olfactory receptors can be detected at pheromone concentrations as low as one attractant molecule per 10^{17} air molecules! The male moth responds by flying rapidly when as few as 40 of the 20 000 receptor cells on an antenna have been stimulated by pheromone molecules. Ants, bees, and wasps may use odour to recognize conspecifics, to identify members of the same hive or nest, or to alert nestmates to danger.

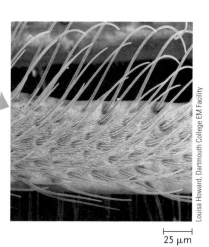

Figure 34.28
The brushlike antennae of a male silkworm moth. Fine sensory bristles containing olfactory receptor cells cover the filaments of the antennae.

© A. Shay/OSF/Animals Animals—Earth Scenes

Louisa Howard, Dartmouth College EM Facility

25 μm

34.5b Vertebrate Animals: More Variations on the Sense of Smell

Taste involves the detection of potential food molecules in objects touched by a receptor. Smell involves the detection of airborne molecules. Taste and smell receptors have hairlike extensions that contain proteins that bind environmental molecules. Hairs of taste receptors are derived from microvilli and contain microfilaments. Hairs of smell receptors are derived from cilia and contain microtubules. Information from taste receptors is typically processed in the parietal lobes, whereas information from smell receptors is processed in olfactory bulbs and temporal lobes.

Taste receptors of most vertebrates form part of a structure called a taste bud, a small, pear-shaped capsule with a pore at the top opening to the exterior **(Figure 34.29)**. Sensory hairs of taste receptors pass through the pore of a taste bud and project to the exterior. The opposite end of the receptor cells synapses with dendrites of an afferent neuron.

Taste receptors of terrestrial vertebrates are concentrated in the mouth. Humans have about 10 000 taste buds, each 30 to 40 μm in diameter, scattered over the tongue, roof of the mouth, and throat. Those on the tongue are embedded in outgrowths called *papillae* (*papula* = pimple), which give the surface of the tongue its rough or furry texture. Taste receptors on the human tongue respond to five basic tastes: sweet, sour, salty, bitter, and umami (savoury). Some receptors for umami respond to the amino acid glutamate (familiar as monosodium glutamate or MSG).

Some taste receptors are stimulated on inhalation (orthonasal) and others on exhalation (retronasal) **(Figure 34.30; Table 34.1)**.

Signals from taste receptors are relayed to the thalamus. From there, some signals lead to gustatory centres in the cerebral cortex, which integrate them into the perception of taste. Others lead to the brain stem and limbic system, which link tastes to involuntary visceral and emotional responses. Through brain stem and limbic connections, a pleasant taste may lead to salivation, secretion of digestive juices, sensations of pleasure, and sexual arousal, whereas an unpleasant taste may produce revulsion, nausea, and vomiting.

Receptors that detect odours are located in the nasal cavities in terrestrial vertebrates. Bloodhounds have more than 200 million olfactory receptors in patches of olfactory epithelium in the upper nasal passages; humans have about 5 million olfactory receptors. On one end, each olfactory receptor cell has 10 to 20 sensory hairs projecting into a layer of mucus covering the olfactory area in the nose. To be detected, airborne molecules must dissolve in the watery mucus solution. At the other end, the olfactory receptor cells make synapses with interneurons in the olfactory bulbs. Olfactory receptors are the only receptor cells that make direct connections with brain interneurons rather than via afferent neurons.

From the olfactory bulbs, nerves conduct signals to the olfactory centres of the cerebral cortex, where they are integrated into the perception of tantalizing or unpleasant odours from a rose to a rotten egg. Most odour perceptions arise from combinations of different olfactory receptors. About 1000 different human genes give rise to an equivalent number of olfactory receptor types, each specific for a different class of chemicals. Recent experiments demonstrate that rats smell in stereo, accurately localizing odours in one or two sniffs. They could do so only with bilateral sampling. Some neurons in the olfactory bulb neurons respond differently to stimuli from the left than from the right. Furthermore, some receptors in the olfactory cortex of mammals fire only upon stimulation by combinations of odourants, perhaps explaining why mixes of odours are perceived as novel by humans.

As in taste, other connections from the olfactory bulbs lead to the limbic system and brain stem, where the signals elicit emotional and visceral responses similar to those caused by pleasant and unpleasant tastes. Olfaction contributes to the sense of taste because vapourized molecules from foods are conducted from the throat to the olfactory receptors in the nasal cavities. Olfactory input is the reason why anything that dulls

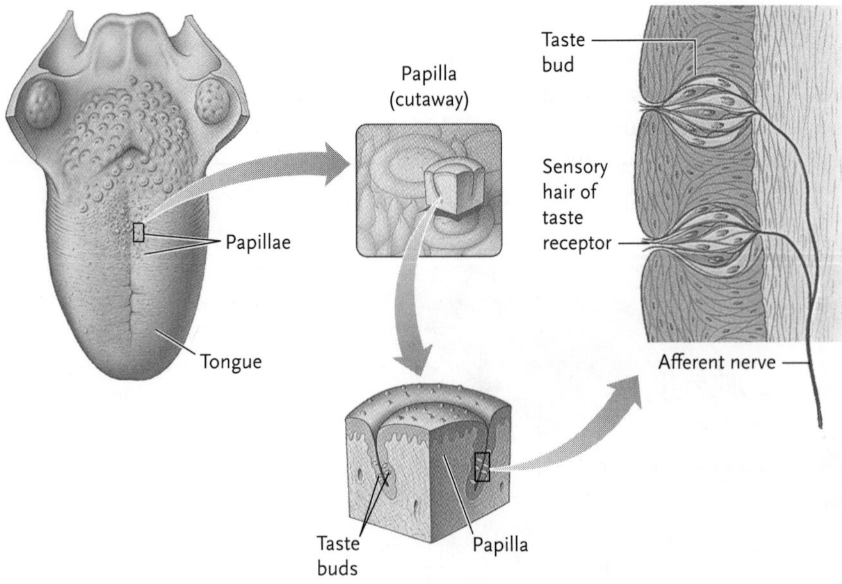

Figure 34.29

Taste receptors in the human tongue. The receptors occur in microscopic taste buds that line the sides of the furry papillae.

Labels: Papilla (cutaway); Taste bud; Papillae; Sensory hair of taste receptor; Tongue; Afferent nerve; Taste buds; Papilla

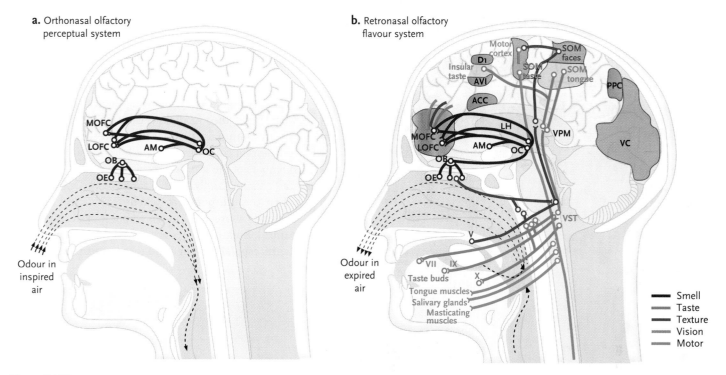

a. Orthonasal olfactory perceptual system

b. Retronasal olfactory flavour system

Odour in inspired air

Odour in expired air

- ——— Smell
- ——— Taste
- ——— Texture
- ——— Vision
- ——— Motor

Figure 34.30

The dual olfactory system (see Table 34.1) means that some smells are perceived during inhalation and activate some parts of the brain **(a)**, whereas others are perceived during exhalation, activating other parts of the brain **(b)**. ACC – accumbens; AM – amygdala; AV1 – anterior ventral insular cortex; D1 – dorsal insular cortex; LH – lateral cortex; MOFC – medial orbitofrontal cortex; NST – nucleus of the solitary tract; OB – olfactory bulb; OC – olfactory cortex; OE – olfactory epithelium; PPC – posterior parietal cortex; SOM – somatosensory cotex; V, VII, IX, and X – cranial nerves; VC – primary visual cortex; VPM – ventral posteriomedial thalamic nucleus.

Table 34.1	The Dual Olfactory System	
Operations	Orthonasal olfaction	Retronasal olfaction
Stimulation route	Through the external nares	From the back of the mouth through the nasopharynx
Stimuli	Floral scents Perfumes Smoke Food aromas Prey/predator smells Social odors Pheromones MHC molecules	Food volatiles
Processed by	Olfactory pathway influences by the visual pathway	Olfactory pathway combined with pathways for taste, touch, sound, and active sensing by proprioception form a "flavour system"

Note the interesting contrast, that orthonasal olfactory perception involves a wide range of types of odors processed through only the olfactory pathway, in comparison with retronasal olfactory perception which involves only food volatiles but processed in combination with many brain pathways.

your sense of smell, such as a head cold or holding your nose, diminishes the apparent flavour of food.

34.5c Communication: Odours as Signals

Like other animals, many mammals communicate with odours, as anyone who has walked a dog knows. Individuals of the same family or colony are identified by their odour; odours are also used to attract mates and to mark territories and trails. In mammals, odourants are detected by ~1000 different odourant receptors. In the olfactory epithelium of mice, trace amine-associated receptors (TAARs) recognize volatile amines found in urine. TAARs were described in 2006, and the genes encoding them are known in humans and fishes, as well as in mice. Humans use the fragrances of perfumes and colognes as artificial sex attractants. The presence of TAARs in humans implies that we make more use of pheromones than previously suspected.

Although olfaction is obviously important to terrestrial animals, it has commonly been believed that, at least in mammals, the olfactory epithelium does not detect odourants in water. Star-nosed moles (*Condylura cristata*) and water shrews (*Sorex palustris*) exhale bubbles while diving. They reinhale the bubbles and, in this way, gain access to airborne olfactory cues **(Figure 34.31, p. 844).**

Figure 34.31
Star-nosed moles (*Condylura cristata*) have papillae around their noses. Shown here, the papillae capture bubbles of air, allowing the submerged mole to smell airborne odours.

Reprinted by permission from Macmillan Publishers Ltd: Nature, Kenneth C. Catania, "OlfactionUnderwater 'sniffing' by semi-aquatic mammals", Vol. 444, pp. 1024-1025, copyright (2006).

STUDY BREAK

1. What is the difference between taste and smell? What are the similarities?
2. What are the five basic tastes to which human receptors respond?
3. To where are taste signals relayed after the thalamus?

34.6 Thermoreceptors and Nociceptors

Thermoreceptors detect changes in the surrounding temperature. Nociceptors respond to stimuli that may potentially damage the surrounding tissues. Both types of receptors consist of free nerve endings formed by the dendrites of afferent neurons, with no specialized receptor structures surrounding them.

34.6a Thermoreception: Heat Detection

Most animals have thermoreceptors. Invertebrates such as mosquitoes and ticks use thermoreceptors to locate warm-blooded prey. Some snakes, including rattlesnakes and pythons, use thermoreceptors to detect the body heat of warm-blooded prey animals. These receptors are located in the pits of some pit vipers **(Figure 34.32)**, whereas those of pythons and boas may not have an opening to the surface. Vampire bats have infrared receptors on their noseleafs **(Figure 34.33)**, allowing them to detect places where blood (their food) flows close to the skin.

In mammals, distinct thermoreceptors respond to heat and cold. Researchers have shown that three members of the *transient receptor potential* (TRP)-gated Ca^{2+} channel family act as heat receptors. One responds when the temperature reaches 33°C and another responds above 43°C, at which point, heat starts to be painful. Both receptors are believed to be involved in **thermoregulation**. The third receptor responds at 52°C

Pit organs

David Hosking/Frank Lane Picture Agency

Figure 34.32
The pit organs of an albino western diamondback rattlesnake (*Crotalus atrox*) are located in depressions on both sides of the head below the eyes. These thermoreceptors detect infrared radiation emitted by warm-blooded prey such as mice and kangaroo rats.

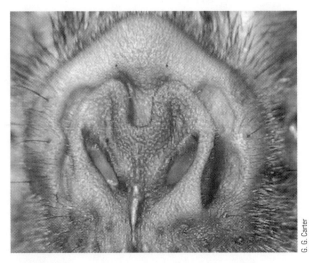

G. G. Carter

Figure 34.33
The noseleaf on the face of a vampire bat (*Desmodus rotundus*) houses an infrared detector, allowing the bat to find places where blood flows close to the skin. The bat then uses razor-sharp teeth to remove a divot of skin and antibleeding chemicals in its saliva to allow it to get a blood meal.

Capsaicin

Biting into a jalapeño pepper (a variety of *Capsicum annuum*) can produce a burning pain in your mouth strong enough to bring tears to your eyes. This painfully hot sensation is due primarily to *capsaicin* **(Figure 1),** a chemical that probably evolved in pepper plants as a defence against foraging animals. The defence is obviously ineffective for the humans who relish peppers and other foods containing capsaicin (such as hot sauce).

David Julius and his coworkers revealed the molecular basis for detection of capsaicin by nociceptors. They designed their experiments to test the hypothesis that the responding nociceptors have a cell surface receptor that binds capsaicin. Binding the chemical opens a membrane channel in the receptor that admits calcium ions and initiates action potentials interpreted as pain.

The Julius team isolated the total complement of messenger RNAs from nociceptors that respond to capsaicin and made complementary DNA (cDNA) clones of the mRNAs. The cDNAs contained thousands of different sequences that encode proteins made in the nociceptors. The team transferred the cDNAs individually into embryonic kidney cells (which do not normally respond to capsaicin), and the transformed cells were screened with capsaicin to identify which took in calcium ions. These would be the cells that had received a cDNA encoding a capsaicin receptor. Messenger RNA transcribed from the identified cDNA clone was injected into both frog oocytes and cultured mammalian cells. Both oocytes and cultured cells responded to capsaicin by admitting calcium ions, confirming that the researchers had found the capsaicin receptor cDNA.

Among the effects noted when the receptor was introduced into oocytes was a response to heat. Increasing the temperature of the solution surrounding the oocytes from 22°C to about 48°C produced a strong calcium inflow. In short, capsaicin and heat produce the same response in cells containing the receptor. Therefore, the feeling that your mouth is on fire when you eat a hot pepper probably results from the fact that, as far as your nociceptors and CNS are concerned, it *is* on fire.

Chili peppers were domesticated in different parts of the New World (from Chile to the Caribbean) by 6000 years b.p., and their use in cooking has spread throughout the world (see Chapter 49).

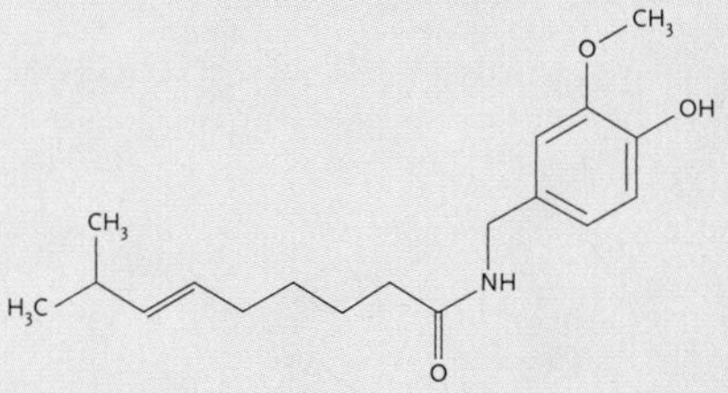

Figure 1
A capsaicin molecule.

and above, in this case producing a pain response rather than being involved in thermoregulation.

Two cold receptors are known in mammals. One responds between 8 and 28°C and is thought to be involved in thermoregulation. The second responds to temperatures below 8°C and appears to be associated with pain rather than thermoregulation. The molecular mechanisms controlling the opening and closing of heat and cold receptor chemical channels are not currently known.

Some neurons in the hypothalamus of mammals function as thermoreceptors, sensing changes in brain temperature and receiving afferent thermal information. They are highly sensitive to shifts from the normal body temperature and trigger involuntary responses such as sweating, panting, or shivering, which restore normal body temperature.

34.6b Nociceptors: Pain

Signals from nociceptors in mammals and possibly other vertebrates detect damaging stimuli that are interpreted by the brain as pain. Pain is a protective mechanism. In humans, pain prompts us to do something immediately to remove or decrease the damaging stimulus. Often pain elicits a reflex response, such as withdrawing the hand from a hot stove, that proceeds before we are consciously aware of the sensation.

Mechanical damage, such as a cut, pin prick, or blow to the body, and temperature extremes can cause pain. Some nociceptors are specific for a particular type of damaging stimulus, whereas others respond to more than one kind. Axons that transmit pain are part of the somatic system of the PNS (see Chapter 33). They synapse with interneurons in the grey matter of the spinal cord and activate neural pathways to the CNS by releasing the neurotransmitters glutamate or substance P (see Chapter 33). Glutamate-releasing axons produce sharp, prickling sensations that can be localized to a specific body part, such as the pain of stepping on a tack. Substance P–releasing axons produce dull, burning, or aching sensations that are not easily localized, such as the pain of tissue damage when you stub your toe.

As part of their protective function, pain receptors adapt very little, if at all. Some pain receptors gradually intensify the rate at which they send out action potentials if the stimulus continues at a constant level. The

Magnetic Sense in Sea Turtles

To determine if loggerhead sea turtles (*Caretta caretta*) use a magnetoreceptor system for orientation, Kenneth Lohmann and colleagues tested the responses of hatchling turtles to magnetic fields. They placed each turtle hatchling they tested in a harness and tethered it to a swivelling, electronic system in the centre of a circular pool of water **(Figure 1a).** The pool was surrounded by a large coil system, allowing the researchers to reverse the direction of the magnetic field **(Figure 1b).** The direction the turtle swam was recorded by the tracking system and relayed to a computer.

The turtles swam under two experimental conditions: half of them in Earth's magnetic field and the other half in a reversed magnetic field. Turtle hatchlings tested in Earth's magnetic field swam, on average, in an east-to-northeast direction, mimicking the direction they follow normally when migrating at sea. The hatchlings tested in the reversed magnetic field swam, on average, in a direction 180° opposite that of the hatchlings swimming in Earth's magnetic field.

The results indicate that loggerhead sea turtle hatchlings have the ability to detect Earth's magnetic field and use it to help them orient their migration. Their direction of migration, east to northeast, matches the inclination of Earth's magnetic field in the Atlantic Ocean where they migrate (see **Figure 1c**).

a.

Kenneth Lohmann/University of North Carolina

Figure 1
(a) Harnessed hatchling loggerhead sea turtles **(b)** were tested in a circular pool in which the magnetic field could be altered. **(c)** Hatchlings swimming in the normal magnetic field of Earth swam in the directions they would travel at sea on migration.

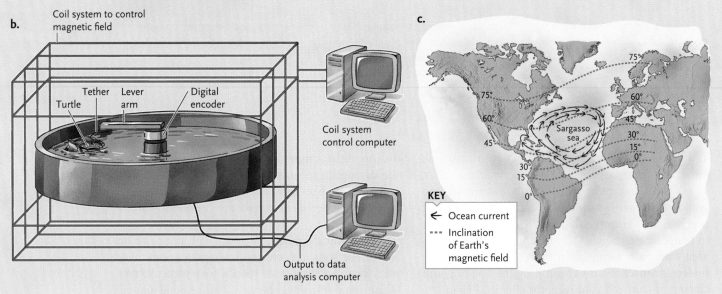

CNS also has a pain-suppressing system. In response to stimuli, such as exercise, hypnosis, and stress, the brain releases *endorphins*, natural painkillers that bind to membrane receptors on substance P neurons, reducing the amount of neurotransmitter released.

Nociceptors contribute to the taste of some spicy foods, particularly those containing capsaicin, the active ingredient in hot peppers. Researchers who study pain use *capsaicin* to identify nociceptors. To some, the burning sensation from capsaicin is addictive because in its presence, nociceptors in the mouth, nose, and throat immediately transmit pain messages to the brain. The brain responds by releasing endorphins that act as a painkiller and create temporary euphoria (see *Molecule Behind Biology*).

STUDY BREAK

1. What do thermoceptors and nociceptors have in common?
2. What are the three heat thermoceptors found in humans?
3. Why do pain receptors not adapt?

34.7 Electroreceptors and Magnetoreceptors

Some animals gain information about their environment by sensing electrical or magnetic fields. In so doing, they directly sense stimuli that humans can detect only with scientific instruments.

34.7a Electroreception

Electroreception is an ancient trait in vertebrates. Although it was lost in ancestral bony fish, it persists today in many sharks and has reappeared in some bony fishes and some amphibians. Mammals such as the star-nosed mole and duck-billed platypus detect electric fields with specialized **electroreceptors**.

Electroreceptors depolarize in an electric field, and the plasma membrane of an electroreceptor cell generates action potentials. The electrical stimuli detected by the receptors are used in different ways. Electrical information can be used to locate prey, to negotiate a way around obstacles in muddy water, or, by some fishes, in communication. Some electroreception systems are passive, detecting electric fields in the environment, not the animal's own electric currents. Passive systems are used mainly to find prey. Sharks and rays use electroreceptors to locate prey buried under sand by detecting electrical currents generated by the prey's heartbeat or by the muscle contractions moving water over the gills.

34.7b Electric Fishes

Fishes in the orders Mormyriformes (elephant fish from Africa) and Gymnotiformes (knifefish from South America; **Figure 34.34a)** emit and receive low-voltage electrical signals, using them to locate prey (electrolocation) and in intraspecific communication. Electric fishes have two kinds of electroreceptors, ampullary and tuberous. Ampullary receptors respond to low-frequency alternating current (AC) fields usually associated with other fishes, vegetation, or electrical stimuli from other electric fishes. Tuberous electroreceptors detect electric organ discharge. Electrical signals are generated by electric organs that are specialized muscle cells.

Some electric fishes can produce discharges of several hundred volts (e.g., *Electrophorus electricus*, the electric eel, **Figure 34.34b**, and *Malapturus electricus*, the electric catfish) that stun or kill prey. The voltage discharged by an electric eel is high enough to stun, but not kill, a human.

34.7b Magnetoreception

Just as the development of a magnetic compass was a pivotal point in humans' ability to navigate, some animals use magnetic compasses in long-distance navigation. **Magnetoreceptors** allow animals to detect and use Earth's magnetic field as a source of directional information. The list includes butterflies, beluga whales, sea turtles (see *Magnetic Sense in Sea Turtles*), homing pigeons, and foraging honeybees (*Apis mellifera*).

The pattern of Earth's magnetic field differs from region to region yet remains almost constant over time, largely unaffected by changing weather or day and night. Animals with magnetic receptors can reliably monitor their location. Although little is known about the receptors that detect magnetic fields, they may depend on the fact that moving a conductor, such as an electroreceptor cell, through a magnetic field generates an electric current. Some magnetoreceptors may depend on the effect of Earth's magnetic field on the mineral *magnetite*, which is found in the bones or teeth of many vertebrates, including humans, and in insects, such as the abdomen of the honeybee and the heads and abdomens of certain ants.

Animals such as homing pigeons (*Columbia livia*), famous for their ability to find their way back to their nests even when released far from home, navigate by detecting their position with reference to both Earth's magnetic field and the Sun. Magnetite is located in the beaks of these birds, which is where magnetoreception likely occurs. Big brown bats (*Eptesicus fuscus*) also have a magnetic sense that influences their navigational abilities.

STUDY BREAK

1. How do animals use electrical information?
2. What are the two types of electroreceptors in electric fishes?
3. Why would a magnetic navigational system be favourable?

a.

b.

Figure 34.34
Two electric fishes from South America. *Eigenmannia eigenmannia* **(a)** is a weakly electric fish that uses electrolocation, whereas *Electrophorus electricus* **(b)**, the electric eel, stuns prey with an electric discharge.

Do humans have a magnetic sense? What is the evidence for a magnetic sense? What is the basis for a magnetic sense—what is the transducer?

Review

Go to CENGAGENOW™ at http://hed.nelson.com/ to access quizzing, animations, exercises, articles, and personalized homework help.

34.1 Overview of Sensory Receptors and Pathways

- Receptors in the sensory system collect information (i.e., stimuli) from internal and external sensors (transducers) and convert (transduce) the information into neural activity. Dendrites of an afferent neuron pick up the stimuli. The axon of the afferent neuron conveys the stimulus to the CNS, providing the organisms with sensory data used to influence behaviour and homeostasis.

- Mechanoreceptors detect mechanical energy (pressure); photoreceptors detect the energy of light; chemoreceptors detect specific molecules or chemical conditions; thermoreceptors detect the flow of heat energy; and nociceptors detect tissue damage or noxious chemicals.

- Some receptors allow the effect of a stimulus to be reduced over time. Otherwise, the receptor could become overloaded and not function properly. Receptors also need a period of rest.

34.2 Mechanoreceptors and the Tactile and Spatial Senses

- You would more likely detect a fruit fly walking on your face than on your leg because touch receptors are more concentrated on a human's face than on a human's leg.

- The vestibular apparatus and cochlea are specialized sensory structures. The vestibular apparatus is responsible for maintaining equilibrium and coordinating head and body movements. The cochlea is used in hearing.

- The strength of the responses of stretch receptors depends on how much and how fast the muscle is stretched.

34.3 Mechanoreceptors and Hearing

- Most invertebrates detect sound through mechanoreceptors in their skin or other surface structures. An example are ears in the common cricket. Crickets detect sound using tympana (ears) on each side of the abdomen or on the first pair of walking legs. The ears are areas of thin exoskeleton (tympanum, singular; tympana, plural).

- Sound waves cause vibrations of the tympanum, which are converted into nerve impulses that travel along the auditory nerve to the CNS.

- Vertebrates use ears to hear sounds. Some vertebrates have pinnae (outer ears) that collect sounds (vibrations) and channel them down the auditory canal to the tympanum (eardrum). There, vibrations in air are converted (transduced) into vibrations of the membrane comprising the tympanum. These vibrations are amplified by vibrations of the malleus, incus, and stapes (auditory ossicles) and conveyed to the oval window, where they are converted to vibrations in the fluid of the coiled cochlea. Vibrations in the fluid inside the cochlea cause vibrations of the basilar membrane, which are detected by cilia and converted to nerve impulses. The nerve impulses move down the auditory nerve to the brain, where they are processed and interpreted.

- The echolocation calls of bats range from about 8 kHz to over 200 kHz. Many are inaudible to humans because they are ultrasonic, above the range of human hearing (20 000 Hz = 20 kHz). Infrasounds (<40 Hz), used by elephants and whales, are below the range of human hearing.

34.4 Photoreceptors and Vision

- The ocellus, the "simplest" eye, lacks a lens and therefore is not image forming. Ocelli (plural) are light receptors that often allow animals to detect differences in the brightness of light. Image-forming eyes (compound eyes and single-lens eyes) are photoreceptors too, but they have lenses that allow light to be focused on the retina, the layer with photosensitive cells.

- Colourblindness is a result of inherited defects in opsin proteins of one or more of the three types of cones. Genes controlling colour vision are located on the X chromosome. Therefore, human males have only one set of genes controlling colour vision, whereas females have two sets. Colourblindness is relatively common in men and relatively rare in women.

- Compound eyes are composed of ommatidia, many individual visual units. Each ommatidium samples a small part of the visual field, and many ommatidia provide the animal with an image that is a mosaic of many individual views. Motion is detected by many ommatidia at once, giving compound eyes special sensitivity to motion.

- Accommodation is the movement of the lens to focus the image on the retina. In cephalopods, muscles move the lens forward and back. In terrestrial vertebrates, muscles change the shape of the lens.

- Rods and cones, the photoreceptor cells in the retina, consist of an outer segment of stacked, flattened membranous disks with photopigments, an inner segment for cellular metabolic activities, and the synaptic terminal for storage and release of neurotransmitter.

34.5 Chemoreceptors

- Taste is the detection of potential food molecules *touched* by a receptor, whereas smell is the detection of *airborne* particles and molecules. Information from taste receptors is processed in the parietal lobes. Information from smell is processed in the olfactory bulb and temporal lobes. Both taste and smell receptors have hairlike extensions that bind molecules.

- The five basic tastes are sweet, sour, bitter, salty, and umami (savoury). Some lead to the gustatory centres of the cerebral cortex, whereas others are linked to the brain stem and limbic system, producing visceral and emotional responses, including physiological responses such as salivation and secretion of gastric juices. The same responses may occur in response to smells.

34.6 Thermoreceptors and Nociception

- Thermoreceptors and nociceptors consist of free nerve endings formed by the dendrites of afferent neurons. No specialized receptor structures surround them. All are members of the *transient receptor potential* (TRP)-gated Ca^{2+} channel family. One responds to temperatures above 33°C, one to temperatures above 43°C, and the third to temperatures above 53°C. The first two are involved in thermoregulation, whereas the last elicits a strong pain response.

- The pain response (nociception) is a protective mechanism. These receptors do not adapt; otherwise, organisms would not withdraw from a prolonged painful stimulus, increasing the level of damage associated with the pain.

34.7 Electroreceptors and Magnetoreceptors

- Electric field information can be used to detect prey, for example, a shark's passive system. Some fishes generate electric signals to detect obstacles and prey and to communicate.

- The two types of electroreceptors are ampullary receptors and tuberous receptors. Ampullary receptors respond to low-frequency alternating current (AC) fields usually associated with other fishes, vegetation, or electrical stimuli from other electric fishes. Tuberous electroreceptors detect electric organ discharge.

- The pattern of the magnetic field of Earth varies from region to region. Animals with a magnetic compass can detect the magnetic field and use the information in navigation (as people use compasses). Earth's magnetic field remains constant over time and so is reliable from year to year. Many animals have magnetic receptors. Other animals use Sun compasses.

Questions

Self-Test Questions

1. Some preying mantises have
   a. two ears.   d. one ear.
   b. two eyes.   e. one eye.
   c. two antennae.

2. Sensory adaptation involves
   a. the loss of eyes in cave-dwelling fish.
   b. the development of ears in insects preyed upon by bats.
   c. a reduction in the effect of stimuli.
   d. the development of an acute sense of smell.
   e. all of the above.

3. Which of the following are examples of proprioceptors?
   a. eyes   d. halteres
   b. statocysts   e. b and d
   c. ears

4. The vestibular system of vertebrates involves the
   a. retina.   d. semicircular canals.
   b. Golgi apparatus. e. c and d.
   c. utricle.

5. The malleus, incus, and stapes occur in
   a. birds.   d. amphibians.
   b. mammals.   e. insects.
   c. bony fish.

6. Golden rice provides _____, which is vital to the development of vision.
   a. vitamin E   d. fatty acids
   b. rhodopsin   e. chlorophyll
   c. vitamin A

7. Accommodation occurs in the eyes of cephalopods when
   a. the shape of the lens is changed.
   b. the retina moves toward the lens.
   c. the retina moves away from the lens.
   d. the lens moves toward or away from the retina.
   e. light is focused on the fovea.

8. Chemoreceptors in *Hydra* are concentrated
   a. in the tentacles.
   b. around the mouth.
   c. in the base.
   d. in the gasteron.
   e. in/around all of the above.

9. Thermoreceptors are widespread in
   a. earthworms.   d. birds.
   b. vampire bats.   e. animals.
   c. pit vipers.

10. Nociceptors are sensitive to
    a. pheromones.   d. light.
    b. pain.   e. vibration.
    c. touch.

Questions For Discussion

1. What is an eye? What are the key elements in the definition? Can robots have eyes?

2. Which are better at evoking memories in humans: visual or olfactory stimuli? What is the evidence supporting either point of view? Why would one kind of stimulus be more effective than the other?

3. Find examples of redundancy in the sensory systems of animals. What are the advantages of redundant systems? What are the disadvantages?

The larva, pupa, and adult moth of the tobacco hornworm, *Manduca sexta,* a model insect that has been used in exploring the hormonal control of metamorphosis.

35 The Endocrine System

WHY IT MATTERS

The larva of the tobacco hornworm, *Manduca sexta,* having reached its critical weight, stops feeding, drops to the ground, and burrows into the soil where it moults into the pupal stage. Within the pupa, nearly all of the old larval tissues are destroyed and replaced by the tissues of the moth, which have been waiting in embryonic form for the signal to develop. When the moth is formed, it wriggles, still enclosed in the pupal cuticle, to the surface of the soil. It begins the behaviour leading to rupture of the pupal cuticle and its emergence as the adult moth as a soft animal with still rumpled wings. Once emergence is complete, it inflates its body and expands its wings. Only then does the cuticle harden. Moths are nocturnal, and the female, feeding on the nectar of several species of flowers, completes the development of its eggs begun in the pupal stage and releases the pheromone that will attract males. After mating, the female takes flight and searches for a suitable host plant in the family Solanaceae, where she lays the 100 or so eggs that she carries.

This carefully timed sequence of developmental and behavioural events is orchestrated by several hormones released in response to internal and external environmental cues. This marvel of communication between the environment and the cells, tissues, and organs of animals involving interactions between the nervous system and endocrine structures is a feature of everyday life in even the simplest of organisms. It is impossible to understand the functioning of any animals in the absence of knowledge of the endocrine system. For more complex animals such as insects and mammals, a galaxy of hormones regulates development, reproduction, and behaviours and helps maintain a stable internal environment.

Hormones are secreted by cells of the **endocrine system** (*endo* = within; *krinein* = separate), so called because it forms a distinct control system within the body. The endocrine system, like the nervous system, regulates and coordinates distant organs. The two systems are structurally, chemically, and functionally related, but they control different types of activities. The nervous system (Chapter 33), through high-speed electrical signals, enables an organism to interact rapidly with the external environment, whereas the endocrine system controls activities that involve slower, longer-acting responses.

The nervous system directs highly specific localized targets: it is a "private" mode of communication. The endocrine system is more "public," often affecting several tissues or organs. Ultimately, the nervous system controls the endocrine system. The mechanisms and functions of the endocrine system are the subjects of this chapter.

35.1 Hormones and Their Secretion

Cells signal other cells using neurotransmitters (see Chapter 33), hormones, and local regulators. Our focus in this chapter is hormones, but we also deal briefly with local regulators, molecules that act locally rather than over long distances.

35.1a The Endocrine System Includes Four Major Types of Cell Signalling

Four types of cell signalling occur in the endocrine system. In *classical endocrine signalling,* hormones are secreted into the blood or extracellular fluid by the cells of ductless secretory organs called **endocrine glands (Figure 35.1a).** In contrast, *exocrine glands,* such as the sweat and **salivary glands,** release their secretions into ducts that lead outside the body or into the cavities of the digestive tract (see Chapter 32). Hormones are circulated throughout the body in the blood or other body fluids, and, as a result, most body cells are constantly exposed to a wide variety of hormones. Only the *target cells* of a hormone, those with *receptor proteins* (Chapter 8) recognizing and binding that hormone, respond to it. Hormones are cleared from the body at a steady rate by enzymatic breakdown in their target cells or blood or organs such as the liver or kidneys, and the breakdown products are excreted.

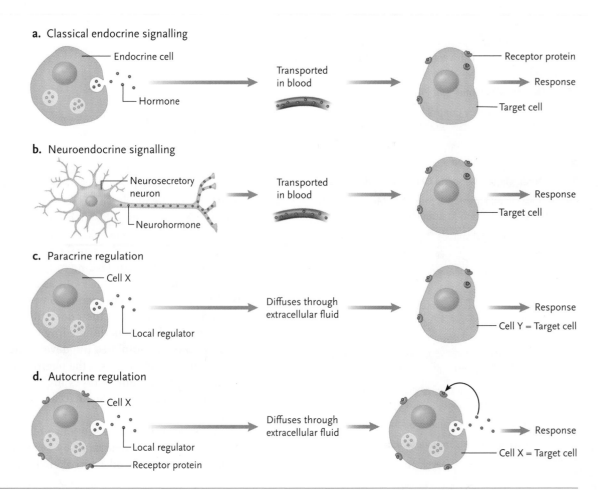

Figure 35.1

The four major types of cell signalling in the endocrine system.

In *neuroendocrine signalling*, specialized **neurosecretory neurons** respond to and conduct electrical signals, but rather than synapsing with target cells, they release a neurohormone into the circulation when appropriately stimulated **(Figure 35.1b)**. The hormone is produced in the cell body and packaged in membrane-bound vesicles that are transported along the axon to the release sites. The neurohormone is usually distributed in blood or other body fluids and elicits a response in target cells that have receptors for the neurohormone. The peptide vasopressin secreted by the pituitary gland acts on the kidney, reducing the water excreted in the urine, and on muscles of blood vessels, increasing blood pressure. It is also released directly into the brain to cause a myriad of social effects, such as pair bonding in some mammals.

Two other sorts of chemical communication between cells are not normally thought of as part of the hormonal signalling system. In *paracrine regulation,* a cell releases a signalling molecule that diffuses through the extracellular fluid and acts on nearby cells. Regulation is *local* **(Figure 35.1 c)** rather than at a distance. In *autocrine regulation,* the local regulator acts on the same cells that release it **(Figure 35.1d)**. Many of the growth factors that regulate cell division and differentiation act in both a paracrine and an autocrine fashion.

35.1b Hormones and Local Regulators Can Be Grouped into Four Classes Based on Their Chemical Structures

More than 60 hormones and local regulators have been identified in humans. Many human hormones are either identical or very similar in structure and function to those in other animals, but other vertebrates, as well as invertebrates, may have hormones not found in humans. Most of these chemicals can be grouped into four molecular classes: amine, peptide, steroid, and fatty acid–derived molecules.

Amine hormones are involved in classical endocrine signalling and neuroendocrine signalling. Most amine hormones are based on tyrosine. With one major exception, they are hydrophilic molecules, which diffuse readily into the blood and extracellular fluids. On reaching a target cell, they bind to receptors at the cell surface. The amines include epinephrine–norepinephrine and, in protostomes, octopamine, already familiar as neurotransmitters released by some neurons (see Chapter 33). The exception is the thyroid hormones secreted by the thyroid gland. These hormones, based on a pair of tyrosines, enter the cell by receptor-mediated endocytosis. Inside the cell, one form of the hormone binds to nuclear receptors in the same way as described for steroids below. Thyroid hormones also act via membrane receptors not only on the surface of the cells but also on mitochondrial membranes.

Peptide hormones consist of amino acid chains, ranging in length from as few as 3 amino acids to more than 200. Some have carbohydrate groups attached. They are involved in classical endocrine signalling and neuroendocrine signalling. Mostly hydrophilic hormones, peptide hormones are released into the blood or extracellular fluid by exocytosis when cytoplasmic vesicles containing the hormones fuse with the plasma membrane. One large group of peptide hormones, the **growth factors**, regulates the division and differentiation of many cell types in the body. Many growth factors act in both a paracrine and an autocrine manner, as well as in classical endocrine signalling.

Steroid hormones are involved in classical endocrine signalling. All are hydrophobic molecules derived from cholesterol and are sparingly soluble in water. They combine with hydrophilic carrier proteins to form water-soluble complexes that diffuse easily in blood or other fluids. On contacting a cell, the hormone is released from its carrier protein, passes through the plasma membrane of the target cell (a process that is sometimes mediated by receptors), and binds to internal receptors in the nucleus or cytoplasm. Steroid hormones include aldosterone, cortisol, the vertebrate sex hormones, and ecdysone, the hormone that governs the formation of new cuticles in ecdyzoan protostomes. Steroid hormones may vary little in structure but produce very different effects. Testosterone and estradiol, two major sex hormones responsible for the development of mammalian male and female characteristics, respectively, differ only in the presence or absence of a methyl group. Steroids can also act via membrane receptors, controlling cellular events such as apoptosis and cell proliferation and more complex events such as behaviour.

Fatty acids represent a very specialized category of hormones. In arthropods and possibly annelids, hormones derived from farnesoic acid include the juvenile hormones that govern metamorphosis and reproduction in arthropods (see Chapter 26, *Molecule Behind Biology*). **Prostaglandins** and their relatives are important local regulators derived from arachidonic acid. They are involved in paracrine and autocrine regulation in all animals. First discovered in semen, they enhance the transport of sperm through the female reproductive tract by increasing the contractions of muscle cells in both vertebrates and insects. In at least some insects, prostaglandins act as endocrines: they are synthesized in the sperm storage organs of mated females and initiate egg laying by acting on the oviducts and possibly the nervous system.

35.1c Many Hormones Are Regulated by Feedback Pathways

The secretion of many hormones is regulated by feedback pathways, some of which operate partially or completely independently of neuronal controls. Most

Figure 35.2

A negative feedback loop regulating secretion of the thyroid hormones. As the concentration of thyroid hormones in the blood increases, the hormones inhibit an earlier step in the pathway (indicated by the negative sign).

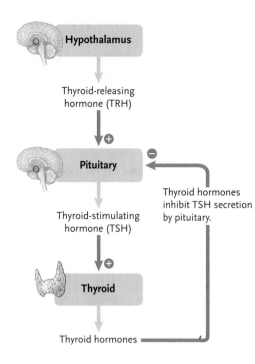

and ghrelin secreted by the stomach and secretin from the intestine; (2) insulin and glucagon, secreted by the pancreas; (3) growth hormone, secreted by the anterior pituitary; (4) epinephrine–norepinephrine, released by the sympathetic nervous system and the adrenal medulla; and (5) glucocorticoid hormones, released by the adrenal cortex.

The entire system of hormones regulating fuel metabolism resembles the fail-safe mechanisms designed by engineers, in which redundancy, overlapping controls, feedback loops, and multiple safety valves ensure that vital functions are maintained at appropriate levels in the face of changing and even extreme circumstances.

STUDY BREAK

1. What are the functions of the endocrine and nervous systems? How are they the same, and how do they differ?
2. What are the four major types of cell signalling that occur in the endocrine system? How do they work?

pathways are controlled by negative feedback in which a product of the pathway inhibits an earlier step in the pathway. In vertebrates, secretion by the thyroid gland is regulated by a negative feedback loop **(Figure 35.2)**. Neurosecretory neurons in the hypothalamus secrete thyroid-releasing hormone (TRH) into a vein connecting the hypothalamus to the pituitary gland. In response, the pituitary releases thyroid-stimulating hormone (TSH) into the blood, which stimulates the thyroid gland to release thyroid hormones. As the thyroid hormone concentration in the blood increases, it begins to inhibit TRH secretion by the hypothalamus. In turn, TSH and secretion of the thyroid hormones are reduced.

35.1d Body Processes Are Regulated by Coordinated Hormone Secretion

Although we mostly discuss individual hormones in the remainder of the chapter, body processes are affected by more than one hormone. The blood concentrations of glucose, fatty acids, and ions such as Ca^{2+}, K^+, and Na^+ are regulated by the coordinated activities of several hormones secreted by different glands. Similarly, body processes such as oxidative metabolism, digestion, growth, sexual development, and reactions to stress are all controlled by multiple hormones.

In many of these systems, negative feedback loops adjust the levels of secretion of hormones that act in antagonistic (opposing) ways, creating a balance in their effects that maintains body homeostasis (see Chapter 43). Consider the regulation of fuel molecules such as glucose, fatty acids, and amino acids in the blood. We usually eat three meals a day and fast to some extent between meals. During these periods of eating and fasting, five hormone systems act in a coordinated fashion to keep the fuel levels in balance: (1) gastrin

35.2 Mechanisms of Hormone Action

Hormones control cell functions by binding to receptor molecules in their target cells. Small quantities of hormones can typically produce profound effects in cells and body functions due to **amplification**. In amplification, an activated receptor activates many proteins, which then activate an even larger number of proteins for the next step in the cellular pathway and so on in each subsequent step (see Chapter 8).

35.2a The Secreted Hormone May Not Be the Active Form

Many hormones are secreted in an inactive or less active form (a "prohormone") and converted by target cells or enzymes in the blood or other tissues to the active form. The best known example is thyroxine, discussed below. Many other hormones are subject to similar processes. **Ecdysone**, a steroid governing the formation of new cuticle in insects, is converted to the much more active functional hormone, 20-OH ecdysone, by the addition in the target cells of a single hydroxyl group. Peptide hormones are commonly synthesized as prohormones that undergo posttranslational conversion to the active forms in the source cell. In some cases, however, further conversion occurs once the hormone has been secreted. Angiotensin is a hormone that governs blood pressure in humans. It is secreted by the liver as angiotensinogen. An inactive form of angiotensin is cleaved from angiotensinogen by an enzyme. This inactive form is converted to the active hormone by angiotensin-converting enzyme (ACE). ACE inhibitors are often prescribed for control of high blood pressure.

35.2b Hormones May Bind to Surface Receptors, Usually Activating Protein Kinases Inside Cells

Hormones that bind to receptor molecules in the plasma membrane produce their responses through signal transduction pathways. In brief, when a surface receptor binds a hormone, it transmits a signal through the plasma membrane. Within the cell, the signal is transduced, changed into a form that causes the cellular response **(Figure 35.3a)**. Typically, the reactions of signal transduction pathways involve protein kinases, enzymes that add phosphate groups to proteins. Adding a phosphate group to a protein may activate or inhibit it, depending on the protein and the reaction. The particular response produced by a hormone depends on the kinds of protein kinases activated in the cell and the types of target proteins they phosphorylate (Chapter 8). The signal transduction pathway may not stop at the cytoplasm: many growth factors and some peptide hormones ultimately affect events in the nucleus. Although action via membrane receptors is character-istic of and most extensively studied in peptide and amine hormones, many steroid and fatty acid hormones also exert some of their actions in this way.

The peptide hormone glucagon illustrates the mechanisms triggered by surface receptors. When glucagon binds to surface receptors on liver cells, it triggers a series of steps leading to the phosphorylation and activation of the enzyme governing the breakdown of glycogen stored in those cells into glucose.

35.2c Hydrophobic Hormones Bind to Receptors Inside Cells, Activating or Inhibiting Genetic Regulatory Proteins

After passing through the plasma membrane, the hydrophobic steroid and thyroid hormones bind to internal receptors in the nucleus or cytoplasm **(Figure 35.3b)**. Binding of the hormone activates the receptor, which then binds to a control sequence of specific genes. Depending on the gene, binding the control sequence either activates or inhibits its transcription, leading to changes in protein synthesis that accomplish the cellular response. The characteristics of the response depend on the specific genes controlled by the activated receptors and on the presence of other proteins that modify the activity of the receptor.

Figure 35.3
The reaction pathways activated by hormones that bind to receptor proteins in plasma membrane **(a)** or inside cells **(b).** In both mechanisms, the signal—the binding of the hormone to its receptor—is transduced to produce the cellular response.

a. Hormone binding to receptor in the plasma membrane

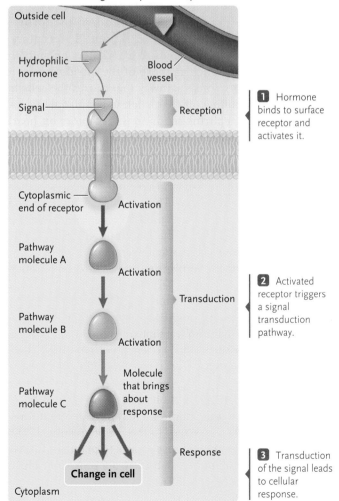

1 Hormone binds to surface receptor and activates it.

2 Activated receptor triggers a signal transduction pathway.

3 Transduction of the signal leads to cellular response.

b. Hormone binding to receptor inside the cell

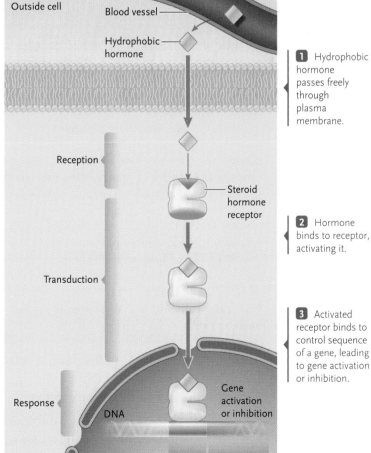

1 Hydrophobic hormone passes freely through plasma membrane.

2 Hormone binds to receptor, activating it.

3 Activated receptor binds to control sequence of a gene, leading to gene activation or inhibition.

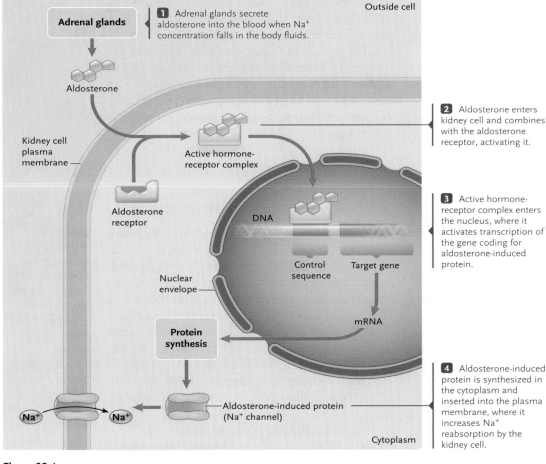

1 Adrenal glands secrete aldosterone into the blood when Na⁺ concentration falls in the body fluids.

2 Aldosterone enters kidney cell and combines with the aldosterone receptor, activating it.

3 Active hormone-receptor complex enters the nucleus, where it activates transcription of the gene coding for aldosterone-induced protein.

4 Aldosterone-induced protein is synthesized in the cytoplasm and inserted into the plasma membrane, where it increases Na⁺ reabsorption by the kidney cell.

Figure 35.4
The action of aldosterone in the increasing Na⁺ reabsorption in the kidneys when concentration of the ion falls in the blood.

One of the actions of the steroid hormone aldosterone illustrates the mechanisms triggered by internal receptors **(Figure 35.4)**. If blood pressure falls below optimal levels, aldosterone is secreted by the adrenal glands. The hormone affects only cells (mostly in the kidney but also in sweat glands and the colon) that contain the aldosterone receptor in their cytoplasm. When activated by aldosterone, the receptor binds to the control sequence of a gene, leading to the synthesis of proteins that increase **reabsorption** of Na⁺ by the kidney cells. The resulting increase in Na⁺ concentration in body fluids increases water retention and, with it, blood volume and pressure.

Many steroid hormones that act at nuclear receptors also use membrane receptors, sometimes in the same cell. Aldosterone can also have rapid effects on Na⁺ reabsorption by activating membrane receptors on kidney cells.

35.2d Target Cells May Respond to More than One Hormone, and Different Target Cells May Respond Differently to the Same Hormone

A single target cell may have receptors for several hormones and respond differently to each hormone. Vertebrate liver cells have receptors for the pancreatic hormones insulin and glucagon. Insulin increases glucose uptake and conversion to glycogen, which decreases blood glucose levels, whereas glucagon stimulates the breakdown of glycogen into glucose, which increases blood glucose levels.

Conversely, particular hormones interact with different types of receptors in or on a range of target cells. Different responses are then triggered in each target cell type because the receptors trigger different transduction pathways. For example, the amine hormone epinephrine prepares the body for handling stress (including dangerous situations) and physical activity. In mammals, epinephrine can bind to three different plasma membrane–embedded receptors: α, β₁, and β₂ receptors. When epinephrine binds to α receptors on smooth muscle cells, such as those of the blood vessels, it triggers a response pathway that causes the cells to constrict, cutting off circulation to peripheral organs. When epinephrine binds to β₁ receptors on heart muscle cells, the contraction rate of the cells increases, which, in turn, enhances blood supply. When epinephrine binds to β₂ receptors on liver cells, it stimulates the breakdown of glycogen to glucose, which is released from the cell. The overall effect of these and a number of other responses to epinephrine secretion is to supply energy to the major muscles responsible for locomotion, preparing the animal for stress or for physical activity. Similar tissue-specific diversification of responses is known for many hormones.

Moreover, the response to a hormone may differ in different animals. For example, melatonin, an amine derived from tryptophan, is important in regulating daily and annual cycles in most animals. However, it also plays a role in regulating the salt gland of marine birds. Thyroxine promotes metamorphosis in amphibians but inhibits metamorphosis in cyclostomes. The same hormone may have different functions at different stages in the life of an animal. The juvenile hormone of insects acts to maintain insects in a larval state but also controls reproduction in the adult.

In summary, the mechanisms by which hormones work have four major features. First, only the cells that contain surface or internal receptors for the hormones respond to them. Second, once bound by their receptors, hormones may produce a response that involves stimulation or inhibition of cellular processes through the specific types of internal molecules triggered by the hormone action. Third, because of the amplification that occurs in both the surface and internal receptor mechanisms, hormones are effective in very small concentrations. Fourth, the response to a hormone differs among target organs and among animals.

In the next two sections, we discuss the major endocrine cells and glands of vertebrates. The locations of these cells and glands in the human body and their functions are summarized in **Figure 35.5** and **Table 35.1, pp. 858–859.** In addition to these major endocrine organs, important hormones are also secreted by organs that have other primary functions, including the kidney, heart, liver, and intestine.

In particular, the digestive system is the source of several peptide hormones, many of which are also produced elsewhere. It is the only known source for peptides such as gastrin, secretin, and ghrelin, which coordinate the digestive secretions of the gut and its associated glands and send signals associated with hunger to the brain. The gut is increasingly recognized as an important endocrine organ in many animals. Among vertebrates, it is a more important source for circulating levels of melatonin than the pineal body with which that hormone is traditionally associated. In insects, several peptide hormones, such as proctolin, produced by neuroendocrine cells in the central nervous system (CNS) are also produced by cells in the intestine.

STUDY BREAK

What are the four major features of a hormone mechanism?

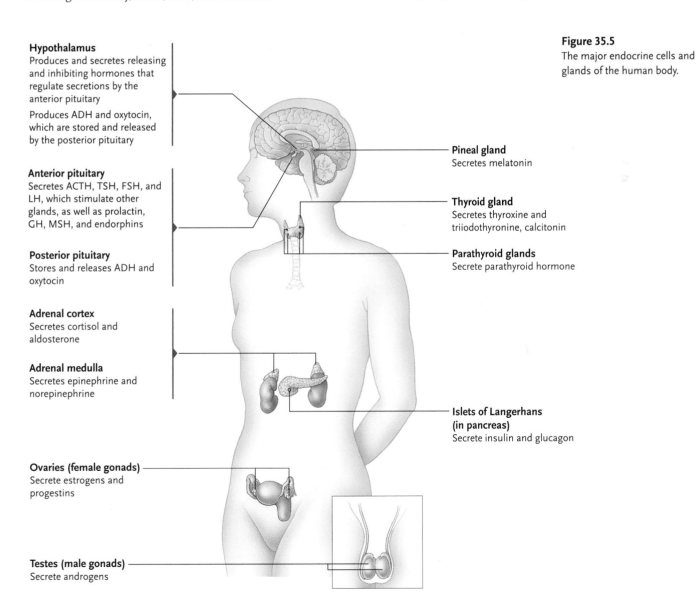

Hypothalamus
Produces and secretes releasing and inhibiting hormones that regulate secretions by the anterior pituitary

Produces ADH and oxytocin, which are stored and released by the posterior pituitary

Anterior pituitary
Secretes ACTH, TSH, FSH, and LH, which stimulate other glands, as well as prolactin, GH, MSH, and endorphins

Posterior pituitary
Stores and releases ADH and oxytocin

Adrenal cortex
Secretes cortisol and aldosterone

Adrenal medulla
Secretes epinephrine and norepinephrine

Ovaries (female gonads)
Secrete estrogens and progestins

Testes (male gonads)
Secrete androgens

Figure 35.5
The major endocrine cells and glands of the human body.

Pineal gland
Secretes melatonin

Thyroid gland
Secretes thyroxine and triiodothyronine, calcitonin

Parathyroid glands
Secrete parathyroid hormone

Islets of Langerhans (in pancreas)
Secrete insulin and glucagon

Table 35.1   The Major Human Endocrine Glands and Hormones

Secretory Tissue or Gland	Hormones	Molecular Class	Target Tissue	Principal Actions
Hypothalamus	Releasing and inhibiting hormones	Peptide	Anterior pituitary	Regulate secretion of anterior pituitary hormones
Anterior pituitary	Thyroid-stimulating hormone (TSH)	Peptide	Thyroid gland	Stimulates secretion of thyroid hormones and growth of thyroid gland
	Adrenocorticotropic hormone (ACTH)	Peptide	Adrenal cortex	Stimulates secretion of glucocorticoids by adrenal cortex
	Follicle-stimulating hormone (FSH)	Peptide	Ovaries in females, testes in males	Stimulates egg growth and development and secretion of sex hormones in females; stimulates sperm production in males
	Luteinizing hormone (LH)	Peptide	Ovaries in females, testes in males	Regulates ovulation in females and secretion of sex hormones in males
	Prolactin (PRL)	Peptide	Mammary glands	Stimulates breast development and milk secretion
	Growth hormone (GH)	Peptide	Bone, soft tissue	Stimulates growth of bones and soft tissues; helps control metabolism of glucose and other fuel molecules
	Melanocyte-stimulating hormone (MSH)	Peptide	Melanocytes in skin of some vertebrates	Promotes darkening of the skin
	Endorphins	Peptide	Pain pathways of PNS	Inhibit perception of pain
Posterior pituitary	Antidiuretic hormone (ADH)	Peptide	Kidneys	Raises blood volume and pressure by increasing water reabsorption in kidneys
	Oxytocin	Peptide	Uterus, mammary glands	Promotes uterine contractions; stimulates milk ejection from breasts
Thyroid gland	Calcitonin	Peptide	Bone	Lowers calcium concentration in blood
	Thyroxine and triiodothyronine	Amine	Most cells	Increase metabolic rate; essential for normal body growth
Parathyroid glands	Parathyroid hormone (PTH)	Peptide	Bone, kidneys, intestine	Raises calcium concentration in blood; stimulates vitamin D activation
Adrenal medulla	Epinephrine and norepinephrine	Amine	Sympathetic receptor sites throughout body	Reinforce sympathetic nervous system; contribute to responses to stress
Adrenal cortex	Aldosterone (mineralocorticoid)	Steroid	Kidney tubules	Helps control body's salt–water balance by increasing Na^+ reabsorption and K^+ excretion in kidneys
	Cortisol (glucocorticoid)	Steroid	Most body cells, particularly muscle, liver, and adipose cells	Increases blood glucose by promoting breakdown of proteins and fats
Testes	Androgens, such as testosterone*	Steroid	Various tissues	Control male reproductive system development and maintenance; most androgens are made by the testes
	Oxytocin	Peptide	Uterus	Promotes uterine contractions when seminal fluid is ejaculated into vagina during sexual intercourse
Ovaries	Estrogens, such as estradiol**	Steroid	Breast, uterus, other tissues	Stimulate maturation of sex organs at puberty and development of secondary sexual characteristics
	Progestins, such as progesterone**	Steroid	Uterus	Prepare and maintain uterus for implantation of fertilized egg and the growth and development of embryo

*Small amounts secreted by ovaries and adrenal cortex.
**Small amounts secreted by testes.

Table 35.1 | The Major Human Endocrine Glands and Hormones (Continued)

Secretory Tissue or Gland	Hormones	Molecular Class	Target Tissue	Principal Actions
Pancreas (islets of Langerhans)	Glucagon (alpha cells)	Peptide	Liver cells	Raises glucose concentration in blood; promotes release of glucose from glycogen stores and production from noncarbohydrates
	Insulin (beta cells)	Peptide	Most cells	Lowers glucose concentration in blood; promotes storage of glucose, fatty acids, and amino acids
Pineal gland	Melatonin	Amine	Brain, anterior pituitary, reproductive organs, immune system, possibly others	Helps synchronize body's biological clock with day length; may inhibit gonadotropins and initiation of puberty
Many cell types	Growth factors	Peptide	Most cells	Regulate cell division and differentiation
	Prostaglandins	Fatty acid	Various tissues	Have many diverse roles

35.3 The Hypothalamus and Pituitary

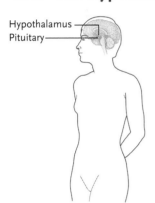

Hypothalamus
Pituitary

Hormones of vertebrates work in coordination with the nervous system. The action of several hormones is closely coordinated by the hypothalamus–pituitary complex.

The hypothalamus is a region of the brain located in the floor of the cerebrum (see Chapter 33). The **pituitary** gland, consisting mostly of two fused lobes, is suspended just below it by a slender stalk of tissue that contains both neurons and blood vessels **(Figure 35.6, p. 860)**. The **posterior pituitary** contains axons and endings of neurosecretory neurons that originate in the hypothalamus. The **anterior pituitary** contains non-neuronal endocrine cells that form a distinct gland. The two lobes are separate in structure and embryonic origins.

35.3a Under Regulatory Control by the Hypothalamus, the Anterior Pituitary Secretes Eight Hormones

The secretions of the anterior pituitary are under the control of peptide neurohormones called **releasing hormones (RHs)** and **inhibiting hormones (IHs)**, produced by the hypothalamus. These neurohormones are carried in the blood to the anterior pituitary in a *portal vein,* a special vein that connects the capillaries of the two glands. The portal vein provides a critical link between the brain and the endocrine system, ensuring that most of the blood reaching the anterior pituitary first passes through the hypothalamus.

RHs and IHs are **tropic hormones** (*tropic* means "stimulating," not to be confused with *trophic,* which means "nourishing") that regulate hormone secretion by another endocrine gland, in this case, the anterior pituitary. The hormones of the anterior pituitary, in turn, control many other endocrine glands of the body and some body processes directly.

Secretion of hypothalamic RHs is controlled by neurons containing receptors that monitor the blood to detect changes in body chemistry and temperature. For example, when body temperature drops, TRH is secreted. Input to the hypothalamus also comes through numerous connections from control centres elsewhere in the brain, including the brain stem and the limbic system. Negative feedback pathways regulate secretion of the releasing hormones, such as the pathway regulating TRH secretion.

Under the control of the hypothalamus, the anterior pituitary secretes six major hormones into the bloodstream (see Figure 35.6): prolactin, growth hormone, thyroid-stimulating hormone, adrenocorticotropic hormone, follicle-stimulating hormone, and luteinizing hormone. **Prolactin (PRL)**, a *nontropic hormone* (a hormone that does not regulate hormone secretion by another endocrine gland), influences reproductive activities and parental care in vertebrates. In mammals, PRL stimulates development of the secretory cells of mammary glands during late pregnancy and milk synthesis after birth. Stimulation of the mammary glands and the nipples, as occurs during suckling, leads to PRL release. PRL occurs in nonmammalian vertebrates, where it has a variety of functions. In fish, for example, it is among the hormones controlling water balance. In all vertebrates, it has a role in promoting both maternal and paternal behaviour.

Growth hormone (GH) stimulates cell division, protein synthesis, and bone growth in children and adolescents, thereby causing body growth. GH also stimulates protein synthesis and cell division in adults. For these actions, GH acts as a tropic hormone by binding to target tissues, mostly liver cells, causing

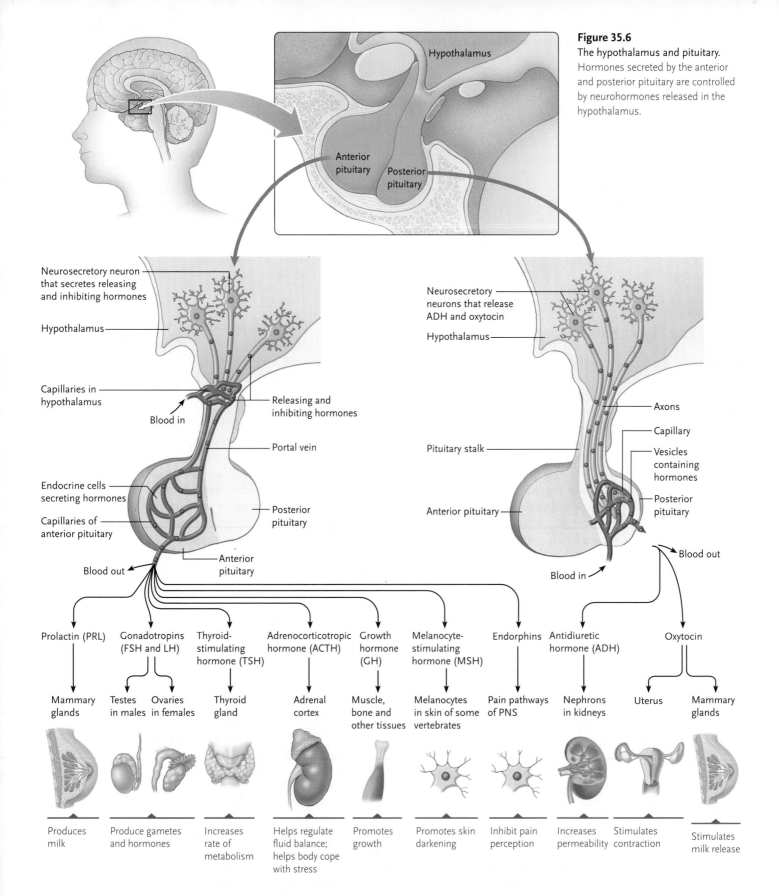

Figure 35.6
The hypothalamus and pituitary.
Hormones secreted by the anterior
and posterior pituitary are controlled
by neurohormones released in the
hypothalamus.

them to release **insulinlike growth factor (IGF)**, a peptide that directly stimulates growth processes. GH also acts as a nontropic hormone to control a number of major metabolic processes in mammals of all ages, including the conversion of glycogen to glucose and fats to fatty acids as a means of regulating their levels in the blood. GH also stimulates body cells to take up fatty acids and amino acids and limits the rate at

Figure 35.7
The results of overproduction and underproduction of growth hormone by the anterior pituitary. The man on the left is of normal height. The man in the centre is a pituitary giant, whose pituitary produced excess GH during childhood and adolescence. The man on the right is a pituitary dwarf, whose pituitary produced too little GH.

which muscle cells take up glucose. These actions help maintain the availability of glucose and fatty acids to tissues and organs between feedings; this is particularly important for the brain. In humans, deficiencies in GH secretion during childhood produce *pituitary dwarfs,* who remain small in stature **(Figure 35.7).** Overproduction of GH during childhood or adolescence, often due to a tumour of the anterior pituitary, produces *pituitary giants,* who may grow to above two metres in height.

Many of the other hormones secreted by the anterior pituitary are tropic hormones that control endocrine glands elsewhere in the body. **Thyroid-stimulating hormone (TSH)** stimulates the thyroid gland to grow in size and secrete thyroid hormones. **Adrenocorticotropic hormone (ACTH)** triggers hormone secretion by cells in the adrenal cortex. **Follicle-stimulating hormone (FSH)** affects egg development in females and sperm production in males. It also has a tropic effect by stimulating the secretion of sex hormones in female mammals. **Luteinizing hormone (LH)** regulates part of the menstrual cycle in human females and the secretion of sex hormones in males. FSH and LH are grouped together as **gonadotropins** because they regulate the activity of the gonads (ovaries and testes). The roles of the gonadotropins and

sex hormones in the reproductive cycle are described in Chapter 38.

Melanocyte-stimulating hormone (MSH) and **endorphins** are nontropic hormones secreted by the anterior pituitary. MSH is named because of its effect in some vertebrates on melanocytes, skin cells that contain the black pigment melanin. An increase in secretion of MSH produces a marked darkening of the skin of fishes, amphibians, and reptiles. The darkening is produced by a dispersal of melanin in melanocytes so that it covers a greater area. In humans, an increase in MSH secretion also causes skin darkening, although the effect is by no means as obvious as in the other vertebrates mentioned. MSH secretion increases in pregnant women. Combined with the effects of increased estrogens, MSH results in increased skin pigmentation. The effects are reversed after the birth of the child.

Endorphins, nontropic peptide hormones produced by the hypothalamus and pituitary, are also released by the intermediate lobe of the pituitary. In the peripheral nervous system (PNS), endorphins act as neurotransmitters in pathways that control pain, thereby inhibiting the perception of pain. Hence, endorphins are often called "natural painkillers."

35.3b The Posterior Pituitary Secretes Two Hormones into the Body Circulation

The neurosecretory neurons in the posterior pituitary secrete two nontropic peptide hormones, antidiuretic hormone and oxytocin, directly into the body circulation (see Figure 35.6).

Antidiuretic hormone (ADH, also known as vasopressin) stimulates kidney cells to absorb more water from urine, thereby increasing the volume of the blood. The hormone is released when sensory receptor cells of the hypothalamus detect an increase in the blood's Na^+ concentration during periods of dehydration or after a salty meal. Ethyl alcohol and caffeine inhibit ADH secretion, explaining in part why alcoholic drinks and coffee increase the volume of urine excreted. Nicotine and emotional stress, in contrast, stimulate ADH secretion and water retention. After severe stress is relieved, the return to normal ADH secretion often makes a trip to the bathroom among our most pressing needs. The hypothalamus also releases a flood of ADH when an injury results in heavy blood loss or some other event triggers a severe drop in blood pressure. ADH helps maintain blood pressure by reducing water loss and by causing small blood vessels in some tissues to constrict.

Hormones with structure and action similar to those of ADH are also secreted in fishes, amphibians, reptiles, and birds. In amphibians, these ADH-like hormones increase the amount of water entering the body through the skin and from the urinary bladder.

Oxytocin stimulates the ejection of milk from the mammary glands of a nursing mother. Stimulation of the nipples in suckling sends neuronal signals to the hypothalamus and leads to the release of oxytocin from the posterior pituitary. The released oxytocin stimulates more oxytocin secretion by a positive feedback mechanism. Oxytocin causes the smooth muscle cells surrounding the mammary glands to contract, forcibly expelling the milk through the nipples. The entire cycle, from the onset of suckling to milk ejection, takes less than a minute in mammals. Oxytocin also plays a key role in childbirth (see Chapter 38).

In males, oxytocin is secreted into the seminal fluid by the testes. When the seminal fluid is ejaculated into the vagina during sexual intercourse, the hormone stimulates contractions of the uterus that aid movement of sperm through the female reproductive tract.

Study Break

1. Distinguish between tropic and nontropic hormones.
2. Distinguish between the anterior and the posterior pituitary. How is the release of hormones from each of these controlled?
3. Name and state the function of the eight hormones released by the anterior pituitary.

35.4 Other Major Endocrine Glands of Vertebrates

In addition to the hypothalamus and pituitary, the body has seven major endocrine glands or tissues, many of them regulated by the hypothalamus–pituitary connection. Included are the thyroid gland, parathyroid glands, adrenal medulla, adrenal cortex, gonads, pancreas, and pineal gland (shown in Figure 35.5 and summarized in Table 35.1).

35.4a The Thyroid Hormones Stimulate Metabolism, Development, and Maturation

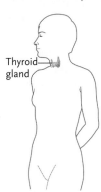

Thyroid gland

The **thyroid gland** is located in the front of the throat in humans and is shaped like a bowtie. It secretes the same hormones in all vertebrates. The thyroid hormones have an extraordinarily wide range of effects. The primary thyroid hormone, **thyroxine**, is known as T_4 because it contains four iodine atoms. The thyroid also

secretes smaller amounts of a closely related hormone, **triiodothyronine** or T_3, which contains three iodine atoms. A supply of iodine in the diet is necessary for production of these hormones. Normally, their concentrations are kept at finely balanced levels in the blood by negative feedback loops such as the loop described in Figure 35.2. Most of the circulating hormone is bound to a transport protein, thyroglobulin, and only the free hormone is available to enter cells.

Thyroid hormones act both in the nucleus and via membrane receptors. Both T_4 and T_3 enter cells, probably via specific uptake receptors. Once inside, the T_4 is deiodinated, forming T_3 and in some cases T_2 (contains two iodine atoms). T_3 is the form that combines with nuclear receptors. Binding of T_3 to receptors alters gene expression, which brings about many of the hormone's effects. In addition, T_2 can act directly on mitochondria. T_3 also acts on receptors on the cell membrane of some cells. It can increase the Ca^{2+}ATPase activity of red blood cells.

Thyroid hormones are vital to growth, development, maturation, and metabolism in all vertebrates. They interact with GH for their effects on growth and development. Thyroid hormones also increase the sensitivity of many body cells to the effects of epinephrine and norepinephrine, hormones released by the adrenal medulla as part of the "fight-or-flight response" (discussed further below).

In amphibians, rising concentrations of thyroid hormones trigger **metamorphosis**, or a change in body form from tadpole to adult **(Figure 35.8).** Teleost fish

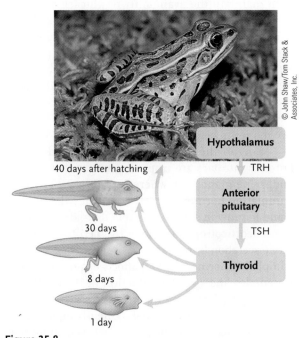

Figure 35.8

Metamorphosis of a tadpole into an adult frog, under the control of thyroid hormones. As part of the metamorphosis, changes in the gene activity lead to a change from an aquatic to a terrestrial habitat. TRH – thyroid-releasing hormone; TSH – thyroid-stimulating hormone.

undergo a form of metamorphosis during their early development, and the transformation from a hatchling "larval" form to a juvenile form is also triggered by rising T_4 concentrations. Curiously, however, the opposite is true in the agnathan lamprey. Their metamorphosis is triggered by decreasing concentrations of T_4. Thyroid hormones also contribute to seasonal moulting, leading to changes in the plumage of birds and coat colour in mammals.

The thyroid also has specialized cells that secrete **calcitonin**, a peptide originally discovered in fish by Harold Copp working at the University of British Columbia. The hormone lowers the level of Ca^{2+} in the blood by inhibiting the ongoing dissolution of calcium from bone. Calcitonin secretion is stimulated when Ca^{2+} levels in blood rise above the normal range and inhibited when Ca^{2+} levels fall below the normal range. Although the specialized cells of the thyroid are the principal source, calcitonin is also synthesized in the lung and intestine. In nonmammalian vertebrates, a separate gland, the ultimobrachial gland, produces calcitonin.

35.4b The Parathyroid Glands Regulate Ca^{2+} Level in the Blood

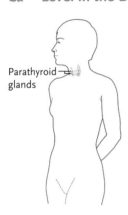

Parathyroid glands

The **parathyroid glands** occur only in tetrapod vertebrates (amphibians, reptiles, birds, and mammals). Each is a spherical structure about the size of a pea. Mammals have four parathyroids located on the posterior surface of the thyroid gland, two on each side. The single hormone they produce, a nontropic hormone called **parathyroid hormone (PTH)**, is secreted in response to a fall in blood Ca^{2+} levels. PTH stimulates bone cells to dissolve the mineral matter of bone tissues, releasing both calcium and phosphate ions into the blood. The released Ca^{2+} is available for enzyme activation, conduction of nerve signals across synapses, muscle contraction, blood clotting, and other uses. How blood Ca^{2+} levels control PTH and calcitonin secretion is shown in **Figure 35.9.**

PTH also stimulates enzymes in the kidneys that convert **vitamin D**, a steroidlike molecule, into its fully active form in the body. The activated vitamin D increases the absorption of Ca^{2+} and phosphates from ingested food by promoting the synthesis of a calcium-binding protein in the intestine. It also increases the release of Ca^{2+} from bone in response to PTH.

PTH underproduction causes Ca^{2+} concentration to fall steadily in the blood, disturbing nerve and muscle function—the muscles twitch and con-

Figure 35.9

Negative feedback control of PTH and calcitonin secretion by blood Ca^{2+} levels.

Stimulus: rising blood Ca^{2+} level

Thyroid gland

Calcitonin

Reduces Ca^{2+} uptake in kidneys

Stimulates Ca^{2+} deposition in bones

Blood Ca^{2+} declines to set point

Homeostasis

Blood Ca^{2+} rises to set point

Increases Ca^{2+} uptake in intestines

Stimulates Ca^{2+} release from bone

Stimulates Ca^{2+} uptake in kidneys

PTH

Parathyroid glands

Stimulus: falling blood Ca^{2+} level

tract uncontrollably, and convulsions and cramps occur. Without treatment, the condition is usually fatal because the severe muscular contractions interfere with breathing. Overproduction of PTH results in the loss of so much calcium from the bones that they become thin and fragile. At the same time, the elevated Ca^{2+} concentration in the blood causes calcium deposits to form in soft tissues, especially in the lungs, arteries, and kidneys (where the deposits form kidney stones).

Although fish do not have a parathyroid gland, they produce PTH, and PTH receptors are known to be present in fish, but the origin of the hormone and its precise function remain uncertain.

35.4c The Adrenal Medulla Releases Two "Fight-or-Flight" Hormones

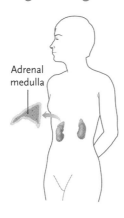

The adrenal glands (*ad* = next to; *renes* = kidneys) of mammals have two distinct regions. The central region, the **adrenal medulla**, contains highly modified neurosecretory neurons that have lost their axons and dendrites. The tissue surrounding it, the **adrenal cortex**, contains non-neural endocrine cells. The two regions secrete hormones with entirely different functions. Nonmammalian vertebrates have glands equivalent to the adrenal medulla and adrenal cortex of mammals, but the two parts are separate entities. Most of the hormones produced by these glands have essentially the same functions in all vertebrates. The only major exception is aldosterone, which is secreted by the adrenal cortex or its equivalent only in tetrapod vertebrates.

In most species, the adrenal medulla secretes two nontropic amine hormones, **epinephrine** and **norepinephrine**, which are **catecholamines**, chemicals derived from tyrosine that can act as hormones or neurotransmitters. They bind to receptors in the plasma membranes of their target cells. Norepinephrine is also released as a neurotransmitter by neurons of the sympathetic nervous system.

Epinephrine and norepinephrine reinforce the action of the sympathetic nervous system and are secreted when the body encounters stresses such as emotional excitement, danger (fight-or-flight situations), anger, fear, infections, injury, and even midterm and final exams. Epinephrine in particular prepares the body for handling stress or physical activity. The heart rate increases. Glycogen and fats break down, releasing glucose and fatty acids into the blood as fuel molecules. In the heart, skeletal muscles, and lungs, the blood vessels dilate to increase blood flow. Elsewhere in the body, the blood vessels constrict, raising blood pressure, reducing blood flow to the intestine and kidneys, and inhibiting smooth muscle contractions, which reduces water loss and slows down the digestive system. Airways in the lungs also dilate, helping to increase the flow of air.

The effects of norepinephrine on heart rate, blood pressure, and blood flow to the heart muscle are similar to those of epinephrine. However, in contrast to epinephrine, norepinephrine causes blood vessels in skeletal muscles to constrict. This contrary effect is largely cancelled out because epinephrine is secreted in much greater quantities.

35.4d The Adrenal Cortex Secretes Two Groups of Steroid Hormones

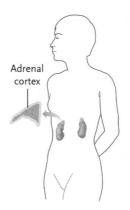

The adrenal cortex of mammals secretes two major classes of steroid hormones: **glucocorticoids** help maintain the blood concentration of glucose and other fuel molecules, and **mineralocorticoids** regulate the levels of Na$^+$ and K$^+$ ions in the blood and extracellular fluid.

The Glucocorticoids. The glucocorticoids help maintain glucose levels in the blood by three major mechanisms: (1) stimulating the synthesis of glucose from noncarbohydrate sources such as fats and proteins, (2) reducing glucose uptake by body cells except those in the CNS, and (3) promoting the breakdown of fats and proteins, which releases fatty acids and amino acids into the blood as alternative fuels when glucose supplies are low. The favouring of glucose uptake in the CNS keeps the brain well supplied with glucose between meals and during periods of extended fasting. **Cortisol** is the major glucocorticoid secreted by the adrenal cortex.

Secretion of glucocorticoids is ultimately under control of the hypothalamus **(Figure 35.10)**. Low glucose concentrations in the blood, or elevated levels of epinephrine secreted by the adrenal medulla in response to stress, are detected in the hypothalamus, leading to secretion of the tropic hormone ACTH by the anterior pituitary. ACTH promotes the secretion of glucocorticoids by the adrenal cortex.

Glucocorticoids also have anti-inflammatory properties; consequently, they are used to treat conditions such as arthritis or dermatitis. They also suppress the immune system and are used in the treatment of autoimmune diseases such as rheumatoid arthritis.

The Mineralocorticoids. In tetrapods, the mineralocorticoids, primarily **aldosterone**, increase the amount of Na$^+$ reabsorbed from the urine in the kidneys and absorbed from foods in the intestine. They also reduce the amount of Na$^+$ secreted by salivary and sweat glands and increase the rate of K$^+$ excretion by the kidneys. The net effect is to keep Na$^+$ and K$^+$ balanced at the levels required for normal cellular functions, including those of the nervous system. Relatedly, secretion of aldosterone is tightly linked to blood volume and indirectly to blood pressure. The adrenal cortex also secretes small amounts of androgens, steroid sex hormones responsible for maintenance of male characteristics, which are synthesized primarily by the gonads.

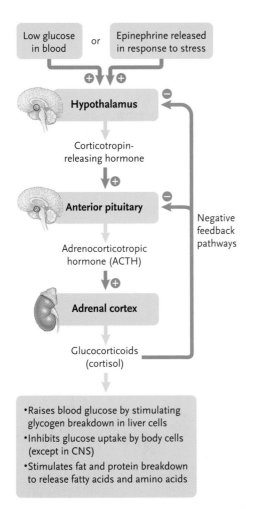

Figure 35.10
Pathways linking secretion of glucocorticoids to low blood sugar and epinephrine secretion in response to stress.

Low glucose in blood or **Epinephrine released in response to stress**

Hypothalamus

Corticotropin-releasing hormone

Anterior pituitary

Adrenocorticotropic hormone (ACTH)

Adrenal cortex

Glucocorticoids (cortisol)

Negative feedback pathways

• Raises blood glucose by stimulating glycogen breakdown in liver cells
• Inhibits glucose uptake by body cells (except in CNS)
• Stimulates fat and protein breakdown to release fatty acids and amino acids

35.4e The Gonadal Sex Hormones Regulate the Development of Reproductive Systems, Sexual Characteristics, and Mating Behaviour

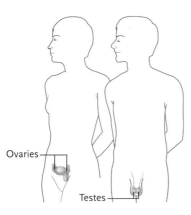

Ovaries

Testes

The **gonads**, the testes and ovaries, are the primary source of sex hormones in vertebrates. The steroid hormones they produce, the **androgens, estrogens,** and **progestins**, have similar functions in regulating the development of male and female reproductive systems, sexual characteristics, and mating behaviour. Both males and females produce all three types of hormones, but in different proportions. Androgen production is predominant in males, whereas estrogen and progestin production is predominant in females. An outline of the actions of these hormones is presented here, and a more complete picture is given in Chapter 38.

The **testes** (singular, testis) of male vertebrates secrete androgens, steroid hormones that stimulate and control the development and maintenance of male reproductive systems. The principal androgen is testosterone, the male sex hormone. In young adult males, a jump in **testosterone** levels stimulates puberty and the development of secondary sexual characteristics, including the growth of facial and body hair, muscle development, changes in vocal cord morphology, and development of normal sex drive. The synthesis and secretion of testosterone by cells in the testes are controlled by the release of LH from the anterior pituitary, which, in turn, is controlled by **gonadotropin-releasing hormone (GnRH)**, a tropic hormone secreted by the hypothalamus.

Androgens are natural types of **anabolic steroids**, hormones that stimulate muscle development. Natural and synthetic anabolic steroids have been in the news over the years because of their use by body builders and other athletes from sports in which muscular strength is important.

The **ovaries** (singular, ovary) of females produce estrogens, steroid hormones that stimulate and control the development and maintenance of female reproductive systems. The principal estrogen is **estradiol**, which stimulates maturation of sex organs at puberty and the development of secondary sexual characteristics. Ovaries also produce progestins, principally **progesterone**, the steroid hormone that prepares and maintains the uterus for implantation of a fertilized egg and the subsequent growth and development of an embryo. The synthesis and secretion of progesterone by cells in the ovaries are controlled by the release of FSH from the anterior pituitary, which, in turn, is controlled by the same GnRH as in males.

35.4f The Pancreatic Islets of Langerhans Hormones Regulate Glucose Metabolism

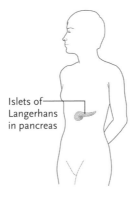

Islets of Langerhans in pancreas

Most of the **pancreas**, a relatively large gland located just behind the stomach, forms an exocrine gland that secretes digestive enzymes into the small intestine (see Chapter 41). About 2% of the cells in the pancreas are endocrine cells that form the **islets of Langerhans.** Found in all vertebrates, the islets secrete the peptide hormones insulin and glucagon into the bloodstream.

MOLECULE BEHIND BIOLOGY

Insulin: More Than Diabetes

Insulin was discovered in 1922 by J.R. Banting and his colleagues, McLeod, Best, and Collip, working at the University of Toronto. Banting and McLeod were awarded the Nobel Prize in 1923. (They were disturbed that their colleagues had not been recognized and shared the prize with them.) In 1955, Frederick Sanger of Cambridge University worked out insulin's complete amino acid sequence (the first protein to be fully sequenced) and was awarded a Nobel Prize in 1958 for this work.

Insulin is a very large peptide of 51 amino acids. Molecular studies show that the insulin gene is present in all vertebrates but is absent from invertebrates. Insulin is a member of a family of genes. Two other members of that family encode two structurally related peptides, IGF I and II. Although they are similar in structure to insulin, they have different but structurally related receptors, and their cellular action is different. They act as growth factors, regulating cell and tissue growth. IGFs are widely dis-

tributed in animals, protists, bacteria, and fungi. In molluscs, insects, and nematodes, these insulinlike peptides (ILPs) are neurohormones, expressed in neurosecretory cells in the brain. Surgical removal of these cells and their reimplantation demonstrate that they control growth, like the IGFs in vertebrates. The impact of these studies on the origin and evolution of the insulin gene is not yet clear.

Insulin and glucagon regulate the metabolism of fuel substances in the body. **Insulin** (see *Molecule Behind Biology*) is secreted by *beta cells* in the islets. It acts mainly on nonworking skeletal muscles, liver cells, and adipose tissue (fat). Brain cells do not require insulin for glucose uptake. Insulin lowers blood glucose, fatty acid, and amino acid levels and promotes their storage. The actions of insulin include stimulation of glucose transport into cells, glycogen synthesis from glucose, uptake of fatty acids by adipose tissue cells, fat synthesis from fatty acids, and protein synthesis from amino acids. Insulin inhibits glycogen degradation to glucose, fat degradation to fatty acids, and protein degradation to amino acids.

Glucagon, secreted by *alpha cells* in the islets, has effects opposite to those of insulin: it stimulates glycogen, fat, and protein degradation. Glucagon also uses amino acids and other noncarbohydrates as the input for glucose synthesis; this aspect of glucagon function operates during fasting. Negative feedback mechanisms keyed to the concentration of glucose in the blood control secretion of both insulin and glucagon to maintain glucose homeostasis **(Figure 35.11)**.

Diabetes mellitus, a disease that afflicts more than 2 million people in Canada, results from problems with insulin production or action. The three classic diabetes symptoms are frequent urination, increased thirst (and consequently increased fluid intake), and increased appetite. Frequent urination occurs because the ability of body cells to take up glucose is impaired in diabetics, leading to abnormally high glucose concentration in the blood. Excretion of the excess glucose in the urine requires water to carry it, which causes increased fluid loss and frequent trips to the bathroom. The need to replace the excreted water causes increased thirst. Increased appetite comes about because cells have low glucose levels as a result of the insulin defect; therefore, proteins and

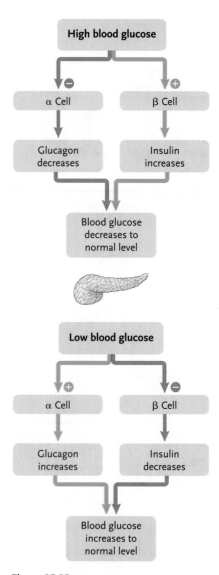

Figure 35.11

The action of insulin and glucagon in maintaining the concentration of blood glucose at an optimal level.

PEOPLE BEHIND BIOLOGY
William Rowan, University of Alberta

In the 1920s, it was generally accepted that birds migrated south from Canada in response to falling temperatures. William Rowan of the University of Alberta had a different view. Recognizing that temperature is an unreliable indicator of seasons, he hypothesized that day length was important. In addition, he recognized that migration was simply part of the annual breeding cycle: birds migrating south were known to have inactive reproductive organs, which became active when they returned in the spring. Rowan conducted two very simple but crucial experiments. In the first, he exposed juncos (*Junco hyemalis*) to artificially lengthened daily light in the autumn; their gonads became active. In the second, he used 500 crows. These were caged in the autumn, and half were exposed to increasing day length, and the others to natural day length. He dyed their tails yellow and released them, alerting surrounding homesteaders to be on the lookout for crows with yellow tails. Those exposed to increased day length headed north, whereas those experiencing normal day length headed south. The association of day length with migration and gonadal development implicated the endocrine system, and these deceptively simple experiments set in motion the entire field of photobiology. Rowan is recognized as among the most influential biologists in the field.

fats are broken down as energy sources. Food intake is necessary to offset the negative energy balance, or weight loss will occur. Two of these classic symptoms gave the disease its name: diabetes is derived from a Greek word meaning "siphon," referring to the frequent urination, and mellitus, a Latin word meaning "sweetened with honey," refers to the sweet taste of a diabetic's urine. (Before modern blood or urine tests were developed, physicians tasted a patient's urine to detect the disease.)

35.4g The Pineal Gland Regulates Some Biological Rhythms

Pineal gland

The **pineal gland** is found at different locations in the brains of vertebrates. In mammals, it is near the centre of the brain, whereas in birds and reptiles, it is on the surface of the brain just under the skull and is directly sensitive to light. The pineal gland regulates some biological rhythms.

The earliest vertebrates had a third, light-sensitive eye at the top of the head, and *Sphenodon* and some lizards have an eyelike structure in this location. In most vertebrates, the third eye became modified into a pineal gland, which in many groups retains some degree of photosensitivity. In mammals, it is too deeply buried in the brain to be affected directly by light; nonetheless, specialized photoreceptors in the eyes make connections to the pineal gland.

The pineal gland secretes the amine hormone **melatonin**, derived from tryptophan, which helps maintain daily biorhythms. Secretion of melatonin is regulated by an inhibitory pathway. Light hitting the eyes generates signals that inhibit melatonin secretion; consequently, the hormone is secreted most actively during periods of darkness. Melatonin targets a part of the hypothalamus called the *suprachiasmatic nucleus,* which is the primary structure coordinating body activity to a daily cycle. The nightly release of melatonin may help synchronize the biological clock with daily cycles of light and darkness. The physical and mental discomfort associated with jet lag may reflect the time required for melatonin secretion to reset a traveller's daily biological clock to match the period of daylight in a new time zone.

Melatonin occurs throughout the animal kingdom, as well as in many plants and fungi. In invertebrates, it is known to be important in the control of diurnal (daily) rhythms.

STUDY BREAK

1. What are the hormones controlling Ca^{2+} levels in the blood of vertebrates?
2. Distinguish between the adrenal medulla and the adrenal cortex.
3. How are levels of glucose in the blood maintained?

35.5 Endocrine Systems in Invertebrates

Some invertebrates have fewer hormones, regulating a narrower range of body processes and responses, than vertebrates. However, in even the simplest animals, such as the cnidarian *Hydra,* hormones produced by neurosecretory neurons control the reproduction, growth, and development of some body features. In

The notion of "stress" may be familiar to all as a vague concept. It was originated by Hans Selye, working with rats at the University of Montreal, as the "General Adaptation Syndrome," in which what he called "nocuous agents" ranging from physical damage to psychological events led to the activation of the fight-or-flight response. This was characterized at the time as the activation of the hypothalamus–pituitary pathway leading to the release of a number of hormones, particularly those from the adrenal glands, which led to increased heart rate, blood pressure, cessation of growth, and, in severe cases, tissue damage and, ultimately, death.

Although we are accustomed to thinking of stress as a human or mammalian phenomenon, it has emerged as a response in a wide range of animals. Any unaccustomed sensory input (whether from external or internal events) leads to release of many hormones, and if the input continues, this can result in pathology or death. Stress, with the release of the corticosteroid hormones, is well studied in most vertebrates. Even in cockroaches, forced activity or forced inactivity will cause the release of many neurohormones, and the insect will die.

annelids, arthropods, and molluscs, endocrine cells and glands produce hormones that regulate development, reproduction, water balance, heart rate, sugar levels, and behaviour.

The known vertebrate hormones, particularly the peptides, also occur in a wide range of organisms. Thus, insulinlike hormones can be found in most invertebrates and receptors are known from insects and nematodes. The protist *Tetrahymena* binds and exhibits responses to insulin and T_4. Whereas some hormones, such as the peptide proctolin and the insect juvenile hormones, do not occur in vertebrates, many of the growing number of peptide hormones identified in invertebrates have structural homologues in the vertebrates, although their functions may be different. Some peptides controlling diuresis in insects are structural homologues of vertebrate corticotropin-releasing factor. Other diuretic peptides in insects are related to calcitonin. The larva of the tapeworm *Spirometra mansonoides* has developed the capacity to secrete vertebrate growth hormone so that its host rat grows larger.

The endocrinology of the more complex invertebrates, formerly thought to be relatively simple, is emerging as very complex: about 200 bioactive peptides have thus far been described in insects, which are probably more closely governed by hormones than any other animals. The development of the eggs and the egg-laying behaviour of insects are controlled by more than a dozen hormones.

35.5a Hormones Regulate Development in Insects

Among the best known invertebrate hormonal systems is the one governing growth and development in insects **(Figure 35.12)**. As insects grow, they undergo a series of moults during which a new cuticle is laid down beneath the old cuticle and the old cuticle is shed (see Chapter 26). The signal to the epidermal cells to begin the process is provided by a steroid hormone, ecdysone, from the prothoracic glands. The prothoracic glands are stimulated to secrete ecdysone by a tropic peptide, prothoracicotropic hormone (PTTH), produced in neuroendocrine cells in the brain and released from the corpus cardiacum. The corpus cardiacum secretes several other hormones and contains both the nerve endings of neurons in the brain and neuroendocrine cells that lack axons and dendrites.

The corpus allatum is an endocrine gland that secretes **juvenile hormone**, a fatty acid derivative (see Chapter 26, *Molecule Behind Biology*). Juvenile hormone controls metamorphosis: when it is present, the insect remains larval. In its absence, the next moult is metamorphic, producing a pupa and then an adult in those insects with a pupal stage or proceeding directly to the adult in those lacking a pupal stage. In the adult of most insects, the corpus allatum becomes active once more, secreting juvenile hormone and stimulating a number of reproductive processes, especially egg development. The secretion of juvenile hormone by the corpus allatum is controlled by both inhibitory and stimulatory tropic peptides from the brain.

The intricate process of shedding the old cuticle involves complex behaviours that are controlled by the interaction of up to five neurohormones, and the hardening of the new cuticle requires a sixth.

Hormones that control moulting have also been detected in crustaceans, including lobsters, crabs, and crayfish. During the period between moults, **moult-inhibiting hormone (MIH)**, a peptide neurohormone secreted by cells in the eyestalks (extensions of the

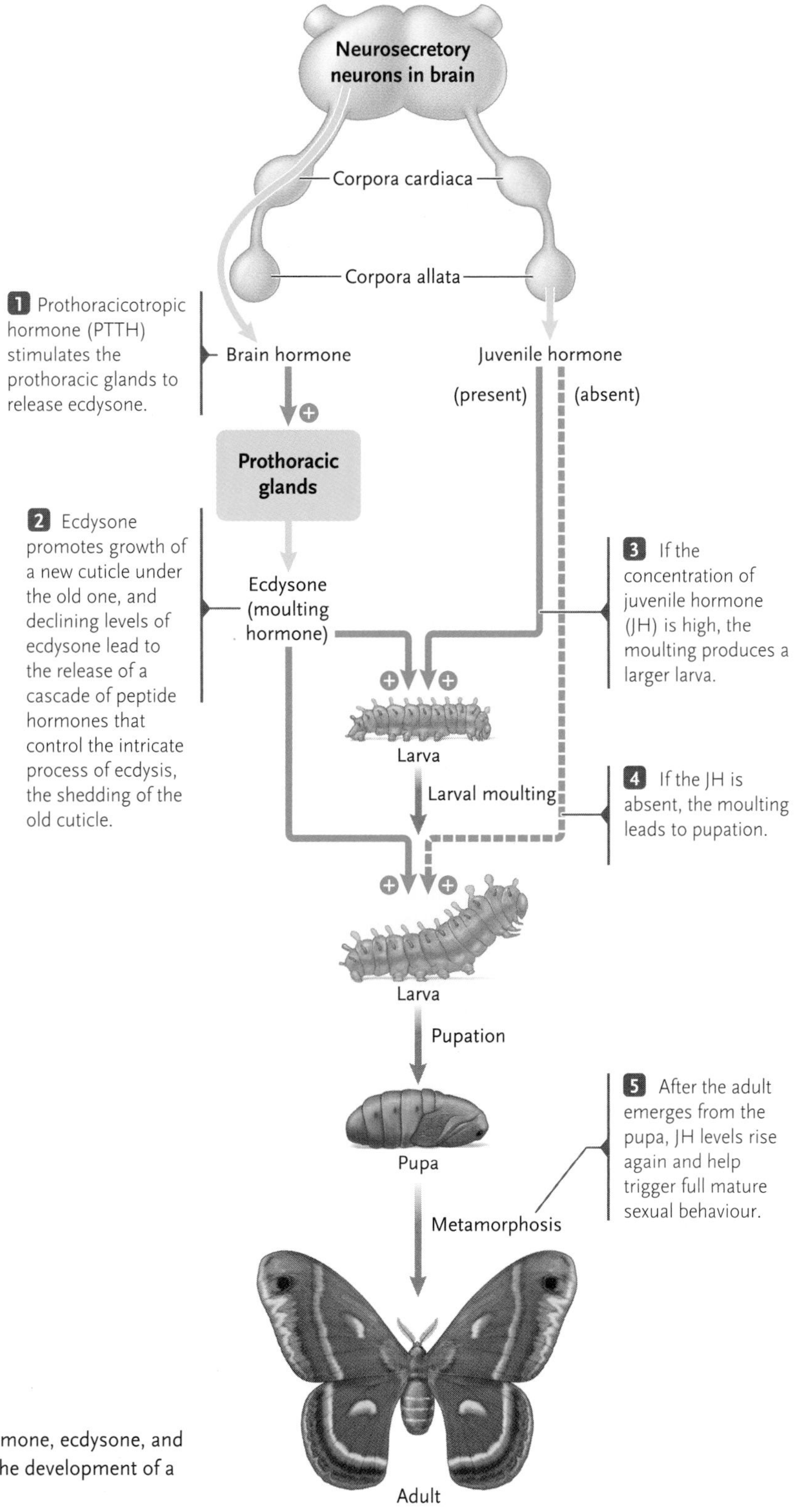
Neurosecretory neurons in brain
Corpora cardiaca
Corpora allata
1 Prothoracicotropic hormone (PTTH) stimulates the prothoracic glands to release ecdysone.
Brain hormone
Juvenile hormone
(present)
(absent)
Prothoracic glands
2 Ecdysone promotes growth of a new cuticle under the old one, and declining levels of ecdysone lead to the release of a cascade of peptide hormones that control the intricate process of ecdysis, the shedding of the old cuticle.
Ecdysone (moulting hormone)
3 If the concentration of juvenile hormone (JH) is high, the moulting produces a larger larva.
Larva
Larval moulting
4 If the JH is absent, the moulting leads to pupation.
Larva
Pupation
5 After the adult emerges from the pupa, JH levels rise again and help trigger full mature sexual behaviour.
Pupa
Metamorphosis
Adult

Figure 35.12
The roles of brain hormone, ecdysone, and juvenile hormone in the development of a silkworm moth.

brain leading to the eyes), inhibits ecdysone secretion. The first step in the moulting process is the inhibition of MIH secretion. Ecdysone secretion increases, and the processes leading to the replacement of the exoskeleton are initiated. As in insects, metamorphosis and reproduction are governed by a hormone different from but structurally related to JH.

Study Break

1. What are the known functions of insect juvenile hormone?
2. How are the pituitary and the corpus cardiacum similar? How do they differ?

The realization that many of the hormones that govern function in vertebrates also occur in protostomes raises the question of the origin of these molecules and their evolution. Some researchers believe that the steroid and fatty acid hormones may have originated as signals in the environment and that animals have simply captured them as internal signals. The sensitivity of protists to some vertebrate hormones supports this view. But this raises the question of the origin and evolution of the receptors for these molecules. It is not immediately obvious, for example, that the sensitivity of protists to thyroid hormones is of evolutionary importance. Research at York University in Toronto suggests that the membrane receptors for thyroxine, present in both vertebrates and insects, may have originated as receptors for CO_2. The origin and evolution of the galaxy of peptide hormones remain a puzzle. Insulinlike peptide (ILP) receptors appear to be active on the surface of some protists. ILPs are present in protists, but their function is not yet clear: are they a means of communication between individuals? Does this suggest that hormones used preexisting receptors that detected environmental signals?

Review

Go to CENGAGENOW™ at http://hed.nelson.com/ to access quizzing, animations, exercises, articles, and personalized homework help.

35.1 Hormones and Their Secretion

- Hormones are molecules secreted by cells of the endocrine system that control the activities of cells elsewhere in the body. The cells that respond to a hormone are its target cells. This contrasts with the nervous system, which controls specific target cells in close proximity to its endings.

- The endocrine system includes four major types of cell signalling: classical endocrine signalling, in which endocrine glands secrete hormones; neuroendocrine signalling, in which neurosecretory neurons release neurohormones into the circulation; paracrine regulation, in which cells release local regulators that diffuse through the extracellular fluid to regulate nearby cells; and autocrine regulation, in which cells release local regulators that regulate the same cells that produced it.

- Most hormones and local regulators fall into one of four molecular classes: amines, peptides, steroids, and fatty acids.

- Neurosecretory neurons secrete hormones under direct control of the central nervous system.

- Many hormones are controlled by negative feedback mechanisms in which a hormone inhibits the reactions that synthesize or release it when its concentration rises in the body.

35.2 Mechanisms of Hormone Action

- Many hormones undergo modification after release that renders them more active

- Only cells that have receptors for the hormone can respond to the hormone. Cells may respond by stimulation or inhibition of a process. Because of amplification involved in receptor mechanisms, hormones are present in body fluids at low concentrations. The response to a hormone may differ among cells and tissues.

- Hormones may bind to receptor proteins in the plasma membrane. When a receptor binds a hormone, its cytoplasmic end is activated, triggering a series of cytoplasmic reactions that may include the activation of protein kinases.

- Hydrophobic hormones bind to receptors in the cytoplasm or nucleus, activating them so that they can bind to the control sequences of specific genes in the cell nucleus. Binding to the control sequence either stimulates or inhibits transcription of the target gene, leading to changes in protein synthesis. They may also bind to membrane receptors.

- The major endocrine cells and glands of vertebrates are the hypothalamus, pituitary gland, thyroid gland, parathyroid gland, adrenal medulla, adrenal cortex, testes, ovaries, islets of Langerhans of the pancreas, and pineal gland.

35.3 The Hypothalamus and Pituitary

- The hypothalamus and pituitary together regulate many other endocrine cells and glands in the body. The posterior pituitary contains the terminals of neurosecretory cells in the hypothalamus.

- The anterior pituitary contains endocrine cells derived from nonnervous tissue.

- The hypothalamus produces tropic hormones (releasing hormones and inhibiting hormones) that control the secretion of eight hormones by the anterior pituitary. Prolactin (PRL), a nontropic hormone, regulates mammary gland development and milk secretion in mammals. As a nontropic hormone, growth hormone (GH) stimulates body growth in children and adolescents, and as a tropic hormone, it stimulates liver cells to make insulinlike growth factor (IGF), which stimulates growth processes. Melanocyte-stimulating hormone (MSH) controls reversible skin darkening. Endorphins can reduce pain. The four other hormones are tropic hormones: thyroid-stimulating hormone (TSH) stimulates secretion by the thyroid gland; adrenocorticotropic hormone (ACTH) regulates hormone secretion by the adrenal cortex; follicle-stimulating hormone (FSH) controls egg development and the secretion of sex hormones by the ovaries in female mammals and the production of sperm cells in males; and luteinizing hormone (LH) regulates part of the menstrual cycle in human females and the secretion of sex hormones in males.

- Antidiuretic hormone (ADH) and oxytocin are secreted by the posterior pituitary. ADH regulates body water balance. In female mammals, oxytocin stimulates the contraction of smooth muscle in the uterus as part of childbirth and triggers milk release from the mammary glands during suckling of the young.

- The intermediate lobe of the pituitary produces the nontropic hormones MSH and endorphins. MSH secretion in some vertebrates produces a darkening of the skin. Endorphins are neurotransmitters that affect pain pathways in the peripheral nervous system, inhibiting the perception of pain.

35.4 Other Major Endocrine Glands of Vertebrates

- The thyroid gland secretes the thyroid hormones and, in mammals, calcitonin. In mammals, the thyroid hormones stimulate the oxidation of carbohydrates and lipids and coordinate with growth hormone to stimulate body growth and development. Calcitonin lowers the Ca^{2+} level in the blood by inhibiting the release of Ca^{2+} from bone. In many vertebrates, thyroid hormones control metamorphosis.

- The parathyroid gland secretes parathyroid hormone (PTH), which stimulates bone cells to release Ca^{2+} into the blood. PTH also stimulates the activation of vitamin D, which promotes Ca^{2+} absorption into the blood from the small intestine.

- The adrenal medulla secretes epinephrine and noreprinephrine, which reinforce the sympathetic nervous system in responding to stress. The adrenal cortex secretes glucocorticoids and mineralocorticoids. Glucocorticoids help maintain glucose at normal levels in the blood; mineralocorticoids regulate Na^+ balance and extracellular fluid volume. The adrenal cortex also secretes small amounts of androgens.

- The gonadal sex hormones, androgens, estrogen, and progestins, regulate the development of reproductive systems, sexual characteristics, and mating behaviour. Both sexes secrete all three, but males primarily produce androgens (secreted by the testes), and females primarily produce estrogens and progestins (secreted by the ovaries).

- The islets of Langerhans of the pancreas secrete insulin and glucagon. Insulin lowers the concentration of glucose in the blood by stimulating glucose uptake by cells, glycogen synthesis from glucose, uptake of fatty acids by adipose tissue cells, fat synthesis from fatty acids, and protein synthesis from amino acids; it inhibits the conversion of noncarbohydrate molecules into glucose. Glucagon raises blood glucose by stimulating glycogen, fat, and protein degradation. The balance of insulin and glucagon regulates the concentration of fuel substances in the blood.

- The pineal gland secretes melatonin, which interacts with the hypothalamus to set the body's daily rhythms.

35.5 Endocrine Systems in Invertebrates

- Even the simplest protostomes use hormones to coordinate growth and reproduction.

- Many of the hormones that occur in vertebrates also occur in invertebrates, although their function may be different.

- Three major hormones, prothoracicotropic hormone (PTTH) from the brain, ecdysone from prothoracic glands, and juvenile hormone (JH) from the corpus allatum, control moulting and metamorphosis in insects. PTTH is a tropic hormone that stimulates the secretion of ecdysone, which initiates and maintains the secretion of the new cuticle. If JH is present, metamorphosis is suppressed, and in its absence, metamorphosis proceeds. The activity of the corpus allatum is governed by stimulatory and inhibitory tropic hormones from the brain. The shedding of the old cuticle and the hardening of the new cuticle are governed by a cascade of neuropeptides.

- JH controls reproduction in the adult insect.

- Similar hormones that control moulting and reproduction are also present in crustaceans, but the secretion of ecdysone is under the control of an inhibitory neurohormone.

Questions

Self-Test Questions

1. When the concentration of thyroid hormone in the blood increases, it
   a. inhibits TRH secretion by the hypothalamus.
   b. stimulates a secretion by the hypothalamus.
   c. stimulates the pituitary to secrete TRH.
   d. stimulates the pituitary to secrete TSH.
   e. activates a positive feedback loop.

2. Blood levels of calcium are regulated directly by
   a. insulin synthesized by the alpha cells of the pancreas.
   b. PTH made by the pituitary.
   c. vitamin D activated in the liver.
   d. prolactin synthesized by the intermediate lobe of the pituitary.
   e. calcitonin secreted by specialized thyroid cells.

3. Proctolin
   a. is secreted by the intestine of mammals.
   b. is a steroid acting on nematodes.
   c. acts on smooth muscle in invertebrates.
   d. is a peptide secreted by insects.
   e. governs egg development in invertebrates.

4. Which of the following functions in mammals is not controlled by a hormone from the anterior pituitary?
   a. growth of muscle
   b. milk production
   c. metabolic rate
   d. contraction of uterine muscles
   e. egg production

5. When blood glucose rises in healthy humans,
   a. the alpha cells of the pancreas increase glucagon secretion.
   b. the beta cells of the pancreas increase insulin production.
   c. the pituitary secretes a tropic hormone controlling the pancreas.
   d. glucagon uses amino acids as an energy source.
   e. target cells decrease their insulin receptors.

Questions for Discussion

1. The occurrence in many invertebrates of peptide hormones that closely resemble those in vertebrates is striking. What are the possible explanations for this in evolutionary terms? What research would you do to help you choose among the possibilities?

2. Stress is commonly regarded as a diseaselike condition that can lead to death. Stresslike phenomena are widely spread in other taxa. If it is pathological, why has evolution not eliminated it? What are the advantages that have led to its retention?

3. A physician sees a patient who complains of a lack of energy and intolerance to cold. What are the possible hormonal causes of these symptoms?

Movement in a long-tailed field mouse (*Apodemus sylvaticus*). Movement of animals occurs as a result of contractions and relaxations of skeletal muscles. When stimulated by the nervous system, actin filaments in the muscles slide over myosin filaments to cause muscle contractions.

G. Delpho/Peter Arnold, Inc.

36 Muscles, Skeletons, and Body Movements

WHY IT MATTERS

A Mexican leaf frog (*Pachymedusa dacnicolor*) sits motionless, its prominent eyes staring into space **(Figure 36.1, p. 874).** But when the frog detects an approaching cricket, it lunges forward at just the right moment, thrusts out its sticky tongue, and captures the prey. This sequence of events, from the beginning of the movement until the frog's mouth closes, sealing the cricket's fate, requires only 260 milliseconds (ms)—about one-quarter of a second. How does the frog move so swiftly, and so surely?

As its prey draws near, the muscles that extend the frog's hindlegs contract and propel the frog forward on its forelimbs toward the cricket. Within 50 ms after the jump begins, the muscles of the lower jaw contract, opening the mouth. Then a muscle on the upper surface of the tongue contracts, which raises the tongue and flips it out of the mouth. As the tongue shoots forward, muscle contractions along the ventral side of the trunk arch the body and direct the head downward toward the prey. Within 80 ms after the lunge begins, the tip of the frog's tongue contacts the cricket. Completion of the lunge folds the tongue—and the cricket—into the frog's mouth, aided by contraction of a muscle on the bottom of the tongue. After the mouth

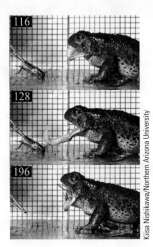

Figure 36.1

A Mexican leaf frog (*Pachymedusa dacnicolor*) capturing a grasshopper.

Kiisa Nishikawa/Northern Arizona University

closes, further muscle contractions pull the legs forward and fold them under the body.

We know this because Kiisa Nishikawa, Lucie Gray, and James O'Reilly of Northern Arizona University recorded the frog's movements using a high-speed video camera linked to a millisecond timer, with a grid in the background that allowed precise measurement of the distances body parts travelled during the capture. Nishikawa's research group uses the camera's record to study movement in frogs in particular, and animals in general.

In Chapter 32, you learned that there are three types of muscle tissue in vertebrates: skeletal, cardiac, and smooth. Skeletal muscle is so named because most muscles of this type are attached by tendons to the skeleton of vertebrates. Cardiac muscle is the contractile muscle of the heart, and smooth muscle is found in the walls of tubes and cavities of the body, including blood vessels and the intestines. In this chapter, we describe the structure and function of skeletal muscles, the skeletal systems found in invertebrates and vertebrates, and how muscles bring about movement.

36.1 Vertebrate Skeletal Muscle: Structure and Function

Vertebrate **skeletal muscles** connect to the bones of the skeleton. The cells forming skeletal muscles are typically long and cylindrical and contain many nuclei (see Chapter 32). Skeletal muscle is controlled by the somatic nervous system (see Chapter 33).

Most skeletal muscles in humans and other vertebrates are attached at both ends across a joint to bones of the skeleton. (Some, such as those that move the lips, are attached to other muscles or connective tissues under skin.) Depending on its points of attachment, contraction of a single skeletal muscle may extend or bend body parts or may rotate one body part with respect to another. The human body has more than 600 skeletal muscles, ranging in size from the small muscles that move the eyeballs to the large muscles that move the legs.

Skeletal muscles are attached to bones by cords of connective tissue called *tendons* (see Chapter 32). Tendons vary in length from a few millimetres to some, such as those that connect the muscles of the forearm to the bones of the fingers, that are 20 to 30 cm long.

36.1a The Striated Appearance of Skeletal Muscle Fibres Results from a Highly Organized Internal Structure

A skeletal muscle consists of bundles of elongated, cylindrical cells called **muscle fibres**, which are 10 to 100 μm in diameter and run the entire length of the muscle (**Figure 36.2**). Muscle fibres contain many nuclei, reflecting their development by fusion of smaller cells. Some very small muscles, such as some of the muscles of the face, contain only a few hundred muscle fibres; others, such as the larger leg muscles, contain hundreds of thousands. In both cases, the muscle fibres are held in parallel bundles by sheaths of connective tissue that surround them in the muscle and merge with the tendons that connect muscles to bones or other structures. Muscle fibres are richly supplied with nutrients and oxygen by an extensive network of blood vessels that penetrates the muscle tissue.

Muscle fibres are packed with **myofibrils**, cylindrical contractile elements about 1 μm in diameter that run lengthwise inside the cells. Each myofibril consists of a regular arrangement of **thick filaments** (13–18 nm in diameter) and **thin filaments** (5–8 nm in diameter) (see Figure 36.2). The thick and thin filaments alternate with one another in a stacked set.

The thick filaments are parallel bundles of myosin molecules; each myosin molecule consists of two protein subunits that together form a *head* connected to a long double helix forming a *tail*. The head is bent toward the adjacent thin filament to form a *crossbridge*. In vertebrates, each thick filament contains some 200 to 300 myosin molecules and forms as many crossbridges. The thin filaments consist mostly of two linear chains of actin molecules twisted into a double helix, which creates a groove running the length of the molecule. Bound to the actin are *tropomyosin* and *troponin* proteins. Tropomyosin molecules are elongated fibrous proteins that are organized end-to-end next to the groove of the actin double helix. Troponin is a three-subunit globular protein that binds to tropomyosin at intervals along the thin filaments.

The arrangement of thick and thin filaments forms a pattern of alternating dark bands and light bands, giving skeletal muscle a striated appearance under the microscope (see Figure 36.2). The dark bands, called *A bands,* consist of stacked thick filaments along with the parts of thin filaments that overlap both ends. The lighter-appearing middle region of an A band, which contains only thick filaments, is the *H zone.* In the centre of the H zone is a disk of proteins called the *M line,* which holds the stack of thick filaments together. The light bands, called *I bands,* consist of the parts of the thin filaments not in the A band. In the centre of each I band is a thin *Z line,* a disk to which the thin filaments are anchored. The region between two adjacent Z lines is a **sarcomere** (*sarco* = flesh; *meros* = segment); sarcomeres are the basic units of contraction in a myofibril.

At each junction of an A band and an I band, the plasma membrane folds into the muscle fibre to form a **T (transverse) tubule (Figure 36.3, p. 876)**. Encircling the sarcomeres is the **sarcoplasmic reticulum**, a complex system of vesicles modified from the smooth endoplasmic reticulum. Segments of the sarcoplasmic retic-

Figure 36.2

Skeletal muscle structure. Muscles are composed of bundles of cells called muscle fibres; within each muscle fibre are longitudinal bundles of myofibrils. The unit of contraction within a myofibril, the sarcomere, consists of overlapping myosin thick filaments and actin thin filaments. The myosin molecules in the thick filaments each consist of two subunits organized into a head and a double-helical tail. The actin subunits in the thin filaments form twisted, double helices, with tropomyosin molecules arranged head to tail in the groove of the helix and troponin bound to the tropomyosin at intervals along the thin filaments.

ulum are wrapped around each A band and I band and are separated from the T tubules in those regions by small gaps.

An axon of an efferent neuron leads to each muscle fibre. The axon terminal makes a single, broad synapse with a muscle fibre called a **neuromuscular junction** (see Figure 36.3). The neuromuscular junction, T tubules, and sarcoplasmic reticulum are key components in the pathway for stimulating skeletal muscle contraction by neural signals—which starts with action potentials travelling down the efferent neuron—as is described next.

36.1b During Muscle Contraction, Thin Filaments on Each Side of a Sarcomere Slide over Thick Filaments

The precise control of body motions depends on an equally precise control of muscle contraction by a signalling pathway that carries information from nerves to muscle fibres. An action potential arriving at the neuromuscular junction leads to an increase in the concentration of Ca^{2+} in the cytosol of the muscle fibre. The increase in Ca^{2+} triggers a process in which the thin filaments on each side of a sarcomere slide over the thick filaments toward the centre of the A band, which brings the Z lines closer together, shortening the sarcomeres and contracting the muscle **(Figure 36.4, p. 876)**. This *sliding filament mechanism* of muscle contraction depends on dynamic interactions between actin and myosin proteins in the two filament types. That is, the myosin crossbridges make and break contact with actin and pull the thin filaments over the thick filaments—the action is similar to rowing, or a ratcheting process. A model for muscle contraction is shown in **Figure 36.5, p. 877.**

Conduction of an Action Potential into a Muscle Fibre. Like neurons, skeletal muscle fibres are *excitable,* meaning that the electrical potential of their plasma membrane can change in response to a stimulus. When an action potential arrives at the neuromuscular junction, the axon terminal releases a neurotransmitter, *acetylcholine,* which triggers an action potential in the muscle fibre (see Figure 36.5, step 1). The action potential travels in all directions over the muscle fibre's surface membrane and penetrates into the interior of the fibre through the T tubules.

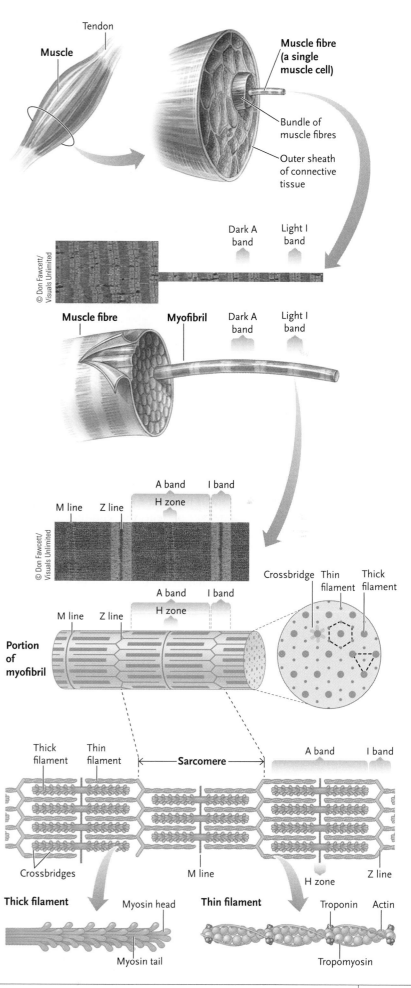

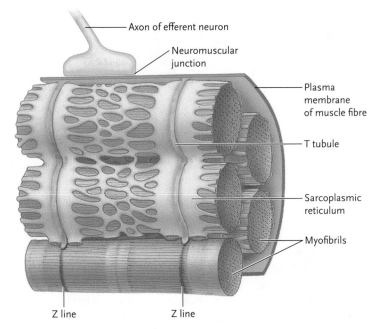

Axon of efferent neuron

Neuromuscular junction

Plasma membrane of muscle fibre

T tubule

Sarcoplasmic reticulum

Myofibrils

Z line   Z line

Figure 36.3
Components in the pathway for the stimulation of skeletal muscle contraction by neural signals. T (transverse) tubules are infoldings of the plasma membrane into the muscle fibre originating at each A band–I band junction in a sarcomere. The sarcoplasmic reticulum encircles the sarcomeres and segments of it end in close proximity to the T tubules.

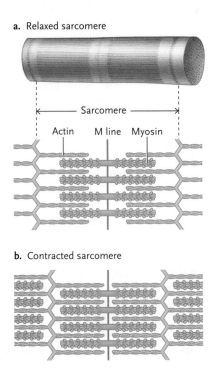

a. Relaxed sarcomere

Sarcomere

Actin   M line   Myosin

b. Contracted sarcomere

Figure 36.4
Shortening of sarcomeres by the sliding filament mechanism, in which the thin filaments are pulled over the thick filaments.

Release of Calcium into the Cytosol of the Muscle Fibre. In the absence of a stimulus, the Ca^{2+} concentration is kept high inside the sarcoplasmic reticulum by active transport proteins that continuously pump Ca^{2+} out of the cytosol and into the sarcoplasmic reticulum. When an action potential reaches the end of a T tubule, it opens ion channels in the sarcoplasmic reticulum that allow Ca^{2+} to flow out into the cytosol (see Figure 36.5, step 2).

When Ca^{2+} flows into the cytosol, the troponin molecules of the thin filament bind the calcium and undergo a conformational change that causes the tropomyosin fibres to slip into the grooves of the actin double helix. The slippage uncovers the actin's binding sites for the myosin crossbridge (see Figure 36.5, step 3). At this point in the process, the myosin crossbridge has a molecule of ATP bound to it, and is not in contact with the thin filament.

The Crossbridge Cycle. Using the energy of ATP hydrolysis, the myosin crossbridge bends away from the tail and binds to a newly exposed myosin crossbridge binding site on an actin molecule (see Figure 36.5, step 4). In effect, this bending compresses a molecular spring in the myosin head. The binding of the crossbridge to actin triggers release of the molecular spring in the crossbridge, which snaps back toward the tail, producing the power stroke (motor) that pulls the thin filament over the thick filament (step 5).

The crossbridge now binds another ATP and myosin detaches from actin (see Figure 36.5, step 6).

The cycle repeats again, starting with ATP hydrolysis (step 4). Contraction ceases when action potentials stop: Ca^{2+} is pumped back into the sarcoplasmic reticulum, and its effect on troponin is reversed, leading to tropomyosin again blocking myosin crossbridge binding sites on actin. Contraction ceases, and the actin thin filaments slide back over the myosin thick filaments to their original relaxed positions (step 7). Crossbridge cycles based on actin and myosin power movements in all living organisms, from cytoplasmic streaming in plant cells and amoebae to muscle contractions in animals.

Although the force produced by a single myosin crossbridge is comparatively small, it is multiplied by the hundreds of crossbridges acting in a single thick filament and by the billions of thin filaments sliding in a contracting sarcomere. The force, multiplied further by the many sarcomeres and myofibrils in a muscle fibre, is transmitted to the plasma membrane of a muscle fibre by the attachment of myofibrils to elements of the cytoskeleton. From the plasma membrane, it is transmitted to bones and other body parts by the connective tissue sheaths surrounding the muscle fibres and by the tendons.

From Contraction to Relaxation. As long as action potentials continue to arrive at the neuromuscular junction, Ca^{2+} is released in response, and ATP is available, the crossbridge cycle continues to run, shortening the sarcomeres and contracting the muscle fibre.

When action potentials stop, excitation of the T tubules ceases, and the Ca^{2+} release channels in

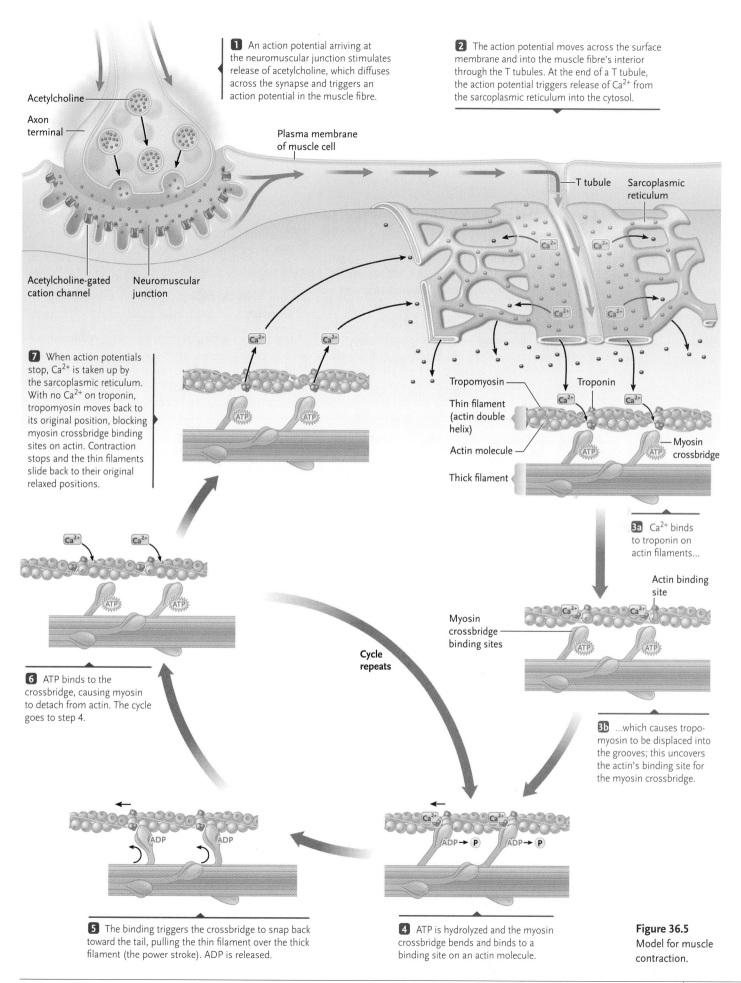

1 An action potential arriving at the neuromuscular junction stimulates release of acetylcholine, which diffuses across the synapse and triggers an action potential in the muscle fibre.

2 The action potential moves across the surface membrane and into the muscle fibre's interior through the T tubules. At the end of a T tubule, the action potential triggers release of Ca^{2+} from the sarcoplasmic reticulum into the cytosol.

Acetylcholine

Axon terminal

Plasma membrane of muscle cell

T tubule

Sarcoplasmic reticulum

Acetylcholine-gated cation channel

Neuromuscular junction

Tropomyosin

Troponin

Thin filament (actin double helix)

Actin molecule

Myosin crossbridge

Thick filament

7 When action potentials stop, Ca^{2+} is taken up by the sarcoplasmic reticulum. With no Ca^{2+} on troponin, tropomyosin moves back to its original position, blocking myosin crossbridge binding sites on actin. Contraction stops and the thin filaments slide back to their original relaxed positions.

3a Ca^{2+} binds to troponin on actin filaments...

Actin binding site

Myosin crossbridge binding sites

Cycle repeats

6 ATP binds to the crossbridge, causing myosin to detach from actin. The cycle goes to step 4.

3b ...which causes tropomyosin to be displaced into the grooves; this uncovers the actin's binding site for the myosin crossbridge.

5 The binding triggers the crossbridge to snap back toward the tail, pulling the thin filament over the thick filament (the power stroke). ADP is released.

4 ATP is hydrolyzed and the myosin crossbridge bends and binds to a binding site on an actin molecule.

Figure 36.5
Model for muscle contraction.

the sarcoplasmic reticulum close. The active transport pumps quickly remove the remaining Ca^{2+} from the cytosol. In response, troponin releases its Ca^{2+} and the tropomyosin fibres are pulled back to cover the myosin binding sites in the thin filaments. The crossbridge cycle stops, and contraction of the muscle fibre ceases. In a muscle fibre that is not contracting, ATP is bound to the myosin head and the crossbridge is not bound to the actin filament (see Figure 36.5, step 7).

36.1c The Response of a Muscle Fibre to Action Potentials Ranges from Twitches to Tetanus

A single action potential arriving at a neuromuscular junction usually causes a single, weak contraction of a muscle fibre called a **muscle twitch (Figure 36.6a).** After a muscle twitch begins, the tension of the muscle fibre increases in magnitude for about 30 to 40 ms and then peaks as the action potential runs its course through the T tubules and the Ca^{2+} channels begin to close. Tension then decreases as the Ca^{2+} ions are pumped back into the sarcoplasmic reticulum, falling to zero in about 50 ms after the peak.

If a muscle fibre is restimulated after it has relaxed completely, a new twitch identical to the first is generated (see Figure 36.6a). However, if a muscle fibre is restimulated before it has relaxed completely, the second twitch is added to the first, producing what is called *twitch summation,* which is basically a summed, stronger contraction **(Figure 36.6b).** And if action potentials arrive so rapidly (about 25 ms apart) that the fibre cannot relax between stimuli, the Ca^{2+} channels remain open continuously and twitch summation produces a peak level of continuous contraction called **tetanus (Figure 36.6c).** Contractile activity will then decrease if either the stimuli cease or the muscle fatigues.

Tetanus is an essential part of muscle fibre function. If we lift a moderately heavy weight, for example, many of the muscle fibres in our arms enter tetanus and remain in that state until the weight is released. Even body movements that require relatively little effort, such as standing still but in balance, involve tetanic contractions of some muscle fibres.

36.1d Muscle Fibres Differ in Their Rate of Contraction and Susceptibility to Fatigue

Muscle fibres differ in their rate of contraction and resistance to fatigue and thus can be classified as slow, fast aerobic, and fast anaerobic muscle fibres. Their properties are summarized in **Table 36.1.** The proportions of the three types of muscle fibres tailor the contractile characteristics of each muscle to suit its function within the body.

Slow muscle fibres contract relatively slowly, and the intensity of contraction is low because their myosin crossbridges hydrolyze ATP relatively slowly. They can

Figure 36.6
The relationship of the tension produced in a muscle fibre to the frequency of action potentials.

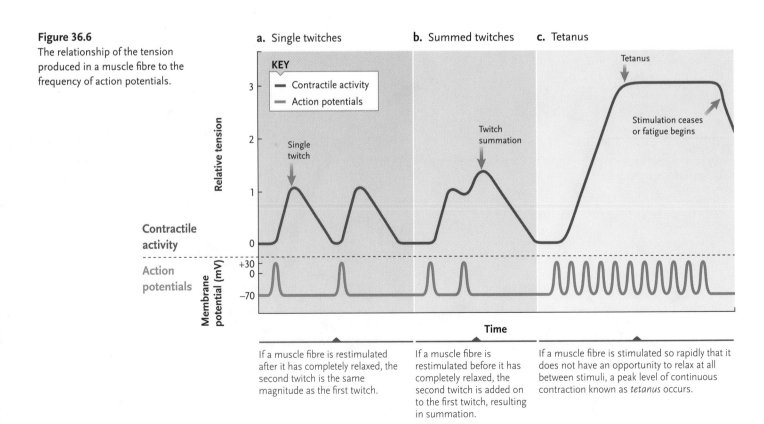

remain contracted for relatively long periods without fatiguing. Slow muscle fibres typically contain many mitochondria and make most of their ATP by oxidative phosphorylation (aerobic respiration). They have a low capacity to make ATP by anaerobic glycolysis. They also contain high concentrations of the oxygen-storing protein **myoglobin**, which greatly enhances their oxygen supplies. Myoglobin is closely related to haemoglobin, the oxygen-carrying protein of red blood cells. Myoglobin gives slow muscle fibres, such as those in the legs of ground birds such as quail, chickens, and ostriches, a deep red colour. In sharks and bony fishes, strips of slow muscles concentrated in a band on either side of the body are used for slow, continuous swimming and maintaining body position.

Fast muscle fibres contract relatively quickly and powerfully because their myosin crossbridges hydrolyze ATP faster than those of slow muscle fibres. Fast aerobic fibres have abundant mitochondria, a rich blood supply, and a high concentration of myoglobin, which makes them red in colour. They have a high capacity for making ATP by oxidative phosphorylation, and an intermediate capacity for making ATP by anaerobic glycolysis. They fatigue more quickly than slow fibres, but not as quickly as fast anaerobic fibres. Fast aerobic muscle fibres are abundant in the flight muscles of migrating birds such as ducks and geese.

Fast anaerobic fibres typically contain high concentrations of glycogen, relatively few mitochondria, and a more limited blood supply than fast aerobic fibres. They generate ATP mostly by anaerobic respiration (glycolysis) and have a low capacity to produce ATP by oxidative respiration. Fast anaerobic fibres produce especially rapid and powerful contractions but are more susceptible to fatigue. Because their myoglobin supply is limited and they contain few mitochondria, they are pale in colour. Some ground birds have flight muscles consisting almost entirely of fast anaerobic muscle fibres. These muscles can produce a short burst of intensive contractions allowing the bird to escape a predator, but they cannot produce sustained flight. Most muscles of lampreys, sharks, fishes, amphibians, and reptiles also contain fast anaerobic muscle fibres, allowing the animals to move quickly to capture prey and avoid danger.

The muscles of most animals are mixed and contain different proportions of slow and fast muscle fibres, depending on their functions. Muscles specialized for prolonged, slow contractions, such as the postural muscles of the back, have a high proportion of slow fibres and are a deep red colour. The muscles of the forearm that move the fingers have a higher proportion of fast fibres and are a paler red than the back muscles. These muscles can contract rapidly and powerfully, but they fatigue much more rapidly than the back muscles.

Table 36.1 | **Characteristics of Slow and Fast Muscle Fibres in Skeletal Muscle**

Property	Fibre Type		
	Slow	Fast Aerobic	Fast Anaerobic
Contraction speed	Slow	Fast	Fast
Contraction intensity	Low	Intermediate	High
Fatigue resistance	High	Intermediate	Low
Myosin–ATPase activity	Low	High	High
Oxidative phosphorylation capacity	High	High	Low
Enzymes for anaerobic glycolysis	Low	Intermediate	High
Mitochondria	Many	Many	Few
Myoglobin content	High	High	Low
Fibre colour	Red	Red	White
Glycogen content	Low	Intermediate	High

36.1e Skeletal Muscle Control Is Divided among Motor Units

The control of muscle contraction extends beyond the simple ability to turn the crossbridge cycle on and off. We can adjust a handshake from a gentle squeeze to a strong grasp or exactly balance a feather or dumbbell in the hand. How are entire muscles controlled in this way? The answer lies in activation of the muscle fibres in blocks called **motor units.**

The muscle fibres in each motor unit are controlled by branches of the axon of a single efferent neuron **(Figure 36.7, p. 880).** As a result, all of those fibres contract each time the neuron fires an action potential. All of the muscle fibres in a motor unit are of the same type—either slow, fast aerobic, or fast anaerobic. When a motor unit contracts, its force is distributed throughout the entire muscle because the fibres are dispersed throughout the muscle rather than being concentrated in one segment.

For a delicate movement, only a few efferent neurons carry action potentials to a muscle, and only a few motor units contract. For more powerful movements, more efferent neurons carry action potentials, and more motor units contract.

Muscles that can be precisely and delicately controlled, such as those moving the fingers in humans, have many motor units in a small area, with only a few muscle fibres—about 10 or so—in each unit. Muscles that produce grosser body movements, such as those moving the legs, have fewer motor units in the same volume of muscle but thousands of muscle fibres in each unit. In the calf muscle that raises the heel, for example, most motor units contain nearly 2000 muscle fibres. Other skeletal muscles fall between these extremes, with an average of about 200 muscle fibres per motor unit.

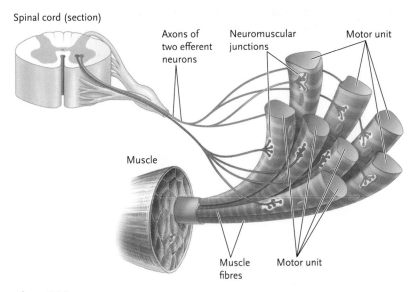

Spinal cord (section)

Axons of two efferent neurons

Neuromuscular junctions

Motor unit

Muscle

Muscle fibres

Motor unit

Figure 36.7

Motor units in vertebrate skeletal muscles. Each motor unit consists of groups of muscle fibres activated by branches of a single efferent (motor) neuron.

36.1f Invertebrates Move Using a Variety of Striated Muscles

Invertebrates also have muscle cells in which actin-based thin filaments and myosin-based thick filaments produce movements by the same sliding mechanism as in vertebrates. In Cnidaria and flatworms, the muscles are not striated, but in most other invertebrates (annelids, molluscs, echinoderms, nematodes, and arthropods), the actin and myosin fibrils are arranged in sarcomeres, forming striated muscle. In general, striated muscle is the dominant muscle type for these invertebrates and functions not only in locomotion but also in movements of the viscera, like the gut and heart. In some invertebrates, muscle cells lacking striations may occur. In the muscles that close the shells of clams and other bivalves (see Chapter 26), smooth muscle cells are present among the striated muscle cells.

In invertebrates, an entire muscle is typically controlled by one or a few motor neurons. Nevertheless,

invertebrate muscles are capable of finely graded contractions because individual neurons make large numbers of synapses with the muscle cells. In arthropods, the muscles may receive up to three types of innervation: fast, slow, and inhibitory. All muscles receive fast innervation in which release of a neuromuscular transmitter produces a twitch. Some also receive slow innervation, in which a graded response results from increased action potentials. As action potentials arrive more frequently, more Ca^{2+} is released into the cells, and they contract more strongly. In addition, there may be inhibitory nerves that prevent the release of Ca^{2+}. The excitatory transmitter for both fast and slow nerves is glutamate, and the inhibitory transmitter is GABA (see Chapter 33).

The muscles responsible for the movement of the wings in insects are highly specialized striated muscles, called fibrillar muscles. They possess a large number of gigantic mitochondria, in some cases about the size of a vertebrate red blood cell, so that the energetic demands of flight can be met. The frequency of wing beat of many flies, bees, and wasps is very high, up to 600 beats per second in mosquitoes. How is this achieved without tetanus being induced? The flight muscles occur in antagonistic pairs **(Figure 36.8)**. When one muscle of the pair contracts, the other is stretched back to its "relaxed" length. Nerve impulses arrive only about three times per second, and this keeps the muscles activated. The frequency with which they contract is determined by the elastic properties of the whole system.

STUDY BREAK

1. Compare thick and thin muscle filaments.
2. What is the role of the sarcoplasmic reticulum in muscle contraction?
3. What are the three types of muscle fibres, and how do they differ?

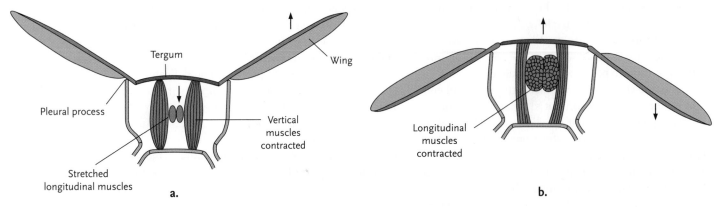

Tergum

Wing

Pleural process

Vertical muscles contracted

Stretched longitudinal muscles

a.

Longitudinal muscles contracted

b.

Figure 36.8

The muscles in a flying insect. When the vertical flight muscles contract, they pull on the cuticle forming the top of the segment, elevating the wings. The segment is constructed so that this action also elongates the cuticle of the thorax from front to rear (at right angles to the plane of the page), extending the longitudinal muscles. This stimulates the activated longitudinal muscles to contract, pushing up the tergum, elevating the wings, and elongating the vertical muscles. Because the cuticle is elastic, the whole system vibrates, producing very rapid wing beats.

36.2 Skeletal Systems

Animal skeletal systems provide physical support for the body and protection for the soft tissues. They also act as a framework against which muscles work to move parts of the body or the entire organism. Three main types of skeletons are found in both invertebrates and vertebrates: hydrostatic skeletons, exoskeletons, and endoskeletons.

36.2a A Hydrostatic Skeleton Consists of Muscles and Fluid

A **hydrostatic skeleton** (*hydro* = water; *statikos* = causing to stand) is a structure consisting of muscles and fluid that, by themselves, provide support for the animal or part of the animal; no rigid support, such as a bone, is involved. A hydrostatic skeleton consists of a body compartment or compartments filled with water or body fluids, which are incompressible liquids. When the muscular walls of the compartment contract, they pressurize the contained fluid. If muscles in one part of the compartment are contracted while muscles in another part are relaxed, the pressurized fluid will move to the relaxed part of the compartment, distending it. In short, the contractions and relaxations of the muscles surrounding the compartments change the shape of the animal.

Hydrostatic skeletons are the primary support systems of cnidarians, flatworms, roundworms, and annelids. In all of these animals, compartments containing fluids under pressure make the body semirigid and provide a mechanical support on which muscles act. For example, sea anemones have a hydrostatic skeleton consisting of several fluid-filled body cavities. The body wall contains longitudinal and circular muscles that work against that skeleton. Between meals, longitudinal muscles are contracted (shortened), whereas the circular ones are relaxed, and the animal looks short and squat **(Figure 36.9a)**. It lengthens into its upright feeding position by contracting the circular muscles and relaxing the longitudinal ones **(Figure 36.9b)**. In flatworms, roundworms, and annelids, striated muscles in the body wall act on the hydrostatic skeleton to produce creeping, burrowing, or swimming movements. Among these animals, annelids have the most highly developed musculoskeletal systems, with an outer layer of circular muscles surrounding the body, and an inner layer of longitudinal muscles **(Figure 36.10)**. Contractions of the circular muscles reduce the diameter of the body and increase the length; contractions of the longitudinal muscles shorten the body and increase its diameter. Because the coelom and musculature are divided into segments, expansion and contraction can be localized to individual segments. Annelids move along a surface or burrow by means of alternating waves of contraction of the two muscle layers that pass along the body, working against the fluid-filled body compartments of the hydrostatic skeleton.

Linda Pitkin/Planet Earth Pictures

Figure 36.9
Sea anemones in **(a)** the resting and **(b)** the feeding position. In **(a)**, longitudinal muscles in the body wall are contracted, and circular muscles are relaxed. In **(b)**, the longitudinal muscles are relaxed, and the circular muscles are contracted. Both sets of muscles work against a hydrostatic skeleton.

a. Resting position   **b.** Feeding position

Some structures of echinoderms are supported by hydrostatic skeletons. The tube feet of sea stars and sea urchins, for example, have muscular walls enclosing the fluid of the water vascular system (see Chapter 27).

In vertebrates, the erectile tissue of the penis is a fluid-filled hydrostatic skeletal structure, although many mammals other than humans also possess a penis bone, the *os penis* or baculum (Figure 36.12d).

Hydrostatic movement may not involve a fluid but may simply depend on the incompressibility of muscles themselves. Although the muscles in the structure may contract, the total body of muscles remains at a constant volume. Our tongues and lips are capable of a range of movements, but no skeletal element supports the movement. The elephant's trunk can lift a large log

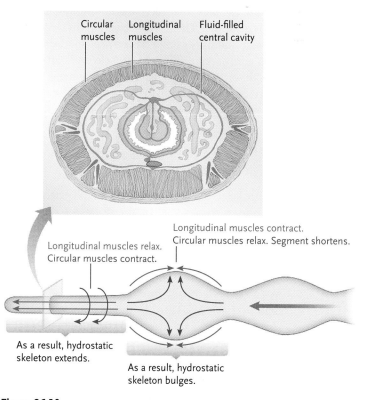

Circular muscles   Longitudinal muscles   Fluid-filled central cavity

Longitudinal muscles contract. Circular muscles relax. Segment shortens.

Longitudinal muscles relax. Circular muscles contract.

As a result, hydrostatic skeleton extends.

As a result, hydrostatic skeleton bulges.

Figure 36.10
Movement of an earthworm, showing how muscles in the body wall act on its hydrostatic skeleton. Contraction of the circular muscles reduces body diameter and increases body length, whereas contraction of the longitudinal muscles decreases body length and increases body diameter.

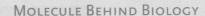
Chitin: An Abundant and Useful Polysaccharide

Chitin is a polymer composed of many repeating units of N-acetylglucosamine **(Figure 1)**. It is a strong, insoluble material that is the principal component of the cuticle of arthropods, where it is crosslinked to proteins. When arthropods moult, the proteins are resorbed, and the cast cuticle is made up almost entirely of chitin. Given the dominance of arthropods in the Earth's biomass, it is thus not surprising that, after cellulose, chitin is the most abundant polymer on Earth, with an annual production of about 10 billion tonnes. Chitin also occurs in molluscs, the eggshells of nematodes, and the cell walls of fungi and algae. Chitin is degraded by bacteria, particularly members of the genus *Vibrio*. Because chitin does not occur in vertebrates, chitin synthesis is a target of some successful insecticides.

Chitin and its closely related compound chitosan, produced by heating chitin in a strongly alkaline solution, are widely used in a variety of fields. Chitosan was found to have healing properties and has been used for medical sutures and as a support for growing skin over severe wounds. The compounds are also said to confer protection against disease in plants, and seeds treated with chitosan have gained some acceptance. Chitin stimulates the growth of soil bacteria that secrete material toxic to nematodes that attack plants: compounds for this purpose are under development. Because the compounds have a strong positive charge, they have the potential to bind negatively charged compounds. Chitosans have been used in water purification plants for many years. More recently, chitosans, because of their ability to bind fats, have been promoted by the natural health products industry as a means to prevent fat absorption, leading to weight loss, but these claims are not supported by good scientific evidence.

Figure 1
Chitin is a polymer made up of many repeating units of N-acetyl-D-glucosamine.

N-acetyl-D-glucosamine

---

but can also pick up objects of a few millimetres. How is this achieved? The trunk is basically an extension of the nose and upper lip. It has no skeleton but consists of an enormous number of muscle units attached to the skin or to one another. The muscle mass remains at a constant volume, and contractions of local muscle groups result in movement.

36.2b An Exoskeleton Is a Rigid External Body Covering

An **exoskeleton** (*exo* = outside) is a rigid external body covering, such as a shell, that provides support. In an exoskeleton, the force of muscle contraction is applied against that covering. An exoskeleton also protects delicate internal tissues such as the brain and respiratory organs.

Many molluscs, such as clams and oysters, have an exoskeleton consisting of a hard calcium carbonate shell secreted by glands in the mantle. Arthropods, such as insects, spiders, and crustaceans, have an external skeleton in the form of a chitinous cuticle, secreted by underlying epidermis, that covers the outside surfaces of the animals. Like a suit of armour, the arthropod exoskeleton has movable joints, flexed and extended by muscles. Most muscles attach directly to the cuticle by extensions of the myofibrils and extend from the inside surface of one section of the cuticle to the inside surface of another section. Since the sections are separated by flexible cuticle, contraction results in movement about the joint **(Figure 36.11)**. The exoskeleton protects against dehydration, serves as armour against predators, and provides the levers against which muscles work. In many flying insects, elastic flexing of the exoskeleton contributes to the movements of the wings.

In vertebrates, the shell of a turtle or tortoise is an exoskeletal structure.

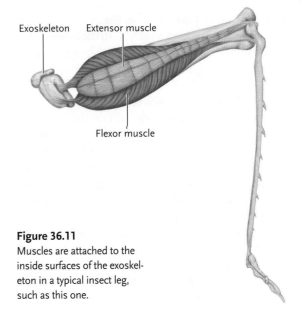

Exoskeleton   Extensor muscle

Flexor muscle

Figure 36.11
Muscles are attached to the inside surfaces of the exoskeleton in a typical insect leg, such as this one.

36.2c An Endoskeleton Consists of Supportive Internal Body Structures Such as Bones

An **endoskeleton** (*endon* = within) consists of internal body structures, such as bones, that provide support. In an endoskeleton, the force of contraction is applied against those structures. Like exoskeletons, endoskeletons also protect delicate internal tissues such as the brain and respiratory organs.

In cephalopod molluscs, the shell has been reduced and internalized. In most squids, it is a long, flat plate of chitin (see *Molecule Behind Biology*), the "pen," that provides support for the mantle. In cuttlefish, it is present as a material such as the shell of snails, but it is divided into air-filled chambers. It may provide some support, but its principal function is to provide buoyancy. The animal has the ability to vary the content of air in the "cuttlebone," providing control of buoyancy. *Nautilus* (see Chapter 26), the only living cephalopod with an external shell, has similar chambers in its shell. Squids also have an internal case of cartilage that surrounds and protects the brain; other segments of cartilage (not chemically identical to vertebrate cartilage) support the gills and siphon in squids and octopuses.

Echinoderms have an endoskeleton consisting of *ossicles* (*ossiculum* = little bone), formed from calcium carbonate crystals. The shells of sand dollars and sea urchins are the endoskeletons of these animals.

The endoskeleton is the primary skeletal system of vertebrates. An adult human, for example, has an endoskeleton consisting of 206 bones arranged in two structural groups **(Figure 36.12, p. 884).** The **axial skeleton**, which includes the skull, vertebral column, sternum, and rib cage, forms the central part of the structure (shaded in red in Figure 36.12). The **appendicular skeleton** (shaded in green) includes the shoulder, hip, leg, and arm bones. The human skeleton is, of course, highly specialized, reflecting our large brain and our upright movement, using only our hindfeet. Three other mammalian skeletons are included in Figure 36.12, illustrating how skeletons are adapted for particular lifestyles.

36.2d Bones of the Vertebrate Endoskeleton Are Organs with Several Functions

The vertebrate endoskeleton supports and maintains the overall shape of the body and protects internal organs. In addition, the bones are a storehouse for calcium and phosphate ions, releasing them as required to maintain optimal levels of these ions in body fluids. Bones are also sites where new blood cells form.

Bones are complex organs built up from multiple tissues, including bone tissue with cells of several kinds, blood vessels, nerves, and, in some, stores of adipose tissue. Bone tissue is distributed between dense, compact bone regions, which have essentially no spaces other than the microscopic canals of the osteons (see Chapter 32), and spongy bone regions, which may open into larger spaces (see Figure 36.12). Compact bone tissue generally forms the outer surfaces of bones and spongy bone tissue the interior. The interior of some flat bones, such as the hip bones and the ribs, are filled with *red marrow*, a tissue that is the primary source of new red blood cells in mammals and birds. The shaft of long bones such as the femur is opened by a large central canal filled with adipose tissue called *yellow marrow*, which is a source of some white blood cells.

Throughout the life of a vertebrate, calcium and phosphate ions are constantly deposited and withdrawn from bones. Hormonal controls maintain the concentration of Ca^{2+} ions at optimal levels in the blood and extracellular fluids (see Chapter 35), ensuring that calcium is available for proper functioning of the nervous system, muscular system, and other physiological processes.

STUDY BREAK

What is a hydrostatic skeleton? How is it different from an exoskeleton or endoskeleton?

36.3 Vertebrate Movement: The Interactions between Muscles and Bones

The skeletal system acts as a framework against which muscles work to move parts of the body or the entire organism. In this section, the muscle–bone interactions that are responsible for the movement of vertebrates are described.

36.3a Joints of the Vertebrate Endoskeleton Allow Bones to Move and Rotate

The bones of the vertebrate skeleton are connected by joints, many of them movable. The most movable joints, including those of the shoulders, elbows, wrists, fingers, knees, ankles, and toes, are *synovial joints,* consisting of the ends of two bones enclosed by a fluid-filled capsule of connective tissue **(Figure 36.13a, p. 885).** Within the joint, the ends of the bones are covered by a smooth layer of cartilage and lubricated by synovial fluid, which makes the bones slide easily as the joint moves. Synovial joints are held together by straps of connective tissue called *ligaments,* which extend across the joints outside the capsule **(Figure 36.13b).** The ligaments restrict the motion of the joint and help prevent it from buckling or twisting under heavy loads.

Figure 36.12

Mammalian skeletons. **(a)** Major bones in the human. Inset shows the structure of the femur (thigh)bone, with the location of red and yellow marrow. Internal spaces lighten the bone's density. At the joints a cartilage layer forms a smooth slippery cushion between bones. Compare the general features of this skeleton with those shown in **(b)**, **(c)**, and **(d)**. Note general and specific resemblances among the skeletons of the four mammals. **(b)** The new world monkey (family Cebidae) lives in trees. **(c)** The gliding lemur (family Cynocephalidae) also is arboreal, but a glider. **(d)** The raccoon (family Procyonidae) is terrestrial. Note the differences in skull shape, limb lengths and feet. The raccoon is a male reflected by the conspicuous baculum or penis bone (see also Fig. 38.14).

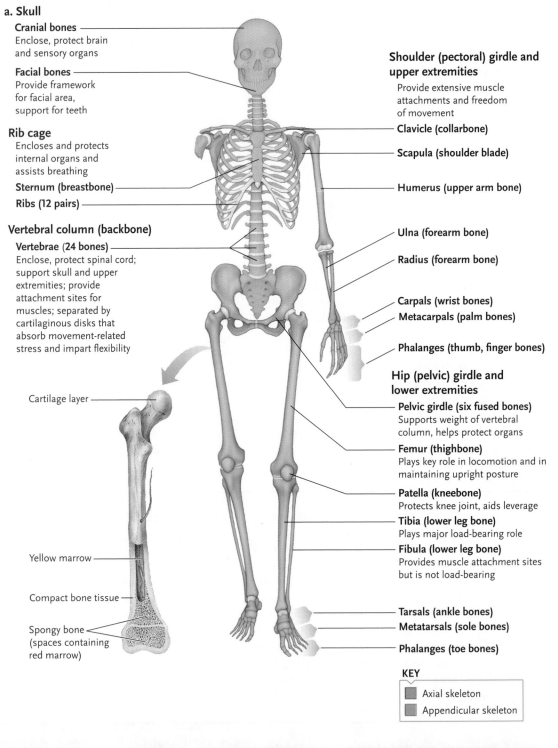

a. Skull

Cranial bones
Enclose, protect brain and sensory organs

Facial bones
Provide framework for facial area, support for teeth

Rib cage
Encloses and protects internal organs and assists breathing

Sternum (breastbone)

Ribs (12 pairs)

Vertebral column (backbone)

Vertebrae (24 bones)
Enclose, protect spinal cord; support skull and upper extremities; provide attachment sites for muscles; separated by cartilaginous disks that absorb movement-related stress and impart flexibility

Cartilage layer

Yellow marrow

Compact bone tissue

Spongy bone (spaces containing red marrow)

Shoulder (pectoral) girdle and upper extremities
Provide extensive muscle attachments and freedom of movement

Clavicle (collarbone)

Scapula (shoulder blade)

Humerus (upper arm bone)

Ulna (forearm bone)

Radius (forearm bone)

Carpals (wrist bones)
Metacarpals (palm bones)

Phalanges (thumb, finger bones)

Hip (pelvic) girdle and lower extremities

Pelvic girdle (six fused bones)
Supports weight of vertebral column, helps protect organs

Femur (thighbone)
Plays key role in locomotion and in maintaining upright posture

Patella (kneebone)
Protects knee joint, aids leverage

Tibia (lower leg bone)
Plays major load-bearing role

Fibula (lower leg bone)
Provides muscle attachment sites but is not load-bearing

Tarsals (ankle bones)
Metatarsals (sole bones)

Phalanges (toe bones)

KEY
Axial skeleton
Appendicular skeleton

b.
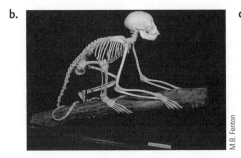
M.B. Fenton

c.

M.B. Fenton

d.
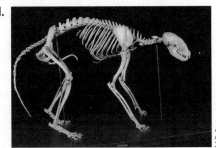
M.B. Fenton

a. Synovial joint cross section

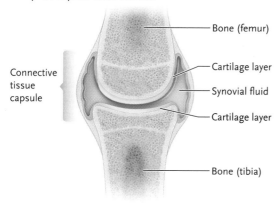

Bone (femur)

Cartilage layer

Synovial fluid

Cartilage layer

Connective tissue capsule

Bone (tibia)

b. Knee joint ligaments

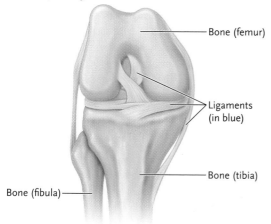

Bone (femur)

Ligaments (in blue)

Bone (tibia)

Bone (fibula)

Figure 36.13
A synovial joint. **(a)** Cross section of a typical synovial joint. **(b)** Ligaments reinforcing the knee joint.

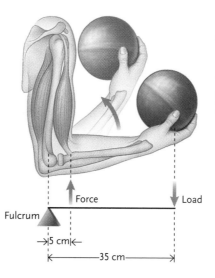

Figure 36.14
A body lever: The lever formed by the bones of the forearm. The fulcrum (the hinge or joint) is at one end of the lever, the load is placed on the opposite end, and the force is exerted at a point on the lever between the fulcrum and the load.

Force

Load

Fulcrum

5 cm

35 cm

In other, less movable joints, called *cartilaginous joints,* the ends of bones are covered with layers of cartilage but have no fluid-filled capsule surrounding them. Fibrous connective tissue covers and connects the bones of these joints, which occur between the vertebrae and some rib bones.

In still other joints, called *fibrous joints,* stiff fibres of connective tissue join the bones and allow little or no movement. Fibrous joints occur between the bones of the skull and hold the teeth in their sockets.

The bones connected by movable joints work like levers. A lever is a rigid structure that can move around a pivot point known as a *fulcrum.* Levers differ with respect to where the fulcrum is located along the lever and where the force is applied. The most common type of lever system in the body—exemplified by the elbow joint—has the fulcrum at one end, the load at the opposite end, and the force applied at a point between the ends **(Figure 36.14)**. For this lever, the force applied must be much greater than the load, but it increases the distance the load moves compared with the distance over which the force is applied. This allows small muscle movements to produce large body movements, as well as allows move-

ments such as running or throwing to be carried out at high speed.

At a joint, a muscle that causes movement in the joint when it contracts is called an **agonist**. In many cases, other muscles that assist the action of an agonist are involved in the movement of a joint. For instance, deltoid and pectoral muscles assist the biceps brachii muscle in lifting a weight.

Most of the bones of vertebrate skeletons are moved by muscles arranged in **antagonistic pairs:** *extensor muscles* extend the joint, meaning increasing the angle between the two bones, whereas *flexor muscles* do the opposite. (Antagonistic muscles are also used in invertebrates for movement of body parts—for example, the limbs of insects and arthropods.) In humans, one such pair is formed by the biceps brachii muscle at the front of the upper arm and the triceps brachii muscle at the back of the upper arm **(Figure 36.15)**. When the biceps muscle contracts, the bone of the lower arm is bent (flexed) around the elbow joint, and the triceps muscle is passively stretched (see **Figure 36.15a**); when the triceps muscle contracts, the lower

a.

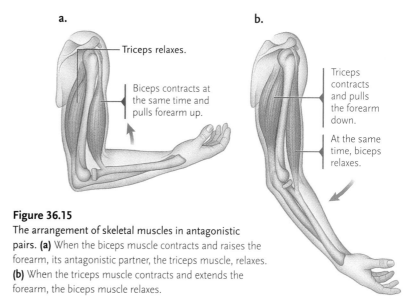

Triceps relaxes.

Biceps contracts at the same time and pulls forearm up.

b.

Triceps contracts and pulls the forearm down.

At the same time, biceps relaxes.

Figure 36.15
The arrangement of skeletal muscles in antagonistic pairs. **(a)** When the biceps muscle contracts and raises the forearm, its antagonistic partner, the triceps muscle, relaxes. **(b)** When the triceps muscle contracts and extends the forearm, the biceps muscle relaxes.

Keir Pearson, University of Alberta

The act of walking is both unconscious and deceptively simple when viewed from a human perspective, involving two legs. But in insects, in which walking on six legs usually involves two legs on one side and one leg on the other being lifted, while the remaining legs form a triangular support, and in nonhuman mammals, in which the order of movement of the legs may differ for various speeds of locomotion, the real complexity of walking becomes more obvious.

Keir G. Pearson of the University of Alberta became interested in this challenging problem while he was a Rhodes Scholar at Oxford University, where he demonstrated the complexity of the patterns of nervous activity controlling a single muscle. He has continued that interest for more than 40 years, progressing to an analysis of walking in insects and vertebrates (usually cats). The techniques he has used are challenging, involving the recording of electrical activity in the leg muscles and the ganglia controlling those muscles. They have revealed that walking in insects is driven by a pattern of rhythmic activity, the central pattern generator (CPG) in the neurons of the ganglia associated with the limbs, influenced by information flowing from proprioceptors (see Chapter 34) in the limbs and by environmental information processed through the brain. These results have influenced the construction of walking robots and physiotherapy for humans with spinal cord damage.

Pearson recently turned his attention to the way in which visual information influences walking. A simple and revealing experiment involved cats **(Figure 1)**. A cat approaching a small barrier steps over the barrier, first with the forelegs and then with the hindlegs. To a less inquiring mind, that might seem simple enough. But a scientist such as Pearson notes that when the hindlegs step over the barrier, the barrier is no longer in the visual field. The CPG in some sense remembers the position of the barrier. Pearson then asked the question: What if there is a delay after the forefeet step over? By offering food, he was able to interrupt the cat before its hindlegs had stepped over the barrier. It turns out that the hindlegs remember the position of the barrier for at least 10 minutes, even if the barrier is removed after the front feet have stepped over it. This is far longer than the memory of the image in the eye. These clear and simple results demonstrate the way in which information processed unconsciously through the brain influences "automatic" events.

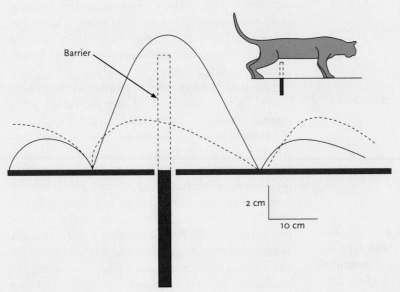

Figure 1

Experiment demonstrating the memory of CPG in the cat. The cat stepped over the barrier with the front feet but was then delayed by offering it food. The barrier, which it could no longer see, was removed. The solid line indicates the path taken by the hindfoot when walking resumed up to 10 minutes later. The dotted line indicates the path taken by the hindfoot when the cat does not encounter a barrier.

arm is straightened (extended), and the biceps muscle is passively stretched (see Figure 36.15b).

36.3b Vertebrates Have Muscle–Bone Interactions Optimized for Specific Movements

The human musculoskeletal system is highly specialized, scarcely revealing its origins from less specialized vertebrates. Vertebrates differ widely in the patterns by which muscles connect to bones and in the length and mechanical advantage of the levers produced by these connections. The segmental muscles of fish, attached to the vertebrae, many ribs, and the skin, are efficient in propelling fish through the water with side-to-side movements. The appearance of limbs and the movement onto land rendered that form of locomotion less useful, although it is often combined with leg movement in some amphibians and reptiles. Crocodiles use a side-to-side movement in the water but can move rapidly on land using their legs. The dinosaurs include bipedal species, able to

a.

b.

Jupiter Images

Petr Kratochvíl/Public Domain Pictures

Figure 36.16

(a) A sea turtle. Note that the forelegs are winglike paddles that enhance swimming. The turtle has limited mobility on land. **(b)** A penguin swimming, using modified forelimbs (wings). Note the resemblance in shape to the paddlelike forelegs of the sea turtle. Locomotion on land depends entirely on the hindlegs.

walk on their hind legs. Other reptiles have lost their legs. Snakes use a wavelike motion to propel themselves on the surface of the ground. Sea turtles have paddlelike forelimbs that they use in swimming, very much like those of penguins, an example of parallel evolution **(Figure 36.16).** Those mammals that have returned to the sea, such as whales, propel themselves by an up-and-down movement of their huge tail. This up-and- down movement is also used by elite human swimmers during some races.

The development of flight, represented by birds and, among the mammals, bats, is accompanied by some common modifications. In both groups, but particularly in birds, the sternum is greatly enlarged to provide attachment for the powerful pectoral muscles. The bones of the forelimb are greatly modified. In bats, the bones of the hand are extended to provide

a framework for the sheets of skin that form the wing. In birds, the bones of the hand are fused and support the feathers that make up the wing. The bones generally contain many cavities or spaces, making the body lighter. Penguins are birds that exploit the marine environment. They do not fly, but they use their reduced wings to "fly" through the water.

Burrowing mammals have short, very stout forelimbs with muscles arranged so that contraction provides relatively little movement but great force.

STUDY BREAK

1. Distinguish synovial joints, cartilaginous joints, and fibrous joints.
2. What are antagonistic muscle pairs?

UNANSWERED QUESTIONS

Is there a genetic predisposition to elite performance?

It is generally accepted that elite performance in some sports is often associated with a general body shape. Sprinters tend to be tall and thin, with long legs. Weight lifters are shorter, and their arms and legs are shorter. These body shapes are the result of the expression of many genes during development. But individual genes have also been associated with performance in specific sports. The *ACTN3* gene is one of a family of genes that code for actin-binding proteins (α-actinins) that are important in attaching the actin (thin filaments) to the Z line. *ACTN3*, producing α-actinin 3, is expressed only in fast muscle fibres, whereas an isoform, *ACTN2*, is expressed in both fast and slow fibres. Moreover, there is a mutant form of *ACTN3* that results in loss of the protein. This appears not to affect health and occurs in about 20% of the population.

In 2003, a group of Australian scientists associated *ACTN3* with power athletes, for whom fast twitch muscles are important (sprinters, weight lifters), whereas endurance athletes (long-distance runners) were more likely to have the mutant form of the *ACTN3* gene, leading to a greater proportion of slow fibres. *ACTN3* has been identified in the media as a "speed gene." It is crucial, however, to recognize that the Australian research simply identified a correlation between the gene and a particular type of athletic activity. The research has not identified a causal relationship: there is no direct evidence that the presence or absence of the functional form of *ACTN3* will result in superior performance in any specific activity. The question remains open, and the issue raises many ethical concerns.

Review

Go to CENGAGENOW at http://hed.nelson.com/ to access quizzing, animations, exercises, articles, and personalized homework help.

36.1 Vertebrate Skeletal Muscle: Structure and Function

- Skeletal muscles move the joints of the body. They are formed from long, cylindrical cells called muscle fibres, which are packed with myofibrils, contractile elements consisting of myosin thick filaments and actin thin filaments. The two types of filaments are arranged in an overlapping pattern of contractile units called sarcomeres.

- Infoldings of the plasma membrane of the muscle fibre form T tubules. The sarcomeres are encircled by the sarcoplasmic reticulum, a system of vesicles with segments separated from T tubules by small gaps.

- In the sliding filament mechanism of muscle contraction, the simultaneous sliding of thin filaments on each side of sarcomeres over the thick filaments shortens the sarcomeres and the muscle fibres, producing the force that contracts the muscle.

- The sliding motion of thin and thick filaments is produced in response to an action potential arriving at the neuromuscular junction. The action potential causes the release of acetylcholine, which triggers an action potential in the muscle fibre that spreads over its plasma membrane and stimulates the sarcoplasmic reticulum to release Ca^{2+} into the cytosol. The Ca^{2+} combines with troponin, inducing a conformational change that moves tropomyosin away from the myosin-binding sites on thin filaments. Exposure of the sites allows myosin crossbridges to bind and initiate the crossbridge cycle in which the myosin heads of thick filaments attach to a thin filament, pull, and release in cyclic reactions powered by ATP hydrolysis.

- When action potentials stop, Ca^{2+} is pumped back into the sarcoplasmic reticulum, leading to Ca^{2+} release from troponin, which allows tropomyosin to cover the myosin-binding sites in the thin filaments, thereby stopping the crossbridge cycle.

- A single action potential arriving at a neuromuscular junction causes a muscle twitch. Restimulation of a muscle fibre before it has relaxed completely causes a second twitch, which is added to the first, causing a summed, stronger contraction. Rapid arrival of action potentials causes the twitches to sum to a peak level of contraction called tetanus. Normally, muscles contract in a tetanic mode.

- Vertebrate muscle fibres occur in three types. Slow muscle fibres contract relatively slowly, but do not fatigue rapidly. Fast aerobic fibres contract relatively quickly and powerfully and fatigue more quickly than slow fibres. Fast anaerobic fibres can contract more rapidly and powerfully than fast aerobic fibres, but fatigue more rapidly. The fibres differ in their number of mitochondria and capacity to produce ATP.

- Skeletal muscles are divided into motor units, consisting of a group of muscle fibres activated by branches of a single motor neuron. The total force produced by a skeletal muscle is determined by the number of motor units that are activated.

- Most invertebrate muscles contain thin and thick filaments arranged in sarcomeres, and contract by the same sliding filament mechanism that operates in vertebrates. In arthropods, fast twitches and slower, graded contractions result from differences in innervation. Insect flight muscle is specialized to contract at high frequency.

36.2 Skeletal Systems

- A hydrostatic skeleton is a structure consisting of a muscle-surrounded compartment or compartments filled with fluid under pressure. Contraction and relaxation of the muscles changes the shape of the animal.

- In an exoskeleton, a rigid external covering provides support for the body. The force of muscle contraction is applied against the covering. An exoskeleton can also protect delicate internal tissues.

- In an endoskeleton, the body is supported by rigid structures within the body, such as bones. The force of muscle contraction is applied against those structures. Endoskeletons also protect delicate internal tissues. In vertebrates, the endoskeleton is the primary skeletal system. The vertebrate axial skeleton consists of the skull, vertebral column, sternum, and rib cage, whereas the appendicular skeleton includes the shoulder bones, the forelimbs, the hip bones, and the hindlimbs.

- Bone tissue is distributed between compact bone, with no spaces except the microscopic canals of the osteons, and spongy bone tissue, which has spaces filled by red or yellow marrow.

- Calcium and phosphate ions are constantly exchanged between the blood and bone tissues. The turnover keeps the Ca^{2+} concentration balanced at optimal levels in body fluids.

36.3 Vertebrate Movement: The Interactions between Muscles and Bones

- The bones of a skeleton are connected by joints. A synovial joint, the most movable type, consists of a fluid-filled capsule surrounding the ends of the bones forming the joint. A cartilaginous joint, which is less movable, has smooth layers of cartilage between the bones, with no surrounding capsule. The bones of a fibrous joint are joined by connective tissue fibres that allow little or no movement.

- The bones moved by skeletal muscles act as levers, with a joint at one end forming the fulcrum of the lever, the load at the opposite end, and the force applied by attachment of a muscle at a point between the ends.

- At a joint, an agonist muscle, perhaps assisted by other muscles, causes movement. Most skeletal muscles are arranged in antagonistic pairs, in which the members of a pair pull a bone in opposite directions. When one member of the pair contracts, the other member relaxes and is stretched.

- Vertebrates have a variety of patterns in which muscles connect to bones, giving different properties to the levers produced. Those properties are specialized for the activities of the animal.

Questions

Self-Test Questions

1. Vertebrate skeletal muscle
   a. is attached to bone by means of ligaments.
   b. may bend but not extend body parts.
   c. may rotate one body part with respect to another.
   d. is found in the walls of blood vessels and intestines.
   e. is usually attached at each end to the same bone.

2. In a resting muscle fibre,
   a. sarcomeres are regions between two H zones.
   b. disks of M line proteins called the A band separate the thick filaments.
   c. I bands are composed of the same thick filaments seen in the A bands.
   d. Z lines are adjacent to H zones, which attach thick filaments.
   e. dark A bands contain overlapping thick and thin filaments with a central thin H zone composed only of thick filaments.

3. The sliding filament contractile mechanism
   a. causes thick and thin filaments to slide toward the centre of the A band, bringing the Z lines closer together.
   b. is inhibited by the influx of Ca^{2+} into the muscle fibre cytosol.
   c. lengthens the sarcomere to separate the I regions.
   d. depends on the isolation of actin and myosin until a contraction is completed.
   e. uses myosin crossbridges to stimulate delivery of Ca^{2+} to the muscle fibre.

4. Which of the following is *not* an example of a hydrostatic skeletal structure?
   a. the tube feet of sea urchins
   b. the body wall of annelids
   c. the body wall of a grasshopper
   d. the body wall of cnidarians
   e. the penis of mammals

5. Endoskeletons
   a. protect internal organs and provide structures against which the force of muscle contraction can work.
   b. differ from exoskeletons in that endoskeletons do not support the external body.
   c. cannot be found in molluscs and echinoderms.
   d. are composed of appendicular structures that form the skull.
   e. compose the arms and legs, which are part of the axial skeleton.

Questions for Discussion

1. A coach must train young athletes for the 100 m sprint. They need muscles specialized for speed and strength rather than for endurance. What kinds of muscle characteristics would the training regimen aim to develop? How would it be altered to train marathoners?

2. Astronauts in zero gravity tend to lose bone mass, particularly in the hip, the long bones of the legs, and the vertebrae. Instruments that measure bone loss cannot be taken on space missions. Can you think of experimental procedures that might be used on earth to explore the phenomenon?

3. If you were a leading researcher in a pharmaceutical company interested in controlling bone loss, what cells and processes would you target?

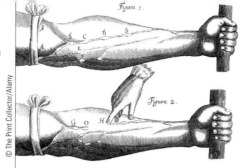

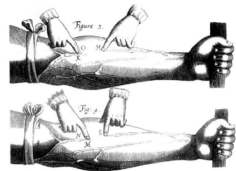

Illustrations from William Harvey's book, published in 1658, in which he demonstrated for the first time that the blood circulates. In one of the experiments, he used a tourniquet to reveal the veins and their valves. Part of the intellectual excitement of the Renaissance period, Harvey was the first experimental biologist.

© The Print Collector/Alamy

37 The Circulatory System

WHY IT MATTERS

Jimmie the bulldog stood on the stage of a demonstration laboratory at a meeting of the Royal Society in London in 1909, with one front paw and one rear paw in laboratory jars containing salt water **(Figure 37.1, p. 892)**. Wires leading from the jars were connected to a galvanometer, a device that can detect electrical currents.

Jimmie's master, Dr. Augustus Waller, a physician at St. Mary's Hospital, was relating his experiments in the emerging field of *electrophysiology*. Among other discoveries, Waller found that his apparatus detected the electrical currents produced each time the dog's heart beat.

Waller had originally experimented on himself. He already knew that the heart produces an electrical current as it beats; other scientists had discovered this by attaching electrodes directly to the heart of experimental animals. Looking for a painless alternative to that procedure, Waller reasoned that because the human body can conduct electricity, his arms and legs might conduct the currents generated by the heart if they were connected to a galvanometer. Accordingly, Waller set up two metal pans containing salt water and connected wires from the pans to a galvanometer. He put his bare left foot in one pan and

a. Jimmie the bulldog

b. Electrocardiogram

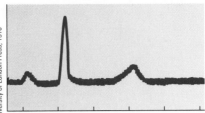

From A. D. Waller, Physiology, The Servant of Medicine, Hitchcock Lectures, University of London Press, 1910

Figure 37.1

The first electrocardiograms. **(a)** Jimmie the bulldog standing in laboratory jars containing salt water, with wires leading to a galvanometer that recorded the electrical currents produced by his heartbeat. **(b)** One of Waller's early electrocardiograms.

his right hand in the other. The technique worked; the indicator on the galvanometer jumped each time his heart beat. And it worked with Jimmie, too.

Waller also invented a method for recording the changes in current, which became the first electrocardiogram (ECG). He constructed a galvanometer by placing a column of mercury in a fine glass tube, with a conducting salt solution layered above the mercury. Changes in the current passing through the tube caused corresponding changes in the **surface tension** of the mercury, which produced movements that could be detected by reflecting a beam of light from the mercury surface. By placing a moving photographic plate behind the mercury tube, Waller could record the movements of the reflected light on the plate (Figure 37.1b shows one of his records). These were the first ECGs.

The beating of Jimmie's heart, recorded as an electrical trace by Augustus Waller, is part of the actions of the **circulatory system**, an organ system consisting of a fluid, a pump (the heart), and vessels for moving important molecules, and often cells, from one tissue to another. Examples of transported molecules are oxygen (O_2), nutrients, hormones, and wastes.

We study these systems in this chapter with emphasis on the circulatory system of humans and other mammals. We also discuss the **lymphatic system**, an accessory system of vessels and organs that helps balance the fluid content of the blood and surrounding tissues and participates in the body's defences against invading disease organisms.

37.1 Animal Circulatory Systems: An Introduction

Protostomes with simple body plans, including sponges, cnidarians, flatworms, and nematodes, function with no specialized circulatory system. Nearly all of these animals are aquatic or, like parasitic flatworms, live surrounded by the body fluids or intestinal contents of a host animal. Their bodies are structured as thin sheets of cells that lie close to the fluids of the surrounding environment. The products of digestion diffuse among the cells via the interstitial fluids, and O_2 and CO_2 are exchanged with the medium through the surface of the animal. Nematodes have a fluid-filled body cavity but no special mechanism for circulating that fluid.

37.1a Animal Circulatory Systems Share Basic Elements

In larger and more complex animals, most cells lie in cell layers too deep within the body to exchange substances directly with the environment via diffusion. Instead, the animals have a circulatory system, composed of tissues and organs, that conducts O_2, CO_2, nutrients, and the products of metabolism among the cells and tissues. The circulatory system may connect with specialized regions of the animal where substances are exchanged with the external environment. For example, oxygen is absorbed from the environment in the gills or lungs of many animals and is carried by the blood to all parts of the body; CO_2 released from body cells is carried by the blood to the lungs or gills, where it is released to the environment. Soluble wastes are conducted from body cells to the kidneys or other excretory organs, which remove wastes from the circulation and excrete them into the environment.

Animal circulatory systems carrying out these roles share certain basic features:

- A specialized fluid medium, usually containing at least some cells. This medium carries nutrients from the digestive system, the products of metabolism, and soluble wastes. With the conspicuous exception of the insects, it also transports O_2 and CO_2.
- In most groups, the fluid is contained in tubular vessels that distribute it to the various organs. Most arthropods and molluscs have an open system that lacks vessels.

a. Open circulatory system: no distinction between haemolymph and interstitial fluid

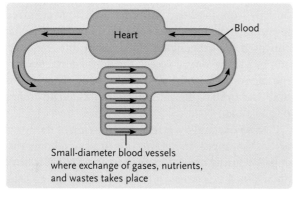

b. Closed circulatory system: blood separated from interstitial fluid

Small-diameter blood vessels where exchange of gases, nutrients, and wastes takes place

Figure 37.2

(a) Open circulatory system: haemolymph bathes the organs and body tissues. **(b)** Closed circulatory system: blood confined in tubes that lie among the cells of all tissues.

- A muscular heart that pumps the fluid through the circulatory system. Words associated with the heart often include *cardio,* from *kardia,* Greek for heart.

Animal circulatory systems take one of two forms, either *open* or *closed* **(Figure 37.2).** In an **open circulatory system,** vessels leaving the heart release fluid, usually termed **haemolymph,** directly into body spaces or into sinuses surrounding organs. Thus, the organs and

tissues are directly bathed in the haemolymph. The haemolymph reenters the heart through valves in the heart wall that close each time the heart pumps, thereby maintaining a unidirectional flow. In a **closed circulatory system,** the blood is confined to blood vessels and is distinct from the interstitial fluid. Substances are exchanged between the blood and the interstitial fluid and then between the interstitial fluid and cells.

37.1b Most Invertebrates Have Open Circulatory Systems

Among the protostomes, arthropods and most molluscs have open circulatory systems with one or more muscular hearts **(Figure 37.3a).** The blood is not conveyed directly to the cells by tubes but bathes the body tissues. In an open system, most of the fluid pressure generated by the heart dissipates when the blood is released into body spaces. Although the pressure remains low, the rate at which the blood circulates can be increased by an increase in the rate of beating of the heart(s). In highly active invertebrates, such as flying insects, the heart rate may rise to three or

a.

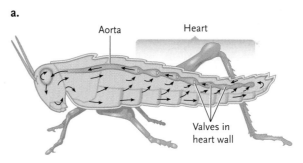

b.

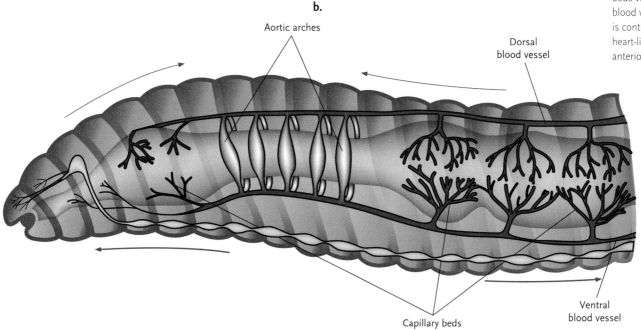

Figure 37.3

(a) Open circulatory system of a grasshopper. **(b)** Closed circulation in an earthworm. Aortic arches help to pump the blood into the ventral vessel that conveys blood to capillaries. The blood returns to the hearts from the capillary beds via the dorsal blood vessel, which is contractile and heart-like at the anterior end.

more times the resting rate, increasing the circulation of the hemolymph among the tissues. In addition, most insects have accessory hearts associated with the wings and each leg, which experience the same increases in rate during periods of high metabolic activity. In insects and molluscs, heart rate is controlled in some cases by nerves but largely by a variety of amine and peptide hormones (see Chapter 35).

37.1c Some Invertebrates and All Vertebrates Have Closed Circulatory Systems

Annelids, cephalopod molluscs such as squids and octopuses, most deuterostome invertebrates, and all vertebrates have closed circulatory systems. In these systems, vessels called **arteries** conduct blood away from the heart at relatively high pressure. From the arteries, the blood enters highly branched networks of microscopic, thin-walled vessels called **capillaries** that are well adapted for diffusion of substances. Nutrients and wastes are exchanged between the blood and body tissues as the blood moves through the capillaries. The blood then flows at relatively low pressure from the capillaries to larger vessels, the **veins**, which carry the blood back to the heart. Typically, the blood is maintained at a higher pressure and moves more rapidly through the body in closed systems than in open systems.

In many animals, closed systems allow precise control of the distribution and rate of blood flow to different body regions by means of muscles that contract or relax to adjust the diameter of the blood vessels. In an earthworm, for example, the anterior part of the dorsal blood vessel and five pairs of aortic arches act as hearts to pump the blood into large ventral blood vessels that extend the length of the body below the gut (Figure 37.3b). In the body segments, lateral branches of the ventral blood vessels lead to capillaries in internal organs and the body wall, where exchange of substances with interstitial fluids takes place. The blood is collected from the capillaries into lateral vessels that lead to a large dorsal vessel running the length of the body above the gut and is carried back to the hearts. The animal's body movements and contractions of the dorsal blood vessel help keep the blood flowing, and one-way valves in the hearts and dorsal blood vessel keep the flow from reversing.

37.1d Vertebrate Circulatory Systems Have Evolved from Single to Double Blood Circuits

A comparison of the different vertebrate groups reveals several evolutionary trends that accompanied the invasion of terrestrial habitats. Among the most striking are the changes that occurred in the major vessels of the body and the heart. These changes converted the single-circuit system of sharks and bony fish, in which the gills are in the same circuit as the rest of the blood vessels, to a double-circuit system in which the circulation to the lungs parallels the circulation to the rest of the body.

There were two major developments. In one, the blood vessels supplying the gills were reorganized to accommodate the appearance of lungs. The second involved developments in the structure of the heart. In the hypothetical chordate ancestor (see Chapter 27), there were six pairs of gills, each supplied by a branch of a single ventral artery, the ventral aorta, that led directly from the heart. After gas exchange occurred in the gills, the blood was collected in two vessels that conveyed the oxygenated blood forward to the brain and in a single dorsal vessel, the dorsal aorta, to carry the blood back to the rest of the body. The vessels that lead to the gills are referred to as aortic arches **(Figure 37.4a)**.

In sharks and their relatives, the first of the aortic arches that appear in the embryos of all vertebrates disappears during development, and only arches II to VI are retained to serve the gills in the adult **(Figure 37.4b)**. In lungfish **(Figure 37.4c)**, the reduction in gill function that accompanies the appearance of lungs is associated with two changes. Arches III and IV no longer run through gills, and, more significantly, arch VI supplies both a gill and the lungs. In a terrestrial amphibian such as a salamander **(Figure 37.4d)**, arch III loses its connection with the other arterial arches to become the carotid arteries. As a result, there is a separate blood supply to the brain directly from the heart. In most reptiles **(Figure 37.4e)**, arch V appears only in the embryo, and the connection between arch VI and the dorsal aorta disappears. Arch VI is exclusively devoted to supplying the lungs as the pulmonary artery. In birds **(Figure 37.4f)** and mammals, arch IV becomes a single rather than a paired vessel, leading to a single dorsal aorta. In birds, the right arch survives, whereas mammals retain the left arch.

These changes are accompanied by changes in the structure of the heart. In a shark or bony fish **(Figure 37.5a, p. 896)**, venous, deoxygenated blood from the tissues enters the first chamber, the atrium. The atrium contracts, forcing open flaplike valves leading into the ventricle and closing valves that prevent backflow into the veins. Contraction of the ventricle propels the blood forward into the ventral aorta leading to the aortic arches, the gill capillaries, and the dorsal aorta, which carries blood to the tissues. In amphibians, the atrium is divided, with one side (left atrium) receiving oxygenated blood from the lungs and skin and the other side (right atrium) receiving deoxygenated blood from the body **(Figure 37.5b)**. In the single ventricle, some separation of the two streams is achieved by one atrium contracting slightly in advance of the other and by

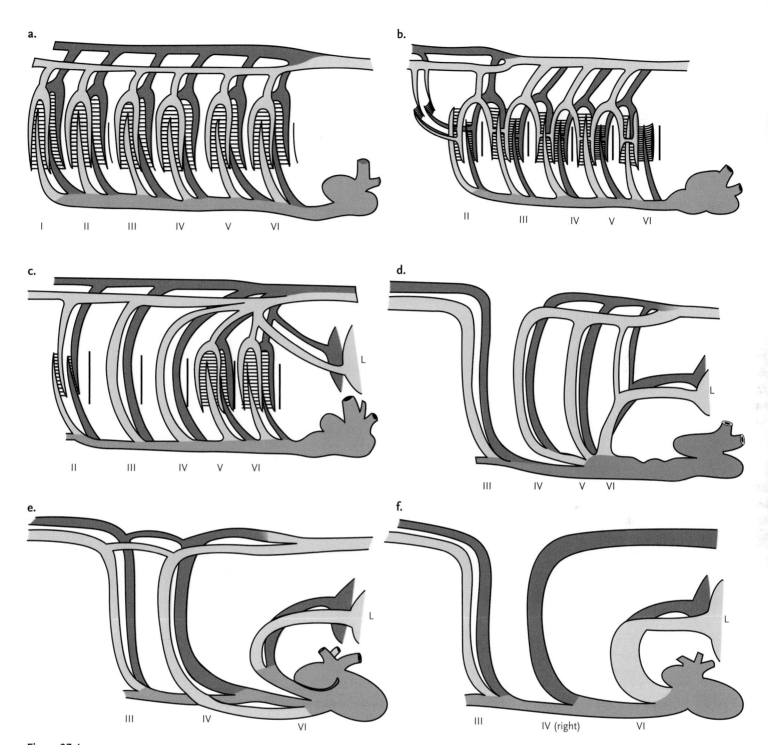

Figure 37.4

Evolution of the aortic arches in vertebrates. The darkest shading indicates the vessels of the right side, the lightest the vessels of the left side, and intermediate shading indicates the single ventral aorta. **(a)** The six arches in a hypothetical ancestor or in embryos. **(b)** In a shark, arch I has disappeared in the adult. **(c)** In lungfish, the lungs and one pair of gills are served from arch VI and the gills are reduced in number. **(d)** In a terrestrial amphibian, the lungs (L) are served by arch VI, which also retains its connection to the general circulation. The supply to the brain via arch III is separate from the general circulation. **(e)** In modern reptiles (snakes and lizards), although arches III and IV are still connected dorsally, arch VI serves only the lungs. **(f)** In birds, the supply to the brain is separated as the carotid arteries. Arch IV is present on the right side only and becomes the dorsal aorta. In mammals, the left side of arch IV remains as the dorsal aorta and the right side disappears.

a. FISH **b.** AMPHIBIAN **c.** REPTILE BIRD **d.** MAMMAL

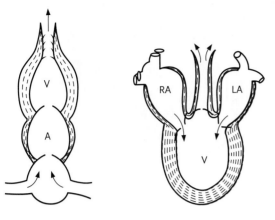 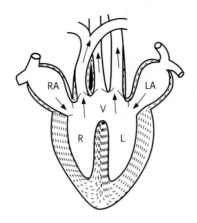 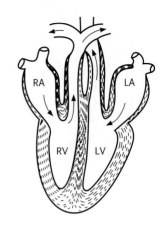

Figure 37.5

Major steps in the evolution of the heart leading to separate circulation for the lungs. **(a)** The two-chambered heart in a shark or a bony fish, **(b)** two atria in an amphibian, **(c)** partial division of the ventricle in modern reptiles, and **(d)** complete separation into two ventricles in crocodilians, birds, and mammals.

a flaplike structure in the vessel leaving the heart that can direct blood into arch VI alternately with the remaining arches. In modern reptiles, such as lizards and snakes, the ventricle is partially divided **(Figure 37.5c).**

Full separation of the blood supply to the lungs occurs in mammals, birds, and crocodilians (alligators and crocodiles share ancestry with birds) by complete division of the ventricle **(Figure 37.5d).** There are thus two separate circuits: one delivering oxygenated blood from the lungs into the left atrium, which then propels it into the left ventricle. The contraction of the left ventricle sends blood to the body circulation via the carotid arteries to the head (remnants of arch III) and the dorsal aorta (derived from arch IV), which supplies the remainder of the body.

Deoxygenated blood from the body and head enters the right atrium and, via the right ventricle, is propelled to the lungs in the pulmonary artery (arch VI). This separation of the body and lung circulation is illustrated in a flow diagram in **Figure 37.6.**

These gradual changes in the heart and aortic arches illustrate two principles:

- the segmental nature of the vertebrate body plan
- although evolution happens by changes in existing structures, changes in one set of structures (the aortic arches) are correlated with changes in other structures (the heart).

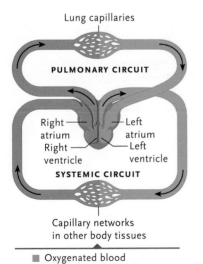

Figure 37.6

The separation of circulation to the lungs (pulmonary circulation) and to the body (systemic circulation) in mammals.

STUDY BREAK

1. Distinguish between open and closed circulatory systems.
2. Which of the following have closed circulatory systems: snails, arthropods, annelids, squids, vertebrates?
3. Which vertebrates have a separate pulmonary circulation?

37.2 Blood and Its Components

In both vertebrates and invertebrates, blood is a complex connective tissue that may contain blood cells suspended in a liquid matrix called the *plasma*. Although the blood of all vertebrates contains blood cells, the blood of some invertebrates may consist exclusively of plasma with few or no suspended cells, as in the Nematoda. In other invertebrates, such as the arthropods, blood cells or haemocytes of various recognizable types may occur in large numbers (up to 275 000 per μL in crickets) that can vary with activity and developmental stage. Whereas some haemocytes circulate with the haemolymph, others may attach temporarily to various tissues. These haemocyctes can be mobilized rapidly and enter the circulation, for example, to take part in wound healing or in defence against disease and parasites. In addition to transporting nutrients, dissolved gases, and metabolic wastes, blood helps stabilize the internal pH and salt composition of body fluids and serves as a highway for cells of the immune system and the antibodies produced by some of these cells.

In vertebrates, blood also helps regulate body temperature by transferring heat between warmer and cooler body regions and between the body and the external environment (see Chapter 43). The total blood volume of an average-sized adult human is about 4 to 5 L and makes up about 8% of body mass. The *plasma,* a clear, straw-coloured fluid, is about 55% of the volume of blood in human males and 58% in human females. Suspended in the plasma are three main types of blood cells, *erythrocytes, leukocytes,* and *platelets,* which account for the remainder of the blood volume. The typical components of human blood are shown in **Figure 37.7.**

37.2a Plasma Is an Aqueous Solution of Proteins, Ions, Nutrient Molecules, and Gases

Plasma is so complex that its complete composition is unknown for any animal. Its known components in humans are given in Figure 37.7. The plasma proteins of vertebrates fall into three classes: the *albumins,* the *globu-lins,* and *fibrinogen.* The **albumins,** the most abundant proteins of the plasma, are important for osmotic balance and pH buffering. They also transport a wide variety of substances through the circulatory system, including hormones, therapeutic drugs, and metabolic wastes. The **globulins** transport lipids (including cholesterol) and fat-soluble vitamins; a specialized subgroup of globulins, the *immunoglobulins,* includes antibodies and other molecules that contribute to the immune response. Some globulins are also enzymes. **Fibrinogen** plays a central role in the mechanism clotting the blood.

The ions of the plasma include Na^+, K^+, Ca^{2+}, Cl^-, and HCO_3^- (bicarbonate) ions. The Na^+ and Cl^- ions are the most abundant ions. Some of the ions, particularly the bicarbonate ion, help maintain arterial blood at its characteristic pH, which in humans is slightly on the basic side at pH 7.4 (see Chapter 43).

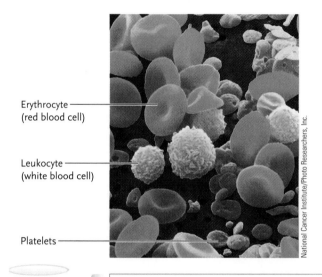

Erythrocyte (red blood cell)

Leukocyte (white blood cell)

Platelets

National Cancer Institute/Photo Researchers, Inc.

Figure 37.7

Typical components of human blood. The colourized scanning electron micrograph shows the three major cellular components. The sketch of the test tube shows what happens when you centrifuge a blood sample. The blood separates into three layers: a thick layer of straw-colored plasma on top, a thin layer containing leukocytes and platelets, and a thick layer of erythrocytes. The table shows the relative amounts and functions of the various components of blood.

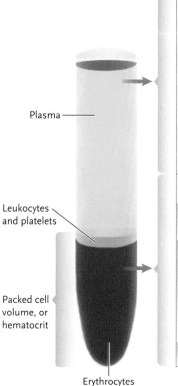

Plasma

Leukocytes and platelets

Packed cell volume, or hematocrit

Erythrocytes

Plasma Portion (55%–58% of total volume):

Components	Relative Amounts	Functions
1. Water	91%–92% of plasma volume	Solvent
2. Plasma proteins (albumin, globulins, fibrinogen, etc.)	7%–8%	Defence, clotting, lipid transport, roles in extracellular fluid volume, etc.
3. Ions, sugars, lipids, amino acids, hormones, vitamins, dissolved gases, urea and uric acid (metabolic wastes)	1%–2%	Roles in extracellular fluid volume, pH, eliminating waste products, etc.

Cellular Portion (45%–42% of total volume):

Components	Relative Amounts	Functions
1. Erythrocytes (red blood cells)	4 800 000–5 400 000 per microliter	Oxygen, carbon dioxide transport
2. Leukocytes (white blood cells)		
Neutrophils	3 000–6 750	Phagocytosis during inflammation
Lymphocytes	1 000–2 700	Immune response
Monocytes/macrophages	150–720	Phagocytosis in all defence responses
Eosinophils	100–360	Defence against parasitic worms
Basophils	25–90	Secrete substances for inflammatory response and for fat removal from blood
3. Platelets	250 000–300 000	Roles in clotting

37.2b Erythrocytes Are the Oxygen Carriers of Vertebrate Blood

Erythrocytes, or red blood cells, carry O_2 from the lungs to body tissues. Each microlitre of human blood normally contains about 5 million erythrocytes, which are small, flattened, and disklike. They measure about 7 μm in diameter and 2 μm in thickness. Microtubules of the cytoskeleton (see Chapter 2) are arranged beneath the surface of the cell so that they are *biconcave*—thinner in the middle than at the edges (see Figure 37.7). The proteins of the cytoskeleton that determine their shape also give them the flexibility to squeeze through narrow capillaries.

Like all blood cells, erythrocytes arise from stem cells in the red bone marrow. As they mature, mammalian erythrocytes lose their nucleus, cytoplasmic organelles, and ribosomes, limiting their metabolic capabilities and life span. The remaining cytoplasm contains enzymes, which carry out glycolysis, and large quantities of *haemoglobin,* the O_2-carrying protein of the blood. The erythrocytes of nearly all other vertebrates retain a nucleus.

Haemoglobin, the molecule that gives erythrocytes, and thus blood, its red colour, consists of four polypeptides, each linked to a nonprotein *haem* group (see *Molecule Behind Biology*) that contains an iron atom in its centre. The iron atom binds O_2 molecules as the blood circulates through the lungs and releases the O_2 as the blood flows through other body tissues.

Some 2 to 3 *million* erythrocytes are produced in the average human each second. The life span of an erythrocyte in the circulatory system is about

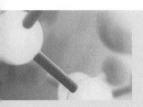

MOLECULE BEHIND BIOLOGY

Haem: Ubiquitous Oxygen Carrier

Haem is a member of a family of complex polycyclic chemicals called porphyrins **(Figure 1)**. Among the properties of porphyrins is their ability to bind metals. In haem, the metal bound is iron. The complex attracts O_2, which binds to the iron. As a constituent of three proteins, haem is an important carrier of O_2 in biological systems. Because these three protein molecules are coloured, they are often referred to as oxygen-carrying pigments.

Haemoglobin is a protein that includes four haem molecules. In the lungs, the higher concentration of O_2 loads the haemoglobin in the erythrocytes with O_2, and in the tissues, the low concentration leads to unloading. The process of unloading the O_2 is assisted by higher concentrations of CO_2 in the tissues.

Myoglobin, a protein that contains one molecule of haem, occurs in muscle. The oxygen that is bound to myoglobin serves as a reservoir that can be called on during periods of high metabolic demand resulting from increased muscle activity.

Cytochrome oxidase, the terminal enzyme in the complex that transfers electrons to molecular O_2 (see Chapter 5), also contains a haem molecule.

Other oxygen-carrying pigments, such as haemocyanin, which occurs in molluscs and some arthropods, and haemerythrin, found in a variety of invertebrates, also function as O_2 carriers. They do not, however, contain haem as the functional group.

Figure 1
The haem molecule.

120 days. At the end of their useful life, erythrocytes are engulfed and destroyed by *macrophages* (*macro* = big; *phagein* = to eat), a type of large leukocyte, in the spleen, liver, and bone marrow.

A negative feedback mechanism keyed to the blood's O_2 content stabilizes the number of erythrocytes in blood. If the O_2 content drops below the normal level, the kidneys synthesize **erythropoietin**, a peptide hormone that stimulates stem cells in bone marrow to increase erythrocyte production. Erythropoietin is also secreted after blood loss and when mammals move to higher altitudes. As new red blood cells enter the bloodstream, the O_2-carrying capacity of the blood rises. If the O_2 content of the blood rises above normal levels, erythropoietin production falls and red blood cell production drops. Erythropoietin has been used in "blood doping" by some athletes to improve their performance.

37.2c Leukocytes Provide the Body's Front Line of Defence against Disease

Leukocytes eliminate dead and dying cells from the body, remove cellular debris, and provide the body's first line of defence against invading organisms. They are called white cells because they are colourless, in contrast to the red blood cells. Because leukocytes retain their nuclei, cytoplasmic organelles, and ribosomes, they are fully functional cells.

Like red blood cells, leukocytes arise from the division of stem cells in red bone marrow. As they mature, they are released into the bloodstream, from which they enter body tissues in large numbers. Some types of leukocytes are capable of continued division in the blood and body tissues. The specific types of leukocytes and their functions in the immune reaction are discussed in Chapter 44.

37.2d Platelets Induce Blood Clots that Seal Breaks in the Circulatory System

Blood **platelets** are oval or rounded cell fragments, 2 to 4 μm in diameter, each enclosed in its own plasma membrane. They are produced in red bone marrow by the division of stem cells. Platelets contain enzymes and other factors that take part in blood clotting. When blood vessels are damaged, collagen fibres in the extracellular matrix are exposed to the leaking blood. Platelets in the blood stick to the collagen fibres and release signalling molecules that induce additional platelets to stick to them. The process continues, forming a plug that helps seal off the damaged site. As the plug forms, the platelets release other factors that convert the soluble plasma protein, fibrinogen, into long, insoluble threads of **fibrin**. Crosslinks between the fibrin threads form a meshlike network that traps blood cells and platelets and further seals the damaged area **(Figure 37.8)**. The entire mass is a blood clot.

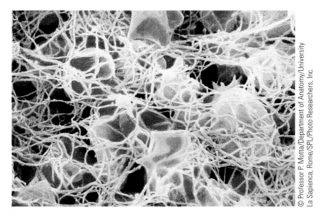

Figure 37.8
Red blood cells caught in a meshlike network of fibrin threads during formation of a blood clot.

© Professor P. Motta/Department of Anatomy/University La Sapienza, Rome/SPL/Photo Researchers, Inc.

STUDY BREAK

> What are the functions of the three main cellular components of vertebrate blood?

37.3 The Heart

The vertebrate heart is composed of cardiac muscle cells (see Chapter 36). In mammals, we have seen that the heart is a four-chambered pump, with two atria (singular, atrium) at the anterior of the heart and two ventricles at the posterior **(Figure 37.9)**. The atria pump blood into the ventricles, and then powerful contractions of the ventricles push the blood at relatively high pressure into arteries leaving the heart. This arterial pressure is responsible for the blood circulation. Valves between the atria and the ventricles, and between the

Figure 37.9
Cutaway view of the human heart showing its internal organization.

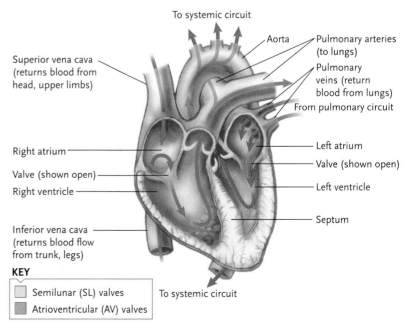

To systemic circuit

Aorta

Pulmonary arteries (to lungs)

Superior vena cava (returns blood from head, upper limbs)

Pulmonary veins (return blood from lungs)

From pulmonary circuit

Right atrium

Left atrium

Valve (shown open)

Valve (shown open)

Right ventricle

Left ventricle

Inferior vena cava (returns blood flow from trunk, legs)

Septum

KEY

☐ Semilunar (SL) valves
☐ Atrioventricular (AV) valves

To systemic circuit

ventricles and the arteries leaving the heart, keep the blood from flowing backward.

The mammalian heart pumps the blood through two completely separate circuits of blood vessels: the **systemic circuit** and the **pulmonary circuit** (**Figure 37.10**). The right atrium (toward the right side of the body) receives blood returning from the entire body, except for the lungs. The *superior vena cava* conveys blood from the head and forelimbs, and the *inferior vena cava* conveys blood from the abdominal organs and hindlimbs. (Most mammals do not have an upright posture, so superior and inferior are anterior and posterior, respectively.) This blood is depleted of O_2

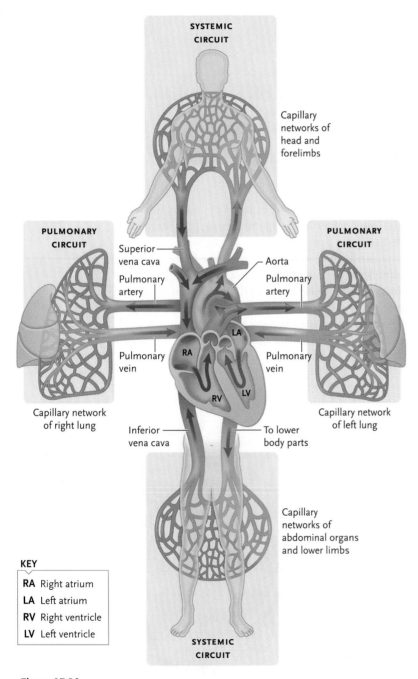

KEY

RA	Right atrium
LA	Left atrium
RV	Right ventricle
LV	Left ventricle

Figure 37.10

The pulmonary and systemic circuits of humans. The right half of the heart pumps blood into the pulmonary circuit, and the left half of the heart pumps blood into the systemic circuit.

and has a high CO_2 content. The right atrium pumps the blood into the right ventricle, which contracts to push the blood into the *pulmonary arteries* (derived from arch VI) leading to the lungs. In the capillaries of the lungs, the blood releases CO_2 and picks up O_2. The oxygenated blood completes this pulmonary circuit by returning in *pulmonary veins* to the heart.

Blood returning from the pulmonary circuit enters the left atrium, which pumps it into the left ventricle. This ventricle, the most thick-walled and powerful chamber, contracts to send the oxygenated blood into a large artery, the **aorta** (the dorsal aorta, derived from arch IV), which branches into arteries leading to all body regions except the lungs.

The arteries divide into smaller and smaller arteries and then into capillary networks, in which the blood releases O_2 and picks up CO_2. The O_2-depleted blood collects in veins, which complete the systemic circuit. The blood from the veins enters the right atrium. The amount of blood pumped by the two halves of the heart is normally balanced so that neither side pumps more than the other.

The heart also has its own circulation, called the *coronary circulation*. Two small *coronary arteries* branch off the aorta and then branch extensively over the heart, leading to dense capillary beds that serve the cardiac muscle cells. The blood from the capillary networks collects into veins that empty into the right atrium. If a coronary artery becomes blocked, the muscle cells it supplies can die and the person can suffer a heart attack (see *People Behind Biology*).

37.3a The Heartbeat Is Produced by a Cycle of Contraction and Relaxation of the Atria and Ventricles

Average heart rates vary among mammals (and among vertebrates generally), depending on body size and the overall level of metabolic activity. A human heart beats 72 times each minute, on average, with each beat lasting about 0.8 second. The heart rate of a trained endurance athlete is typically much lower. The heart of a flying bat may beat 1200 times a minute, whereas that of an elephant beats only 30 times a minute. **Systole** is the period of contraction and emptying of the heart, and **diastole** is the period of relaxation and filling of the heart between contractions. The systole–diastole sequence of the heart is called the **cardiac cycle** (**Figure 37.11**). The following discussion goes through one cardiac cycle.

Starting when both atria and ventricles are relaxed in diastole, the atria begin to fill with blood (step 1 in Figure 37.11). At this point, the **atrioventricular valves** (**AV valves**) between each atrium and ventricle and the **semilunar valves** (**SL valves**) between the ventricles and the aorta and pulmonary arteries are closed. As the atria fill, the pressure pushes open the AV valves and begins to fill the relaxed ventricles (step 2). When the ventricles are about 80% full, the atria contract

PEOPLE BEHIND BIOLOGY

Lorrie Kirshenbaum, University of Manitoba

Cardiac muscle cells do not divide after birth; the growth of the heart is the result of an increase in the size of the muscle cells. When the cells are damaged in a heart attack, they cannot repair themselves, nor can they be replaced.

Lorrie Kirshenbaum, Canada Research Chair in Molecular Cardiology at the University of Manitoba, explores the molecular events controlling the growth and death of cardiac muscle. *BNIP3* is one of a family of genes that can initiate cell death. It is particularly associated with a lack of oxygen. His lab has been active in describing the pathways leading to activation of the *BNIP3* gene and the intracellular pathways that the protein product of the gene uses to kill the cells. Dr. Kirshenbaum is also exploring how this information might be used to prevent the activation of the gene or its effects and thus prevent the death of cells when deprived of oxygen. Conversely, could the gene be activated to kill cancer cells in other tissues?

A different approach involves exploring ways to replace damaged heart cells. Cardiac muscle cells, which do not divide, have the normal cell division pathway blocked. Dr Kirshenbaum's lab is using growth factors, delivered as genes in viruses, to activate the cell division pathway. However, it turns out that adult myocytes that have the cell division pathway turned on enter a pathway leading to death. To get around this difficulty, he also delivers growth factors that block the self-destructive pathway together with the growth factors that stimulate cell division. This is a promising alternative to using stem cells to replace damaged heart cells.

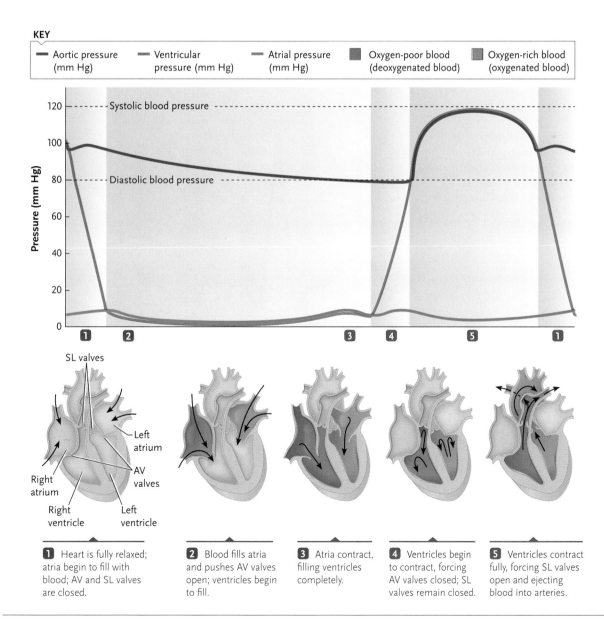

Figure 37.11
The cardiac cycle.

1 Heart is fully relaxed; atria begin to fill with blood; AV and SL valves are closed.

2 Blood fills atria and pushes AV valves open; ventricles begin to fill.

3 Atria contract, filling ventricles completely.

4 Ventricles begin to contract, forcing AV valves closed; SL valves remain closed.

5 Ventricles contract fully, forcing SL valves open and ejecting blood into arteries.

and completely fill the ventricles with blood (step 3). Although there are no valves where the veins open into the atria, the atrial contraction compresses the openings, sealing them so that little backflow occurs into the veins.

As the ventricular muscles begin to contract, rising pressure in the ventricular chambers forces the AV valves shut (step 4). As they continue to contract, the pressure in the ventricular chambers rises above that in the arteries leading away from the heart, forcing open the SL valves. Blood now rushes from the ventricles into the aorta and the pulmonary arteries (step 5).

Completion of the contraction squeezes about two-thirds of the blood in the ventricles into the arteries. Now the ventricles relax, lowering pressure in ventricular chambers below that in arteries. This reversal of the pressure gradient reverses the direction of blood flow in the regions of the SL valves, causing them to close. For about half a second, both atria and ventricles remain in diastole and blood flows into the atria and ventricles. Then the blood-filled atria contract, and the cycle repeats.

In an adult human at rest, each ventricle pumps roughly 5 L of blood per minute, an amount roughly equivalent to the entire volume of blood in the body. At maximum rate and strength, the human heart pumps about five times the resting amount, or more than 25 L/min.

37.3b The Cardiac Cycle Is Initiated within the Heart

Contraction of cardiac muscle cells is triggered by action potentials that spread across the muscle cell membranes. Crustaceans, such as crabs and lobsters, have **neurogenic hearts**, that is, hearts that beat under the control of signals from the nervous system. Each contraction is initiated by signals from a cardiac ganglion located in the heart, and the heart will continue to beat and respond to some environmental signals in isolation from the central nervous system, as long as the ganglion is intact. Other animals, including all insects and all vertebrates, have **myogenic hearts** that maintain their contraction rhythm with no requirement for signals from the nervous system. Isolated cardiac myocytes contract rhythmically when grown in a suitable medium. Both neurogenic and myogenic hearts can also be influenced by signals from the central nervous system and by hormones.

The rate and timing of the contraction of individual cardiac muscle cells in a mammalian myogenic heart are coordinated by a region of the heart called the **sinoatrial node (SA node)**. The SA node consists of **pacemaker cells**, which are specialized cardiac muscle cells in the upper wall of the right atrium **(Figure 37.12, step 1)**. Ion channels in these cells open in a cyclic,

Figure 37.12

The electrical control of the cardiac cycle. The bottom part of the figure shows how a signal originating at the SA node leads to ventricular contraction. The top part of the figure shows the electrical activity for each of the stages as seen in an ECG. The colours in the hearts show the location of the signal at each step and correspond to the colours in the ECG.

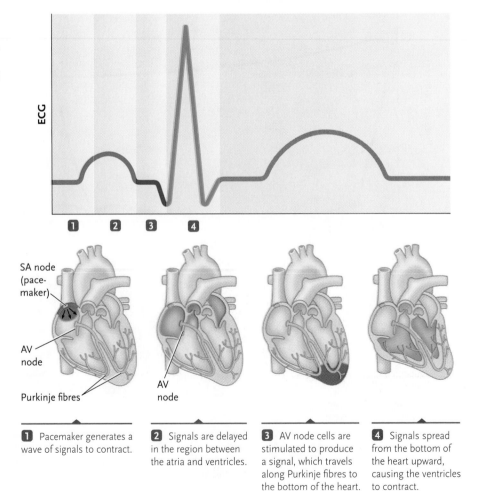

1 Pacemaker generates a wave of signals to contract.

2 Signals are delayed in the region between the atria and ventricles.

3 AV node cells are stimulated to produce a signal, which travels along Purkinje fibres to the bottom of the heart.

4 Signals spread from the bottom of the heart upward, causing the ventricles to contract.

self-sustaining pattern that alternately depolarizes and repolarizes their plasma membranes. The regularly timed depolarizations initiate waves of contraction that travel over the heart, causing the atria to contract.

A layer of connective tissue separates the atria from the ventricles, acting as a layer of electrical insulation between the top and the bottom of the heart. The insulating layer keeps a contraction signal from the SA node from spreading directly from the atria to the ventricles (step 2). Instead, the atrial wave of contraction excites cells of the **atrioventricular node (AV node)**, located in the heart wall between the right atrium and right ventricle, just above the insulating layer of connective tissue. The signal produced travels from the AV node to the bottom of the heart via *Purkinje fibres* (step 3). These fibres follow a path downward, through the insulating layer, to the bottom of the heart, where they branch through the walls of the ventricles. The signal carried by the Purkinje fibres induces a wave of contraction that begins at the bottom of the heart and proceeds upward, squeezing the blood from the ventricles into the aorta and pulmonary arteries (step 4). The transmission of a signal from the AV node to the ventricles takes about 0.1 second; this delay gives the atria time to finish their contraction before the ventricles contract.

As Augustus Waller found in experiments with Jimmie the bulldog, the electrical signals passing through the heart can be detected by attaching electrodes to different points on the surface of the body. The signals change in a regular pattern corresponding to the electrical signals that trigger the cardiac cycle, producing what is known as an **electrocardiogram** (**ECG**; also EKG, from German, *Elektrocardiogramm*). The highlighted region of the ECG above each stage of the cardiac cycle in Figure 37.12 indicates the electrical activity measured in those stages.

37.3c Arterial Blood Pressure Cycles between a High Systolic and a Low Diastolic Pressure

The pressure exerted by a fluid in a confined space is called *hydrostatic pressure*. That is, fluid in a container exerts some pressure on the wall of the container. Blood vessels are essentially tubular containers that are part of a closed system filled with fluid. Hence, the blood in vessels exerts hydrostatic pressure against the walls of the vessels. *Blood pressure* is the measurement of that hydrostatic pressure on the walls of the arteries as the heart pumps blood through the body. Blood pressure is determined by the force and amount of blood pumped by the heart and the size and flexibility of the arteries. In any animal, blood pressure changes in response to activity, temperature, body position, behaviour, time of day, and diet.

As the ventricles contract, a surge of high-pressure blood moves outward through the arteries leading from the heart. This peak of high pressure, called the *systolic blood pressure,* can be felt as a *pulse* by pressing

a finger against an artery that lies near the skin, such as the arteries of the neck or the artery that runs along the inside of the wrist. Between ventricular contractions, the arterial blood pressure reaches a low point called the *diastolic blood pressure*. In healthy humans at rest, the systolic pressure, measured in the large artery in the forearm, is equivalent to between 90 and 120 mm of mercury and the diastolic to between 60 and 80 mm. These pressures are illustrated in Figure 37.11.

The blood pressure in the systemic and pulmonary circuits is highest in the arteries leaving the heart and drops as the blood passes from the arteries into the capillaries. By the time the blood returns to the heart, its pressure has dropped to 2 to 5 mm Hg, with no differentiation between systolic and diastolic pressures. The reduction in pressure occurs because the blood encounters resistance as it moves through the vessels, primarily due to the friction created when blood cells and plasma proteins move over each other and over vessel walls.

STUDY BREAK

1. Distinguish between systolic and diastolic blood pressure.
2. What is the function of the sinoatrial node?

37.4 Blood Vessels of the Circulatory System

Both systemic and pulmonary circuits consist of a continuum of different blood vessel types that begin and end at the heart **(Figure 37.13)**. From the heart, large arteries carry blood and branch into progressively

Figure 37.13
The structure of arteries, capillaries, and veins and their relationship in blood circuits.

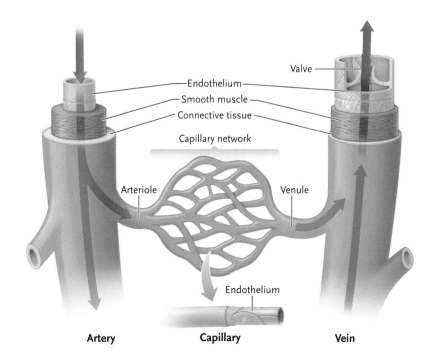

Valve
Endothelium
Smooth muscle
Connective tissue
Capillary network
Arteriole
Venule
Endothelium

Artery   **Capillary**   **Vein**

smaller arteries, delivering blood to the various parts of the body. When a small artery reaches the organ it supplies, it branches into yet smaller vessels, the **arterioles**. Within the organ, arterioles branch into capillaries, the smallest vessels of the circulatory system. Capillaries form a network in the organ, where they exchange substances between the blood and the surrounding interstitial fluid. Capillaries rejoin to form small **venules**, which merge into the small veins that leave the organ. The small veins progressively join to form larger veins that eventually become the large veins that enter the heart.

37.4a Arteries Transport Blood Rapidly to the Tissues and Serve as a Pressure Reservoir

Arteries have relatively large diameters and therefore provide little resistance to blood flow. They are structurally adapted to the relatively high pressure of the blood passing through them. The walls of arteries consist of three major tissue layers (see Figure 37.13):

- an outer layer of connective tissue containing collagen fibres mixed with fibres of the protein elastin, giving the vessel recoil ability,
- a relatively thick middle layer of vascular smooth muscle cells also mixed with elastin fibres, and
- an inner layer of flattened cells only one cell in thickness, forming an endothelium.

In addition to being conduits for blood travelling to the tissues, arteries also act as a pressure reservoir for blood movement when the heart is relaxing. When contraction of the ventricles pumps blood into arteries, a greater volume of blood enters the arteries than leaves them to flow into the smaller vessels downstream because of the higher resistance to blood flow in those smaller vessels. Arteries accommodate the excess volume of blood because of their elastic walls, which allow the arteries to expand in diameter. When the heart relaxes and blood is no longer being pumped into the arteries, the arterial walls recoil passively back to their original state. The recoil pushes the excess blood from the arteries into the smaller downstream vessels. As a result, blood flow to tissues is continuous during systole and diastole.

37.4b Capillaries Are the Sites of Exchange between the Blood and Interstitial Fluid

Capillaries thread through nearly every tissue in the body and are arranged in networks bringing them within 0.01 mm of most body cells. In humans, they are estimated to have a surface area of about 2600 km^2 for the exchange of gases, nutrients, and wastes with the interstitial fluid. Capillary walls consist of a single layer of endothelial cells, resting on a thin basement membrane (see Chapter 32).

Control of Blood Flow through Capillaries. Blood flow through capillary networks is controlled by contraction of smooth muscle in arterioles **(Figure 37.14)**. In addition to the normal layer of smooth muscle, some arterioles may have circular rings of smooth muscle at the entrance to a capillary, called *precapillary sphincter muscles*. When the arteriole and sphincter smooth muscles are relaxed, blood flows readily through the arterioles and capillary networks. In the most contracted state, the blood flow is limited through the arterioles and capillary networks. Variation in the contraction of the arteriole and sphincter smooth muscles adjusts the rate of flow through the capillary networks

a. Relaxed

Precapillary sphincters   Capillaries

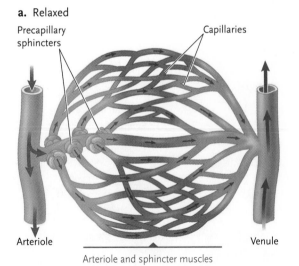

Arteriole   Venule

Arteriole and sphincter muscles fully relaxed—maximal blood flow

b. Contracted

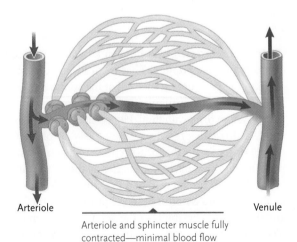

Arteriole   Venule

Arteriole and sphincter muscle fully contracted—minimal blood flow

Figure 37.14

Control of blood flow through capillary networks. **(a)** Maximal blood flow when arteriole and sphincter muscles are fully relaxed. **(b)** Minimal blood flow when the arteriole and sphincter muscles are fully contracted.

for particular organs. For example, the flow of blood through the capillary networks of the intestines is increased after a meal.

The Velocity of Blood Flow through Capillaries.

Although their total surface area is astoundingly large, the diameter of individual capillaries is so small that red blood cells must squeeze through most of them in single file **(Figure 37.15)**. As a result, each capillary presents a high resistance to blood flow. In addition, there are so many billions of capillaries in the networks that their combined diameter is about 1300 times greater than the cross-sectional area of the aorta. As a result of the resistance and the vastly increased diameter of the combined tubes, blood slows considerably as it moves through capillaries, maximizing opportunities for exchange between the blood and the interstitial fluids. As they leave the tissues, capillaries rejoin to form venules and veins. Veins have a total cross-sectional area that is smaller than the total cross-sectional area of capillaries, so the velocity of flow increases as blood returns to the heart.

The Exchange of Substances across Capillary Walls.

In most body tissues, narrow spaces between the capillary endothelial cells allow water, ions, and small molecules such as glucose to pass freely between blood and interstitial fluid. Erythrocytes, platelets, and most plasma proteins are too large to pass between the cells and are retained inside capillaries, except for molecules that are transported through epithelial cells by specific carriers. Leukocytes, however, can squeeze actively between the cells and pass from the blood to the interstitial fluid.

There are exceptions to these general properties. In the brain, endothelial cells are tightly sealed together, preventing all molecules and ions from passing between them. The tight seals set up the *blood–brain barrier* (Chapter 33). This limits the exchange between capillaries and brain tissues to molecules and ions that are specifically transported through the capillary endothelial cells. At the other extreme are capillaries in the liver, in which the spaces between endothelial cells are wide enough to admit most plasma proteins (most plasma proteins are synthesized in the liver), and in the small intestines, where wide spaces between capillary endothelial cells allow many nutrient molecules to pass into the bloodstream. In bone marrow and other sites of erythrocyte production, spaces are large enough to admit red blood cells.

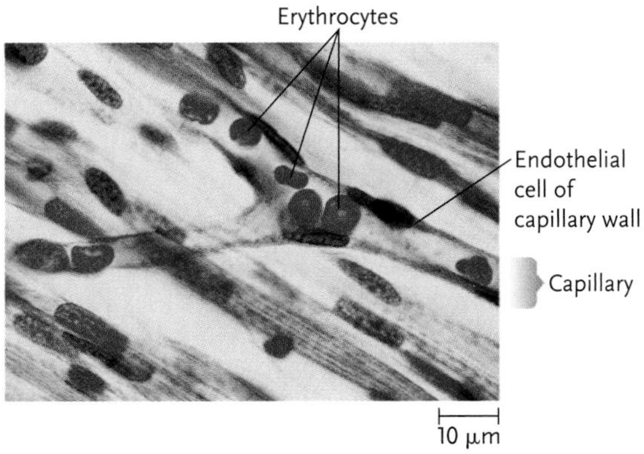

37.4c Venules and Veins Serve as Blood Reservoirs in Addition to Conduits to the Heart

The walls of venules and veins are thinner than those of arteries and contain little elastin. Many veins have flaps of connective tissue that extend inward from their walls. These flaps form one-way valves that keep blood flowing toward the heart (see Figure 37.13).

Rather than stretching and contracting elastically, like arteries, the relatively thin walls of venules and veins can expand and contract over a relatively wide range, allowing them to act as blood reservoirs as well as conduits. At times, venules and veins may contain from 60 to 80% of the total blood volume of the body. The stored volume is adjusted by skeletal muscle contraction and the valves, in response to metabolic conditions and signals carried by hormones and neurotransmitters.

Although blood pressure in the venous system is relatively low, several mechanisms assist the movement of blood back to the heart. The contraction of skeletal muscles compresses nearby veins, increasing their internal pressure **(Figure 37.16)**. The one-way valves in the veins, especially numerous in the larger veins of the limbs, keep the blood from flowing backward when the muscles relax. Respiratory movements also force blood from the abdomen toward the chest cavity.

Figure 37.16
How skeletal muscle contraction and the valves inside veins help move blood toward the heart.

Erythrocytes

Endothelial cell of capillary wall

Capillary

10 μm

Lennart Nilsson from Behold Man © 1974 by Albert Bonniers Forlag and Little, Brown and Company

Figure 37.15
Erythrocytes moving through a capillary that is just wide enough to admit the cells in single file.

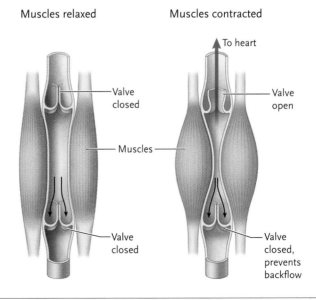

Muscles relaxed   Muscles contracted

To heart

Valve closed   Valve open

Muscles

Valve closed   Valve closed, prevents backflow

37.5 Maintaining Blood Flow and Pressure

Arterial blood pressure is the principal force moving blood to the tissues. Blood pressure must be regulated carefully so that the brain and other tissues receive adequate blood flow, but not so high that the heart is overburdened, risking damage to blood vessels. The three main regulators of blood pressure are

- *cardiac output* (the pressure and amount of blood pumped by the left and right ventricles),
- the degree of constriction of the blood vessels (primarily the arterioles), and
- the total blood volume.

The autonomic nervous system and the endocrine system interact to coordinate the mechanisms controlling these factors. The system effectively counteracts the effects of constantly changing internal and external conditions, such as movement from rest to physical activity or ending a period of fasting by eating a large meal. In humans, for example, moderate physical activity results in an increase in blood flow to the heart itself by 360%, to the muscles of the skin by 370% (increases loss of heat), and to the skeletal muscles by 1060%. Flow is decreased to the digestive tract and liver by 60%, to the kidneys by 40%, and to the bone and most other tissues by 30%. Only the blood flow to the brain remains unchanged. These changes are the result of cardiac output together with adjustments to the muscles in the arterioles supplying the various organs.

37.5a Cardiac Output Is Controlled by Regulating the Rate and Strength of the Heartbeat

Regulation of the strength and rate of the heartbeat starts at stretch receptors called *baroreceptors* (a type of mechanoreceptor; see Chapter 34), located in the walls of blood vessels. The baroreceptors in the cardiac muscle, aorta, and carotid arteries (which supply blood to the brain) are the most crucial. By detecting the amount of stretch of the vessel walls, baroreceptors constantly provide information about blood pressure, sending signals to the medulla within the brain stem. In response, the brain stem sends signals to the heart (primarily the SA node) and muscles of the blood vessels via the autonomic nervous system (see Chapter 33). The sympathetic system, using norepinephrine, stimulates the heart, whereas the parasympathetic system uses acetylcholine to slow the rate. These signals adjust the rate and force of the heartbeat: the heart beats more slowly and contracts less forcefully when arterial pressure is above normal levels, and it beats more rapidly and contracts more forcefully when arterial pressure is below normal levels.

The O_2 content of the blood, detected by chemoreceptors in the aorta and carotid arteries, also influences cardiac output. If O_2 concentration falls below normal levels, the brain stem integrates this information with the baroreceptor signals and issues signals that increase the rate and force of the heartbeat. Too much O_2 in the blood has the opposite effect, reducing cardiac output.

37.5b Hormones Regulate Both Cardiac Output and Arteriole Diameter

Hormones secreted by several glands contribute to the regulation of blood pressure and flow. For example, as part of the stress response, the adrenal medulla reinforces the action of the sympathetic nervous system by secreting epinephrine and norepinephrine into the bloodstream (see Chapter 35). Epinephrine in particular raises blood pressure by increasing the strength and rate of the heartbeat and stimulating vasoconstriction (decrease in diameter) of arterioles in some parts of the body, including the skin, gut, and kidneys. At the same time, by inducing vasodilation (increase in diameter) of arterioles that deliver blood to the heart, skeletal muscles, and lungs, epinephrine increases blood flow to these structures.

37.5c Local Controls Also Regulate Arteriole Diameter

Several automated mechanisms also operate locally to increase the flow of blood to body regions engaged in increased metabolic activity. Repeated contraction of the muscles of your legs during an extended uphill bike ride produces local low O_2 and high CO_2 concentrations because of the increased oxidation of glucose and other fuels. This increases vasodilation of the arterioles and hence the blood supply serving the muscles. At least part of the vasodilation is caused by nitric oxide (NO) produced by arterial endothelial cells. NO is broken down quickly after its release, ensuring that its effects are local.

Life on the Edge discusses aquatic animals and their specialized circulatory systems that allow deep and prolonged dives.

1. What are the three factors by which blood pressure and flow are controlled?
2. How do O_2 and CO_2 concentrations affect local blood flow?
3. What are baroreceptors, and how do they affect circulation?

Taking a Dive

Reptiles, birds, and mammals evolved as land animals, but many species (e.g., snakes, turtles, penguins, loons, otters, seals, and whales) have taken up life in freshwater or marine environments and feed under water. This may involve prolonged dives, sometimes to considerable depths and for extended times. One of the physiological responses to diving is a slowing of the heart rate, which comes into play as soon as the face or equivalent of the animal is wet. This "diving reflex" is a characteristic of all birds and mammals, including humans, but it is more prominent among those animals that dive for their food. Many of these species do not dive to great depths and remain submerged for a relatively brief period that still exceeds the abilities of land dwellers.

Whales dive to very great depths and may remain submerged for more than an hour. The champion divers are sperm whales, *Physeter macrocephalus* **(Figure 1).** These gigantic

Figure 1
A sperm whale resting between dives.

animals (up to 18 m in length and weighing up to 40 t [t = tonne, or 1000 kg]) feed on schools of deep water squid and other animals that live near the bottom of the ocean. The whales can dive to depths of 2000 m (typically less than 1000 m) and remain submerged for as long as 90 minutes. Of course, air-breathing animals swimming to and at such depths encounter several problems. There is the enormous pressure: for each 10 m of depth, the pressure increases by about 1 atmosphere, so these animals are experiencing pressure changes of up to 200 atmospheres! They require a great deal of muscular effort to swim to such depths. The supply of oxygen to the muscles, the brain, and the other tissues needs to be maintained throughout the dive. How does a whale "hold its breath" for more than an hour while swimming actively?

The diving reflex is particularly well developed in marine mammals. It involves a slowing of the heart rate by about 90% and a redirection of blood flow, by adjustment of arteriole muscles, away from skin and digestive organs and toward the heart, brain, and skeletal muscles. The reduction in blood flow to the skin in particular permits the temperature of that organ to drop and reduces its metabolic rate.

Sperm whales have gigantic heads (macrocephalus means large head), which accommodate large quantities of an oil, ambergris, that may help maintain neutral buoyancy, reducing the muscular effort required for swimming. As a whale descends and the pressure increases, the specially adapted rib cage collapses; the lungs, which are not large in whales, shrink; and the air contained in them is compressed. Thus, whales do not "hold their breath": the oxygen that they require is stored in the blood and muscles. Whales, particularly sperm whales, have more erythrocytes per unit of blood than other animals, and these erythrocytes are larger. Moreover, whale muscles contain far more myoglobin (see *Molecule Behind Biology*) than terrestrial mammals. Thus, they are able to store oxygen in the blood and muscles. More than 40% of the oxygen taken in at the surface is stored in muscle, whereas a human diver is able to store only about 13% in muscle.

When the whale returns to the surface, the air in the lungs expands and circulation returns to the surface condition. CO_2 accumulated during the dive is exhaled during the familiar "blow," and the animal breathes on the surface for about 10 minutes before diving again to feed.

37.6 The Lymphatic System

Under normal conditions, a little more fluid from the blood plasma in the capillaries enters the interstitial fluid than is reabsorbed into the plasma. The **lymphatic system** is an extensive network of vessels that collect excess interstitial fluid and return it to the venous blood **(Figure 37.17, p. 908).** Interstitial fluid picked up by the lymphatic system is called **lymph.** This system also collects fats that have been absorbed from the small intestine and delivers them to the blood circulation. The lymphatic system is also a key component of the immune system (see Chapter 44).

37.6a Vessels of the Lymphatic System Extend throughout Most of the Body

Vessels of the lymphatic system collect lymph and transport it to *lymph ducts* that empty into the veins of the circulatory system. *Lymph capillaries,* the smallest vessels of the system, are distributed throughout the body, intermixed intimately with the capillaries of the circulatory system. Although they are several times larger in diameter than the blood capillaries, the walls of lymph capillaries also consist of a single layer of endothelial cells resting on a basement membrane. Interstitial fluid becomes lymph when it enters the lymph capillaries at sites in their walls where the endothelial cells overlap, forming a flap that is forced

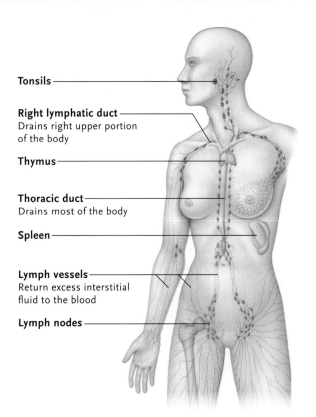

Tonsils

Right lymphatic duct
Drains right upper portion
of the body

Thymus

Thoracic duct
Drains most of the body

Spleen

Lymph vessels
Return excess interstitial
fluid to the blood

Lymph nodes

Figure 37.17
The human lymphatic system.
Patches of lymphoid tissue in the small intestine and in the appendix are also part of the lymphatic system.

through veins. Over a day, the human lymphatic system returns about 3 to 4 L of fluid to the bloodstream. In fishes, amphibians, and reptiles, the lymphatic vessels have lymphatic hearts, regions of the ducts equipped with striated muscle that propels the lymph through the vessels.

37.6b Lymphoid Tissues and Organs Act as Filters and Participate in the Immune Response

Tissues and organs of the lymphatic system include the *lymph nodes,* the *spleen,* the *thymus,* and the *tonsils.* They play primary roles in filtering viruses, bacteria, damaged cells, and cellular debris from the lymph and bloodstream and in defending the body against infection and cancer. Lymphoid tissue also occurs in other regions of the body, particularly the digestive tract, where patches of lymph cells can be found beneath the epithelium of the intestine, in the colon, and in the appendix.

Lymph nodes are small, bean-shaped organs spaced along the lymph vessels and clustered along the sides of the neck, in the armpits and groin, and in the centre of the abdomen and chest cavity (see Figure 37.17). Spaces in nodes contain macrophages, a type of leukocyte that engulfs and destroys cellular debris and infecting bacteria and viruses in the lymph. The lymph nodes also contain other leukocytes that produce antibodies that aid in the destruction of invading pathogens (see Chapter 44).

open by the higher pressure of the interstitial fluid. The openings are wide enough to admit all components of the interstitial fluid, including bacteria, damaged cells, cellular debris, and **lymphocytes.**

Lymph capillaries merge into *lymph vessels,* which contain one-way valves that prevent the lymph from flowing backward. Lymph vessels lead to the thoracic duct and the right lymphatic duct (see Figure 37.17), which empty the lymph into a vein beneath the clavicles (collarbones).

Breathing movements and movements of skeletal muscles adjacent to lymph vessels help move lymph through the vessels, just as they help move the blood

STUDY BREAK

1. What are the three functions of the lymphatic system?
2. What are the main organs of the lymphatic system?

UNANSWERED QUESTIONS

Can gene therapy be used to treat genetic diseases? Haemophilia is a disease in which normal blood clotting is inhibited because the sufferer lacks one of the proteins important in the pathway leading to the formation of clots. It is a sex-linked genetic disease associated with the X chromosome (see Chapter 12). Therapies in the past have included regular transfusions of normal blood concentrates, but that process carries with it the risk of transmitting viruses, including hepatitis and AIDS. People with haemophilia lack either factor VIII or factor IX of the blood-clotting proteins. The cloning of the genes for these two proteins has made recombinant proteins available for injection, but these are expensive. An ideal solution would have the haemophiliac

produce his own clotting factors as the result of some form of genetic technology.

Dr. David Lillicrap at Queen's University is leading two approaches. The first involves the delivery of the gene to the haemophiliac's cells using a vector virus. The virus is stripped of most of its genes, the clotting factor genes are added, and the virus is injected. The results from mice have been promising. In the second approach, Dr. Lillicrap leads a team in the Stem Cell Network, a large international group of researchers with headquarters at the University of Ottawa. They will insert the gene for factor VIII into stem cells from haemophiliac dogs. The cells will then be cultured to greatly increase the numbers before returning them to the dog.

Review

Go to CENGAGENOW™ at http://hed.nelson.com/ to access quizzing, animations, exercises, articles, and personalized homework help.

37.1 Animal Circulatory Systems: An Introduction

- Those invertebrates with a simple internal structure (sponges, cnidarians, flatworms, and nematodes) have no specialized circulatory systems.

- Animals with circulatory systems have a muscular heart that pumps a specialized fluid, such as blood, from one body region to another through tubular vessels. The blood carries O_2 and nutrients to body tissues and carries away CO_2 and wastes.

- Animal circulatory systems are either open or closed. In an open system, the heart pumps haemolymph into vessels that empty into body spaces. The haemolymph collects in other vessels to be returned to the heart. In a closed system, the blood is confined in blood vessels throughout the body and does not mix directly with the interstitial fluid. Closed systems circulate the blood at higher pressures and allow more rapid distribution of O_2 and nutrients and clearance of CO_2 and wastes.

- In the protostome invertebrates, open circulatory systems occur in arthropods and most molluscs, whereas closed circulatory systems occur in annelids and in cephalopod molluscs (squids and octopuses).

- In vertebrates, the circulatory system has evolved from a heart with a single series of chambers, pumping blood through a single circuit to the gills and then on to the brain and other organs, to a double heart in birds and mammals that pumps blood through separate pulmonary and systemic circuits.

- The changes in the structure of the heart are accompanied by alterations in the six segmental aortic arches that originally served the gills of the chordate ancestor.

37.2 Blood and Its Components

- Mammalian blood is a fluid connective tissue consisting of erythrocytes, leukocytes, and platelets suspended in a fluid matrix, the plasma.

- Plasma contains water, ions, dissolved gases such as O_2 and CO_2, glucose, amino acids, lipids, vitamins, hormones, and plasma proteins. The plasma proteins include albumins, which transport substances through the blood or act as enzymes; globulins, which transport lipids and include antibodies; and fibrinogen, which takes part in the clotting reaction.

- Erythrocytes contain haemoglobin, which transports O_2 between the lungs and all body regions. Erythrocytes also transport some CO_2 from interstitial fluid to the lungs and contribute to the reactions maintaining blood pH. They are formed by division of stem cells in the red bone marrow.

- Leukocytes engulf cellular debris and dead and diseased cells, and they defend the body against infecting pathogens. Leukocytes are produced by division of stem cells in the red bone marrow and by division of existing leukocytes.

- Platelets are functional cell fragments that trigger clotting reactions at sites of damage to the circulatory system.

37.3 The Heart

- The mammalian heart is a four-chambered pump. Two atria at the anterior of the heart pump the blood into two ventricles at the posterior of the heart, which pump blood into two separate pulmonary and systemic circuits of blood vessels.

- In both circuits, the blood leaves the heart in large arteries, which branch into smaller arteries, the arterioles. The arterioles deliver the blood to capillary networks, where substances are exchanged between the blood and the interstitial fluid. Blood is collected from the capillaries in small veins, the venules, which join into larger veins that return the blood to the heart.

- Contraction of the ventricles pushes blood into the arteries at a peak pressure, the systolic pressure. Between contractions, the blood pressure in the arteries falls to a minimum pressure, the diastolic pressure. The systole–diastole sequence is the cardiac cycle.

- Contraction of the atria and ventricles is initiated by signals from the sinoatrial (SA) node (pacemaker) of the heart. The signals move over the atria and then activate cells in the atrioventricular (AV) node. From there, the signals move along Purkinje fibres to trigger contraction of the ventricles starting at the bottom of the heart and moving upward.

37.4 Blood Vessels of the Circulatory System

- The walls of arteries consist of an inner endothelial layer, a middle layer of smooth muscle, and an outer layer of elastic fibres. By expanding during systole and rebounding during diastole, the elastic arteries ensure that blood flow is continuous.

- Capillary walls consist of a single layer of endothelial cells and their basement membrane. Blood flow through capillaries is controlled by contraction and relaxation of the smooth muscles of arterioles and precapillary sphincters.

- In the capillary networks, the rate of blood flow is considerably slower than in arteries and veins. This maximizes the time for exchange of substances between blood and tissues. Two major mechanisms drive the exchange of substances: diffusion along concentration gradients and bulk flow.

- Venules and veins have thinner walls than arteries, allowing the vessels to expand and contract over a wide range. As a result, they act both as blood reservoirs and conduits.

- The return of blood to the heart is aided by pressure exerted on the veins when surrounding skeletal muscles contract and by respiratory movements, which force blood from the abdomen toward the chest cavity. One-way valves in the veins prevent the blood from flowing backward.

37.5 Maintaining Blood Flow and Pressure

- Blood pressure and flow are regulated by controlling cardiac output, the degree of constriction of blood vessels (primarily arterioles), and the total blood volume. The autonomic nervous system and the endocrine system interact to coordinate these mechanisms.

- Regulation of cardiac output starts with baroreceptors, which detect blood pressure changes in the large arteries and veins and send signals to the medulla of the brain stem. In response, the brain stem sends signals via the autonomic nervous system that alter the rate and force of the heartbeat.

- Hormones secreted by several glands contribute to the regulation of blood pressure and flow. Epinephrine increases blood pressure by increasing the strength and rate of the heartbeat and stimulating vasoconstriction of arterioles in some areas of the body.

- Local controls respond primarily to O_2 and CO_2 concentrations in tissues. Low O_2 and high CO_2 concentration cause dilation of arteriole walls, increasing the arteriole diameter and blood flow. High O_2 and low CO_2 concentrations have the opposite effects. Nitric oxide released by arterial endothelial cells acts locally to increase arteriole diameter and blood flow.

37.6 The Lymphatic System

- The lymphatic system is an extensive network of vessels that collect excess interstitial fluid, which becomes lymph, and return it to the venous blood. The system also collects fats absorbed from the small intestine and delivers them to the blood circulation, and it is a key component of the immune system.

- The tissues and organs of the lymphatic system include the lymph nodes, the spleen, the thymus, and the tonsils. They remove viruses, bacteria, damaged cells, and cellular debris from the lymph and bloodstream and defend the body against infection and cancer.

Questions

Self-Test Questions

1. Which circulatory system best describes the animal?
   a. Squids and octopuses have open circulatory systems with ventricles that pump blood away from the heart.
   b. Fishes have a single-chambered heart with an atrium that pumps blood through gills for oxygen exchange.
   c. Amphibians have the most oxygenated blood in the pulmocutaneous (leading to and from the lungs and skin) circuit and the most deoxygenated blood in the systemic circuit.
   d. Amphibians and reptiles use a two-chambered heart to separate oxygenated and deoxygenated blood.
   e. Birds and mammals pump blood to separate pulmonary and systemic systems from two separate ventricles in a four-chambered heart.

2. A characteristic of blood circulation through or to the mammalian heart is that
   a. the superior vena cava conveys blood to the head.
   b. the inferior vena cava conveys blood to the right atrium.
   c. the pulmonary arteries convey blood from the lungs to the left atrium.
   d. the pulmonary veins convey blood into the left ventricle.
   e. the aorta branches into two coronary arteries that convey blood from heart muscle.

3. Which of the following is correct? (Any number of answers from a–e may be correct.)
   a. Carotid arteries supply the brain with oxygenated blood.
   b. Vertebrate hearts are neurogenic.
   c. Red blood cells in vertebrates contain no nuclei.
   d. The mammalian aorta is derived from aortic arch VI on the left side.
   e. Erythropoietin is secreted by red blood cells.

4. Characteristics of veins and venules are
   a. thick walls.
   b. large muscle mass in walls.
   c. a large quantity of elastin in the walls.
   d. low blood volume compared with arteries.
   e. one-way valves to prevent backflow of blood.

5. To increase cardiac output,
   a. the adrenal medulla and sympathetic nervous system secrete epinephrine and norepinephrine.
   b. baroreceptors in the brain signal the sympathetic nerves.
   c. the brain stem signals the baroreceptors, causing the heart to beat faster.
   d. the autonomic nervous system responds to low oxygen detected by chemoreceptors and decreases the force of the heartbeat.
   e. chemoreceptors, stimulated by excessive blood oxygen, increase the rate of the heartbeat.

Questions for Discussion

1. *Aplastic anaemia* develops when certain drugs or radiation destroy red bone marrow, including the stem cells that give rise to erythrocytes, leukocytes, and platelets. Predict some symptoms a person with aplastic anaemia would be likely to develop. Include at least one symptom related to each type of blood cell.

2. Haemoglobin occurs in the blood of many animals, but in many invertebrates, it occurs free in the plasma. What advantages can you think of that might have led to its incorporation in erythrocytes during evolution of vertebrates? The nonvertebrate chordates do not have haemoglobin in their blood, but they do have cells in their blood that have a role in defence, suggesting that erythrocytes may have originated from such cells. What evidence could be used to examine this hypothesis?

3. In some people, the pressure of the blood pooling in the legs leads to a condition called *varicose veins*, in which the veins stand out like swollen, purple knots. Explain why this might happen and why veins closer to the leg surface are more susceptible to the condition than those in deeper leg tissues.

Acropora millepora.

ZEOvit.com

38 Animal Reproduction

WHY IT MATTERS

Over a few nights each year along the Great Barrier Reef off the east coast of Australia, many species of reef-building corals synchronously spawn, releasing eggs and sperm into the water. Timing is important for the corals because they rely on external fertilization. Male and female gametes must be released at the same time to maximize the chances of fertilization.

The circadian clocks of the corals control the reproductive event, which is synchronized to the lunar cycle. The massed spawning occurs over several nights after the full moon. Spawning is triggered by changes in lunar irradiance intensity. *Acropora millepora* (as seen in the opening photograph) is one coral involved in the mass reproductive event, providing a spectacular example of photosensitive responses.

A. millepora and other corals are sensitive to blue light (see Chapter 1) that entrains the circadian clocks of insects and mammals. The light sensors, called cryptochromes (CRYs), are DNA photolyaselike receptor proteins (see *Molecule Behind Biology*). Until 2007, cryptochromes had been reported only in higher animals (vertebrates and insects), and related but different proteins have now been found in plants and eubacteria. In *A. millepora*, the gene *cry2* showed increased

Cryptochromes

In 1881, Charles Darwin reported that the growth of plants toward the sun (heliotropism) could be eliminated by filtering blue wavelengths (380 to 500 nm) from the light reaching the plant. Cryptochromes **(Figure 1)** is the name given to the photoreceptors sensitive to blue light—blue, ultraviolet-A receptors. The prefix "crypto-" was used because the identity of the pigments was unknown for some time.

We now know that many organisms respond to blue light and that the absorption spectrum of flavins is similar to the action spectrum of blue light. Repair of DNA is mediated by photolyases, which are flavoproteins. Cryptochromes also regulate the circadian clocks in animals and plants. The operation of cryptochromes depends on both the *cry1* and *cry2* genes for normal expression of circadian behavioural rhythms in organisms ranging from corals to *Arabidopsis* (a small plant related to mustard) and from *Drosophila* to mammals.

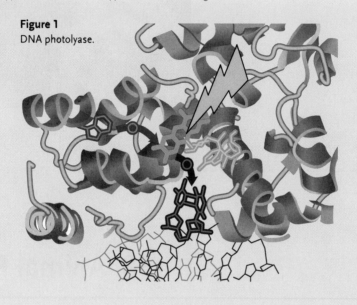

Figure 1
DNA photolyase.

expression on full moon versus new moon nights. These findings suggest that cryptochromes underlie synchronization of mass spawning events in many other invertebrates.

Many species of corals release eggs and sperm at the same time, but chemicals in the egg coatings and in the sperm **acrosomes** interact to ensure fertilization of eggs only by sperm of the same species.

38.1 The Drive to Reproduce

Organisms have a strong drive to reproduce, and most will go to considerable lengths to ensure that their genes are represented in future generations. The behaviour of some parasitic worms provides a telling example.

Acanthocephalan worms (*Moniliformis dubius*; **Figure 38.1**) are dioecious parasites that live in the alimentary systems of vertebrates. When laboratory rats are experimentally infected with juvenile acanthocephalan worms, the worms occupy different parts of a rat's gastrointestinal tract. Females live in the carbohydrate-rich anterior region just posterior to the stomach, whereas males migrate anteriorly. Females reach sexual maturity more or less synchronously in a 20-cm long portion of intestine. In the experimental system, males had access to many females, and each individual inseminated, on average, eight females.

Acanthocephalan worms are long-lived and presumably have many opportunities to mate. The density of these worms can be quite high, perhaps forcing males to compete for access to females. The male testes empty into the **vas deferens** and then into the *cirrus,* which terminates in a bursa or eversible cup. When copulating, a male wraps his everted bursa around the posterior end of the female so that the bursa can enter the female's gonopore (vagina). After mating, a male uses cement from glands in front of his testes to seal closed

Robert Poulin

Figure 38.1
An acanthocephalan worm, *Moniliformis dubius.*

(or cap) the female's gonopore. This behaviour protects the male's investment in the female, perhaps ensuring that his sperm fertilizes her eggs and effectively outcompetes other males. Presumably, the plug is lost when the female releases her eggs. This is an example of the role that accessory glands (in this case cement glands) can play in reproduction.

Male acanthocephalan worms take other steps to outcompete other males for access to females. In addition to attempting to copulate with females, male *M. dubius* attempt to mate with other males, this time using the cement gland secretions to render the other males' genitalia inoperable.

Some variations in the mating and reproductive behaviour and biology of animals are best understood in the context of individuals protecting their investments in mate selection and production of young.

STUDY BREAK

How do acanthocephalan worms protect their investment in mate choice? Why?

38.2 Asexual and Sexual Reproduction

Reproduction is the means of passing on an individual's genes to a new generation, making it the most vital function of living organisms. In **asexual reproduction**, a single individual gives rise to offspring with no genetic input from another individual. In **sexual reproduction**, male and female parents produce zygotes (fertilized eggs) through the union of egg and sperm.

38.2a Asexual Reproduction: Reproduction without Recombination

Many aquatic invertebrates and some terrestrial annelids and insects reproduce asexually. This mode of reproduction is much less common among vertebrates. In asexual reproduction, from one to many cells of a parent's body develop directly into a new individual. Cells involved in asexual reproduction in animals are usually produced by mitosis, less commonly by meiosis. When cells involved in asexual reproduction are produced by mitosis, the resulting offspring are genetically identical to one another and to the parent (clonal reproduction).

Genetic uniformity of offspring can be advantageous in stable, uniform environments. In these cases, successful individuals with the "best" combinations of genes perpetuate the most competitive genotypes through asexual reproduction. Individuals do not have to expend energy to produce gametes or find a mate. Asexual reproduction is also advantageous to individuals

living in sparsely settled populations or to sessile animals.

In animals, asexual reproduction involving mitosis occurs by three basic mechanisms: *fission, budding,* and *fragmentation.* In **fission**, the parent splits into two or more offspring of approximately equal size. Some species of planarians (Platyhelminthes) reproduce asexually by fission, dividing transversely or longitudinally. In **budding**, a new individual grows and develops while attached to the parent. Sponges, tunicates, and some cnidarians reproduce asexually by budding, and offspring may break free from the parent or remain attached to form a *colony.* In the cnidarian *Hydra,* an offspring buds and grows from one side of the parent's body and then detaches to become a separate individual **(Figure 38.2).** Often in corals, buds remain attached when their growth is complete, forming colonies of thousands of interconnected individuals. In **fragmentation**, pieces separate from a parent's body and develop (*regenerate*) into new individuals. Many species of cnidarians, flatworms, annelids, and some echinoderms can reproduce by fragmentation.

Some animals produce offspring by **parthenogenesis** (*parthenos* = virgin; *genesis* = birth), which is the growth and development of an unfertilized egg. Offspring produced by parthenogenesis may be haploid or diploid, depending on the species. When the eggs involved in parthenogenesis are produced by meiosis, the offspring are not genetically identical to the parent or to each other.

Parthenogenesis occurs in some invertebrates, including certain aphids, water fleas, bees, and crustaceans. In bees, haploid drones (males) are produced parthenogenetically from unfertilized eggs produced by reproductive females (queens). New queens and sterile workers develop from fertilized eggs. Parthenogenesis also occurs in some vertebrates (e.g., certain fishes, salamanders, amphibians, lizards, and

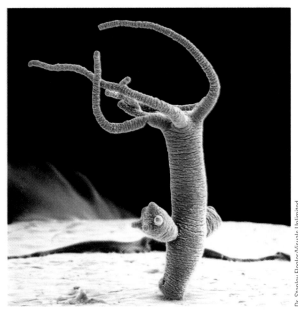

Figure 38.2
Asexual reproduction by budding in *Hydra* species.

Dr. Stanley Flegler/Visuals Unlimited

turkeys). In these animals, an egg, produced by meiosis, typically doubles its chromosomes to produce a diploid cell that begins development. In species in which females have two identical sex chromosomes, the offspring are female, whereas in single-sex species, in which males have two identical sex chromosomes, the offspring are males. All whip-tail lizard (*Cnemidophorus* spp.; **Figure 38.3**) species consist of females produced by parthenogenesis. These females still go through the motions of mating with each other.

In 2008, Eugene A. Gladyshev and two colleagues reported that bdelloid rotifers (see Chapter 26) contain many genes that apparently originated in other organisms (bacteria, fungi, and plants). Horizontal gene transfer—the capture and functional assimilation of exogenous genes—may be important in the evolution of bdelloid rotifers and be a factor allowing them to circumvent overt recombination during reproduction.

38.2b Sexual Reproduction: Add Recombination

Animals reproduce sexually through the union of **sperm** (motile gamete) and **eggs** (nonmotile gametes), both produced by **meiosis**. The overriding advantage of sexual reproduction is the generation of genetic diversity among offspring. Genetic diversity increases the chances that at least some offspring will be better adapted to local conditions and be more likely to survive and reproduce. Genetic recombination and independent assortment of chromosomes are two mechanisms of meiosis that give rise to genetic diversity in eggs and sperm (see Chapter 9). Genetic recombination mixes the alleles of parents into new combinations within chromosomes. Independent assortment randomly combines maternal and paternal chromosomes in the gamete nuclei. Additional variability is generated at fertilization when eggs and sperm from genetically different individuals fuse together at random to initiate the development of new individuals. Adding to the effects of genetic recombination and independent assortment, random mutations in DNA are the ultimate source of variability for both sexual and asexual reproduction.

Sexual reproduction can be a disadvantage because of the costs in energy and raw materials associated with producing gametes and finding mates. Finding mates can expose animals to predation and may conflict with the need to find food and shelter and caring for existing offspring.

STUDY BREAK

What are the advantages of sexual reproduction over asexual reproduction?

38.3 Cellular Mechanisms of Sexual Reproduction

The cellular mechanisms of sexual reproduction include **gametogenesis**, the formation of male and female gametes, and **fertilization**, the union of gametes that initiate development of a new individual. Mating is the pairing of a male and a female for sexual reproduction.

38.3a Gametogenesis: Production of Gametes

Gametes in most animals are formed from **germ cells**, cell lines set aside early in embryonic development that remain distinct from **somatic cells** of the body. During development, germ cells collect in gonads, specialized gamete-producing organs: **testes** (singular, testis) in males and **ovaries** in females. Mitotic division of germ cells produces **spermatogonia** (singular, spermatogonium) in males and **oogonia** (singular, oogonium) in females. These cells then undergo meiosis to produce gametes **(Figure 38.4)**. In some animals, germ cells give rise to families of cells that assist gamete development.

Meiosis reduces the number of chromosomes from diploid to haploid. Somatic cells have two copies of each chromosome; gametes have one. Fertilization, the fusion of a haploid sperm and a haploid egg, restores the diploid condition and produces a **zygote** or fertilized egg, the first cell of a new individual.

At the beginning of meiosis, developing gametes start as **spermatocytes** or **oocytes** and by the end are *spermatids* or *ootids*. When meiosis is complete, haploid cells develop into mature sperm cells (**spermatozoa**; singular, spermatozoon = sperm) or eggs (**ova**; singular, ovum). Spermatogenesis is the process of producing sperm; oogenesis is the process of producing eggs. Sperm are specialized to move toward, contact, and penetrate eggs. Sperm must be numerous, so each spermatid produces four sperm. Mobility means small size, a flagellum **(Figure 38.5, p. 916)** for propulsion, and mitochondria for energy. Sperm have nuclei that carry the male's chromosomal contributions to the zygote. The *acrosome* allows sperm to enter the egg.

Although sperm are usually smaller than eggs, they show considerable range in size, for example, from

Figure 38.3
A whip-tail lizard (*Cnemidophorus deppei*), in which all individuals are female.

Nature's Images, Inc./Photo Researchers, Inc.

a. Spermatogenesis

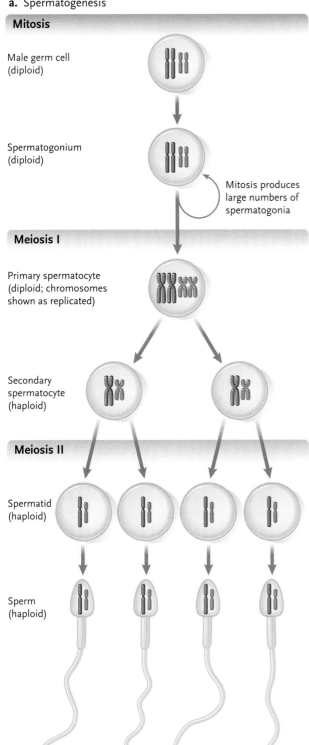

b. Oogenesis

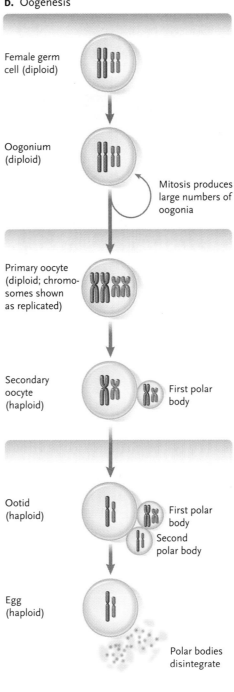

Figure 38.4
The mitotic and meiotic divisions that produce eggs and sperm from germ cells. **(a)** Spermatogenesis. **(b)** Oogenesis. The first polar body may or may not divide, depending on the species, so that either two or three polar bodies may be present at the end of meiosis. Two are shown in this diagram.

4.5 to 16.5 μm long among *Drosophila* species. The size of individual sperm influences the speed at which they swim (longer sperm move more quickly than shorter sperm). But producing longer sperm may reduce total sperm production and limit a male's reproductive success because sperm must be numerous to maximize the chances of fertilization. In mammals, variations in sperm size, volume of ejaculate, and sperm density often reflect mating behaviour (see Chapter 40).

Spermatogenesis produces haploid cells specialized to deliver their nuclei to conspecific eggs. Two meiotic divisions produce four haploid spermatids (see Figure 38.4a) that each develop into a mature sperm (see Figure 38.5). During sperm maturation, most of the cytoplasm is lost. Mitochondria are in the cytoplasm around the base of the flagellum. These mitochondria produce ATP, the energy source for beating of the flagellum. At the opposite end of the sperm, the acrosome is a specialized secretory vesicle forming a cap over the nucleus. The acrosome contains enzymes and other

Figure 38.5

Spermatozoa.
(a) Photomicrograph of human sperm and **(b)** the structure of a sperm.

a. Human sperm

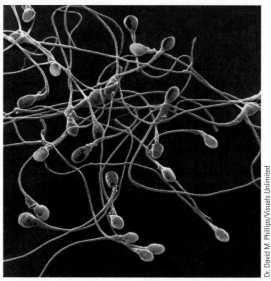

Dr. David M. Phillips/Visuals Unlimited

b. Sperm structure

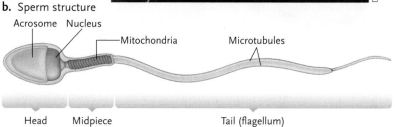

Acrosome   Nucleus   Mitochondria   Microtubules

Head   Midpiece   Tail (flagellum)

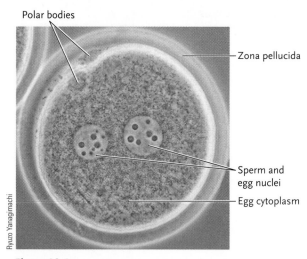

Polar bodies

Zona pellucida

Sperm and egg nuclei

Egg cytoplasm

Ryuzo Yanagimachi

Figure 38.6

A mature hamster egg that has been fertilized. The sperm and egg nuclei are about to fuse together.

proteins that help the sperm attach to and penetrate the surface coatings of an egg of the same species.

Oogenesis produces eggs. Only one of the cell products of meiosis develops into a functional egg, with that cell retaining almost all of the cytoplasm of the parent. The other products form polar bodies (see Figure 38.4b). Unequal cytoplasmic divisions concentrate nutrients and other molecules required for development in the egg. In most species, polar bodies eventually disintegrate and do not contribute to fertilization or embryonic development.

Meiosis in the oogonia of most animals can be very protracted, taking as long as from embryonic development until fertilization of the next generation's zygote. Within a few weeks after a female mammal is born, her oocytes stop developing at the end of the first meiotic prophase. Oocytes remain in the ovary in this stage until the female is sexually mature. Then one to several oocytes advance to metaphase of the second meiotic division and are released from the ovary. The timing of egg release varies among mammals according to species and season. As in other animals, mammalian eggs complete meiosis at fertilization to produce the fully mature egg **(Figure 38.6)**. In humans, some oocytes may remain in prophase of the first meiotic division for 50 years.

The egg typically has specialized features, including stored nutrients required for at least the early stages of embryonic development, as well as one or more kinds of coatings that protect the egg from mechanical injury and infection. In some species, egg coats protect the embryo immediately after fertilization and prevent penetration by more than one sperm.

Egg coats are surface layers added during oocyte development or fertilization. The **vitelline coat** (the **zona pellucida** in mammals; see Figure 38.6) is a gel-like matrix of proteins, glycoproteins, and/or polysaccharides lying immediately outside the plasma membrane of the egg cell. Insect eggs have additional outer protein coats forming a hard, water-impermeable layer that prevents desiccation. In amphibians and some echinoderms, egg jelly forms the outer coat protecting the egg from desiccation.

In birds, reptiles, and monotremes (see Chapter 27), egg white, a thick solution of proteins, surrounds the vitelline coat. Outside the white is the *shell* of the egg, which is flexible and leathery in reptiles and mineralized and brittle in birds. Both egg white and shell are added while the egg (fertilized or not) moves along the **oviduct**, the tube connecting the ovary to the outside of the body.

In the mammalian ovary, the egg is surrounded by **follicle cells** during its development. Follicle cells grow from ovarian tissue and nourish the developing egg. They also make up part of the zona pellucida while the egg is in the ovary and remain as a protective layer after it is released.

Mature eggs can be the largest cells in an animal (see Figure 27.44). Mammalian eggs are microscopic, with few stored nutrients, because the embryo develops inside the mother and is supplied with nutrients by her body. The eggs of birds are huge because they contain all of the nutrients required for complete embryonic development. The bird egg includes the "yolk," which contains the nutrients for the developing egg, the ovum or egg cell, and the white. Regardless of size, cytoplasm makes up most of the volume of an animal egg and the nucleus of the egg is usually microscopic or nearly so.

38.3b Fertilization: Union of Egg and Sperm

Eggs and sperm are delivered from the ovaries and testes to the site of fertilization by oviducts (females) and sperm ducts (males). In many species, external accessory sex organs participate in the delivery of gametes. The basic design of vertebrate and invertebrate reproductive systems **(Figure 38.7)** is similar. Nonmotile eggs move through oviducts on currents generated by the beating of cilia that line the oviducts or by contractions of the oviducts or the body wall.

Fertilization may be external (outside the body of either parent) in a watery medium or internal in a watery fluid inside a female's body. **External fertilization** occurs in most aquatic invertebrates, bony fishes, and amphibians. Sperm and eggs are shed into the surrounding water. Sperm swim until they collide with an egg of the same species. The process is helped by synchronization of the release of eggs and sperm and by the enormous numbers of gametes released (see *Why It Matters*). In animals such as sea urchins and amphibians, sperm are attracted to eggs by diffusible attractant molecules released by the egg.

Most amphibians, even terrestrial species such as toads, mate in an aquatic environment. Frogs typically mate by a reflex response called *amplexus,* in which the male clasps the female tightly around the body with his forelimbs **(Figure 38.8, p. 918).** Amplexus stimulates the female to shed a mass of eggs into the water through the *cloaca.* The cloaca is the cavity into which intestinal, urinary, and genital tracts empty in reptiles, birds, amphibians, and many fishes. As the eggs are released, they are fertilized by sperm released by the male.

Internal fertilization is widespread in animals, and is not restricted to mammals. Internal fertilization occurs in invertebrates such as annelids, some arthropods, and some molluscs and in vertebrates from fishes and salamanders to reptiles, birds, and mammals. In internal fertilization, sperm are released by the male close to or inside the entrance to the female's reproductive tract. Sperm swim through fluids in the reproductive tract until one reaches and fertilizes an egg. In some species, molecules released by the egg attract the sperm. Internal fertilization involves copulation, which occurs when a male's accessory sex organ (e.g., a penis) is inserted into a female's accessory sex organ (e.g., a vagina). Internal fertilization makes terrestrial life possible because the female's body provides the aquatic medium required for fertilization without the danger of gametes drying when exposed to the air. Effecting internal fertilization means close contact between individuals.

Male sharks and rays use a pair of modified pelvic fins as accessory sex organs that channel sperm directly into the female's cloaca. Male reptiles, birds, and mammals also use accessory sex organs to place sperm directly inside the reproductive tract of females, where fertilization takes place. In reptiles and birds, sperm fertilize eggs as they are released from the ovary and

a. *Drosophila* (fruit fly)

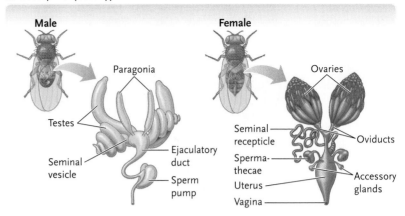

b. Amphibian (frog)

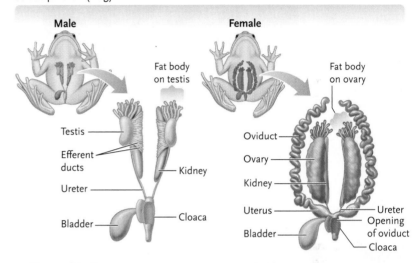

c. Mammal (cat)

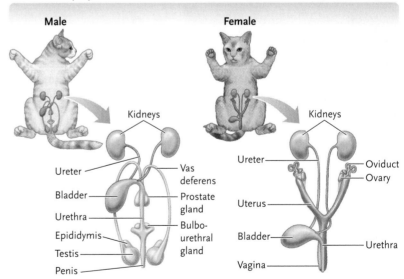

Figure 38.7
Some reproductive systems. **(a)** An insect, *Drosophila* (fruit fly). **(b)** An amphibian, a frog. **(c)** A mammal, a cat. Female systems are shown in blue and male systems in yellow.

travel through the oviducts, before the shell is added. In mammals, the penis delivers sperm into the female's vagina, which is a specialized structure for

Figure 38.8

A male leopard frog (*Rana pipiens*) clasping a female during a mating embrace known as **amplexus.** The tight squeeze by the male frog stimulates the female to release her eggs, which can be seen streaming from her body, embedded in a mass of egg jelly. Sperm released by the male fertilize the eggs as they pass from the female.

Hans Pfletschinger

— Eggs

reproduction. Fertilization takes place when a sperm meets an egg in the oviducts.

Once a sperm touches the outer surface of an egg of the same species **(Figure 38.9a),** receptor proteins in the sperm plasma membrane bind the sperm to the vitelline coat or zona pellucida **(Figure 38.9b,** step 4).

In most animals, only a conspecific sperm is recognized and binds to the egg surface. Species recognition between sperm and eggs is particularly important in animals using external fertilization because water surrounding the egg may contain sperm from many different species. This aspect is less important in species using internal fertilization where structural adaptations and behavioural patterns usually limit sperm transfer from males to females of the same species (see *Variations in Internal Fertilization*).

After initial attachment of sperm to egg, the events of fertilization proceed in rapid succession (see Figure 38.9b). The acrosome of the sperm releases its contents, including enzymes that dissolve a path through the egg coats. The sperm, with its tail still beating, follows the path until its plasma membrane touches and fuses with the egg's plasma membrane (step 5). Fusion introduces the sperm nucleus into the egg cytoplasm and activates the egg to complete meiosis and begin development.

38.3c   Polyspermy: Keeping Sperm Out of the Fertilized Egg

Protection against polyspermy (more than one sperm fertilizing an egg) is widespread in the animal kingdom. Two mechanisms help prevent polyspermy: a

a. Sperm adhering to egg

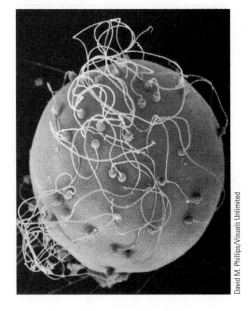

David M. Phillips/Visuals Unlimited

b. Steps in fertilization

1 A sperm contacts the jelly layer of the egg.

2 The acrosomal reaction begins.

3 Acrosomal enzymes dissolve a path through the jelly layer.

4 Proteins in its plasma membrane bind the sperm to the vitelline coat.

5 The sperm lyses a hole in the vitelline coat. The sperm and egg plasma membranes fuse.

6 Membrane depolarization produces the fast block to polyspermy.

7 The sperm nucleus and centriole enter the egg. The sperm nucleus then fuses wtih the egg nucleus.

8 Cortical granules discharge their contents, producing the slow block to polyspermy.

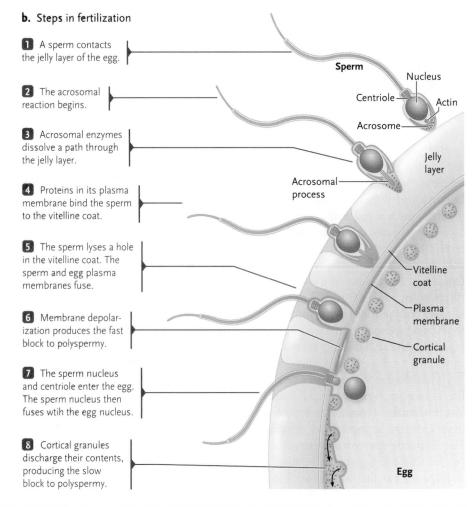

Sperm

Nucleus
Centriole
Actin
Acrosome

Acrosomal process

Jelly layer

Vitelline coat

Plasma membrane

Cortical granule

Egg

Figure 38.9

Fertilization. **(a)** Sperm adhering to the surface coat of a sea urchin egg. Of the many sperm that may initially adhere to the outer surface of the egg, usually only one accomplishes fertilization. **(b)** Steps of fertilization in a sea urchin.

Variations in Internal Fertilization

Internal fertilization benefits animals in several ways. The placement of sperm in the female's reproductive system maximizes the chances of fertilization. Specializations of the male and female genitalia increase the precision with which sperm is delivered, whereas a combination of behavioural and structural (often referred to as "lock-and-key") arrangements minimizes the chances of mistaken identity (cross-species matings). Internal fertilization can give both males and females more control over which sperm fertilize which eggs.

Invertebrate animals such as bedbugs, insects in the family Cimicidae, have well-developed genitalia but practise "traumatic insemination." This involves the male bedbug (e.g., *Cimex lectularius*) piercing the female's abdominal wall with his external genitalia and delivering sperm directly into her coelom (body cavity). The male always pierces the female's body at the spermalege, consisting of an ectospermalege and a mesospermalege, where the sperm are deposited **(Figure 1).**

Females are wounded during traumatic insemination, and the spermalege appears to be an anatomical counterstrategy to minimize the damage associated with this form of mating. "Last male precedence," the fact that the sperm of the last male to mate fertilizes the eggs, appears to be the main benefit of this approach to mating. The spermalege also reduces the chances of infection by pathogens transferred in the mating process.

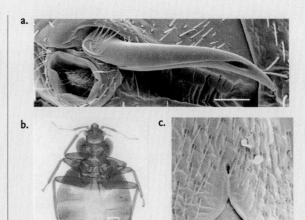

Figure 1

(a) The intromittant organ (penis) of *Cimex lectularius* (scale bar 0.1 μm) is shown with **(b)** a ventral view of the whole animals and **(c)** the ectospermalege, an incurving in one of the female's sternites that guides the male's intromittant organ (scale bar 1.5 μm).

fast block that works within seconds of fertilization and a **slow block** that works in minutes. In invertebrate species such as the sea urchin, fusion of egg and sperm opens ion channels in the egg's plasma membrane, spreading a wave of electrical depolarization over the egg surface (much like a nerve impulse travelling along a neuron). Depolarization alters the egg plasma membrane so that it cannot fuse with any additional sperm, eliminating the possibility that more than one set of paternal chromosomes enters the egg. This fast block occurs within a few seconds of fertilization.

The fast block depends on a change in the egg's membrane potential from negative to positive. It is not established when the membrane potential of a sea urchin egg is experimentally kept at a negative value. In this case, additional sperm fuse with the plasma membrane. Fertilization was entirely blocked if the membrane was kept positive before sperm contact.

In vertebrates, the wave of membrane depolarization following sperm–egg fusion is not as pronounced as it is in sea urchins and does not prevent additional sperm from fusing with the egg. However, additional sperm nuclei that enter the egg cytoplasm usually break down and disappear, so only the first sperm nucleus to enter fuses with the egg nucleus.

In both invertebrates and vertebrates, fusion of egg and sperm also triggers the release of stored calcium (Ca^{2+}) ions from the endoplasmic reticulum into the cytosol. Ca^{2+} ions activate control proteins and enzymes that initiate intense metabolic activity in the fertilized egg, including a rapid increase in cellular oxidations and synthesis of proteins and other molecules.

Ca^{2+} ions also cause **cortical granules** to fuse with the egg's plasma membrane and release their contents to the outside (see Figure 38.9b, step 8). Enzymes released from cortical granules alter the egg coats within minutes of fertilization, so no further sperm can attach to or penetrate the egg. This process is the slow block to polyspermy.

The importance of Ca^{2+} to cortical granule release has been demonstrated experimentally. Granules are released in unfertilized eggs if Ca^{2+} is added experimentally to the cytoplasm. Conversely, if chemicals that bind Ca^{2+} are added to the cytoplasm of unfertilized eggs, the concentration of Ca^{2+} cannot rise and cortical granule release does not occur after fertilization.

After the sperm nucleus enters the egg cytoplasm, microtubules move the sperm and egg nuclei together in the egg cytoplasm until they fuse. The chromosomes of egg and sperm nuclei then assemble together and enter mitosis. The subsequent, highly programmed events of embryonic development convert the fertilized egg into an individual capable of independent existence.

The paternal chromosomes, the microtubule organizing centre, and one or two centrioles (see Chapter 9) are the only components of sperm to survive in the egg. Therefore, almost all cytoplasmic structures of the embryo and of the new individual are maternal in origin.

On the Road to Vivipary

Although mammals could be considered the archetypical viviparous animals, not all living mammals are viviparous. Furthermore, living therian mammals (marsupials and placentals) do not all have the same kind of placenta. We have no data about whether early mammals such as *Castorocauda* (see Figures 19.8 and 19.9) were viviparous. At least 50 gene loci regulate the development of the placenta in "placental" mammals such as mice and rats. Included are gene families that produce protein hormones and hemoglobin. Some of these genes and gene families are adapted to fetal development.

Guppies, familiar aquarium fish, (see Ch. 45 *Life Histories of Guppies*)and other related fish in the genus *Poeciliopsis* exhibit a range of reproductive patterns. Some species are oviparous. Females in other species retain eggs in their bodies after fertilization and give birth to live young with no further provisioning, whereas the females of still other species develop a "follicular pseudoplacenta" that functions much like the placenta of a mouse (or a human). The level of maternal investment (matrotrophic index) ranges from none to extensive **(Figure 1)**. A phylogeny of *Poeciliopsis* and other species **(Figure 2)** shows repeated evolution of placentalike structures in these fish.

Vivipary also appears in different evolutionary lines of elasmobranchs (sharks, rays, and holocephalians; see Chapter 27). The young of various rays and tiger sharks (*Galaeocerdo cuvier*) develop inside the mother and are nourished by

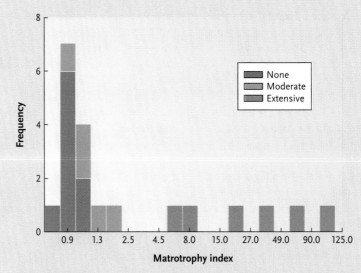

Figure 1

A matrotrophy index (MI) results when values are assigned to the levels of investment that females make in their young. Matrotrophy indices range from low (ovipary) to high (some ovovivipary and vivipary), and there are intermediate conditions. There is a range of MI indices for species in the genus *Poeciliopsis*, illustrating the range of conditions. Low MI values are the most common (highest frequency values), but there are significant differences among species. In this figure, species with the same colour do not differ significantly in MI values.

her, but there is no evidence of a placenta. Species such as grey nurse sharks (*Ginglymostoma cirratum*), mako sharks (*Isurus oxyrinchus*), porbeagle sharks (*Lamna nasus*), and thresher sharks (*Alopias* species) are oophagous. This means that while pregnant, the females of these species of sharks continue to produce large numbers of small eggs that are consumed by the developing embryos. At least one species of fossil holocephalan chondrichthyian fish (*Delphyodontos dacriformes*) appears to have been oophagous. The

evidence comes from the well-developed slashing and piercing teeth of this lower Carboniferous fossil **(Figure 3)**.

By retaining developing young in their bodies, adults (usually females) better protect them from predators, provide an appropriate environment for development, and, in many cases, ensure an adequate food supply. As a recurring theme in animals, viviparity is of little value in determining phylogeny (evolutionary relationships). We explore this subject in more detail when we consider development in Chapter 39.

The centrioles of the new individual are normally of paternal origin.

38.3d Patterns of Development: Moving from Zygote to Complete Organism

Ovipary, vivipary, and ovovivipary are three patterns of embryonic support. **Oviparous** animals lay eggs, whereas viviparous and ovoviviparous animals bear live young. The eggs of oviparous animals contain all of the nutrients necessary for development of the embryo outside the mother's body. In **viviparous** animals, the mother's body provides nutrients and oxygen to and

removes wastes from the developing embryo. In **ovoviviparous** animals, eggs with yolk are retained in the mother's body, but embryonic development and growth are supported by the yolk (see *On the Road to Vivipary*).

Viviparous animals (*vivi* = alive) retain the embryo within the mother's body and nourish it during at least early embryo development. All living mammals except monotremes are viviparous. Vivipary occurs in other vertebrate groups except crocodilians, turtles, and birds. An exceptionally well-preserved Devonian fossil placoderm (*Materpiscis attenboroughi*) reveals that vivipary is an ancient trait in vertebrates. This fossilized female was connected by an umbilical cord to one embryo in

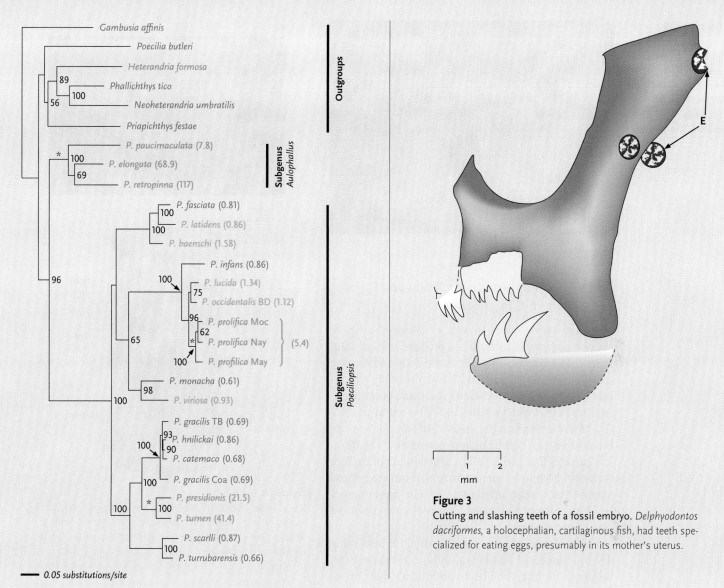

Figure 2

Matrotrophy index values are shown for species in a phylogeny (phylogram) that illustrates proposed evolutionary relationships among species in the genus *Poeciliopsis* and other species. This phylogeny demonstrates that different evolutionary lineages of guppylike fish have independently achieved different levels of association between mothers and their developing young. The colour codes as per MI indices in Figure 1 in *Variations in Internal Fertilization*.

Figure 3

Cutting and slashing teeth of a fossil embryo. *Delphyodontos dacriformes*, a holocephalian, cartilaginous fish, had teeth specialized for eating eggs, presumably in its mother's uterus.

her uterus. Another fossilized female placoderm from the same deposit contains three embryos.

In viviparous animals, embryonic development takes place in the **uterus** (womb), a specialized, saclike organ. There are two basic groups of viviparous mammals. In placental mammals (Eutheria), a specialized temporary structure called the placenta facilitates the transfer of nutrients from the mother's blood to the embryo and of wastes in the opposite direction. The young are born with fully developed limbs. In marsupials (Metatheria), the young are born at an earlier stage before their hindlimbs have developed. Metatherians have a placenta, but it is derived from a different tissue than that of eutherians.

The metatherian placenta provides nutrients to the embryo from an attached membranous sac containing yolk, but only for the early stages of its development. In marsupials such as kangaroos, koalas, wombats, and opossums, the tiny newborn crawls from the opening of the birth canal to its mother's pouch (marsupium), where it attaches to a nipple and completes its development **(Figure 38.10, p. 922).**

Ovovivipary (adjective, **ovoviviparous**) is common in some fishes, lizards, and amphibians; many snakes; and many invertebrates. No uterus or placenta is involved. When development is complete, the eggs hatch inside the mother and the young are released to the exterior.

Figure 38.10
Developing offspring of a marsupial mammal, an opossum (*Didelphis virginiana*), attached to a nipple in the marsupium (pouch) of its mother.

John Cancalosi/Peter Arnold, Inc.

38.3e Hermaphroditism: Producing Eggs and Sperm in One Individual

Hermaphroditic (from *Hermes + Aphrodite,* a Greek god and goddess) individuals can produce both eggs and sperm. **Hermaphroditism** is more common among sponges (Porifera), cnidaria, flatworms (Platyhelminthes), earthworms, land snails, and some other invertebrates than it is in humans and other mammals.

Most hermaphroditic individuals do not fertilize themselves. Self-fertilization is prevented by anatomical barriers that preclude introduction of sperm into their own body or by mechanisms that cause eggs and sperm to mature at different times. The prevention of self-fertilization maintains the genetic variability of sexual reproduction.

Simultaneous hermaphrodites are individuals that develop functional ovaries and testes at the same time. Sequential hermaphrodites are individuals that change from one sex to the other. Earthworms, as shown in **Figure 38.11**, are a good example of **simultaneous hermaphroditism**. The only known vertebrate simultaneous hermaphrodites are hamlets (genus *Hypoplectrus*), a group of predatory sea basses **(Figure 38.12)**. **Sequential hermaphroditism** occurs in many invertebrates (for example, some crustaceans) and some ectothermic vertebrates. Well-known examples include the genus *Amphiprion,* in which, in some species, the initial sex is male (as in clownfish), whereas in others, it is female. In still other species, individuals start as males, turn into females, and then turn back into males.

STUDY BREAK

1. When does meiosis occur in animal life cycles? How does this compare with the situation in plants?
2. Why do the eggs of animals differ so much in size? What are the implications for development?
3. What are the benefits of internal fertilization?

a.

Robin Chittenden; Frank Lane Picture Agency/Corbis

b. Sex organs

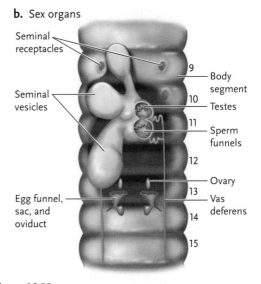

Seminal receptacles

Seminal vesicles

Egg funnel, sac, and oviduct

9 — Body segment
10 — Testes
11 — Sperm funnels
12
13 — Ovary
14 — Vas deferens
15

Figure 38.11
Simultaneous hermarphroditism in the earthworm. **(a)** Copulation by a mating pair of earthworms, in which each individual releases sperm that fertilizes the eggs in its partner. **(b)** The sex organs in the earthworm.

Andrew J. Martinez/Photo Researchers, Inc.

Figure 38.12
Hypoplectrus gummigutta, a predatory sea bass that is a simultaneous hermaphrodite.

38.4 Sexual Reproduction in Humans

Reproductively, humans are typical eutherian (placental) mammals. Males and females each have a pair of gonads (testes or ovaries). As in other vertebrates, gonads serve a dual function, producing gametes and secreting hormones responsible for sexual development and mating behaviour (see Chapters 35, 39, and 40).

38.4a Females: Produce Eggs, Get Pregnant, Lactate

Human females have a pair of ovaries suspended in the abdominal cavity. An oviduct leads from each ovary to the uterus, which is hollow with walls that contain smooth muscle. The uterus is lined by the endometrium, formed by layers of connective tissue with embedded glands and richly supplied with blood vessels. If an egg is fertilized and begins development, it must implant in the endometrium to continue developing. The lower end of the uterus, the cervix, opens into a muscular canal, the vagina, which leads to the exterior. Sperm enter the female reproductive tract via the vagina, and at birth, the baby passes from the uterus to the vagina to the outside.

The **vulva**, external female sex organs (genitalia), surround the opening of the vagina **(Figure 38.13).** Two folds of tissue, the **labia minora**, run from front to rear on either side of the opening. Labia minora are partially covered by **labia majora**, a pair of fleshy, fat-padded folds that also run from front to rear on either side of the vagina. At the anterior end of the vulva, the labia minora join to partly cover the **clitoris**, a bulblike erectile organ with the same embryonic origins as the penis. Two **greater vestibular glands** open near the entrance to the vagina and secrete a mucus-rich fluid that lubricates the vulva. The urethra that conducts urine from the bladder to the outside opens between the clitoris and the vaginal opening. Most nerve endings associated with erotic sensations are concentrated in the clitoris and the labia minora and around the opening of the vagina. When a human female is born,

a thin flap of tissue, the **hymen**, partially covers the opening of the vagina. This membrane, if it has not already been ruptured by physical exercise or other disturbances, is broken during the first sexual intercourse.

At birth, each ovary contains about 1 million oocytes whose development is arrested at the end of the first meiotic prophase. Although 200 000 to 380 000 oocytes survive until a female reaches sexual maturity, only about 380 are actually ovulated, released as immature eggs into the abdominal cavity and pulled into the nearby oviduct by the current produced by the beating of the cilia that line the oviduct. The cilia propel the egg along the oviduct and into the uterus. Fertilization usually occurs in the oviduct.

In most vertebrates, **ovulation**, the release of the egg from the ovary, usually occurs during a well-defined mating season, the time of year when males and females are *fertile*—physiologically and behaviourally ready to reproduce. The timing of mating seasons is usually under the general control of day length (photoperiod), with some adjustment for local weather. This pattern ensures that the young are born at a time of year when food is plentiful. Humans, however, do not show any evidence of a mating season. Mating and fertilization can occur at any time of the year. Furthermore, ovulation appears to be cryptic in humans, meaning that women do not know when they are ovulating, nor do their partners.

Reproduction in human females is under neuroendocrine control, involving complex interactions between the hypothalamus, pituitary, ovaries, and uterus. The *ovarian cycle* occurs from puberty to menopause and involves the events in the ovaries leading to the release of a mature egg approximately every 28 days. The ovarian cycle is coordinated with the **uterine cycle** or **menstrual cycle** (*menses* = month), events in

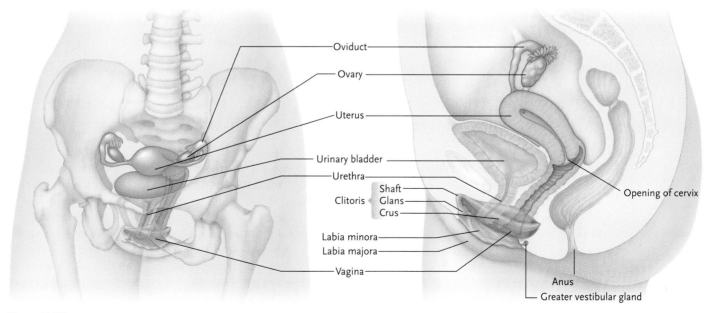

Figure 38.13
The reproductive organs of a human female.

Delaying Reproduction

Many animals have a distinct reproductive season. Its timing is often triggered by changes in photoperiod that herald the changing seasons of the year or by lunar events (e.g., *Acoropora millepora*). Other animals are more opportunistic. Sea turtles and seals are two examples of animals that separate the acts of mating from the timing of egg laying or birth. Both sea turtles and seals must go ashore to lay eggs or give birth to their young, but for the rest of the year, the animals lead largely pelagic lives. In sea turtles **(Figure 1)**, males gather off the beaches where females come to lay their eggs and mate with females en route to the beach. But males and females also mate during chance meetings in the open ocean. Females can store sperm for up to five years so that they can be ready to lay eggs and fertilize them (from their supply of stored sperm) at any time.

Many species of seals also disassociate the act of mating from, in their case, birth (also known as parturition). Males defend territories on beaches where females haul out to give birth. Females undergo a postpartum estrus, so they are ready to mate (fertile) immediately after giving birth. The seals mate, the egg is fertilized, and the zygote is formed, but implantation is delayed for several months, and the young are born a year later. The time between mating and birth is considerably longer than the gestation period (the time needed for growth and development of the fetus). This approach to reproduction maximizes the chances of males and females finding mates.

Bats in the families Rhinolophidae and Vespertilionidae separate the acts of copulation and ovulation. The gestation periods in these species are about 60 days. In temperate regions, species in both families mate in late summer and early autumn **(Figure 2)**. Females store the sperm **(Figure 3)** in their uteri and then enter hibernation. Ovulation and fertilization occur when females leave hibernation in the spring, and the young are born when spring is well advanced.

Fertilization followed by delays in development or implantation can allow males and females more control over mate choice through interactions among sperm or between sperm and the females' reproductive tract.

Figure 2
A pair of little brown bats, *Myotis lucifugus*, mate in an abandoned mine in southern Ontario. The male is on the female's back.

Delays achieved by sperm storage and postponement of ovulation raise possibilities of competition between sperm or other mechanisms for selecting the sperm that fertilizes the egg.

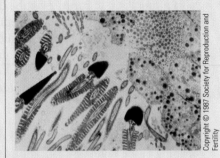

Figure 3
Sperm stored in the uterus of a female *M. lucifugus.*

Figure 1
A pair of green turtles (*Chelonia mydas*) mate in the waters of Tortuguero in Costa Rica.

---

the uterus that prepare it to implant the egg if fertilization occurs.

The beginning of the ovarian cycle **(Figure 38.14)** is stimulated by the release of gonadotropin-releasing hormone (GnRH) by the hypothalamus. GnRH stimulates the pituitary to release follicle-stimulating hormone (FSH) and luteinizing hormone (LH) into the bloodstream **(Figure 38.15a, p. 926)**. FSH stimulates 6 to 20 oocytes in the ovaries to begin meiosis. As oocytes develop, they become surrounded by cells that form a follicle (the ovum and follicle cells) (see Figure 38.14, step 1, and Figure 38.15a, day 2). During this phase, the follicle grows and develops and, at its largest size, becomes filled with fluid and may be 12 to 15 mm in diameter. Usually, only one follicle develops

to maturity with release of the egg (secondary ooctye) by ovulation. Multiple births can result if two or more follicles develop and their eggs ovulate in one cycle.

As the follicle enlarges, FSH and LH interact to stimulate estrogen (female sex hormone, primarily estradiol) secretion by follicular cells. Initially, estrogens are secreted in low amounts and have a negative feedback effect on the pituitary, inhibiting secretion of FSH. As a result, FSH secretion declines briefly. But estrogen secretion increases steadily, and its level peaks about 12 days after the beginning of follicle development **(Figure 38.15c**, day 12). High estrogen level has a positive feedback effect on the hypothalamus and pituitary, increasing secretion of GnRH and stimulating the pituitary to release a burst of FSH and LH.

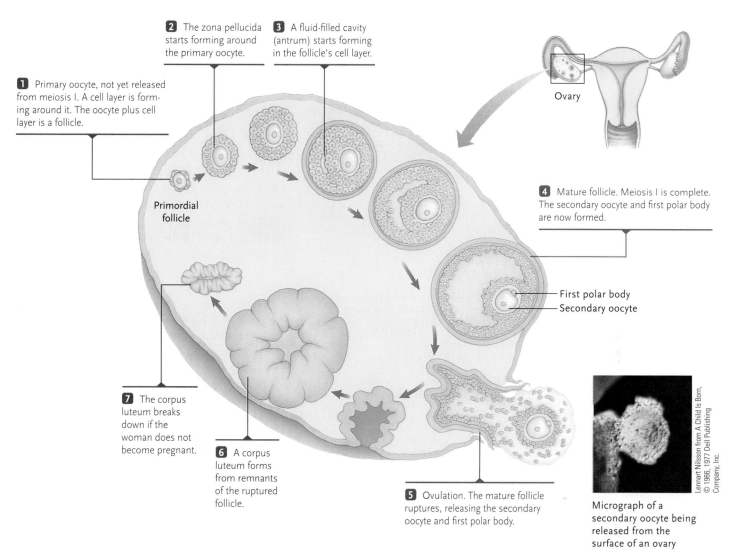

2 The zona pellucida starts forming around the primary oocyte.

3 A fluid-filled cavity (antrum) starts forming in the follicle's cell layer.

1 Primary oocyte, not yet released from meiosis I. A cell layer is forming around it. The oocyte plus cell layer is a follicle.

Ovary

Primordial follicle

4 Mature follicle. Meiosis I is complete. The secondary oocyte and first polar body are now formed.

First polar body
Secondary oocyte

7 The corpus luteum breaks down if the woman does not become pregnant.

6 A corpus luteum forms from remnants of the ruptured follicle.

5 Ovulation. The mature follicle ruptures, releasing the secondary oocyte and first polar body.

Lennart Nilsson from A Child Is Born, © 1966, 1977 Dell Publishing Company, Inc.

Micrograph of a secondary oocyte being released from the surface of an ovary

Figure 38.14
The growth of a follicle, ovulation, and the formation of the corpus luteum in a human ovary.

Increased estrogen levels convert the mucus secreted by the uterus to a thin and watery consistency, making it easier for sperm to swim through the uterus.

Ovulation occurs after the burst in LH secretion stimulates the follicle cells to release enzymes that digest away the wall of the follicle, causing it to rupture and release the egg (see Figure 38.14, step 5). LH also causes the follicle cells remaining at the surface of the ovary to grow into the **corpus luteum**, an enlarged, yellowish structure (*corpus* = body; *luteum* = yellow), initiating the luteal phase. The corpus luteum (see Figure 38.14, step 6) acts as an endocrine gland that secretes estrogens, as well as large quantities of progesterone, a second female sex hormone, and **inhibin**, another hormone. Progesterone stimulates growth of the uterine lining and inhibits contractions of the uterus. Progesterone and inhibin have a negative feedback effect on the hypothalamus and pituitary. Progesterone inhibits secretion of GnRH and, in turn, secretion of FSH and LH by the pituitary. Inhibin specifically inhibits FSH secretion. The fall in FSH and LH levels diminishes the signal for follicular growth, and no new follicles begin to grow in the ovary.

If fertilization does not occur, the corpus luteum gradually shrinks, perhaps because of the low levels of LH. About 10 days after ovulation, the shrinkage has inhibited secretion of estrogen, progesterone, and inhibin. In the absence of progesterone, *menstruation* begins. As progesterone and inhibin levels decrease, FSH and LH secretion is no longer inhibited, and a new monthly cycle begins.

The uterine (menstrual) cycle includes the changes in the uterus over one ovarian cycle. The hormones that control the ovarian cycle also control the menstrual cycle **(Figure 38.15d)**, physiologically connecting the two processes. Day 0 of the monthly cycle is the beginning of follicular development in the ovary **(Figure 38.15b)**, and in the uterus, menstrual flow begins.

Menstrual flow results from the breakdown of the endometrium, which releases blood and tissue breakdown products from the uterus to the outside through the vagina. When the flow ceases (at day 4 to 5 of the

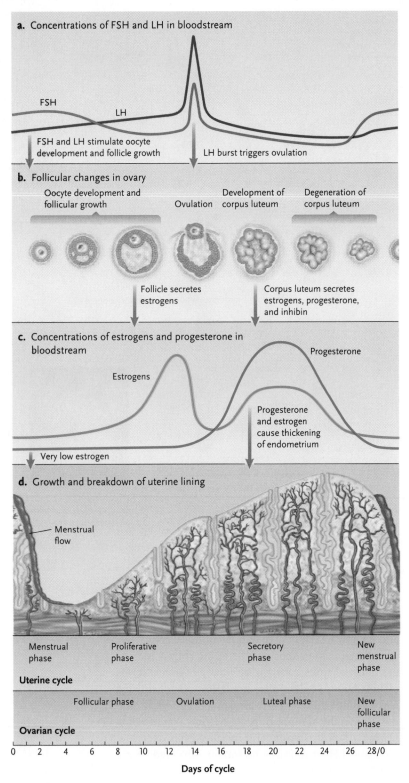

a. Concentrations of FSH and LH in bloodstream

FSH

LH

FSH and LH stimulate oocyte
development and follicle growth

LH burst triggers ovulation

b. Follicular changes in ovary

Oocyte development and
follicular growth

Ovulation

Development of
corpus luteum

Degeneration of
corpus luteum

Follicle secretes
estrogens

Corpus luteum secretes
estrogens, progesterone,
and inhibin

c. Concentrations of estrogens and progesterone in
bloodstream

Estrogens

Progesterone

Progesterone
and estrogen
cause thickening
of endometrium

Very low estrogen

d. Growth and breakdown of uterine lining

Menstrual
flow

Menstrual phase	Proliferative phase	Secretory phase	New menstrual phase

Uterine cycle

Follicular phase	Ovulation	Luteal phase	New follicular phase

Ovarian cycle

0 2 4 6 8 10 12 14 16 18 20 22 24 26 28/0

Days of cycle

Figure 38.15
The ovarian and uterine (menstrual) cycles of a human female. The days of the monthly cycle are given in the scale at the bottom of the diagram. **(a)** The changing concentrations of FSH and LH in the bloodstream, triggered by GnRH secretion by the hypothalamus. **(b)** The cycle of follicle development, ovulation, and formation of the corpus luteum in the ovary. **(c)** The concentrations of estrogens and progesterone in the bloodstream. **(d)** The growth and breakdown of the uterine lining.

cycle), the proliferation phase begins as the endometrium begins to grow again. As the endometrium gradually thickens, oocytes in both ovaries begin to develop further, eventually leading to ovulation (usually a single egg from one ovary) at about 14 days into the cycle. The uterine lining continues to grow for another 14 days after ovulation. This is the secretory phase. At that time, if fertilization has not taken place, the absence of progesterone results in contraction of the arteries supplying blood to the uterine lining, shutting down the blood supply and causing the lining to disintegrate. The menstrual flow begins. Contractions of the uterus, no longer inhibited by progesterone, help expel the debris. Prostaglandins released by the degenerating endometrium add to uterine contractions, making them severe enough to be felt as the pain of "cramps" and sometimes producing other effects, such as nausea, vomiting, and headaches.

Menstruation occurs only in human females and our closest primate relatives, gorillas and chimpanzees. In other mammals, the uterine lining is completely reabsorbed if a fertilized egg does not implant during the period of reproductive activity. The uterine cycle in those mammals is called the *estrous* cycle, and females are said to be *in estrus* when fertile.

38.4b Males: Produce and Deliver Sperm

Organs that produce and deliver sperm comprise the male reproductive system **(Figure 38.16)**. Human males have a pair of testes (singular, testis), suspended in a baglike **scrotum**. Keeping the testes at cooler temperatures than the body core provides an optimal environment for sperm development. Some land mammals such as elephants and monotremes with relatively low body temperatures have internal (cryptic) testes carried within the body. Marine mammals such as whales and dolphins also have internal testes despite relatively high body temperatures. In these animals, countercurrent exchange between cool blood flowing from the tail flukes is delivered to the testes, cooling them enough to allow the production of fertile sperm. In many mammals (e.g., grey squirrels, *Sciurus carolinensis*), the testes descend into the scrotum only during the mating season. Otherwise, they are cryptic, kept in the body captivity, where temperatures are too warm to produce fertile sperm.

In human males, each testicle is packed with about 125 metres of **seminiferous tubules**, in which sperm proceed through all stages of spermatogenesis **(Figure 38.17, p. 928)**. The entire process, from spermatogonium to sperm, takes 9 to 10 weeks, and the testes produce about 130 million fertile sperm each day.

Sertoli cells are supportive cells, completely surrounding the developing spermatocytes in the seminiferous tubules. Sertoli cells supply nutrients to the spermatocytes and seal them off from the body's blood

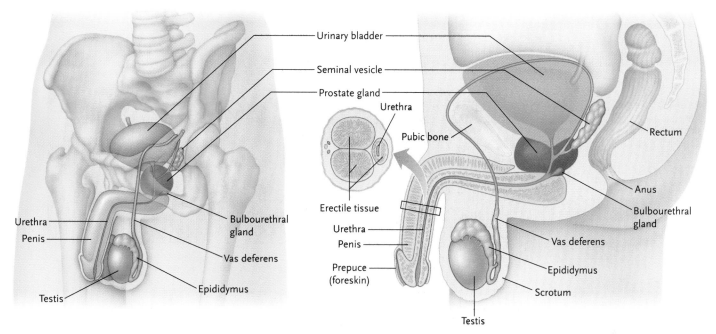

Figure 38.16

The reproductive organs of a human male.

supply. **Leydig cells**, located in the tissue surrounding the developing spermatocytes, produce the male sex hormones (**androgens**), particularly **testosterone.**

Mature sperm flow from seminiferous tubules into the **epididymis,** a coiled storage tubule attached to the surface of each testis. Rhythmic muscular contractions of the epididymis move sperm into a thick-walled, muscular tube, the **vas deferens** (plural, vasa deferentia), which extends through the abdominal cavity. Just below the bladder, the vasa deferentia join the urethra. During ejaculation, muscular contractions force the sperm into the urethra and out of the penis. At this time, the sperm are activated and become motile when they come in contact with alkaline secretions added to the ejaculated fluid by accessory glands.

About 150 to 350 million sperm are released in a single ejaculation. Semen, the ejaculate, is a mixture of sperm and the secretions of several accessory glands. In humans, about two-thirds of the volume is produced by a pair of **seminal vesicles** that secrete seminal fluid, a thick, viscous liquid, into the vasa deferentia near the point where they join with the urethra. Seminal fluid contains prostaglandins that, when ejaculated into the female, trigger contractions of the female reproductive tract that help move the sperm into and through the uterus.

The **prostate gland**, which surrounds the region where the vasa deferentia empty into the urethra, adds a thin, milky fluid to the semen. The alkaline prostate secretion makes up about one-third of the volume of semen, raising its pH (and that of the vagina) to about pH 6, the level of acidity best tolerated by sperm. This pH level also fosters sperm motility. As part of the prostate secretion, a fast-acting enzyme converts the

semen to a thick gel at ejaculation. The thickened consistency helps keep the semen from draining from the vagina when the penis is withdrawn. A second, slower-acting enzyme in the prostate secretion gradually breaks down the semen clot and releases the sperm to swim freely in the female reproductive tract.

Finally, a pair of **bulbourethral glands** secretes a clear, mucus-rich fluid into the urethra before and during ejaculation. This fluid lubricates the tip of the penis and neutralizes the acidity of any residual urine in the urethra. In total, the secretions of the accessory glands make up more than 95% of the volume of semen; less than 5% is sperm.

Most of the interior of the penis is filled with three cylinders of spongelike tissue (corpora cavernosa) that become filled with blood and cause erection during sexual arousal. Although the human penis depends solely on engorgement of spongy tissue for erection, the males of many mammals, including bats, rodents, carnivores, and most other primates, have a baculum or penis bone **(Figure 38.18, p. 928)** that helps maintain the penis in an erect state. The presence of bacula in a species usually coincides with the presence of a baubellum (clitoris bone) in females.

The penis ends in the **glans**, a soft, caplike structure. Most nerve endings producing erotic sensations are crowded into the glans and the region of the penile shaft just behind the glans. The **prepuce** or foreskin is a loose fold of skin that covers the glans (see Figure 38.16). In many human cultures, the foreskin is removed for hygienic, religious, or other ritualistic reasons by **circumcision** ("around cut"). In 2007, the World Health Organization stated that male circumcision is an important strategy to

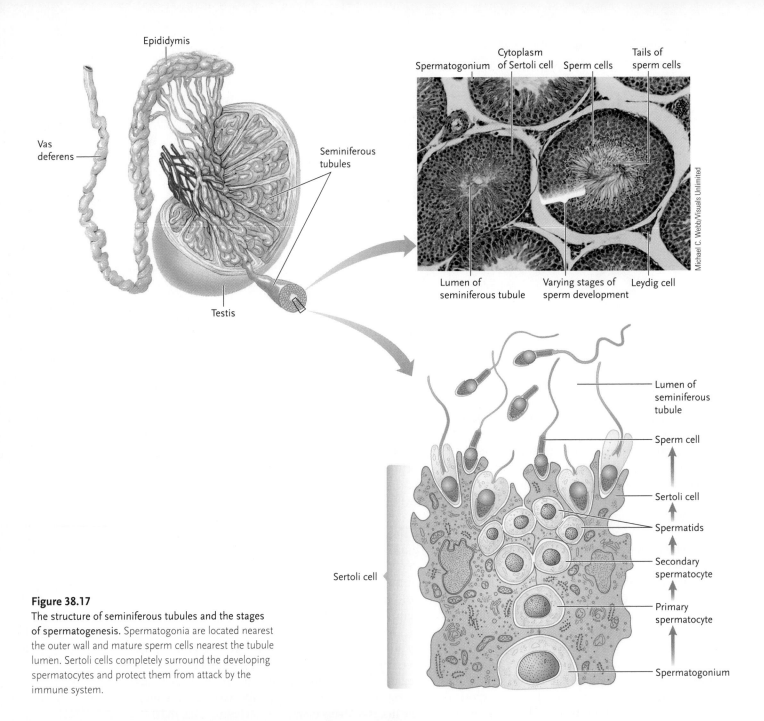

Figure 38.17

The structure of seminiferous tubules and the stages of spermatogenesis. Spermatogonia are located nearest the outer wall and mature sperm cells nearest the tubule lumen. Sertoli cells completely surround the developing spermatocytes and protect them from attack by the immune system.

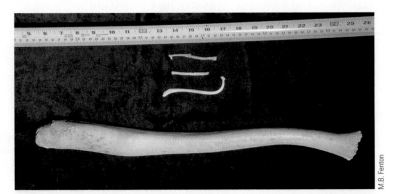

Figure 38.18

The bacula of a wolverine (*Gulo luscus*), a red fox (*Vulpes fulva*), a raccoon (*Procyon lotor*), and a walrus (*Odobenus rosmarus*) (top to bottom).

prevent heterosexually acquired HIV infection in males. Female circumcision, the removal of the labia minora and the clitoris, is often called female genital mutilation or FGM.

Many of the hormones regulating the menstrual cycle, including GnRH, FSH, LH, and inhibin, also regulate male reproductive functions. Testosterone, secreted by the Leydig cells in the testes, also plays a key role **(Figure 38.19)**. In sexually mature males, the hypothalamus secretes GnRH in brief pulses every 1 to 2 hours. GnRH stimulates the pituitary to secrete LH and FSH. LH stimulates the Leydig cells to secrete testosterone, which stimulates sperm production and controls the

John P. Wiebe, Professor Emeritus, University of Western Ontario

What happens to hormones after they are produced? John P. Wiebe, professor emeritus in the Department of Biology at the University of Western Ontario in London, and his graduate students study two steroid hormones derived from progesterone. Specifically, 5α-pregnane-3,20-dione (5αP) and 3α-hydroxy-4-pregen-20-one (3αP) are metabolites of progesterone. 5αP is cancer promoting, stimulating cells to proliferate (tumour growth) and detach (metastasis). 3αP is cancer inhibiting because it suppresses cell proliferation and metastasis.

Wiebe and his colleagues have demonstrated that tumorigenic cells produce higher levels of 5αP and lower levels of 3αP. When cells become tumorigenic, there are strong increases in the mRNA expression of the enzyme that catalyzes conversion of progesterone to 5αP, specifically 5α-reductase. This increase is paralleled by a decrease in 3α-hydroxysteroidoxidoreductase that catalyzes conversion to 3αP.

It is clear that this work has potential in the control and treatment of breast cancer. It also alerts us to the impact of exposure to hormones, whether in the food we eat or in the water we drink (see Box 45.4).

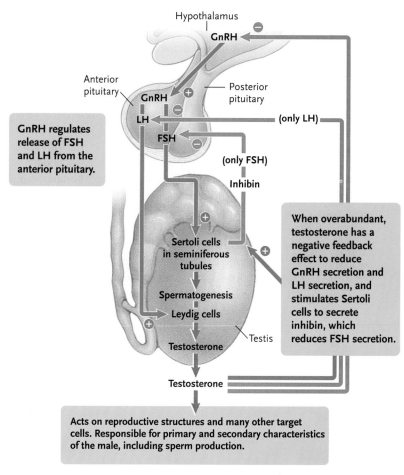

Figure 38.19

Hormonal regulation of reproduction in the male and the negative feedback systems controlling hormone levels.

the concentration of testosterone falls in the bloodstream, the hypothalamus responds by increasing GnRH secretion. If testosterone levels rise too high, the overabundance inhibits LH secretion. An overabundance of testosterone also stimulates Sertoli cells to secrete inhibin, which inhibits FSH secretion by the pituitary. As a result, testosterone secretion by the Leydig cells drops off, returning the concentration to optimal levels in the bloodstream.

When the male is sexually aroused, sphincter muscles controlling the flow of blood to the spongy erectile tissue of the penis relax, allowing the tissue to become engorged with blood (the penis is a hydrostatic skeleton structure; see Chapter 37). As the spongy tissue swells, it maintains the pressure by compressing and almost shutting off the veins draining blood from the penis. The engorgement produces an erection in which the penis lengthens, stiffens, and enlarges. During continued sexual arousal, lubricating fluid secreted by the bulbourethral glands may be released from the tip of the penis.

Female sexual arousal results in enlargement and erection of the clitoris, in a process analogous to erection of the penis. The labia minora become engorged with blood and swell in size, and lubricating fluid is secreted onto the surfaces of the vulva by the vestibular glands. In addition to these changes, the nipples become erect by contraction of smooth muscle

growth and function of male reproductive structures. FSH stimulates Sertoli cells to secrete a protein and other molecules required for spermatogenesis.

Concentrations of male reproductive hormones are maintained by negative feedback mechanisms. If

cells, and the breasts swell in size due to engorgement with blood.

Insertion of the penis into the vagina and the thrusting movements of copulation lead to the reflex actions of ejaculation, including spasmodic contractions of muscles surrounding the vasa deferentia, accessory glands, and urethra. During ejaculation, the sphincter muscles controlling the exit from the bladder close tightly, preventing urine from mixing with the ejaculate. Ejaculation is usually accompanied by *orgasm,* a sensation of intense physical pleasure that is the peak (climax) of excitement for sexual intercourse, followed by feelings of relaxation and gratification.

The motions of copulation stretch the vagina and stimulate the clitoris, sometimes inducing orgasm in females. Vaginal stretching also stimulates the hypothalamus to secrete oxytocin, which induces contractions of the uterus. The contractions keep the sperm in suspension and aid their movement through the reproductive tract. Uterine contractions are also induced by the prostaglandins in the semen.

Sperm reach the site of fertilization in the oviducts within 30 minutes of being ejaculated. Of the millions of sperm released in a single ejaculation, only a few hundred actually reach the oviducts. After orgasm, the penis, clitoris, and labia minora gradually return to their unstimulated size. Females can experience additional orgasms within minutes or even seconds of a first orgasm, but most males enter a *refractory period* lasting 15 minutes or longer before they can regain an erection and have another orgasm.

38.4c Fertilization of Human Eggs: Producing Zygotes

A human egg can be fertilized only during its passage through the third of the oviduct nearest the ovary. If the egg is not fertilized during the 12 to 24 hours that it is in this location, it disintegrates and dies. Sperm do not swim randomly for a chance encounter with the egg. Rather, they first swim up the cervical canal to reach the oviduct and then are propelled up the oviduct by contractions of the oviduct's smooth muscles. There is evidence that eggs release chemical attractant molecules that the sperm recognize, causing them to swim directly toward the egg.

Sperm must first penetrate the layer of follicle cells surrounding the egg, aided by enzymes in the sperm plasma membrane **(Figure 38.20)**. Then the sperm adhere to receptor molecules on the surface of the zona pellucida. This contact triggers the **acrosome reaction**, in which enzymes contained in the acrosome are released from the sperm onto the zona pellucida, where they digest a path to the plasma membrane of the egg. As soon as the first sperm cell reaches the egg, sperm and egg plasma membranes fuse, and the sperm cell is engulfed by the cytoplasm of the egg. Although only one sperm fertilizes the egg, the combined release of acrosomal enzymes from many sperm greatly increases the chance that a complete channel will be opened through the zona pellucida. This is in part why a low sperm count is often a source of male infertility. Low sperm counts can be caused by infection, heat, frequent intercourse, smoking, and excess alcohol consumption.

Membrane fusion activates the egg. The sperm that enters the egg releases nitric oxide that stimulates the release of stored Ca^{2+} in the egg. Ca^{2+} triggers cortical granule release to the outside of the egg. Enzymes from the cortical granules crosslink molecules in the zona pellucida, hardening it and sealing the channels opened by acrosomal enzymes. The enzymes also destroy receptors that bind sperm to the surface of the zona pellucida. As a result, no further sperm can bind to the zona pellucida or reach the plasma membrane of the egg. The Ca^{2+} also triggers the completion of meiosis of the egg. The sperm and egg nuclei then fuse, and the cell is now considered the zygote. Mitotic divisions of the zygote soon initiate embryonic development.

The first cell divisions of embryonic development take place while the fertilized egg is still in the oviduct. About seven days after ovulation, the embryo passes from the oviduct and implants in the uterine lining. During and after implantation, cells associated with the embryo secrete **human chorionic gonadotropin (hCG)**, a hormone that keeps the corpus luteum in the ovary from breaking down. Excess hCG is excreted in the urine; its presence in urine or blood provides the basis of pregnancy tests.

Continued activity of the corpus luteum keeps estrogen and progesterone secretion at high levels, which maintain the uterine lining and prevent menstruation. The high progesterone level also thickens the mucus secreted by the uterus, forming a plug that seals the opening of the cervix from the vagina. The plug keeps bacteria, viruses, and sperm cells from further copulations from entering the uterus.

About 10 weeks after implantation, the placenta takes over the secretion of progesterone, hCG secretion drops off, and the corpus luteum regresses. However, the corpus luteum continues to secrete the hormone *relaxin,* which inhibits contraction of the uterus until near the time of birth.

STUDY BREAK

1. How does the reproductive pattern of human females differ from that of other mammals? (See also Chapters 27, 46, and 49.)
2. Which hormones are good predictors of ovulation?
3. What role does follicle-stimulating hormone play?

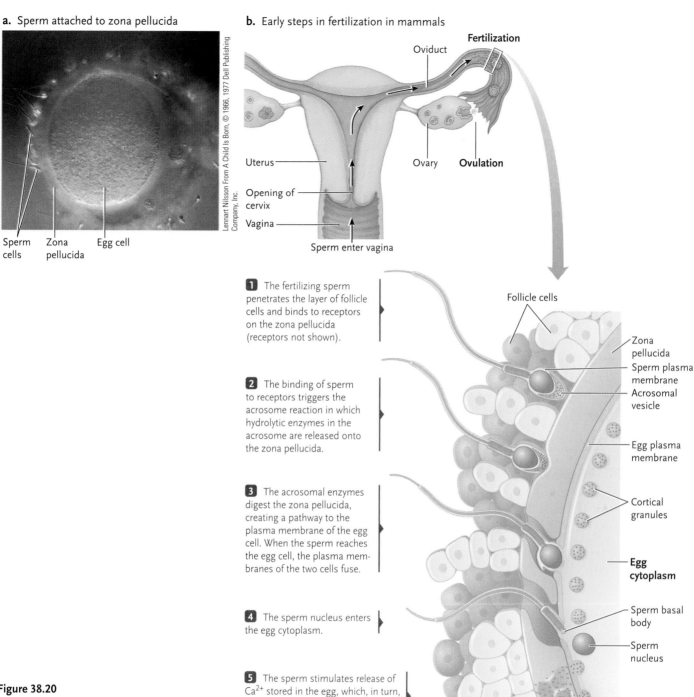

a. Sperm attached to zona pellucida

b. Early steps in fertilization in mammals

Fertilization

Oviduct

Ovulation

Uterus

Opening of cervix

Vagina

Ovary

Sperm enter vagina

Lennart Nilsson From A Child Is Born, © 1966, 1977 Dell Publishing Company, Inc.

Sperm cells

Zona pellucida

Egg cell

1 The fertilizing sperm penetrates the layer of follicle cells and binds to receptors on the zona pellucida (receptors not shown).

2 The binding of sperm to receptors triggers the acrosome reaction in which hydrolytic enzymes in the acrosome are released onto the zona pellucida.

3 The acrosomal enzymes digest the zona pellucida, creating a pathway to the plasma membrane of the egg cell. When the sperm reaches the egg cell, the plasma membranes of the two cells fuse.

4 The sperm nucleus enters the egg cytoplasm.

5 The sperm stimulates release of Ca^{2+} stored in the egg, which, in turn, triggers the cortical reaction, leading to the slow block in polyspermy.

Follicle cells

Zona pellucida

Sperm plasma membrane

Acrosomal vesicle

Egg plasma membrane

Cortical granules

Egg cytoplasm

Sperm basal body

Sperm nucleus

Figure 38.20
Fertilization in mammals. **(a)** Sperm attached to the zona pellucida of a human egg cell. **(b)** Early steps in the fertilization process.

38.5 Controlling Reproduction

Knowledge about the details of reproduction can allow us to control fertility. In some cases, this means increasing the chances of reproducing, whereas in other cases, it means minimizing them. In human society, pregnancy can be a blessing or a disaster, depending on the situation. Statistics describing the effectiveness of different means of limiting human reproductive output **(Table 38.1, p. 932)** illustrate our progress in the area of family planning. Knowledge about the timing of ovulation, for example (see Figure 38.16), can provide the means to maximize or minimize the chances of pregnancy.

Biologists working to conserve biodiversity often attempt to control reproduction. When a species is on the brink of extinction, the goal is to maximize reproductive output. Techniques can range from the use of foster parents to raise young to using reproductive technologies such as in vitro fertilization and implantation of embryos. For example, biologists

Table 38.1   Pregnancy Rates for Birth Control Methods

Method	Lowest Expected Rate of Pregnancy[a]	Typical-Use Rate of Pregnancy[b]
Rhythm method	1%–9%	25%
Withdrawal	4%	19%
Condom (male)	3%	14%
Condom (female)	5%	21%
Diaphragm and spermicidal jelly	6%	20%
Vasectomy (male sterilization)	0.1%	0.15%
Tubal ligation (female sterilization)	0.5%	0.5%
Contraceptive pill (combination estrogen–progestin)	0.1%	5%
Contraceptive pill (progestin only)	0.5%	5%
Implant (progestin)	0.09%	0.09%
Intrauterine device (IUD) (copper T)	0.6%	0.8%

[a]Rate of pregnancy when the birth control method was used correctly every time.
[b]Rate of pregnancy when the method was used typically, meaning that it may not have been always used correctly every time.
Source: U.S. Food and Drug Administration, http://www.fda.gov/fdac/features/1997/conceptbl.html. Data reported in 1997 for effectiveness of methods in a 1-year period.

working with black-footed ferrets (see Box 45.2), which are highly endangered, strive to maximize reproductive output to increase the population.

Similarly, when an increasing population of one species threatens to overwhelm other species or an ecosystem, the goal is to prevent reproduction. Biologists faced with growing populations of African elephants try to reduce reproductive output using techniques ranging from the application of contraceptives to females (see Box 45.4) to "culling" (killing) individuals in the population. Culling usually targets females because they produce young (see Chapter 45). The same principles of controlling reproductive output have been central to humans' domestication of other organisms (see Chapter 49).

STUDY BREAK

1. How can biologists control reproduction? Why is access to control of reproduction important?
2. What are the advantages and disadvantages of different approaches to contraception for mammals?

UNANSWERED QUESTIONS

What factors could have led to the evolution of sexual reproduction? Does the evidence suggest that sexual reproduction evolved once, or did it evolve several times? What are the advantages of sexual versus asexual reproduction?

Review

Go to CENGAGENOW at http://hed.nelson.com/ to access quizzing, animations, exercises, articles, and personalized homework help.

38.1 The Drive to Reproduce

- After mating, a male acanthocephalan worm (*Moniliformis dubius*) uses a cement gland to seal the vagina of the female, preventing her from mating with other males. This protects his investment. To reduce competition from other males, a male *M. dubius* may "rape" other males and use the product of his cement gland to prevent the other male from mating.

38.2 Asexual and Sexual Reproduction

- One important advantage of sexual reproduction over asexual reproduction is the generation of genetic diversity in offspring. Genetic recombination and independent assortment of chromosomes during meiosis give rise to diversity and reduce vulnerability to deleterious effects carried on recessive alleles.

- Meiosis in animals occurs during the production of gametes, although the actual timing of meiosis varies between males and females and among species. In plants and fungi, meiosis is not always used to produce gametes.

38.3 Cellular Mechanisms of Sexual Reproduction

- The amount of yolk in eggs varies enormously, influencing egg size and the time of incubation and development. Development in eggs with large amounts of yolk often involves a small embryonic disk floating on top of the yolk (e.g., bird) as opposed to development around the yolk (e.g., frog).

- Internal fertilization can reduce the amounts of sperm necessary to achieve fertilization. Internal fertilization also can increase the certainty of mate choice where only the sperm of a male that mated come in contact with the egg(s) of a female.

- Polyspermy, more than one sperm entering an egg, is prevented in two ways. The fast block depends on a change in the egg's membrane potential from negative to positive, whereas the slow block involves Ca^{2+} ions that cause cortical granules to fuse with the egg's plasma membrane and release their enzyme contents to the outside.

- Viviparous animals give birth to live young. Before birth, the mother provides the developing embryos with food and oxygen and removes its metabolic wastes. Usually, but not always, this occurs in the body of the female. Ovoviviparous animals may also give birth to live young, but here the eggs develop and grow inside the mother using yolk as the source of energy. Oviparous animals lay eggs that develop outside the mother's body.

- Most mammals show a well-defined mating season, the time when copulation, fertilization, and development take place. The onset of the mating season is often triggered by changes in photoperiod (day length). Mammalian females typically display behavioural and physiological signs of fertility (estrus). Humans, on the other hand, do not have a mating season, and both sexes are receptive to mating at any time.

- Follicle-stimulating hormone (FSH) is released by the pituitary in response to gonadotropin-releasing hormone. FSH stimulates the development of oocytes in the ovaries. Developing oocytes are surrounded by cells that form a follicle. A peak in the level of luteinizing hormone is a clear indicator of ovulation (the release of the secondary oocyte).

- Cryptic testes (e.g., elephants and whales) are housed inside the body. In many mammals, testes are housed in a scrotum outside the body because sperm develop best at temperatures below the core temperature of a mammal. A baculum is a penis bone, a feature of many mammals but not humans. The baubellum (clitoris bone) is the equivalent in females.

38.4 Sexual Reproduction in Humans

- Alkaline prostate secretions raise the pH of semen and the vagina to about pH 6, activating sperm motility. When enzymes contained within the acrosome of a sperm are released into the zona pellucida, they digest a path to the plasma membrane of the egg. This process is initiated by contact between and then adherence of the sperm to receptor molecules on the surface of the zona pellucida.

38.5 Controlling Reproduction

- Controlling (promoting or reducing) reproductive output is important for our species in social and political contexts. It is also important in conservation.

Questions

Self-Test Questions

1. Asexual reproduction can involve
   a. fission.
   b. budding.
   c. copulation.
   d. parthenogenesis.
   e. All of the above are correct.

2. Germ cells give rise to
   a. eggs.
   b. somatic cells.
   c. sperm.
   d. muscles.
   e. Only a and c are correct.

3. Internal fertilization is rarely seen in
   a. platyhelminthes.
   b. bedbugs.
   c. mammals.
   d. frogs.
   e. bony fishes.

4. In slow blocks, more than one sperm is prevented from fertilizing an egg by changes in
   a. Ca^{2+} ions.
   b. Cl^- ions.
   c. Na^+ ions.
   d. K^+ ions.
   e. cortical granules.

5. Ovulation in women is signalled by a rise in
   a. luteinizing hormone.
   b. follicle-stimulating hormone.
   c. estrogen.
   d. testosterone.
   e. relaxin.

6. In mammals, cryptic testes are typical of
   a. humans.
   b. elephants.
   c. dogs.
   d. bulls.
   e. hermaphrodites.

7. The reproductive cycle of some bats and turtles involves
   a. delayed fertilization.
   b. delayed implantation.
   c. delayed development.
   d. postpartum estrus.
   e. All of the above are correct.

8. Cryptochromes are sensitive to
   a. red light.
   b. white light.
   c. blue light.
   d. green light.
   e. All of the above are correct.

9. Amplexus is the term used to describe mating behaviour in
   a. birds.
   b. frogs and toads.
   c. salmon.
   d. mammals.
   e. sharks.

Questions for Discussion

1. How do plants exploit animals to effect pollination and dispersal of seeds? Are there examples of animals exploiting plants to achieve reproduction?

2. How does variation in the pattern of fertilization (internal versus external) differ among the animal phyla? Do these variations indicate different ancestral conditions? How do these patterns differ between plants and animals?

3. What methods do zoos and botanical gardens use to control reproduction of captive organisms? Is this appropriate?

4. What steps could conservation biologists take to increase reproductive output of rare and endangered species?

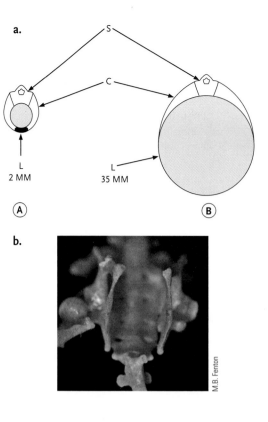

(a) A diagrammatic view of the bony birth canal of (A) a nonpregnant Brazilian free-tailed bat (*Tadarida brasiliensis*) and (B) a female in the process of giving birth. The sacrum (S), coxal bone (C), and interpubic ligament (L) are shown. **(b)** The ventral view of the pelvis of another nonpregnant free-tailed bat. There is no connection between the pubic bones, making it possible for the animal to achieve the expansion of the birth canal required during birth.

39 Animal Development

WHY IT MATTERS

The size of the mother and of the young can be a challenging aspect of the process of giving birth (parturition). Bats are an extreme example when it comes to offspring size. Typically, a female bat bears a pup that is 25 to 30% of her normal body mass—the equivalent of a 60-kg woman bearing a 15-to 20-kg baby. The birth canal of placental mammals (such as bats and people) passes between the two halves of the pelvic girdle. Not surprisingly, the pelvis of adult female placental mammals differs from that of males.

The magnitude of the challenge of parturition to a female Brazilian free-tailed bat is shown above. In a nonpregnant female, the bony birth canal is 2 mm in diameter, but it expands to 35 mm in diameter during birth. In these bats, birth takes about 90 seconds.

The elasticity of the interpubic ligament is the key to the birth process in female mammals that give birth to large young. This ligament has an abundance of elastic fibres, which intermingle with collagen fibres at the ligament's core. The situation in bats appears to be the same as in other mammals: elastic fibres stretch, whereas collagen fibres slide in relation to each other. The hormone relaxin plays a fundamental role in this process, promoting the stretchability

MOLECULE BEHIND BIOLOGY

Relaxin

The hormone relaxin (**Figure 1**) is a polypeptide produced by the ovaries during pregnancy. In humans, relaxin occurs at higher levels earlier in pregnancy than closer to parturition. Relaxin promotes angiogenesis, the growth of new blood vessels, and influences the interface between the uterus and the placenta. Relaxin inhibits muscular contractions of the uterus that could terminate pregnancy and stimulates

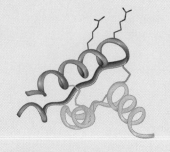

Figure 1
Relaxin.

the growth of glands that produce milk in breast tissue.

Near parturition, relaxin causes relaxation of the pubic ligaments and softens and enlarges the cervical opening.

---

of interpubic ligaments at the time of parturition (see *Molecule Behind Biology*).

39.1 Housing and Fuelling Developing Young

Some animal parents invest significant energy in housing and feeding their developing young. This is one aspect of the genetically selfish drive to ensure that their genes are represented in future generations.

39.1a Housing: Providing a Place in Which the Embryo Can Develop

There is a recurring tendency across phyla for parents to put eggs and developing young in situations that minimize their exposure to predators and parasites while maximizing favourable conditions for growth and development. Many species of birds use nests to house their eggs and unfledged young. Parents of other species, such as some species of scorpions (see Figure 3.22a), frogs, and insects, carry their young with them, often on their backs. This allows the parent (parents) to avoid or actively deter would-be predators.

An escalation in parental investment is moving eggs and young inside the parent's body (vivipary and ovovivipary; see Chapter 38). This approach to parental care has several different stages (see Chapter 38, *On the Road to Vivipary*). Although we associate vivipary with mammals, many species of fish are mouth-breeders, keeping eggs and, for a time, developing young in their mouths. Other fish, such as sea horses and pipefish (family Syngnathidae, order Gasterosteiformes; **Figure 39.1**), keep eggs and developing young in specialized incubation areas, called brood pouches, located on the tail or trunk of the male. "Pregnancy" in male sea horses represents an increase in parental investment. It also allows males to be confident about the paternity of the young they raise.

Some amphibians also show high levels of parental care. In Australia, female frogs, *Rheobatrachus silus,* use their stomachs as brood pouches. While the young are developing, they secrete prostaglandin E_2, which inhibits the secretion of gastric acid in the stomach and saves the developing young from being digested. On Mount Nimba in west Africa, female toads *Nectophrynoides occidentalis* harbour developing young in their uterus, where the young feed on uterine secretions in the absence of a placenta. The gestation period for these toads is nine months, and newborns are 7 to 8 mm long and weigh 30 to 60 mg. Retention of developing embryos in the oviducts has evolved independently in each of the three living groups of Amphibia: Anura, Urodela, and Gymnophiona (see Chapter 27).

39.1b Feeding: Aiding and Abetting Developing Young

Almost everyone has seen pictures of parent birds feeding their young (see Figures 40.2 and 40.4). In many species, both males and females deliver food to the nestlings. Some fruit-eating adult birds feed insects to their young

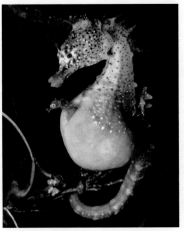

Figure 39.1
A male sea horse gives birth.

Joseph G. Kunkel

Figure 39.2

Diploptera punctata, a cockroach that produces live young. Starting from the left, an adult female and male, egg, last stage fetus, larval instars.

because a higher protein diet promotes rapid growth of the young. Producing high-quality food is the next level of parental investment, and "milk" is a prime example.

The term "milk" is usually applied to secretions of the mammary glands of mammals, and it is the quintessentially mammalian food. However, other animals also make "milk." Female cockroaches **(Figure 39.2)** house developing embryos in a brood sac and give birth to them as first-instar (first stage) larvae. The brood sac is an infolding of a ventral intersegmental membrane, and its epithelium produces the milk, a blend of water-soluble proteins encoded by a multigene family.

Both male and female discus fish **(Figure 39.3)** feed their hatchling young (known as "fry") skin secretions, the first and only food eaten by the fry. This fish "milk" appears as a slight mucus coating on the adults' bodies, particularly above the lateral lines. Skin feeding also occurs in a caecilian (an amphibian). The skin in brooding females **(Figures 39.4 and 39.5)** is transformed

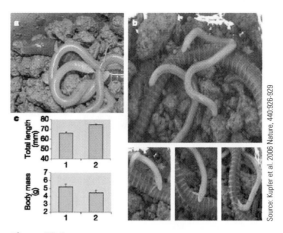

Source: Kupfer et al. 2006 Nature, 440:926-929

Figure 39.4

A female caecilian (*Boulengerula taitanus*) feeding her young skin secretions.

to provide a rich supply of nutrients, and the young have specialized teeth for peeling and eating the outer layer of their mother's skin **(Figure 39.6, p. 938)**. In some other caecilians, young develop in the uterus and feed on the lining of the oviduct. "Milk" also has been reported in birds, where the crop milk of pigeons (*Columba livia domestica*) is fed to young (known as squabs) from hatching to about age 19 days. In some cases, pigeon lactation continues to day 28. Pigeon crop milk is composed mainly of proteins and lipids and is highly nutritious. Obviously, female mammals have not cornered the "milk market." There are records of male mammals lactating, the most notable being *Dyacopterus spaediacus,* a fruit bat from Indonesia. Some male *D. spaediacus* produce milk, although not as much as females, but the behavioural significance of male lactation in this species remains unknown.

M.B. Fenton

Figure 39.3

Symphysodon discus, a fish that produces "'milk'" to feed its young.

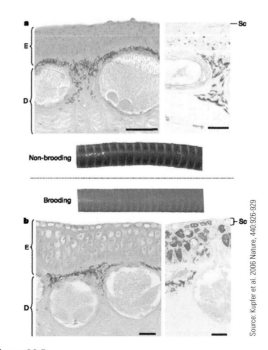

Source: Kupfer et al. 2006 Nature, 440:926-929

Figure 39.5

Details of the skin of **(a)** a nonbrooding and **(b)** a brooding female caecilian.

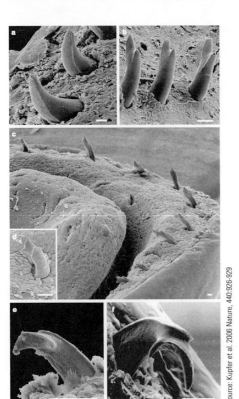

Source: Kupfer et al. 2006 Nature, 440:926–929

Figure 39.6
Scanning electron micrographs of the specialized teeth of young caecilians.

Study Break

1. What is relaxin? What role does it play?
2. Where can embryos develop in a parent?
3. What is milk? What animals produce it?

39.2 Mechanisms of Embryonic Development

When a sperm fertilizes an egg, a zygote is produced. At this point, embryonic development begins, ultimately producing a free-living individual. All of the instructions required for development are packed into the zygote. Mitotic divisions of the zygote are the beginning of developmental activity (see Chapter 9).

Information that directs the initiation of development is stored in two locations in the zygote. The nucleus houses the DNA derived from egg and sperm nuclei. This DNA directs development as individual genes are activated or turned off in a regulated and ordered manner. The balance of the information is stored in the zygote's cytoplasm.

Because sperm contribute essentially no cytoplasm to the zygote, its cytoplasm is maternal in origin. The mRNA and proteins stored in the egg cytoplasm are known as *cytoplasmic determinants*, which direct the first stages of animal development before genes become active. Depending on the animal group, control of early development by cytoplasmic determinants may be limited to the first few divisions of the zygote (e.g., in mammals), or it may last until the actual tissues of the embryo are formed (e.g., in most invertebrates).

The zygote's cytoplasm also contains ribosomes and other cytoplasmic components required for protein synthesis and early divisions of embryonic cells. Zygote cytoplasm contains the tubulin molecules required to form spindles for early cell divisions, as well as mitochondria and nutrients stored in granules in the yolk and in lipid droplets. In many animals, zygotes contain pigments that colour the egg or regions of it.

Yolk contains nutrients. In the eggs of typical insects, reptiles, and birds, large amounts of yolk supply all of the nutrients for development of the embryo. In contrast, the eggs of placental mammals contain very little yolk, which is used only to support the earliest stages of development.

Depending on the species, yolk may be concentrated at one end or in the centre or distributed evenly throughout the egg. Yolk distribution influences the rate and location of cell division during early embryonic development. Typically, cell division proceeds more slowly in the region of the egg containing the yolk. In the large, yolky eggs of birds and reptiles, cell division takes place only in a small, yolk-free patch at the egg's surface.

Unequal distribution of yolk and other components in the egg is termed **polarity.** In most species, the egg's nucleus is located toward one end, called the **animal pole.** The animal pole typically gives rise to surface structures and the anterior end of the embryo. The opposite end of the egg, the **vegetal pole,** typically gives rise to internal structures such as the gut, along with the posterior end of the embryo. When yolk is unequally distributed in the egg cytoplasm, it is usually concentrated in the vegetal half of the egg. Egg polarity plays a role in setting the three body axes of bilaterally symmetrical animals, namely the anterior–posterior axis, the dorsal–ventral (back-front) axis, and the left–right axis **(Figure 39.7).**

39.2a Cleavage and Gastrulation: Zygote to Multicellular Embryo

Soon after fertilization, the zygote begins a series of mitotic cleavage divisions in which cycles of DNA replication and division occur without the production of new cytoplasm. Thus, the cytoplasm of the zygote is partitioned into successively smaller cells without increasing the size or mass of the embryo **(Figure 39.8).** In the frog *Xenopus laevis,* 12 cleavage divisions produce an embryo of about 4000 cells that collectively occupy about the same volume and mass as the original zygote.

Cleavage is the first of three major developmental stages that, with modifications, are common to the early development of most animals. **Gastrulation,** the

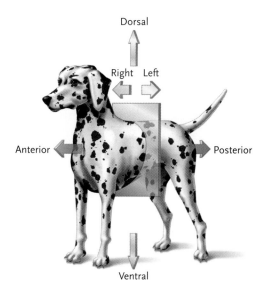

Dorsal

Right | Left

Anterior

Posterior

Ventral

Figure 39.7
Body axes: anterior–posterior, dorsal–ventral, and left–right.

stage following cleavage, produces an embryo with three distinct primary tissue layers. **Organogenesis** follows gastrulation and gives rise to development of major organ systems. At the end of organogenesis, the embryo has the body organization characteristic of its species. Cell division, cell movements, and cell rearrangements occur during gastrulation and in organogenesis. **Figure 39.9, p. 940,** shows these stages as part of the life cycle of a frog.

In frogs, cleavage divisions form two different structures in succession. The **morula** (*morula* = mulberry) is a solid ball or layer of cells. As cleavage divisions continue, the ball or layer hollows out to form the **blastula** (*blast* = bud or offshoot; *ula* = small), the second structure in which cells, now called **blastomeres** (*mere* = part or division), enclose a fluid-filled cavity, the **blastocoel** (*coel* = hollow).

After cleavage is complete, cells of the blastula migrate and divide to produce the **gastrula** (*gaster* = gut or belly). Gastrulation, a morphogenetic process, dramatically rearranges the cells of the blastula into the three **primary cell layers** of the embryo: **ectoderm**, the outer layer (*ecto* = outside; *derm* = skin); **endoderm**, the inner layer (*endo* = inside); and **mesoderm** (*meso* = middle), the middle layer between ectoderm and endoderm. Gastrulation establishes the body pattern. Each tissue and organ of the adult animal originates from one of the three primary cell layers of the gastrula **(Table 39.1, p. 940)**. Cell movements also contribute to the formation of the **archenteron** (*arch* = beginning; *enteron* = intestine or gut), a new cavity within the embryo that is lined with endoderm.

As the blastula develops into the gastrula, embryonic cells begin to differentiate, becoming recognizably different in biochemistry, structure, and function. The developmental potential of each cell becomes more limited than that of the zygote from which it originated. Although a fertilized egg could develop

into a complete embryo, a mesoderm cell may develop into muscle or bone but does not normally develop into outside skin or brain. Scientists once thought that this restriction of developmental potential resulted because cells lost all of their genes except those required for the structure and function of the cell type they would become. However, differentiating cells each contain the complete genome of the organism, but each type of cell has a different program of gene expression.

Although development in all animals is accomplished by mechanisms under genetic control, the mechanisms are influenced to some extent by environmental factors such as temperature. The six mechanisms are as follows:

- Mitotic cell divisions
- Cell movements
- **Selective cell adhesions**, in which cells make and break specific connections to other cells or to the extracellular matrix (ECM)
- **Induction**, in which one group of cells (inducing cells) causes or influences another nearby group of cells (responding cells) to follow a particular developmental pathway. The key to induction is that only certain cells can respond to the signal from the inducing cells. Induction typically involves signal transduction events (see Chapter 35). Some induction events are triggered by direct cell–cell contact involving interaction between a membrane-embedded protein on the inducing cell and a receptor protein on the responding cell's surface. Others are triggered by

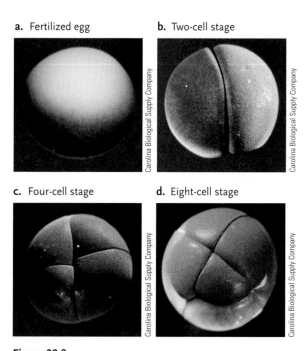

a. Fertilized egg

b. Two-cell stage

c. Four-cell stage

d. Eight-cell stage

Carolina Biological Supply Company

Figure 39.8
The first three cleavage divisions of a frog embryo, which convert the fertilized egg into the eight-cell stage. Note that the cleavage divisions cut the volume of the fertilized egg into successively smaller cells.

Figure 39.9
Stages of animal
development shown
in a frog.

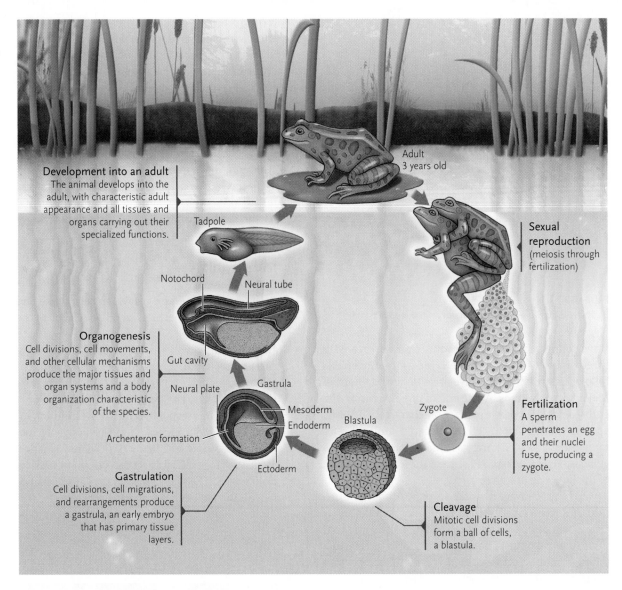

Development into an adult
The animal develops into the
adult, with characteristic adult
appearance and all tissues and
organs carrying out their
specialized functions.

Adult
3 years old

Sexual
reproduction
(meiosis through
fertilization)

Tadpole

Notochord

Neural tube

Organogenesis
Cell divisions, cell movements,
and other cellular mechanisms
produce the major tissues and
organ systems and a body
organization characteristic
of the species.

Gut cavity

Neural plate

Gastrula

Mesoderm

Endoderm

Zygote

Blastula

Fertilization
A sperm
penetrates an egg
and their nuclei
fuse, producing a
zygote.

Archenteron formation

Ectoderm

Gastrulation
Cell divisions, cell migrations,
and rearrangements produce
a gastrula, an early embryo
that has primary tissue
layers.

Cleavage
Mitotic cell divisions
form a ball of cells,
a blastula.

a signal molecule released by the inducing cell that interacts with a receptor on the responding cell (e.g., paracrine regulation; see Chapter 35).

- **Determination** sets the developmental fate of a cell. Before determination, a cell has the potential to become any cell type of the adult. Afterward, the cell is committed to becoming a particular cell type. Typically, determination results from induction, although in some cases, it results from the asymmetric segregation of cellular determinants.
- **Differentiation** follows determination and involves the establishment of a cell-specific developmental program in the cells. Differentiation results in cell types with clearly defined structures and functions. These features are derived from specific patterns of gene expression in cells.

Table 39.1	Origins of Adult Tissues and Organs in the Three Primary Tissue Layers
Primary Tissue Layer	**Adult Tissues and Organs**
Ectoderm	Skin and its elaborations, including hair, feathers, scales, and nails; nervous system, including brain, spinal cord, and peripheral nerves; lens, retina, and cornea of eye; lining of mouth and anus; sweat glands, mammary glands, adrenal medulla, and tooth enamel
Mesoderm	Muscles; most of skeletal system, including bones and cartilage; circulatory system, including heart, blood vessels, and blood cells; internal reproductive organs; kidneys and outer walls of digestive tract
Endoderm	Lining of digestive tract, liver, pancreas, lining of respiratory tract, thyroid gland, lining of urethra, and urinary bladder

STUDY BREAK

1. What is yolk? What role does it play?
2. What mechanisms are involved in animal development from the zygote?

39.3 Major Patterns of Cleavage and Gastrulation

39.3a Sea Urchin

Cleavage divisions proceed at approximately the same rate in all regions of a sea urchin embryo (**Figure 39.10, step 1**), reflecting uniform distribution of yolk in the egg. These divisions continue until a blastula containing about a thousand cells is formed (step 2).

Gastrulation begins at the vegetal pole of the blastula. Through induction, some cells in the middle of the vegetal pole become elongated and cylindrical, causing the region to flatten and thicken. Then some cells (primary mesenchyme: *mesen* = middle; *chyme* = juice) break loose and migrate into the blastocoel (step 3), making and breaking adhesions until eventually they attach along the ventral sides of the blastocoel. These cells form the future mesoderm (see Figure 39.10, step 7), which give rise to skeletal elements of the embryo. Next, the flattened vegetal pole of the blastula invaginates, pushing gradually into the interior (steps 4 and 5). The cells that invaginate will become endoderm cells. The inward movement, much like pushing in the side of a hollow rubber ball, generates the archenteron, a new cavity that opens through the blastopore.

As the archenteron forms, extensions of cells of the invaginated layer stretch across the blastocoel and contact the inside of the ectoderm (step 6). These extensions make tight adhesions and then contract, pulling the invaginated cell layer inward with them, eliminating most of the blastocoel.

Now the embryo has two complete cell layers. The outer layer—the original blastula surface—forms embryonic ectoderm. Cells of the second, inner layer are derived from the archenteron and become endoderm. Mesodermal cells, which begin to form a third layer, are derived from the primary mesenchyme cells and from *secondary mesenchyme* cells that migrated into the space between the ectoderm and endoderm (step 7). After the formation of the three primary cell layers, cells begin to differentiate based on synthesis of different proteins in each layer.

As ectoderm, mesoderm, and endoderm layers develop, the embryo lengthens into an ellipsoidal shape, with the blastopore marking the posterior end of the embryo. From here on, further cell divisions, combined with cell movements, selective cell adhesions, induction, and differentiation, lead to differentiation of organ systems. In sea urchins and other deuterostomes, the blastopore forms the anus, and the mouth will form at the opposite, anterior end of the gut.

39.3b Amphibians

In the eggs of amphibians such as frogs, yolk is concentrated in the vegetal half, giving it a pale colour. The animal half is darkly coloured because of a layer of pigment granules just below the surface. A sperm normally

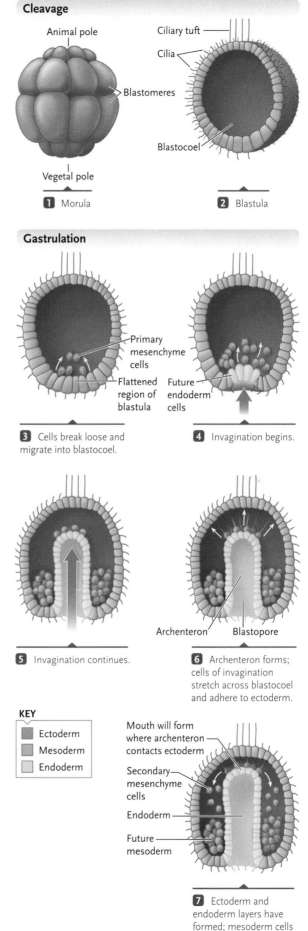

Cleavage

Animal pole
Blastomeres
Vegetal pole

Ciliary tuft
Cilia
Blastocoel

1 Morula

2 Blastula

Gastrulation

Primary mesenchyme cells
Flattened region of blastula
Future endoderm cells

3 Cells break loose and migrate into blastocoel.

4 Invagination begins.

Archenteron   Blastopore

5 Invagination continues.

6 Archenteron forms; cells of invagination stretch across blastocoel and adhere to ectoderm.

KEY
Ectoderm
Mesoderm
Endoderm

Mouth will form where archenteron contacts ectoderm
Secondary mesenchyme cells
Endoderm
Future mesoderm

7 Ectoderm and endoderm layers have formed; mesoderm cells are between them.

Figure 39.10
Cleavage and gastrulation in the sea urchin.

Figure 39.11
Rotation of the pigment layer and development of the grey crescent after fertilization in a frog egg. The grey crescent marks the site where gastrulation of the embryo will begin.

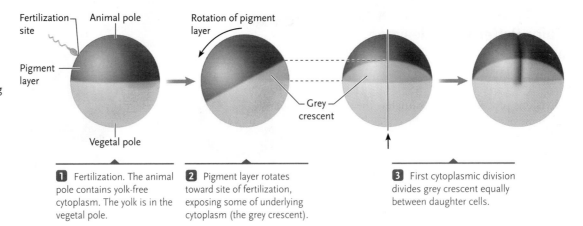

1 Fertilization. The animal pole contains yolk-free cytoplasm. The yolk is in the vegetal pole.

2 Pigment layer rotates toward site of fertilization, exposing some of underlying cytoplasm (the grey crescent).

3 First cytoplasmic division divides grey crescent equally between daughter cells.

fertilizes the egg in the animal half (**Figure 39.11,** step 1). After fertilization, the pigmented layer of cytoplasm rotates toward the site of sperm entry, exposing a crescent-shaped region of underlying cytoplasm at the side opposite the point of sperm entry (step 2). This region, the **grey crescent,** establishes the dorsal–ventral axis of the embryo and marks the future dorsal side of the animal.

Normally, the first cleavage division runs perpendicular to the long axis of the grey crescent and divides the crescent equally between resulting cells (step 3). If the first two blastomeres are experimentally divided so that one does not receive grey crescent material, and the two cells are separated, the blastomere without grey crescent material divides but ends up in a disordered mass that stops developing. The blastomere receiving grey crescent produces a normal embryo. Cytoplasmic material localized in the grey crescent is essential to normal development in frog embryos.

As cleavage of the frog embryo continues, cell divisions proceed more rapidly in the animal half, producing smaller and more numerous cells there than in the yolky vegetal half. By the time cleavage has produced an embryo with 15 000 cells, the animal half has hollowed out, forming the blastula (**Figures 39.12,** step 1, and **39.13a**).

Gastrulation begins when cells from the animal pole begin to migrate across the embryo surface to reach the region derived from the grey crescent. This site is marked by a crescent-shaped depression rotated clockwise 90° and called the **dorsal lip of the blastopore** (see Figure 39.12, step 2, and **Figure 39.13b**). These cells invaginate, changing shape and pushing inward from the surface to produce the depression. The depression eventually forms a complete circle (the blastopore) after further inward movement of additional cells (see Figure 39.12, step 3, and **Figure 39.13c**).

By **involution,** cells migrate into the blastopore, and the pigmented cell layer of the animal half expands to cover the entire embryo surface (see Figures 39.12, step 4, and 39.13c). Cells of the vegetal half are enclosed by this cell migration, becoming visible on the outside as a yolk plug in the blastopore. The blastopore gives rise to the anus.

Continuing involution moves cells into the interior and upward (see Figure 39.12, steps 3 and 4), forming two layers that line the inside top half of the embryo. Dorsal mesoderm (shown in red) is the uppermost of these induced layers. Beneath it is the endoderm

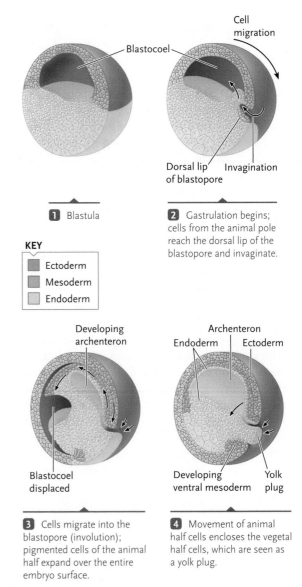

KEY

▨	Ectoderm
▨	Mesoderm
▨	Endoderm

1 Blastula

2 Gastrulation begins; cells from the animal pole reach the dorsal lip of the blastopore and invaginate.

3 Cells migrate into the blastopore (involution); pigmented cells of the animal half expand over the entire embryo surface.

4 Movement of animal half cells encloses the vegetal half cells, which are seen as a yolk plug.

Figure 39.12
Gastrulation in a frog embryo. Yolk cells are shown in paler yellow.

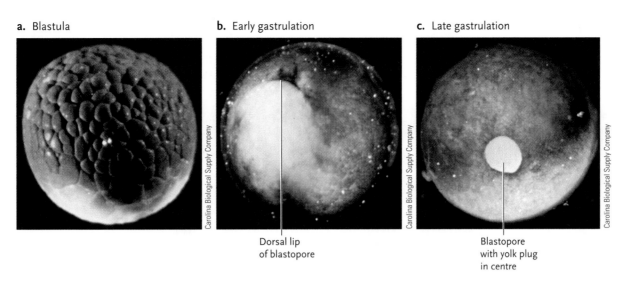

a. Blastula   **b.** Early gastrulation   **c.** Late gastrulation

Dorsal lip
of blastopore

Blastopore
with yolk plug
in centre

Figure 39.13
Photomicrographs of a frog embryo. **(a)** Blastula. **(b)** Early gastrulation and the formation of the dorsal lip of the blastopore. **(c)** Late gastrulation, showing the completed blastopore, closed by the yolk plug.

(shown in yellow), containing cells originating from both the outer surface of the embryo and the yolky interior. Ectoderm (shown in blue) forms from pigmented cells remaining at the surface of the embryo. Induction of the ventral mesoderm begins near the vegetal pole.

As mesoderm and endoderm form, the depression created by inward cell movements gradually deepens and extends inward as the archenteron (see Figure 39.12, steps 3 and 4), displacing the blastocoel. Cells of the three primary cell layers continue to increase in number by further migrations and divisions as development proceeds.

The major induction centre during frog gastrulation is the dorsal lip of the blastopore. If cells are removed from the dorsal lip and transplanted elsewhere in the egg, they form a second blastopore (and a second embryo).

39.3c Birds

Gastrulation in amniotes (see Chapter 27) such as birds and reptiles is modified by the distribution of the yolk, which occupies almost the entire volume of the egg. A thin layer of cytoplasm at the egg's surface gives rise to primary tissues of the embryo. Although mammalian eggs have relatively little yolk, gastrulation in them follows a similar pattern.

Early cleavage divisions in birds produce the **blastodisk**, a thin layer of cells at the yolk's surface **(Figure 39.14, p. 944,** step 1). The complete blastodisk is a layer with about 20 000 cells. Cells of the blastodisk then separate into two layers, the **epiblast** (top layer) and the **hypoblast** (bottom layer). The blastocoel is the flattened cavity between them (step 2).

Gastrulation begins as cells in the epiblast stream toward the midline of the blastodisk, thickening it in this region. The thickened layer (or **primitive streak**) is first evident in the posterior end of the embryo and extends toward the anterior end as more cells of the epiblast move into it (step 3). A thickening at the anterior end of the primitive streak (the *primitive knot*) is the functional equivalent of the amphibian dorsal lip of the blastopore. The primitive streak initially marks

the future posterior end of the embryo, and by the time it has elongated fully, it has established the left and right sides of the embryo. The streak forms on what will become the dorsal side of the embryo, with the ventral side below.

As the primitive streak forms, its midline sinks, forming the **primitive groove**, a conduit for migrating cells to move into the blastocoel. Epiblasts are the first cells to migrate through the primitive groove (see Figure 39.14, step 4) and produce the endoderm. Mesoderm is formed from cells migrating laterally between the epiblast and endoderm. Epiblast cells remaining at the surface of the blastodisk form ectoderm (step 4).

In the chick embryo, all primary tissue layers arise from the epiblast. Only a few of the hypoblast cells—near the posterior end of the embryo—contribute directly to the embryo. These form *germ cells* that later migrate to developing gonads, founding cell lines leading to eggs and sperm (see Chapter 9).

Initially, ectoderm, mesoderm, and endoderm are located in three more or less horizontal layers in the chick embryo. During gastrulation, the endoderm pushes upward along its midline and its left and right sides fold downward, forming a tube that is oriented parallel to the primitive streak (see Figure 39.14, step 5). The archenteron is the central cavity of the tube, the primitive gut. Mesoderm separates into two layers, forming the coelom, a fluid-filled body cavity lined with mesoderm. These movements complete the formation of the gastrula.

39.3d Extraembryonic Membranes: Amnion, Chorion, Allantois

Each primary tissue layer of a bird embryo extends outside the embryo to form **extraembryonic membranes (Figure 39.15, p. 944)** that conduct nutrients from yolk to embryo, exchange gases with the environment outside the egg, or store metabolic wastes removed from the embryo. The **yolk sac** is an extension of mesoderm and endoderm enclosing the yolk.

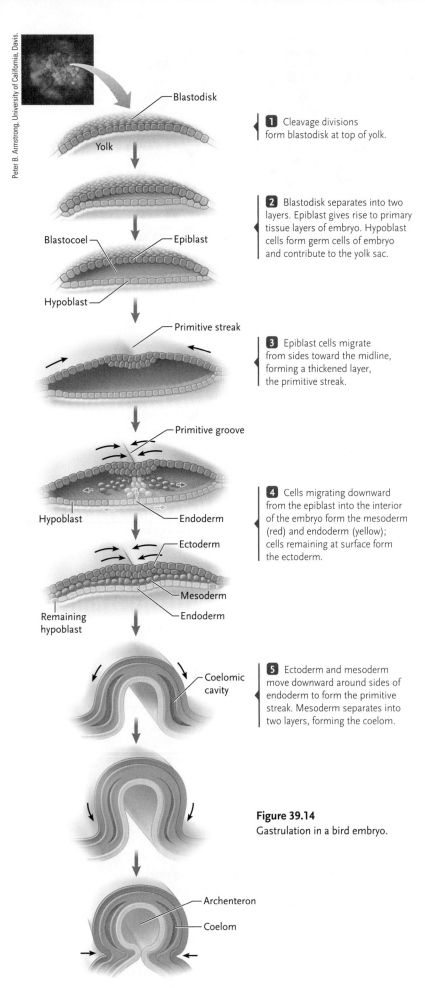

Peter B. Armstrong, University of California, Davis.

1 Cleavage divisions form blastodisk at top of yolk.

Blastodisk

Yolk

2 Blastodisk separates into two layers. Epiblast gives rise to primary tissue layers of embryo. Hypoblast cells form germ cells of embryo and contribute to the yolk sac.

Blastocoel — Epiblast

Hypoblast

Primitive streak

3 Epiblast cells migrate from sides toward the midline, forming a thickened layer, the primitive streak.

Primitive groove

4 Cells migrating downward from the epiblast into the interior of the embryo form the mesoderm (red) and endoderm (yellow); cells remaining at surface form the ectoderm.

Hypoblast — Endoderm

Ectoderm

Remaining hypoblast — Mesoderm — Endoderm

5 Ectoderm and mesoderm move downward around sides of endoderm to form the primitive streak. Mesoderm separates into two layers, forming the coelom.

Coelomic cavity

Figure 39.14
Gastrulation in a bird embryo.

Archenteron

Coelom

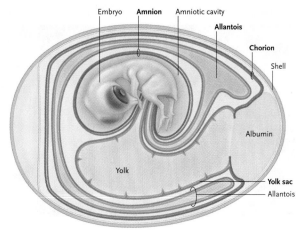

Embryo   **Amnion**   Amniotic cavity
                                    Allantois
                                        Chorion
                                            Shell

Albumin

Yolk

Yolk sac
Allantois

Figure 39.15
The four extraembryonic membranes in a bird embryo (in bold).

Although the yolk sac remains connected to the gut of the embryo by a stalk, yolk does not directly enter the embryo by this route. Rather, it is absorbed by blood in vessels in the membrane, which then transport the nutrients to the embryo.

The **chorion**, produced from ectoderm and mesoderm, completely surrounds the embryo and yolk sac and lines the inside of the shell. The chorion exchanges oxygen and carbon dioxide with the environment through the egg's shell. The **amnion** closes over the embryo to form the *amniotic cavity*. Cells of the amnion secrete *amniotic fluid* into the cavity, which bathes the embryo and provides an aquatic environment in which it can develop. Reptilian and mammalian embryos are also surrounded by an amnion and amniotic fluid. Providing the embryo with an aquatic environment is presumed to have been a key factor in the evolution of fully terrestrial vertebrates, the **Amniota**.

The **allantoic membrane** forms from mesoderm and endoderm that has bulged outward from the gut. It encloses the allantois, a sac that closely lines the chorion and fills much of the space between chorion and yolk sac. The **allantois** stores nitrogenous wastes (primarily uric acid) removed from the embryo. The part of the allantoic membrane lining the chorion forms a rich bed of capillaries connected to the embryo by arteries and veins. This circulatory system delivers carbon dioxide to the chorion and picks up the oxygen that is absorbed through the shell and chorion. At hatching, part of the allantoic membrane becomes the lining of the bladder.

STUDY BREAK

1. What are the main developmental differences between protostomes and deuterostomes?
2. What are important differences between typical patterns of development in birds and in mammals?
3. What are the features of an amniote egg? What is its evolutionary significance?

39.4 Organogenesis: Gastrulation to Adult Body Structures

Following gastrulation, organogenesis gives rise to the body organization characteristic of the species. Organogenesis involves the same mechanisms used in gastrulation, namely cell division, cell movements, selective cell adhesion, induction, and differentiation. Organogenesis also involves an additional mechanism, **apoptosis**, in which certain cells are programmed to die (see Chapter 9). To illustrate how cellular mechanisms of development interact in organogenesis, we follow the formation of major organ systems in the bird embryo. Then we describe the generation of one organ, the eye, which follows a pathway typical of eye development in all vertebrates.

39.4a Ectoderm and the Nervous System: Neural Tube and Neural Crest Cells

In vertebrates, organogenesis begins with **neurulation**, which is the development of nervous tissue from ectoderm. As a preliminary to neurulation, cells of the mesoderm form the notochord, a solid rod of tissue extending the length of the embryo under the dorsal ectoderm. Notochord cells carry out a major induction, causing the overlying ectoderm to form the **neural plate**, a thickened and flattened longitudinal band of cells (**Figure 39.16,** steps 1 and 2). The neural plate does not form if the notochord is removed.

Once induced, the neural plate sinks downward along its midline (steps 2 and 3), creating a deep longitudinal groove and ridges (neural crests) that rise along the sides of the neural plate. The neural tube forms when the neural crests move together and close over the centre of the groove along the length of the developing embryo (steps 4 and 5). The neural tube then pinches off from the overlying ectoderm, which closes over the tube (step 6). The central nervous system, including the brain and spinal cord, develops directly from the neural tube.

During formation of the neural tube, **neural crest** cells (see Figure 39.16) migrate into the mesoderm and follow specific routes to reach distant points in the developing embryo, where they contribute to the formation of a variety of organ systems. Some cells develop into cranial nerves in the head, whereas others contribute to the bones of the inner ear and skull, cartilage of facial structures, and teeth. Still others form ganglia of the autonomic nervous system, peripheral nerves leading from the spinal cord to body structures, and nerves of the developing gut. Neural crest cells also move to the skin, where they form pigment cells, and to the adrenal glands, where they form the medulla of the kidney. The migration of neural crest cells contributes to development in all vertebrates (see *People Behind Biology*).

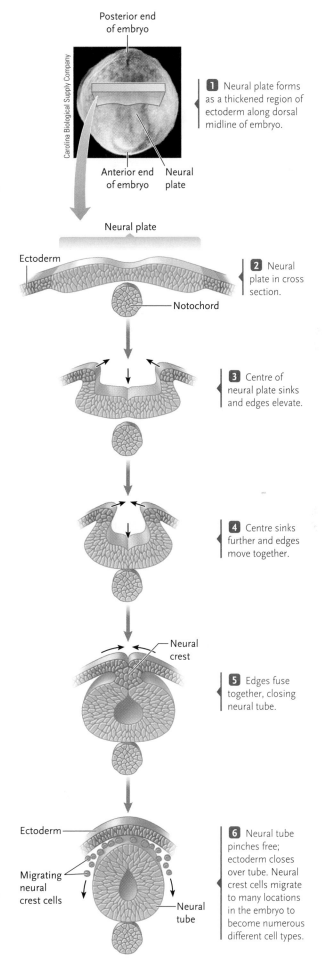

Posterior end of embryo

Anterior end of embryo / Neural plate

Carolina Biological Supply Company

1 Neural plate forms as a thickened region of ectoderm along dorsal midline of embryo.

Neural plate

Ectoderm

Notochord

2 Neural plate in cross section.

3 Centre of neural plate sinks and edges elevate.

4 Centre sinks further and edges move together.

Neural crest

5 Edges fuse together, closing neural tube.

Ectoderm

Migrating neural crest cells

Neural tube

6 Neural tube pinches free; ectoderm closes over tube. Neural crest cells migrate to many locations in the embryo to become numerous different cell types.

Figure 39.16
Development of the neural tube and neural crest cells in vertebrates. Photo is of an amphibian embryo; drawings show steps in a bird embryo.

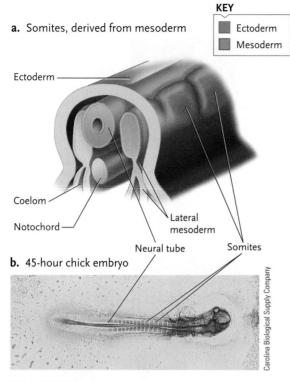

a. Somites, derived from mesoderm

Ectoderm

Coelom

Notochord

Lateral mesoderm

Neural tube

Somites

b. 45-hour chick embryo

Carolina Biological Supply Company

Figure 39.17

Later development of the mesoderm. **(a)** Somites develop into segmented structures such as the vertebrae, the ribs, and the musculature between the ribs. The lateral mesoderm gives rise to other structures, such as the heart and blood vessels and the linings of internal body cavities. **(b)** The somites in a 45-hour chick embryo.

Other structures differentiate in the embryo while the neural tube is forming. On each side of the notochord, mesoderm separates into **somites**, blocks of cells spaced one after the other **(Figure 39.17)**. Somites give rise to the vertebral column, ribs, repeating sets of muscles associated with the ribs and vertebral column, and limb muscles. Mesoderm outside the somites extends around the primitive gut (lateral mesoderm in Figure 39.17) and

splits into two layers, one covering the surface of the gut and the other lining the body wall. The space between the layers is the adult coelom.

39.4b Development of the Eye: Interactions between Cells

Eyes develop by the same basic five-step pathway in all vertebrates **(Figure 39.18)**. The brain forms at the anterior end of the neural tube from a cluster of hollow vesicles that swell outward from the neural tube (see Figure 39.18, step 1). Each of the two optic vesicle develops into an eye. Figure 39.18 depicts the development of a frog eye; the development of the rest of the brain is not shown.

The optic vesicles grow outward until they contact the overlying ectoderm, inducing a series of developmental responses in both tissues. The optic cup, a double-walled structure, forms when the outer surface of the optic vesicle thickens and flattens at the region of contact and then pushes inward. The optic cup ultimately becomes the retina. The lens forms from the lens placode, a disklike swelling that arises when the optic cup induces thickening of overlying ectoderm (step 2). The centre of the lens placode sinks inward toward the optic cup, and its edges eventually fuse together, forming the lens vesicle, a ball of cells (step 3).

The developing lens cells begin to synthesize crystallin, a fibrous protein that collects into clear, glassy deposits. Lens cells finally lose their nuclei and form the elastic, crystal-clear lens.

As the lens develops, it contacts the overlying ectoderm that has closed over it. In response, the ectoderm cells form the cornea by losing their pigment granules and becoming clear. Eventually, the developing cornea joins with the edges of the optic cup to complete the primary structure of the eye (step 4). Other cells contribute to accessory structures of the eye. Mesoderm and neural crest cells contribute to reinforcing tissues

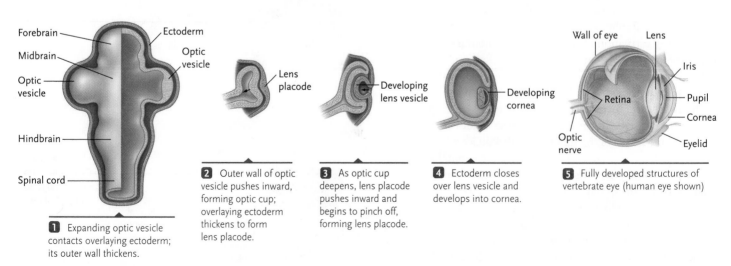

1 Expanding optic vesicle contacts overlaying ectoderm; its outer wall thickens.

2 Outer wall of optic vesicle pushes inward, forming optic cup; overlaying ectoderm thickens to form lens placode.

3 As optic cup deepens, lens placode pushes inward and begins to pinch off, forming lens placode.

4 Ectoderm closes over lens vesicle and develops into cornea.

5 Fully developed structures of vertebrate eye (human eye shown)

Figure 39.18

Stages in the development of the vertebrate eye from the optic vesicle of the brain and the overlying ectoderm.

PEOPLE BEHIND BIOLOGY

Brian K. Hall, Dalhousie University

Brian K. Hall, a professor of biology at Dalhousie University in Halifax, Nova Scotia, and his students study developmental patterns in the embryos of fishes, frogs, and chicks, in particular the development of the skeleton. Considerable work has focused on the embryonic origins of the cells that form the cartilages of the head and shows that skeletal elements in vertebrates often originate in the developing nervous system, specifically in neural crest cells. The research produces insight into the evolution of structures by showing how they are derived and from where (see Chapter 20).

Honours thesis students working with Professor Hall have studied a range of topics, from looking at asymmetry between the left and right limbs of mice to the impact of thyroxine on metamorphosis in a South African frog (*Hymenochirus*). Other research on this frog has explored its capacity to regenerate hindlimbs and the patterns of migration of neural crest cells and pigment cells. This work makes Professor Hall one of the founding fathers of evolutionary developmental biology, or "evo-devo," which is really an extension of traditional developmental biology.

His work on development has led Professor Hall to consider how biologists use words to describe what they see in the natural world. For example, how does one distinguish "rudimentary" and "vestigial" features in animals? It is well known that snakes evolved from limbed ancestors and that whales evolved from ancestors that had teeth. Studies on the development of snakes clearly show that they have rudimentary limbs, partly formed or incomplete transformations of developing features. Vestiges, evolutionary remnants of ancestral features, also occur in animals. The tooth buds (fetal teeth) of baleen whales are examples of vestigial features, as are the rudimentary clavicles that appear in the developing embryos of toothed whales. Once again, the problem of imposing specific terms and definitions on a natural continuum can make the study of biology unnecessarily complex.

Professor Hall's detailed studies of development make it easier to understand the patterns that emerge, which also means appreciating the shortcomings of the labels we use.

---

in the wall of the eye and the muscles that move the eye. Figure 39.18, step 5, shows a fully developed vertebrate eye.

Initial induction by optic vesicles is necessary for development of the eye. If an optic vesicle is removed before lens formation, ectoderm fails to develop into the lens placode and vesicle. Moreover, placing a removed optic vesicle under the ectoderm in other regions of the head causes a lens to form in the new location. If ectoderm over an optic vesicle is removed and ectoderm from elsewhere in the embryo is grafted in its place, a normal lens develops in the grafted ectoderm. This occurs even though in its former location it would not differentiate into lens tissue.

Eye development also demonstrates differentiation. Ectoderm cells induced to form the lens synthesize crystallin. In other locations, ectoderm cells synthesize mainly keratin, a different protein. Keratin is a component of surface structures such as skin, hair, feathers, scales, and horns. In response to induction by the optic vesicle, genes of ectoderm cells coding for crystallin are activated, but genes coding for keratin are not.

39.4c Apoptosis: Programmed Cell Death

Induction and differentiation build complex, specialized organs from three fundamental tissue types. Apoptosis (see Chapter 9), programmed cell death, complements these processes by removing tissues needed during development but not present in the fully formed organ. Apoptosis plays an important role in the development of both invertebrates and vertebrates. The best example of apoptosis in frog development occurs during metamorphosis, when the tadpole changes into an adult frog. The tadpole's tail becomes progressively smaller and finally disappears because its cells disintegrate and their components are absorbed and recycled by other cells. Cells eliminated by apoptosis, like those of a tadpole's tail, are parts of structures required at one stage of development but not later.

STUDY BREAK

1. What processes and embryonic layers are involved in the development of the eye?
2. How does the neural tube develop?

39.5 Embryonic Development of Humans and Other Mammals

The embryonic development of humans is representative of placental mammals. In the uterus, the embryo is nourished by the placenta, which supplies oxygen and nutrients to the embryo and carries carbon dioxide and nitrogenous wastes away from it.

Pregnancy or gestation, the period of mammalian in utero development, varies among species. Larger mammals bearing larger young tend to have longer gestation periods. From fertilization to birth, pregnancy

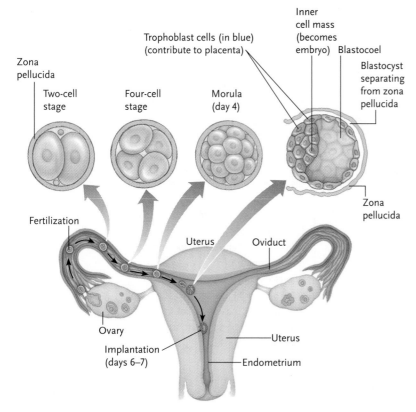

Figure 39.19
Early stages in the development of the human embryo.

Labels in figure:

Inner cell mass (becomes embryo)

Trophoblast cells (in blue) (contribute to placenta)

Blastocoel

Blastocyst separating from zona pellucida

Zona pellucida

Two-cell stage

Four-cell stage

Morula (day 4)

Zona pellucida

Fertilization

Uterus

Oviduct

Ovary

Implantation (days 6–7)

Uterus

Endometrium

lasts about 600 days in elephants, about 365 days in blue whales, and 21 days in hamsters.

In humans, gestation takes an average of 266 days, about 38 weeks. Because the date of fertilization can be difficult to establish, human gestation is usually calculated from the beginning of the menstrual cycle in which fertilization took place. The nine-month period is divided into three **trimesters**, each three months long.

Major developmental events in human gestation—cleavage, gastrulation, and organogenesis—take place during the first trimester. By week 4, the embryo's heart is beating, and by the end of week 8, the major organs and organ systems have formed. From this point until birth, the developing human is called a **fetus**. Only 5 cm long by the end of the first trimester, the fetus grows during the second and third trimesters to an average length of 50 cm and an average mass of 3.5 kg.

Cleavage occurs during the passage of the developing embryo down the fallopian tube and while it is still enclosed in the zona pellucida, the original coat of the egg **(Figure 39.19).**

By day 4, the morula, a ball of 16 to 32 cells, has been produced. By the time the endometrium (uterine lining) is ready for implantation (about seven days after ovulation), the morula has reached the uterus and, through further cell divisions and differentiation, has become a blastocyst. The **blastocyst** is a single-cell-layered hollow ball of about 120 cells with a fluid-filled cavity, the *blastocoel*, which has a dense mass of cells localized on one side. This **inner cell mass** will become the embryo

itself, whereas the outer single layer of cells of the blastocyst, the **trophoblast**, will become tissues that support development of the embryo in the uterus.

When ready to implant, the blastocyst breaks out of the zona pellucida and sticks to the endometrium on its inner cell mass side **(Figure 39.20a)**. Implantation begins when the trophoblast cells that overlie the inner cell mass secrete proteases that digest pathways between endometrial cells. Dividing trophoblast cells fill in the digested spaces, appearing as fingerlike projections into the endometrium. These cells continue to digest nutrient-rich endometrial cells, producing a hole in the endometrium for the blastocyst and releasing nutrients for the developing embryo after it has consumed the small amount of yolk contained in egg cytoplasm. While the blastocyst burrows into the endometrium, the inner cell mass separates into the *embryonic disk*, which consists of two distinct cell layers (see Figure 39.20a). The epiblast, the layer farther from the blastocoel, gives rise to the embryo proper. The hypoblast, the layer nearer the blastocoel, generates part of the extraembryonic membranes. When implantation is complete, the blastocyst has completely burrowed into the endometrium and is covered by a layer of endometrial cells **(Figure 39.20b)**.

Gastrulation proceeds as in birds (see Figure 39.14), with the formation of a primitive streak in the epiblast. Soon after the inner cell mass separates into epiblast and hypoblast, a layer of cells separates from the epiblast along its top margin (see Figure 39.20b). The amniotic cavity is the fluid-filled space created by the separation. The layer of cells forming its roof becomes the amnion, which expands until it completely surrounds the embryo, suspending it in amniotic fluid.

Also as in birds, the hypoblast develops into the yolk sac. In mammals, the mesoderm of the yolk sac gives rise to the blood vessels in the embryonic portion of the placenta. The allantois stores nitrogenous wastes in birds, but it is a small, vestigial sac in human embryos because most nitrogenous wastes are transferred across the placenta to the mother via blood vessels in the placenta and the umbilical cord.

While the amnion is expanding around the embryo, blood-filled spaces form in maternal tissue, and trophoblast cells grow rapidly around both the embryo and amnion to form the chorion **(Figure 39. 20c)**, the membrane that forms most of the embryonic portion of the placenta. Next, a connecting stalk forms between the embryonic disk and the chorion, which begins to grow into the endometrium as fingerlike extensions called **chorionic villi** (singular, villus) **(Figure 39.20d)**. Chorionic villi increase the surface area of the chorion. The placenta forms in the area where these villi grow into the endometrium. As the chorion develops, mesodermal cells of the yolk sac grow into it and form a rich network of blood vessels, the embryonic circulation of the placenta. At the same time, the expanding chorion stimulates the blood vessels of the endometrium to

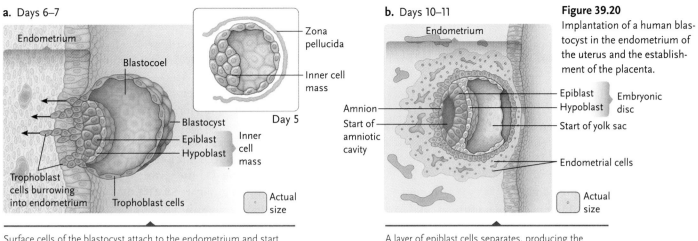

a. Days 6–7

Endometrium

Blastocoel

Blastocyst

Epiblast

Hypoblast

} Inner cell mass

Trophoblast cells burrowing into endometrium

Trophoblast cells

Zona pellucida

Inner cell mass

Day 5

Actual size

Surface cells of the blastocyst attach to the endometrium and start to burrow into it. Implantation is under way.

b. Days 10–11

Endometrium

Amnion

Start of amniotic cavity

Epiblast

Hypoblast

} Embryonic disc

Start of yolk sac

Endometrial cells

Actual size

Figure 39.20
Implantation of a human blastocyst in the endometrium of the uterus and the establishment of the placenta.

A layer of epiblast cells separates, producing the amniotic cavity. The cells above the cavity become the amnion, which eventually surrounds the embryo. The hypoblast begins to form around the yolk sac.

c. Day 12

Blood-filled spaces

Chorion

Yolk sac

Amniotic cavity

Start of chorionic cavity

Actual size

Blood-filled spaces form in maternal tissue. The chorion forms, derived from trophoblast cells, and encloses the chorionic cavity.

d. Day 14

Chorionic villi

Chorion

Chorionic cavity

Yolk sac

Connecting stalk

Actual size

A connecting stalk has formed between the embryonic disk and chorion. Chorionic villi, which will be features of a placenta, start to form.

e. Day 25

Chorion

Chorionic cavity

Embryo

Amniotic cavity

Amnion

Yolk sac

Chorionic villi

Maternal blood

The chorion continues to grow into the endometrium, producing the chorionic villi. The chorion growth stimulates blood vessels of the endometrium to grow into the maternal circulation of the placenta.

f. Day 45

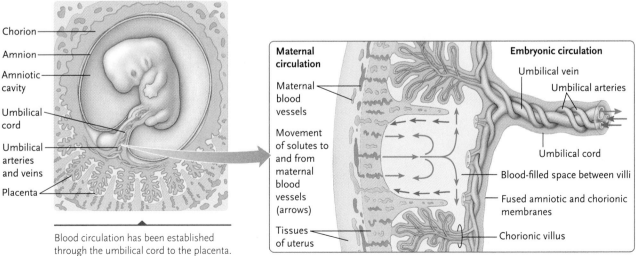

Chorion

Amnion

Amniotic cavity

Umbilical cord

Umbilical arteries and veins

Placenta

Blood circulation has been established through the umbilical cord to the placenta.

Maternal circulation

Maternal blood vessels

Movement of solutes to and from maternal blood vessels (arrows)

Tissues of uterus

Embryonic circulation

Umbilical vein

Umbilical arteries

Umbilical cord

Blood-filled space between villi

Fused amniotic and chorionic membranes

Chorionic villus

grow into the maternal circulation of the placenta **(Figure 39.20e)**.

Within the placenta of humans, apes, monkeys, and rodents, the maternal circulation opens into spaces where maternal blood directly bathes capillaries coming to the placenta from the embryo **(Figure 39.20f)**. (Other mammals have different types of placentae.) Embryonic circulation remains closed so that the embryonic blood and the maternal blood do not mix directly. This isolation prevents the mother from developing an immune reaction against cells of the embryo, which may be recognized as foreign. Eventually, the placenta and its blood circulation grow to cover about a quarter of the inner surface of the enlarged uterus and reach the size of a dinner plate.

When the amnion forms (see Figure 39.20e), the embryo remains connected to the developing placenta through the **umbilicus**, a cord of tissue. Blood vessels in the umbilical cord conduct blood between the embryo and the placenta (see Figure 39.20f, inset). Within the placenta, nutrients and oxygen pass from the mother's circulation into the circulation of the embryo. Besides nutrients and oxygen, many other substances taken in by the mother, such as alcohol, caffeine, drugs, pesticide residues, and toxins in smog and cigarette smoke, can pass from mother to embryo. Carbon dioxide and nitrogenous wastes pass from the embryo to the mother and are disposed of by the mother's lungs and kidneys.

Cells from the embryonic portion of the placenta or from the amniotic fluid are derived from the embryo. To test for the presence of genetic diseases such as cystic fibrosis or Down syndrome, these cells can be obtained by chorionic villus sampling or by amniocentesis (*centesis* = puncture, referring to the use of a needle, which is pushed through the abdominal wall to obtain fluid from the amniotic cavity). Chorionic villus sampling can be carried out as early as the eighth week of pregnancy, compared with 14 weeks for amniocentesis.

39.5a Birth: The Fetus Leaves the Mother

By the end of its fourth week, a human embryo is 3 to 5 mm long, 250 to 500 times the size of the zygote **(Figure 39.21a)**. It has a tail and gill arches, embryonic features of all vertebrates (see Chapter 27). Gill arches contribute to the formation of the face, neck, mouth, nasal cavities, larynx, and pharynx. After five to six weeks, most of the tail has disappeared, and the embryo begins to be a recognizable human form **(Figure 39.21b)**. At eight weeks, the embryo, now a fetus, is about 2.5 cm long **(Figure 39.21c)**. Its organ systems have formed, and its limbs, with fingers or toes at their ends, have developed.

After about 38 weeks, fetal growth comes to a close, the cervix of the uterus softens, and the fetus typically turns so that its head is downward, pressed against the cervix. At this time, a steep rise in levels of estrogen secreted by the placenta cause uterine cells to express the gene for the receptor of the hormone *oxytocin* (secreted by the pituitary gland). Receptors become inserted into the plasma membranes of uterine cells. Oxytocin binds to its receptors, triggering contractions of smooth muscle cells of the uterine wall, beginning the rhythmic contractions of labour. These contractions mark the beginning of the three steps culminating in birth or **parturition** (*parturire* = to be in labour).

Contractions push the fetus against the cervix and stretch its walls **(Figure 39.22,** step 1). In response, stretch receptors in the walls send nerve signals to the hypothalamus, which responds by stimulating the pituitary to secrete more oxytocin. Oxytocin stimulates

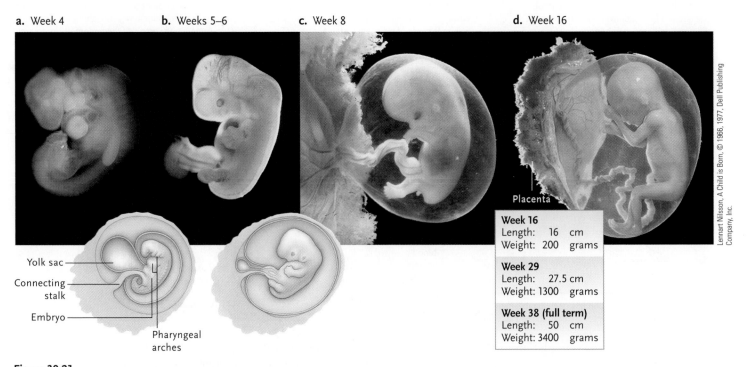

a. Week 4   **b.** Weeks 5–6   **c.** Week 8   **d.** Week 16

Yolk sac
Connecting stalk
Embryo
Pharyngeal arches

Placenta

Week 16		
Length:	16	cm
Weight:	200	grams
Week 29		
Length:	27.5 cm	
Weight:	1300	grams
Week 38 (full term)		
Length:	50	cm
Weight:	3400	grams

Lennart Nilsson, A Child is Born, © 1966, 1977, Dell Publishing Company, Inc.

Figure 39.21

The human embryo at various stages of development, beginning at week 4. The chorion has been moved aside to reveal the embryo in the amnion at week 8 and week 16. By week 16, movements begin as nerves make functional connections with the forming muscles.

Figure 39.22

Birth of the fetus. Hormonal events of birth are at the top, and physical events are at the bottom.

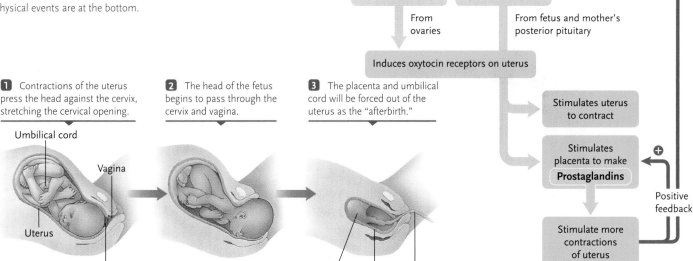

1 Contractions of the uterus press the head against the cervix, stretching the cervical opening.

2 The head of the fetus begins to pass through the cervix and vagina.

3 The placenta and umbilical cord will be forced out of the uterus as the "afterbirth."

Umbilical cord
Vagina
Uterus
Partially dilated cervix

Placenta Uterus Umbilical cord

Estrogen — From ovaries

Oxytocin — From fetus and mother's posterior pituitary

Induces oxytocin receptors on uterus

Stimulates uterus to contract

Stimulates placenta to make **Prostaglandins**

Positive feedback

Stimulate more contractions of uterus

more forceful contractions of the uterus, pressing the fetus more strongly against the cervix and further stretching its walls. The positive feedback cycle continues, steadily increasing the strength of the uterine contractions.

As the contractions force the head of the fetus through the cervix (step 2), the amniotic membrane bursts, releasing the amniotic fluid. Usually after 12 to 15 hours from the onset of uterine contractions, the head passes entirely through the cervix. Once the head is through, the rest of the body follows quickly and the entire fetus is forced out through the vagina, still connected to the placenta by the umbilical cord (step 3).

After the baby takes its first breath, the umbilical cord is cut and tied off by the birth attendant. Uterine contractions continue expelling the placenta and any remnants of the umbilical cord and embryonic membranes as the afterbirth, usually within 15 to 60 minutes after the infant's birth. The short length of umbilical cord still attached to the infant dries and shrivels within a few days. Eventually, it separates entirely and leaves a scar, the umbilicus or navel, to mark its former site of attachment during embryonic development. Immediately after birth, some mammals (e.g., *Gazella* species) can stand and are soon able to run. The newborns of these precocial species contrast with those of altricial species, which are immobile and helpless for some considerable time after birth. The same terms, precocial and altricial, also apply to other animals.

39.5b Milk: Food for the Young

Before birth, estrogen and progesterone secreted by the placenta stimulate the growth of the mammary glands in the mother's breasts. But high levels of these hormones prevent mammary glands from responding to *prolactin*, the hormone secreted by the pituitary that

stimulates the glands to produce milk. After birth and the release of the placenta, levels of estrogen and progesterone in the mother's bloodstream fall steeply, and the breasts begin to produce milk (stimulated by prolactin) and secrete it (stimulated by oxytocin).

Continued milk secretion depends on whether the infant suckles. Stimulation of the nipples sends nerve impulses to the hypothalamus, which responds by signalling the pituitary to release a burst of prolactin and oxytocin. Hormonal stimulation of milk production and secretion continues as long as the infant is breast-fed.

39.5c Gonadal Development: The Gender of the Fetus

Gonads and their ducts begin to develop in the fetus during week 4 of gestation. Until week 7, male and female embryos have the same set of internal structures derived from mesoderm, including a pair of gonads **(Figure 39.23a, p. 953)**. Each gonad is associated with two primitive ducts, the **Wolffian duct** and the **Müllerian duct**, that lead to a cloaca. These internal structures are *bipotential* because they can develop into either male or female sexual organs.

The presence or absence of a Y chromosome determines whether the internal structures develop into male or female sexual organs. In a fetus with XY combination of sex chromosomes, *SRY* (the sex-determining region of the Y), a single gene on the Y chromosome, becomes active in week 7. The protein encoded by *SRY* induces a molecular switch that causes primitive gonads to develop into testes. Fetal testes secrete two hormones, testosterone and the *anti-Müllerian hormone* (*AMH*). Testosterone stimulates development of Wolffian ducts into a male reproductive tract, including the epididymis, vas deferens, and seminal vesicles **(Figure 39.23b)**.

Hormones and External Genitalia

Figure 1
Crocuta crocuta, a spotted hyena.

Many Africans believe that spotted hyenas (*Crocuta crocuta*, **Figure 1**) are hermaphroditic because the females have external genitalia resembling those of the males. Specifically, the clitoris is peniform, and there is a pseudoscrotum **(Figure 2)**. The combined effect means that anyone looking at a spotted hyena easily confuses males and females. Confusion turns to puzzlement when the apparent male is obviously lactating, nursing young. Spotted hyenas are not hermaphroditic. Males mate with females, and fertilized eggs develop into fetuses that are born as in other placental mammals.

How do females come to look like males? In mammals, the neutral condition for genitalia is the arrangement typical of females. The derived condition is what we associate with males. Genetically, female embryos exposed to testosterone during gestation develop malelike genitalia. Blood samples obtained from pregnant female spotted hyenas had relatively high levels of circulating testosterone, 5α-dihydrotestosterone, and androstenedione, albeit not as high as in males. Male and female fetuses experience the same levels of maternal androgens; therefore, females have malelike genitalia.

What selective advantage would female spotted hyenas gain from looking like males? Spotted hyenas are social animals that live in clans. As adults, females are larger and more aggressive than males. During greeting ceremonies, when members of a clan meet after a separation, individuals sniff

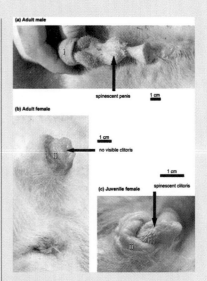

Figure 3
In their external genitalia, specifically their clitorises, **(c)** subadult female fossas (*Cryptoprocta ferox*) resemble **(a)** males, whereas **(b)** adult females do not.

one another's genitals. During these encounters, males erect their penises and females their peniform clitorises. Females appear to dominate spotted hyena societies even though their levels of testosterone and 5α-dihydrotestosterone are lower than in adult males. This does not support the proposal that females are more aggressive because of levels of circulating male hormones. Spotted hyena cubs are precocial and aggressive to the point of siblicide, perhaps reflecting hormone levels at birth. Masculinization of female genitalia could be a side effect of selection for aggressive neonates.

Masculinization of female genitalia occurs in some other members of the order Carnivora, for example, the fossa (*Cryptoprocta ferox*) from Madagascar **(Figure 3)**. Transient masculinization of young female (but not adult female) fossas could allow them to avoid sexual harassment by adult males and to escape from aggression by adult females.

Figure 2
In spotted hyenas (*C. crocuta*), the external genitalia of a nulliparous female (left) resembles that of a male (right). Note the peniform clitoris and pseudoscrotum on the female.

AMH causes the Müllerian ducts to degenerate and disappear. Testosterone also stimulates development of male genitalia.

In a fetus with XX chromosomes, no *SRY* protein is produced. The primitive gonads, under the influence of estrogens and progesterone secreted by the placenta, develop into ovaries. Müllerian ducts develop into oviducts, the uterus, and part of the vagina, and the Wolffian ducts degenerate and disappear **(Figure 39.23c)**. Female sex hormones also stimulate the development

Figure 39.23
Development of the internal sexual organs of males and females from common bipotential origins.

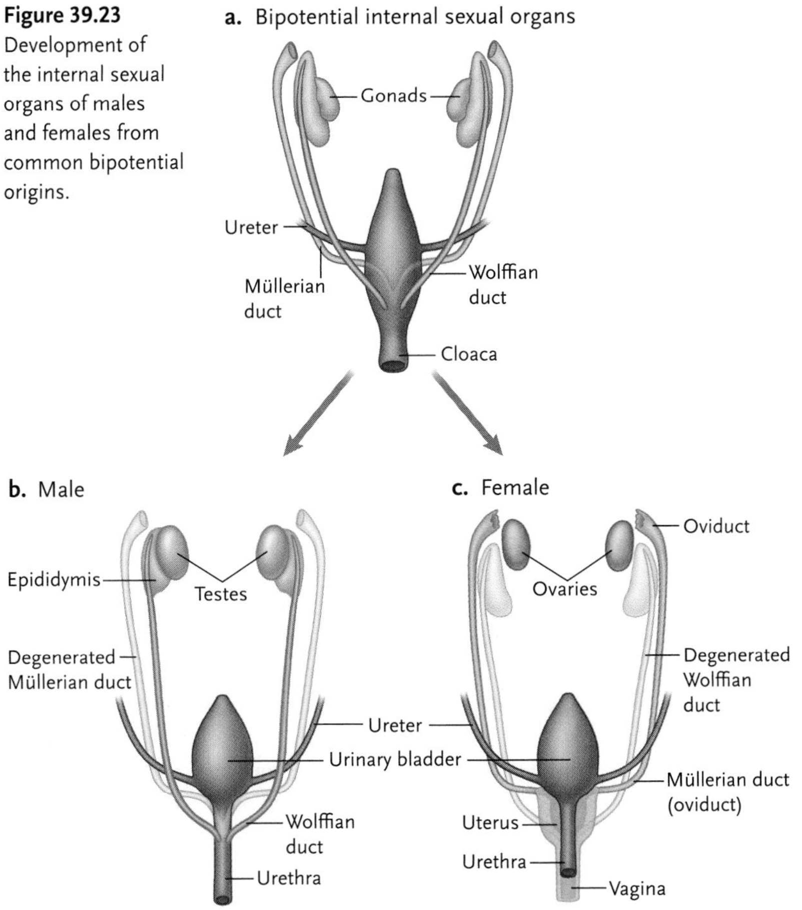

a. Bipotential internal sexual organs

Gonads
Ureter
Müllerian duct
Wolffian duct
Cloaca

b. Male

Epididymis
Testes
Degenerated Müllerian duct
Ureter
Urinary bladder
Wolffian duct
Urethra

c. Female

Oviduct
Ovaries
Degenerated Wolffian duct
Müllerian duct (oviduct)
Uterus
Urethra
Vagina

of the external female genitalia (see *Hormones and External Genitalia*).

39.5d Further Development

Once fetal development is over, the newborn animal follows a prescribed course of further growth and development leading to the mature adult. In humans, internal and external sexual organs mature and secondary sexual characteristics appear at puberty (see Figure 20.18). Similar changes occur in most mammals.

There are many examples among different animal groups of developmental changes that take place after hatching or birth. In some cases, offspring hatch in forms distinctly different in structure from the adult. Examples among invertebrates include insects such as *Drosophila* and butterflies, in which eggs hatch to produce larvae that undergo metamorphosis into adults. Some frogs hatch as tadpoles, which undergo metamorphosis to produce adults.

STUDY BREAK

1. Define the terms blastocyst, chorionic villi, and parturition.
2. What is a placenta? Name some animals with placentae.
3. What *is* SRY? What role does it play?

39.6 Cellular Basis of Development

Orientation and *rate* of mitotic cell division have special significance in the development of the shape, size, and location of organ systems of the embryo. Regulation of the orientation and rate occurs at all stages of development.

"Orientation of cell division" refers to the angles at which daughter cells are added to older cells as development proceeds. Orientation is determined by the location of the furrow separating the cytoplasm after mitotic division of the nucleus (see Chapter 9). The furrow forms in alignment with the spindle midpoint so that when the spindle is centrally positioned in the cell, the furrow leads to symmetrical division of the cell. When the spindle is displaced to one end of the cell, the furrow leads to asymmetrical division into a smaller cell and a larger cell. Little is known about how spindle positioning is regulated.

The rate of cell division primarily reflects the time spent in the G_1 period of interphase (see Chapter 9). Once DNA replication begins, the rest of the cell cycle takes the same time in all cells of a species. As an embryo develops and cells differentiate, the time spent in interphase increases and varies in different cell types. Therefore, different cell types proliferate at various rates as they differentiate, giving rise to tissues and organs with different cell numbers. When fully differentiated, some cells remain fixed in interphase and stop replicating DNA or dividing. Nerve cells in the mammalian brain and spinal cord stop dividing once the nervous system is fully formed. Ultimately, the rate of cell division is under genetic control.

Frog egg cleavage provides examples of how both changes in orientation and rate of mitotic division affect development. The first two cleavages start at the animal pole and extend to the vegetal pole, producing four equal blastomeres (see Figure 39.8). The third cleavage occurs equatorially, but because of yolk in the vegetal region of the embryo, this cleavage furrow forms toward the animal pole. This cleavage produces an eight-cell embryo with four small blastomeres near the animal pole and four large blastomeres in the vegetal region. Blastomeres in the animal region of the embryo proceed to divide rapidly, whereas blastomeres in the vegetal region divide more slowly because division is inhibited by yolk. The resulting morula consists of an animal region with many small cells and a vegetal region with relatively few but larger blastomeres.

39.6a Microtubules and Microfilaments: Movements of Cells

Embryonic cells undergo changes in shape that generate movements, such as the infolding of surface layers to produce endoderm or mesoderm. Entire cells also move during embryonic growth, both singly and in groups. Movements are also produced by changes in rates of growth or by breakdowns of microtubules and microfilaments. Changes in both cell shape and cell movement play important roles in cleavage, gastrulation, and organogenesis.

39.6b Change in Cell Shape: Adjusting to New Roles

Changes in cell shape typically result from reorganization of the cytoskeleton. During development of the neural plate in frogs, the ectoderm flattens and thickens and cells in the ectoderm layer change from cubelike to columnar in shape (Figure 39.24a).

Sinking of the neural plate downward along its midline reflects changes in cell shape from columnar to wedgelike (Figure 39.24b). As one end of each cell narrows, the entire cell layer invaginates (is forced inward). How does this occur? Each wedge-shaped cell contains a group of microfilaments arranged in a circle at the top. Microfilaments slide over each other, tightening the ring like a drawstring and narrowing the top of the cell. If an experimenter adds cytochalasin, a chemical that interferes with microfilament

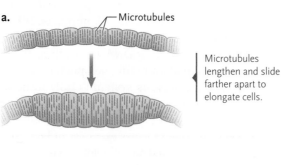

Microtubules lengthen and slide farther apart to elongate cells.

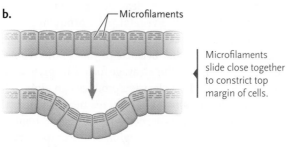

Microfilaments slide close together to constrict top margin of cells.

Figure 39.24
The roles of **(a)** microtubules and **(b)** microfilaments in the changes in cell shape that produce developmental movements.

assembly, to the cells, the microfilament circle disperses, and invagination does not occur.

39.6c Movements of Whole Cells to New Positions in the Embryo

Cell movements during gastrulation and long-distance migrations of neural crest cells are striking examples of movements of whole cells during embryonic development. These movements involve coordinated activity by microtubules and microfilaments. The typical pattern of movement is a repeating cycle of extension, anchoring, and contracting. First, a cell attaches to the substrate (**Figure 39.25,** step 1), and then it moves forward by elongating from the point of attachment (step 2). The cell then makes a new attachment at the advancing tip (step 3) and contracts until the rearmost attachment breaks (step 4). The front attachment now serves as the base for another movement.

How do the cells know where to go? Typically, cells migrate over the surfaces of stationary cells in one of the embryo's layers. In many developmental systems, migrating cells follow tracks formed by molecules of the ECM, secreted by cells along the route over which they travel. Fibronectin, an important track molecule, is a fibrous, elongated protein of the ECM. Migrating cells recognize and adhere to fibronectin, and in response, internal changes in cells trigger their movement in a direction based on alignment of fibronectin molecules.

Some migrating cells follow concentration gradients rather than molecular tracks. Gradients are created by diffusion of molecules (often proteins) released by cells in one part of an embryo. Cells with receptors for the diffusing molecule follow the gradient toward or away from the source.

Selective cell adhesion—the ability of an embryonic cell to make and break specific connections to other cells—is closely related to cell movement. As development proceeds, many cells break initial adhesions, move, and form new adhesions in different locations. Final cell adhesions hold the embryo in its correct shape and form. Junctions of various kinds, including tight, anchoring, and gap junctions, reinforce final adhesions. Selective cell adhesions were first demonstrated in a classic experiment by Johannes Holtfreter and P.L. Townes (see *Making Cell Connections*).

Many cell surface proteins are responsible for selective cell adhesions, including **cell adhesion molecules** (CAMs) and **cadherins** (*calcium-dependent adhesion molecules*). Cadherins require calcium ions to set up adhesions—hence their name. As cells develop, different types of CAMs or cadherins appear or disappear from their surfaces as they make and break cell adhesions. The changes reflect alterations in gene activity, often in response to molecular signals arriving from other cells. During early development in the chick, cells of the ectoderm are held together by E- and N-cadherins. As the neural plate appears, cells destined to form the neural

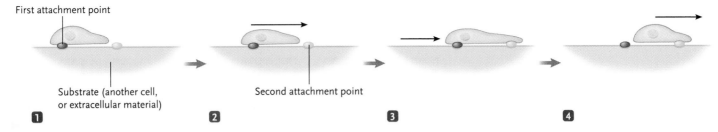

Figure 39.25

The cycle of attachment, stretching, and contraction by which a cell moves over other cells or extracellular materials in embryos.

tube lose their cadherins, and N-CAMs appear on their surfaces. As these surface molecules appear, the neural tube cells break loose from the ectoderm and adhere to each other to form the neural tube. If N-cadherin is added experimentally to ectoderm cells, the neural tube stays anchored and never separates.

39.6d Induction: Interactions between Cells

Induction is the process in which a group of inducing cells causes or influences a nearby group of responding cells to follow a particular developmental pathway. Induction is the major process responsible for *determination*, in which the developmental fate of a cell is set. Induction occurs through the combination of signal molecules with surface receptors on the responding cells. The signal molecules may be located on the surface of the inducing cells or released by inducing cells. The surface receptors are activated by binding the signal molecules. In the activated form, signal molecules trigger internal response pathways that produce the developmental changes (see Chapter 15). Often responses include changes in gene activity.

In the 1920s, Hans Spemann and Hilde Mangold conducted the first experiments identifying induction in embryos. They found that if the dorsal lip of a newt embryo was removed and grafted into a different position of another embryo, cells moving inward from the dorsal lip induced a neural plate, a neural tube, and eventually an entire embryo at the new location (see *Spemann and Mangold's Experiment Demonstrating Induction in Embryos*). They proposed that the dorsal lip is an *organizer*, acting on other cells to alter the course of development. This action is now known as *induction*. Spemann received the Nobel Prize in 1935 for his research. (Mangold had died in a tragic accident in 1924 when her kitchen gasoline heater exploded, the year their research paper was published. She would likely have also received the Nobel Prize, but it is never awarded posthumously.)

Spemann and Mangold's findings touched off a search for inducing molecules that must pass from inducing cells to responding cells. In 1992, researchers constructed a DNA library from *Xenopus* gastrulas. By isolating and cloning cellular DNA in gene-sized pieces, they made mRNA transcripts of cloned genes and injected them into early *Xenopus* embryos in which the induc-

ing ability of mesoderm had been destroyed by exposure to ultraviolet light. Some injected mRNAs, translated into proteins in the embryos, were able to induce formation of a neural plate and tube, leading to a normal embryo. More than 10 proteins that act as inducing molecules have been identified in the *Xenopus* system.

Differentiation produces specialized cells without the loss of genes. By this process, cells that have committed to a particular developmental fate by the determination process (see Section 39.2) develop into specialized cell types with distinct structures and functions. As part of differentiation, cells concentrate on the production of molecules characteristic of the specific types. For example, 80 to 90% of the total protein synthesized by lens cells is crystallin.

Research into differentiation confirmed that as cells specialize, they retain all of the genes of the original egg cell. Except in rare instances, differentiation does not occur through selective gene loss. Robert Briggs and Thomas King, and later John B. Gurdon, used ultraviolet light to destroy the nucleus of a fertilized frog egg. A micropipette was then used to transfer a nucleus from a fully differentiated tissue, intestinal epithelium, to the enucleated egg. Some eggs receiving transplanted nuclei subsequently developed into normal tadpoles and adult frogs. This outcome is possible only if the differentiated intestinal cells retained the full complement of genes. This conclusion was extended to mammals in 1997 when Ian Wilmut and his colleagues successfully cloned a sheep (named Dolly) starting with an adult cell nucleus.

From the early days of studying development, embryologists focused on describing not only how embryos form and develop but also exactly how adult tissues and organs are produced from embryonic cells. An important goal of embryology was to trace cell lineages from embryo to adult. For most organisms, it is not possible to trace lineages at the individual cell level, primarily because of the complexity of the developmental process and the opacity of embryos. Experimenters developed *fate mapping*, mapping adult or larval structures onto the region of the embryo from which each structure developed. Fate mapping is done by following the development of living embryos under the microscope, either using species in which the embryo is transparent or by marking cells so they can be followed. Cells may be marked with vital dyes

Making Cell Connections: Selective Adhesion Properties of Cells

1. Holtfreter and Townes separated ectoderm, mesoderm, and endoderm tissue from amphibian embryos soon after the neural tube had formed. They used embryos from amphibian species that had cells of different colours and sizes, so they could follow under the microscope where each cell type ended up. (The colours shown here are for illustrative purposes only.)

2. The researchers placed the tissues individually in alkaline solutions, which caused the tissues to break down into single cells.

3. Holtfreter and Townes then combined suspensions of single cells in various ways. Shown here are ectoderm + mesoderm and ectoderm + mesoderm + endoderm. When the pH was returned to neutrality, the cells formed aggregates. Through a microscope, the researchers followed what happened to the aggregates on agar-filled petri dishes.

RESULTS: In time, the reaggregated cells sorted themselves with respect to cell type; that is, instead of the cell types remaining mixed, each cell type became separated spatially. That is, in the ectoderm + mesoderm mixture, the ectoderm moved to the periphery of the aggregate, surrounding mesoderm cells in the centre. In no case did the two cell types remain randomly mixed. The ectoderm + mesoderm + endoderm aggregate showed further that cell sorting in the aggregates generated cell positions reflecting the positions of the cell types in the embryo. That is, the endoderm cells separated from the ectoderm and mesoderm cells and became surrounded by them. In the end, the ectoderm cells were located on the periphery, the endoderm cells were internal, and the mesoderm cells were between the other two cell types.

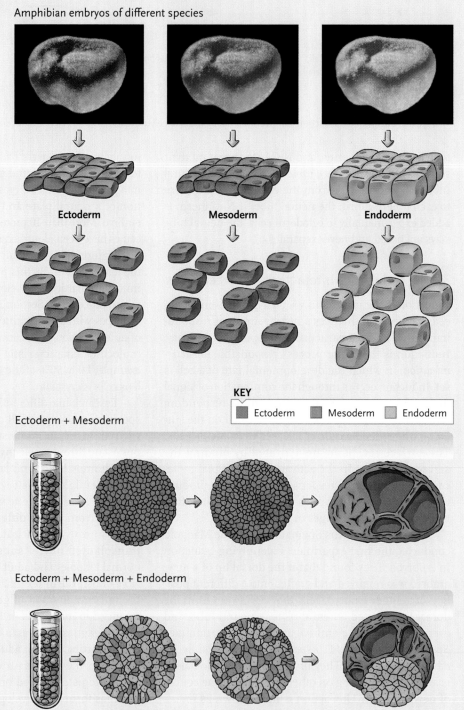

Amphibian embryos of different species

Ectoderm   Mesoderm   Endoderm

KEY

Ectoderm   Mesoderm   Endoderm

Ectoderm + Mesoderm

Ectoderm + Mesoderm + Endoderm

CONCLUSION: Holtfreter interpreted the results to mean that cells have selective affinity for each other; that is, cells have selective adhesion properties. Specifically, he proposed that ectoderm cells have positive affinity for mesoderm cells but negative affinity for endoderm cells, whereas mesoderm cells have positive affinity for both ectoderm cells and endoderm cells. In modern terms, these properties result from cell surface molecules that give cells specific adhesion properties.

(that do not kill cells), fluorescent dyes, or radioactive labels. Fate maps have been produced for *Xenopus*, the chick, and *Drosophila*.

In most cases, a fate map is not detailed enough to show how particular cells in the embryo gave rise to cells of the adult. The exception is the fate map of the

Spemann and Mangold's Experiment Demonstrating Induction in Embryos

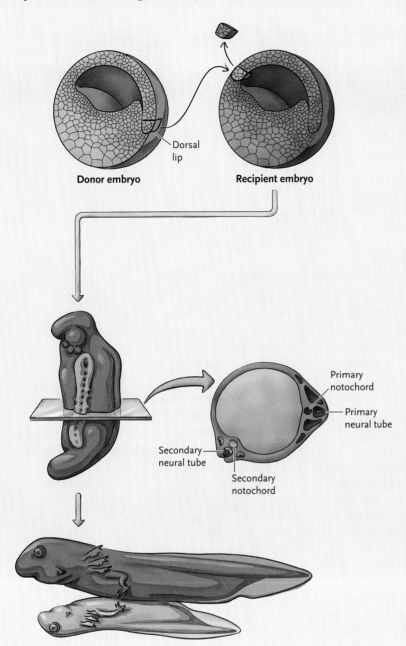

Donor embryo

Recipient embryo

Dorsal lip

Primary notochord

Primary neural tube

Secondary neural tube

Secondary notochord

QUESTION: Does induction occur in embryonic development?

EXPERIMENT: Hans Spemann and Hilde Mangold performed transplantation experiments with newt embryos, the results of which demonstrated that specific induction of development occurs in the embryos. The researchers removed the dorsal lip of the blastopore from one newt embryo and grafted it onto a different position—the ventral side—of another embryo. The two embryos were from different newt species that differed in pigmentation, allowing them to follow the fate of the tissue easily. The embryo with the transplant was allowed to develop.

CONCLUSION: The grafted dorsal lip of the blastopore induced a second gastrulation and subsequent development in the ventral region of the recipient embryo. The result demonstrated the ability of particular cells to induce the development of other cells.

nematode *Caenorhabditis elegans*, an organism with a fixed, reproducible developmental pattern. *C. elegans* has a transparent body, and scientists have mapped the fate (traced the **cell lineage**) of every cell as the zygote divides and the resulting embryo differentiates into a 959-cell adult hermaphrodite or 1031-cell adult male **(Figure 39.26, p. 958)**. All somatic cells of the adult can be traced from five somatic *founder cells* produced during early development. Knowing the cell lineages of *C. elegans* has been a valuable tool for research into the genetic and molecular control of development because mutants affecting development can be easily visualized.

STUDY BREAK

1. What role do microtubules and microfilaments play in development?
2. What happens during induction? Where does it occur?
3. How has work with *Caenorhabditis elegans* advanced our knowledge of developmental biology? What about *Xenopus*?

a. Founder cells

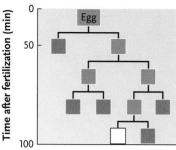

b. Cell lineage for intestinal cells

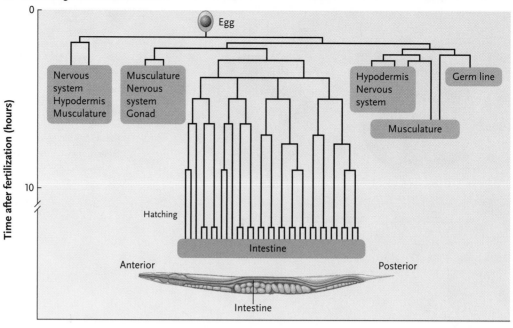

Figure 39.26

Cell lineages of *C. elegans*. **(a)** The founder cells (blue) produced in early cell divisions from which all adult somatic cells are produced. The white cell gives rise to germ-line cells. **(b)** The cell lineage for cells that form the intestine. The detailed lineages for the other parts of the adult are not shown.

39.7 Genetic and Molecular Control of Development

Developmental biologists are interested in identifying and characterizing the genes involved in development and defining how gene products regulate and bring about elaborate events. One productive research approach has been to isolate mutants that affect developmental processes and then identify the genes involved, clone these genes, and analyze them in detail to build models for molecular functions of gene products in development. Model organisms used for these studies include the fruit fly (*Drosophila melanogaster*) and *C. elegans* among invertebrates, and the zebrafish (*Danio rerio*) and the house mouse (*Mus musculus*) among vertebrates.

39.7a Genetic Control of Development

Gene expression regulates changes that occur through determination and differentiation. One well-studied example of the genetic control of these processes is the production of skeletal muscle cells from somites in mammals **(Figure 39.27)**. Somites are blocks of mesoderm cells that form along both sides of the notochord (see Figure 39.17). Under genetic control, some cells of each somite differentiate into skeletal muscle cells. First, paracrine signalling from nearby cells induces somite cells to express the master regulatory gene, *myoD*. The product of *myoD* is the transcription factor MyoD. By turning on specific muscle-determining genes, the action of MyoD brings about determination of those cells, converting them to undifferentiated

muscle cells, called **myoblasts**. Among genes that MyoD regulates are *myogenin* and *MEF* genes. Both are regulatory genes, expressing transcription factors in myoblasts that turn on yet another set of genes. The products

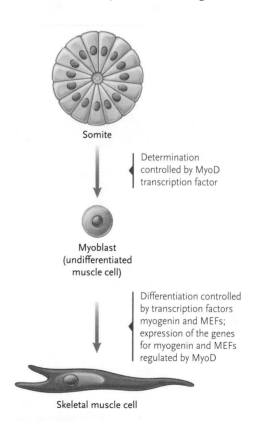

Somite

Determination controlled by MyoD transcription factor

Myoblast (undifferentiated muscle cell)

Differentiation controlled by transcription factors myogenin and MEFs; expression of the genes for myogenin and MEFs regulated by MyoD

Skeletal muscle cell

Figure 39.27

The genetic control of determination and differentiation involved in mammalian skeletal muscle cell formation.

of those genes, including myosin (a major protein involved in muscle contraction), promote differentiation of myoblasts into specific types of muscle cells, such as skeletal muscle cells or cardiac muscle cells.

Molecular mechanisms involved in determination and differentiation usually depend on regulatory genes that encode regulatory proteins that control the expression of other genes. Regulatory genes act as master regulators, and in most cases, the expression of the regulatory genes is controlled by induction.

39.7b Gene Control of Pattern Formation: Developing a Body

As part of the signals guiding differentiation, cells receive positional information that tells them where they are in the embryo. Positional information is vital to **pattern formation**, the arrangement of organs and body structures in their proper three-dimensional relationships. Positional information is laid down primarily as concentration gradients of regulatory molecules produced by genetic control. In most cases, gradients of several different regulatory molecules interact to tell a cell, or a cell nucleus, where it is in the embryo. Genetic control of pattern formation is well documented in *Drosophila melanogaster*. Developmental principles discovered in *D. melanogaster* also apply to many other animal species, including humans.

39.7c Embryogenesis in *Drosophila*: Fruit Fly Model

Production of an adult fruit fly from a fertilized egg occurs in a sequence of genetically controlled development events. Following fertilization, division of the nucleus begins by mitosis. This produces a multinucleate blastoderm because the cytoplasm does not divide in the early embryo (cytokinesis does not occur) **(Figure 39.28)**. At the tenth nuclear division, the nuclei migrate to the periphery of the embryo, where, three divisions later, the 6000 or so nuclei are organized into separate cells. At this stage, the embryo is a *cellular blastoderm*, corresponding to the late blastula stage in the animals discussed above. Ten hours after fertilization, the cellular blastoderm develops into a segmented embryo (an embryo with distinct segments). About 24 hours after fertilization, the egg hatches into a larva that will undergo three moults before becoming a pupa. The adult fly emerges after metamorphosis 10 to 12 days after fertilization. The colours in Figure 39.28 illustrate how segments of the embryo can be mapped to the segments of the adult fly.

The study of developmental mutants has provided important information about *Drosophila* development. Three researchers performed key, pioneering research with developmental mutants: Edward B. Lewis, Christiane Nüsslein-Volhard, and Eric Wieschaus. The three shared a Nobel Prize in 1995 "for their

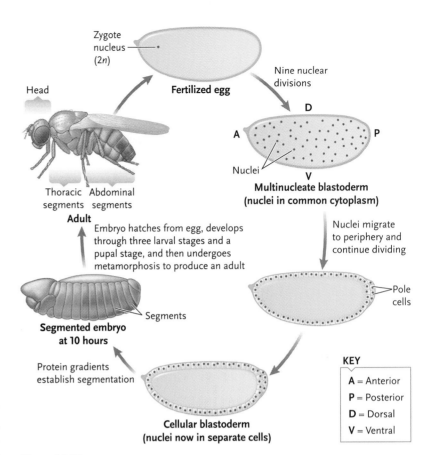

Figure 39.28

Embryogenesis in *Drosophila* and the relationship between segments of the embryo and segments of the adult.

discoveries about the genetic control of early embryonic development."

Nüsslein-Volhard and Wieschaus studied early embryogenesis. They searched for *every* gene required for early pattern formation in the embryo by looking for recessive, *embryonic lethal* mutations. When homozygous, these mutations result in embryo death during development. By determining the stage at which the embryo died and how development was disrupted, they gained insights into the role of the particular genes in embryogenesis.

Lewis studied mutants that changed the fates of cells in particular embryonic regions, producing structures in the adult that normally were produced by other regions. His work was the foundation of research identifying master regulatory genes that control the development of body regions in a wide range of organisms.

39.7d Maternal-Effect Genes and Segmentation Genes: Segmenting the Body

A number of genes control the establishment of the embryo's body plan. These genes regulate the expression of other genes. Two classes, *maternal-effect genes* and *segmentation genes*, work sequentially **(Figure 39.29, p. 960)**. Many **maternal-effect genes** are expressed by the mother during oogenesis. These genes control egg polarity and thus embryo polarity. Some of these genes

Figure 39.29
Maternal-effect genes and segmentation genes and their role in *Drosophila* embryogenesis.

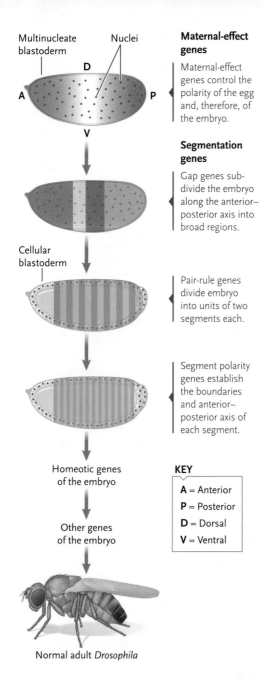

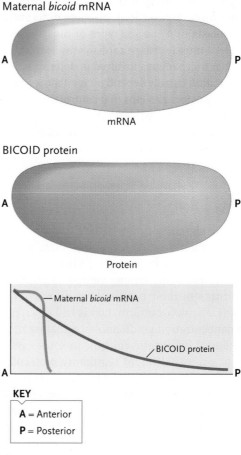

Figure 39.30
Gradients of *bicoid* mRNA and BICOID protein in the *Drosophila* egg.

control formation of embryonic anterior structures, whereas others control formation of posterior structures. Still others control formation of the terminal end.

The *bicoid* gene is the key maternal-effect gene responsible for development of the head and thorax. This gene is transcribed in the mother during oogenesis, and the resulting mRNAs are deposited in the egg, localizing near the anterior pole **(Figure 39.30)**. After fertilization, translation of mRNAs produces BICOID protein, which diffuses through the zygote to form a gradient with highest concentration at the anterior end and fading to none at the posterior end. BICOID is a transcription factor that activates some genes and represses others along the anterior–posterior axis of the embryo. Embryos with mutations in the *bicoid* gene lack thoracic structures but have posterior structures at each end. In normal embryos, the *bicoid* gene is a master regulator gene controlling the expression of

genes for the development of anterior structures (head and thorax).

The activities of products of other maternal-effect genes in gradients in the embryo are also involved in axis formation. The *nanos* gene is the key maternal-effect gene for the posterior structures. When the *nanos* gene is mutated, embryos lack abdominal segments.

Once the axis of the embryo is set, expression of at least 24 **segmentation genes** progressively subdivides the embryo into regions, determining the segments of the embryo and the adult (see Figure 39.29). Gradients of BICOID and other proteins encoded by maternal-effect genes regulate expression of the embryo's segmentation genes differentially. So each segmentation gene is expressed at a particular time and in a particular location during embryogenesis.

Three sets of segmentation genes are regulated in a cascade of gene activations. **Gap genes** such as *hunchback* and *tailless* are the first to be expressed. These genes are activated based on their positions in the maternally directed anterior–posterior axis of the zygote by reacting to the concentrations of BICOID and other proteins. Products of gap genes control subdivision of the embryo into several broad regions along the anterior–posterior axis. Mutations in gap genes

a. Gap genes   **b.** Pair-rule genes   **c.** Segment polarity genes

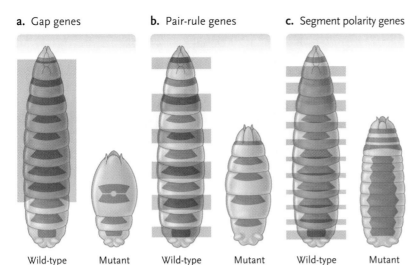

Wild-type   Mutant   Wild-type   Mutant   Wild-type   Mutant

Figure 39.31

Examples of mutations in the different types of segmentation genes of *Drosophila*. Orange highlights indicate wild-type segments that are mutated. **(a)** Gap gene mutants lack one or more segments. **(b)** Pair-rule gene mutants are missing every other segment. **(c)** Segment polarity genes have segments with one part missing and the other part duplicated as a mirror image.

result in the loss of one or more body segments in the embryo **(Figure 39.31a).**

Products of gap genes are transcription factors that activate **pair-rule genes**, such as *even-skipped* and *fushi tarazu*. The products of pair-rule genes divide the embryo into units of two segments each. Mutations in pair-rule genes lead to the deletion of every other segment of the embryo **(Figure 39.31b).**

Expression of **segment polarity genes**, *engrailed* and *gooseberry*, is regulated by products of pair-rule genes. The actions of the products of segment polarity genes set boundaries and the anterior–posterior axis of each segment in the embryo. Mutations in segment polarity genes produce segments in which one part is missing and the other part is duplicated as a mirror image **(Figure 39.31c).** The products of segment polarity genes are transcription factors and other molecules that regulate other genes involved in laying down the pattern of the embryo.

39.7c   Homeotic Genes: Structure and Determining Outcomes

Once the segmentation pattern has been set, **homeotic** (structure-determining) **genes** of the embryo specify what each segment will become after metamorphosis. In normal flies, homeotic genes are master regulatory genes controlling development of structures such as eyes, antennae, legs, and wings on particular segments **(Figure 39.32).** Mutations of these genes allowed researchers to discover the role of homeotic genes. In the *Antennapedia* mutant fly, legs develop instead of antennae (see Figure 39.32).

How do homeotic genes regulate development? Homeotic genes encode transcription factors that regulate expression of genes responsible for the development of adult structures. Each homeotic gene has a common region called a homeobox that is key to its function. A homeobox corresponds to an amino acid section of the encoded transcription factor called the **homeodomain**. The homeodomain of each protein binds to a

region in the promoters of the genes whose transcription it regulates.

Eight homeobox (*Hox*) genes in *Drosophila* are organized along a chromosome in the same order as they are expressed along the anterior–posterior body axis **(Figure 39.33, p. 962).** The discovery of *Hox* genes in *Drosophila* led to a search for equivalent genes in other organisms. *Hox* genes are present in all major animal phyla, where they control the development of segments or regions of the body and are arranged in order in the genome. Homeobox sequences in *Hox* genes are highly conserved, indicating common function in the wide range of animals in which they occur. Homeobox sequences of mammals are the same as or similar to those of the fruit fly (see Figure 39.33). Homeotic genes also are found in plants, where they affect flower development. Homeobox genes have been identified and analyzed in *Arabidopsis* (see Chapter 31).

Normal

Oliver Meckes/Nicole Ottawa/Photo Researchers, Inc.

Antennapedia mutant

UCSF Computer Graphics Laboratory, National Institutes, NCRR Grant 01081

Figure 39.32

Antennapedia, a homeotic mutant of *Drosophila*, in which legs develop in place of antennae.

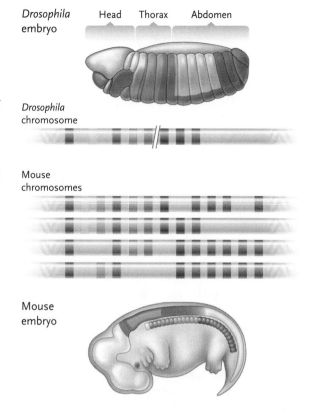

a. Weeks 5–6

b. Week 18

Carolina Biological Supply Company

*Lennart Nilsson from A Child Is Born, © 1966, 1977
Dell Publishing Company, Inc.*

Figure 39.34
An illustration of apoptosis in humans: the removal of tissue between developing fingers and toes to produce the free fingers and toes.

39.7f Cell-Death Genes: Apoptosis

Apoptosis plays a role in the breakdown of a tadpole's tail and in many other patterns of development in vertebrates and invertebrates. In humans, developing fingers and toes are initially connected by tissue, forming paddle-shaped structures. Later in development, cells of this tissue die by apoptosis, resulting in separated fingers and toes **(Figure 39.34)**. Like many other mammals, kittens and puppies are born with their eyes sealed shut by an unbroken layer of skin. Just after birth, cells die in a thin line across the middle of each eyelid, freeing the eyelids and allowing them to open. During pupation from caterpillar to butterfly, many tissues of the larva break down by apoptosis to be replaced by newly formed adult tissues.

Apoptosis results from gene activation in response to molecular signals from receptors on the surfaces of marked cells. In effect, the signals are death notices, delivered at a specific time during embryonic development. In the nematode *C. elegans*, division of the zygote produces 1090 cells. Of these, exactly 131 die at prescribed times to produce a total of 959 cells in the adult hermaphrodite.

In *C. elegans*, a "death signal" molecule that binds to a receptor in the plasma membrane of the target cell results in apoptosis. When the receptor is activated, it leads to activation of proteins that kill the cell. The killing proteins remain inactive in the absence of the death signal.

In the absence of a death signal, the membrane receptor is inactive **(Figure 39.35a)**. This allows CED-9, a protein associated with the outer mitochondrial membrane (encoded by the *ced-9* cell-death gene), to inhibit CED-4 (encoded by the *ced-4* gene) and CED-3 (encoded by the *ced-3* gene). These two proteins are needed to turn on the cell-death program. Cells with the *ced-9* gene expressed and its product CED-9 active normally survive in the adult nematode. When a death signal binds to and activates a receptor, the resulting events are typical of signal transduction pathways (see Figure 39.37). Now the activated receptor leads to inactivation of CED-9. In the absence of CED-9, CED-4 is activated, which, in turn, activates CED-3. Activated CED-3 triggers a cascade of reactions, including activation of proteases and nucleases that degrade cell structures and chromosomes.

Studies of mutants have helped us understand the role of cell-death genes in *C. elegans*. In mutants lacking normal *ced-3* or *ced-4* genes, the 131 marked cells fail to die, producing a disorganized embryo. In the nervous system, the 103 cells that normally die by apoptosis live to form extra neurons in mutants. These extra neurons are inserted at random in the embryo, leading to a disorganized, nonfunctional nervous system.

Genes related to *ced-3* and *ced-4* occur in all animals tested for their presence. In humans and other mammals, the *caspase-9* gene is equivalent to *ced-3*. The *caspase-9* gene, which encodes a protease that degrades cell structures, is activated in cells that form the webbing between the fingers and toes in a human embryo, causing it to break down. The equivalent of *ced-4* is the *Apaf* gene (for *a*poptotic *p*rotease-*a*ctivating *f*actor). Mammalian cells are saved from death by the *Bcl* family of genes, which are the equivalent of *ced-9* in *C. elegans*. The genes are so closely related that they retain their effects if they are exchanged between *C. elegans* and human cells.

STUDY BREAK

1. Give two examples of genetic control of development.
2. What are maternal-effect genes?
3. What are homeotic genes, and what do they do?

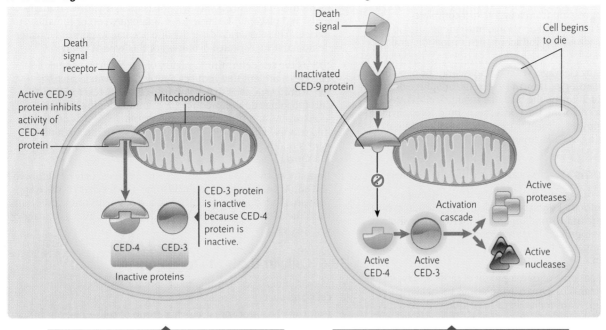

a. No death signal

Death signal receptor

Active CED-9 protein inhibits activity of CED-4 protein

Mitochondrion

CED-4   CED-3

Inactive proteins

CED-3 protein is inactive because CED-4 protein is inactive.

Apoptosis is inhibited as long as CED-9 protein is active; cell remains alive.

b. Death signal

Death signal

Cell begins to die

Inactivated CED-9 protein

Active CED-4   Active CED-3

Activation cascade

Active proteases

Active nucleases

When a death signal binds to the death signal receptor, it activates the receptor, which leads to inactivation of CED-9 protein. As a result, CED-4 protein is no longer inhibited and becomes active, activating CED-3 protein. Active CED-3 triggers a cascade of activations producing active proteases and nucleases, which cause the changes seen in apoptotic cells and eventually to cell death.

Figure 39.35

The molecular basis of apoptosis in C. *elegans*. **(a)** In the absence of a death signal, no apoptosis occurs. **(b)** In the presence of a death signal, activation of CED-4 and CED-3 proteins triggers a pathway that leads to the cell's death.

Unanswered Question

How can the use of stem cells in the development of human medical procedures be informed by changes in development from zygote to gastrula?

Review

Go to CENGAGENOW™ at http://hed.nelson.com/ to access quizzing, animations, exercises, articles, and personalized homework help.

39.1 Housing and Fuelling Developing Young

- In mammals, relaxin is a polypeptide hormone produced by the ovaries during pregnancy. Relaxin inhibits muscular contractions of the uterus and promotes the growth of glands that produce milk. As the time for parturition approaches, relaxin causes relaxation of the pubic ligaments and softens and enlarges the opening to the cervix.

- Vivaparous animal embryos often develop in the uterus (or a uteruslike structure in the reproductive tract). Ovoviviparous animal embryos may develop in the oviduct, the stomach (at least one species of frog), the mouth (several species of fish), or a brood pouch (other fish, such as sea horses). Ovoviviparous animals lay eggs, and the embryos develop outside the body.

- Strictly defined, milk is produced by female mammals to feed their young. Milklike substances (milk analogues) are produced by many other animals: the "crop milk" of pigeons, secretions of the skin in some fish and caecilians, or the uterine milk of a variety of animals, from some cockroaches to elasmobranchs.

- Yolk is food housed within the egg and used to support the growth and development of the embryo. The eggs of birds, insects, reptiles, and many other animals have large deposits of yolk that support the complete development of the young. Other animals have varying amounts of yolk corresponding to shorter periods of in-egg development.

39.2 Mechanisms of Embryonic Development

- The process of progressing from a zygote to a complete organism involves mitotic cell divisions, movements of cells, selective cell adhesions, induction, determination, and differentiation.
- The zygote divides by cleavage, without increasing the overall size or mass of the embryo. It forms a morula, or sphere of blastomeres (cells), which forms the blastula, a single-cell-layered hollow ball of cells enclosing the blastocoel, a fluid-filled cavity. The blastula invaginates to form the gastrula, a hollow ball of cells with an opening, the blastopore.

39.3 Major Patterns of Cleavage and Gastrulation

- Protostomes show determinant development in the progression from egg to animal. From the first divisions, blastomeres develop into specific tissues and organs. Deuterostomes show indeterminant development. Blastomeres produced by the first several divisions remain totipotent, capable of developing into a complete organism.
- Differences in the patterns of development of birds and typical mammals are functions of oviparity (egg-laying) versus viviparity (bearing live young). Mammalian eggs have little yolk, and the developing embryo depends on its mother for food and oxygen, as well as the collection and removal of wastes. The eggs of birds have large amounts of yolk, and the complete development of the embryo takes place independent of the mother's body. Bird embryo development begins with the formation of the blastodisk, a layer of cells on the surface of the yolk.

39.4 Organogenesis: Gastrulation to Adult Body Structures

- In the amniote egg, the amnion (an extraembryonic membrane) encloses the developing embryo in a pool of amniotic fluid. The chorion, another extraembryonic membrane, is produced from ectoderm and mesoderm and lines the inside of the egg shell. It is the site of oxygen exchange for the developing embryo. The allantois is enclosed by the allantoic membrane, which forms from mesoderm and endoderm. The allantois is the storage site for metabolic end products such as uric acid. The amniote egg allows development outside of water and was a fundamental breakthrough in the evolution of fully terrestrial vertebrates.
- In vertebrates, the eye forms when two sides of the anterior end of the neural tube (optic vesicles) swell outward until they contact the ectoderm. The optic cup, a double-walled structure, forms when the outer surface of the optic vesicle thickens and flattens at the region of contact and then pushes inward. The optic cup becomes the retina. The lens forms from the lens placode, a swelling arising when the optic cup induces thickening of the overlying ectoderm. Developing lens cells synthesize crystallin, a fibrous protein that collects into glassy deposits. Lens cells lose their nuclei and form the lens.

39.5 Embryonic Development of Humans and Other Mammals

- The chorion of the mammalian embryo forms chorionic villi, fingerlike extensions into the endometrium that increase the surface area of the chorion where the placenta will form.

- The placenta is the interface between the developing embryo and its mother. There are a variety of placental structures in mammals, and placentalike structures (analogues) occur in fishes and insects. There is a rich diversity of placentae in guppylike fishes, demonstrating repeated evolution of this type of structure.
- *SRY* is the sex-determining region of the Y chromosome. *SRY* encodes a protein that induces a molecular switch causing the primitive gonads to develop into testes. Fetuses with XX chromosomes do not produce *SRY*, and under the influence of estrogens and progesterone, the primitive gonads develop into ovaries. High levels of circulating testosterone in a pregnant female cause masculinization of the genitalia in some mammals.
- In many animals, a larval stage is intermediate between embryo and adult. Larval stages occur in some species that produce eggs with small amounts of yolk. Larvae are often strikingly different from the adults and may be the feeding and/or dispersal stage of the species.

39.6 Cellular Basis of Development

- Microtubules and microfilaments produce movements of whole cells and changes in cell shape
- Induction is a process by which a group of cells causes or influences changes in a nearby group of cells, leading them to follow a particular pathway. Induction works through interactions between signal molecules from inducing cells and surface receptors on responding cells. When signal molecules bind to surface receptors, the responding cells are activated or inactivated.

39.7 Genetic and Molecular Control of Development

- Work with the nematode worm *Caenorhabditis elegans* has allowed biologists to follow development patterns and trace the fate of every cell produced from the zygote. *C. elegans* is transparent, making it even more appropriate for this kind of work.
- Skeletal muscles in mammals develop from somites, blocks of mesoderm along either side of the notochord. Paracrine signalling by nearby cells induces somite cells to express *myoD*, which turns on specific muscle-determining genes, converting them to myoblasts, undifferentiated muscle cells.
- Maternal-effect genes expressed by the mother during oogenesis control egg polarity and thus embryo polarity. The *bicoid* gene is a key maternal-effect gene responsible for the development of the head and thorax. Segmentation genes subdivide the embryo into regions, determining the segmentation in the body plan.
- Homeotic genes determine structure. In *Drosophila*, they are master regulatory genes controlling the development of body parts such as eyes, antennae, legs, and wings. In each homeotic gene is a common region called a homeobox. Homeobox-containing genes are called *Hox* genes.
- Apoptosis is programmed cell death, a process central to the development of some parts of the body. The process results from gene activation in response to molecular signals from receptors on the surfaces of marked cells.

Questions

Self-Test Questions

1. Vivipary occurs in
   a. class Osteichthyes.
   b. class Amphibia.
   c. class Mammalia.
   d. class Chondrihthyes.
   e. All of the above are correct.

2. Gastrulation occurs in
   a. all chordates.
   b. only in amniotes.
   c. only in echinoderms.
   d. only in amphibians.
   e. all metazoa.

3. Large amounts of yolk occur in the eggs of all
   a. bony fish.
   b. amphibians.
   c. birds.
   d. mammals.
   e. none of the above.

4. In vertebrates, the blastopore becomes
   a. the mouth.
   b. the nostrils.
   c. the anus.
   d. gill slits.
   e. the auditory meatus.

5. Apoptosis occurs in the development of
   a. mammals.
   b. birds.
   c. reptiles.
   d. bony fish.
   e. all of the above.

6. *Pax*-6 genes control the development of eyes in
   a. chordates.
   b. arthropods.
   c. mammals.
   d. cnidarians.
   e. all of the above.

7. Milk produced by mammary glands is a characteristic of
   a. discus fish.
   b. pigeons.
   c. caecilians.
   d. mammals.
   e. birds.

8. Induction results from interactions between cells and occurs in the development of
   a. eggs.
   b. tissues.
   c. embryos.
   d. nervous systems.
   e. b, c, and d.

9. Segmentation of developing embryos is controlled by
   a. *Pax*-6 genes.
   b. gap genes.
   c. pair-rule genes.
   d. homeotic genes.
   e. b and c.

10. Relaxin is important in the process of birth in
    a. sea horses.
    b. sharks.
    c. mammals.
    d. birds.
    e. cockroaches.

Questions for Discussion

1. How can apoptosis be used in the treatment of cancer?

2. How does the process of development differ between fraternal and identical twins? In humans, what is the incidence of fraternal twins with two fathers? How does this compare with other species of mammals?

3. How does the pattern of development of compound eyes differ from that of the eyes of vertebrates and molluscs?

4. What is the role of ectoderm in the development of the nervous system?

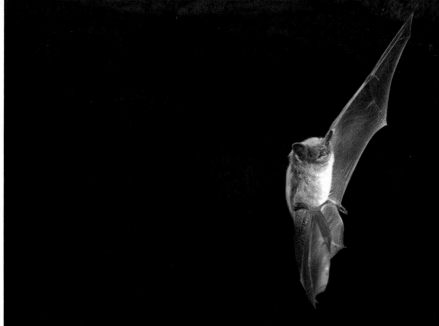

A little brown bat (*Myotis lucifugus*) flies through an abandoned mine. Mouth open, the animal produces echolocation calls that, in this setting, allow it to orient through the underground space.

M.B. Fenton

40 Animal Behaviour

WHY IT MATTERS

When it comes to food, many animals quickly learn to take advantage of new opportunities and show great versatility in behaviour from hunting to planning. Here are four examples.

During the Vietnam War (1959–1975), tigers (*Panthera tigris*) learned to associate the sound of gunfire with an opportunity to eat. The tigers' behaviour meant that some wounded soldiers waiting for treatment received a different kind of attention than what they expected. During the Second World War, wolves (*Canis lupis*) showed the same behaviour in some areas of Poland. A food reward is a strong reinforcer of behaviour.

In the 1970s, Kim McCleneghan and Jack Ames were studying sea otters (*Enhydra lutris*) in California waters. These otters dive and collect food (sea urchins, *Pisaster brevispinnis*, and clams, *Saxidomus nuttallii*) from the bottom and bring their catch to the surface to eat it. The observers were surprised to see some otters resurfacing with empty beverage cans. These otters would lie on their backs in the ocean swells, take a can, bite it open, and, in some cases, remove and eat something before discarding the can. Some cans appeared to be empty and were discarded after opening. The biologists collected their own beverage cans and discovered that many harboured young octopods (*Octopus* species). Populations of these cephalopods are limited by the number of shelters available. Young octopods were exploiting new opportunities for shelter, and the sea otters, in turn, were taking advantage of the molluscs' behaviour.

Meanwhile, in **savannah** woodlands in Senegal (West Africa), Jill Pruetz and Paco Bertolani observed chimpanzees (*Pan troglodytes*) hunting bushbabies (*Galago senegalensis*). The fact that the chimps were not vegetarians was no surprise because they had been reported using grass stalks to fish for termites and working in gangs to hunt young baboons (*Papio ursinus*). The discovery that savannah chimps in Senegal used "spears" to impale bushbabies hidden in tree hollows extended the repertoire of chimps. Pruetz and Bertolani watched chimps modifying branches they had broken off by biting to sharpen them prior to using them against bushbabies. The chimps that Pruetz and Bertolani studied appeared to plan their hunts in advance.

Other experiments have revealed how Western Scrub Jays (*Aphelocoma californica*) cache food in preparation for the next day's breakfast. Proving that animals plan ahead means that the experiments have to demonstrate that the animal executes a novel action or combination of actions and anticipates an emotional state different from the one at the time of planning. These two conditions rule out behaviours associated with migration and hibernation or those associated with meeting an immediate need for food.

In foraging behaviour, animals exhibit an array of opportunism and adaptation that we often believe is the exclusive domain of *Homo sapiens*. The purpose of this chapter is to introduce you to the topic of **animal behaviour.**

40.1 Genes, Environment, and Behaviour

Learning, as demonstrated by the foraging animals introduced above, illustrates how some behaviour patterns are acquired rather than inherited. But animal behaviourists had long debated whether animals are born with the ability to perform most behaviours completely or whether experience is necessary to shape their actions. Today, the emerging picture is that no behaviour is determined entirely by genetics or entirely by environmental factors. Rather, behaviours develop through complex gene–environment interactions.

Why do adult male White-crowned Sparrows sing a song that no other species sings **(Figure 40.1)**? They could have an innate (inborn) ability to produce their particular song, an ability so reliable that young males sing the "right" song the first time they try. According to this hypothesis, their distinctive song would be an **instinctive behaviour**, one genetically or developmentally "programmed" that appears in complete and functional form the first time it is used. An alternative hypothesis is that they acquire the song as a result of certain experiences, such as hearing the songs of adult male White-crowned Sparrows that live nearby. If so, this species' distinctive song might be an example of a **learned behaviour**, one that depends on having a particular kind of experience during development.

How can we determine which of these two hypotheses is correct? If the White-crowned Sparrow's song is instinctive, isolated male nestlings that have never heard other members of their species should be able to sing their species' song when they mature. If the learning hypothesis is correct, young birds deprived

Figure 40.1
Songbirds and their songs. Sound spectrograms (visual representations of sound graphed as frequency versus time) illustrate differences in the songs of the White-crowned Sparrow (*Zonotricha leucophrys*), Song Sparrow (*Melospiza melodia*), and Swamp Sparrow (*Melospiza georgiana*).

White-crowned Sparrow

Song Sparrow

Swamp Sparrow

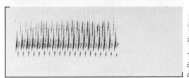

A. & J. Binns/Vireo

© J. Schumacher/Vireo

© G. Mc Elroy/Vireo

Dr. Stephen Yezerinac, Bishop's University, Lennoxville, Quebec, Canada

Frequency (kHz)

Time

Time

Time

of certain essential experiences should not sing "properly" when they become adults.

Peter Marler tested these two hypotheses. He took newly hatched White-crowned Sparrows from nests in the wild and reared them individually in sound-proof cages in his laboratory. Some of the chicks heard recordings of a male White-crowned Sparrow's song when they were 10 to 50 days of age, whereas others did not. Juvenile males in both groups first started to vocalize at about 150 days of age. For many days, the birds produced whistles and twitters that only vaguely resembled the songs of adults. Gradually, the young males that had listened to tapes of their species' song began to sing better and better approximations of that song. At about 200 days of age, these males were right on target, producing a song that was nearly indistinguishable from the one they had heard months before. Males that had not heard recordings of White-crowned Sparrow songs never sang anything close to the songs typical of wild males.

These results show that learning is essential for a young male White-crowned Sparrow to acquire the full song of its species. Although birds isolated as nestlings sang instinctively, they needed the acoustical experience of listening to their species' song early in life if they were to reproduce it months later. These data allow us to reject the hypothesis that White-crowned Sparrows hatch from their eggs with the ability to produce the "right" song. Their species-specific song, and perhaps the songs of many other songbirds, have both instinctive and learned components.

Early researchers generally classified behaviours as either instinctive or learned, but we now know that most behaviours include both instinctive and learned components. Nevertheless, some behaviours have a stronger instinctive component than others and vice versa.

Study Break

1. How can the study of bird song lend itself to understanding the influence of genes on behaviour?
2. How is behaviour learned?

40.2 Instinct

Instinctive behaviours presumably can be performed without the benefit of previous experience. They can be grouped into functional categories, such as feeding, defence, mating, and parental care. We assume that they have a strong genetic basis and that natural selection has preserved them as adaptive behaviours.

Many instinctive behaviours are highly stereotyped. When an animal is triggered by a specific cue,

a. Herring Gulls (*Larus argentatus*)

Marie Read Natural History Photography

b.

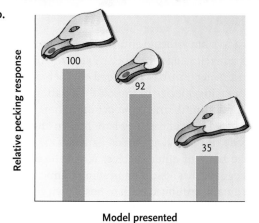

Figure 40.2

(a) A Herring Gull (*Larus argentatus*) chick begs its parent for food. **(b)** Nestling Herring Gulls also begged when presented with various models of an adult gull. In the experiment, nestlings pecked at a model with a red spot on the lower jaw almost as often as they did to a real gull. A model lacking the jaw spot elicited much fewer begging pecks from the nestling.

it performs the same response over and over in almost exactly the same way. These **fixed action patterns** are triggered by **sign stimuli**. Very young Herring Gull chicks use a begging response **(Figure 40.2)**, a fixed action pattern, to secure food from their parents. Begging chicks peck at the red spot on the parent's bill, and the tactile stimulus serves as a sign stimulus inducing the adult to regurgitate food from its crop. Baby gulls eat the chunks of fish, clams, or other food that have been regurgitated for them. We know that the spot on the parent's bill releases the begging response of the young gull because the same response is triggered by an artificial bill that looks only vaguely like an adult bill, provided that it has a dark contrasting spot near the tip (see Figure 40.2). Simple cues can activate fixed action patterns.

Human infants often respond innately to the facial expressions of adults **(Figure 40.3, p. 970)**. Researchers can trigger smiling in even very young babies simply by moving a mask toward the infant, as long as the mask has two simple, diagrammatic eyes. Clearly, the infant, like a nestling Herring Gull, is not reacting to every

Figure 40.3

Instinctive responses in humans. The smiling face of an adult is a sign stimulus that triggers smiling in very young infants.

Evan Cerasoli

which stimuli elicit its begging behaviour. During their early performances, instinctive behaviours can be modified in response to particular experiences.

Behavioural differences between individuals may reflect genetic differences because performance of instinctive behaviours does not depend on previous experience. Stevan Arnold studied innate responses of captive newborn garter snakes to olfactory stimuli provided by potential food items they had never before encountered. Arnold measured the snakes' responses to cotton swabs that had been dipped in a smelly extract of banana slug, a shell-less mollusc. Young snakes born to a mother captured in coastal California, where adult garter snakes regularly eat banana slugs, almost always began tongue-flicking at slug-scented cotton swabs (Figure 40.5). Newborn snakes whose parents came from central California, where banana slugs do not occur, rarely tongue-flicked at the swabs. Although the coastal and inland snakes belong to the same species, their instinctive responses to banana slug chemicals differed markedly.

feature of a face but rather to simple cues that function as sign stimuli releasing a fixed behavioural response.

Natural selection has moulded the behaviour of some parasitic species to exploit the relationship between sign stimuli and fixed action patterns for their own benefit. In effect, they have broken another species' code. Birds that are brood parasites lay their eggs in the nests of other species. When the brood parasite's egg hatches, the alien nestling mimics sign stimuli ordinarily exhibited by their hosts' own chicks. The parasitic chick begs for food by opening its mouth, bobbing its head, and calling more vigorously than the host's chicks. These exaggerated behaviours elicit feeding by the foster parents, and the young brood parasite often receives more food than the hosts' own young (Figure 40.4).

Although instinctive behaviours are often performed completely the first time an animal responds to a stimulus, they can be modified by an individual's experiences. The fixed action patterns of a young Herring Gull change over time. Although the youngster initially begs by pecking at almost anything remotely similar to an adult gull's bill, it eventually learns to recognize the distinctive visual and vocal features associated with its parents. The chick uses this information to become increasingly selective about

a. Banana slug

Eugene Kozloff

b. Adult coastal garter snake eating a banana slug

Stevan Arnold

c. Newborn coastal garter snake "smelling" slug extra

Stevan Arnold

Figure 40.5
Genetic control of food preference. (a) Banana slugs (*Ariolimax columbianus*) are a preferred food of (b) an adult garter snake (*Thamnophis elegans*) from coastal California. (c) A newborn snake from a coastal population of garter snakes flicks its tongue at a cotton swab drenched with tissue fluids from a banana slug.

Figure 40.4
This European Cuckoo Hedge Sparrow (*Cuculus canorus*) is a brood parasite that stimulates food delivery by its foster parent, a Hedge Sparrow (*Prunella modularis*). The cuckoo elicits food delivery by displaying exaggerated versions of the sign stimuli used by the host offspring. The exaggerated stimuli are "releasers" initiating the appropriate behaviour from a parent with food.

© Stephen Dalton/Photo Researchers, Inc.

Knockouts: Genes and Behaviour

Almost all eukaryotic organisms share a series of developmental interactions called the *wingless/Wnt* pathway. The name comes from the original discovery of the pathway in the fruit fly *Drosophila melanogaster,* in which mutant genes of the pathway cause alterations in the wings and other segmental structures. Recently, three genes closely related to *dishevelled,* one of the genes of the *Drosophila wingless/Wnt* pathway, were isolated from and identified in mice. No functions have yet been identified for the proteins encoded in the three mouse *dishevelled* genes, but they are highly active in both embryos and adults. Their function must be important, but what could it be?

Nardos Lijam and his coworkers sought an answer to this question by developing a line of mice that totally lacked one of the *dishevelled* genes, *Dvl1,* in genetic shorthand. First, they constructed an artificial copy of the *Dvl1* gene with the central section scrambled so that no functional proteins could be made from its encoded directions. Next, they introduced the artificial gene into embryonic mouse cells. Cells that successfully incorpo-

rated the gene were injected into very early mouse embryos. Some mice grown from these embryos were heterozygotes, with one normal copy of the *Dvl1* gene and one nonfunctional copy. Interbreeding of the heterozygotes produced some individuals that carried two copies of the altered *Dvl1* gene and no normal copies. Individuals lacking the normal gene are called "knockout" mice for the missing gene.

Surprisingly, knockout mice grew to maturity with no apparent morphological defects in any tissue examined, including the brain. Their motor skills, sensitivity to pain, cognition, and memory all appeared to be normal. However, their social behaviour was different. In cages with normal mice, the knockouts failed to take part in the common activities of mouse social groups: social grooming, tail pulling, mounting, and sniffing. Although normal mice build nests and sleep in huddled groups, knockouts tended to sleep alone, without constructing full nests from cage materials. Mice heterozygous for the *Dvl1* gene (those with one normal and one altered copy of the gene) behaved normally in all of these social activities.

The knockout mice also jumped around wildly in response to an abrupt, startling sound, whereas the response of normal mice was less extreme. A neural circuit of the brain inhibits the startle response of normal mice, so the reaction of knockout mice suggested that this inhibitory circuit was probably altered. Humans with schizophrenia, obsessive–compulsive disorders, Huntington disease, and some other brain dysfunctions also show an intensified startle reflex similar to that of the *Dvl1* knockout mice.

The researchers' analysis revealed that the *Dvl1* gene modifies developmental pathways affecting complex social behaviour in mice and probably in other mammals. It is one of the first genes identified that affects mammalian behaviour. The similarity in startle reflex intensity between the knockout mice and humans with neurological or psychiatric disorders suggests that mutations in the *Dvl* genes and the *wingless* developmental pathway may underlie some human mental illnesses. If so, further studies of the *Dvl* genes may give us clues to the molecular basis of these diseases and a possible means to their cure.

Arnold then tested whether newborn snakes would eat bite-sized chunks of slug. After a brief flick of the tongue, 85% of newborn snakes from a coastal population routinely struck at the piece of slug and swallowed it even though they had no previous experience with this food. Even when no other food was available, only 17% of newborn snakes from the inland population consistently tongue-flicked at or ate pieces of slug. Arnold hypothesized that coastal and inland garter snakes have different alleles at one or more gene loci controlling their odour-detection mechanisms and leading to differences in their behaviour. Arnold crossbred coastal and inland snakes. If genetic differences contribute to the food preferences of snakes from the two populations, hybrid offspring receiving genetic information from each parent should behave in an intermediate fashion. The results of the experiment confirmed the prediction. When presented with bite-sized chunks of slug, 29% of the newborn snakes of mixed parentage ate them every time.

Many other experiments have confirmed that genetic differences between individuals can translate into behavioural differences between them (see *Knockouts: Genes and Behaviour*). Bear in mind, however, that single genes do not control complex behaviour patterns directly. Rather, the alleles determine the kinds of enzymes that cells can produce, influencing biochemical pathways involved in the development of an animal's nervous system. The resulting neurological differences can translate into a behavioural difference between individuals that have certain alleles and those that do not.

STUDY BREAK

1. What are the differences between instinctive and learned behaviours?
2. What are fixed action patterns and sign stimuli?
3. How did experiments with garter snakes demonstrate the influence of genetics on behaviour?

40.3 Learning

Unlike instinctive behaviours, learned behaviours are not performed completely the first time an animal responds to a specific stimulus. They change in response to environmental stimuli that an individual experiences as it develops. Behavioural scientists generally define learning as a process in which experiences change an animal's behavioural responses. Different types of learning occur under different environmental circumstances.

Imprinting occurs when animals learn the identity of a caretaker or the key features of a suitable mate during a **critical period**, a stage of development early in life. Newly hatched geese imprint on their mother's appearance and identity, staying near her for months. When they reach sexual maturity, young geese try to mate with other geese exhibiting the visual and behavioural stimuli on which they had imprinted as youngsters. When Konrad Lorenz, a founder of **ethology** (the study of animal behaviour), tended a group of newly hatched Greylag Geese, they imprinted on him rather than on an adult of their own species **(Figure 40.6)**. Male geese not only followed Lorenz, but at sexual maturity, they also courted humans.

Other forms of learning can occur throughout an animal's lifetime. Ivan Pavlov, a Russian physiologist, demonstrated **classical conditioning** in experiments with dogs. Like many other animals, dogs developed a mental association between two phenomena that are usually unrelated. Dogs typically salivate when they eat. Food is an *unconditioned stimulus* because the dogs instinctively respond to it and do not need to learn to salivate when presented with food. Pavlov rang a bell just before offering food to dogs. After about 30 trials in which dogs received food immediately after the bell rang, the dogs associated the bell with feeding time and drooled profusely whenever it rang, even when no food was forthcoming. The bell had become a *conditioned stimulus*, one that elicited a particular learned response. In classical conditioning, an animal learns to respond to a conditioned stimulus (e.g., the bell) when it precedes an unconditioned stimulus (e.g., food) that normally triggers the response (e.g., salivation). If your pet cat becomes exceptionally friendly

whenever it hears the sound of a can opener, its behaviour is the result of classical conditioning.

Operant conditioning, trial-and-error learning, is another form of associative learning. Here animals learn to link a voluntary activity, an *operant*, with its favourable consequences, a *reinforcement*. A laboratory rat will explore a new cage randomly. If the cage is equipped with a bar that releases food when it is pressed, the rat eventually leans on the bar by accident (the operant) and immediately receives a morsel of food (the reinforcement). After a few such experiences, a hungry rat learns to press the bar in its cage more frequently, provided that the bar-pressing behaviour is followed by access to food. Laboratory rats also have learned to press bars to turn off disturbing stimuli, such as bright lights.

Insight learning occurs when an animal can abruptly learn to solve problems without apparent trial-and-error attempts at the solution. Captive chimpanzees solved a novel problem: how to get bananas hung far out of reach. The chimps studied the situation and then stacked several boxes, stood on them, and used a stick to knock the fruit to the floor.

Habituation occurs when animals lose their responsiveness to frequent stimuli not quickly followed by the usual reinforcement. Habituation can save the animal the time and energy of responding to stimuli that are no longer important. Sea hares (*Aplysia* species) are shell-less molluscs that typically retract their gills when touched on the side. Gill retraction helps protect sea hares from approaching predators. But a sea hare stops retracting its gills when it is touched repeatedly over a short period of time with no harmful consequences.

Cross-fostering experiments with song birds, Blue Tits (*Cyanistes caeruleus*), and Great Tits (*Parus major*) revealed that early learning is essential to the realization of ecological niches. In this work, Tore Slagsvold and Karen Wiebe transferred fertilized eggs from the nests of Great Tits to those of Blue Tits and vice versa. Compared with their genetic parents, the fostered young shifted their feeding niches, and the shift was lifelong. The changes in foraging behaviour were greater for fostered Great Tits, the species with more specialized foraging behaviour.

STUDY BREAK

1. How do the examples of feeding behaviour (see also *Why It Matters*) inform us about learning?
2. What are the differences between instinctive and learned behaviours? Give examples of each.

Figure 40.6

Imprinting. Having imprinted on him shortly after hatching, young Greylag Geese (*Anser anser*) frequently joined Karl Lorenz for a swim.

© Nina Leen/Time and Life Pictures/Getty Images

40.4 Neurophysiology and Behaviour

Research in **neuroscience** has shown that all behavioural responses, whether mostly instinctive or mostly learned, depend on an elaborate physiological foundation provided by the biochemistry and structure of the nerve cells. Nerve cells that regulate an innate response and make it possible for an animal to learn something are products of a complex developmental process. Here genetic information and environmental contributions are intertwined. Although the anatomical and physiological basis for some behaviours is present at birth, an individual's experiences alter the cells of its nervous system in ways that produce particular patterns of behaviour.

Marler's experiments helped explain the physiological underpinnings of singing behaviour in male White-crowned Sparrows. If acoustical experience shapes singing, a sparrow chick's brain must be able to acquire and store information present in the songs of other males. Then, months later, when the young male starts to sing, its nervous system must have special features enabling the bird to match its vocal output to the stored memory of the song that it heard earlier. Eventually, when it achieves a good match, the sparrow's brain must "lock" on the now complete song and continue to produce it when the bird sings.

Additional experiments demonstrated that when young birds did not hear a taped song during their critical period (10–50 days of age), they never produced the full song of their species, even if they heard it later in life. In addition, young birds that heard recordings of *other* bird species' songs during the critical period never generated replicas of those songs as they matured. These and other findings suggested that certain nerve cells in the young male's brain are influenced only by appropriate stimuli, in this case, acoustical signals from individuals of its own species, and only during the critical period. Neuroscientists have identified nuclei (singular, *nucleus*), clusters of nerve cells that make song learning and song production possible.

Every behavioural trait appears to have its own neural basis. Another songbird, a male Zebra Finch **(Figure 40.7)**, can discriminate between the songs of strangers and those of established neighbours. These finches live in territories, plots of land defended by individual males or breeding pairs. Defence of the territory ensures that the residents have exclusive access to food and other necessary resources.

Zebra Finches' ability to discriminate between the songs of neighbours and those of strangers involves a nucleus in the forebrain. Cells in this nucleus fire frequently the first time the Zebra Finch hears the song of a new conspecific. As the song is played again and again, the cells of this nucleus cease to respond, indicating that the bird has become habituated to a now familiar song. The same bird still reacts to the songs of strangers. Neurophysiological networks that make this selective learning possible enable male Zebra Finches to behave differently toward familiar neighbours that they largely ignore and unfamiliar singers they attack and drive away.

Figure 40.7
Zebra Finches (*Taeniopygia guttata*) are native to Indonesia. They have played an important role in studies of the physiological basis of song learning. The male has the striped throat.

Molecular and cellular techniques have been used to identify the role of genes in learning. When a bird is exposed to relevant acoustical stimuli, such as songs of potential rivals, certain genes are "turned on" within neurons in the song-controlling nuclei of the bird's brain. When a Zebra Finch hears the elements of its species' song, a gene called *zenk* becomes active in the brain, producing an enzyme that changes the structure and function of neurons. The ZENK enzyme programs nerve cells of the bird's brain to "anticipate" key acoustical events of potential biological importance. When they occur, these events trigger additional changes in the bird's brain, affecting its actions. In this way, a **territory** owner learns to ignore (= habituates to) a singing neighbour with which it shares an established territorial boundary. The same bird retains the ability to detect and respond to new intruding conspecifics because they are a real threat to its continued control of its territory.

STUDY BREAK

1. What role does the ZENK enzyme play?
2. How do nerve connections influence behaviour?

40.5 Hormones and Behaviour

Hormones are chemical signals that can trigger the performance of specific behaviours. Hormones often work by regulating the development of neurons and neural networks or by stimulating cells within endocrine organs to release chemical signals.

How did the neurons in an adult Zebra Finch acquire the remarkable capacity to change in response to specific stimuli? In Zebra Finches, only males produce courtship songs. Very early in its life, certain cells in the brain of a male songbird produce estrogen, which affects target neurons in the higher vocal centre, an area of the developing brain. Estrogen leads to a complex series of biochemical changes resulting in the production of more nerve cells in the parts of the brain that regulate singing. Brains of developing females do not produce estrogen. In the absence of estrogen, the number of neurons in the higher vocal centre of females *declines* over time **(Figure 40.8).** If young female Zebra Finches are given estrogen, they produce more nerve cells in the higher vocal centre and are capable of singing. Specific stimuli, such as the songs of familiar or unfamiliar males, can alter the genetic activity of the nerve cells that control the behaviour of adult birds.

Just as estrogen influences the development of singing ability in Zebra Finches, other hormones mediate the development of the nervous system in other species. A change in the concentration of a certain hormone can be the physiological trigger that induces important changes in an animal's behaviour as it matures.

As they age, worker honeybees perform different tasks. Bees less than 15 days old after emerging from the pupa tend to care for larvae and maintain the hive. Bees older than 15 days often make foraging excursions from the hive to collect food, nectar and pollen **(Figure 40.9).** These behavioural changes are induced by rising concentrations of juvenile hormone (see Chapter 35) released by a gland near the bee's brain. Despite its name, circulating levels of juvenile hormone actually increase as a honeybee ages.

Juvenile hormone may exert its effect on bee behaviour by stimulating genes in certain brain cells to produce proteins that affect nervous system function. Octopamine, for example, stimulates neural transmissions and reinforces memories. Octopamine is con-

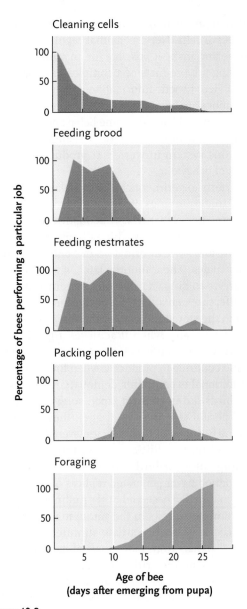

Figure 40.9
Age and task specialization in honeybee (*Apis mellifera*) workers. Newly emerged bees typically clean cells and feed the brood, whereas older workers leave the hive to forage for food.

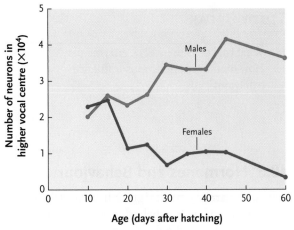

Figure 40.8
Hormonally induced changes in brain structure. The brains of young male Zebra Finches secrete estrogen, which stimulates production of additional neurons in the higher vocal centre. Lacking estrogen, young female Zebra Finches have fewer neurons in this region of the brain.

centrated in the antennal lobes, parts of the bee's brain that contribute to the analysis of chemical scents in the external environment. Octopamine is found at higher concentrations in older, foraging bees that have higher levels of juvenile hormone. When extra juvenile hormone is experimentally administered to bees, their production of octopamine increases. Increased octopamine levels in the antennal lobes may help a foraging bee home in on the odours of flowers from which it can collect nectar and pollen.

The honeybee example illustrates how genes and hormones interact in the development of behaviour. Genes code for the production of hormones that become part of the intracellular environment of assorted target cells. Hormones then directly or indirectly change genetic activity and enzymatic biochemistry in their targets. When the target cells are neurons, changes in biochemistry translate into changes in the animal's behaviour.

The African cichlid fish illustrates how hormones regulate reproductive behaviour. Some adult males maintain nesting territories on the bottom of Lake Tanganyika in East Africa **(Figure 40.10)**. Territory holders are relatively brightly coloured and exhibit elaborate behavioural displays to attract egg-laden females. These males defend their real estate aggressively against neighbouring territory holders and against incursions by males without territories of their own. Nonterritorial males (called "drifters") are much less colourful and aggressive and do not control a patch of suitable nesting habitat. They make no effort to court females.

Differences in levels of GnRH (gonadotropin-releasing hormone; see Chapter 38) cause behavioural differences between the two types of males. In the hypothalamus of the brain of territorial males, large, biochemically active cells produce GnRH. The same cells in the brains of drifters are small and inactive. GnRH stimulates the testes to produce testosterone and sperm. When circulating sex hormones are carried to the brain of the fish, they modulate the activity of nerve cells that regulate sexual and aggressive behaviour. In the absence of GnRH, male fish do not court females or attack other males.

What causes the differences in the neuronal and hormonal physiology of the two types of male fish? Russell Fernald and his students manipulated the territorial status of males. Some territorial males were changed into nonterritorial males and vice versa, whereas the territorial status of other males was left unchanged as a control **(Figure 40.11)**. Four weeks after the changes, Fernald and his students compared experimental and control fish. They considered colouration and behaviour, as well as the size of the GnRH-producing cells in the brains. Territorial males that had been changed to

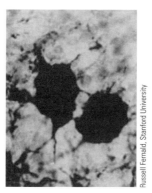
Territorial control

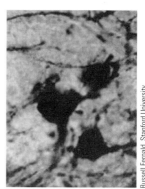
Territorial to nonterritorial experimentals

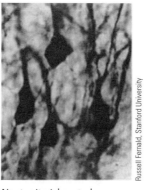
Nonterritorial control

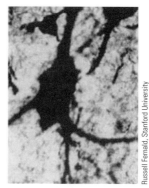
Nonterritorial to territorial experimentals

Figure 40.11
Gonadotropin-releasing hormone (GnRH) cells in male *Haplochromis burtoni*.

nonterritorial males quickly lost their bright colours and stopped being combative. Moreover, their GnRH-producing cells were smaller than those of the territory-holding controls. Conversely, males that gained a territory in the experiment quickly developed bright colours and displayed aggressive behaviours toward other males. GnRH-producing cells in their brains were larger than those of fish that had maintained their status as non-territory-holding controls.

This example shows how what is happening inside a fish affects its environmental situation—its success or failure at gaining and holding a territory. Fish can detect and store information about their aggressive interactions. Neurons that process this information transmit their input to the hypothalamus, where it affects the size of cells producing GnRH, in turn dictating the hormonal state of the male. A decrease in GnRH production can turn a feisty territorial male into a subdued drifter. Drifters bide their time and build energy reserves for a future attempt at defeating a weaker male and taking over his territory. If successful in regaining territorial status, the male's GnRH levels will increase again. The once peaceful male reverts to vigorous sexual and aggressive behaviour.

Note the general similarity of these processes to those described for the White-crowned Sparrow's song learning. The fish's brain has cells that can change their biochemistry, structure, and function in response

African cichlid fish (*Haplochromis burtoni*)

Nonterritorial male

Territorial male

Figure 40.10
A comparison of a territorial and a nonterritorial male *Haplochromis burtoni*.

to well-defined social stimuli. These physiological changes make it possible for the fish to modify its behaviour depending on its social circumstances.

STUDY BREAK

1. How has research on White-crowned Sparrows and Zebra Finches advanced our knowledge of the impact of neurobiology on behaviour?
2. How does juvenile hormone affect the behaviour of adult bees?
3. How does gonadotropin-releasing hormone affect the behaviour of fish?

40.6 Neural Anatomy and Behaviour

Some specific behaviours are produced by anatomical structures in an animal's nervous system. The nervous systems of many animal species allow them to respond rapidly to key stimuli. Often sensory systems are structured to acquire a disproportionately large amount of information about the stimuli that are most important to survival and reproductive success.

Important information acquired by the senses can be relayed directly to motor neurons, for example, providing prey animals with behaviour that can save them from a predator's attack. Insects such as crickets that fly mainly at night avoid day-flying, predatory birds. But flying at night exposes them to attacks by insectivorous bats.

Insectivorous bats hunting at night use echolocation to detect and track flying prey (see Chapter 34). The echolocation calls of bats hunting flying insects are usually intense, measuring about 130 decibels sound pressure level at 10 cm, making the calls stronger than the sound of a smoke detector alarm. The bats' calls can cover frequencies from ~10 kHz to >200 kHz, beyond the 20 kHz upper limit of human hearing. By comparing its calls with the echoes from its calls, the bat uses echolocation to detect, assess, and track its flying prey. However, a bat's echolocation calls give crickets (and other prey; see Chapter 34) warning of their approach (see *Echolocation: Communication*).

With ears on their front legs, black field crickets hear bat echolocation calls (see Figure 34.14, and the anatomical structure of the cricket's nervous system produces a behavioural response that takes the cricket out of harm's way. Sensory neurons connected to the ears fire in response to the bat's calls, and the information is immediately translated into evasive action. When a bat attacks from the cricket's right side, the right ear receives a stronger stimulation than the left ear. The cricket's nervous system relays incoming messages from the *right* ear to the motor neurons controlling the *left* hindleg. Sufficient stimulation on the right side induces

firing by motor neurons for the left hindleg, causing the leg to jerk up. This, in turn, blocks the movement of the left hindwing and reduces the flight power generated on the left side of the cricket's body. These changes cause the flying cricket to swerve sharply to the left and lose altitude, effectively diving down and away from the approaching bat **(Figure 40.12)**.

The structure and neural connections of sensory systems allow some animals to distinguish potentially life-threatening situations from more mundane stimuli. Fiddler crabs live and feed on mud flats, where they dig burrows that provide safe refuge from predators such as crab-hunting shorebirds. To use its burrow to advantage, a crab must distinguish between predatory gulls and other fiddler crabs. Otherwise, it would dash for cover whenever anything moved in its field of vision.

Fiddler crabs have long-stalked eyes held above their carapaces and perpendicular to the ground **(Figure 40.13)**. John Layne wondered whether a crab might use a divided field of view to distinguish dangerous predators from fellow crabs. An approaching large gull would stimulate receptors on the upper part of the eye, whereas another crab's movements would be slightly below the midpoint of the eyes. A split field of view would allow the crab to distinguish between the two kinds of stimuli. To achieve this, receptors above and below the retinal equator must relay signals to different groups of neurons, effectively wiring the crab's nervous system to distinguish for a split field of view. If this were the case, stimulation of receptors

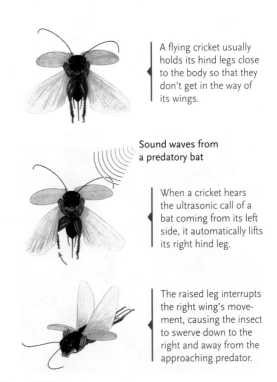

A flying cricket usually holds its hind legs close to the body so that they don't get in the way of its wings.

Sound waves from a predatory bat

When a cricket hears the ultrasonic call of a bat coming from its left side, it automatically lifts its right hind leg.

The raised leg interrupts the right wing's movement, causing the insect to swerve down to the right and away from the approaching predator.

Figure 40.12

A neural mechanism for escape behaviour in the black field cricket (*Teleogryllus oceanicus*).

Echolocation: Communication

In echolocation, the echolocator stores the outgoing signal in its brain for comparison with returning echoes. The difference between what the animal says and what it hears is the data used in echolocation. However, when an echolocating bat (see Figure 40.1) or dolphin produces echolocation signals, the signals can also be heard by other animals.

When the bat or dolphin is foraging, potential prey (some insects for the bat, some fish for the dolphin) hear the signals and flee from the sound source (= negative phonotaxis) in an effort to evade the approaching predator. When the bat is close (strong echolocation signals), moths with ears dive to the ground or go into erratic flight to evade the bats. Moths with ears sensitive to bat echolocation calls avoid bat attacks 40% of the time. Insects lacking bat detectors are caught at much higher rates, sometimes >90% of the time. Acoustic warfare between bats and insects entertains biologists, involving measures and countermeasures by both predator and prey.

The same echolocation calls that alert potential prey are also available to any other animals within earshot, provided that their ears are sensitive to the frequencies in the signals. Little brown bats may use feeding buzzes (signals associated with attacks on prey; see Figure 34.3) to locate concentrations of prey. Spotted bats (*Euderma maculatum*) either approach a calling conspecific, apparently to chase it away, or turn and leave the area. Resident killer whales (*Orcinus orca*) in the Pacific Ocean off the west coast of Canada typically use echolocation to detect, track, and locate the salmon they eat. Transient killer whales in the same area feed mainly on marine mammals. These killer whales rarely echolocate. Local marine mammals, such as seals, quickly leave the water when they hear killer whales approaching.

The study of echolocation is a rich source of information about signals, signal design, hearing systems, and behaviour.

Figure 40.13
A fiddler crab, *Uca pugilator*.

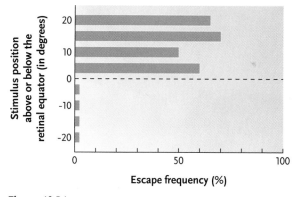

Figure 40.14
Stimuli that activated the upper part of the retinas of *Uca pugilator* elicited escape behaviour much more often than those activating the lower retinas.

above the midline of the eye would activate neurons controlling an escape response, triggering a dash for the burrow. A moving stimulus at or below eye level would stimulate a different response. Responses to other crabs are likely to be gender-dependent.

To explore this, Layne placed crabs one at a time in a glass jar on an elevated platform. He presented a black square to each crab from two different heights. Sometimes the stimulus circled the jar above the crab's eyes; sometimes it circled below them. Stimuli activating the upper part of the retina induced escape behaviour, whereas those below the retinal equator were usually ignored **(Figure 40.14).** Specific nervous system connections between a fiddler crab's eyes and brain provide appropriate responses to different specific stimuli.

The match between the structure of an animal's nervous system and the real-world challenges it faces extends beyond the ability to avoid predators. Star-nosed moles live in wet tunnels in North American marshlands and spend almost all of their lives in complete darkness. Like nocturnal insect-eating bats, star-nosed moles must find food without the benefit of visual cues. Like the bats, its receptor–perceptual system enables it to feed effectively. A star-nosed mole eats mainly earthworms it locates with its nose, but not by smell. As the mole proceeds down its tunnel, 22 fingerlike tentacles on its nose sweep the area directly ahead of it. Each tentacle is covered with thousands of Eimer's organs (touch receptors; **Figure 40.15, p. 978).** Sensory nerve terminals in Eimer's organs generate complex and detailed patterns of signals about the objects they contact. These messages are relayed by neurons to the cortex of the mole's brain,

a. Sensory organs on the tentacle of a star-nosed mole

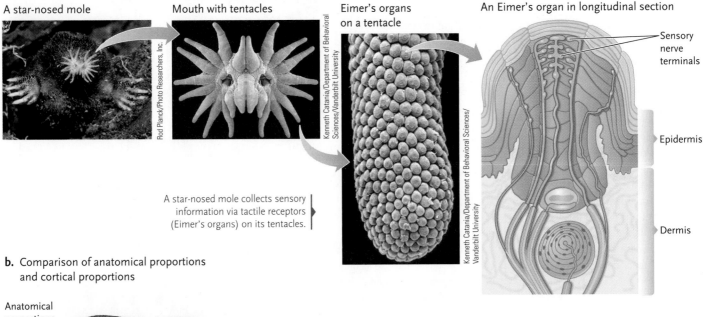

A star-nosed mole

Rod Planck/Photo Researchers, Inc.

Mouth with tentacles

Kenneth Catania/Department of Behavioral Sciences/Vanderbilt University

Eimer's organs on a tentacle

Kenneth Catania/Department of Behavioral Sciences/Vanderbilt University

An Eimer's organ in longitudinal section

Sensory nerve terminals

Epidermis

Dermis

A star-nosed mole collects sensory information via tactile receptors (Eimer's organs) on its tentacles.

b. Comparison of anatomical proportions and cortical proportions

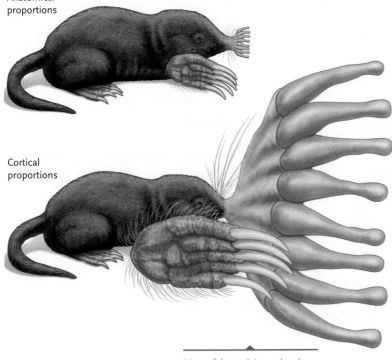

Anatomical proportions

Cortical proportions

Most of the mole's cereberal cortex is devoted to the tentacles and front, digging feet.

Figure 40.15

The collection and analysis of sensory information by the star-nosed mole (*Condylura cristata*). **(a)** The mole's nose has 22 fleshy tentacles covered with cylindrical tactile receptors called Eimer's organs. Each Eimer's organ contains sensory nerve terminals. **(b)** The mole's cerebral cortex devotes far more space and neurons to analysis of input from the tentacles than from elsewhere on the body. These drawings compare the relative amounts of sensory information coming from different parts of a mole's body.

much of which is devoted to the analysis of information received from the nose's touch receptors.

The structural basis of the mole's sensory analysis is reflected by the amount of brain tissue responding to signals from its nose. The mole's brain contains many more cells decoding input from Eimer's glands than the combined input received from all other parts

of the animal's body (see Figure 40.15b). Moreover, the brain does not treat inputs from all 22 of the mole's "nose fingers" equally. Instead, the brain devotes more cells to input from tentacles closest to the mouth. Fewer cells analyze messages from those farther away. Processing tactile information by star-nosed moles is related to the importance of finding food in dark, underground tunnels. Moreover, the extra attention given to signals from certain tentacles helps the star-nosed mole locate prey that are close to its mouth, in turn allowing it to feed more efficiently (see Figure 34.31).

Animals' nervous systems do not offer neutral and complete pictures of the environment. Instead, the pictures are distorted, but the unbalanced perceptions of the world are advantageous because certain types of information are far more important than others for the animals' survival and reproductive success.

STUDY BREAK

1. How do crickets hear the echolocation calls of bats?
2. How does what they see influence the behaviour of fiddler crabs?

40.7 Communication

In animal communication, one individual produces a signal that is received by another, changing the behaviour of one or both individuals in a way that benefits signaller and/or signal receiver. The signaller is the individual transmitting information (the signal) and the signal *receiver* **(Figure 40.16)**, the one receiving the signal. Some animals have broken the signal codes of others and exploited them to their advantage. Some people who study animal communication consider that only signals intended to communicate should be called communication (think back to problems with the definition of species or genes).

Animals use a variety of sensory modalities when producing signals, including acoustical, chemical, electrical, vibrational, and visual. Some signals combine modalities. Sometimes the animal itself is a signal; in other situations, the animal's excretory or eliminated products are signals.

Bird songs are acoustical signals, heard by the signal receivers. The song of a male Whippoorwill (*Caprimulgis vociferous*) advertises his presence to females and may help him secure a mate. The same song is heard by other males, who recognize it as a territorial display. After the eggs have been laid, the same song is heard by the young developing in the eggs. Other birds, male Club-winged Manakins (e.g., *Machaeropterus deliciosus*), use sounds produced by feather stridulations as their acoustic courting signal. Sounds are used as signals by many other animals, such as insects and rattlesnakes. Pacific herring (*Clupea pallasi*) communicate with conspecifics through the noise generated with little bursts of gas (known as fast repetitive transient signals [f*rts]) passed from the anus.

A striped skunk's (*Mephitis mephitis*) black and white stripes constitute a visual signal. Other examples are humans' facial expressions and body language. These **visual signals** are available to anyone viewing them. Visual signals can be enhanced by morphological features, such as the erectile crest of a Royal Kingbird (*Tyrannus melancholicus*), or semaphore flags used by people. In darkness, some animals use bioluminescent signals (e.g., Figure 41.29). In many animals, visual signals are *ritualized*—they have become exaggerated and stereotyped, enhancing their function as signals **(Figure 40.17)**.

Many species produce chemical signals, well known to anyone who has walked a dog. **Pheromones** are distinctive volatile chemicals released in minute amounts to influence the behaviour of conspecifics. The body of a worker ant contains a battery of glands, each releasing a different pheromone **(Figure 40.18, p. 980)**. One set of pheromones recruits fellow workers to battle colony invaders, whereas another stimulates workers to collect food that has been discovered outside the colony. Pheromones are used by some animals to attract mates. Female silkworm moths (*Bombyx mori*) produce the pheromone bombykol (see *Molecule Behind Biology*).

Figure 40.16
Song birds, such as this Grey Vireo (*Vireo vincinior*), use songs to advertise their presence to unmated females and to other males.

Figure 40.17
Visual display. The courtship display of a male Albatross (*Diomedea exulans*) includes ritualized postures and movements of the wings and body.

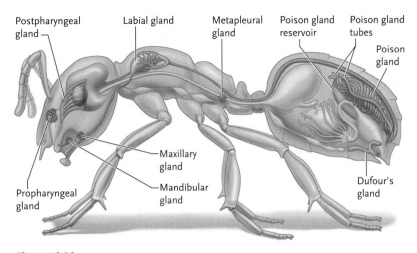

Figure 40.18

Chemical signals. An ant's body contains a host of pheromone-producing glands, each of which manufactures and releases its own volatile chemical or chemicals.

A single molecule of bombykol can generate a message in specialized receptors on the antennae of any male silkworm moth that is downwind (see *Knockouts: Genes and Behaviour*). Chemicals used as signals are often exploited by predators.

In many species, touch conveys important messages from a signaller to a receiver. **Tactile signals** can operate only over very short distances, but for social animals living in close company, they play a significant role in the development of friendly bonds between individuals **(Figure 40.19)**.

a. Round dance

b. Waggle dance

c. Coding direction in the waggle dance

When the bee moves straight down the comb, other bees fly to the source directly away from the sun.

When the bee moves 45° to the right of vertical, other bees fly at a 45° angle to the right of the sun.

When the bee moves straight up the comb, other bees fly straight toward the sun.

Figure 40.20

Dance communication by honeybees (*Apis mellifera*). Foraging honeybees transmit information about the location and quality of a food source by dancing on vertical honeycomb. **(a)** If the food source is close to the hive, the forager performs a "round dance." **(b)** When food is farther from the hive, the honeybee performs a "waggle dance." **(c)** The dancing bee indicates the direction to the distant food source by the angle of the waggle run.

Figure 40.19

Tactile signals. Grooming by Hyacinth Macaws (*Ano dorhynchus hyacinthinus*) removes ectoparasites and dirt from feathers. The close physical contact promotes friendly relationships between groomer and groomee.

Some freshwater fish species, especially those that occupy murky tropical rivers where visual signals cannot be seen, use weak **electrical signalling** to communicate (see Figure 34.34). These fish have electric organs that can release charges of variable intensity, duration, and frequency, allowing substantial modulation of the message that a signaller sends. Among the New World knifefish (order Gymnotiformes), including the electric eel (Figure 34.34b), electrical discharges can signal threats, submission, or a readiness to breed.

Animals often use several channels of communication simultaneously. Karl von Frisch demonstrated that the famous dance of the honeybee involves tactile, acoustical, and chemical modes **(Figure 40.20)**. When a foraging honeybee discovers a source of pollen or nectar, it returns to its colony. There, in the darkness of the hive, it performs a dance on the vertical surface of the honeycomb. The dancer moves in a circle, attracting a crowd of workers. Some workers follow and maintain physical contact with the dancer. The dance delivers information about food source, its quality, and the distance and direction observers will need to fly to locate it.

Bombykol

Male silkworm moths (*Bombyx mori*) respond to the pheromone bombykol (Figure 1) produced and released by females. Bombykol is a pheromone designed to function in communication. Animals use pheromones to bring males and females together. Male *B. mori* detect bombykol using specialized receptors on their antennae (see Chapter 34).

Not surprisingly, predators exploit the powerful attractiveness of pheromones to lure prey. Female bolas spiders (*Mastophora cornigera* and other species in the genus) use a sticky ball of web impregnated with a chemical that mimics the odour of sex pheromones secreted by female moths. Male moths respond to the lure of these odours, approach the pheromone-soaked web, and are captured by the spiders.

Bolas spiders do not prey on just one species of moth. Adult female

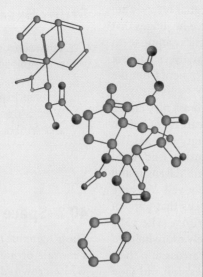

Figure 1
Bombykol.

M. cornigera produce three sex pheromone compounds, (Z)-9-tetradecenyl acetate, (Z)-9-tetradecenal, and

(Z)-11-hexadecenal. These pheromones attract several different species of moths. Moth attractants produced by *Mastiphora hutchinsoni* are effective on several moths that use quite different pheromones. *M. hutchinsoni* adjust the production of pheromone mimic to match the times of maximum activity by *Tetanolita mynesalis* (smoky tetanolita) and *Lacinipolia renigera* (bristly cutworm). The pheromone blend for the early-flying *L. renigera* interferes with attraction of the late-flying *T. mynesalis*, so the spider adjusts the blend of pheromone it uses in its lure. This spider lures the early-flying moth with one blend and the late-flying one with another.

When the food source is less than 75 m from the hive, the bee performs a "round dance" (see Figure 40.20a). Here the bee moves in tight circles, swinging its abdomen back and forth. Bees surrounding the dancer produce a brief acoustical signal that stimulates the dancer to regurgitate a sample of the food it discovered. The regurgitated sample serves as a chemical cue for other workers that search for the food.

When the food source is farther away, the forager performs the "waggle dance": a half-circle in one direction, then a straight line while waggling its abdomen, and then a half-circle in the other direction (see Figure 40.20b). With each waggle, the dancer produces a brief buzzing sound. The angle of the waggle run relative to the vertical honeycomb indicates the direction of the food source relative to the position of the Sun (see Figure 40.20c). The duration of the waggles and buzzes carries information about distance to the food. The more time spent waggling and buzzing, the farther the food is from the hive.

Signal receivers often respond to communication from signallers in predictable ways. A male White-crowned Sparrow generally avoids entering a neighbouring territory simply because it hears the song of the resident male. Similarly, young male baboons and mandrills often retreat without a fight when they see an older male's visual threat display

(Figure 40.21), even with the loss of a chance to mate with a female. Why do these receivers behave in ways that appear to be beneficial to their rivals but not to themselves?

Explaining behavioural interactions often means considering how an animal's actions affect its reproductive output. The retreating White-crowned Sparrow avoids wasting time and energy on a battle he is likely to lose. By retreating, the would-be intruder minimizes the chances of being injured or killed by a resident male. Moreover, ousting the current resident might be more tiring and risky than finding a suitable unoccupied breeding site. Resident males usually win physical contests, and intruders typically succeed in gaining a territory from a resident only after a prolonged series of exhausting clashes. Observations of territorial species, such as birds, lizards, frogs, fish, or insects, generally support these predictions.

Tom and Pat Leeson

Figure 40.21
Threat display. Exposed canines epitomize the threat display of a dominant male mandrill (*Mandrillus sphinx*) that is used to drive away rival males.

Applying a similar argument to competition among male mandrills, we can predict that smaller or younger males will concede females to threatening older rivals without fighting. The signal receiver retreats after receiving the threat because he judges that he would not win, and a male mandrill's canine teeth are not just for show. Evolutionary analyses suggest that the signaller and signal receiver benefit from the exchange of signals.

In winter, some Ravens (*Corvus corax*) may emit a strange "yell" call when they find a carcass of a deer. The loud yell attracts a crowd of hungry Ravens. The calling behaviour puzzled Bernd Heinrich, who noted that when paired, territory-holding adult Ravens found a carcass, they fed quietly and did not yell. Yells are produced by young, wandering Ravens that happen on a carcass in another bird's territory. The yells attracted other Ravens that collectively overwhelmed the residents' efforts to defend the carcass and their territory. Wanderers used yells to exploit the food supply, whereas residents just ate. Bernd Heinrich concluded that the reproductive benefit of resident Ravens was enhanced by uninterrupted feeding. Wandering Ravens succeeded in their trespassing only when they attracted others.

40.7a Language: Syntax and Symbols

Although language is communication, not all communication is language. Many people believe that language is the exclusive domain of humans, but the distinction is not clear and crisp. The round and waggle dances of honeybees contain both syntax (the order in which information is presented) and symbols (a display that represents something else) and are considered by many to meet the criteria for language. Furthermore, by blackening the dancer's ocelli (see Chapter 34), James L. Gould was able to get a dancing bee to lie to other bees. When there is a light in the hive, the dancer orients the waggle dance to the light as if it were the Sun. Dancers with blackened ocelli do not see the light as do other bees—thus, the lie.

Vervet monkeys (see Figure 18.2) have a repertoire of signals to alert conspecifics to different predators. Vervet monkeys use one signal for snakes, another for leopards, and still another for raptors and show different predator-specific defensive behaviours. Chickadees (*Parus atricapillus*) also use different alarm calls to alert others to approaching danger. Captive, trained chimpanzees and gorillas (*Gorilla gorilla*) have been reported to use AMSLAN, American Sign Language.

In the area of communication, humans are not as distinct from other animals as some people would like to believe. To appreciate redundancy in animal communication, observe the body language and facial expressions of someone talking on a telephone. The eloquence of these signals is not conveyed to the signal receiver at the other end of the phone!

STUDY BREAK

1. What sensory modalities do animals use in communication?
2. How do "yells" influence the behaviour of Ravens? Explain.
3. What is the meaning and importance of syntax and symbols in signalling?

40.8 Space

The geographic range of many animal species includes a mosaic of habitat types. The breeding ranges of White-crowned Sparrows can encompass forests, meadows, housing developments, and city dumps. Other animals have a limited range, for example, a Kirtland's Warbler (*Dendroica kirtlandii*) is found only in young jackpine forests. An animal's choice of habitat is critically important because the habitat provides food, shelter, nesting sites, and the other organisms with which it interacts. If an animal chooses a habitat that does not provide appropriate resources, it will not survive and reproduce.

On a large spatial scale, animals almost certainly use multiple criteria to select the habitats they occupy, but no research has yet established any general principles about how animals make these choices. When a migrating bird arrives at its breeding range, it probably cues on large-scale geographic features, such as a pond or a patch of large trees. If the bird does not find the food or nesting resources it needs, or if other individuals have already occupied the space and perhaps depleted those resources, it may move to another habitat patch.

On a very fine spatial scale, basic responses to physical factors enable some animals to find suitable habitats. Kinesis (kine = movement; es = inward) is a change in the rate of movement or the frequency of turning movements in response to environmental stimuli. Wood lice (terrestrial crustaceans in the order Isopoda) typically live under rocks and logs or in other damp places. Although these arthropods are not attracted to moisture per se, when a wood louse encounters dry soil, it scrambles around, turning frequently. When it reaches a patch of moist soil, it moves much less. This kinesis results in wood lice accumulating in moist habitats. Wood lice exposed to dry soil quickly dehydrate and die, so those that move to moister habitats are more likely to survive.

A **taxis** (taxis = ordered movement) is a response directed either toward or away from a specific stimulus. Cockroaches (order Blattodea) exhibit negative phototaxis, meaning that they actively avoid light and seek darkness. Negative phototaxis makes cockroaches less vulnerable to predators that use vision to find their food.

Biologists generally assume that habitat selection is adaptive and has been shaped by natural selection. Some animals instinctively select habitats where they are well camouflaged and less detectable by predators. Predators would discover and eliminate individuals that do not select a matching background, along with any alleles responsible for the mismatch. Many insects have genetically determined preferences for the plants they eat as larvae (e.g., caterpillars). Adults often lay their eggs only on appropriate food plants, effectively selecting the habitats where their offspring will live and feed.

Vertebrates sometimes exhibit innate preferences, as demonstrated by two closely related species of European birds, Blue Tits (*Cyanistes caeruleus*) and Coal Tits (*Parus ater*). Adult Blue Tits forage mainly in oak trees and Coal Tits in pines. When researchers reared the young of both species in cages without any vegetation and then offered them a choice between oak branches and pine branches, Coal Tits immediately gravitated toward pines and Blue Tits toward oaks, suggesting an innate preference **(Figure 40.22)**. Each species feeds most efficiently in the tree species it prefers.

Habitat preferences can also be moulded by experiences early in life. Tadpoles of red-legged frogs (*Rana aurora*) usually live in aquatic habitats cluttered with sticks, strands of algae, and plant stems. In the laboratory, these tadpoles prefer striped backgrounds to plain ones. In contrast, tadpoles of the closely related cascade frog (*Rana cascadae*) live over gravel bottoms and prefer plain substrates over striped ones. These habitat preferences do not appear when red-legged frogs are reared over plain substrates and cascade frogs over striped substrates and are later given a choice of substrate.

40.8a Home Range and Territory: Occupied and Defended Areas

Space is an important resource for animals. Although many animals are motile, moving about in space, others are sessile. Sessile species such as barnacles (see Figure 3.3g and Chapter 26) anchor themselves to the substrate but are motile as larvae. Barnacles that live on whales or the hulls of ships are sessile but mobile because of the substrate they selected. Motile animals have a home range, the space they regularly traverse during their lives. Home ranges or parts of home ranges become territories when they are defended. In species such as pronghorned antelopes **(Figure 40.23)**, some males hold territories, but others do not. Females are not usually territorial. There is a direct connection between territory quality and male reproductive success. Male pronghorn antelopes defending the "best" territories (those with the best food resources) attract the most females, offering the male the most opportunities to mate with the most females.

Male Jarrow's spiny lizards **(Figure 40.24, p. 984)** are normally territorial during the autumnal mating season, when they have elevated levels of testosterone in their blood. Catherine Marler and Michael Moore implanted small doses of testosterone or a placebo under the skins of experimental animals during the nonmating season (June and July). Testosterone-enhanced males were more active and displayed more

Blue Tit   Coal Tit

Percent of time

100   100

50   50

0   0

Wild Hand-reared   Wild Hand-reared

KEY

Pine   Oak

Figure 40.22

Habitat selection by birds. Wild Blue Tits (*Cyanistes caeruleus*) show a strong preference for oak trees; Coal Tits (*Parus ater*) show a strong preference for pines. Hand-reared birds raised in a vegetation-free environment showed identical but slightly weaker responses.

Alan Williams/Alamy

Mark Hamblin/Oxford Scientific/Index Stock

www.amazilia.net

Figure 40.23

Pronghorn antelopes, *Antilocapra americana*.

Figure 40.24

Jarrow's spiny lizard, *Sceloporus jarrovi*.

often than control males. Experimental males spent less time feeding, even though they used about 30% more energy per day than control males. In one seven-week period, testosterone-enhanced males suffered significantly higher mortality than placebo males. It can be expensive to be territorial.

Territorial defence is always a costly activity. Patrolling territory borders, performing displays hundreds of times per day, and chasing intruders take time and energy. Moreover, territorial displays increase an animal's like-

Figure 40.25

Long-distance migration. Arctic Terns (*Sterna paradisaea*) migrate from the high Arctic to Antarctica each year, a round-trip journey of 40 000 km. This species' summer breeding range is shaded on the map.

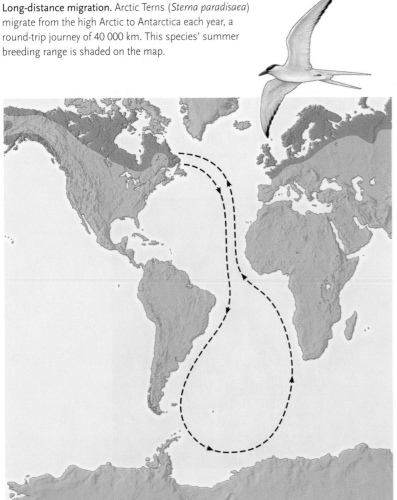

lihood of being injured or detected and captured by a predator.

But territorial behaviour has its benefits, such as access to females. Territorial surgeonfish (*Acanthurus lineatus*) living in coral reefs around American Samoa may engage in as many as 1900 chases per day, defending their small territories from incursions by other algae-eating fish. Territorial surgeonfish eat five times as much food as nonterritory holders because they have more exclusive access to the food in their territories. It also costs the territory holders more to patrol and defend their realm.

STUDY BREAK

1. Define kinesis and taxis.
2. What is the difference between a home range and a territory?
3. How are home ranges and territories different from a species' range?

40.9 Migration

Many animal species make a seasonal **migration**, travelling from the area where they were born or hatched to a distant and initially unfamiliar destination. The migration is complete when they later return to their natal site. The Arctic Tern, a seabird, makes an annual round-trip migration of 40 000 km **(Figure 40.25).** Other vertebrate species, such as grey whales (*Eschrichtius robustus*) and salmon (*Salmo* species), undertake long and predictable journeys. The same is true of arthropods such as spiny lobsters that form long conga lines and move seasonally between coral reefs and the open ocean floor **(Figure 40.26).**

Moving animals use various mechanisms to find their way during migration. There are three categories of way-finding mechanisms: **piloting, compass orientation**, and **navigation**. Many species probably use some combination of these mechanisms to guide their movements.

40.9a Piloting: Finding the Way

Piloting is the simplest way-finding mechanism, involving the use of familiar landmarks to guide a journey. Grey whales migrate from Alaska to Baja California and back using visual cues provided by the Pacific coastline of North America. When it is time to breed and lay eggs, Pacific salmon use olfactory cues to pilot their way from the ocean back to the stream in which they hatched.

Animals that do not migrate also use specific landmarks to identify their nest site or places where they have stored food. Female digger wasps (*Philanthus*

Howard Hall/Oxford Scientific/Index Stock

Figure 40.26

Migrating arthropods. Spiny lobsters (*Panulirus argus*) make seasonal migrations between coral reefs and the open ocean floor. As many as 50 individuals march in single file for several days.

triangulum) nest in soil. In 1938, Niko Tinbergen showed that after foraging flights, these wasps used visual landmarks to find their nests **(Figure 40.27)**. While the female wasp was in the nest, Tinbergen arranged pinecones in a circle around it. As she left, the wasp flew around the area, apparently noting nearby landmarks. Tinbergen then moved the circle of pinecones a short distance away. Each time the female returned, she searched for her nest within the pinecone circle. She never once found her nest unless the pinecones were returned to their original position. Later Tinbergen rearranged the pinecones into a triangle after females left their nests and added a ring of stones nearby. The returning females looked for their nest in the stone circle. Tinbergen concluded that digger wasps respond to the general outline or geometry of landmarks around their nests and not to the specific objects comprising the landmarks.

40.9b Compass: Which Way Is North?

Animals using compass orientation move in a particular direction, often over a specific distance or for a prescribed length of time. Some day-flying migratory birds orient themselves using the Sun's position in the sky in conjunction with an internal biological clock (see Chapter 1). The internal clock allows the bird to use the Sun as a compass, compensating for changes in its position through the day. The clock also may allow some birds to estimate how far they have travelled since beginning their journey. Other animals, including birds, mammals, reptiles, amphibians, fish, crustaceans, and insects, use Earth's magnetic field as a compass. This requires detection of weak magnetic fields (~50 μT – micro-Teslas). In 2008, some biologists suggested that magnetically sensitive free radical reactions could be the basis of a magnetic sense, but the search for the transducer(s) (see Chapter 34) is ongoing.

40.9c Stars: Celestial Navigation

Some birds that migrate at night determine their direction by using the positions of stars. The Indigo Bunting flies about 3500 km from the northeastern United States to the Caribbean or Central America each fall and makes the return journey each spring. Stephen Emlen demonstrated that Indigo Buntings direct their migration

Wasp's flight pattern on leaving nest

Wasp's return, looking for nest

Figure 40.27

Female digger wasps find their nets. A ring of pinecones serves as a landmark for a female digger wasp (*Philanthus triangulum*). By manipulating the location of landmarks, Nikko Tinbergen demonstrated the role they serve in the wasp's orientation behaviour.

using celestial cues **(Figure 40.28)**. Emlen confined individual buntings in cone-shaped test cages whose sides were lined with blotting paper. He placed inkpads on the cage bottoms and kept the cages in an outdoor enclosure so that the birds had a full view of the night sky. Whenever a bird made a directed movement, its inky footprints indicated the direction in which it was trying to move. On clear nights in fall, the footprints pointed to the south, but in spring, they pointed north. On cloudy nights, when the buntings could not see the stars, Emlen recorded that their footprints were evenly distributed in all directions. The data indicated that the compass of indigo buntings required a view of the stars.

40.9d Navigation: A Complex Challenge

Navigation is the most complex way-finding mechanism. It occurs when an animal moves toward a specific destination, using both a compass and a "mental map" of where it is in relation to the destination. Hikers in unfamiliar surroundings routinely use navigation to find their way home. They use a map to determine their current position and the necessary direction of movement and a compass to orient themselves in that direction. Scientists have documented true navigation in a few animal species, notably the Homing Pigeon (*Columba livia*). These birds can navigate to their home coops from any direction, probably using the Sun's position as their compass and olfactory cues as their map.

40.9e Reasons for Migratory Behaviour

Migrations by White-crowned Sparrows and many other species are triggered by changes in day length. Shortening day length indicates approaching fall and winter; lengthening day length indicates spring. Day length changes the anterior pituitary of the bird's brain to generate a series of hormonal changes. In response, birds feed heavily and accumulate the fat reserves necessary to fuel their long journey. Sparrows also become increasingly restless at night, until one evening, they begin their nocturnal migration. Their ability to adopt and maintain a southerly orientation in autumn (and a northerly one in spring) rests in part on their capacity to use the positions of stars to provide directional information.

Migratory behaviour entails obvious costs, such as the time and energy devoted to the journey and the risk of death from exhaustion or predator attack. Migratory behaviour is not universal—many animals never migrate, spending their lives in one location. Why do some species migrate? What ecological pressures give migrating individuals higher fitness than individuals that do not migrate? Remember that many species of terrestrial animals migrate, such as wildebeest and caribou.

For migratory birds, seasonal changes in food supply are the most widely accepted hypothesis to explain migratory behaviour. Insects can be abundant in higher latitude (>50° N or S) habitats during the warm spring and summer, providing excellent resources for birds to raise offspring. As summer wanes and fall and winter approach, insects all but disappear. Bird species that remain in temperate habitats over winter eat mainly seeds and dormant insects. When it is winter at higher latitudes, energy supplies are more predictably available in the tropical grounds used by overwintering migratory birds.

Two-way migratory journeys may provide other benefits. Avoiding the northern winter is probably adaptive because endotherms must increase their metabolic rates just to stay warm in cold climates (see Chapter 43). Moreover, summer days are longer at high latitudes than they are in the tropics (see Chapter 3), giving adult birds more time to feed and rear a brood.

Seasonal changes in food supply also underlie the migration of monarch butterflies

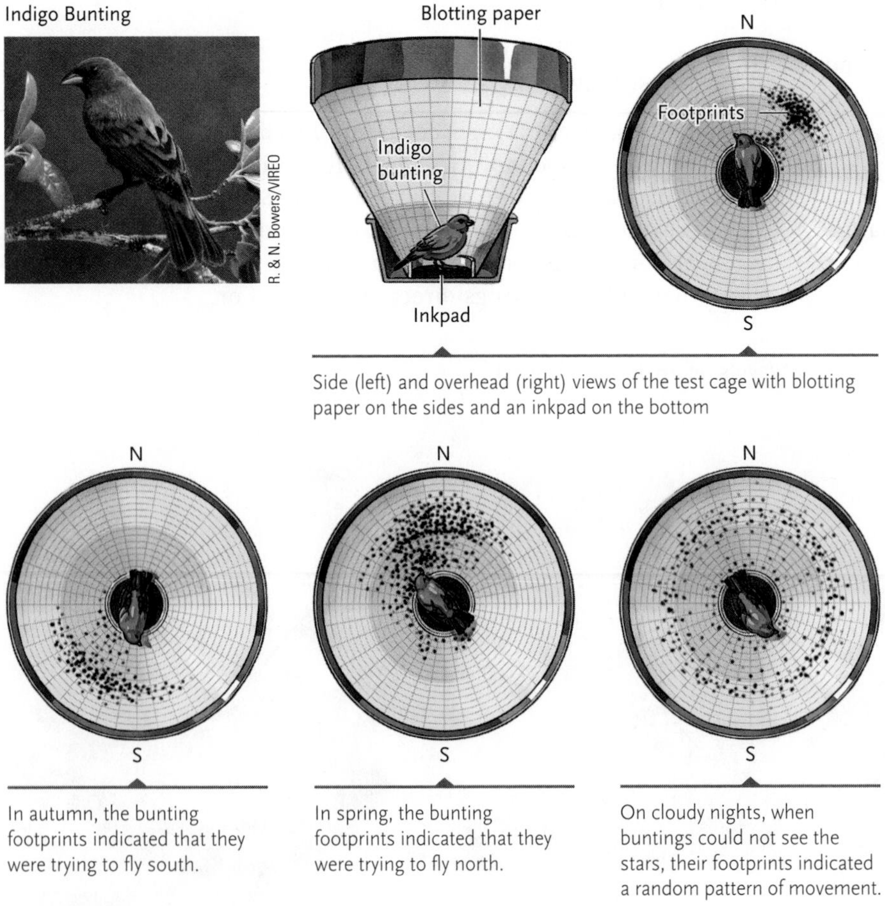

Indigo Bunting

R. & N. Bowers/VIREO

Blotting paper

Indigo bunting

Inkpad

N

Footprints

S

Side (left) and overhead (right) views of the test cage with blotting paper on the sides and an inkpad on the bottom

N

S

In autumn, the bunting footprints indicated that they were trying to fly south.

N

S

In spring, the bunting footprints indicated that they were trying to fly north.

N

S

On cloudy nights, when buntings could not see the stars, their footprints indicated a random pattern of movement.

Figure 40.28
Orientation by Indigo Buntings. The footprints of inked Indigo Buntings (*Passerina cyanea*) demonstrated migrating birds' responses to celestial cues.

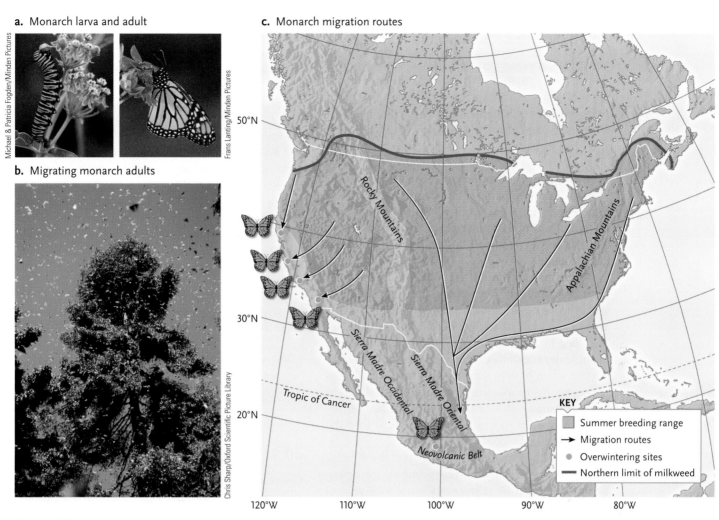

a. Monarch larva and adult

c. Monarch migration routes

b. Migrating monarch adults

KEY

☐ Summer breeding range
→ Migration routes
● Overwintering sites
━ Northern limit of milkweed

Figure 40.29

Migrating monarch butterflies. **(a)** Monarch butterflies (*Danaus plexippus*) eat milkweed plants as caterpillars. **(b)** When milkweed plants in their breeding range die back at the end of summer, monarchs migrate south. The following spring, after passing the winter in a semi-dormant state, they migrate north. **(c)** Monarchs that live and breed east of the Rocky Mountains migrate to Mexico. Those living west of the Rocky Mountains overwinter in coastal California.

that eat milkweed leaves as caterpillars and milkweed nectar as adults. In eastern North America, milkweed plants grow only during spring and summer. Many adult monarchs head south in late summer, when the plants begin to die. Some migrate as much as 4000 km from eastern and central North America to central Mexico, where they cluster in spectacular numbers **(Figure 40.29b),** apparently using olfactory cues to find preferred resting places. Unlike migrant birds, these insects do not feed while at their overwintering grounds. Instead, their metabolic rate decreases in the cool mountain air, and the butterflies become inactive for months, conserving precious energy reserves. When spring arrives, the butterflies become active again and begin the return migration to northern breeding habitats. The northward migration is slow, however, and many individuals stop along the way to feed and lay eggs. These offspring, and their offspring, continue the northward migration through the summer. Some descendants of these migrants

eventually reach Canada for a final round of breeding. The summer's last generation then returns south to the spot where their ancestors, two to five generations removed, spent the previous winter.

For other animals, migration to breeding grounds may provide the special conditions necessary for reproduction. Grey whales migrate south, where females give birth to their young in quiet, shallow lagoons where predators are rare and warm water temperatures are more conducive to the growth of their calves.

STUDY BREAK

1. Define migration. Give examples of migratory animals, including some not mentioned in the text. Do any humans migrate?
2. How do migrating animals find their way? Distinguish between navigation and compass orientation.

40.10 Mates as Resources

Mating systems have evolved to maximize reproductive success, partly in response to the amount of parental care that offspring require and partly in response to other aspects of a species' ecology. **Monogamy** describes the situation in which a male and a female form a pair bond for a mating season or, in some cases, for the individuals' reproductive lives. **Polygamy** occurs when one male has active pair bonds with more than one female (**polygyny**) or one female has active pair bonds with more than one male (**polyandry**). **Promiscuity** occurs when males and females have no pair bonds beyond the time it takes to mate.

When young require a great deal of care that both parents can provide, monogamy often prevails. Songbirds, such as the White-crowned Sparrow **(Figure 40.30)**, are altricial (naked and helpless) when they hatch. They beg for food, and both parents can bring it to them. Males and females achieve higher rates of reproduction when both parents are actively involved with raising young. In mammals, the situation is different because females provide the food (milk). Monogamy occurs in species in which males indirectly feed the young by bringing food to the mother.

If males have high-quality territories, the females living there may be able to raise young on their own. These males may be polygynous (mate with several females). The male's role is that of sperm donor and protector of the space rather than that of an active parent to all of his young. In birds such as Red-winged Blackbirds

Figure 40.30

Reproductive success. Parental care is just one of the many behaviours required for successful reproduction in White-crowned Sparrows and in many other animal species. The number of surviving nestlings will determine the reproductive success of their parents and the representation of their genes in the next generation.

(*Agelaius phoenecius*), some males hold large, resource-filled territories that support several females. These males will be attractive to females even if a female (or females) already lives on the territory. Polygyny is prevalent among mammals because, compared with males, females make a much larger investment in raising young (through egg development and care of the young).

Promiscuous mating systems occur when females are only with males long enough to receive sperm and there is no pair bond. These males make no contribution to raising young. Sage Grouse **(Figure 40.31)** and hammer-headed bats (*Hysignathus monstrosus*) are examples of this approach. Both species form leks, congregations of displaying males, where females come only to mate. There are more details about Sage Grouse below.

STUDY BREAK

What do the terms monogamy, polygamy, and promiscuity mean?

40.11 Sexual Selection

Given the drive to reproduce (see Chapter 38), competition for access to mates coupled with mate choice sets the stage for sexual selection. Sexual dimorphism, in which one gender is larger or more colourful than the other, can be an outcome of sexual selection. When males compete for females, males are often larger than females and may have ornaments and weapons, such as horns and antlers, useful for attracting females and for butting, stabbing, or intimidating rival males. Displays of adornments or weapons can simultaneously warn off other males and attract the attention of females. Peacocks strut in front of female peahens while spreading a gigantic fan of tail feathers, which they shake, rattle, and roll.

Why should females choose males with exaggerated structures conspicuously displayed? A male's large size, bright feathers, or large horns might indicate that he is particularly healthy. His appearance could indicate that he can harvest resources efficiently or simply that he has managed to survive to an advanced age. The features are, in effect, signals of male quality, and if they reflect a male's genetic makeup, he is likely to fertilize a female's eggs with sperm containing successful alleles. Large showy males may hold large, rich territories. Females that choose these males can gain access to the resources their territories contain.

The degree to which females *actively* choose genetically superior mates varies among species. In northern elephant seals, female choice is more or less passive.

Ray Richardson/Animals, Animals–Earth Scenes

Figure 40.31

Lekking behaviour. Male Sage Grouse (*Centrocercus urophasianus*) use their ornamental feathers in visual courtship displays performed at a lek. There each male has his own small territory. The smaller brown females observe the performing males before picking a mate.

Large numbers of females gather on beaches to give birth to their pups before becoming sexually receptive again (see Chapter 38, *Delaying Reproduction*). Males locate clusters of females and fight to keep other males away (see Figure 17.8). Males that win have exceptional reproductive success because they mate with many females, but only after engaging in violent and relentless combat with rival males. In this mating system, the females struggle during a male's attempts to mate with them. A female's struggles attract other males, who try to interrupt the attempted mating. Only the largest and most powerful males are not interrupted in their copulations, and they inseminate the most females. These attributes may be associated with alleles that will increase their offspring's chances of living long enough to reproduce.

In other species, females exercise more active mate choice, mating only after inspecting several potential partners. Among birds, active female mate choice is most apparent at **leks**, display grounds where each male holds a small territory from which it courts attentive females. The male is the only resource on the territory. Male Sage Grouse in western North America gather in open areas among stands of sagebrush. Each male defends a few square metres, where it struts in circles while emitting booming calls and showing off its elegant tail feathers and big neck pouches (see Figure 40.31). Females wander among displaying males, presumably observing the males' visual and acoustical displays. Eventually, each female selects one mate from among the dozens of males that are present. Females repeatedly favour males that come to the lek daily, defend their small area vigorously, and display more frequently than the average lek participant. Males preferred by females sustain their territorial defence and high display rate over long periods, abilities that may correlate with other useful genetic traits. Ultimately, the male holding the "best" position in the lek mates with the most females.

The results of experiments with peafowl suggest that the top Peacocks (*Pavo cristatus*) supply advantageous alleles to their offspring. In nature, peahens prefer males whose tails have many ornamental eyespots **(Figure 40.32).** In an experiment on captive birds, some peahens were mated to peacocks with highly attractive tails, but others were paired with males whose tails were less impressive. The offspring of both groups were reared under uniform conditions for several months and then released into an English woodland. After three months on their own, the offspring of fathers with impressive tails survived better and weighed significantly more than did those whose fathers had less attractive tails. The evidence demonstrates that a peahen's mate choice influences her offsprings' chances of survival.

According to the handicap hypothesis, females select males that are successful—the ones with ornate structures. These structures may impede their locomotion, and their elaborate displays may attract the attention of predators. Females select ornate males because they have survived *despite* carrying such a handicap. Successful alleles responsible for the ornamental handicap are passed to the female's offspring.

© Ashley Cooper/Corbis

Figure 40.32

Sexual selection for ornamentation. The attractiveness of a peacock to peahens depends in part on the number of eyespots in his extraordinary tail. The offspring of males with elaborate tails are more successful than the offspring of males with plainer tails.

What are the distinguishing features of a lek?

40.12 Social Behaviour

Social behaviour, the interactions that animals have with other members of their species, has profound effects on an individual's reproductive success. Some animals are solitary, getting together only briefly to mate (house flies and leopards). Others spend most of their lives in small family groups (gorillas). Still others live in groups with thousands of relatives (termites and honeybees). Some species, such as caribou and humans, live in large social units composed primarily of nonrelatives. In many species, the level of social interactions varies seasonally, usually reflecting the timing of reproduction, which, in turn, is influenced by changes in day length.

40.12a African Lions: Infanticide

African lions (*Panthera leo*) usually live in prides, one adult male with several females and their young. Males typically sire the young born to the females in their pride, achieving a high reproductive output. Females benefit from the support of the others in the group, which includes caring for young and cooperating in foraging. Female lions living in prides wean more young per litter than those living alone. The females in a pride are often genetically related, and their estrus cycles are usually synchronized. Male lions are bigger (~200 kg) than females (~150 kg), and males fight vigourously for the position of pride male. Males protect their females from incursions by other males.

When a new male takes over a pride, he kills all nursing young, bringing the females into estrus. At first, this infanticide seems counterproductive. However, it benefits the male because it increases the chances of his succeeding at reproducing. Were he to wait until the females had raised their dependent young, his reproductive contributions could be delayed for some time, perhaps as much as a year or more.

Females are not large enough to protect their young from the male. If a female takes her nursing young and leaves the pride, her efficiency as hunter declines, and she is less able to protect her young. Her reproductive success plummets. Females can be more productive (measured by output of young) when they are part of a pride.

But why live in a group in the first place? By hunting together, lions are more efficient foragers than when they hunt alone; therefore, they raise more young. Perhaps more important is the threat posed by spotted hyenas, which live in large groups (clans). Although individually smaller (~60 kg), when spotted

hyenas outnumber lions, they can chase lions from their kills. Furthermore, many of the lion's main prey also live in groups, and group defences affect lions' hunting success.

The situation in lions exemplifies some biological realities. Males and females do not have the same strategies when it comes to reproduction. Understanding behaviour means considering genetic relatedness and production of offspring, as well as the setting in which the animals live.

40.12b Group Living: Costs and Benefits

Social Behaviours. Ecological factors have a large impact on the reproductive benefits and costs of social living. Groups of cooperating predators frequently capture prey more effectively than they would on their own. White Pelicans (*Pelecanus erythrorhynchos*) often encircle a school of fish before attacking, so being part of a group provides a better yield to individuals than working alone. On the other hand, prey subject to intense predation may benefit from group defence. This can mean more pairs of watchful eyes or ears to detect an approaching danger. It may also translate into multiple lures so that when a predator attacks, it is more difficult to focus on an individual. When you are part of a group that is attacked, it may be someone other than you that is captured, diluting the risk to any one group member.

When attacked by wolves, adult muskoxen form a circle around the young, so attackers are always confronted by horns and hooves **(Figure 40.33)**. Insects such as Australian sawfly caterpillars also show cooperative defensive behaviour **(Figure 40.34)**. When predators disturb the caterpillars, all group members rear up, writhe about, and regurgitate sticky, pungent oils. The caterpillars collect the oils from the eucalyptus leaves they eat. The oils do not harm the caterpillars but are toxic and repellent to birds.

Living in groups can also be expensive. One cost can be increased competition for food. When thousands of royal penguins crowd together in huge colonies **(Figure 40.35)**, the pressure on local food supplies is great, increasing the risk of starvation. Communal living may facilitate the spread of contagious diseases and parasites. Nestlings in large colonies of Cliff Swallows (*Petrochelidon pyrrhonota*) are often stunted in growth because the nests swarm with blood-feeding, bedbuglike parasites, *Oeciacus vicarious*, **(Figure 40.36, p. 992)**. The parasites move readily from nest to nest in crowded conditions. Some social animals learn to recognize and avoid diseased group members. Caribbean spiny lobsters live in groups but avoid conspecifics infected by a lethal virus (PaV1). It is no surprise that most animals live alone.

Group living brings both costs and benefits. Not all animals that live in groups are "social," a term implying some organization of the group. The 10 million

Figure 40.33
Muskoxen.

Figure 40.34
Social defensive behaviour. Australian sawfly (*Perga dosalis*) caterpillars clump together on tree branches. They regurgitate yellow blobs of sticky aromatic fluid that repels birds. The accumulation of regurgitate from a group of caterpillars is an effective defence.

Brazilian free-tailed bats emerging from a cave roost near San Antonio, Texas, are no more a social group than the dozens of people leaving a high-rise apartment or university residence. Within the aggregation, there may be social units, but the aggregation itself is not necessarily a social unit.

Social animals usually live in groups characterized by some form of structure. Some individuals may dominate others (a **dominance hierarchy**), manifested in access to resources. Dominant (alpha or α) individuals get priority access to food (or mates or sleeping sites). In some situations, only dominant individuals (a male and a female) reproduce. Dominance hierarchies may be absolute, such as when the same individual always

has priority access to any resource. In relative dominance hierarchies, an individual's status depends on the circumstance.

Dominance brings its costs. In animals such as wild dogs (*Lycaeon pictus*) or grey wolves, dominant animals must constantly defend their status. Dominants often have high levels of cortisol and other stress-related hormones in their blood (see Chapter 35) compared with subordinates. Elevated cortisol levels may induce high blood pressure, the disruption of sugar metabolism, and other pathological conditions.

Subordinance brings its benefits. Subordinate group members, like all members of the group, gain protection from predators. They may also gain experience by

Figure 40.35
Colonial living. Royal Penguins (*Eudyptes schlegeli*) on Macquarie Island between New Zealand and Antarctica experience benefits and costs from living together in huge groups.

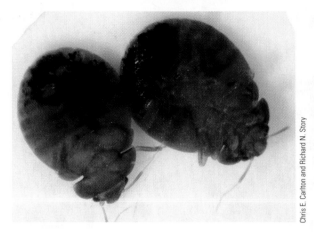

Figure 40.36
Dorsal and ventral views of a bedbug that specializes on birds (*Ornithocoris pallidus*).

Chris E. Carlton and Richard N. Story

John Dominis/Time & Life Pictures/Getty Images

Figure 40.37
When attacked by a leopard (*Panthera pardus*), a solitary baboon (*Papio anubis*) is defiant but unlikely to survive.

helping dominant individuals raise young. Over time, subordinate individuals can rise in a dominance hierarchy and avoid some of the side effects of dominance. Many social animals cannot survive on their own **(Figure 40.37)**.

STUDY BREAK

1. What is infanticide? Why does it occur?
2. Gives some examples of advantages and disadvantages of group living.

40.13 Kin Selection and Altruism

Behavioural ecologist William D. Hamilton recognized that helping genetic relatives effectively propagates the helper's genes because family members share alleles inherited from their ancestors. By calculating the degree of relatedness, we can quantify the average percentage of alleles shared by relatives **(Figure 40.38)**. Half-siblings, by definition, share one genetic parent, so they share, on average, 0.25 of their alleles by inheritance from their shared parent. Their degree of relatedness is 0.25. Full siblings share both parents' share, 0.25 of their alleles through the mother and 0.25 through the father, for a total, on average, of $0.25 + 0.25 = 0.5$ of their alleles. The degree of relatedness between a nephew or niece and an aunt or uncle is 0.25 and between first cousins is 0.125. Individuals should be more likely to help close relatives because increasing a close relative's fitness means that the individual is helping to propagate some of its own alleles. This is **kin selection**.

A male grey wolf helps his parents rear four pups to adulthood, pups that would have died without the extra assistance he provided. The pups are his younger full siblings, sharing 0.5 of his genes, so, on average, the helper has created "by proxy" two ($0.50 \times 4 = 2$) copies of any allele they shared. However, the costs of his helping must be measured against this indirect reproductive success. If he had found a mate, sired offspring, and raised two of them, each would have

carried half of his alleles, preserving only one ($0.50 \times 2 = 1$) copy of a given allele. In this situation, reproducing on his own would have produced fewer copies of his alleles in the next generation than helping to raise his siblings. This example is hypothetical, but sibling helpers have been documented in many species of birds and mammals. The phenomenon is especially common among animals in which inexperienced parents are not very successful at reproducing on their own offspring. By helping, they gain experience and realize some genetic benefit.

Altruism involves doing something that enhances the situation of another individual, but Hamilton's kin selection theory demonstrates why parental behaviour (or helping parents raise siblings) is genetically selfish, not altruistic. Therefore, the behaviour of the young wolf (above) is not altruistic. Robert Trivers proposed that individuals will help nonrelatives if they are likely to return the favour in the future. Trivers called this **reciprocal altruism** because each member of the partnership can potentially benefit from the relationship. Trivers hypothesized that reciprocal altruism would be favoured by natural selection as long as individuals that do *not* reciprocate (cheaters) are denied future aid.

Among the many features of social animals, the evolution of cooperative behaviour can be one of the most challenging to

Half Siblings

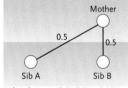

Relatedness = (0.5)(0.5) = 0.25

Full Siblings

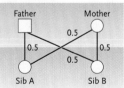

Relatedness
Through mother = (0.5)(0.5) = 0.25
Through father = (0.5)(0.5) = 0.25
Total relatedness = 0.25 + 0.25 = 0.5

First Cousins

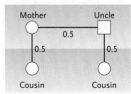

Relatedness = (0.5)(0.5)(0.5) = 0.125

Figure 40.38
Calculating degrees of relatedness.

understand. Why has cooperative behaviour arisen in populations of animals? How does it arise? And how is it maintained in populations?

John M. McNamara and three colleagues wrote about the coevolution of choosiness and cooperation in a paper published in *Nature* in 2007. Using modelling and simulation experiments, these authors examined the consequences arising in situations in which one individual's cooperativeness influences the decisions about actions by other individuals toward group members. They postulated a situation of "competitive altruism" in which individuals actually compete with one another to be more cooperative.

The results of their analysis suggest that longer-lived species are more likely to develop cooperative behaviour than short-lived ones. This is important because the model does not require intermediate situations involving negotiation behaviour. The model helps us understand the appearance of cooperative behaviour in all animals, including in *Homo sapiens*.

Reciprocal altruism is one element of cooperative behaviour, and dolphins may be reciprocal altruists. Many species of dolphins are long-lived and social, living in groups. Dolphins and other cetaceans show many forms of aid-giving behaviour, from attending injured group members ("standing by" in naval parlance) to assisting with difficult births. They also use group behaviour to protect themselves from the attacks of sharks. Richard Connor and Kenneth Norris proposed that the persistent threat of attacks by sharks and the perils of living in the ocean combined to provide dolphins with many opportunities to help one another, even members of other species. Connor and Norris did not have specific details of genetic relationships among group members, but they proposed that dolphins are reciprocal altruists.

STUDY BREAK

1. What is the main argument in Hamilton's kin selection theory?
2. Imagine that four of your first cousins, two siblings, and two half-siblings are about to fall from a cliff and die. You have the option of taking their place. In terms of kin selection, which is more beneficial to you, your life or the life of your genetic relatives?
3. Which of the following behaviours is altruistic: parental care, mate selection, courtship feeding, self-defence, and/or helping nonrelatives? Why? Why not?

40.14 Eusocial Animals

Hamilton's insights led to the prediction that self-sacrificing behaviour should be directed to kin. Evidence from many species of animals, particularly bees, ants, termites, and wasps, overwhelmingly supports this prediction. In a colony of **eusocial** insects, thousands of genetically related individuals, most of them sterile workers, live and work together for the reproductive benefit of one individual, a single queen and her mate(s). The workers may even die in defence of their colonies.

How did this social behaviour evolve, and why does it persist over time? A colony of honeybees may contain 30 000 to 50 000 related individuals, but only the queen bee is fertile. All of the workers are her daughters **(Figure 40.39)**. The queen's role in the colony is to reproduce. The workers perform all of the other tasks in maintaining the hive, from feeding the queen and her larvae to constructing new honeycomb and foraging for nectar and pollen. They also transfer food to one another and sometimes guard the entrance to the hive. Some pay the ultimate sacrifice when they sting intruders because stinging tears open the bee's abdomen, leaving the stinger and the poison sac behind in the intruder's skin but killing the bee.

In bees and other eusocial insects, sex is determined genetically through **haplodiploidy (Figure 40.40, p. 994).** Female bees are diploid because they receive a set of chromosomes from each parent. Male bees are haploid because they hatch from unfertilized eggs. When a queen bee mates with a drone (a male), all of the sperm he delivers are genetically identical because males have only one set of chromosomes. When a queen bee mates with just one male, all of her worker offspring will inherit exactly the same set of alleles from their male parent, ensuring at least a 50% degree of relatedness among them. Like other diploid organisms, workers are related to each other by an average of 25% through their female parent. Adding these two components of relatedness, workers are related to each other by an average of 75%, a higher degree of relatedness than they would have to any offspring they would have produced had they been fertile.

The high degree of relatedness among workers in some colonies of eusocial insects may explain their exceptional level of cooperation. When Hamilton first worked out this explanation of eusocial behaviour, he suggested that workers devote their lives to caring for their siblings (the queen's other offspring) because a few of those siblings, those carrying 75% of the workers' alleles, may become future queens and produce enormous numbers of offspring themselves.

Naked mole rats are eusocial mammals with nonbreeding workers. In East Africa, these small, almost hairless animals live in underground colonies of 70 to

a. Queen with sterile workers

Kenneth Lorenzen

b. Workers sharing food and passing pheromones

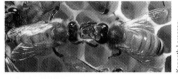

Kenneth Lorenzen

Figure 40.39
(a) In a hive of honeybees, a court of sterile workers (daughters) surround their mother (the queen). **(b)** Worker bees routinely share food and transfer pheromones to one another.

Advantage Social Behaviour

The McNamara et al. model of competitive altruism helps explain the evolution of blood-sharing behaviour in vampire bats (see Figure 3.19). Vampire bats are the only euthermic blood-feeders, and like many other bats, they are long-lived in the wild (recorded to at least 19 years of age). The three living species of vampire bats, common vampire bat (*Desmodus rotundus*), white-winged vampire bat (*Diaemus youngi*), and hairy-legged vampire bat (*Diphylla ecaudata*), all practise food sharing. An individual unsuccessful in foraging can return to its roost and beg blood from a successful forager among its roost mates. The donor bat regurgitates some of its blood meal to the recipient. G.S. Wilkinson's work with common vampire bats demonstrated that individuals roost together with both genetic relatives and nonrelatives. Familiarity, not relatedness, was the key to food sharing by these bats.

The selection process for the behaviour can be placed in context by evidence about a bat's success. Adult common vampire bats typically are unsuccessful in obtaining blood one night per month. An adult can survive two days (daytime periods) without feeding but not three. This means that on any night in any month in a colony of 30 adult vampire bats, one individual will benefit from the cooperativeness of a roost mate.

Even more importantly, young bats may be unsuccessful three or four times a week. Blood-feeding bats thus live on the edge of survival and likely depend on a network of cooperation by roost mates. The social network demonstrated for common vampire bats probably applies to white-winged and hairy-legged species as well. The network is based on cooperation and may be the key to being a successful euthermic blood-feeder.

The drone (male parent) has only one set of chromosomes (symbolized by a red circle), which he contributes to the genome of every female worker. Thus, the workers are related to each other by 50% through their male parent.

Drone
♂

The queen (female parent) has two sets of chromosomes (symbolized by a green triangle and a blue square). She contributes half of her alleles to each female offspring (either a triangle or a square) in this simplified presentation.

Queen
♀

Workers that receive different alleles from the queen share no genetic relationship through their female parent. Thus, they are related to each other only by 50% (the alleles inherited from their male parent).

Workers that receive the same alleles from the queen are related to each other by 50% through their male parent plus an additional 50% through their female parent.

Figure 40.40

Haplodiploidy. The genetic system of eusocial insects produces full siblings with exceptionally high degrees of relatedness. Although this simplified model ignores recombination between the queen's two sets of chromosomes, it demonstrates how half of the workers are related to each other by 50% and half are related to each other by 100%. On average, the relatedness between workers is 75%.

80 individuals. Like eusocial insects, naked mole rats share an exceptionally high proportion of alleles (see *People Behind Biology*).

Animals living in groups, whether they are aggregations or social units, may be at greater risk of inbreeding than those living alone. Dispersal is a mechanism that can reduce the chances of incestuous matings and inbreeding. Although spotted hyenas live in clans, the males tend to disperse from their natal units, minimizing the risk of inbreeding. Using microsatellite profiling, O.P. Höner and colleagues showed that a female preferred mates that had been born into or immigrated into the clan after she was born.

STUDY BREAK

1. What is haplodiploidy? How does it relate to Hamilton's prediction (above)?

40.15 Human Social Behaviour

Humans and chimpanzees share 95% of their genomes. In Chapter 27, we considered the relative importance of the 5% difference. Compared with humans, both chimpanzees and bonobos (*Pan paniscus*) live in relatively unstructured social groups. The brains of *Homo sapiens* are approximately three times larger than those of great apes such as the chimps and bonobos. Brain tissue has a very high metabolic rate, so growing and operating a large brain imposes significant costs.

a. Physical domain

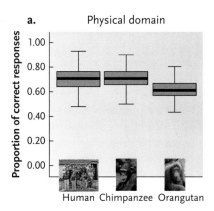

b. Social domain

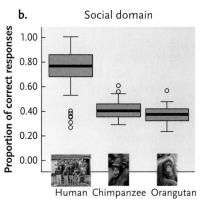

Figure 40.41
Humans, chimpanzees, orangutans. Box plots showing the proportion of correct responses to survey questions in the physical and the social domains. In the social domain, outlying data points (circles) were at least 1.5 times the interquartile distances (shown by the error bars).

The cultural intelligence hypothesis proposes that large brain size in humans reflects cognitive skill sets absent from great apes. Large brains allow humans to perform many cognitive tasks more rapidly and efficiently than other species with smaller brains. The tasks include those associated with memory, learning time, long-range planning, and complexity of interindividual interactions.

To test this, Esther Herrmann and her colleagues administered a large battery of cognitive tests to chimpanzees, orangutans (*Pongo pygmaeus*), and 2½-year-old human children. The children in the experiment were preschool and preliteracy. Although the children, chimpanzees, and orangutans had similar cognitive skills for dealing with the physical world, the children had more sophisticated cognitive skills for dealing with the social world **(Figure 40.41)**. The data support the hypothesis that cultural intelligence is an important way to distinguish humans from their closest living relatives.

The ultimatum game is an economic decision-making tool for assessing the responses of individuals to opportunities and the behaviour of others. Responses allow researchers to distinguish between players on the basis of sensitivity and sense of fairness. Keith Jensen and his colleagues used the ultimatum game to compare humans and chimpanzees. Two anonymous individuals can play a round of this game. One, the proposer, is offered a sum of money (or a food reward) and can decide whether to share it with the other, the responder. The responder can accept or reject the proposer's offer. If the responder accepts the offer, then both receive their share of the reward. If the responder rejects the offer, then neither gets any reward. The economic model predicts that the proposer will offer the responder the minimum award.

When humans play the ultimatum game, proposers typically offer 40 to 50% of the reward, and responders typically reject offers of <20%, making them "rational maximizers." When chimpanzees play the game, they are rational maximizers **(Figure 40.42)**.

They follow the economic model and show little sensitivity to fairness or the interests of others.

Together, the cultural intelligence hypothesis and the results of the ultimatum game suggest social differences between humans and their closest relatives. These findings support the views of people who believe that humans are "not animals."

In other ways, humans behave like other animals. In the area of reproduction and genetic selfishness, some humans show little difference from their mammalian cousins. Kin selection predicts that humans (and other animals) that are genetic relatives will benefit from assisting the members of their family. What happens when there is no close genetic tie between parents and children?

Margo Wilson and Martin Daly wondered if child abuse might be more common in families with stepparents who are not genetically related to all of the children in their care. They examined data on criminal child abuse within families, made available by the police department of a Canadian city. They found that the chance that a young child would be subject to criminal abuse was 40 times higher when children lived with one stepparent and one genetic parent

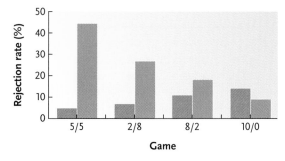

Figure 40.42
The ultimatum game. Data from chimpanzees (orange bars) and humans (green bars) show rejection rates (percentage of offers) indicating fundamental differences in the way that humans and chimps approach issues of fairness. The chimps are rational maximizers, whereas the humans are not.

Naked Mole Rats

Naked mole rats are sightless and essentially hairless burrowing mammals **(Figure 1)** that live in mazes of subterranean tunnels in parts of Ethiopia, Somalia, and Kenya. Colonies of naked mole rats may number from 25 to several hundred individuals. In each colony, a single "queen" and one to three males are the breeders. All of the others, males and females, are nonbreeding workers that, like worker bees, ants, and termites in insect colonies, do all of the labour, including digging and defending the tunnels and caring for the queen and her mates. H. Kern Reeve and his colleagues set out to determine if close kinship could explain the behaviour of worker naked mole rats. They used molecular techniques resembling DNA fingerprinting analysis (see Chapter 16) to obtain data about relatedness. The technique

Figure 1
Naked mole rats (*Heterocepahlus glaber*) live in colonies containing many workers that are effectively sterile.

depends on a group of repeated DNA sequences that vary to a greater or lesser extent among individuals (e.g., they are polymorphic). No two individuals (except identical twins) are likely to have exactly the same combination of sequences. Brothers and sisters with the same parents have the most closely related sequences, and differences increase as genetic relationships become more distant.

Reeve and his colleagues captured mole rats living in four colonies in Kenya. Individuals from the same colony were placed together in a system of artificial tunnels. Samples of the entire DNA complement were extracted from individuals that died naturally in the artificial colonies. The extracted DNA was then "probed" with radioactively labelled DNA sequences that paired with and marked the three distinct groups of polymorphic sequences in the mole rat DNA.

Naked mole rat sequences were then fragmented by treatment with a restriction endonuclease. This procedure produced a group of fragments that, reflecting the variations in polymorphic sequences, is unique for each individual. As a final experimental step, the fragments for each individual were separated into a pattern of bands by gel electrophoresis. The pattern of

bands, different for each individual, is the DNA fingerprint.

Reeve and his colleagues compared the DNA fingerprint of each mole rat with those of other members of the same and other colonies. In the comparisons, bands that were the same in two individuals were scored as "hits." The number of hits was then analyzed to assign relatedness by noting which individuals shared the greatest number of bands.

Individuals in the same mole rat colony were found to be closely related. They shared an unusually high number of bands, higher than human siblings, and approaching the kin similarity of identical twins. The number of bands shared between individuals of different colonies was significantly lower but still higher than that noted between unrelated individuals of other vertebrate species. Close relatedness of even separate colonies may be due to similar selection pressures or to recent common ancestry among colonies in the same geographic region.

In naked mole rats, close genetic relatedness among individuals in a colony could explain the altruistic behaviour of workers. The persistence of the social organization reflects its importance to the survival of individual naked mole rats, rather like the situation in lion prides.

---

compared with children living with both genetic parents **(Figure 40.43)**.

This example illustrates the insights that an evolutionary analysis of human behaviour can provide. Wilson and Daly made the point that humans may have some genetic characteristic making it more difficult to invest in children they know are not their own, particularly if they also care for their own genetic children. They did not excuse child abusers or claim that abusive stepparenting is acceptable. These results are not just academic. Most stepparents cope well with the difficulties of their role, but a few do not. Knowing the familial circumstances under which child abuse is more likely to occur may allow us to provide social assistance that could prevent some children from being abused in the future.

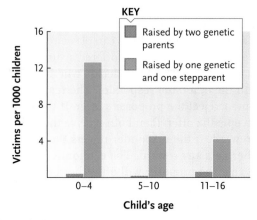

Figure 40.43
Children raised by one genetic parent and one stepparent were 40 times more likely to suffer criminal abuse at home than children living with two genetic parents.

R.F. (Griff) Ewer, University of Ghana

Born Rosalie Griffith, Griff Ewer was an outstanding leader in the study of animal behaviour. She married Denis William (Jakes) Ewer, and with their two children, they moved to the University of Natal in 1946 and from there to Rhodes University. In 1963, the family moved to the University of Ghana. To the unsuspecting person meeting her, the fact that she wore her hair short, dressed in slacks and a suit jacket, and smoked a pipe meant that some visitors were at first confused. On one paleontological expedition, she used the newspaper in which meat had been wrapped as a tablecloth. Her young daughter started to read the paper, drawing Dr. Ewer's atten-

tion to the fact that she was "in the tablecloth"—a story about the findings of a "woman paleontologist." Later, when she met the reporter who had written the story, she proceeded to express her annoyance about being identified as a "woman paleontologist." As she and her daughter walked away and were out of earshot, her daughter observed that she must have been annoyed with the man. Dr. Ewer said that she was not annoyed and was then told that her foot must have been. Dr. Ewer observed that from a child's perspective, her foot stamping was more obvious than it had been to the reporter (see Chapter 34, *Good Vibrations*)!

Dr. Griff Ewer had an insatiable curiosity about animal behaviour. In her classic book *Ethology of Mammals*, published in 1968, she encouraged would-be students of animal behaviour to take animals into their homes and live with them. Animals had free run of her house. She observed that if there was something about their behaviour that you did not understand, the animal would patiently demonstrate the behaviour time and time again. A visitor to her house was usually quickly scent-marked by the resident mongooses. The book was dedicated to her pet meerkats. *Ethology of Mammals* stands as a classic book, a tribute to Dr. Ewer and her contributions.

In recent years, the application of evolutionary thinking to human behaviour has produced research on many kinds of questions. Some questions are interesting or even profound. Why do some tightly knit ethnic groups discourage intermarriage with members of other groups? At other times, the issues may seem frivolous. Why do men often find women with certain physical characteristics attractive? Although evolutionary hypotheses about the adaptive value of behaviour can be tested, helping us understand why we behave as we do, the hypotheses should never be used to justify behaviour that is harmful to other individuals. Understanding why

we get along or fail to get along with each other and the ability to make moral judgments about our behaviour are uniquely human characteristics that set us apart from other animals.

STUDY BREAK

1. What is the ultimatum game? How does it help us understand behaviour?
2. What genetic reason helps explain the domestic risks to foster children and stepchildren?

UNANSWERED QUESTION

How did echolocation behaviour evolve? Compare echolocation with electrolocation, radar, and sonar.

Review

Go to CENGAGENOW™ at http://hed.nelson.com/ to access quizzing, animations, exercises, articles, and personalized homework help.

40.1 Genes, Environment, and Behaviour

- Instinctive behaviours are genetically or developmentally "programmed." They appear in complete and functional form the first time they are used. Examples include eating, defence, mating, and parental care. Learned behaviours depend on having a particular kind of experience during development. Examples include language, mobility, and foraging. Marler's work with White-crowned Sparrows demonstrated some aspects of learned and instinctive behaviours.

40.2 Instinct

- Fixed action patterns are triggered by specific cues (sign stimuli). Fixed action patterns are repeated over and over in almost exactly the same way. The begging behaviour of Herring Gull chicks is a good example.

- Garter snakes from some areas of California often eat slugs, whereas those from other areas do not. Snakes that regularly eat slugs flick their tongues in response to the odour of slugs. Cross-breeding snakes from populations that eat slugs with those that do not demonstrate that the response to slugs was partly under genetic control.

- The feeding situations described in *Why It Matters* demonstrate how animals learn to adjust their foraging behaviour according to the availability of prey. The examples also demonstrate how some animals plan their meals ahead. The examples demonstrate the flexibility of animal behaviour.

40.3 Learning

- Research with White-crowned Sparrows and Zebra Finches demonstrated how the brains of these birds are involved in learning and in matching song outputs. The work also has revealed the neurophysiological networks involved in selective learning, including how individual birds learn to ignore (habituate to) familiar signals.

40.4 Neurophysiology and Behaviour

- Studies of bird brains have demonstrated how specific enzymes are turned on to activate different patterns of behaviour.

- Changes in the level of factors controlling synapses can clearly influence animal behaviour.

40.5 Hormones and Behaviour

- In honeybees, juvenile hormone stimulates genes in certain brain cells to produce proteins that affect functions of the nervous system. Octopamine is a product that stimulates neural transmissions and reinforces memories—it occurs in higher concentrations in older bees. This example shows how hormones and genes interact to affect behaviour.

- Gonadotropin-releasing hormone (GnRH) causes differences in the behaviour of fish such as *Haplochromis buroni*. Levels of GnRH are useful predictors of territorial behaviour.

40.6 Neural Anatomy and Behaviour

- Some crickets use ears (tympanic membranes) on their front legs to detect the echolocation calls of bats. Differential stimulation of left and right ears allows crickets to rapidly change their flight behaviour in response to approaching bats.

- Fiddler crabs (*Uca pugilator*) have compound eyes that give them a split field of view, which allows them to distinguish between the movements of other crabs and the approach of potential predators.

40.7 Communication

- Animals use at least five classes of signals in communication: acoustic signals such as songs used to attract a mate or warn intruders; visual signals such as the bioluminescent display of fireflies; chemical signals such as pheromones like bombykol, produced by female silkworm moths (and predatory spiders); tactile signals such as grooming in group-living primates; and electrical signals such as the pulses produced by knifefish or elephant fish.

- The "yells" of Ravens that have found a patch of food attract others. This behaviour allows nonresidents to gang up on resident ravens and obtain food that would otherwise not be available to them.

- Syntax (or grammar) and symbols are characteristic of language. Humans use words to represent objects (chair, table, dog), and the order of words (syntax) affects meaning (she was bitten by a dog versus she bit a dog). Dancing honeybees use symbols and syntax, and animals such as vervet monkeys use different calls to represent different predators.

40.8 Space

- Kinesis is a change in the rate of movement or the frequency of turning movements in response to environmental stimuli. A taxis is a response directed either toward or away from a specific stimulus.

- An animal's home range is the area it regularly uses, typically to move from where it sleeps to where it feeds. A territory is space that is defended to allow exclusive use by an individual or group of individuals. Home ranges and territories are features of individuals, whereas a range is a feature of a species (a population of individuals).

40.9 Migration

- Migration is a seasonal movement to and from an area. Migrations can be lengthy or short. Some birds, bats, insects, whales, and caribou migrate between different areas according to the season. Some nomadic populations of humans (and some retired people) move to and from habitats, also according to the season. Migrations are usually triggered by some combination of changes in day length and weather.

- Animals may find their way by piloting, landmarks, compass, or celestial (stars and the Sun) cues in navigation. Compass orientation involves movement in a specific direction as indicated by an external source such as the Sun, stars, or Earth's magnetic field. Animals continue to move in the same direction until they have completed their journey. Navigation is more complicated. Although still using compass orientation, navigation also requires the use of a "mental map," some independent indication of the animal's location. The "mental map" enables the animal to determine its relative position to the target location, and the compass provides the direction back to the target location.

40.10 Mates as Resources

- In many species, the behaviour of individuals of one sex is determined by the distribution of the other sex. In polygynous species, the distribution of males is influenced by the distribution of females, which are defended by the males. In other cases, albeit less common, males are defended by females. Be careful to distinguish situations in which the individuals are the resource that is defended as opposed to access to food or roosts (for example).

40.11 Sexual Selection

- Monogamy describes the situation when a male and a female form a pair bond and do not mate with others.
- Polygamy occurs when a male (polygyny) or a female (polyandry) mates with multiple partners. Polygamy usually involves pair bonds between the mating individuals. In promiscuity, there are no pair bonds, and males and females may mate with multiple partners.
- Lek mating systems are promiscuous. Males congregate in a display area (an arena), where they are visited by females. Females mate with the most attractive male (the one with the best display area). Males are the only resource at the display site. In leks, females typically mate with one male and males with multiple females.

40.12 Social Behaviour

- Infanticide is the killing of conspecific young. Male African lions practise infanticide when they take over a pride (group of females) because stopping nursing brings females back into heat, allowing the male to increase his direct fitness.
- Group living can provide significant advantages to animals such as easier-to-locate food resources either through cooperative hunting or simply an increased searching capacity (many eyes), ease in finding a mate, more efficient raising of young, more effective defence against predators, and increased vigilance. Group living also can present disadvantages, including increased competition for food, increased spread of parasites and disease, and increased conspicuousness to predators.

40.13 Kin Selection and Altruism

- Hamilton's kin selection theory states that helping genetic relatives effectively propagates the helper's genes because family members share alleles inherited from their ancestors. Individuals should be more likely to help close relatives because increasing a close relative's fitness means that the individual is helping to propagate some of its own alleles.
- Degree of relatedness between individuals

$$= (4 \times \text{first cousin relatedness})$$
$$+ (2 \times \text{sibling relatedness})$$
$$+ (2 \times \text{half-sibling relatedness})$$
$$= (4 \times 0.125) + (2 \times 0.5) + (2 \times 0.25)$$
$$= 0.5 + 1 + 0.5$$
$$= 2$$

- Helping only nonrelatives is altruistic because each of the other examples involves genetically selfish behaviour (= getting the actor's genes into the next generation).

40.14 Eusocial Animals

- Haplodiploidy occurs when the females in a colony are diploid and the male are haploid. Females receive a set of chromosomes from each parent, but males hatch from unfertilized eggs (one set of chromosomes). This affects the level of genetic relationship among females. Worker bees hatching from eggs laid by one queen bee could all have the same father if the queen mated with only one male. We can use Hamilton's kinship theory to understand the behaviour of eusocial animals.

40.15 Human Behaviour

- In the ultimatum game, players have the opportunity to demonstrate their sense of fairness and their sensitivity to others. When people and chimps play the ultimatum game, the results demonstrate a fundamental difference in their levels of social behaviour.

Questions

Self-Test Questions

1. Peter Marler concluded that White-crowned Sparrows can learn their species' song only
   a. after receiving hormone treatments.
   b. during a critical period of their development.
   c. under natural conditions.
   d. from their genetic father.
   e. if they are reared in isolation cages.

2. In cichlid fish, high levels of the hormone GnRH
   a. make females more receptive to male attention.
   b. cause males to be sexually aggressive but not territorial.
   c. stimulate a male to defend its territory.
   d. cause males to abandon their territories.
   e. cause males to lose their bright colours.

3. Sensory bias in the nervous system of a cricket ensures that ultrasound perceived on one side of the body causes
   a. a movement in a leg on the same side of the body.
   b. a movement in a leg on the opposite side of the body.
   c. the cricket to respond with a vocalization.
   d. the cricket to stop vocalizing.
   e. the cricket to fly toward the sound.

4. In the brain of a star-nosed mole, more cells decode
   a. tactile information from its feet than from all other parts of its body.
   b. tactile information from the tentacles on its nose than from all other parts of its body.
   c. tactile information from its mouth than from all other parts of its body.
   d. visual information from the top part of its visual field than the bottom part.
   e. visual information from the bottom part of its visual field than the top part.

5. Which of the following statements about animal migration is true?
   a. Piloting animals use the position of the Sun to acquire information about their direction of travel.
   b. Animals migrating by compass orientation use mental maps of their position in space.
   c. Navigating animals use familiar landmarks to guide their journey.
   d. Navigating animals use a compass and a mental map of their position to reach a destination.
   e. Most migrating birds use olfactory cues to return to the place where they hatched from eggs.

6. Squashing an ant on a picnic blanket often attracts many other ants to its "funeral." What kind of signal did squashing the ant likely produce?
   a. an electrical signal
   b. a visual signal
   c. an acoustical signal
   d. a chemical signal
   e. a tactile signal

7. Compared with males, the females of many animal species
   a. compete for mates.
   b. choose mates that are well camouflaged in their habitats.
   c. choose to mate with many partners.
   d. are always monogamous.
   e. choose their mates carefully.

8. Social behaviour
   a. is exhibited *only* by animals that live in groups with close relatives.
   b. cannot evolve in animals that maintain territories.
   c. evolved because group living provides benefits to individuals in the group.
   d. is never observed in insects and other invertebrate animals.
   e. can only be explained by the hypothesis of kin selection.

9. Altruism is a behaviour that
   a. cannot evolve.
   b. advances the welfare of the entire species.
   c. increases the number of offspring an individual produces.
   d. can indirectly spread the altruist's alleles.
   e. can only evolve in animals with a haplodiploid genetic system.

Questions for Discussion

1. When can communication behaviour be called "language"? What is language?

2. Using an example from your own experience, explain why habituation to a frequent stimulus might be beneficial. Describe an example in which habituation might be harmful or even dangerous.

3. Is learning always superior to instinctive behaviour? If you think so, why do so many animals react instinctively to certain stimuli? Are there environmental circumstances in which being able to respond "correctly" the first time would have a big payoff?

4. What effects might global warming have on animal species that undertake seasonal migrations?

5. Develop three evolutionary hypotheses to explain why male birds are likely to involve themselves in caring for their young.

Foods that are high in starch include potatoes (*Solanum tuberosum*), corn (*Zea mays*), wild rice (*Zizania* species), and domestic rice (*Oryza* species).

M.B. Fenton

41 Plant and Animal Nutrition

WHY IT MATTERS

Salivary amylase **(Figure 41.1, p. 1002)**, an enzyme produced by the saliva glands, initiates digestion of starch during chewing and can break down significant amounts of starch even before the food is swallowed. The action of **salivary amylase** explains why this happens. Blood glucose has been found to be significantly higher when high-starch foods (e.g., corn, rice, and potatoes) are first chewed and then swallowed rather than swallowed without chewing. The same is not true of low-starch foods such as apples. Immediate oral access to energy is especially important during episodes of diarrhea. Furthermore, in the stomach and intestines, salivary amylase augments the activity of pancreatic amylase and may further buffer the body against the impact of digestive disorders.

In humans, the salivary amylase gene (*AMY1*) varies in the numbers of copies according to the amount of starch in the diet of different populations. More copies of *AMY1* coincides with higher levels of salivary amylase in saliva **(Figure 41.2, p. 1002)** among seven populations of people, three with high-starch diets and four with low-starch diets. Humans living in agricultural societies and those in hunter–gatherer societies in arid environments have high-starch diets. Hunter–gatherers

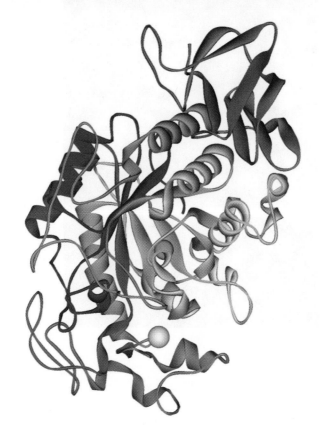

Figure 41.1
Salivary amylase molecule.

Figure 41.2
Salivary amylase.
(a) In humans, more copies of the gene *AMY1* correlates with higher levels of salivary amylase **(b)** and more starch in the diet.

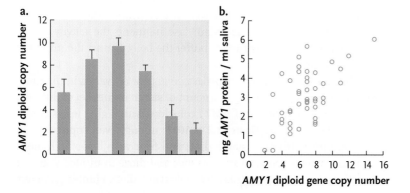

in rainforest and circumarctic habitats, as well as some pastoralists, have low-starch diets.

Although most digestion in mammals occurs beyond the esophagus, particularly in omnivores, salivary amylase presents an interesting exception. By hydrolyzing starch into the disaccharide maltose, salivary amylase can have an immediate impact on levels of blood sugar. The same impact on blood sugar is not achieved when people eat (and carefully chew) low-starch foods such as apples.

41.1 Nutrition: Essential Materials

All organisms require sources of matter and energy for metabolism and homeostasis (see Chapter 43), as well as growth and reproduction. No organism grows normally when deprived of a chemical element essential for its metabolism. In the latter half of the nine-

teenth century, plant physiologists exploited rapid advances in chemistry to explore the chemical composition of plants and the essential nutrients they needed to survive. Because living organisms require some nutrients in only trace amounts, biologists developed more sophisticated methods to understand the roles that **trace elements** play in organisms' nutrition and well-being.

41.1a Macronutrients and Micronutrients in Plant Metabolism

Most of a plant's nutrition is derived through photosynthesis, the combining of water and carbon dioxide to produce sugars and starches. Plants also require much smaller amounts of other nutrients, which they obtain from soil or water.

By weight, the tissues of most plants are >90% water. Burning a plant (or other organism) and analyzing the resulting ash was one way to obtain a rough estimate of the composition of a plant's dry weight. Although this method yielded a long list of elements, the results were flawed because chemical reactions during burning often dissipated quantities of some important elements (e.g., hydrogen, oxygen, and nitrogen). Furthermore, plants take up a variety of ions they do not use, and the exact combination of these ions depends on the minerals present in the soil where a plant grows. The tissues of organisms may contain non-nutritive elements such as gold, lead, arsenic, and uranium.

41.1b Hydroponics: A New Way to Study Plant Nutrition

In 1860, German plant physiologist Julius von Sachs pioneered hydroponics (*hydro* = water; *ponos* = work), an experimental method for identifying minerals absorbed into plant tissues that were essential for growth. Sachs carefully measured amounts of compounds containing specific minerals and mixed them in different combinations with pure water. He then grew plants in the solutions and studied their growth after eliminating one element at a time. In this way, Sachs deduced a list of six essential plant nutrients, in descending order of the amount required: nitrogen, potassium, calcium, magnesium, phosphorus, and sulphur.

Sachs's innovative research paved the way for decades of increasingly sophisticated studies of plant nutrition. One variation on his approach involved growing a plant in a solution containing a complete spectrum of known and possible essential nutrients **(Figure 41.3).** The healthy plant is then transferred to a solution that is identical except that it lacks one element with an unknown nutritional role. Abnormal growth of the plant in this solution is evidence that the missing element is essential. Normal growth indicates that the missing element may not be essential, but only further experimentation can confirm this hypothesis.

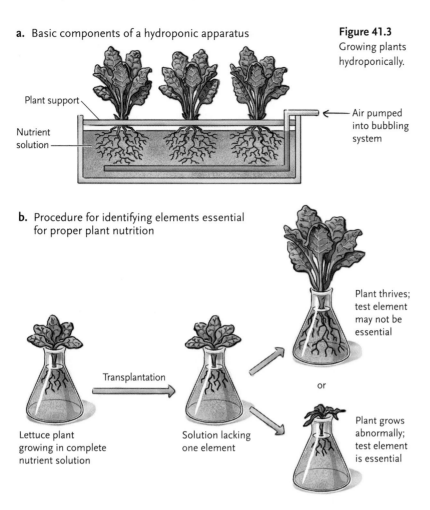

a. Basic components of a hydroponic apparatus

Plant support

Nutrient solution

Air pumped into bubbling system

b. Procedure for identifying elements essential for proper plant nutrition

Transplantation

Lettuce plant growing in complete nutrient solution

Solution lacking one element

or

Plant thrives; test element may not be essential

Plant grows abnormally; test element is essential

Figure 41.3
Growing plants hydroponically.

proteins and nucleic acids. Plants also add phosphorus to C, H, and O to construct nucleic acids, ATP, and phospholipids and potassium for functions ranging from enzyme activation to mechanisms that control the opening and closing of stomata. Rounding out the list of macronutrients are calcium, sulphur, and magnesium.

Carbon, hydrogen, and oxygen are the only plant nutrients not considered to be minerals. Plants obtain C, H, and O from the air and water. The other six macronutrients are mineral nutrients, inorganic substances available to plants through the soil as ions dissolved in water.

Micronutrients are mineral elements required in trace amounts, but they are as vital as macronutrients to a plant's health and survival. Although 2.5 tonnes of potatoes contain roughly the amount of copper in a single, copper-plated, Canadian penny, without it, potato plants are sickly and do not produce normal tubers.

Chlorine was identified as a micronutrient nearly a century after Sachs's experiments. Using **hydroponic culture** in a California laboratory near the Pacific Ocean, researchers discovered the vital role of Cl^- because the coastal air contains sodium chloride. The test plants obtained tiny but sufficient quantities of chlorine from the air or from sweat (which also contains NaCl) on the researchers' hands. Denying the plants access to Cl^- required careful control of the test environments in which they grew.

In some cases, plant seeds contain enough of certain trace minerals to sustain the adult plant. Nickel (Ni^{2+}) is a component of urease, the enzyme required to hydrolyze urea. Urea is a toxic by-product of the breakdown of nitrogenous compounds, and it will kill the cells in which it accumulates. In the late 1980s, investigators found that barley seeds contain enough Ni^{2+} to sustain two complete generations of barley plants. Plants grown in the absence of Ni^{2+} did not begin to show signs of Ni^{2+} deficiency until the third generation. Besides the 17 essential elements, some plant species may require additional micronutrients. Many, perhaps most, plants adapted to hot, dry conditions require sodium ions, including many plants using the C_4 pathway (see Chapter 7). A few plant species require selenium, also an essential micronutrient for animals (see Table 41.1). Horsetails (*Equisetum*) require silicon, as may some grasses (such as wheat). Scientists continue to discover additional micronutrients for specific plant groups.

In a typical modern hydroponic apparatus (see Figure 41.3b), the nutrient solution is refreshed regularly and air is bubbled into it to supply oxygen to the roots. In the absence of sufficient oxygen for respiration, plants' roots do not absorb nutrients efficiently. The same thing occurs in poorly aerated soil. Hydroponics are used on a commercial scale to grow vegetables such as lettuce and tomatoes.

41.1c Essential Macro- and Micronutrients for Plants

Hydroponics research has revealed that plants generally require 17 essential elements **(Table 41.1, p. 1004–1005).** An **essential element** is defined as one that is necessary for normal growth and reproduction and cannot be functionally replaced by a different element. Essential elements can have one or more roles in plant metabolism. With the combination of enough sunlight and the 17 essential elements, plants can synthesize all of the compounds they need.

Nine of the essential elements are **macronutrients** because plants incorporate relatively large amounts of them into their tissues. Together, carbon (C), hydrogen (H), and oxygen (O) account for about 96% of a plant's dry mass and are the key components of lipids and of carbohydrates such as cellulose. With the addition of nitrogen, C, H, and O are the basic building blocks of

Table 41.1	Essential Elements and Their Functions in Plants and Animals			
Element	Commonly Absorbed Forms	Some Known Functions, Macronutrient or Micronutrient	Some Deficiency Symptoms	Sources for Humans
Carbon*	CO_2	Essential elements for life, including raw materials for photosynthesis	Rarely deficient	
Hydrogen*	H_2O			
Oxygen*	O_2, H_2O, CO_2			
Boron	H_3BO_3	**Plants, micro:** roles in germination, flowering, fruiting, cell division, nitrogen metabolism	**Plants only:** terminal buds and lateral branches die; leaves thicken, curl, and become brittle	
Calcium	Ca^{2+}	**Plants, macro:** formation and maintenance of cell walls and membrane permeability, enzyme cofactor **Animals, macro:** bone and tooth formation, blood clotting, action of nerves and muscles	**Plants:** leaves deformed, terminal buds die, poor root growth **Animals:** stunted growth, diminished bone mass (osteoporosis in humans)	Dairy products, leafy green vegetables, legumes, whole grains, and nuts
Chlorine	Cl^-	**Plants, micro:** role in root and shoot growth and photosynthesis **Animals, macro:** formation of HCl (in stomach), contributes to acid–base balance, neural function, water balance	**Plants:** wilting, chlorosis, some leaves die (deficiency not seen in nature) **Animals:** muscle cramps, impaired growth, poor appetite	Table salt, meat, eggs, dairy products
Chromium	Cr	**Animals, macro:** roles in carbohydrate metabolism	**Animals only:** impaired responses to insulin (increased risk of type 2 diabetes mellitus in humans)	Meat, liver, cheese, whole grains, brewer's yeast, peanuts
Cobalt	Co	**Animals, macro:** constituent of vitamin B$_{12}$ (required for normal red blood cell maturation)	**Animals only:** same as for vitamin B$_{12}$ (see Table 41.2)	Meat, liver, fish, milk
Copper	Cu^+, Cu^{2+}	**Plants and animals, micro:** component of several enzymes **Animals:** used in synthesis of melanin, hemoglobin, and in some electron transport chain components in mitchondria	**Plants:** chlorosis, dead spots in leaves, stunted growth **Animals:** anemia, changes in bone and blood vessels	Nuts, legumes, seafood, drinking water, whole grains
Fluorine	F	**Animals, macro:** bone and tooth maintenance	**Animals only:** tooth decay	Fluoridated water, seafood, tea
Iodine	I	**Animals, macro:** thyroid hormone formation	**Animals only:** goitre (enlarged thyroid) and metabolic disorders	Marine fish, shellfish, iodized salt
Iron	Fe^{2+}, Fe^{3+}	**Plants and animals:** roles in electron transport, component of cytochrome **Plants micro:** role in chlorophyll synthesis **Animals, macro:** component of hemoglobin and myoglobin	**Plants:** chlorosis, yellow and green striping in grasses **Animals:** iron-deficiency anemia	Liver, whole grains, green leafy vegetables, legumes, nuts, eggs, lean meat, molasses, dried fruit, shellfish
Magnesium	Mg^{2+}	**Plants, macro:** component of chlorophyll, activation of enzymes **Animals, macro:** activation of enzymes, roles in functioning of nerves and muscles	**Plants:** chlorosis, drooping leaves **Animals:** weak and sore muscles, impaired neural function	Whole grains, green vegetables, legumes, nuts, dairy products

Table 41.1 | *(continued)*

Element	Commonly Absorbed Forms	Some Known Functions, Macronutrient or Micronutrient	Some Deficiency Symptoms	Sources for Humans
Manganese	Mn^{2+}	**Plants and animals, macro:** role is activation of enzymes, coenzyme action **Plants:** involved in chlorophyll synthesis **Animals:** plays a role in synthesis of urea and fatty acids	**Plants:** dark veins, but leaves whiten and fall off **Animals:** abnormal bone and cartilage	Whole grains, nuts, legumes, many fruits
Molybdenum	MoO_4^{2-}	**Plants and animals, macro:** component of enzyme used in nitrogen metabolism **Animals:** components of some enzymes	**Plants:** pale green, rolled or cupped leaves **Animals:** impaired nitrogen excretion	Dairy products, whole grains, green vegetables, legumes
Nickel	Ni^{2+}	**Plants, micro:** component of enzyme required to break down urea generated during nitrogen metabolism	**Plants only:** dead spots on leaf tips (deficiency not seen in nature)	
Nitrogen	NO_3^-, NH_4^+	**Plants and animals, macro:** component of proteins, nucleic acids, coenzymes **Plants:** component of chlorophylls	**Plants:** stunted growth, light-green older leaves, older leaves yellow and die (chlorosis)	
Phosphorus	$H_2PO_4^-$, HPO_4^{2+}	**Plants and animals, macro:** component of nucleic acids, phospholipids, ATP, several coenzymes **Animals:** component of bones	**Plants:** purplish veins, stunted growth, fewer seeds, fruits **Animals:** muscular weakness, loss of minerals from bone	Whole grains, legumes, poultry, red meat, dairy products
Potassium	K^+	**Plants and animals, macro:** activation of enzymes, key role in maintaining water–solute balance and so influences osmosis **Animals:** involved in actions of nerves and muscles	**Plants:** reduced growth; curled, mottled, or spotted older leaves; burned leaf edges; weakened plant **Animals:** muscular weakness	Meat, many fruits and vegetables
Selenium	Se	**Animals, macro:** constituent of several enzymes, antioxidant	**Animals only:** muscle pain	Meat, seafood, cereal grains, poultry, garlic
Sodium	Na^+	**Animals, macro:** acid–base balance, roles in functioning of nerves and muscles	**Animals only:** muscle cramps	Table salt, dairy products, meats, eggs
Sulphur	SO_4^{2-}	**Plants and animals, macro:** component of most proteins, coenzyme A	**Plants:** light-green or yellowed leaves, reduced growth **Animals:** same symptoms as those associated with protein deficiencies	Meat, eggs, dairy products
Zinc	Zn^{2+}	**Plants, micro:** plays a role in formation of auxin, chloroplasts, and starch; enzyme component **Animals, micro:** component of digestive enzymes and transcription factors; role in normal growth, wound healing, sperm formation, as well as taste and smell	**Plants:** chlorosis, mottled or bronzed leaves, abnormal roots **Animals:** impaired growth, scaly skin, impaired immune function	Whole grains, legumes, nuts, meats, seafood

Micronutrients and macronutrients play vital roles in plant metabolism. Many function as cofactors or coenzymes in protein synthesis, starch synthesis, photosynthesis, and aerobic respiration. Some also have a role in creating solute concentration gradients across plasma membranes, which are responsible for the osmotic movement of water.

41.1d Consequences of Nutrient Deficiencies for Plants

Plant species differ in the quantity of each nutrient they require, so the amount of an essential element adequate for one plant species may be insufficient for another. Lettuce and other leafy plants require more N and magnesium than many other plants. Alfalfa requires significantly more K^+ than lawn grasses. The amount of an essential element adequate for one species may be toxic for another. The amount of boron required for normal growth of sugar beets is toxic for soybeans. Thus, the nutrient content of soils is an important factor determining which plants grow well in a given location.

Plants deficient in one or more of the essential elements develop characteristic abnormalities **(Figure 41.4;** see also Table 41.1), and the symptoms give an indication of the metabolic roles the missing elements play in plant growth and development (see Chapter 31). Deficiency symptoms typically include stunted growth, abnormal leaf colour, dead spots on leaves, or abnormally formed stems (see Figure 41.4). Iron is a compo-

nent of cytochromes essential to the cellular electron transfer system and plays a role in reactions that synthesize chlorophyll. **Chlorosis**, a symptom of Fe deficiency, is a yellowing of plant tissues resulting from lack of chlorophyll (see Figure 41.4b). Because ionic iron (Fe^{3+}) is relatively insoluble in water, gardeners often fertilize plants with chelated Fe, a soluble Fe compound that staves off or cures chlorosis. Chelating agents are also used in *phytoremediation*, the use of plants to clean up contaminated soils (see Phytoremediation). Mg^{+2} is also a necessary component of chlorophyll, so plants deficient in it have fewer chloroplasts than normal. Plants that are Mg^{+2} deficient appear paler green than normal, and their growth is stunted because of reduced photosynthesis (see Figure 41.4c).

Plants that lack adequate nitrogen may also become chlorotic (see Figure 41.4d), with older leaves yellowing first because the N is preferentially shunted to younger, actively growing plant parts. This adaptation is not surprising, given N's central role in the synthesis of amino acids, chlorophylls, and other compounds vital to plant metabolism. However, young leaves are the first to show symptoms of some other mineral deficiencies. These observations underscore the point that plants use different nutrients in specific, often metabolically complex ways.

Soils often have too little rather than too much N, P, K, or some other essential mineral, so farmers and gardeners typically add nutrients to suit the types of plants they wish to cultivate. Farmers and gardeners can assess deficiency symptoms of plants grown in their

Figure 41.4
Leaves and stems of tomato plants showing visual symptoms of seven different mineral deficiencies. The plants were grown in the laboratory, where the experimenter could control which nutrients were available. (Photos by E. Epstein, University of California, Davis.)

a. Plant grown using a complete growth solution

b. Iron deficiency

c. Magnesium deficiency

d. Nitrogen deficiency

e. Phosphorus deficiency

f. Potassium deficiency

g. Sulphur deficiency

h. Calcium deficiency

Phytoremediation

The pollution of soils contaminated by industrial waste is a global environmental problem. The contamination includes a range of heavy metals as well as numerous toxic organic compounds. One high-profile target for cleanup is the highly toxic organic compound methylmercury (MeHg) because industrial and agricultural activities have released several hundred thousand tonnes of mercury into the biosphere during the past century. MeHg is present in coastal soils and wetlands that are contaminated by industrial wastes containing an ionic form of the element mercury called Hg(II). This builds up when bacteria in contaminated sediments metabolize Hg(II) and generate MeHg as a metabolic by-product. Once MeHg forms, it enters the food web and eventually becomes concentrated in the tissues of fishes and other animals. MeHg is particularly toxic in humans as it can lead to degeneration of the nervous system and is the cause of most cases of mercury poisoning due to consuming contaminated fish (see Chapter 47).

To date, the remediation of mercury-polluted sites has been slow because chemical engineering methods currently employed to remove mercury-containing compounds are expensive and highly disruptive to the environment. Solutions such as "concrete capping" are designed to stabilize the mercury but render sites uninhabitable for plants, insects, and other organisms.

Phytoremediation, the use of plants as a natural means of removing contaminants from contaminated soils, wetlands, and aquatic habitats, may be a viable alternative for remediating contaminated sites. The procedure relies on the ability of a range of plant species to take up and sequester toxic compounds. In the 1990s, a team of scientists including Scott Bizily and Richard Meagher used genetic engineering to try to develop plants capable of detoxifying mercury-contaminated soil and wetlands. It was already known that bacteria in contaminated sediments possess two genes, *merA* and *merB*, that encode enzymes that convert MeHg into elemental mercury (Hg), a relatively inert substance that is much less dangerous to organisms. Using standard molecular biology techniques (see Chapter 16), the research team expressed both *merA* and *merB* in *Arabidopsis thaliana*, a model plant system for plant biology research **(Figure 1).**

The group produced three transgenic lines: one in which plants contained only the *merA* gene, some that contained only the *merB* gene, and some that contained both *merA* and *merB*. Seeds from each group were grown (along with wild-type controls) in five different growth media, one containing no mercury and the other four containing increasing concentrations of methylmercury. Wild-type and *merA* seeds germinated and grew only in the mercury-free growth medium. The *merB* seedlings fared somewhat better. They germinated and grew briefly even at the highest concentrations of MeHg but soon became chlorotic and died. Not only did the seeds with the *merA/merB* genotype germinate, but the resulting seedlings also grew into robust plants with healthy root and shoot systems. In later tests, *merA/merB* plants were grown in chambers in which the chemical composition of the air could be monitored. This study revealed that the doubly transgenic plants were transpiring large amounts of Hg. The results showed that *A. thaliana* with *merA* and *merB* genes could take up toxic methylmercury with no ill effects and convert it to a harmless form. Meagher and his colleagues are now experimenting with ways of increasing the efficiency of phytore-mediating enzymes when plant cells express *merA* and *merB*. They are also studying the mechanisms by which ionic mercury taken up by roots may be transported via the xylem to leaves and other shoot parts. The goal is to engineer plants that accumulate large quantities of mercury in aboveground tissues that can be harvested, leaving the living plant to continue its "work" of detoxifying a contaminated landscape.

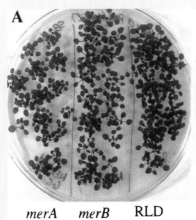
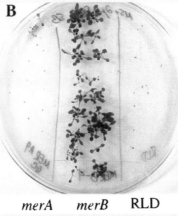
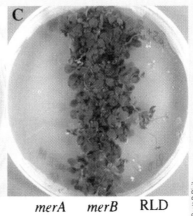

Figure 1
Plates of parental (RLD, a wild-type ecotype of *Arabidopsis*) and transgenic *Arabidopsis thaliana* (*merA* and *merB*) grown on media with no mercury **(a)** or 2 µM phenylmercuric acetate **(b** and **c)**; a and b were photographed after 4 weeks and c after 6 weeks. RLD and *merA* seeds did not germinate in the presence of mercury, whereas *merB* seeds did.

locale or have soil tested in a laboratory and then select a fertilizer with the appropriate balance of nutrients to compensate for deficiencies. Packages of commercial fertilizers use a numerical shorthand (e.g., 15-30-15) to indicate the percentages of N, P, and K they contain.

41.1e Essential Elements for Animals

Plants and other photosynthesizers use sunlight as an energy source and a supply of simple inorganic precursors such as water, carbon dioxide, and minerals to make all of the organic molecules they require. Animals require a diet of organic molecules as a source of energy and nutrients that they cannot make for themselves. Animals can be classified according to the sources of organic molecules they use (see Chapters 3 and 48). Primary consumers eat plants, whereas **secondary consumers** primarily eat other animals. "Primarily" is an appropriate modifier because many herbivores eat animal matter sometimes (e.g., insects on the plants), and secondary consumers often eat plant material (e.g., a cat eating grass). Animals that regularly take food from different **trophic levels** (see Chapter 48) are omnivores.

Organic molecules are the basis of two of the most fundamental processes of life, namely fuels for oxidative reactions supplying energy and as building blocks for making complex biological molecules.

Fuel. Animals must acquire enough fuel in their diets to cover their basic costs of operation. Carbohydrates and fats are primary organic fuel molecules used in cellular respiration (see Chapter 6). Undernourished animals suffer from inadequate intake of organic fuels or abnormal assimilation of these fuels. **Undernutrition** is commonly referred to as **malnutrition**, a condition resulting from an improper diet. **Overnutrition** is caused by excessive intake of specific nutrients and is another form of malnutrition.

An undernourished animal can be starving for one or more nutrients or just be eating fewer calories than needed for daily activities. Animals with chronic undernutrition lose weight because they use molecules of their own bodies as fuels. In times when food is abundant, some animals accumulate stores of fat for use in lean times. Birds on long migratory flights metabolize fat, as well as other tissues, to meet their energy needs. Relatively short-term use of an animal's own proteins as fuels leads to the wasting of muscles and other tissues and cannot be sustained over long periods of time.

Building Blocks. As in plants, organic molecules serve as building blocks for carbohydrates, lipids, proteins, and nucleic acids. Animals can synthesize many of the organic molecules that they do not obtain directly in their diet by converting one type of building block into another. But there are some amino acids and fatty acids that most animals cannot synthesize and must obtain from organic molecules in their food. Lack of these essential amino acids and **essential fatty acids** can

have serious consequences. Protein synthesis cannot continue unless all 20 amino acids are present. In the absence of essential amino acids in the diet, an animal would have to break down its own proteins to obtain the necessary building blocks for new protein synthesis.

Vitamins and Coenzymes. Animals also must ingest **vitamins**, organic molecules that are required in small quantities. Many animals cannot synthesize these for themselves. Many vitamins are coenzymes, nonprotein organic subunits associated with enzymes that assist in enzymatic catalysis (see Chapter 4). Individual species differ in the vitamins and essential amino acids and fatty acids they require in their diets. Various species also have different dietary requirements for inorganic elements such as calcium, iron, and magnesium. Required inorganic elements are known collectively as **essential minerals**. **Essential nutrients** include amino acids, fatty acids, vitamins, and minerals, but the precise list varies from species to species, even from individual to individual.

As in plants, elemental macronutrients and micronutrients are essential in the animal diet (see Table 41.1). Humans require macronutrients in amounts ranging from 50 mg to more than 1 g per day and micronutrients (trace elements) such as zinc in small amounts, some <1 mg per day. All of the minerals, although listed as elements, are ingested by animals as compounds or as ions in solution.

41.1f Essential Elements for Humans

Adult humans require eight essential amino acids: lysine, tryptophan, phenylalanine, threonine, valine, methionine, leucine, and isoleucine. Infants and young children also require histidine. Proteins in fish, meat, egg whites, milk, and cheese supply all of the essential amino acids as long as they are eaten in adequate quantities. In contrast, the proteins of many plants are deficient in one or more of the amino acids essential to humans. Corn contains inadequate amounts of lysine, and beans contain little methionine. Vegetarians, especially vegans, whose diet includes no animal-derived nutrients, must choose their foods carefully to obtain all of the essential amino acids **(Figure 41.5).**

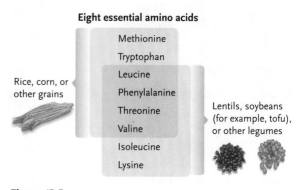

Eight essential amino acids

Methionine
Tryptophan
Leucine
Phenylalanine
Threonine
Valine
Isoleucine
Lysine

Rice, corn, or other grains

Lentils, soybeans (for example, tofu), or other legumes

Figure 41.5
Obtaining essential amino acids in a human vegetarian diet.

Protein deficiency occurs when essential amino acids are not part of the diet. Consequently, many enzymes and other proteins cannot be synthesized in sufficient quantities. Protein deficiency is most damaging to the young because of their need for proteins for normal development and growth. Even mild protein starvation during pregnancy or for some months after birth can retard a child's growth and have negative effects on mental and physical development.

Two fatty acids, linoleic acid and linolenic acid, are essential because they are used in the synthesis of phospholipids that form parts of biological membranes and certain hormones. Because almost all foods contain these fatty acids, most people have no problem obtaining them. However, people on a low-fat diet deficient in linoleic acid and linolenic acid are at serious risk for developing coronary heart disease. There is an inverse correlation between the concentration of these essential fatty acids in the diet and the incidence of coronary heart disease. Hindu vegetarians from India eat mainly low-fat grains and legumes, a low-fat diet. Yet their rate of coronary heart disease is higher than rates in the United States and Europe, where dietary fat content is higher.

Humans require 13 known vitamins in their diet **(Table 41.2)**. Many metabolic reactions depend on vitamins, and the absence of one vitamin can affect the functions of the others. Vitamins fall into two classes: **water-soluble** (hydrophilic) **vitamins** and **fat-soluble** (hydrophobic) **vitamins**. The body stores excess fat-soluble vitamins in adipose tissues (fat), but any

Table 41.2	Vitamins: Sources, Functions, and Effects of Deficiencies in Humans		
Vitamin	Common Sources	Main Functions	Effects of Chronic Deficiency
Fat-Soluble Vitamins			
A (retinol)	Yellow fruits, yellow or green leafy vegetables; also in fortified milk, egg yolk, fish liver	Used in synthesis of visual pigments, bone, teeth; maintains epithelial tissues	Dry, scaly skin; lowered resistance to infections; night blindness
D (calciferol)	Fish liver oils, egg yolk, fortified milk; manufactured when body exposed to sunshine	Promotes bone growth and mineralization; enhances calcium absorption from gut	Bone deformities (rickets) in children; bone softening in adults
E (tocopherol)	Whole grains, leafy green vegetables, vegetable oils	Antioxidant; helps maintain cell membrane and red blood cells	Lysis of red blood cells; nerve damage
K (napthoquinone)	Intestinal bacteria; also in green leafy vegetables, cabbage	Promotes synthesis of blood clotting protein by liver	Abnormal blood clotting, severe bleeding (hemorrhaging)
Water-Soluble Vitamins			
B_1 (thiamine)	Whole grains, green leafy vegetables, legumes, lean meats, eggs, nuts	Connective tissue formation; folate utilization; coenzyme forming part of enzyme in oxidative reactions	Beriberi; water retention in tissues; tingling sensations; heart changes; poor coordination
B_2 (riboflavin)	Whole grains, poultry, fish, egg white, milk, lean meat	Coenzyme	Skin lesions
Niacin	Green leafy vegetables, potatoes, peanuts, poultry, fish, pork, beef	Coenzyme of oxidative phosphorylation	Sensitivity to light; contributes to pellagra (damage to skin, gut, nervous system, etc.)
B_6 (pyridoxine)	Spinach, whole grains, tomatoes, potatoes, meats	Coenzyme in amino acid and fatty acid metabolism	Skin, muscle, and nerve damage
Pantothenic acid	In many foods (meats, yeast, egg yolk especially)	Coenzyme in carbohydrate and fat oxidation; fatty acid and steroid synthesis	Fatigue, tingling in hands, headaches, nausea
Folic acid	Dark green vegetables, whole grains, yeast, lean meats; intestinal bacteria produce some folate	Coenzyme in nucleic acid and amino acid metabolism; promotes red blood cell formation	Anemia; inflamed tongue; diarrhea; impaired growth; mental disorders; neural tube defects and low birth weight in newborns
B_{12} (cobalamin)	Poultry, fish, eggs, red meat, dairy foods (not butter)	Coenzyme in nucleic acid metabolism; necessary for red blood cell formation	Pernicious anemia; impaired nerve function
Biotin	Legumes, egg yolk; colon bacteria produce some	Coenzyme in fat and glycogen formation and amino acid metabolism	Scaly skin (dermatitis), sore tongue, brittle hair, depression, weakness
C (ascorbic acid)	Fruits and vegetables, especially citrus, berries, cantaloupe, cabbage, broccoli, green pepper	Vital for collagen synthesis; antioxidant	Scurvy, delayed wound healing, impaired immunity

amount of water-soluble vitamins above daily nutritional requirements is passed in urine. Thus, meeting the daily minimum requirements of water-soluble vitamins is critical. The body can tap its stores of fat-soluble vitamins to meet daily requirements; however, these stores can be quickly depleted, and prolonged deficiencies of the fat-soluble vitamins may also become critical to health.

Humans can synthesize vitamin D (calciferol) through the action of ultraviolet light on lipids in the skin. People who are not exposed to enough sunlight to make sufficient quantities of the vitamin must rely on dietary sources. Although we cannot make vitamin K, much of what we require is supplied through the metabolic activity of bacteria living in our large intestine. Vitamin K deficiency is rare in healthy people. Vitamin K plays a role in blood clotting, so individuals with vitamin K deficiency will bruise easily and show increased blood clotting times. Vitamin K deficiency can be caused in people on long-term antibiotic therapy because the antibiotics kill intestinal bacteria.

Other mammals have basically the same vitamin requirements as humans, with some differences. Most mammals can synthesize vitamin C, but not primates, guinea pigs, and fruit bats. To date, no mammals are known to synthesize B vitamins, but **ruminants** such as cattle and deer are supplied with these vitamins by microorganisms living in the digestive tract.

STUDY BREAK

1. What elements are essential to plants and to humans?
2. Why is iron essential for plants? For animals? What are the symptoms of iron deficiency in plants? In animals?
3. What do vitamins do for animals?

41.2 Soil

What is soil? Although we tend to think of soil as inert matter, just "dirt," it is really the living skin of the Earth: a complex mixture of mineral particles, organic matter, air, and water, combined with a great diversity of living organisms (e.g., Figure 19.3), many as yet unnamed. If you went outside and scooped up a handful of soil, you would be holding millions of prokaryotic and **unicellular organisms**, several kilometres of fungal hyphal filaments, and numerous invertebrates. If you took a larger volume of soil, you would also find a wide diversity of larger animals, some of which may never have been seen before. These organisms, along with plant roots, live in the soil and have a major influence on its composition and characteristics. Bacteria and fungi decompose organic matter; burrowing creatures such as earthworms aerate the soil; and when plant roots die, they contribute their organic matter to the soil.

Soil is also the source of water for most plants and of O_2 for cellular respiration in root cells. The physical texture of soil determines whether root systems have access to sufficient water and dissolved oxygen. The physical and chemical properties of soils in different habitats have a major impact on the ability of plant species to grow, survive, and reproduce there.

41.2a Soil Components and Particle Sizes Determine Properties of Soil

Most soils initially develop from the physical or chemical weathering of rock, which also liberates mineral ions. Different kinds of soil particles range in size from sand (2.0–0.02 mm in diameter) to silt (0.02–0.002 mm) and clay (<0.002 mm). The relative proportions of different sizes of mineral particles help determine the number and volume of pores—air spaces—soil contains. A soil that is sticky when wet with few air spaces is mostly clay, whereas one that dries quickly and may wash or blow away is mostly sand. The mineral particles alone do not make a soil; it also depends on the presence of living organisms and the accumulation of their products and decomposing bodies. In most soils, mineral particles are intermixed with various organic components, including **humus**, the decomposing parts of plants and animals, animal droppings, and other organic matter. Humus can absorb a great deal of water, contributing to the soil's capacity to hold water. Humus is a reservoir of nutrients vital to living plants, including N, P, and S. Organic material in humus feeds decomposers, whose metabolic activities, in turn, release minerals that plant roots can take up.

As soils develop naturally, they tend to take on a characteristic vertical profile, with a series of layers or **horizons (Figure 41.6)**. Each horizon has a distinct texture and composition that varies with soil type. The *A horizon*, or topsoil, is the region with the most biological activity and usually is the most fertile layer. Most plant roots are concentrated in this layer, although some roots can extend much deeper in the soil (see Figure 46.14). Below the topsoil is the *B horizon*, or **subsoil**, characterized by the accumulation of mineral ions, including those needed by plants, washed down from above. Mature tree roots generally extend into this layer. The *C horizon*, the lowest layer, is the parent material. Regions where the topsoil is deep and rich in humus are ideal for agriculture, such as the vast grasslands of the Canadian prairies and Ukraine. Without soil management and intensive irrigation, crops usually cannot be grown in deserts because of the lack of rainfall and low humus in the soil—nor can agriculture flourish for long on land cleared of a tropical rainforest because rainforest soil is surprisingly poor and has very little humus.

41.2b Soil Characteristics Affect Nutrient Uptake by Plants

Some nutrients that plants require are present as mineral ions dissolved in the soil water. Soil composition influences the ability of plant roots to obtain nutrients, as outlined below.

Water Availability. As water flows through soil, gravity pulls much of it down through the spaces between soil particles into deeper soil layers. This available water is part of the **soil solution (Figure 41.7),** a combination of water and dissolved substances that coats soil particles and partially fills pore spaces. The solution develops through ionic interactions between water molecules and soil particles. The surfaces of clay particles and humus particles are often negatively charged and so attract polar water molecules, which form hydrogen bonds with the soil particles.

The proportion of the water entering a soil that is actually available to plants depends on the soil's composition—the size of the air spaces that allow water to flow and the proportions of water-attracting particles of clay and organic matter. By volume, soil is about one-half solid particles and one-half air space.

The size of the particles in a soil has a major effect on how well plants will grow there. Sandy soil has relatively large air spaces, so water drains rapidly down from the top two soil horizons, where most plant roots are located. Soils rich in clay or humus are often high in water content, but in the case of clay, ample water is not necessarily an advantage for plants. Whereas a humus-rich soil contains lots of air spaces, the closely spaced particles in clay allow few air spaces, and existing spaces tend to hold on to the water that enters them. "Air" spaces in clay soils filled with water also severely limit the supplies of oxygen to roots for cellular respiration, and the plant's metabolic activity suffers. Thus, few plants can grow well in clay soils, even when water content is high. Overwatered houseplants die because their roots are similarly "smothered" by water.

Mineral Availability. Most mineral ions enter plant roots passively along with the water in which they are dissolved. Plant cell membranes have ion-specific transport proteins by which they selectively absorb ions from soil (see Chapter 29). Some mineral nutrients enter plant roots as cations and others as anions. Although both cations and anions may be present in soil solutions, they are not equally available to plants.

Cations such as Mg^{2+}, Ca^{2+}, and K^+ cannot easily enter roots because they are attracted by the net negative charges on the surfaces of soil particles. To varying degrees, they become reversibly bound to negative ions on the surfaces. Attraction in this form is called

adsorption. Cations are made available to plant roots through **cation exchange**, a mechanism in which one cation, usually H^+, replaces a soil cation **(Figure 41.8, p. 1012).** There are two main sources of H^+ ions. Respiring root cells release carbon dioxide, which dissolves in the soil solution, yielding carbonic acid (H_2CO_3). Subsequent reactions ionize H_2CO_3 to produce bicarbonate (HCO_3^-) and hydrogen ions (H^+). Reactions involving organic acids inside roots also

William Ferguson

Fallen leaves and other organic material littering the surface of mineral soil

A horizon
Topsoil, which contains some percentage of decomposed organic material and which is of variable depth; here it extends about 30 cm below the soil surface

B horizon
Subsoil; larger soil particles than the A horizon, not much organic material, but greater accumulation of minerals; here it extends about 60 cm below the A horizon

C horizon
No organic material, but partially weathered fragments and grains of rock from which soil forms; extends to underlying bedrock

Bedrock

Figure 41.6
A representative profile of soil horizons.

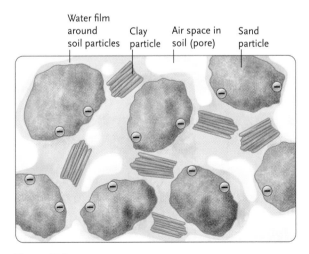

Water film around soil particles | Clay particle | Air space in soil (pore) | Sand particle

Figure 41.7
Location of the soil solution. Negatively charged ions on the surfaces of soil particles attract water molecules, which coat the particles and fill spaces between them (blue). Hydrogen bonds between water and soil components counteract the pull of gravity and help hold some water in soil spaces.

produce H$^+$, which is excreted. As H$^+$ enters the soil solution, it displaces adsorbed mineral cations attached to clay and humus, freeing them to move into roots. Other types of cations may also participate in this type of exchange (see Figure 41.8).

Meanwhile, anions in the soil solution, such as nitrate (NO$_3^-$), sulphate (SO$_4^{2-}$), and phosphate (PO$_4^-$), are only weakly bound to soil particles, so they generally move fairly freely into root hairs. However, because they are so weakly bound compared with cations, anions are more subject to loss from soil by leaching (draining from the soil, carried by excess water).

Soil pH also affects the availability of some mineral ions. Soil pH is a function of the balance between cation exchange and other processes that raise or lower the concentration of H$^+$ in soil. Areas receiving heavy rainfall tend to have acidic soils because moisture promotes the rapid decay of organic material in humus. As the material decomposes, the organic acids it contains are released. **Acid precipitation**, which results from the release of S and N oxides into the air, also contributes to soil acidification. Soils in arid regions are often alkaline.

Although most plants are not directly sensitive to soil pH, chemical reactions in very acid (pH <5.5) or very alkaline (pH >9.5) soils can have a major impact on whether plant roots can take up certain mineral cations. In the presence of OH$^-$ ions in alkaline soil, Ca^{2+} and P^{3+} ions react to form insoluble calcium phosphates. Phosphate captured in these compounds is as unavailable to roots as if it were completely absent from the soil.

For a soil to sustain plant life over long periods, mineral ions that the plants take up must be replenished naturally or artificially. In the long run, some mineral nutrients enter the soil from ongoing weathering of rocks and smaller bits of minerals. Over the shorter term, nutrients are returned to the soil by the decomposition of organisms and their parts or wastes. Other inputs occur when airborne compounds, such as S in volcanic and industrial emissions, become dissolved in rain and fall to Earth. Still others, including compounds of N and P, enter soil in fertilizers.

Although using commercial fertilizers maintains high crop yields (see Chapter 49), it does not add humus to the soil. Fertilizers can also cause serious problems, as when N-rich runoff from agricultural fields promotes the serious overgrowth of algae in lakes and ocean shorelines.

41.3 Obtaining and Absorbing Nutrients

Natural habitats show a wide range of soil conditions (minerals, humus, pH, biota, and other factors) that influence the growth and health of plants. Soil managed for agriculture can be plowed, precisely irrigated, and chemically adjusted to provide air, water, and nutrients in optimal quantities for a particular crop. Natural habitats, on the other hand, show wide variations in soil minerals, humus, pH, the presence of other organisms, and other factors that influence the availability of essential elements. Some nutrients, particularly N, P, and K, are often relatively scarce in soil. The evolutionary solutions to these challenges include an array of adaptations in the structure and functioning of plant roots.

41.3a Acquisition of Nutrients by Plants

Immobile organisms such as plants must locate nutrients in their immediate environment. For plants, root systems are the adaptive solution to this problem. Roots make up 20 to 50% of the dry weight of many plants and even more in species that grow where water or nutrients are especially scarce, such as the Arctic tundra. As long as a plant lives, its root system continues to grow, branching out through the surrounding soil. Roots do not necessarily grow *deeper* as a root system branches out, however. In arid regions, a shallow but broad root system may be better positioned than a deep one to take up water from occasional rains that may never penetrate below the first few centimetres of soil.

Figure 41.8

Cation exchange on the surface of a clay particle. When cations come into contact with the negatively charged surface of a clay particle, they become adsorbed. As one type of cation (e.g., H$^+$) is adsorbed, other ions are liberated and can be taken up by the roots of plants.

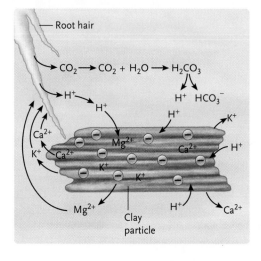

A root system grows most extensively in soil where water and mineral ions are abundant (see Chapter 29). Roots take up ions in regions just above the root tips. Over successive growing seasons, long-lived plants such as trees can develop millions, even billions, of root tips, each one a potential absorption site. Root hairs are diminutive absorptive structures (see Figure 28.20a) that are also associated with the uptake of mineral ions and water. In a plant such as a mature red oak (*Quercus rubrum*) with a vast root system, the total number of root hairs is astonishing. Even in young plants, root hairs greatly increase the root surface area available for absorbing water and ions.

Plant cell membranes (see Chapter 29) have ion-specific transport proteins by which they selectively absorb ions from soil. Studies of *Arabidopsis thaliana*, a plant that has become a model organism for plant research, showed that transport channels for K^+ are embedded in the cell membranes of root cortical cells. These ion transporters absorb more or less of a particular ion depending on chemical conditions in the surrounding soil.

Plants can only take up inorganic or mineral forms of nutrients. A plant growing in a soil rich in organic forms of N such as proteins cannot access the N until decomposers have converted the protein into mineral forms such as ammonium (NH_4^+) or nitrate (NO_3^-). Associations with fungi and bacteria allow many plants to shortcut the decomposition process and gain access to organic and other forms of nutrients. **Mycorrhizas**, symbiotic associations between a fungus and the roots of a plant (see Chapter 24), increase the uptake of nutrients, especially P and N, by most species of vascular plants. The fungal partner in the association grows inside the plant's roots and in the soil beyond the roots, forming an extensive network of hyphal filaments. This network provides a very large surface area for absorbing ions from a large volume of soil. Furthermore, many mycorrhizal fungi can access forms of nutrients not available to plants. For instance, many fungi obtain nutrients from organic forms of nutrients (e.g., proteins) and even obtain other nutrients directly from rocks!

41.3b Movements of Nutrients in Plants

Most mineral ions enter plant roots passively along with the water in which they are dissolved. Some mineral ions enter root cells immediately, whereas others travel in solution *between* cells until they meet the endodermis sheathing the root's **stele** (see Figure 29.6). Here the ions are actively transported into the endodermal cells and then into the xylem for transport throughout the plant.

Inside cells, most mineral ions enter vacuoles or cell cytoplasm, where they become available for metabolic reactions. Nutrients, such as N-containing ions,

move in phloem from site to site in the plant, as dictated by growth and seasonal needs. In plants that shed their leaves in autumn (= deciduous), significant amounts of N, P, K, and Mn move out of them and into twigs and branches before the leaves age and fall. This adaptation conserves the nutrients, which will be used in new growth the next season. Likewise, in late summer, mineral ions move to the roots and lower stem tissues of perennial range grasses that typically die back during the winter. These activities are regulated by hormonal signals (see Chapter 31).

41.3c Plants Can Be Limited by the Availability of Nitrogen

A lack of N is the single most common limit to plant growth. Although air contains plenty of gaseous N (~80% by volume), plants lack the enzyme necessary to break the three covalent bonds in each N_2 molecule ($N \equiv N$). Some N from the atmosphere reaches the soil in the form of nitrate, NO_3^-, and ammonium ions, NH_4^+. Plants can absorb both of these inorganic nitrogen compounds, but usually there is not nearly enough of them to meet plants' ongoing needs for N.

Organic N also enters the soil through decomposition of dead organisms and animal wastes. Dried blood is about 12% N by weight and chicken manure about 5%. But this N is bound up in complex organic molecules such as proteins that are not available to plants. The actions of bacteria are the main natural processes that replenish soil N and convert it to the absorbable form (see Figure 47.7). They are part of the *nitrogen cycle* (see Figure 47.15), the global movement of N in its various chemical forms from the environment to organisms and back to the environment.

N fixation is the incorporation of atmospheric N into compounds accessible to plants. Metabolic pathways of *nitrogen-fixing bacteria*, which live in the soil or in mutualistic association with plant roots, add H to atmospheric N_2, producing two molecules of NH_3 and one H_2 for each N_2 molecule **(Figure 41.9, p. 1014)**. The process requires a substantial input of ATP and is catalyzed by the enzyme nitrogenase. In a final step, H_2O and NH_3 react, forming NH_4^+ and OH^-. Ammonification is another bacterial process producing NH_4^+ when ammonifying soil bacteria break down decaying organic matter. This process recycles nitrogen previously incorporated in plants and other organisms. N fixation and ammonification form NH_4^+ that plant cells can use to synthesize organic compounds.

Most plants absorb N as nitrate, NO_3^-, which is produced in the soil by **nitrification** when NH_4^+ is oxidized to NO_3^-. Soils generally teem with *nitrifying bacteria* that carry out this process. Because of ongoing nitrification, nitrate is far more abundant than ammonium in most soils. Usually, plants take

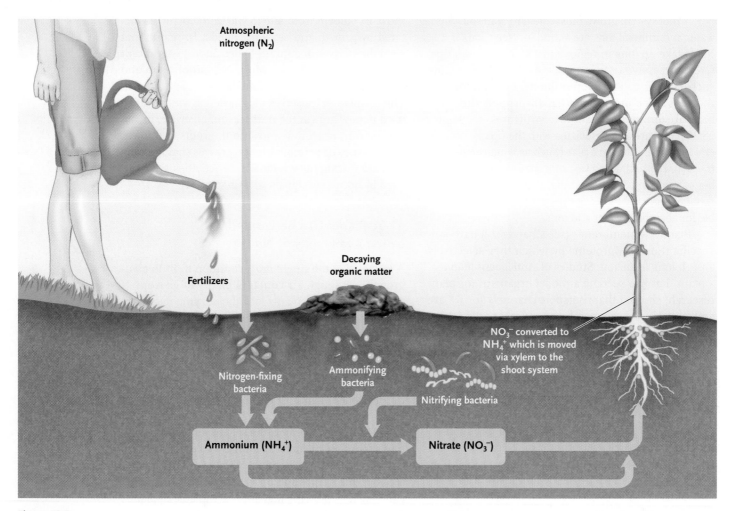

Figure 41.9

How plants obtain nitrogen from soil. Many commercial N fertilizers are nitrates that plant roots readily take up. Others are ammonium (NH$_4$), which nitrifying bacteria convert to nitrate.

Labels in figure:

Atmospheric nitrogen (N$_2$)

Fertilizers

Decaying organic matter

NO$_3^-$ converted to NH$_4^+$ which is moved via xylem to the shoot system

Nitrogen-fixing bacteria

Ammonifying bacteria

Nitrifying bacteria

Ammonium (NH$_4^+$)

Nitrate (NO$_3^-$)

up ammonium directly only in highly acidic soils, such as in bogs, where the low pH is toxic to nitrifying bacteria.

Within root cells, absorbed NO$_3^-$ is converted by a multistep process back to NH$_4^+$, which is rapidly used to synthesize organic molecules, mainly amino acids. Amino acids pass into the xylem, which transports them throughout the plant. In some plants, nitrogen-rich precursors travel in xylem to leaves, where different organic molecules are synthesized, and, in turn, travel to other plant cells in the phloem.

Although some N-fixing bacteria live free in the soil (see Figure 41.9), by far the largest percentage of nitrogen is fixed by species of *Rhizobium* and *Bradyrhizobium* that form mutualistic associations with the roots of plants in the legume family (e.g., peas, beans, clover, alfalfa). The host plant supplies organic molecules used by the bacteria for cellular respiration, and the bacteria supply NH$_4^+$, which the plant uses to produce proteins and other nitrogenous molecules. The N-fixing bacteria reside in **root nodules**, localized swellings on legume roots **(Figure 41.10)**. Rotating legumes with other crops (e.g., corn) allows farmers to increase soil N. When the legume crop is harvested, root nodules and other tissues remaining in the soil enrich its N content.

Decades of research have revealed the details of how the remarkable relationship between plant and bacteria unfolds. Usually, a single species of N-fixing bacteria colonizes a single legume species, drawn to the plant's roots by flavenoids, chemical attractants secreted by the roots. Through a sequence of exchanged molecular signals, bacteria penetrate a root hair and form a colony inside the root cortex **(Figure 41.11)**.

The soybean (*Glycine max*) and its bacterial partner (*Bradyrhizobium japonicum*) illustrate the process. In response to a specific flavenoid released by the soybean's roots, *nod* (for nodule) genes in the bacteria begin to be expressed (see Figure 41.11a). Products of the *nod* gene cause the tip of the root hair to curl toward the bacteria and trigger the bacteria to release enzymes that break down the cell wall of the root hair (see Figure 41.11b). As bacteria enter the cell and multiply, the plasma membrane forms an *infection thread*, a tube that extends into the root cortex, allowing the bacteria to invade cells of the cortex (see Figure 41.11c). The enclosed bacteria, now called **bacteroids**, enlarge and become immobile. Stimulated by still other *nod* gene products, cells of the root cortex begin to divide, and this region of proliferating cortex cells forms the root nodule (see Figure 41.11d). Typically, each cell in a root nodule contains several thousand bacteroids. The plant takes up some of

a. Root nodules

b. Field experiment with soybeans *(Glycine max)* and *Rhizobium*

c. Bacteroids

Root nodule

Figure 41.10

(a) Root nodules. **(b)** When soybeans *(Glycine max)* are grown in nitrogen-poor soil (left), they do not thrive compared with their growth in soils innoculated with *Rhizobium* cells (right). Also, the plants on the right developed root nodules. **(c)** A false colour transmission electron micrograph shows membrane-bound bacteroids as red in a root nodule cell. Membranes that enclose the bacteroids appear blue. The large yellow-green structure is the nucleus.

the N fixed by the bacteroids, whereas the bacteroids use some compounds produced by the plant.

Inside bacteroids, nitrogenase catalyzes the reduction of N_2 to NH_4^+ using ATP produced by cellular respiration. NH_4^+ can be highly toxic to cells if it accumulates, so it is immediately moved out of the bacteroids and into the surrounding nodule cells and converted, for example, to the amino acids glutamine and asparagine.

The protein *leghemoglobin* ("legume hemoglobin") also results from stimulation of plant nodule cells by a product of the bacterial *nod* genes. Like the hemoglobin of animal red blood cells, leghemoglobin contains a reddish, iron-containing heme group that binds oxygen. Its colour gives root nodules a pinkish cast (see Figure 41.11). Leghemoglobin picks up oxygen at the cell surface and shuttles it inward to bacteroids. This method of oxygen delivery is vital because nitrogenase,

the enzyme responsible for nitrogen fixation, is irreversibly inhibited by excess O_2. Leghemoglobin delivers just enough oxygen to maintain bacteroid respiration without shutting down the action of nitrogenase.

41.3d How Animals Eat to Acquire Nutrients

Animals are adapted to obtain the food they need (see Chapters 3 and 40). There are four basic groups of overall feeding methods and the physical states of the organic molecules they eat: fluid feeders, suspension feeders, deposit feeders, and bulk feeders **(Figure 41.12, p. 1016)**.

Fluid feeders ingest liquids containing organic molecules in solution. Among invertebrates, aphids, mosquitoes, leeches, butterflies, and spiders are fluid feeders. Among vertebrates, lamprey eels, hummingbirds, nectar-feeding bats, and vampire bats are examples of fluid feeders (see Figure 41.12a). Many fluid

Figure 41.11

Root nodule formation in legumes, which interact mutualistically with the nitrogen-fixing bacteria *Rhizobium* and *Bradyrhizobium*.

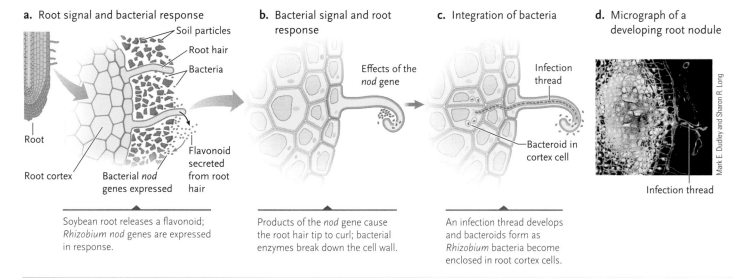

a. Root signal and bacterial response

Soil particles
Root hair
Bacteria
Root
Root cortex
Bacterial *nod* genes expressed
Flavonoid secreted from root hair

Soybean root releases a flavonoid; *Rhizobium nod* genes are expressed in response.

b. Bacterial signal and root response

Effects of the *nod* gene

Products of the *nod* gene cause the root hair tip to curl; bacterial enzymes break down the cell wall.

c. Integration of bacteria

Infection thread
Bacteroid in cortex cell

An infection thread develops and bacteroids form as *Rhizobium* bacteria become enclosed in root cortex cells.

d. Micrograph of a developing root nodule

Infection thread

a. Fluid feeder

Sanford/Angliolo/Corbis

b. Suspension feeder

Baleen

James Watt/Animals, Animals—Earth Scenes

c. Deposit feeder

Joe McDonald/Corbis

d. Bulk feeder

Gunter Ziesler/Bruce Coleman, Inc.

Figure 41.12

General feeding methods in animals. **(a)** Fluid feeders, for example, the hummingbird that eats nectar. **(b)** Suspension feeders, for example, the northern right whale (*Balaena glacialis*) that gulps tonnes of water and filters out plankton. **(c)** Deposit feeders, for example, the fiddler crab (*Uca* species) that sifts edible material from detritus. **(d)** Bulk feeders, for example, this python that can open its mouth very wide and take in large objects such as a gazelle.

feeders have mouthparts specialized for reaching the source of their nourishment. Mosquitoes, bedbugs, and aphids have needlelike mouthparts that pierce body surfaces. Nectar-feeding butterflies, birds, and bats have long tongues that can extend deep within flowers. Some fluid feeders use enzymes or other chemicals to liquefy their food or to keep it liquid during feeding. Spiders inject digestive enzymes that liquefy tissues inside their victim, providing a nutrient soup they can ingest. The saliva of mosquitoes, leeches, and vampire bats includes chemicals that keep blood from clotting.

Suspension feeders ingest small organisms suspended in water, such as bacteria, protozoa, algae, and small crustaceans, or fragments of these organisms. Suspension feeders include aquatic invertebrates, such as clams, mussels, and barnacles, and vertebrates, such as many species of fishes, as well as some birds, pterosaurs, and whales (see Figure 41.12b). Suspension feeders strain (filter) food particles suspended in water through a body structure covered with sticky mucus or through a filtering network of bristles, hairs, or other body parts. Trapped particles are funnelled into the animal's mouth, and the water is pushed out. Bits of organic matter are trapped by the gills of bivalves such as clams and oysters, and plankton is filtered from water by the sievelike fringes of horny fibre, called baleen, hanging in the mouths of baleen whales (see Figure 41.12b).

Deposit feeders pick up or scrape particles of organic matter from solid material they live in or on. Earthworms are deposit feeders that eat their way through soil, taking the soil into their mouth and digesting and absorbing any organic material it contains. Some burrowing molluscs and tube-dwelling polychaete worms use body appendages to gather organic deposits from the sand or mud around them. Mucus on the appendages traps the organic material, and cilia move it to the mouth. The fiddler crab (*Uca* species) is a deposit feeder (see Figure 41.12c) with front claws differing dramatically in size. The small claw picks up sediment and moves it to the mouth, where the contents are sifted (the large claw is used in signalling). The edible parts of the sediment are ingested, and the rest is put back on the sediment as a small ball. The feeding-related movement of the small claw over the larger claw looks as if the crab is playing the large claw like a fiddle, giving the crab its name.

Bulk feeders consume sizable food items whole or in large chunks. Most mammals eat this way, as do reptiles, most birds and fishes, and adult amphibians. Depending on the animal, adaptations for bulk feeding include teeth for tearing or chewing, as well as claws and beaks for holding large food items. Some bulk feeders have flexible jaws allowing them to ingest objects that are larger in diameter than their head (see Figure 41.12d).

Study Break

1. What role do roots and root hairs play in the lives of plants?
2. What do mycorrhizae do for plants?
3. Describe the four basic feeding types in animals.

41.4 Digestive Processes in Animals

Most invertebrates and all vertebrates have a tubelike digestive system with two openings, a mouth for ingesting food and an anus for eliminating unused material. In these animals, contents move in one direction along the tube. The lumen of a digestive tube (also known as a gut, alimentary canal, digestive tract, or gastrointestinal tract) is external to all body tissues.

Mechanical and chemical digestive processes break food into its component parts, eventually breaking molecules into molecular subunits that can be absorbed into body fluids and transported to and moved into cells. Mechanical breakdown often involves grinding, sometimes with teeth or sometimes in a muscular gizzard. Chemical breakdown occurs by **enzymatic hydrolysis**, in which chemical bonds are broken by the addition of H^+ and OH^-, the components of water (see Chapter 3). Specific enzymes speed these reactions: *amylases* catalyze the hydrolysis of starches, *lipases* break down fats and other lipids, *proteases* hydrolyze proteins, and *nucleases* digest nucleic acids. Enzymatic hydrolysis of food molecules may take place inside or outside the body cells, depending on the animal.

41.4a Intracellular Digestion

In intracellular digestion, primarily in sponges and some cnidarians, cells take in food particles by endocytosis. Inside the cell, endocytic vesicles containing food particles fuse with a lysosome, a vesicle containing hydrolytic enzymes. The molecular subunits produced by the hydrolysis pass from the vesicle to the cytosol. Any undigested material remaining in the vesicle is released to the outside of the cell by exocytosis.

In sponges, water-containing particles of organic matter and microorganisms enter the body through pores in the body wall (see Figure 26.8). In the body cavity, individual *choanocytes* (collar cells) lining the body wall trap food particles, take them in by endocytosis, and transport them to amoeboid cells, where intracellular digestion takes place.

41.4b Extracellular Digestion

Extracellular digestion takes place in a pouch or tube enclosed within the body but outside the body cells—the digestive tract. Epithelial cells lining the pouch or tube secrete enzymes that digest the food. Processing food in this specialized compartment prevents self-digestion of the body tissues of the animal itself. Extracellular digestion, which occurs in most invertebrates and all vertebrates, greatly expands the range of available food sources by allowing animals to deal with much larger food items than those that can be engulfed by single cells. Extracellular digestion also allows animals to eat large batches of food that can be stored and digested while the animal continues other activities.

Some animals, including flatworms and cnidarians such as hydras, corals, and sea anemones, have a saclike digestive system with a single opening that serves as both the entrance for food and the exit for undigested material. In some of these animals, such as the flatworm *Dugesia* **(Figure 41.13),** the digestive cavity is called a gastrovascular cavity because it circulates and digests food. Food is brought to the mouth by a protrusible pharynx (a throat that can be everted) and then enters the gastrovascular cavity (see Figure 41.13), where glands in the cavity wall secrete enzymes that begin the digestive process. Cells lining the cavity then take up the partially digested material by endocytosis and complete digestion intracellularly. Undigested matter is released to the outside through the pharynx and mouth.

41.4c Gastrointestinal Tracts

In most animals with a digestive tube, digestion occurs in five successive steps, each taking place in a specialized region of the tube. The tube is a biological disassembly line, with food entering at one end and leftovers leaving from the other. Five main processes occur from ingestion of food to expulsion of wastes:

1. *Mechanical processing*. Chewing, grinding, and tearing food chunks into smaller pieces, making them easier to move through the tract and increasing the surface area exposed to digestive enzymes.

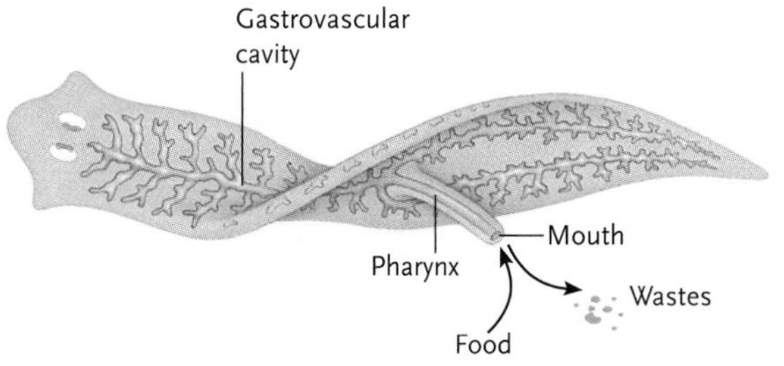

Figure 41.13
The digestive system of a flatworm (*Dugesia* species). The gastrovascular cavity (in blue) is a blind sac with one opening to the exterior through which food is ingested and wastes are expelled.

2. *Secretion of enzymes and other digestive aids.* Enzymes and other substances that aid the process of digestion, such as acids, emulsifiers, and lubricating mucus, are released into the tube.
3. *Enzymatic hydrolysis.* Food molecules are broken down through enzyme-catalyzed reactions into absorbable molecular subunits.
4. *Absorption.* The molecular subunits are absorbed from the digestive contents into body fluids and cells.
5. *Elimination.* Undigested materials are expelled through the anus.

Material being digested is pushed through the digestive tube by peristalsis, muscular contractions of its walls. During its progress through the tube, the digestive contents may be stored temporarily at one or more locations. Storage allows animals to take in larger quantities of food than they can process immediately, so feedings can be spaced in time rather than continuous.

Digestion in an Annelid. The earthworm (*Lumbricus* species, **Figure 41.14a**) is a deposit feeder that ingests a great deal of material, only some of which is edible. As it burrows, it pushes soil particles into its mouth. The particles pass from the mouth, through the esophagus, and into the crop (an enlargement of the digestive tube), where contents are stored and mixed with lubricating mucus. This mixture enters the muscular gizzard, where muscular contractions and

abrasion by sand grains grind the food mixture into fine particles. The pulverized mixture then enters a long intestine, where organic matter is hydrolyzed by enzymes secreted into the digestive tube. As muscular contractions of the intestinal wall move the mixture along, cells lining the intestine absorb the molecular subunits produced by digestion. The absorptive surface of the intestine is increased by folds of the wall called *typhlosoles*. At the end of the intestine, the undigested residue is expelled through the anus.

Digestion in an Insect. Herbivorous insects such as grasshoppers **(Figure 41.14b)** are more selective in what they ingest. When eating, grasshoppers tear leaves and other plant parts into small particles with their mandibles, the hard external mouthparts. From the mouth, food particles pass through the pharynx, where salivary secretions moisten the mixture before it enters the esophagus and passes into the crop and begins the process of chemical digestion. From the crop, the food mass enters the muscular gizzard, where it is ground into smaller pieces. Food particles then enter the stomach, where food is stored and digestion continues. In gastric cecae (saclike outgrowths of the stomach; *cecum* = blind), enzymes hydrolyze food, and the products of digestion are absorbed through the walls of the ceca. Undigested food moves into the intestine for further digestion and absorption. At the end of the intestine, water is absorbed from undigested matter and the remnants (frass) are expelled through the anus. The digestive systems of other arthropods are similar to the insect system.

Figure 41.14
The digestive systems of **(a)** an annelid (earthworm), **(b)** an insect (grasshopper), and **(c)** a bird (pigeon).

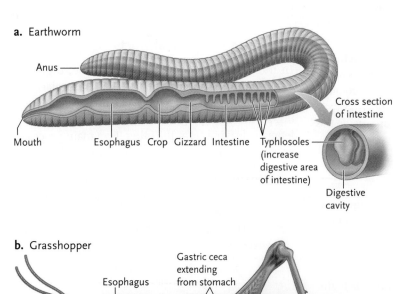

a. Earthworm

Anus

Mouth Esophagus Crop Gizzard Intestine Typhlosoles (increase digestive area of intestine)

Cross section of intestine

Digestive cavity

b. Grasshopper

Gastric ceca extending from stomach

Esophagus

Pharynx

Mouth

Crop Gizzard Intestine

Anus

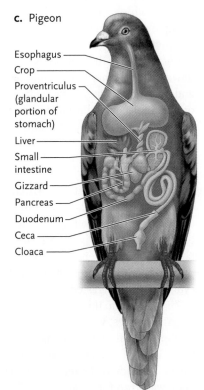

c. Pigeon

Esophagus
Crop
Proventriculus (glandular portion of stomach)
Liver
Small intestine
Gizzard
Pancreas
Duodenum
Ceca
Cloaca

Digestion in a Bird. A pigeon **(Figure 41.14c)** is also selective in what it ingests. A pigeon picks up seeds with its bill and uses its tongue to move them into its mouth, where they are moistened by mucus-filled saliva and swallowed whole. Seeds pass through the pharynx and into the esophagus. Birds such as parrots or cardinals use their bills to crack open seeds, but ingestion occurs as it does in pigeons. The anterior end of the esophagus is tubelike, but it opens into a crop, where food can be stored. The food moves to the proventriculus, the anterior glandular portion of the stomach that secretes digestive enzymes and acids. The food then passes to the gizzard, where muscular action grinds the seeds into fine particles, aided by ingested bits of sand and rock. Food particles then enter the intestine, where secretions from the liver **bile** and pancreas (digestive enzymes) are added. Molecular subunits produced by enzymatic digestion are absorbed as the mixture passes along the intestine, and the undigested residues are expelled through the anus, which opens to the cloaca. Structures such as the mouth, pharynx, esophagus, stomach, intestine, liver, and pancreas occur in almost all vertebrates.

STUDY BREAK

1. What is a gizzard? What does it do? Which animals have one?
2. What are choanocytes? What do they do?
3. Is the diet of *Dugesia* typical for flatworms? Why?

41.5 Digestion in Mammals

The mammalian digestive system is a series of specialized regions including the mouth, pharynx, esophagus, stomach, small and large intestines, rectum, and anus, that perform the five steps listed above **(Figure 41.15)**. These regions are under the control of the nervous and endocrine systems, allowing mammals to meet basic needs for fuel molecules and for a wide range of nutrients, including the molecular building blocks of carbohydrates, lipids, proteins, and nucleic acids. If the diet is adequate, the digestive system also absorbs the essential nutrients (the amino acids, fatty acids, vitamins, and minerals that cannot

Figure 41.15
The human digestive system.

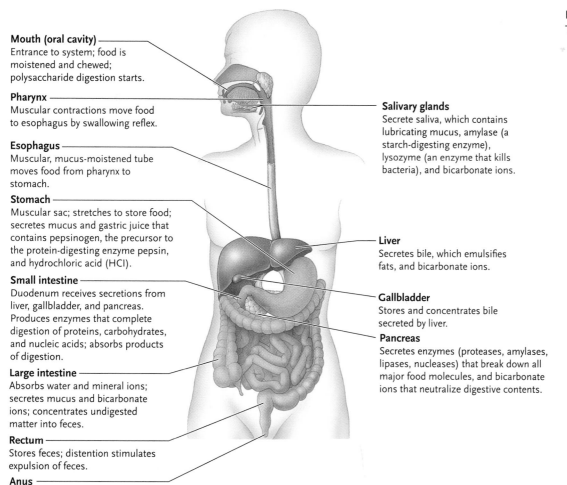

Mouth (oral cavity)
Entrance to system; food is moistened and chewed; polysaccharide digestion starts.

Pharynx
Muscular contractions move food to esophagus by swallowing reflex.

Esophagus
Muscular, mucus-moistened tube moves food from pharynx to stomach.

Stomach
Muscular sac; stretches to store food; secretes mucus and gastric juice that contains pepsinogen, the precursor to the protein-digesting enzyme pepsin, and hydrochloric acid (HCl).

Small intestine
Duodenum receives secretions from liver, gallbladder, and pancreas. Produces enzymes that complete digestion of proteins, carbohydrates, and nucleic acids; absorbs products of digestion.

Large intestine
Absorbs water and mineral ions; secretes mucus and bicarbonate ions; concentrates undigested matter into feces.

Rectum
Stores feces; distention stimulates expulsion of feces.

Anus
End of system; opening through which feces are expelled.

Salivary glands
Secrete saliva, which contains lubricating mucus, amylase (a starch-digesting enzyme), lysozyme (an enzyme that kills bacteria), and bicarbonate ions.

Liver
Secretes bile, which emulsifies fats, and bicarbonate ions.

Gallbladder
Stores and concentrates bile secreted by liver.

Pancreas
Secretes enzymes (proteases, amylases, lipases, nucleases) that break down all major food molecules, and bicarbonate ions that neutralize digestive contents.

Carnivore (a dog)

Fred Bruemmer

Herbivore (a rabbit)

Jane Burton/Bruce Coleman

especially the cell walls, is particularly difficult to digest—hence the longer intestine. The rabbit cecum houses symbiotic, plant-digesting microorganisms that help extract nutrients from plants (see *Digesting Cellulose: Fermentation*).

41.5a Gut Layers

The wall of the gut in mammals and other vertebrates contains four major layers, each with specialized functions **(Figure 41.17),** from the inner surface outward:

1. The **mucosa** contains epithelial and glandular cells and lines the inside of the gut. Epithelial cells absorb digested nutrients and seal off the digestive contents from body fluids. The glandular cells secrete enzymes, aids to digestion such as lubricating mucus, and substances that adjust the pH of the digestive contents.

2. The **submucosa** is a thick layer of elastic connective tissue containing neuron networks and blood and lymph vessels. Neuron networks provide local control of digestive activity and carry signals between the gut and the central nervous system. Lymph vessels carry absorbed lipids to other parts of the body.

3. In most regions of the gut, the **muscularis** is formed by two smooth muscle layers, a *circular layer* that constricts the diameter of the gut when it contracts and a *longitudinal layer* that shortens and widens the gut. The stomach also has an *oblique layer* running diagonally around its wall. Peristalsis occurs when circular and longitudinal muscle layers of the muscularis coordinate their activities to push the digestive contents through the gut **(Figure 41.18, p. 1022).** In **peristalsis**, the circular muscle layer contracts in a wave that passes along the gut, constricting the gut and pushing the digestive contents onward. Just in front of the advancing constriction, the longitudinal layer contracts, shortening and expanding the tube and making space for the contents to advance.

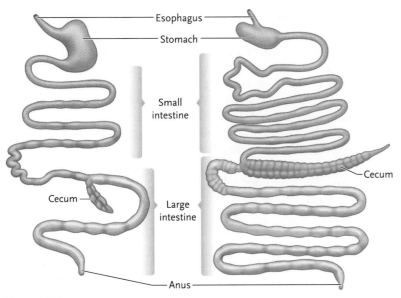

Figure 41.16

Comparison of the lengths of the digestive system of a carnivore (*Canis familiaris*) and a herbivore (*Oryctolagus cuniculus*). Note differences in length and the development of the cecae.

be synthesized within our bodies). In other mammals, differences in diet are reflected by the structure of the digestive tract **(Figure 41.16).** A carnivore's diet is relatively easy to digest; therefore, it does not require as long an intestine as a herbivore does. Plant matter,

Figure 41.17

Layers of the gut wall in vertebrates as seen in the stomach wall.

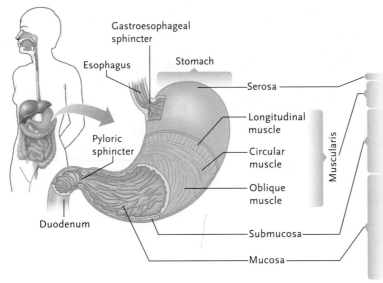

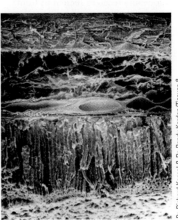

Dr. Richard Kessel & Dr. Randy Kardon/Tissues & Organs/Visuals Unlimited

Digesting Cellulose: Fermentation

The digestive tracts of mammals vary in length (see Figure 41.16). Most animals cannot digest cellulose because they lack cellulase, which hydrolyzes cellulose into glucose subunits. Many herbivorous animals (primary consumers) use the hydrolytic capabilities of microorganisms that do produce cellulase. In this way, bacteria, protists, and fungi aid other animals to digest plant material.

Herbivores using microorganisms house these symbionts in specialized structures along the alimentary canal. These structures occur in the esophagus, stomach, or cecae, depending on the species. Ruminant mammals (Bovidae, Cervidae) and termites are well-known examples of animals that use symbionts to digest cellulose.

Ruminants use their teeth to crop and chew plant material. They swallow the masticated material, moving it to a complex, four-chambered rumen **(Figure 1).** The first three chambers of the rumen are derived from the esophagus, whereas the fourth, the abomasum, is the stomach. Swallowed food material arrives in the reticulum and then moves to the rumen. Then ruminants "chew their cuds," regurgitating material from the reticulum and rumen, rechewing it, and macerating it into smaller fragments before swallowing it again. This exposes more surface area to microbial enzymes, giving them more time to act.

Fermentation by the microorganisms occurs in the reticulum and in the rumen. Oxygen levels in the chambers are too low to support mitochondrial reactions (see Chapter 7). Matter digested and liquefied by microorganisms moves to the *omasum*, where water is absorbed from the mass. In the *abomasum* (the ruminant's true stomach), acids and pepsin are added to the food mass, killing the microorganisms and starting the process of "typical" vertebrate digestion. As the food mass moves to the small intestine, dead microorganisms, themselves a rich source of proteins, vitamins, and other nutrients, are digested and absorbed along with other hydrolyzable molecules.

Fermentation generates products such as alcohols and amino acids that are used as nutrients. It also produces volatile fatty acids that move from the rumen to the blood and are used as sources of carbon and energy. Microorganisms use 40 to 60% of the food protein produced by fermentation, and, in turn, their bodies are protein for the host. Methane, another product, collects in the fermentation chambers, so ruminants belch the gas in huge quantities. One cow can release more than 400 L of methane per day. Cattle are estimated to contribute 20% of the methane polluting our atmosphere. A 500-kg cow with a 70-L rumen produces about 60 L of saliva a day and ingests 40 L of water. Fermentation takes time: the leaves eaten by a cow take about 55 hours to move through its digestive system.

Many other mammals have esophageal or gastric chambers that house plant-digesting symbiotic microorganisms. Biologists had long thought that the weight of the fermentation chamber made digestion by fermentation inaccessible to birds. But at least one species of bird, the South American hoatzin, uses fermentation **(Figure 2).** Freshly caught hoatzins smell like cattle dung, perhaps giving a clue to their use of fermentation. Hoatzins eat young leaves that are fermented in the enlarged forestomach. This fermentation centre takes up some space occupied by flight muscles in "normal" birds. Hoatzins are weak fliers, probably because of reduction in the mass of their flight muscles to accommodate the enlarged forestomach.

D. Robert Franz/Plant Earth Pictures

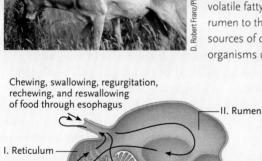

Chewing, swallowing, regurgitation, rechewing, and reswallowing of food through esophagus

II. Rumen

I. Reticulum

III. Omasum

IV. Abomasum (true stomach)

To small intestine

Figure 1
Ruminants, such as the pronghorn (*Antilocapra americanus*), have a four-chambered stomach system for digesting plant material (cellulose) by fermentation. Some other mammalian herbivores use the same approach, but others, such as the rabbit *O. cuniculus*, achieve fermentation in other parts of the digestive tract (see Figure 41.16).

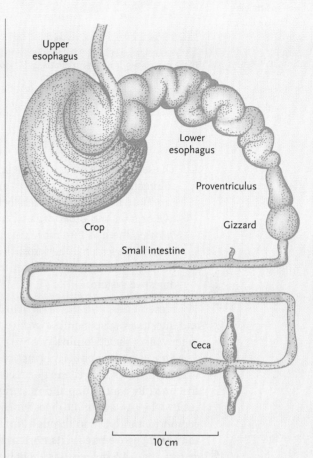

Upper esophagus

Lower esophagus

Proventriculus

Crop

Gizzard

Small intestine

Ceca

10 cm

Figure 2
The hoatzin (*Opisthocomus hoatzin*) uses foregut fermentation to digest cellulose to meet a high percentage of its energy requirements. The contents of the crop and lower esophagus account for over 15% of adult mass. Deep ridges in the lining of the crop increase its surface area.

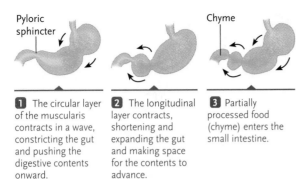

1 The circular layer of the muscularis contracts in a wave, constricting the gut and pushing the digestive contents onward.

2 The longitudinal layer contracts, shortening and expanding the gut and making space for the contents to advance.

3 Partially processed food (chyme) enters the small intestine.

4. The **serosa** is the outermost gut layer. It consists of connective tissue that secretes an aqueous, slippery fluid that lubricates areas between the digestive organs and other organs, reducing friction between them as they move together as a result of muscle movement. The serosa is continuous with the mesentery along much of the length of the digestive system.

Mesenteries, thin tissues attached to the stomach and intestines, suspend the digestive system from the inner wall of the abdominal cavity.

Sphincters are powerful rings of smooth muscle that form valves between major regions of the digestive tract. By contracting and relaxing, the sphincters control the passage of the digestive contents from one region to another and through the anus. The adult human digestive tract in its normal living contracted state is about 4.5 m long. It is about twice as long when fully extended, as in a cadaver (all muscles relaxed).

41.5b Down the Tube

Food begins its travel through the gastrointestinal tract in the mouth, where the teeth cut, tear, and crush food items. During chewing, three pairs of **salivary glands** secrete saliva through ducts that open on the inside of the cheeks and under the tongue. Saliva, which is more than 99% water, moistens the food and, as we have seen, begins digestion (salivary amylase; see *Why It Matters*). Saliva also contains mucus to lubricate the food mass and bicarbonate ions (HCO_3^-) to neutralize acids in the food and keep the pH of the mouth between 6.5 and 7.5, the optimal range for salivary amylase to function. Saliva also contains a *lysozyme*, an enzyme that kills bacteria by breaking open their cell walls.

After a suitable period of chewing, the food mass, called a **bolus**, is pushed by the tongue to the back of the mouth, where touch receptors detect the pressure and trigger the *swallowing reflex* (**Figure 41.19**). This reflex is an involuntary action produced by contractions of muscles in the walls of the pharynx that direct food into the esophagus. Peristaltic contractions of the esophagus, aided by mucus secreted by the esophagus, propel the bolus toward the stomach. The passage down the esophagus stimulates the gastroesophageal sphincter at the junction between the esophagus and the stomach to open and admit the bolus to the stomach. After the bolus enters the stomach, the sphincter closes tightly. If the closure is imperfect, the acidic stomach contents can enter the esophagus, in humans causing *acid reflux* or heartburn.

Mammals consciously initiate the swallowing reflex, but once it has begun, they cannot stop it because whereas the muscles of the pharynx and upper esophagus are skeletal muscles under voluntary control, the muscles below are smooth muscles under involuntary control.

Involuntary movements of the tongue and soft palate at the back of the mouth prevent food from backing into the mouth or nasal cavities. The glottis (space between vocal cords) and the epiglottis, a flap-like valve, prevent entry of food into the tracheae.

41.5c Stomach

The stomach is a muscular, elastic sac that stores food and adds secretions, furthering digestion. The stomach lining, the mucosa, is covered with tiny *gastric pits*, entrances to millions of *gastric glands*. These glands extend deep into the stomach wall and contain cells that secrete some of the products needed to digest food. Entry of food into the stomach activates stretch receptors in its wall. Signals from stretch receptors stimulate the secretion of gastric juice (**Figure 41.20**), which contains the digestive enzyme pepsin, hydrochloric acid (HCl), and lubricating mucus. The stomach secretes about 2 L of gastric juice each day.

Pepsin begins the digestion of proteins by

Figure 41.19
The swallowing reflex.

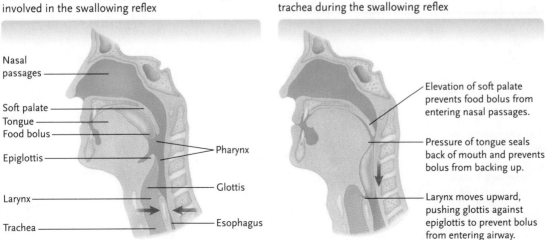

Structures of the mouth, pharynx, and esophagus involved in the swallowing reflex

Nasal passages
Soft palate
Tongue
Food bolus
Epiglottis
Pharynx
Glottis
Larynx
Trachea
Esophagus

Motions that seal the nasal passages, mouth, and trachea during the swallowing reflex

Elevation of soft palate prevents food bolus from entering nasal passages.

Pressure of tongue seals back of mouth and prevents bolus from backing up.

Larynx moves upward, pushing glottis against epiglottis to prevent bolus from entering airway.

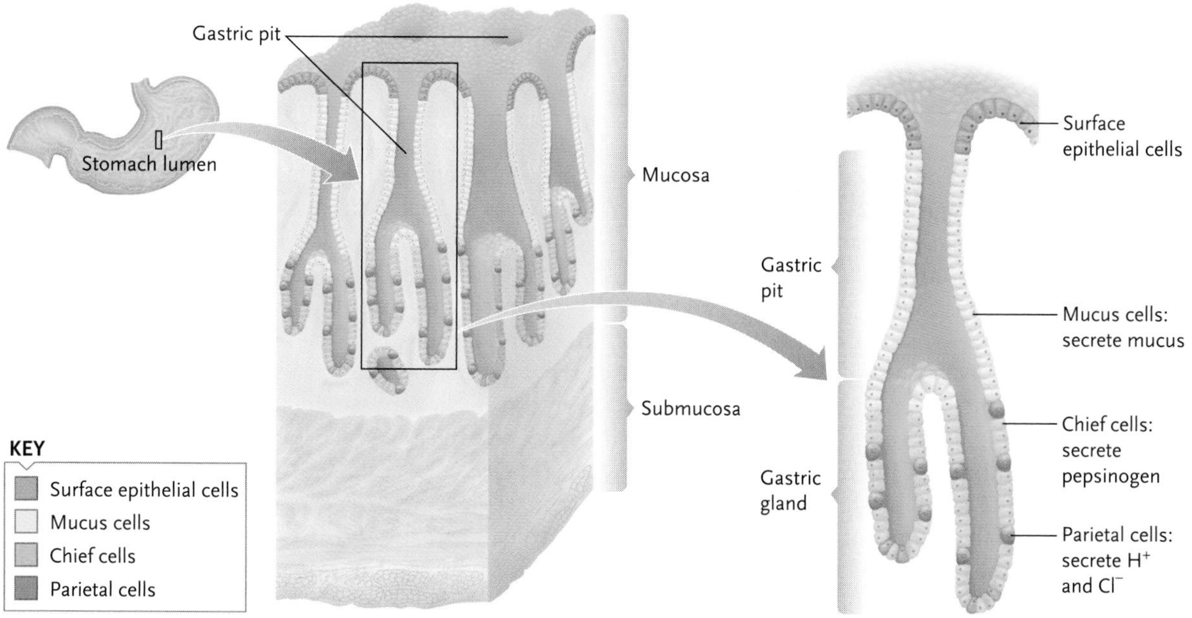

Figure 41.20
Cells that secrete mucus, pepsin, and HCl in the stomach lining.

introducing breaks in polypeptide chains. Pepsin is secreted in the form of an inactive precursor molecule, pepsinogen, by cells called *chief cells*. Pepsinogen is converted to pepsin by the highly acid conditions of the stomach. Once produced, pepsin itself can catalyze the reaction, converting more pepsinogen to pepsin. The activation of pepsin illustrates a common theme in digestion. Powerful hydrolytic enzymes such as pepsin would be dangerous to the cells that secrete them. However, enzymes are synthesized as inactive precursors and not converted into active form until exposed to digestive contents.

Parietal cells secrete H^+ and Cl^- that combine to form HCl in the lumen of the stomach. The HCl lowers the pH of the digestive contents to pH $\leq$ 2, the level at which pepsin reaches optimal activity. To put this pH in perspective, lemon juice is pH 2.4, and sulphuric acid or battery acid is about pH 1. The acidity of the stomach helps break up food particles and causes proteins in the stomach contents to unfold, exposing their peptide linkages to hydrolysis by pepsin. The acid also kills most bacteria that reach the stomach and stops the action of salivary amylase. Some nectar-feeding bats that digest pollen drink their own urine to make their stomach more acid.

A thick coating of alkaline mucus is secreted by *mucus cells* and protects the stomach lining from attack by pepsin and HCl. Behind the mucus barrier, tight junctions between cells prevent gastric juice from seeping into the stomach wall. Even so, there is some breakdown of the stomach lining. The damage is normally repaired by rapid division of mucosal cells, which replaces the entire stomach lining about every three days. Most bacteria cannot survive the highly acid environment of the stomach, but one, *Helicobacter pylori*, thrives there. Ulcers result when *H. pylori* breaks down the mucus barrier and exposes the stomach wall to

attack by HCl and pepsin (see Chapter 21, *People Behind Biology*).

Contractions of the stomach walls continually mix and churn the contents. Peristaltic contractions move the digestive contents toward the *pyloric sphincter* (*pylorus* = gatekeeper) at the junction between the stomach and the small intestine. The arrival of a strong stomach contraction relaxes and opens the valve briefly, releasing a pulse of the stomach contents, **chyme**, into the small intestine.

Feedback controls regulate the rate of gastric emptying, matching it to the rate of digestion, so that food is not moved along more quickly than it can be chemically processed. In particular, chyme with high fat content and high acidity stimulates the secretion of hormones by cells in the mucosal layer of the **duodenum**. These hormones slow the process of stomach emptying. Fat is digested in the lumen of the small intestine and more slowly than other nutrients, so further emptying of the stomach is prevented until fat processing has been completed in the small intestine. This is why a fatty meal, such as a greasy pizza, feels so heavy in the stomach. Highly acidic chyme must be neutralized by bicarbonate in the small intestine. Unneutralized stomach acid inactivates digestive enzymes secreted in the small intestine and inhibits further emptying of the stomach until it is neutralized.

41.5d Small Intestine

The small intestine completes digestion and begins the absorption of nutrients. Nutrients are not absorbed in the mouth, pharynx, or esophagus. Substances such as alcohol, aspirin, caffeine, and water are absorbed in the stomach, but most absorption occurs in the small intestine, where digestion is completed. The small intestine is smaller in diameter than the large intestine. The lining of the small intestine is folded into ridges densely covered by microscopic, fingerlike extensions, the intestinal villi (singular, *villus*). In addition, the epithelial cells covering the villi have microvilli, a *brush border* consisting of fingerlike projections of the plasma membrane **(Figure 41.21)**. The intestinal villi and microvilli in humans increase the absorptive surface area of the small intestine to 300 m², about the size of a doubles tennis court.

Digestion in the small intestine depends on enzymes and other substances secreted by the intestine itself and by the pancreas and liver. Secretions from the pancreas and liver enter a common duct that empties into the lumen of the dudodenum, a 20-cm long segment of the small intestine **(Figure 41.22)**.

About 95% of the volume of material leaving the stomach is absorbed as water and nutrients as digestive contents travel along the small intestine. Movement of the contents from the duodenum to the end of the

small intestine takes three to five hours. By the time the digestive contents reach the large intestine, almost all nutrients have been hydrolyzed and absorbed.

In humans, the pancreas is an elongated, flattened gland located between the stomach and duodenum (Figure 41.22; see also Figure 41.15). Exocrine cells in the pancreas secrete bicarbonate ions ($H_2CO_3^-$) and pancreatic enzymes into ducts that empty into the lumen of the duodenum. The bicarbonate ions neutralize the acid in chyme, bringing the digestive contents to a slightly alkaline pH. Alkaline pH allows optimal activity of the enzymes secreted by the pancreas, including proteases, an amylase, nucleases, and lipases. All of these enzymes act in the lumen of the small intestine. Like pepsin, the proteases released by the pancreas are secreted in an inactive precursor form and are activated by contact with the digestive contents. The enzyme mixture includes trypsin, which hydrolyzes bonds within polypeptide chains, and carboxypeptidase, which cuts amino acids from polypeptide chains one at a time.

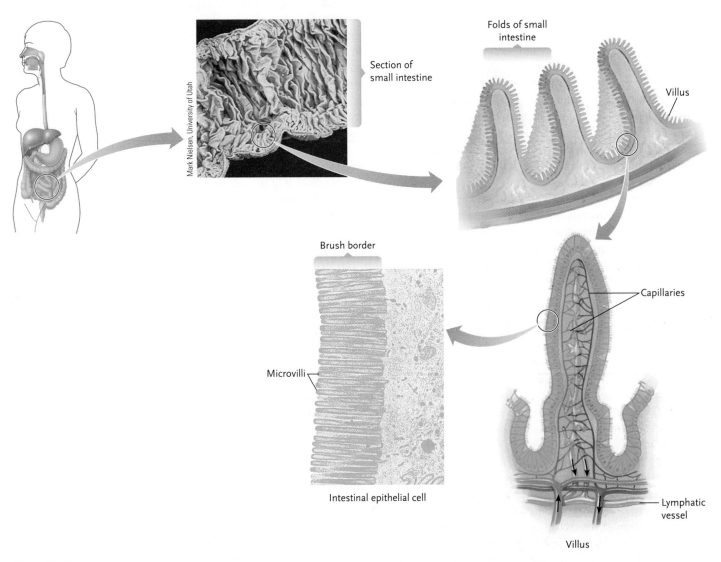

Figure 41.21

The structure of villi in the small intestine. The plasma membrane of individual epithelial cells of the villi extends into fingerlike projections, the microvilli, which greatly expand the absorptive surface of the small intestine. Collectively, the microvilli form the brush border of an epithelial cell of the intestinal mucosa.

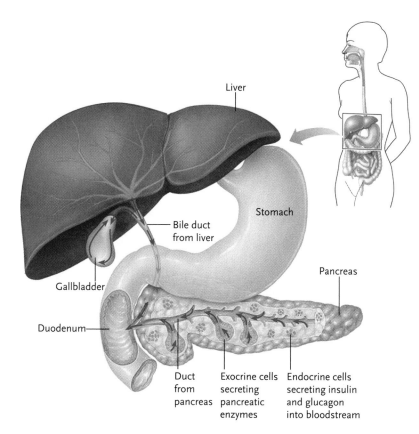

Liver

Bile duct
from liver

Stomach

Gallbladder

Pancreas

Duodenum

Duct
from
pancreas

Exocrine cells
secreting
pancreatic
enzymes

Endocrine cells
secreting insulin
and glucagon
into bloodstream

dase cuts amino acids from the end of a polypeptide, and a dipeptidase splits dipeptides into individual amino acids. Nucleases and other enzymes complete digestion of nucleic acids into five-carbon sugars and nitrogenous bases (**Figure 41.23, p. 1026**).

Water-soluble products of digestion enter the intestinal mucosa cells by active transport or facilitated diffusion (**Figure 41.24a, p. 1027**), and water follows by osmosis. The nutrients are then transported from the mucosal cells into the extracellular fluids, from where they enter the bloodstream in the capillary networks of the submucosa. The absorption of fatty acids, monoglycerides, fat-soluble vitamins, and cholesterol and other products of lipid breakdown by lipase occurs with the assistance of the micelles formed by bile salts (Figure 41.24b). When a micelle contacts the plasma membrane of a mucosal cell, the hydrophobic molecules within the droplet penetrate through the membrane and enter the cytoplasm.

In mucosal cells, fatty acids and monoglycerides are combined into fats (**triglycerides**) and packaged into **chylomicrons**, small droplets covered by a protein coat. Cholesterol absorbed in the small intestine is also packed into the chylomicrons. The protein coat of the chylomicrons provides a hydrophilic surface that keeps the droplets suspended in the cytosol. After travelling across the mucosal cells, the chylomicrons are secreted into the interstitial fluid of the submucosa, where they are taken up by lymph vessels. Eventually, they are transferred by lymph into the blood circulation.

Many nutrients absorbed by the small intestine are processed by the liver. Capillaries absorbing nutrient molecules in the small intestine collect into veins that join to form the **hepatic portal vein**, a larger blood vessel that leads to capillary networks in the liver. In the liver, some nutrients leave the bloodstream and enter liver cells for chemical processing. Among the reactions taking place in the liver is the combination of excess glucose units into glycogen that is stored in liver cells. This reaction reduces the glucose concentration in the blood exiting the liver to about 0.1%. If the glucose concentration in the blood entering the liver falls below 0.1% between meals, the reaction reverses. The reversal adds glucose to return the blood concentration to the 0.1% level before it exits the liver.

The liver secretes bicarbonate ions and bile, a mixture of substances including bile salts, cholesterol, and bilirubin. Bile salts are derivatives of cholesterol and amino acids that aid fat digestion through their detergent action. They form a hydrophilic coating around fats and other lipids, allowing the churning motions of the small intestine to emulsify fats. During emulsification, fats are broken into tiny droplets called micelles, much the same effect as mixing oil and vinegar in a salad dressing. Lipase, a pancreatic enzyme, can then hydrolyze fats in the micelles to produce monoglycerides and free fatty acids. Bilirubin, a waste product derived from worn-out red blood cells, is yellow and gives the bile its colour. Bacterial enzymes in the intestines modify the pigment, resulting in the characteristic brown colour of feces.

The liver secretes bile continuously. Between meals, when no digestion is occurring, bile is stored in the gallbladder, where it is concentrated by the removal of water. After a meal, entry of chyme into the small intestine stimulates the gallbladder to release the stored bile into the small intestine.

Microvilli on the villi of the small intestine secrete water and mucus into the intestinal contents. They also carry out intracellular digestion by transporting products of earlier digestion, including disaccharides, peptides, and nucleotides, across their plasma membranes and producing enzymes to complete hydrolysis of these nutrients. Different disaccharidases break maltose, lactose, and sucrose into individual monosaccharides. Two proteases complete protein digestion: an aminopepti-

Figure 41.23
Enzymatic digestion of carbohydrates, proteins, fats, and nucleic acids in the human digestive system.

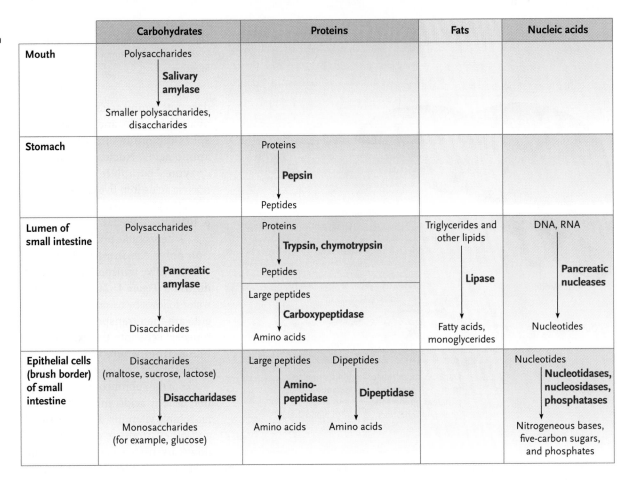

	Carbohydrates	Proteins	Fats	Nucleic acids
Mouth	Polysaccharides ↓ **Salivary amylase** Smaller polysaccharides, disaccharides			
Stomach		Proteins ↓ **Pepsin** Peptides		
Lumen of small intestine	Polysaccharides ↓ **Pancreatic amylase** Disaccharides	Proteins ↓ **Trypsin, chymotrypsin** Peptides Large peptides ↓ **Carboxypeptidase** Amino acids	Triglycerides and other lipids ↓ **Lipase** Fatty acids, monoglycerides	DNA, RNA ↓ **Pancreatic nucleases** Nucleotides
Epithelial cells (brush border) of small intestine	Disaccharides (maltose, sucrose, lactose) ↓ **Disaccharidases** Monosaccharides (for example, glucose)	Large peptides Dipeptides ↓ **Amino-peptidase** ↓ **Dipeptidase** Amino acids Amino acids		Nucleotides ↓ **Nucleotidases, nucleosidases, phosphatases** Nitrogeneous bases, five-carbon sugars, and phosphates

The liver also synthesizes lipoproteins that transport cholesterol and fats in the bloodstream, detoxifies ethyl alcohol and other toxic molecules, and inactivates steroid hormones and many types of drugs. As a result of the liver's activities, the blood leaving it has a markedly different concentration of nutrients than the blood carried into the liver by the hepatic portal vein. From the liver, blood goes to the heart and is then pumped to deliver nutrients to all parts of the body.

41.5e Large Intestine

From the small intestine, the contents move on to the large intestine, or colon. A sphincter at the junction between the small and large intestines controls the passage of material and prevents backward movement of contents. The inner surface of the large intestine is relatively smooth and contains no villi.

The large intestine has several distinct regions. At the junction with the small intestine, a part of the large intestine forms the **cecum**, a blind pouch. A fingerlike sac, the **appendix**, extends from the cecum. The cecum merges with the colon, which forms an inverted U, finally connecting with the **rectum**, the terminal part of the large intestine.

The large intestine secretes mucus and bicarbonate ions and absorbs water and other ions, primarily Na^+ and Cl^-. The absorption of water condenses and compacts the digestive contents into solid masses, the feces. Normally, fecal matter reaching the rectum contains less than 200 mL of the fluid that entered the digestive tract each day. Animals suffering from diarrhea produce liquid fecal matter. Diarrhea is a higher-than-normal rate of movement of materials through the small intestine, which does not leave adequate time for absorption of water. Diarrhea can be caused by irritation of the small intestine wall, caused by infection, emotional stress, or irritation of the small intestine wall.

As many as 500 species of bacteria comprise 30 to 50% of the dry matter of feces in humans and other vertebrates. Most of these bacteria live as essentially permanent residents in the large intestine. *Escherichia coli* is the most common in humans and other mammals. Intestinal bacteria metabolize sugars and other nutrients remaining in the digestive residue. They produce useful fatty acids and vitamins (such as vitamin K, the B vitamins, folic acid, and biotin), some of which are absorbed in the large intestine. Bacterial activity in the large intestines produces large quantities of gas (*flatus*), primarily CO_2, methane, and hydrogen sulphide. Most of the gas is absorbed through the intestinal mucosa, and the rest is expelled through the anus in the process of *flatulence*. The amount and composition of flatus depend on the type of food ingested and the particular population of bacteria present in the large intestine.

a. Absorption of water-soluble products of digestion by intestinal mucosa cells

b. Absorption of fat-soluble products of digestion by intestinal mucosa cells

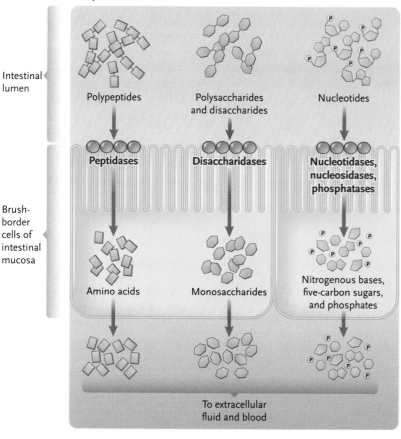

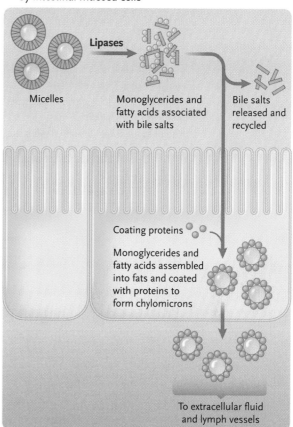

Water-soluble molecules are broken into absorbable subunits at brush borders of mucosal cells and transported inside; the subunits are transported on the other side to extracellular fluid and blood.

Micelles (fats coated with bile salts) are digested to monoglycerides and fatty acids, which penetrate into cells and are assembled into fats. The fats are coated with proteins to form chylomicrons, which are released by exocytosis to extracellular fluids, where they are picked up by lymph vessels.

Figure 41.24
Absorption of digestive products by the epithelial cells of the intestinal mucosa.

Foods such as beans contain carbohydrates that humans cannot digest. These carbohydrates can be metabolized by gas-producing intestinal bacteria, explaining the folkloric connection between beans and flatulence.

Feces entering the rectum stretch its walls, at some point triggering a *defecation reflex* that opens the *anal sphincter* and expels the feces through the anus. Because the anal sphincter contains rings of voluntary skeletal muscle as well as involuntary smooth muscle, animals can resist the defecation reflex by voluntarily tightening the striated muscle ring—to a point.

STUDY BREAK

1. Where and why is methane produced along the gastrointestinal tract?
2. What is peristalsis? What does it do? Is it unique to mammals?
3. How many species of bacteria can occur in a mammalian gastrointestinal tract?

41.6 Regulation of the Digestive Processes

Most of the digestive process is regulated and coordinated by largely automated controls, most of which originate in the neuron networks of the submucosa. Other controls, particularly those regulating appetite and oxidative metabolism, originate in the brain, in control centres that form part of the hypothalamus (see *People Behind Biology*).

Movement of food through the digestive system is controlled by receptors in and hormones secreted by various parts of the system **(Figure 41.25, p. 1029)**. Control starts with the mouth, where the presence of food activates receptors that increase the rate of salivary secretion by as much as 10-fold over the resting state.

Swallowed food expands the stomach and sets off signals from stretch receptors in the stomach walls. Chemoreceptors in the stomach respond to the presence of food molecules, particularly proteins. Signals from

PEOPLE BEHIND BIOLOGY

Richard E. Peter, University of Alberta

Our knowledge about the neuroendocrine regulation of reproduction and growth in fish was strongly influenced by the work of Dick Peter (1943–2007). Dr. Peter studied the physiological role of nuclei in the hypothalamus. He and his colleagues were among the first to measure the secretion of fish pituitary hormones, contributing to both our knowledge of fish and the broader field of comparative endocrinology.

In vertebrates, a complex set of interactions governs regulation of appetite and body mass. By combining techniques from animal behaviour, molecular biology, and immunohistochemistry, Dr. Peter explored neuroendocrine regulation of feeding behaviour and food intake in fish. Key factors from the hypothalamus stimulate food intake (orexigenic factors), whereas other factors inhibit it (anorexigenic factors). Pathways of interactions of various factors are shown in **Figure 1.**

Dr. Peter also contributed to the advancement of comparative endocrinology by initiating regional, national, and international meetings and organizations. At these meetings, colleagues and their students could interact and share ideas and data. His leadership extended into other areas, serving as chair of the Department of Zoology at the University of Alberta and later as the dean of science there.

Dr. Peter is particularly remembered for the excellence of his mentorship. During his career, he supervised the work of 17 Ph.D. students.

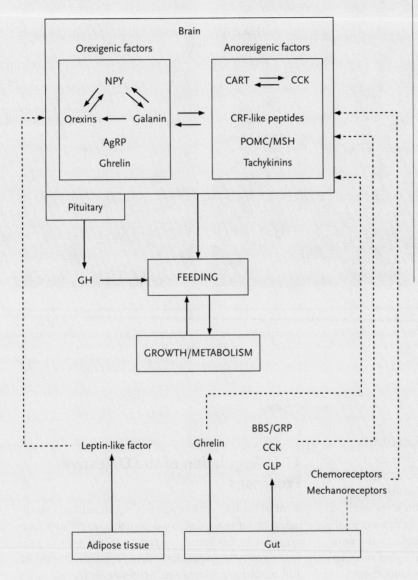

Figure 1
General schematic diagram showing how central and peripheral peptides regulate food intake in fish. AgRP – agouti-related protein; BBS/GRP – bombesin/gastrin releasing protein; CART – cocaine and amphetamine regulated transcript; CCK – cholecystokinin; CRF – corticotropin-releasing factor; GH – growth hormone; GLP – glucagonlike peptide: MSH – melanocyte-stimulating hormone; NPY – neuropeptide Y; POMCD – proopiomelanocortin.

these receptors are integrated in neuron networks in the stomach and autonomic nervous system to produce several reflex responses. One response is an increase in the rate and strength of stomach contractions. Another is secretion of a hormone, *gastrin*, into the blood leaving the stomach. After travelling through the circulatory system, gastrin returns to the stomach, where it stimulates the secretion of HCl and pepsinogen. These molecules are used in the digestion of the protein in the food that was responsible for their secretion. Gastrin also stimulates stomach and intestinal contractions, activities serving to keep the digestive contents moving through the digestive system when a new meal arrives.

Three hormones secreted into the lumen of the duodenum participate in regulating digestive processes. When chyme is emptied into the duodenum, its acidic nature stimulates the release of the hormone *secretin*. Secretin inhibits further gastric emptying to prevent more acid from entering the duodenum until the newly arrived chyme is neutralized. Secretin also

LIFE ON THE EDGE

Fuelling Hovering Flight

Hovering flight is extremely expensive in terms of fuel consumption whether the hoverer is a Harrier Jump Jet, a helicopter, or a hummingbird. In a hovering hummingbird, >90% of the animal's metabolic rate (overall energy consumption) is accounted for by the flight muscles. Hovering allows hummingbirds and some nectar-feeding bats to feed at flowers and tank up with nectar, which is essentially sugar-water. How do these animals pay the costs of hovering?

Some humans eat a lot of sugar before exercising, but only 25 to 30% of the energy they burn while running comes from the sugar ingested just before or during exercise. It is a mistake to think that eating a bar of chocolate as you run will immediately increase the energy available to you.

In contrast, hummingbirds fuel ~95% of the cost of hovering from the sugar they are ingesting as they hover.

High levels of sucrase activity in the birds' intestines translate into rapid hydrolysis of sucrose, explaining the rapid mobilization of this fuel. Nectar produced by the flowers visited by hummingbirds is high in sucrose.

What happens with flower-visiting bats? The Pallas' long-tongued bat (*Glossophaga soricina*, family Phyllostomidae) appears to be a nocturnal version of a hummingbird. These bats hover in front of flowers and use long, extensible tongues to extract nectar. Kenneth Welch and two colleagues used measures of oxygen consumption to indicate metabolism and the ratios of $^{13}\Delta/^{12}\Delta$ carbon to identify sources of energy for hovering Pallas' long-tongued bats. Their measurements indicated that these bats mobilized ~78% of the energy needed for hovering from sucrose ingested during hovering. Sucrase levels in the

intestines of the bats are about half of those recorded for hummingbirds, probably accounting for the difference in immediate access to fuel. These results suggest convergent evolution between flower-visiting bats and birds. It remains to be determined if the flower-visiting bats of the Old World tropics (family Pteropodidae) have evolved the same adaptations.

The nectar content of flowers pollinated by hummingbirds is about 60% sucrose, whereas that of flowers pollinated by bats is 20% sucrose. These differences may make the bats less efficient at immediately covering the costs of hovering than the hummingbirds. Hovering to extract high-energy food (nectar) amounts to living on the edge, and animals that do so are highly adapted to pay the costs of this specialized flight behaviour (see *The Puzzling Biology of Weight Control*).

---

inhibits gastric secretion to reduce acid production in the stomach and stimulates HCO_3^- secretion into the lumen of the duodenum to neutralize the acid. If the acid is not neutralized, the duodenal wall can be damaged.

Fat and, to a lesser extent, protein in the chyme entering the duodenum stimulate the release of the hormone *cholecystokinin* (*CCK*). CCK inhibits gastric activity, allowing time for nutrients in the duodenum to be digested and absorbed. CCK also stimulates the secretion of pancreatic enzymes to digest macromolecules in chyme.

The hormone glucose-dependent insulinotrophic peptide (*GIP*) acts primarily to stimulate insulin release into the blood by the pancreas. Insulin changes the metabolic state of the body after a meal is ingested so that new nutrients, particularly glucose, are used and stored. Glucose in the duodenum increases GIP secretion, triggering the release of insulin.

Two interneuron centres in the hypothalamus work in opposition to control appetite and oxidative metabolism. One centre stimulates appetite and reduces oxidative metabolism; the other stimulates

Hormone controls

Acidic chyme stimulates release of the hormone secretin in the small intestine. Secretin inhibits gastric emptying and gastric secretion and stimulates HCO_3^- secretion into the duodenum.

Fat (mostly) in chyme stimulates release of the hormone cholecystokinin (CCK). CCK inhibits gastric activity and stimulates secretion of pancreatic enzymes.

A meal entering the small intestine stimulates GIP secretion, which triggers insulin release. Insulin stimulates the uptake and storage of glucose from the digested food.

Receptor controls

Receptors in the mouth respond to food by increasing salivary secretion.

Stretch receptors in the stomach respond to food, signalling neuron networks to increase stomach contractions.

Chemoreceptors in the stomach respond to food, signalling neuron networks to stimulate the stomach to secrete the hormone gastrin, which, in turn, stimulates the stomach to secrete HCl and pepsinogen.

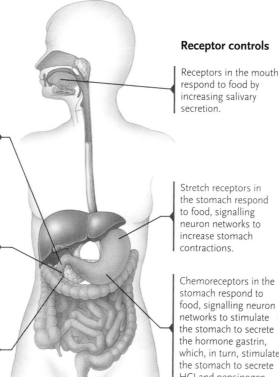

Figure 41.25

Control of digestion by receptors and hormones in the digestive system.

the release of α-melanocyte-stimulating hormone (α-MSH), a peptide hormone inihibiting appetite. Leptin (*leptos* = thin), a peptide hormone, is a major link between these two pathways. Leptin was discovered in mice by Jeffrey Friedman and his coworkers. Fat-storing cells secrete leptin when deposition of fat increases in the body. Leptin travels in the bloodstream and binds to receptors in both centres in the hypothalamus. Binding stimulates the centre that reduces appetite and inhibits the centre that stimulates appetite. Leptin also binds to receptors on body cells, triggering reactions that oxidize fatty acids rather than converting them to fats. When fat storage is reduced, leptin secretion drops off, and signals from other pathways activate the appetite-stimulating centre in the hypothalamus and turn off the appetite-inhibiting centre. These controls closely match the activity of the digestive system to the amount and types of foods ingested and coordinate appetite and oxidative metabolism with the body's needs for stored fats (see *The Puzzling Biology of Weight Control* and *Molecule Behind Biology*).

STUDY BREAK

What is leptin? What role does it play?

41.7 Variations in Obtaining Nutrients

The leaves of plants may be specialized to catch insect prey (see Figure 3.22f), whereas the teeth of vertebrates, particularly those of mammals and fish, indicate the range of diets these animals can exploit (see Chapter 27). This is also reflected in the general lengths of the gastrointestinal tracts and the size and shape of cecae (see Figure 41.16). Some species of fungi lasso nematodes (see Chapter 24), whereas some sponges (see Chapter 26) use specialized spicules to impale prey. Many interesting adaptations and specializations are involved with feeding and digestion.

41.7a Bladderworts: Vacuum Feeding

Worldwide, about 200 species of aquatic bladderworts (*Utricularia* species) use underwater bladders to capture aquatic animals **(Figure 41.26a)**. Bladderworts lack roots and get their name from the small bladderlike organs (utricles), thin-walled suction traps that draw in small animals. Bladderworts use active water transport to generate negative pressure inside the bladder, achieved by a two-step ATP-driven pump. Bladderworts have modified cytochrome *c* oxidase, a rate-limiting enzyme in cellular respiration (see Chapter 6), to set

a. Bladderwort

Perennou Nuridsany / Photo Researchers, Inc.

b. Dodder *(Cuscuta)*

© Grant Heilman Photography

d. Lady-of-the-night orchid *(Brassavola nodosa)*

© Prem Subrahmanyam / www.premdesign.com

c. Snow plant *(Sarcodes sanguinea)*

Beverly McMillan

Figure 41.26

(a) A bladderwort (*Utricularia* sp.) has bladders that serve as insect traps. **(b)** About 150 species of dodder (*Cuscuta* species) have slender yellow to orange stems that twine around host plants before producing haustorial roots that absorb nutrients and water from the host's xylem and phloem. **(c)** A snow plant (*Sarcodes sanguinea*) pops up in the deep humus of shady coniferous forests after snow melt in spring. Snow plants lack chlorophyll and do not photosynthesize, but their roots intertwine with the hyphae of soil fungi that are also associated with the roots of conifers. **(d)** The lady-of-the-night orchid (*Brassavola nodosa*) is a tropical epiphyte.

their traps. Opening the portal to the bladder is triggered by release of elastic instability.

41.7b Dodders: Parasitic Plants

Dodders **(Figure 41.26b)** and several thousand other species of flowering plants are parasites that obtain some or all of their nutrients from the tissues of other plants. Parasitic species develop haustorial roots (similar to the **haustoria** of fungi) that penetrate deep into the host plant and tap into its vascular tissues. Although some parasitic plants, such as mistletoe, contain chlorophyll and thus can photosynthesize, dodders and other nonphotosynthesizers rob the host of sugars as well as water and minerals (see also Figure 3.17).

41.7c Mycoheterotrophic Plants

Coralroot (*Coralloriza* species) is another variation on this theme. As its deep red colour suggests **(Figure 41.26c),** coralroot lacks chlorophyll and does not have haustorial roots. Rather, these plants are mycoheterotrophs, which obtain carbon from other plants via shared mycorrhizal fungi (see Chapter 24).

41.7d Epiphytes: Plant Hangers-On

Epiphytes, such as some tropical orchids **(Figure 41.26d)** or Spanish moss **(Figure 41.27)**, are not parasitic even though they grow on other plants. Spanish moss is not a moss but a vascular plant, despite its common name. Some epiphytes trap falling debris and rainwater among their leaves, whereas their roots (including mycorrhizas, in the case of the orchid) invade the moist leaf litter and absorb nutrients from it as the litter decomposes. In temperate forests, many mosses and lichens are epiphytes.

41.7c Gutless Animals

Most molluscs have a prominent alimentary tract, from mouth to stomach to intestine to anus, whether the animal is a bivalve, a gastropod, or a cephalopod. But several species in the bivalve genus *Solemya* have smaller guts. At least one benthic species (*Solemya borealis*) from the northeastern Pacific is gutless, lacking any evidence of a digestive tract. This burrow-dwelling species lives in areas rich in nutrients and appears to use secretions from the pedal gland to effect extraorganism digestion. Ctenidial lamellae on the gills are probably used to absorb dissolved organic molecules. The lamellae are well serviced by circulating blood, and cilia clean the sediment from them. Gutlessness is also known from species in the phylum Pogonophora (beardworms), as well as in tapeworms (Playthelminthes, Cestoda).

41.7f A Termite-Eating Flatworm: Unusual Lifestyles

At night in a garden in Harare, Zimbabwe, flatworms, *Microplana termitophaga*, gather around the vents of termite mounds where they harvest termites, *Odontotermes transvaalensis* **(Figure 41.28, p. 1033).** *M. termitophaga* are most numerous at termite mounds one to two hours after dawn, with peaks of activity after a rain. When hunting termites, *M. termitophaga* attach the posterior third of their bodies to the ground at the entrance to the termite mound, leaving the anterior end of the body mobile. The planarians appear to visually detect termites at ranges of 5 mm. Their eyes consist of groups of photoreceptors (see Figure 1.15), and although not image-forming, their eyes allow them to detect movement.

The flatworm uses its head, apparently with mucus and perhaps suction, to capture a termite, which is

a.

b.

Figure 41.27

(a) A forest in the southeastern United States has lush azaleas (*Rhododendron* species) as well as a southern oak (*Quercus virginiana*) draped with "Spanish moss" (*Tillandsia usnoides*), an unusual flowering plant. The roots of trees and shrubs and most other plants take up water and minerals from the soil. **(b)** Spanish moss is an epiphyte because it lives independently on other plants and obtains nutrients via absorptive hairs on its leaves and stems (background scale is in millimetres).

The Puzzling Biology of Weight Control

Obesity can lead to elevated blood pressure, heart disease, stroke, diabetes, and other ailments. Obese humans are $\geq 20\%$ heavier than an optimal body weight. The body mass index (BMI), a measure of an individual's body fat, is a standard way to estimate obesity: BMI = weight in kilograms $\div$ (height in metres)^2.

People with a BMI between 18.5 and 24.9 have a normal weight, whereas those with a BMI of 25 to 29.9 are considered overweight. People with a BMI of ≥ 30.0 are considered obese. People with a BMI >27 have a moderately increased risk for developing type 2 diabetes, high blood pressure, and heart disease. Those with a BMI >30 have a greatly increased risk for these conditions.

Conventional wisdom asserts that eating less is the way to lose weight. The relationship between eating and excessive body weight, however, is much more complex. One complicating factor is the probable existence of a genetically determined, homeostatic set point for body weight. If our body weight varies from the set point, compensating mechanisms adjust metabolism and eating behaviour to return body weight to the set point. Thus, if we diet and eat fewer calories, compensating mechanisms reduce the number of calories we use and make the diet less effective. Research shows that the metabolic activity of most body cells in a person on a crash diet decreases by about 15. Each person has a different set point: people of the same height who eat the same number of daily calories vary widely in body weight. Some remain thin, whereas others grow fatter every day on the same amount of food.

If a genetically determined set point governs human body weight, then why is the incidence of obesity increasing in the population? Some researchers speculate that our set points are changing toward greater deposition of fat because, on average, we are less active physically and have greater access to food, especially fatty foods and sugars. The average male Mennonite farmer takes >20 000 steps a day. Wear a pedometer and see how your activity compares. Our evolutionary history may also be a factor. Humans evolved under conditions in which food was occasionally scarce, so our built-in physiological mechanisms may actually favour raising the set point and thus storing extra nutrients when food is available. This would provide some protection against starvation if food becomes unavailable.

The search for factors governing the set point and general concern over growing obesity have sparked intensive research into the genetic mechanisms that might control fat deposition and maintain body weight. In this area, Jeffrey Friedman discovered leptin, and Louis Tartaglia and coworkers found leptin receptors. Leptin is a circulating hormone derived from adipocytes (fat cells). It informs the brain about energy stores. Mice with mutant forms of the genes encoding leptin or leptin receptors become morbidly fat.

The discovery of leptin seemed to offer a "magic bullet" to control human obesity: administer leptin to people and they will lose weight. Further studies revealed, however, that the genetics of weight control are considerably more complex in humans than they are in mice. In addition to genes controlling the production of leptin and the formation of receptors for it, in humans, at least four other genes are involved in appetite control and weight gain, complicating the effects of leptin compared with the situation in mice. In trials with humans, obese patients with mutant leptin genes benefitted from leptin injections, but some with normal leptin genes unexpectedly gained weight. Furthermore, none of the obese patients with normal leptin genes had any deficiency in leptin production. Obese individuals produced more leptin than people of normal weight.

Inconclusive results of leptin trials have turned attention to the development of other drugs to control obesity. PYY (pancreatic polypeptide YY), a recently discovered hormone, stimulates the appetite-suppressive centre in the hypothalamus and inhibits the appetite-stimulating centre. Trial injections of PYY in mice, rats, and humans have led to a significant decrease in appetite and eating.

Rimonabant is an experimental drug that shows promise in producing weight loss in obese patients. This drug works by blocking one of the receptors for the active ingredient in cannabis (marijuana). This receptor is widely distributed in the brain and other organs and helps regulate fat and sugar metabolism, energy balance, and appetite. The receptor is stimulated by cannabislike neurotransmitters produced in the body. When an individual smokes cannabis, the receptor is stimulated, increasing the appetite of the smoker. Blocking the receptor inhibits the stimulation of appetite by natural cannabislike neurotransmitters. In clinical trials, 363 obese Europeans who took rimonabant daily for a year lost an average of 8.6 kg and showed healthy changes in cholesterol triglycerides in the blood. Interestingly, rimonabant appears to be effective in controlling diabetes and helping people stop smoking tobacco.

The search for a magic pill for weight loss, which has gone on for decades, continues.

Figure 41.28
Microplana termitophaga hunting termites at a termite mound in Harare, Zimbabwe. One planarian (centre) has caught a termite. The termites include both soldiers (large head and jaws) and workers.

Figure 41.29
An anglerfish with a bioluminescent lure.

subdued and held against the pharyngeal region. In this position, *M. termitophaga* digests its prey, ingests the juices produced by digestion, and then expels the undigestable remains. In 135 captures, *M. termitophaga* took mainly worker termites rather than soldiers. On average, a flatworm ate a termite in 5 minutes 58 seconds. In 3 hours 14 minutes, one *M. termitophaga* ate 13 termites.

This pattern of feeding appears to be typical of predatory planarians. They may release poisonous secretions (to immobilize prey), adhesive mucus (to hold it), and copious amounts of digestive fluid from their extensible pharynges. Other species of flatworms eat earthworms (see Chapter 49), and some make group attacks on giant African land snails (*Achatina* species).

41.7g Fish Predation: More than You Imagined

Invisible in the inky darkness, a deep-sea anglerfish (*Chaenophryne longiceps*; order Lophiiformes) lies in wait for prey, its gaping mouth lined with sharp teeth. Just above the mouth dangles a glowing lure suspended from a fishing rod–like structure that is a spine of the fish's dorsal fin **(Figure 41.29)**. The lure resembles a tiny fish, wiggling back and forth. The lure's glow is produced by bioluminescent bacteria that live symbiotically in the fish's lure.

Attracted to the lure, a hapless fish comes within range. The oral cavity of the anglerfish expands suddenly, and powerful suction draws the prey into the gaping mouth. The angler's backward-angling teeth keep the prey from escaping, and it is swallowed. The strike takes 6 milliseconds, among the fastest of any known fish. Contractions of throat muscles send the prey to the anglerfish's stomach, which can expand to accommodate a meal as large as the fish itself. The anglerfish now digests, sits, and waits. Some cephalo-

pods, some other fish, and the siphonophore *Erenna* species are examples of other animals using bioluminescent lures.

Other predatory fish hunt from ambush. Moray eels, such as *Muraena retifera*, hide and lie in wait in holes. Their eel-like shape makes it easy for them to fit into small openings. Like the anglerfish, moray eels can swallow items (fish and cephalopods) larger than their heads. But although moray eels lack the effective suction mechanisms of anglerfish, they have two sets of jaws **(Figure 41.30, p. 1034)**. Like other gnathostomes, the moray eel's mouth is bordered by upper (maxilla and premaxilla) and lower (mandibular) jaws bearing teeth. Unlike other gnathostomes, moray eels also have pharyngeal jaws. In the two upper pharyngeal jaws, pharyngobranchial bones bear teeth and connect to the lower pharyngeal jaws, which also bear teeth. Upper and lower pharyngeal jaws are connected by the epibranchial bones. The jaw arrangement of moray eels allows them to grab and transport (swallow) prey in a system similar to the mechanisms in snakes. The combination of hunting from ambush and quick swallowing of prey makes moray eels efficient predators.

Vandellia cirrhosa, the dreaded candiru **(Figure 40.31, p. 1035)**, is a specialized catfish that takes a completely different approach to feeding. Normally living as a gill parasite on larger fish, this small and slender predator lodges itself in the gills of a fish. There, like a leech or mosquito, tick, or lamprey eel, candirus drink the host's blood. To understand why the candiru is "dreaded," consider other aspects of its behaviour. Candirus locate hosts by swimming into currents, especially those bearing metabolic end products. Freshwater fish often pass metabolic end products out across their gills. The behaviour can also lead it into the urethra of an animal urinating while submerged in the water. In South American waters inhabited by candirus, local humans hold their water when swimming or bathing and may wear protective covers cut from coconuts to foil invasive candirus.

MOLECULE BEHIND BIOLOGY

Sorbitol

Based on the number of OH groups, sorbitol contains more energy than glucose (**Figure 1**). Yet a glance at the label of many "sugar-free" products reveals sorbitol as the major sweetener. How can the sugar-free product contain more calories than the "real" thing? The answer lies in taste. Sorbitol, like other sugar-free sweeteners, is much sweeter in flavour than glucose. This means that people who use sorbitol-based products to avoid the calories of sugar (glucose) are taking advantage of its sweetness. A little sorbitol makes that cup of coffee or tea or that sugar-free drink sweet with fewer calories (molecules of sorbitol). If the consumer

CH_2OH

Figure 1
Sorbitol.

used as much sorbitol as glucose, the sweetness would be overwhelming, rather like the calorie count.

The safety of food substitutes is often a concern for people who watch what they eat. In 2008, Juergen Bauditz and four colleagues reported two cases of severe weight loss and digestive upset associated with excessive consumption of chewing gum flavoured with sorbitol. One patient, a 21-year-old woman, had lost 11 kg and weighed 40.8 kg (body mass index 16.6) and suffered from severe diarrhea. At first, she was diagnosed as having infectious colitis, but further investigation revealed that her daily consumption of 18 to 20 g of sorbitol by chewing gum (1.25 g of sorbitol per stick of gum) was responsible for the condition. One year after going on a sorbitol-free diet, this patient had gained 7 kg, and her body mass index was 19.5. Another patient, a 46-year-old man, weighed 79.9 kg (body mass index 25.8) and had lost 22 kg before being examined

by the physicians. He reported having eaten about 30 g of sorbitol every day. Six months after going on a sorbitol-free diet, he had gained 5 kg (body mass index 27.4). Both patients had normal bowel movements after going on sorbitol-free diets.

Two main ingredients of sorbitol are mannitol and xylitol, polyalcohol sugars that are common ingredients in laxatives. People often eat sugar substitutes such as sorbitol to minimize their caloric intake and to reduce the incidence of cavities in their teeth. Sorbitol is poorly absorbed by the small intestine and acts as an osmotic agent. Ingestion of 5 to 20 g of sorbitol daily can cause bloating and gas, as well as abdominal pain. Higher doses, such as those reported for the two patients in the case studies, cause osmotic diarrhea. It is likely that reducing caloric intake by eating less is a healthier alternative than eating large amounts of artificial sweeteners such as sorbitol.

41.7h Egg-Eating Snakes: Caviar of a Different Sort

Eggs well provisioned with yolk may be the ultimate food, rich in proteins and carbohydrates—everything a predator could want. Mammals such as mongooses break ostrich eggs by throwing rocks against them. Egyptian vultures (*Neophron percnopterus*) break these

large eggs by dropping sticks or stones on them. But egg-eating snakes cannot throw stones or sticks.

Dasypeltus scabra is one of the best studied species of egg-eating snakes. Widespread in Africa, these snakes find birds' nests and help themselves to eggs. Like many other snakes, the jaws of *D. scabra* are loosely connected with elastic tendons and ligaments, allowing them to

Figure 41.30

(a) A moray eel with its mouth open does not reveal its pharyngeal jaws, the second part of of its bite. (b) The skeleton and teeth of the pharyngeal jaws. (c) A scanning electron micrograph view of the upper pharyngeal recurved teeth. Scale bars 1 cm (b) and 500 μm (c).

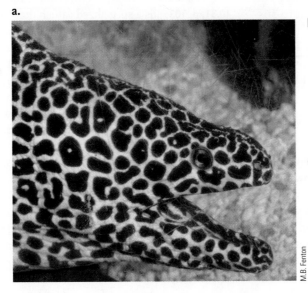

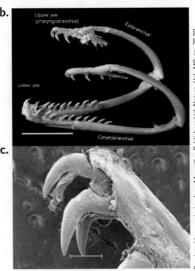

Reprinted by permission from Macmillan Publishers Ltd: Nature, Vol. 449: pp. 79-82, Raptorial jaws in the throat help moray eels swallow large prey by Rita S. Mehta and Peter C. Wainwright, copyright (2007). 1035: bottom, Based on illustration from Gans 1974 Biomechanics Fig. 2-22

M.B. Fenton

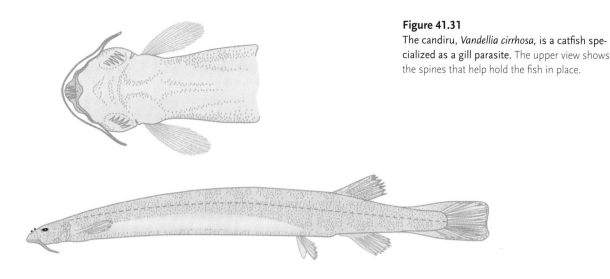

Figure 41.31
The candiru, *Vandellia cirrhosa*, is a catfish specialized as a gill parasite. The upper view shows the spines that help hold the fish in place.

open their mouths extremely wide **(Figure 41.32)**. In a bird's nest, the snake can push an egg against the rim of the nest and swallow it whole. A wide gape is important, but so are an extensible epiglottis and ribbed tracheae, which together allow the snake to breathe while it swallows an egg.

After the egg has passed out of the mouth and into the esophagus, the snake makes a coil in front of the swallowed egg. By moving the coil backward along its body, the snake moves the egg down its digestive tract. You can watch the egg move down the snake, but then, with a cracking noise, the outline of the egg

disappears. The snake has pushed the egg against ventral hypophyses, anteriorly pointed extensions of specialized vertebrae **(Figure 41.33, p. 1036)** that protrude into the lumen of the gastrointestinal tract. *D. scabra* has a built-in egg-cracker.

STUDY BREAK

1. What do anglerfish and bladderworts have in common?
2. How do you define "carnivorous"?

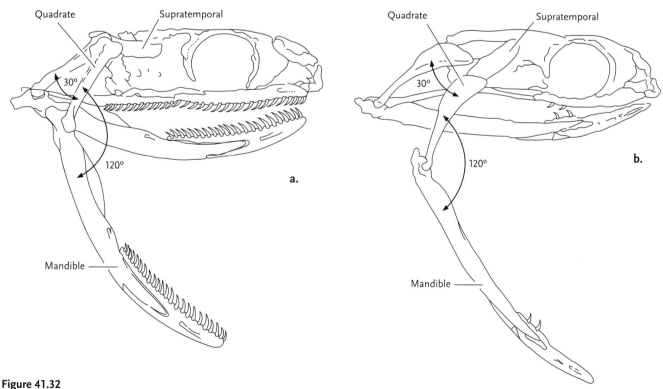

Figure 41.32
The articulation between the supratemporal bone and the braincase (more than between the quadrate and the supratemporal bones) is responsible for a great increase in the gape of a snake. Combined with the position of the quadrate, these features allow one snake **(b)** a 20% larger gape than the other **(a)**.

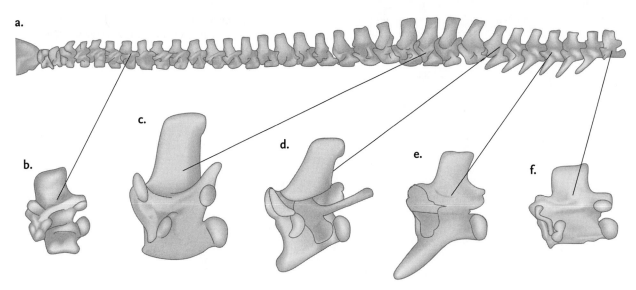

Figure 41.33

Dasypeltus scabrus has **(a)** a specialized vertebral column. **(b)** The anterior vertebrae are not especially modified, **(c)** whereas high neural arches make more posterior vertebrae highly modified. **(d)** Other vertebrae are transitional to **(e)** the highly modified, egg-breaking vertebrae. **(f)** Posterior, the vertebrae are more typical of snakes. The spines on the egg-breaking vertebrae are ventral hypophyses.

Unanswered Questions

Under what circumstances have achlorophyllous plants evolved? What opportunity are they exploiting?

Review

Go to CENGAGENOW™ at http://hed.nelson.com/ to access quizzing, animations, exercises, articles, and personalized homework help.

41.1 Nutrition: Essential Materials

- Living organisms require a variety of materials to survive. Use Table 41.1 to review and compare nutrients essential to plants and animals.

- Malnutrition can manifest itself as undernutrition or overnutrition, involving inadequate intake of organic fuels or abnormal ingestion of fuels, respectively. Undernutrition is commonly referred to as malnutrition and may involve ingestion of too little energy to fuel daily activities or failure to ingest essential nutrients. Malnutrition kills thousands of people annually (see Chapter 49). Overnutrition can result in excessive gain in body mass and other health problems.

- Hydroponics allowed biologists to explore the details of nutrients essential to plants and recognize symptoms of deficiencies of different nutrients in plants.

- Animals such as birds on long migratory flights may mobilize energy from their own bodies during prolonged periods of not feeding.

- In animals, vitamins are essential for different metabolic operations.

41.2 Soil

- Soil is the living skin of much terrestrial habitat on Earth. It provides essential materials for plants and a habitat for animals. Soil consists of particles of different sizes (<0.002−2 mm in diameter). Three main horizons (A, B, and C) develop in many soils.

- Soil composition influences plants' access to water and nutrients in soil. Roots anchor plants and also serve as connections between plants and soil.

41.3 Obtaining and Absorbing Nutrients

- Nutrients such as N, P, and K are often scarce in soil, so they are supplied in fertilizers used in agricultural operations.

- Mycorrhizas are symbiotic associations between fungi and the roots of plants.

- Specialized structures in plant roots, stems, and leaves conduct materials up and down the plant.

- Nitrogen can be an important limiting factor for many plants because atmospheric N is not available to them. Some carnivorous plants obtain N from animals they catch and digest. Other plants use associations with nitrogen-fixing bacteria to obtain N.

- Deposit feeders such as some burrowing molluscs and tube-dwelling polychaete worms (see Chapter 26) pick up or scrape particles of organic matter from their surroundings. Arthropods such as fiddler crabs sift food (organic matter) from sediments.

41.4 Digestive Processes in Animals

- The gizzard is a muscular structure that grinds food into small particles. In annelids such as earthworms, the gizzard uses sand to increase abrasive action. Birds and reptiles often pick up stones for the gizzard to achieve the same effect. Insects often have gizzards.

- Sponge chanocytes (collar cells) trap food particles and take them in by endocytosis. Amoeboid cells transfer food throughout the sponge.

- The diet of *Dugesia* is not typical of free-living flatworms (see Chapter 26). Whereas *Dugesia* mainly feed on detritis, other flatworms are predatory, taking earthworms, snails, or termites.

- Along the gastrointestinal tract, food is first mechanically processed, breaking it into smaller pieces. Then enzymes and other digestive aids, such as acids and mucus, are secreted and commence chemical breakdown of food. Enzymatic hydrolysis breaks food molecules into absorbable molecular subunits. This involves reactions catalyzed by enzymes. Then the food molecules are absorbed from the gastrointestinal tract. Finally, undigested materials are expelled through the anus.

- The hoatzin is a flying bird known to use fermentation to digest cellulose. Just-caught hoatzins often smell like cow dung. Their digestive systems include a fermentation centre that uses space normally occupied by flight muscles. Hoatzins are not strong fliers. Other animals from termites to some mammals use fermentation to digest cellulose.

41.5 Digestion in Mammals

- Living birds lack teeth and use the gizzard to break food into finer particles. Many (but not all) mammals use teeth to mechanically break up food.

- Mammals such as humans require eight essential amino acids: lysine, tryptophan, phenylalanine, threonine, valine, methionine, leucine, and isoleucine.

- The layers in the gut of a mammal are the mucosa, submucosa, muscularis, and serosa. Each plays a different role in the operation of the gastrointestinal tract. Sphincter muscles control the movement of material through the gastrointestinal tract.

- Peristalsis is rhythmic contractions of circular bands of smooth muscle. Peristalsis constricts the gut and moves food along the gastrointestinal tract. Peristalsis usually occurs in waves.

- Gastric juices are produced in gastric glands that line the stomach. Production and release of gastric juices are stimulated by output of stretch receptors. The 2 L of gastic juices secreted daily by the average human includes pepsin, hydrochloric acid, and lubricating mucus.

- The liver secretes bile salts, cholesterol, and bilirubin. Bile salts are derivates of cholesterol and amino acids and aid the digestion of fat. They operate by forming a hydrophilic coating around fats and other lipids, allowing the churning of the intestine to emulsify fats. Bilirubin is a waste product derived from worn-out erythrocytes.

- As many as 500 species of bacteria may be living in the gastrointestinal tract of mammals. Bacterial activity is partly responsible for flatus in the gastrointestinal tract, the production of CO_2, methane, and hydrogen sulphide. Bacteria may comprise 20 to 50% of the dry matter in feces.

41.6 Regulation of the Digestive Processes

- Leptin is a peptide hormone that links two interneuron centres in the hypothalamus. One centre stimulates appetite, whereas the other reduces oxidative metabolism. Fat-storing cells secrete leptin when deposition of fat increases in the body. Leptin moves through the bloodstream and binds to receptors in both centres in the hypothalamus. Binding stimulates the centre reducing appetite and inhibits the centre increasing appetite.

41.7 Variations in Obtaining Nutrients

- Bladderworts are vacuum-feeding carnivorous plants.
- Dodders are parasitic plants.
- Some plants are mycoheterotrophs.
- Epiphytes grow on other plants.
- Tapeworms and some species of molluscs lack digestive tracts.
- Some moray eels have two functional sets of jaws, each armed with teeth. In addition to their "normal" jaws, moray eels have pharyngeal jaws that make it easier for them to seize and then swallow prey. The mechanism works like a rachet and is reminiscent of similar systems in some snakes.
- Some egg-eating snakes swallow birds' eggs whole. They use hypophyses, ventral, anterior-pointing processes from some vertebrae, as egg-crackers.

Questions

Self-Test Questions

1. Vitamins are
   a. coenzymes.
   b. fatty acids.
   c. organic molecules.
   d. amino acids.
   e. carbohydrates.

2. Some plants can gain access to nitrogen (N) through
   a. direct absorption of atmospheric N.
   b. carnivory.
   c. nitrogen-fixing bacteria.
   d. fertilizers.
   e. b, c, and d.

3. Mycorrhizas are associations between
   a. fungi and mammals.
   b. roots of terrestrial plants and fungi.
   c. solar slugs and fungi.
   d. fungi and algae.
   e. all of the above.

4. The crop is part of the digestive tract in
   a. birds.
   b. mammals.
   c. reptiles.
   d. earthworms.
   e. a and d.

5. Humans require __ essential amino acids.
   a. 0
   b. 2
   c. 5
   d. 8
   e. 10

6. In humans, pepsin reaches optimal levels of activity at
   a. pH 1.
   b. pH 2.
   c. pH 4.
   d. pH 6.
   e. pH 8.

7. Epiphytes grow on other plants, usually trees. They include
   a. lichens.
   b. Spanish moss.
   c. some orchids.
   d. dodders.
   e. a, b, and c.

8. Cellulose is digested by fermentation in
   a. humans.
   b. ruminants.
   c. the hoatzin.
   d. carnivores.
   e. b and c.

9. Animals that eat blood and body fluids include
   a. leeches, vampire bats, and candirus.
   b. tapeworms, flukes, and mosquitoes.
   c. blackflies, lamprey eels, and bedbugs.
   d. wolves, cats, and vampire bats.
   e. a and c.

10. Sorbitol is an effective substitute for glucose because
    a. it has less energy.
    b. digesting it burns more energy than it produces.
    c. it is much sweeter than glucose.
    d. it is not toxic.
    e. All of the above are correct.

Questions for Discussion

1. How does secondment of chloroplasts benefit animals (see Chapter 3)? In what animals does it occur? Do these animals have genetic control over the chloroplasts? If so, how do they acquire this control?

2. Although humans evolved as omnivores, some eat only meat, others are vegetarian, and many (perhaps most) eat a combination of plant and animal material. What should a balanced diet include?

3. If you are planting a vegetable garden, what fertilizers would you plan to apply? For the vegetables to qualify as "organic" produce, what kinds of fertilizers would you be allowed to use? What constitutes "organic" meat?

4. Do animals that are usually vegetarian readily eat animal protein? Find an example of a disease associated with this behaviour.

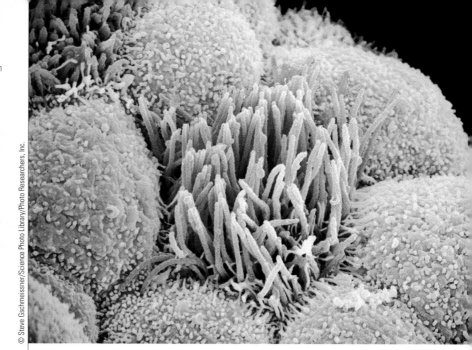

Lining of the trachea (windpipe) shown in a colourized scanning electron micrograph, with mucus-secreting cells (white) and epithelial cells with cilia (pink). The trachea is positioned between the larynx and the lungs, providing a conduit for air entering and leaving the body.

© Steve Gschmeissner/Science Photo Library/Photo Researchers, Inc.

42 Gas Exchange: The Respiratory System

WHY IT MATTERS

In Africa, a huge swarm of the desert locust, *Schistocerca gregaria* **(Figure 42.1, p. 1040)**, composed of millions of individuals, takes off, escaping the lack of food caused by their voracious feeding (a swarm may consume the equivalent of food for 2500 people in a single day). Driven by the wind, the swarm normally descends in the evening to take off again in the morning but may remain airborne for more than 24 hours. A swarm may continue to fly during the day for several days until the wind delivers it to food. The wings of each locust, which weighs about 2.5 grams, will beat about 20 times per second. Their relatively large flight muscles require very large amounts of oxygen to provide the energy derived from fat stores. Insects in flight have the highest O_2 consumption per gram ever recorded for animals.

Like these locusts, all organisms with active metabolism need to exchange gases with their surroundings. The chloroplasts of plant cells need CO_2 for the Calvin cycle, and in the light, these chloroplasts release O_2 by the action of photosystem II. Likewise, the mitochondria of eukaryotic cells need a constant supply of O_2 as it is required as the terminal electron acceptor of respiratory electron transport. In addition, respiration also produces CO_2, which needs to be rapidly

Figure 42.1
(a) Migratory locust. **(b)** A small portion of a locust swarm in Mauritania in 2004.

removed from cells because in animals, high cellular CO_2 is a narcotic poison, damaging nerve function. In this chapter, we introduce the physical basis for gas exchange and how evolution has produced a range of adaptations that maximizes the rate of gas exchange both into and out of animal tissues.

42.1 General Principles of Gas Exchange

Air normally contains about 78% diatomic nitrogen (N_2), 21% diatomic oxygen (O_2), and less than 1% carbon dioxide (CO_2) and other gases.

The percentage composition of air does not change with the total amount of air. Thus, as you climb a mountain, the total amount of air falls, and there are fewer molecules of O_2 (and any other gas) in the environment. The density of the air in the atmosphere is measured as the atmospheric pressure, which is greater the closer you are to sea level and falls with increasing altitude. The unit of measurement is often millimetres of mercury (mm Hg). At sea level, the atmospheric pressure is 760 mm Hg: the pressure is sufficient to support a vertical column of mercury 760 mm high. This pressure is the sum of the pressures of all of the gases in a mixture. The individual pressure exerted by each gas within a mixture of gases such as air is defined as its **partial pressure**. For any one gas, the partial pressure is calculated by multiplying the fractional composition of that gas by the atmospheric pressure. Given that O_2 is 0.21 of air, the partial pressure of O_2, abbreviated P_{O_2}, at sea level is 0.21×760 mm Hg or 160 mm Hg.

42.1a Fick's Equation of Diffusion

The partial pressure of a gas is the key factor in determining the direction in which a gas will move. You can think of the partial pressure of a gas as being similar

to the concentration of a solute in solution. Just as a solute will move by simple diffusion from an area of high concentration to an area of low concentration, a gas will move down a partial pressure gradient, from a region of high partial pressure to an area of low partial pressure.

But the rate (amount per unit time) at which a gas will diffuse is dependent on a set of factors, only one of which is the difference in partial pressure between two regions. Fick's equation, which is important to understanding the diffusion of gases, recognizes the importance of these other factors. Fick's equation can be stated in a number of ways, but for the diffusion of gas across a membrane or other surface, it can be stated as follows:

$$Q = \frac{DA \times (P1 - P2)}{L}$$

where

- Q is the rate of diffusion between the two sides of the membrane.
- D is the diffusion coefficient for the gas involved. This is simply a factor specific to the gas molecule that recognizes its size and vibrational activity, the medium (gas, solid, liquid) in which the diffusion occurs, and the temperature.
- A is the area across which diffusion takes place.
- P1 and P2 are the partial pressures of the gases at the two locations.
- L is the path length or distance between the two locations (the thickness of the membrane).

42.1b Gas Exchange by Simple Diffusion

The additional variables introduced by Fick's equation are the area across which diffusion takes place and the distance over which diffusion occurs. Relying on diffusion alone for gas exchange limits both the size and to some degree the shape of the organism. The importance of these factors is made obvious by a consideration of the surface to volume ratio. Bacteria, with a volume of about 10^{-18} m^3 and a surface area of about 6×10^{-12} m^2, have a surface to volume ratio of 6 000 000:1. They can clearly rely on diffusion alone for gas exchange because the surface is large with respect to the volume, and the distance that the gases must diffuse is relatively small. The same is true of protists, with a surface to volume ratio of about 60 000:1. Among multicellular organisms, however, an increase in size can be accommodated only if the distance over which diffusion must occur is minimized. Gas exchange by diffusion can occur only if the organisms are thin and flat.

Flatworms **(Figure 42.2a)** represent an example of a multicellular organism that relies on simple diffusion for gas exchange. Most free-living flatworms are small but may range up to 10 cm or more in length, and parasitic forms such as tapeworms may be as long as 3 m or more (see Chapter 26). But all are thin, so L in Fick's equation is minimized and A is maximized.

42.1c The Plant Leaf Is a Special Case

Like flatworms, a plant leaf has a large surface area over which gas exchange can take place. However, unlike the flatworm, the external surface of a leaf is not available for gas exchange: leaves are covered by a waxy cuticle that is impermeable to gases. We learned in Chapter 28 that gases gain access to the interior of the leaf through pores called stomata. Inside the leaf, there is gas-filled space where gases can move such that the diffusion distance (L in Fick's equation) between the external environment (the air within the leaf) and the individual cells is very short **(Figure 42.3)**. Moreover, the total area (A in Fick's equation) of the cell walls inside the leaf is large. These two factors maximize the rate of diffusion of gases between the cells and the environment.

42.1d Adaptations That Increase the Surface Available for Gas Exchange in Animals

In animals, gas exchange occurs across a **respiratory surface**, which may consist of the external surface of the organism. We have seen, however, that Fick's equation limits the size and shape of such organisms. The evolution of larger specialized respiratory surfaces and/or some means of transporting gases to and from the surfaces of cells within the organism has permitted the development of larger and more complex organisms. The specialized respiratory surfaces are often very large, increasing A in Fick's equation. The total area of the lungs in humans is about 100 m^2. In addition, the cells that make up the

a. Extended body surface: flatworm

b. External gills: mudpuppy

c. Lungs: human

Figure 42.2
Adaptations increasing the area of the respiratory surface. **(a)** The flattened and elongated body surface of a flatworm. **(b)** The highly branched, feathery structure of the external gills in an amphibian, the mudpuppy (*Necturus*). **(c)** The many branches and pockets expanding the respiratory surface in the human lung.

specialized respiratory layer are squamous epithelium, decreasing L. If a circulatory system is bathing the inside of the respiratory surface, it will maintain a large difference between P1 and P2, increasing the value of Q. Both of these adaptations, the development of specialized respiratory surfaces and the development of circulatory systems that transport the gases, have reduced the limitation on size and shape imposed by reliance on simple diffusion

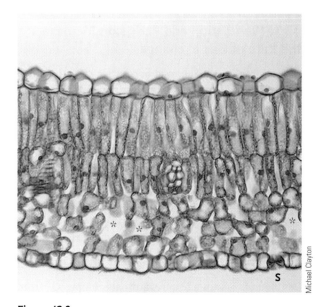

Figure 42.3
A cross section of a leaf of a lilac bush. The network of interconnecting spaces, marked by an asterisk, brings most cells in direct contact with gases. S marks a stomata through which gases gain access to the spaces inside the leaf.

through the animal's surface. These are among the adaptations that have permitted animals to penetrate a wider range of environments. We will see that insects, like the plant leaf, represent a special case. In insects, tubes called tracheae penetrate from the surface of the insect to every cell and carry respiratory gases to and from the surfaces of the cells.

For the gases to pass across the epithelium, they must be in solution. For aquatic and marine animals, for which the **respiratory medium** is water, that is already the case, and gills, outward extensions of the body surface (see Figure 42.2 b), are the site of gas exchange. For terrestrial animals, the respiratory medium is air and the respiratory epithelium is covered by a thin film of fluid. Loss of water is minimized by the location of the expanded respiratory epithelium, called lungs, deep within the body of an animal (see Figure 42.2c).

Some of the CO_2 produced by the cells remains in solution as a gas, but in many organisms, significant amounts may combine with water to produce carbonic acid (H_2CO_3), which dissociates into bicarbonate (HCO_3^-) and H^+ ions. This reaction maintains a maximal concentration gradient of CO_2 between the cells and the blood. Of course, the capacity of the blood or body fluids for CO_2 is limited. It is a means of storing the gas in a harmless form temporarily until it can be transported to the respiratory surface of the animal for release once more as a gas.

Most of the H^+ ions produced by the **dissociation** of carbonic acid combine with haemoglobin or with proteins in the blood. The combination, by removing excess H^+ from the blood solution, *buffers* the pH of the blood, helping to maintain it at the set point appropriate for the species, usually about 7.4.

42.1e Ventilation and Perfusion Increase the Rate of Gas Exchange

Although all gas exchange occurs by diffusion, two adaptations help most animals maintain the difference in concentration between gases outside and inside the respiratory surface. In Fick's equation, they maximize the value of P1-P2, maintaining a steep gradient of the partial pressure of the gas across the respiratory surface and enhancing the rate of diffusion. One is **ventilation**, the flow of the respiratory medium (air or water, depending on the animal) over the respiratory surface. The second is **perfusion**, the flow of blood or other body fluids on the internal side of the respiratory surface.

Ventilation. As they respire, animals remove O_2 from the respiratory medium and replace it with CO_2. Without ventilation, the concentration of O_2 would fall in the respiratory medium close to the respiratory surface, and the concentration of CO_2 would rise, gradually reducing the value of P1-P2 in the Fick

equation for both gases and reducing the rate of diffusion below that necessary to sustain life. Examples of ventilation include the one-way flow of water over the gills in fishes and many other aquatic animals and the in-and-out flow of air in the lungs of most vertebrates and in the tracheal system of insects at rest.

Perfusion. The rate at which blood or other fluids are replaced on the internal side of the respiratory surface similarly helps keep P1-P2 at an acceptable level. In animals with a circulatory system, the circulatory system brings blood to the internal side of the respiratory surface, transporting CO_2 (often in the form of bicarbonate) from all cells of the body. At the surface, CO_2 is released into the medium, and a fresh supply of O_2 is picked up. Insects, as we will see, do not use blood to transport these gases.

42.1f Water and Air Have Advantages and Disadvantages as Respiratory Media

Because their respiratory surfaces are exposed directly to the environment, aquatic and marine animals have no problem keeping the respiratory surface wet. However, aquatic animals face challenges in obtaining O_2 from water compared with terrestrial animals. An important variable in Fick's equation is the diffusion coefficient, D. This factor varies with the gas, the medium, and the temperature. The difference between D in gas and water is large: for O_2 at 20°C, the value in water is 1.97×10^{-5}, whereas in air, it is 0.219—approximately 10 000 times greater. The rate of diffusion in air is thus 10 000 times faster than in water. In addition, for the same volume, there is approximately 30 times less O_2 in water than in air at 15°C, reducing the value of P1-P2. These two factors require animals that rely on water for their gas exchange to pass a vastly greater volume of water over the respiratory surface in order to be exposed to the same volume of O_2 as an animal relying on air. Moreover, the density of water is about 1000 times that of air, and its viscosity is about 50 times that of air. Therefore, it takes significantly more energy to move water than air over a respiratory surface. Ventilation in most aquatic animals takes place in a one-way direction, compensating to some degree for these two effects. In bony fishes, for instance, water enters the mouth, flows over the gills, and exits through the gill covers, all in one direction.

In addition, temperature and solutes affect the O_2 content of water. That is, as either the temperature or the amount of solutes increases, the amount of gas that can dissolve in water decreases. Therefore, with respect to obtaining O_2, aquatic animals that live in warm water are at a disadvantage compared with those that live in cold water. And because solutes (such as sodium chloride) are higher in seawater compared with freshwater, animals living in a marine environment are at a disadvantage compared with those living in an aquatic environment.

The relatively high O_2 content, low density, and low viscosity of air greatly reduce the energy required to ventilate the respiratory surface. These advantages allow animals with lungs to breathe in and out, reversing the direction of flow of the respiratory medium, without a large energy penalty.

A major disadvantage of air is that it constantly evaporates water from the respiratory surface unless it is saturated with water vapour. Therefore, except in an environment with 100% humidity, animals lose water by evaporation during breathing and must replace the water to keep the respiratory surface from drying.

We next turn to the adaptations that allow water-breathing and air-breathing animals to obtain O_2 and release CO_2 in aquatic and terrestrial environments. These adaptations allow animals to exploit the advantages and circumvent the disadvantages of water and air as respiratory media.

STUDY BREAK

1. What variables in Fick's equation affect the rate of diffusion across a membrane?
2. What are the disadvantages of water as a respiratory medium?

42.2 Adaptations for Gas Exchange

Although most animals that live in marine or aquatic environments exchange gases through the skin or gills, a few use lungs and breathe air. Whales, seals, and dolphins are mammals that have returned to the sea and have special adaptations that permit them to remain submerged for long periods, surfacing to breathe (see Chapter 37, *Life on the Edge*). Many amphibians respire through the skin as well as the lungs, and their larvae have gills. Some terrestrial arthropods, such as land crabs and some spiders, have internalized gills into lunglike structures.

42.2a External and Internal Gills

Gills are respiratory surfaces that are branched and folded evaginations (outward extensions) of the body. They increase the area over which diffusion can take place. **External gills (Figure 42.4a)** extend out from the body and do not have protective coverings. They occur in some molluscs, some annelids, the larvae of some aquatic insects, the larvae of some fishes, and the larvae of amphibians. **Internal gills (Figure 42.4c, d)** are located within chambers of the body. This not only provides protection for delicate structures but also allows currents of water to be directed over the gills.

a. External gills: nudibranch

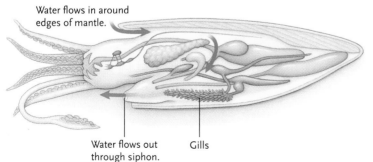

b. Internal gills: clam

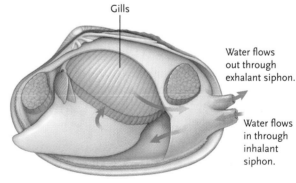

Gills

Water flows out through exhalant siphon.

Water flows in through inhalant siphon.

c. Internal gills: cuttlefish

Water flows in around edges of mantle.

Water flows out through siphon. Gills

d. Internal gills: fish

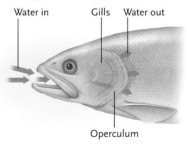

Water in   Gills   Water out

Operculum

Figure 42.4

External and internal gills. **(a)** The external gills of a nudibranch (*Flabellina iodinea*). **(b)** The internal gills in a clam. **(c)** The internal gills of a cuttlefish. **(d)** The internal gills of a bony fish. Water enters through the mouth and passes over the filaments of the gills before exiting through an opening at the edges of the flaplike protective covering, the operculum.

Most crustaceans, molluscs, sharks, and bony fishes have internal gills. Some invertebrates, such as clams and oysters, use beating cilia to circulate water over their internal gills (Figure 42.4b). Others, such as the cuttlefish, use contractions of the muscular mantle to pump water over their gills (see Figure 42.4c). In adult bony fishes, the gills extend into a chamber covered by gill flaps or *opercula* (singular, *operculum* = little lid) on either side of the head. The operculum serves as part of a one-way pumping system that ventilates the gills (Figure 42.4d).

42.2b Many Animals with Internal Gills Use Countercurrent Flow to Maximize Gas Exchange

Sharks, fishes, and some Crustacea take advantage of one-way flow of water over the gills to maximize the amounts of O_2 and CO_2 exchanged with water. In this mechanism, called **countercurrent exchange**, the water flowing over the gills moves in a direction opposite to the flow of blood under the respiratory surface.

Figure 42.5 illustrates countercurrent exchange in the uptake of O_2. At the point where fully oxygenated water first passes over a gill filament in countercurrent flow, the blood flowing beneath it in the opposite direction is also almost fully oxygenated. However, the water still contains O_2 at a higher concentration than the blood, and the gas diffuses from the water into the blood, raising the concentration of O_2 in the blood almost to the level of the fully oxygenated water. At the opposite end of the filament, much of the O_2 has been removed from the water, but the blood flowing under the filament, which has just arrived from body tissues and is fully deoxygenated, contains even less O_2. As a result, O_2 also diffuses from the water to the blood at this end of the filament. All along the gill filament, the same relationship exists, so that at any point,

the water is more highly oxygenated than the blood. P1-P2 is maximized, and O_2 diffuses at a high rate from the water and into the blood across the respiratory surface.

The overall effect of countercurrent exchange is the removal of 80 to 90% of the O_2 content of water as it flows over the gills. In comparison, by breathing in and out and constantly reversing the direction of air flow, mammals manage to remove only about 25% of the O_2 content of air. Efficient removal of O_2 from water is important because of the much lower O_2 content of water compared with air.

42.2c Insects Use a Tracheal System for Gas Exchange

Insects breathe air by a unique respiratory system consisting of air-conducting tubes called tracheae (*trachea* = windpipe) (Figure 42.6). The tracheae are invaginations of the outer epidermis of the animal

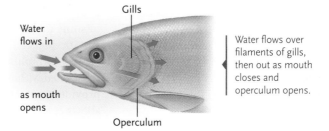

a. The flow of water around the gill filaments

b. Countercurrent flow in fish gills, in which the blood and water move in opposite directions

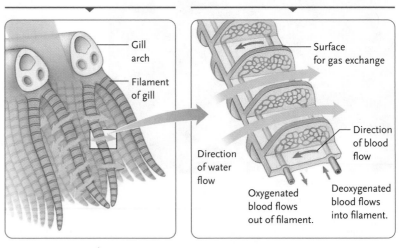

c. In countercurrent exchange, blood leaving the capillaries has the same O_2 content as fully oxygenated water entering the gills

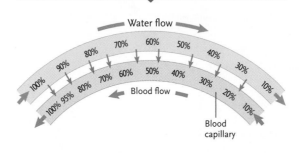

Figure 42.5
Ventilation and countercurrent exchange in bony fishes.
(a) Water flows around the gill filaments. **(b)** Water and blood flow in opposite directions through the gill filaments.
(c) Countercurrent exchange: oxygen from the water diffuses into the blood, raising its oxygen content. The percentages indicate the degree of oxygenation of water (blue) and blood (red).

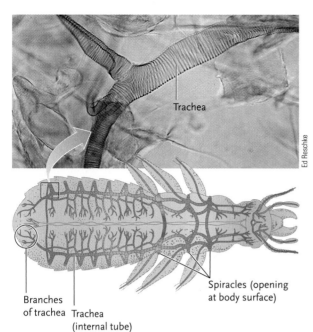

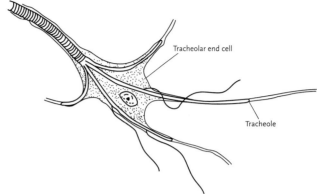

Figure 42.6

The tracheal system of insects. The photograph shows the chitinous rings that reinforce the tracheae, keeping them from collapsing. The tracheal system terminates in many tracheolar end cells that have branches with a diameter of less than 1 μM.

Trachea

Branches of trachea

Trachea (internal tube)

Spiracles (opening at body surface)

Tracheolar end cell

Tracheole

Ed Reschke

and as such consist of the epithelial cells and the cuticle secreted by those cells. They are lined with a thin layer of the same cuticle as the exoskeleton and are reinforced by rings of cuticle. They lead from the body surface and branch repeatedly. With each branching, the diameter of the tracheae is reduced, and the tracheae ultimately end as tracheoles less than 1 μM in diameter. Every cell except the hemocytes in an insect's body makes contact with at least one tracheole. In the case of large, metabolically active cells, such as the flight muscles, tracheoles may penetrate inside the cell via invaginations of the cell membrane. Tracheoles are dead-end tubes with very small tips filled with fluid that are in contact with cells of the body. Air is transported by the tracheal system to those tips, and gas exchange occurs directly across the very thin cuticle and epithelium of the tracheoles and the plasma membranes of the body cells. At places within the body, the tracheae may expand into internal air sacs that act as reservoirs to increase the volume of air in the system.

Air enters and leaves the tracheal system at openings in the insect's chitinous exoskeleton called **spiracles** (spiraculum = airhole). The spiracles are located in a row on either side of the thorax and abdomen, typically one pair per body segment. Each spiracle incorporates a muscle that allows the spiracles to open and close. For many insects at rest, the spiracles are minimally open, allowing a very small current of air to enter. O_2 is consumed by the tissues, and CO_2 is taken up by the bicarbonate buffering system, resulting in a small negative pressure inside the tracheal system. The small inward current of air flowing through the reduced opening of the spiracle prevents water vapour from escaping. As the

bicarbonate buffering system becomes saturated, free CO_2 builds up inside the insect, causing the spiracles to open briefly, allowing CO_2 (and water vapour) to escape. In periods of greater activity, as in flight, this mechanism is replaced by one in which alternating compression and expansion of the thorax by the flight muscles also pumps air through the tracheal system. The spiracles open and close in synchrony with this rhythm.

42.2d Lungs Allow Animals to Live in Completely Terrestrial Environments

Lungs are one of the primary adaptations that allowed vertebrates to fully invade terrestrial environments. Many authorities believe that the bony fish evolved from a freshwater ancestor that had both fins and lungs, which arose as invaginations of the upper digestive tract. Two lines evolved from this ancestor. In one line, the lung lost its connection to the digestive system and became the swim bladder that controls buoyancy in the modern teleosts. The other line (Sarcopterygii), represented by only a few living species, retained the lung, enabling them to survive in O_2-poor water or periods when pools dried up. This line gave rise to the tetrapod vertebrates. In these fish, air is obtained by **positive pressure breathing**, a gulping or swallowing motion that forces air into the lungs (see Chapter 27).

The lungs of mature amphibians such as frogs and salamanders are also thin-walled sacs with relatively little folding or pocketing. Amphibians fill their lungs by positive pressure breathing **(Figure 42.7, p. 1046)**. A breathing cycle begins by opening the nostrils and lowering the floor of the mouth cavity

Figure 42.7
Positive pressure breathing in an amphibian (frog).

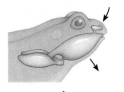 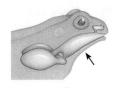 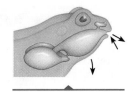 

1 The frog lowers the floor of its mouth and inhales through its nostrils.

2 It closes its nostrils, opens the glottis, and elevates the floor of the mouth, forcing air into the lungs.

3 Rhythmic ventilation assists in gas exchange.

4 Air is forced out when muscles in the body wall above the lungs contract and the lungs recoil elastically.

with the entrance into the lungs constricted by the glottis. The nostrils are closed, the glottis is opened, and the floor of the mouth is raised, forcing air into the lungs. During this period, the floor of the mouth moves up and down, ensuring mixing of the gases. The nostrils open, and the gases in the lungs are expelled by contraction of muscles on the sides of the frog and the rebound elasticity of the lungs. Rhythmic motions of the floor of the mouth with the nostrils open ensure that the buccal cavity contains fresh air for the beginning of the next cycle. The efficiency of the system is increased because much of the CO_2 is lost through the skin. (Remember from Chapter 37 that frogs have a pulmonary-cutaneous circulation.)

In reptiles, birds, and mammals, the lungs become more folded, with many pockets, increasing the surface for gas exchange. Mammalian lungs consist of millions of tiny air pockets, the **alveoli** (singular, *alveolus*), each surrounded by dense capillary networks. Reptiles and mammals fill their lungs by **negative pressure breathing**, in which muscular contractions expand the lungs, lowering the pressure of the air in the lungs and causing air to be pulled inward. In crocodilians, for example, the contraction of a muscle connecting the liver to the pelvis pulls the liver back, causing the lungs to expand, and another muscle pulls the liver forward, forcing gases out of the lungs. The mechanism in mammals is described in detail in the next section.

In birds, a countercurrent exchange system provides the most complex and efficient vertebrate lungs **(Figure 42.8)**. In addition to paired lungs, birds have nine pairs of air sacs that branch off the respiratory tract. The air sacs, which collectively contain several times as much air as the lungs, are not respiratory surfaces. They set up a pathway that allows air to flow in one direction through the lungs rather than in and out, as in other vertebrates. Within the lungs, air flows through an array of fine, parallel tubes that are surrounded by a capillary network. The blood flows in the direction opposite to the air flow, setting up a countercurrent exchange. The countercurrent exchange allows bird lungs to extract about one-third of the O_2 from the air compared with about one-fourth in the lungs of mammals.

STUDY BREAK

1. What variable in Fick's equation is most affected by the countercurrent mechanism in gas exchange in teleost fish?
2. What is the difference between positive pressure breathing and negative pressure breathing?

42.3 The Mammalian Respiratory System

All mammals have a pair of lungs and a diaphragm in the chest cavity that plays an important role in negative pressure breathing. Rapid ventilation of the respiratory surface and perfusion by blood flow through dense capillary networks maximizes gas exchange.

42.3a The Airways Leading from the Exterior to the Lungs Filter, Moisten, and Warm the Entering Air

The human respiratory system is typical for a terrestrial mammal **(Figure 42.9, p. 1048)**. Air enters and leaves the respiratory system through the nostrils and mouth. Hairs in the nostrils and mucus covering the surface of the airways filter out and trap dust and other large particles. Inhaled air is moistened and warmed as it moves through the mouth and nasal passages.

Next, air moves into the throat or **pharynx**, which forms a common pathway for air entering the **larynx** or "voice box" and food entering the esophagus, which leads to the stomach. The airway through the larynx is open except during swallowing.

From the larynx, air moves into the **trachea**, which branches into two airways, the **bronchi** (singular, *bronchus*). The bronchi lead to the two elastic, cone-shaped lungs, one on each side of the chest cavity. Inside the lungs, the bronchi narrow and branch repeatedly, becoming progressively narrower and more numerous. The terminal airways, the **bronchioles**, lead into cup-shaped pockets, the alveoli; shown in Figure 42.9 insets).

a. Lungs and air sacs of a bird

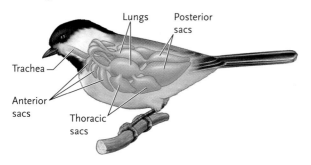

b. Countercurrent exchange

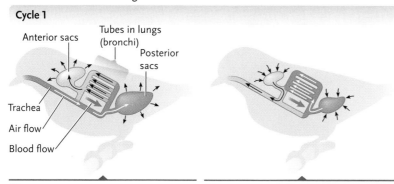

Cycle 1

1 During the first inhalation, most of the oxygen flows directly to the posterior air sacs. The anterior air sacs also expand but do not receive any of the newly inhaled oxygen.

2 During the following exhalation, both anterior and posterior air sacs contract. Oxygen from the posterior sacs flows into the gas-exchanging tubes (bronchi) of the lungs.

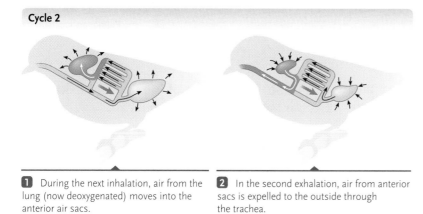

Cycle 2

1 During the next inhalation, air from the lung (now deoxygenated) moves into the anterior air sacs.

2 In the second exhalation, air from anterior sacs is expelled to the outside through the trachea.

Figure 42.8

Countercurrent exchange in bird lungs. **(a)** Unlike mammalian lungs, bird lungs do not expand and contract. Changes in pressure in the expandable air sacs move air in and out. **(b)** Air flows in one direction through the tubes of the lungs; blood flows in the opposite direction in the surrounding capillary network. Two cycles of inhalation and exhalation are needed to move a specific volume of air through the bird respiratory system.

Each of the 150 million alveoli in each lung is surrounded by a dense network of capillaries. By the time inhaled air reaches the alveoli, it has been moistened to the saturation point and brought to body temperature. The many alveoli provide an enormous area for gas exchange. If the alveoli of an adult human were flattened out in a single layer, they would cover an area approaching 100 square metres, about the size of a tennis court! The epithelium of the alveoli is composed of very thin squamous cells. In terms of Fick's law, A is very large and D is minimized.

The tracheae and larger bronchi are nonmuscular tubes encircled by rings of cartilage that prevent the tubes from compressing (recall the analogous but not homologous arrangement in the tracheae of insects). The largest of the rings, which reinforces the larynx, stands out at the front of the throat as the Adam's apple, more prominent in males. The walls of the smaller bronchi and the bronchioles contain smooth muscle cells that contract or relax to control the diameter of these passages and with it the amount of air flowing to and from the alveoli.

The epithelium lining each bronchus contains cilia and mucus-secreting cells. Bacteria and airborne particles such as dust and pollen are trapped in the mucus (see *Molecule Behind Biology*) and then moved upward and into the throat by the beating of the cilia lining the airways. Infection-fighting macrophages (see Chapter 44) also patrol the respiratory epithelium.

42.3b Contractions of the Diaphragm and Muscles between the Ribs Ventilate the Lungs

The lungs are located in the rib cage above the *diaphragm,* a dome-shaped sheet of skeletal muscle separating the chest cavity from the abdominal cavity. The lungs are covered by a double layer of epithelial tissue called the **pleura.** The inner pleural layer is attached to the surface of the lungs, and the outer layer is attached to the surface of the chest cavity. A narrow space between the inner and outer layers is filled with slippery fluid, which allows the lungs to move within the chest cavity without rubbing or abrasion as they expand and contract.

Contraction of the diaphragm and the intercostal muscles between the ribs brings air into the lungs by a negative pressure mechanism. As an inhalation begins, the diaphragm contracts and

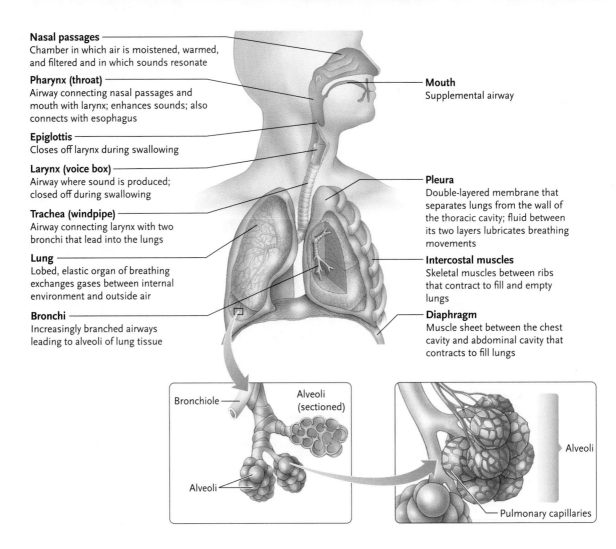

Figure 42.9
The human respiratory system, which is typical for a terrestrial mammal.

Nasal passages
Chamber in which air is moistened, warmed, and filtered and in which sounds resonate

Pharynx (throat)
Airway connecting nasal passages and mouth with larynx; enhances sounds; also connects with esophagus

Epiglottis
Closes off larynx during swallowing

Larynx (voice box)
Airway where sound is produced; closed off during swallowing

Trachea (windpipe)
Airway connecting larynx with two bronchi that lead into the lungs

Lung
Lobed, elastic organ of breathing exchanges gases between internal environment and outside air

Bronchi
Increasingly branched airways leading to alveoli of lung tissue

Mouth
Supplemental airway

Pleura
Double-layered membrane that separates lungs from the wall of the thoracic cavity; fluid between its two layers lubricates breathing movements

Intercostal muscles
Skeletal muscles between ribs that contract to fill and empty lungs

Diaphragm
Muscle sheet between the chest cavity and abdominal cavity that contracts to fill lungs

Bronchiole

Alveoli (sectioned)

Alveoli

Alveoli

Pulmonary capillaries

flattens, and one set of muscles between the ribs, the external intercostal muscles, contracts, pulling the ribs upward and outward **(Figure 42.10)**. These movements expand the chest cavity and lungs, lowering the air pressure in the lungs below that of the atmosphere. As a result, air is drawn into the lungs, expanding and filling them.

The expansion of the lungs is much like filling two rubber balloons. Like balloons, the lungs are elastic and resist stretching as they are filled. And also like balloons, the stretching stores energy that can be released to expel air from the lungs. During an exhalation by a person at rest, the diaphragm and muscles between the ribs relax, and the elastic recoil of the lungs expels the air.

When physical activity increases the body's demand for O_2, contractions of other muscles help expel the air by forcefully reducing the volume of the chest cavity. That is, abdominal wall muscles contract, which increases abdominal pressure. That pressure exerts an upward-directed force on the diaphragm, which is pushed upward. In addition, internal intercostal muscles contract, pulling the chest wall inward and downward, causing it to flatten. As a result, the dimensions of the chest cavity decrease.

42.3c The Volume of Inhaled and Exhaled Air Varies over Wide Limits

The volume of air entering and leaving the lungs during inhalation and exhalation is called the **tidal volume.** In a person at rest, the tidal volume amounts to about 500 mL. As physical activity increases, the tidal volume increases to match the body's demands for O_2; at maximal levels, the tidal volume reaches about 3400 mL in females and 4800 mL in males. This maximum tidal volume is called the **vital capacity** of an individual.

Even after the most forceful exhalation, about 1200 mL of air remains in the lungs in males and about 1000 mL in females; this is the **residual volume** of the lungs. In fact, the lungs cannot be deflated completely because small airways collapse during forced exhalation, blocking further outflow of air. Because air cannot be removed from the lungs completely, some gas exchange can always occur between blood flowing through the lungs and the air in the alveoli.

The respiratory movements are controlled by centres in the medulla and pons, part of the brain stem (see Chapter 33). Nerve signals from these centres to the muscles involved in breathing can vary the intake of air from as little as 5 to 6 L per minute to as much as

MOLECULE BEHIND BIOLOGY

Mucin: Sticky Lubricant

The airways leading to the lungs have a surface coating of mucus secreted by specialized epithelial cells. The mucus traps dust particles, bacteria, and other foreign bodies and is swept upward by the action of cilia to be swallowed into the digestive tract or expelled by the act of blowing your nose or spitting. Mucus is composed of a rodlike protein, mucin, that is very heavily glycosylated (the addition of sugars) after translation and that forms giant polymers up to 10 million Da. The very dense coating of sugar provides a lot of water-holding capacity. This gives mucin its slippery character, which makes it useful as a lubricant. Mucins constitute a family of proteins, and at least 19 genes are known for humans.

Three of these are expressed in airway epithelium. Mucins are also important in the mucus that coats the intestinal epithelium, where mucus serves as a lubricant and protects the epithelium from gastric acids and enzymes. Mucus also lines the reproductive tract. Mucins are important constituents of saliva, keeping the membranes of the oral cavity moist and lubricating the food as it is chewed. Tears contain mucins, continuously washing the eye free of dust. All of these mucins are produced continuously, often in large quantities.

Other mucins, the "tethered" or membrane-associated mucins, are attached to cell membranes, where they serve a variety of functions. In addition to the mucins in tears,

there are also mucins attached to the membranes of the surface of the eye, a slippery surface over which tears can move more easily. The function of the several membrane-associated mucins is far from clear. They may have a role in preventing infection or in cell-to-cell attachment, or in some cases, they may be part of a cell-signalling mechanism.

Mucins are ancient molecules and occur in most metazoan phyla as well as in protists. In insects, a mucin gene is expressed in the intestine, and mucins are known from nematodes. In molluscs, mucins are found in the matrix that contains calcite to form the shell.

---

150 L per minute (for very brief periods). These centres integrate information about O_2 and CO_2 in the blood from O_2 and CO_2 receptors located in special sense organs (the carotid bodies) in the carotid arteries that supply the brain and in the aorta (the **aortic body**) that supplies blood to the rest of the body. These receptors are more sensitive to changes in CO_2: the P_{O_2} must drop below about 100 mm Hg before they are activated. The medulla integrates this information with information coming from its own receptors that monitor the pH of the cerebrospinal fluid. The pH of this fluid is determined mostly by the CO_2 concentration in the

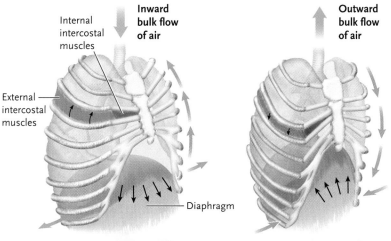

Internal intercostal muscles

External intercostal muscles

Inward bulk flow of air

Outward bulk flow of air

Diaphragm

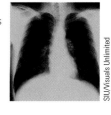

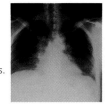

Inhalation. Diaphragm contracts and moves down. The external intercostal muscles contract and lift rib cage upward and outward. The lung volume expands.

SIU/Visuals Unlimited

Exhalation during breathing or rest. Diaphragm and external intercostal muscles return to the resting positions. Rib cage moves down. Lungs recoil passively.

SIU/Visuals Unlimited

Figure 42.10

The respiratory movements of humans during breathing at rest. The movements of the rib cage and diaphragm fill and empty the lungs. Inhalation is powered by contractions of the external intercostal muscles and diaphragm, and exhalation is passive. During exercise or other activities characterized by deeper and more rapid breathing, contractions of the internal intercostal muscles and the abdominal muscles add force to exhalation. The X-ray images show how the volume of the lungs increases during inhalation and exhalation.

blood. (Remember that the pH decreases as CO_2 levels increase.) In general, the CO_2 level is most closely monitored. The O_2 receptors act as a backup system that comes into play only when blood O_2 concentration falls to critically low levels. The level of CO_2 in the blood and body fluids is much more closely monitored and has a much greater effect on breathing than the O_2 level. This reflects the fact that small fluctuations in blood pH have much greater effects on the ability of haemoglobin to carry oxygen and on enzyme activity in the blood and interstitial fluid than fluctuations in the O_2 level.

STUDY BREAK

1. What is the site of gas exchange in human lungs?
2. What is tidal volume?

42.4 Mechanisms of Gas Exchange and Transport

In both the lungs and body tissues, gas exchange occurs when the gas diffuses from an area of higher concentration to an area of lower concentration. In this section, we consider the mechanics of gas exchange between air and the blood in mammals and the means by which gases are transported between the lungs and other body tissues. A major part of this story involves haemoglobin, the vertebrate respiratory pigment.

In the lungs, even though the P_{O_2} is reduced by mixing with the air in the residual volume, it is still much higher than the P_{O_2} in deoxygenated blood entering the network of capillaries in the lungs (Figure 42.11). As a result, O_2 readily diffuses from the alveolar air into the plasma solution in the capillaries.

42.4 a Haemoglobin Greatly Increases the O_2-Carrying Capacity of the Blood

After entering the plasma, O_2 diffuses into erythrocytes, where it combines with haemoglobin. The combination with haemoglobin removes O_2 from the plasma,

lowering the P_{O_2} of the plasma and increasing P1-P2 between alveolar air and the blood. This increases the rate of diffusion of O_2 across the alveoli and into the plasma.

Recall from Chapter 37 that a mammalian haemoglobin molecule has four haem groups, each containing an iron atom that can combine reversibly with an O_2 molecule. A haemoglobin molecule can therefore bind a total of four molecules of O_2. The combination of O_2 with haemoglobin allows blood to carry about 60 times more O_2 (about 200 mL per litre) than it could if the O_2 simply dissolved in the plasma (about 3 mL per litre). About 98.5% of the O_2 in blood is carried by haemoglobin, and about 1.5% is carried in solution in the blood plasma.

The reversible combination of haemoglobin with O_2 is related to the P_{O_2} in a pattern shown by the *haemoglobin–O_2 dissociation curve* in **Figure 42.12**. (The curve is generated by measuring the amount of haemoglobin saturated at a given P_{O_2}.) The curve is S-shaped, with a plateau region, rather than linear. The top, plateau, part of the curve above 60 mm Hg is in the blood P_{O_2} range found in the pulmonary capillaries, where O_2 is binding to haemoglobin. For this part of the curve, the blood remains highly saturated with O_2 over a relatively large range of P_{O_2}. Even at P_{O_2} levels much higher than shown on the graph, only a

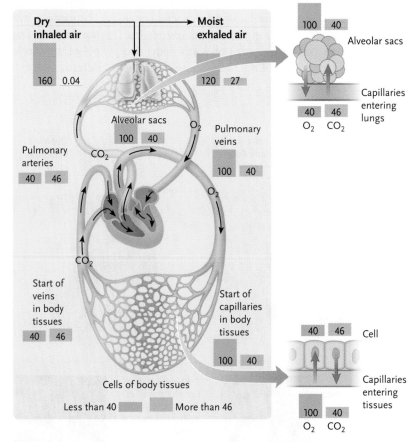

Figure 42.11

The partial pressures of O_2 (pink) and CO_2 (blue) in various locations in the body.

a. Haemoglobin saturation level in lungs

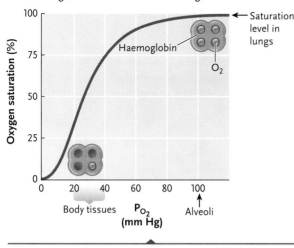

In the alveoli, in which the P_{O_2} is about 100 mm Hg and the pH is 7.4, most haemoglobin molecules are 100% saturated, meaning that almost all have bound four O_2 molecules.

b. Haemoglobin saturation range in body tissues

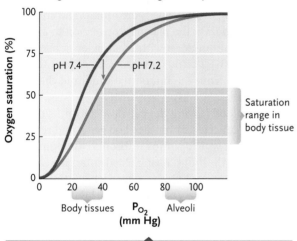

In the capillaries of body tissues, where the P_{O_2} varies between about 20 and 40 mm Hg depending on the level of metabolic activity and the pH is about 7.2, haemoglobin can hold less O_2. As a result, most haemoglobin molecules release two or three of their O_2 molecules to become between 25% and 50% saturated. Note that the drop in pH to 7.2 (red line) in active body tissues reduces the amount of O_2 haemoglobin can hold as a compared with pH 7.4. The reduction in binding affinity at lower pH increases the amount of O_2 released in active tissues.

Figure 42.12
Haemoglobin–O_2 dissociation curves, which show the degree to which haemoglobin is saturated with O_2 at increasing P_{O_2}.

small extra amount of O_2 will bind to haemoglobin. The steep part of the curve between 0 and 60 mm Hg is in the range found in the capillaries in the rest of the body. For this part of the curve, small changes in P_{O_2} result in large changes in the amount of O_2 bound to haemoglobin.

Because the P_{O_2} in alveolar air is about 100 mm Hg, most of the haemoglobin molecules in the blood

leaving the alveolar networks are fully saturated, meaning that most of the haemoglobin molecules have bound four O_2 molecules (see Figure 42.12a). The P_{O_2} of the O_2 in solution in the blood plasma has risen to approximately the same level as in the alveolar air, about 100 mm Hg. The blood has also changed colour, reflecting the bright red colour of oxygenated haemoglobin compared with the darker red colour of deoxygenated haemoglobin.

The oxygenated blood exiting from the alveoli collects in venules, which merge to form the pulmonary veins leaving the lungs. These veins carry the blood to the heart, which pumps the blood through the systemic circulation to all parts of the body.

As the oxygenated blood enters the capillary networks of body tissues, it encounters regions in which the P_{O_2} in the interstitial fluid and body cells is lower than that in the blood, ranging from about 40 mm Hg downward to 20 mm Hg or less (see Figure 42.11b). As a result, O_2 diffuses from the blood plasma into the interstitial fluid and from the fluid into body cells. As O_2 diffuses from the blood plasma into body tissues, it is replaced by O_2 released from haemoglobin.

Several factors contribute to the release of O_2 from haemoglobin, including increased acidity (lower pH) in active tissues. The acidity increases because oxidative reactions release CO_2, which combines with water to form carbonic acid (H_2CO_3). The lowered pH reduces the affinity of haemoglobin for O_2, which is released and used in cellular respiration.

The net diffusion of O_2 from blood to body cells continues until by the time the blood leaves the capillary networks in the body tissues, much of the O_2 has been removed from haemoglobin. The blood, now with a P_{O_2} of 40 mm Hg or less, returns in veins to the heart, which pumps it through the pulmonary arteries to the lungs for oxygenation.

42.4b Carbon Dioxide Diffuses Down Concentration Gradients from Body Tissues and into the Blood and Alveolar Air

The CO_2 produced by cellular oxidations diffuses from active cells into the interstitial fluid, where it reaches a partial pressure of about 46 mm Hg. Because this P_{CO_2} is higher than the 40 mm Hg P_{CO_2} in the blood entering the capillary networks of body tissues, CO_2 diffuses from the interstitial fluid into the blood plasma **(Figure 42.13a, p. 1052)**.

Some of the CO_2 remains in solution as a gas in the plasma. Remember, however, that most of the free CO_2, about 70%, combines with water to produce carbonic acid (H_2CO_3), which dissociates into bicarbonate (HCO_3^-) and H^+ ions. In the erythrocyte, the enzyme carbonic anhydrase accelerates the reaction.

PEOPLE BEHIND BIOLOGY

Peter Hochachka, University of British Columbia

Dr. Peter Hochachka (1937–2002), of the University of British Columbia, spent his career exploring the various biochemical mechanisms that allow animals to exploit extreme environments. Among his interests were the metabolic characteristics of animals in environments low in O_2. He brought Sherpas from the Himalaya and Quechuas from the high Andes as volunteer research subjects to his lab and used positron emission tomography (PET) and magnetic resonance spectroscopy, two techniques that permit the noninvasive characterisation of metabolic activity, as well as magnetic resonance imaging (MRI) to understand their metabolism. He and his collaborators showed that the brains of the volunteers living at high altitudes metabolized O_2 at lower rates. Their hearts relied more on glucose as a fuel, an arrangement that produces more work per O_2 molecule than the greater reliance on fatty acids characteristic of the hearts of people living nearer sea level. Thus, although the increased O_2 binding capacity of haemoglobin is important in animals that live at high altitudes, it is only one of a suite of genetic adaptations to living at high altitudes.

a. Body tissues

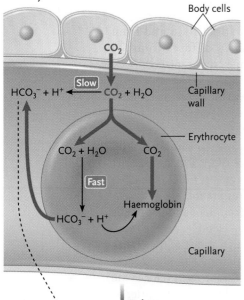

b. Lungs

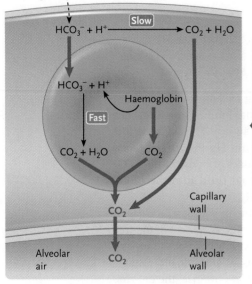

Figure 42.13

The reactions occurring during the transfer of CO_2 from body tissues to alveolar air.

In body tissues, some of the CO_2 released into the blood combines with water in the blood plasma to form HCO_3^- and H^+. However, most of the CO_2 diffuses into erythrocytes, where some combines directly with haemoglobin and some combines with water to form HCO_3^- and H^+. The H^+ formed by this reaction combines with haemoglobin; the HCO_3^- is transported out of erythrocytes to add to the HCO_3^- in the blood plasma.

In the lungs, the reactions are reversed. Some of the HCO_3^- in the blood plasma combines with H^+ to form CO_2 and water. However, most of the HCO_3^- is transported into erythrocytes, where it combines with H^+ released from haemoglobin to form CO_2 and water. CO_2 is released from haemoglobin. The CO_2 diffuses from the erythrocytes and, with the CO_2 in the blood plasma, diffuses from the blood into the alveolar air.

Most of the H^+ ions produced by the dissociation of carbonic acid combine with haemoglobin or with proteins in the plasma, so that the pH is maintained. Note, however, that if CO_2 levels are high, pH will fall, resulting in changes in breathing. The combination of solution in the plasma, conversion to bicarbonate, and combination with haemoglobin operate to maximize P1-P2 of the gaseous CO_2 so that the rate of diffusion from the interstitial fluid into the blood is optimal.

The blood leaving the capillary networks of body tissues is collected in venules and veins and returned to the heart, which pumps it through the pulmonary arteries into the lungs. As the blood enters the capillary networks surrounding the alveoli, the entire process of CO_2 uptake is reversed **(Figure 42.13b)**. The P_{CO_2} in the blood, now about 46 mm Hg, is higher than the P_{CO_2} in the alveolar air, about 40 mm Hg (shown in Figure 42.11). As a result, CO_2 diffuses from the blood and into the air. The diminishing CO_2 concentrations in the plasma, along with the lower pH encountered in the lungs, promote the release of CO_2 from haemoglobin. As CO_2 diffuses away, bicarbonate ions in the blood combine with H^+ ions, forming carbonic acid molecules that break down into water and additional CO_2. This CO_2 adds to the quantities diffusing from the blood into the alveolar air. By the time the blood leaves the capillary networks in the lungs, its P_{CO_2} has been reduced to the same level as that of the alveolar air, about 40 mm Hg.

STUDY BREAK

1. In mammals, which lung tissue serves as a major site of gas exchange?
2. What is the role of pH in gas exchange?

LIFE ON THE EDGE

Prospering in Thin Air

With increasing altitude, atmospheric pressure decreases, and with it, the P_{O_2} also decreases. At an elevation of 5000 M, the atmospheric pressure is about half that at sea level, and the P_{O_2} is thus $380 \times 21/100$ or 80 mm Hg, about half that at sea level. This reduces P1-P2 between the alveolar air and the blood, and, in turn, the supply of O_2 to the tissues is reduced. Humans who normally live at or near sea level and move to higher elevations above about 2000 M experience fatigue, dizziness, and nausea until their systems produce additional erythrocytes, a physiological response to the stress of reduced O_2.

However, there are animals that live at high altitudes, such as the llama (*Lama glama*), from the Andes at about 4000 M, or the bar-headed goose (*Anser indicus*), which migrates over the Himalaya mountains at elevations in excess of 8000 M. The major factor that permits these animals to exploit what is a marginal environment for other animals is a genetic difference in the haemoglobin molecule that produces a higher affinity for O_2. The haemoglobin in these animals shifts the haemoglobin–oxygen dissociation curve in Figure 42.10 to the left so that the haemoglobin is closer to saturation at lower P_{O_2} levels.

Haemoglobin is particularly polymorphic: the gene has a number of alleles, and the alleles present in the animals that can live at very high altitudes produce the appropriate forms of haemoglobin. This is well illustrated by the deer mouse, *Peromyscus maniculatus,* which occupies an extreme range of altitudes from below sea level in Death Valley to above 4300 M in the Sierra Nevada mountains. The populations of deer mice at higher altitudes have alleles of the haemoglobin genes with higher affinities for O_2 than the alleles of mice at low altitudes.

UNANSWERED QUESTIONS

We have become accustomed to thinking exclusively of physiological adaptation as the principal means for humans functioning at high altitudes. The discovery by Hochachka and others that there is a strong genetic component in humans and other animals and that this genetic component is not limited to the haemoglobin gene raises interesting questions. Which genes govern the improved performance at high altitudes? Did these genetic adaptations permit the astonishing feat of Rheinhold Messner, who, together with Peter Habelen in 1978, was the first human to climb to the pinnacle of Mount Everest without the use of supplementary O_2? A few others have matched this heroic achievement, and, indeed, Messner repeated his feat alone in 1980. Messner was born and raised in the mountainous Tyrol region of northern Italy. Although in no way diminishing this extraordinary accomplishment, he may have had some genetic predisposition that permitted his survival, however agonizing, at an altitude where the atmospheric pressure is reduced from 760 mm Hg at sea level to 250 mm and the P_{O_2} is only 53 mm **(Figure 1)**.

The questions surrounding the genetic adaptations that permit survival in extreme environments involve more than scientific curiosity.

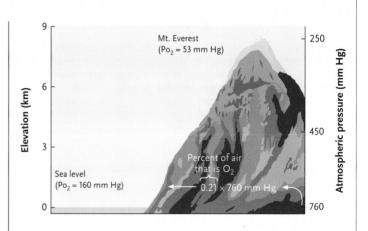

Figure 1
Changes in partial pressure of oxygen with height above sea level.

Humans have begun the exploration of space, and knowledge of the genetic control of those processes that are involved in our ability to tolerate extreme environments will be important.

Review

Go to CENGAGENOW™ at http://hed.nelson.com/ to access quizzing, animations, exercises, articles, and personalized homework help.

42.1 General Principles of Gas Exchange

- The percentage of a gas in a mixture of gases times the atmospheric pressure yields the partial pressure.
- Fick's equation describes the rate of diffusion across a biological membrane. The rate of diffusion is proportional to the product of the area over which diffusion occurs times the difference in partial pressures of the gas on either side of a membrane and is inversely proportional to the distance over which diffusion must occur.
- Gas exchange by simple diffusion is limited to small or flattened organisms.
- In larger animals, respiratory surfaces are increased, and the difference in partial pressures across a membrane is optimized by ventilation and perfusion.
- Water and air, as respiratory media, have different advantages and challenges. Water contains less oxygen, the rate of diffusion is greatly reduced, and its density requires greater energy for ventilation.

42.2 Adaptations for Gas Exchange

- Gills are evaginations of the body surface. Water moves over gills by the beating of cilia or by muscular pumping.
- Water moves over the gills of sharks, bony fishes, and some arthropods, allowing countercurrent exchange to maximize the difference in partial pressure across the respiratory surface.
- Insects have a tracheal system that brings air directly to every cell.
- Lungs are invaginated body surfaces with greatly expanded respiratory surfaces. They may be ventilated by positive pressure breathing, in which air is forced into the lungs, or by negative pressure breathing. Negative pressure breathing relies on muscles that alternately increase or decrease the pressure within the body cavity.

42.3 The Mammalian Respiratory System

- Air enters and leaves the lungs via the nostrils and mouth leading to the trachea, which branch into two bronchi. The bronchi branch many times into broncioles leading to alveoli, which are surrounded by blood capillaries.
- Mammals rely on a negative pressure mechanism. The tidal volume is the volume of air moved in and out of the lungs during normal breathing. The vital capacity is the volume that can be moved in and out by breathing as deeply as possible. The residual volume is the volume remaining in the lungs after exhaling as much as possible.
- Breathing is controlled by centres in the brain reacting to sensors for CO_2 and O_2 in the carotid arteries. The concentration of CO_2 has the greatest influence.

42.4 Mechanisms of Gas Exchange and Transport

- The site of gas exchange in mammals is at the alveolar surface. The P_{O_2} in the alveolar air is greater than that in the blood, causing O_2 to diffuse into the blood and enter the erythrocytes, where it is bound by haemoglobin, thus maximizing the difference in partial pressure of that gas between the air and the blood.
- In the tissues, the reverse is true, and O_2 leaves the blood for the tissues
- The P_{CO_2} is higher in the tissues than in the blood; P_{CO_2} leaves the tissues and dissolves in the plasma and enters the erythrocytes, where most of it is converted into H^+ and HCO_3^- and released back into the plasma. This causes a slight drop in pH, promoting the release of O_2 from haemoglobin. The remaining CO_2 combines with haemoglobin. At the alveolar surface, the P_{CO_2} in the alveolar cells and the air in the lungs is lower than that in the blood, the HCO_3^- releases its CO_2, and CO_2 flows down the gradient.

Questions

Self-Test Questions

1. Which statement is not true?
   a. The partial pressure of O_2 increases with altitude.
   b. The rate of diffusion of a gas increases with an increase in the difference of partial pressures on either side of a membrane.
   c. Countercurrent flow optimizes the difference in partial pressures of a gas across a respiratory surface.
   d. Gases are in aqueous solution at respiratory surfaces.
   e. The solubility of O_2 decreases with an increase in temperature.

2. Tracheal systems are characterized by
   a. closed tubes that circulate gases.
   b. uncontrolled diffusion of gases between the atmosphere and the tissues.
   c. the transport of respiratory gases directly to every cell.
   d. positive pressure breathing.
   e. CO_2 sensors in the segmental ganglia.

3. Which of the following statements is correct? (More than one may be correct.)
   a. The concentration of O_2 in water rises with increasing temperature.
   b An advantage of breathing in air is the reduced energy required to move the gases over the respiratory surface.
   c Birds use a one-way countercurrent flow through their lungs for gas exchange.
   d. O_2 receptors in the medulla have the greatest influence on mammalian breathing.
   e. CO_2 receptors in the carotid body influence mammalian breathing.

4. The Olympic speed skating champion Catriona LeMay Doan is finishing her last lap. At this time,
   a. the diaphragm and rib muscles contract when she exhales.
   b. positive pressure brings air into her lungs.

    c.   her lungs undergo an elastic recoil when she inhales.
    d.   her tidal volume is at vital capacity.
    e.   her residual volume momentarily reaches zero.

5.   The haemoglobin–O_2 dissociation curve
    a.   reflects about 50% dissociation in the alveoli.
    b.   shifts to the left when pH rises.
    c.   shows that haemoglobin holds less O_2 when the pH rises.
    d.   shows a lack of dependence on CO_2 levels.
    e.   explains how haemoglobin can bind more pH in the lungs and release it at the tissues where the pH is lower.

Questions for Discussion

1.   The ability to live at high elevations appears to have genetic components beyond the properties of the blood in many animals. What experiments can you devise to explore this possibility? (Hint: some animals have high and low elevation populations.)

2.   Hospital patients frequently have a small ring on the end of a finger that shines a red light from the pad of tissue to a detector on the fingernail. What do you think this apparatus measures, and why is it important to measure it continuously? What factors could change the value of this measurement?

3.   The control of the spiracular opening in insects is assumed to be important in avoiding water loss during gas exchange. Suggest an experiment to test this hypothesis. How does this mechanism differ from that controlling the opening of the stomata in plants?

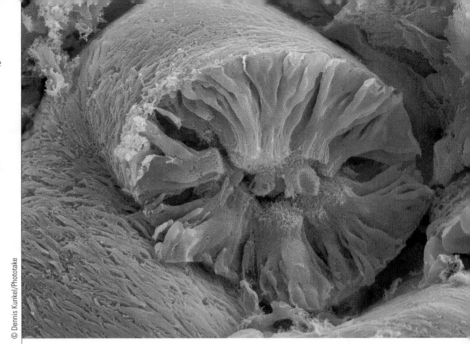

A nephron in a human kidney (colourized scanning electron micrograph). Nephrons are the specialized tubules in kidneys that filter the blood to conserve nutrients and water, balance salts in the body, and concentrate wastes for excretion from the body.

© Dennis Kunkel/Phototake

43 Regulating the Internal Environment

WHY IT MATTERS

In the Miramichi River of New Brunswick, an Atlantic salmon (*Salmo salar*) has spent two or three years growing from an egg to a fish about 10 to 25 cm in length. In the spring, as day length increases and the water temperature begins to rise, it undergoes a number of physiological, morphological, and behavioural changes. It loses some of its mottled coloration, becoming silvery in appearance, and joins other similar fish in migrating downstream. It may pause at the **estuary** of the river for a day or two, but then it abandons its freshwater environment and enters the sea, where it will remain for two or more years, feeding and growing to maturity. Eventually, it will return to the Miramichi to spawn **(Figure 43.1, p. 1058)**.

When the salmon leaves the freshwater of the river and enters the salt water of the North Atlantic Ocean, it moves from an environment in which the total concentration of solutes is about 0.1% to one with a concentration of about 3.5%. Its own body fluids contain about 1% of solutes. The salmon moves, over the course of a few days, to an

Figure 43.1
An adult Atlantic salmon, ready to return to its river of birth.

environment in which the solute concentration is more than three times greater than the concentration of its body fluids. It thus faces a continuous loss of water. However, the physiological changes that accompany its morphological changes allow it to prosper in an environment that is potentially hostile.

Organisms, particularly those on land, are subject to seasonal and shorter-term fluctuations in their external environment that present challenges for them to maintain not only the integrity of their internal fluids but also the functioning of all the systems that sustain life. The maintenance of a steady internal environment, called **homeostasis** (introduced in Chapter 32), is the subject of this chapter. During evolution, a variety of physiological and behavioural mechanisms have appeared that have permitted organisms to exploit environments that may be highly variable.

In the background are those mechanisms that accompanied the emergence from aqueous to terrestrial environments. There are some important differences between aqueous and terrestrial environments. Water, essential to life, is obviously more abundant in aqueous environments. But aqueous environments may contain greater or lesser amounts of solutes than the body fluids do, posing different problems for homeostasis of the body fluids. Terrestrial environments require mechanisms to conserve water, but they are also subject to much greater variation in temperature. The temperature of aqueous environments is seldom greater than about 25°C, and the lower limit is a little above −2°C (salt water freezes at about this temperature). By contrast, organisms in the northern temperate zone of Canada, for example, encounter temperatures that may range between approximately 40°C and −40°C.

43.1 Introduction to Osmoregulation and Excretion

Living cells contain water, are surrounded by water, and constantly exchange water with their environment. The water of the external environment directly surrounds the cells of the simplest animals. For more complex animals, an aqueous extracellular fluid surrounds the cells and is separated from the external environment by a body covering. In animals with a circulatory system, the extracellular fluid includes both the blood and the interstitial fluid immediately surrounding the cells.

In this section, we review the mechanisms cells use to exchange water and solutes with the surrounding fluid through *osmosis*. We also look at how animals harness osmosis to regulate their internal *water balance,* the equilibrium between the inward and the outward flow of water.

43.1a Osmosis: Passive Diffusion

In osmosis (see Chapter 5), water molecules move across a selectively permeable membrane (one that lets water through but excludes most solutes) from a region where they are more highly concentrated to a region where their concentration is lower. The difference in water concentration is produced by different numbers of solute molecules or ions on each side of the membrane. The side of the membrane with a *lower* solute concentration has a *higher* concentration of water molecules, so water moves osmotically to the other side, where water concentration is *lower.* Selective permeability is a key factor in osmosis because it helps maintain differences in solute concentration on either side of biological membranes. Proteins are among the most important solutes in establishing the conditions producing osmosis.

The osmotic concentration of a solution, measured in *osmoles,* is determined by the total molal concentration of solute particles, both molecules and ions. A nonionic solute such as glucose contributes one osmole of osmotic concentration per mole. A fully dissociated ionic solute, such as NaCl, however, contributes two osmoles per mole of the salt. The osmotic concentration (**osmotic pressure**) of a solution, called its **osmolality**, is the number of osmoles per kilogram of solute. The molal standard is used rather than the molar (moles per litre of solution) partly because the volume of water changes with temperature.

Because of the complexity of biological fluids, their osmotic concentration is normally measured rather than calculated. The principles that determine osmotic concentration also determine the depression of the freezing point (or elevation of the boiling point) and lowering of the vapour pressure of a solution. Freezing point and vapour pressure of biological fluids are easily measured and calibrated to standard solutions of known osmotic concentration. Because the total solute concentration in the body fluids of most animals is less than 1 osmole, osmolality is usually expressed in thousandths of an osmole, or *milliosmoles* (mOsm) per kilogram. As shown in **Figure 43.2,** the osmolality of body fluids in mammals (including humans) is about 300 mOsm/kg; osmolality in a flounder, a marine teleost (bony fish), is about 330 mOsm/kg, and in a goldfish, a freshwater teleost, it is about 290 mOsm/kg. The relatively low osmotic concentration in marine teleosts reflects their evolutionary history. Early marine teleosts invaded freshwater and prospered there. During the extensive radiation of the group in freshwater, many have reinvaded the marine habitat. By contrast, sharks and many marine invertebrates such as lobsters have osmolalities close to that of seawater, about 1000 mOsm/kg, whereas

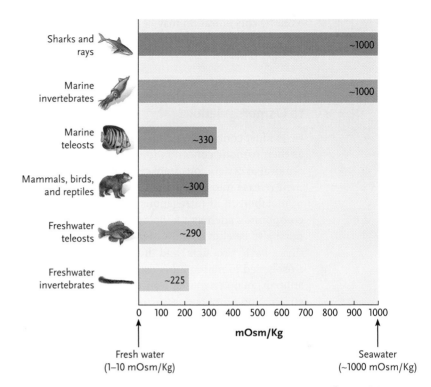

Figure 43.2
Osmolality of body fluids in some animal groups.

freshwater invertebrates have an osmolality of about 225 mOsm/kg.

A solution of higher osmolality on one side of a selectively permeable membrane is said to be *hyperosmotic* to a solution of lower osmolality on the other side, and a solution of lower osmolality is said to be *hypoosmotic* to a solution of higher osmolality. If the solutions on either side of a membrane have the same osmotic concentrations, they are *isoosmotic.* Water moves across the membrane between solutions that differ in osmolality, whereas when two solutions are isoosmotic, no *net* water movement occurs, although water exchanges from one side to the other.

43.1b Animals Use Different Approaches to Regulate Osmosis

Because even small differences in osmotic concentration can cause cells to swell or shrink, animals must keep their cellular and extracellular fluids isoosmotic. In some animals, called **osmoconformers**, the osmotic concentrations of the cellular and extracellular solutions simply match that of the environment.

Many marine invertebrates are osmoconformers: when placed in dilute seawater, the osmotic concentration of their body fluids decreases, and their weight increases as a result of the osmotic influx of water. Other animals, called **osmoregulators**, use control mechanisms to keep the osmolality of cellular and extracellular fluids constant but at levels that may differ from the osmolality of the surroundings. Most freshwater and terrestrial invertebrates, and almost all vertebrates, are osmoregulators. It is important to recognize that the various solutes contributing to the osmotic concentration may be at different concentrations inside the cell, in the extracellular fluids, and in the environment.

43.1c Excretion Is Closely Tied to Osmoregulation

Cells must control their ionic and pH balance as well as their osmotic concentration. This may require the removal of certain ions from cells and body fluids and their **release** into the environment. The end products of metabolism of nitrogenous (nitrogen-containing) compounds such as amino acids and nucleic acids must also be eliminated. Water serves as a **solvent** for these waste products, and their elimination is thus closely tied to maintaining osmolality. For terrestrial animals, maintenance of osmotic concentration while eliminating nitrogenous wastes is a challenge, particularly since many animals may confront wet and dry seasonal conditions.

43.1d Animals Excrete Nitrogen Compounds as Metabolic Wastes

The metabolism of ingested food is a source of both energy and molecules for the biosynthetic activities of an animal. Importantly, metabolism of ingested food produces water, *metabolic water,* that is used in chemical reactions as well as being involved in physiological processes such as the excretion of wastes.

The proteins, amino acids, and nucleic acids in food are continually broken down as part of digestion (see Chapter 41) and from the constant turnover and replacement of these molecules in body cells. The nitrogenous products of this breakdown are excreted by most animals as *ammonia, urea,* or *uric acid* or a combination of these substances **(Figure 43.3).** The particular molecule or combination of molecules depends on a balance among toxicity, water conservation, and energy requirements.

Ammonia. Ammonia (NH_3) results from the metabolism of amino acids and proteins and is highly toxic: it can be safely transported and excreted from the body only in dilute solutions. Those animals with a plentiful supply of water, such as aquatic or marine invertebrates, teleost fish, and larval amphibians, excrete ammonia as their primary nitrogenous waste. Other animals detoxify ammonia by converting it to urea or uric acid.

Urea. All mammals, most amphibians, some reptiles, some marine fishes, and some terrestrial invertebrates combine ammonia with HCO_3^- and convert the product in a series of steps to *urea,* a soluble substance that is less toxic than ammonia. Although producing urea requires more energy than forming ammonia, excreting urea instead of ammonia requires much less water.

Uric Acid. Water is conserved further in some animals, including many terrestrial invertebrates, reptiles, and birds, by the formation of uric acid instead of ammonia or urea. Uric acid is nontoxic, but its great advantage is its low solubility. During the concentration of the urine in the final stages of its formation, the uric acid precipitates as crystals that can be expelled with minimal water. (The white substance in bird droppings is uric acid.) The embryos of reptiles and birds, which develop within leathery or hard-shelled eggs that are impermeable to liquids, also conserve water by forming uric acid, which is stored as a waste product inside the shell. Similarly, the pupae of insects store uric acid in the rectum.

Many animals have the capacity to form all three products of nitrogen metabolism. Mammalian urine, for example, contains small amounts of uric acid, although urea predominates. Some tree frogs, such as *Phyllomedusa sauvagei,* have uric acid as their principal excretory product. This has enabled them to exploit the woodlands of South America, where the dry season is extremely arid. Conversely, the American cockroach, *Periplaneta americana,* an insect that normally lives

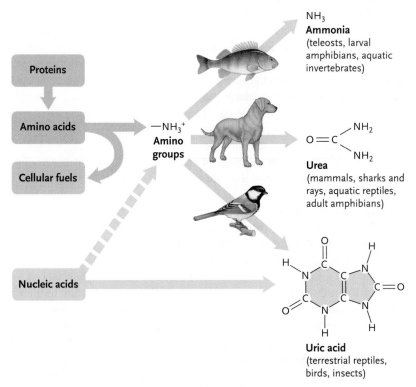

Figure 43.3

Nitrogenous wastes excreted by different animal groups. Although humans and other mammals primarily excrete urea, they also excrete small amounts of ammonia and uric acid.

in damp environments, uses ammonia as its primary excretory product and stores uric acid in special cells during periods when water is less available.

Sharks and rays maintain their osmotic concentration near that of their marine environment by retaining urea in their tissues and blood.

In the following sections, we look at the specifics of **osmoregulation** and excretion in different animal groups, beginning with the invertebrates.

STUDY BREAK

1. Define osmolality.
2. What are the sources of ammonia, urea, and uric acid in excretory products?

43.2 Osmoregulation and Excretion in Invertebrates

Both osmoconformers and osmoregulators occur among the invertebrates, and most carry out excretion by specialized excretory structures.

43.2a Osmoconformers and Osmoregulators

Many marine invertebrates (sponges, cnidarians, some molluscs, and echinoderms) are osmoconformers. If placed in dilute solutions of seawater, they increase in weight because of the entry of water. They release nitrogenous wastes, usually in the form of ammonia, directly from body cells to the surrounding seawater. The cells of these animals do not normally swell or shrink because the osmotic concentrations of their intracellular and extracellular fluids and the surrounding seawater are the same, about 1000 mOsm/kg. Although they do not expend energy to maintain their osmolality, osmoconformers do expend energy to keep some ions, such as Na^+, at concentrations different from the concentration in seawater.

In general, the invertebrates that spend their entire lives in the open sea, where the environment is osmotically stable, have very little capacity for osmoregulation. Thus, many marine molluscs, such as squid and octopus, are osmoconformers, as are most marine arthropods, such as lobsters.

Other marine invertebrates are more diverse in their responses to variations in the osmotic concentration of the environment. Invertebrates living in the intertidal zone or at the mouths of tidal rivers experience regular changes in the osmotic concentration of their environment. Some marine annelids and arthropods that live in such environments are capable of short-term osmoregulation, slowing or delaying the changes in osmotic concentration of their body fluids that result from dilution of seawater by the outflow from rivers **(Figure 43.4)**. Some animals that live in the intertidal zone

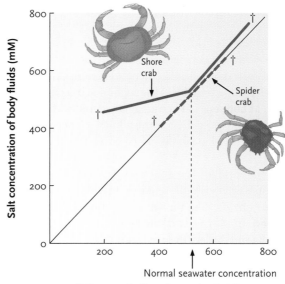

Figure 43.4
The concentration of the body fluids of two crabs when immersed in seawater of different concentrations. The spider crab lives in the sea, where it is not exposed to variation in the concentration of the water. The concentration of its fluids follows the 45° line that marks equivalency between the seawater and the body fluid concentration. By contrast, the shore crab, which lives in the intertidal zone and in estuaries, can regulate its body fluids to some degree. The crosses at the end of each line indicate the concentration at which each crab dies. Note that osmotic concentrations were not measured in this experiment.

use behavioural responses, such as closing their shells (mussels and clams) or retreating into burrows (some annelids) to avoid desiccation when the tide is out.

By contrast, all freshwater and terrestrial invertebrates are osmoregulators. Those that live in aquatic environments are faced with a potential influx of water, diluting their body fluids. Although osmoregulation is energetically expensive, these invertebrates can live in more varied habitats than osmoconformers can.

The internal hyperosmocity of freshwater osmoregulators such as flatworms and mussels causes water to move constantly from the surroundings into their bodies. This excess water must be excreted, at a considerable cost in energy, to maintain this hyperosmocity. These animals obtain the salts they need from foods and by actively transporting salt ions from the water into their bodies (even freshwater contains some dissolved salts). This active ion transport occurs through the body surface or gills.

Terrestrial osmoregulators include annelids (earthworms), arthropods (insects, spiders and mites, millipedes, and centipedes), and molluscs (land snails and slugs). Although they do not have to excrete water entering by osmosis, they must constantly replace water lost from their bodies by evaporation and by excretion. Most obtain water from their food, and some drink water. Like their freshwater relatives, these invertebrates must obtain salts from their surroundings, usually in their foods.

43.2b Specialized Excretory Tubules Participate in Osmoregulation

Most invertebrates (except marine osmoconformers) use specialized tubular structures for carrying out excretion. These include *protonephridia* in flatworms and larval molluscs, *metanephridia* in annelids and most adult molluscs, and *Malpighian tubules* in insects and other arthropods. In protonephridia, the excretory tubules are open only at one end. Body fluids do not enter protonephridia directly. An *ultrafiltrate* enters the tubule through narrow extracellular spaces that permit only small molecules to enter and exclude larger molecules such as proteins. Metanephridia, by contrast, are open at both ends. They are characteristic of animals with coeloms, and the coelomic fluid is already an ultrafiltrate of the blood in the closed circulatory system.

Protonephridia. The flatworm *Dugesia* provides an example of the simplest form of invertebrate excretory tubule, the *protonephridium* (*proto* = before; *nephros* = kidney). In *Dugesia*, two branching networks of protonephridia run the length of the body **(Figure 43.5).** The cell at the blind end of each tubule has a bundle of cilia on its inner surface. The synchronous beating of the cilia resembles the flickering of a flame, and these cells are called *flame cells*. The cilia help draw a filtrate of body fluids through very small spaces between the cell membranes of the flame cell and those of the adjacent tubule cell and propel the filtrate along the tubule. As the fluids pass along the tubule, some molecules and ions are reabsorbed, whereas others are secreted into the tubules. The urine resulting from this filtration system is released through pores that connect the network of protonephridia to the body surface. The principal nitrogenous excretory product is ammonia. Although some ammonia passes out in the urine, most of it passes through the body wall.

Metanephridia. Animals with metanephridia have coelomic cavities (see Chapter 26) that are separate from the circulatory system. The fluid in the coelom is a filtrate of the haemolymph or blood. Coelomic fluid enters the proximal end of the excretory tubule, and ions and other solutes are reabsorbed or secreted as the fluid moves

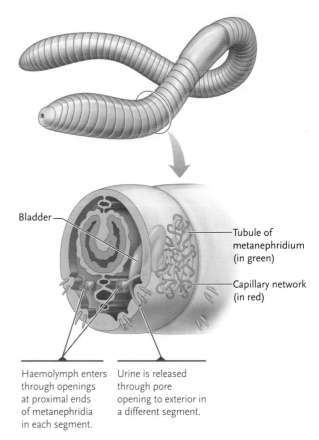

Bladder

Tubule of metanephridium (in green)

Capillary network (in red)

Haemolymph enters through openings at proximal ends of metanephridia in each segment.

Urine is released through pore opening to exterior in a different segment.

Figure 43.6
A metanephridium of an earthworm.

along the tubule. In annelids, the **metanephridium (Figure 43.6)** is a segmental structure. The proximal ends of a pair of metanephridia are located in each body segment, one on each side of the animal. A funnel-like opening surrounded by cilia admits coelomic fluid. Each tubule of the pair extends into the following segment, where it bends and folds into a convoluted arrangement surrounded by a network of blood vessels. Reabsorption and secretion of specific molecules and ions take place in the convoluted section. Urine from the distal end of the tubule collects in a saclike storage organ, the *bladder*, from where it is released through a pore in the surface of the segment.

Malpighian Tubules. The excretory tubules of insects, the *Malpighian tubules*, have a closed proximal end that is immersed in the haemolymph **(Figure 43.7).** The distal ends of the tubules empty into the gut. The fluid in the tubules results primarily from secretion, although in some insects, an ultrafiltrate of the haemolymph may enter the upper part of the tubule through extracellular spaces. In particular, uric acid and several ions, including Na^+ and K^+, are actively secreted into the tubules. As the concentration of these substances rises, water moves osmotically from the haemolymph into the tubule. The fluid then passes into the hindgut (intestine and rectum) of the insect as dilute urine. Cells in the hindgut wall actively reabsorb most of the Na^+ and K^+ back into the haemolymph, and water

Figure 43.5
The protonephridia of the planarian *Dugesia*, showing the flame cells.

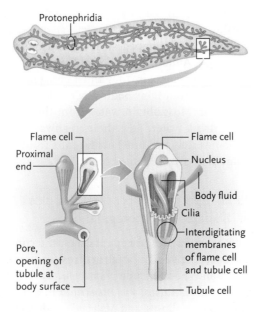

Protonephridia

Flame cell

Proximal end

Flame cell

Nucleus

Body fluid

Cilia

Interdigitating membranes of flame cell and tubule cell

Pore, opening of tubule at body surface

Tubule cell

LIFE ON THE EDGE

Surviving Drought

Some organisms live in temporary aquatic environments, consisting of ponds that dry up completely during periods of prolonged drought. One survival strategy involves complete desiccation of the animal, leading to anhydrobiosis (life without water). For example, the aquatic larvae of a midge (a small dipteran), *Polypedilum vanderplanki* **(Figure 1)**, inhabit pools in Africa that can dry up completely for long periods. The larvae construct nests of mud, but within these nests, their water content is almost completely eliminated, and signs of life are absent. These desiccated larvae can withstand exposure to temperatures as low as −270°C and as high as +106°C. Immersed in water, the larvae recover within less than an hour, even after as long as 17 years of life without water. The precise mechanisms are not fully understood, but the animals accumulate high concentrations of the disaccharide trehalose as they enter the anhydrobiotic state. This sugar is thought to form a glasslike structure that prevents the formation of crystals that would damage the cells. Similar mechanisms are known in yeasts and other microorganisms.

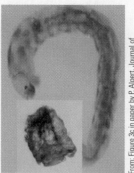

From: Figure 3c in paper by P. Alpert, Journal of Experimental Biology vol 209 p 1579, 2006.

Figure 1
The larva of *Polypedilum vanderplanki* in its active state and (inset) fully desiccated.

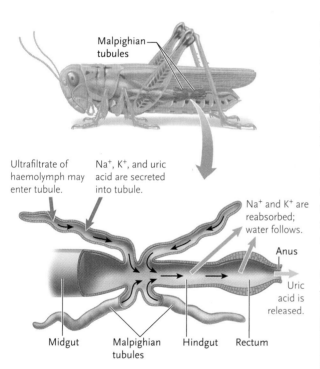

Figure 43.7
Excretion through Malpighian tubules in a grasshopper.

follows by osmosis. The uric acid left in the gut precipitates into crystals, which mix with the undigested matter in the rectum and are released with the faeces. This arrangement is important in conserving water.

Hydrostatic Skeleton. Nematodes have a hydrostatic skeleton that requires the maintenance of hydrostatic pressure in the body fluids against which the muscles can contract (see Chapter 26). Nematodes live in a wide range of environments, including marine, freshwater, terrestrial, and parasitic. They must maintain the osmotic concentration of their body fluids: a net entry of water would increase hydrostatic pressure, and a net loss of water would reduce the pressure. Either of these would make it difficult for the animal to move. Most nematodes have a system of two or three cells with ducts that run the length of the animal and open to the outside. These function in osmoregulation in some nematodes but not in others.

The body wall is important in osmoregulation. Researchers have cut parasitic cod worms and made sausagelike sacs by removing the intestine and closing the cut ends with ligatures. The sacs are capable of maintaining the internal osmotic concentration in environments of different osmotic concentrations.

STUDY BREAK

1. How does a protonephridium differ from a metanephridium?
2. What is the excretory product of most insects? How does it get into the urine?

43.3 Osmoregulation and Excretion in Mammals

In all vertebrates, specialized excretory tubules contribute to osmoregulation and excretion. The excretory tubules, called **nephrons**, are located in a specialized organ, the kidney. We begin our survey of vertebrate osmoregulation and excretion with a description of the structure and function of the mammalian kidney.

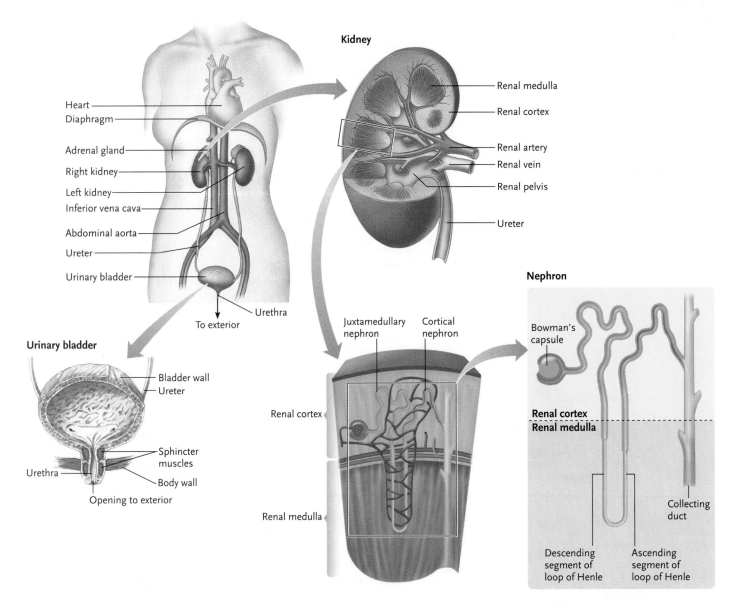

Figure 43.8
Human kidneys and urinary system in a female.

Labels in figure:

Body diagram:
Heart
Diaphragm
Adrenal gland
Right kidney
Left kidney
Inferior vena cava
Abdominal aorta
Ureter
Urinary bladder
Urethra
To exterior

Kidney:
Renal medulla
Renal cortex
Renal artery
Renal vein
Renal pelvis
Ureter

Urinary bladder:
Bladder wall
Ureter
Sphincter muscles
Urethra
Body wall
Opening to exterior

Nephron cross-section:
Juxtamedullary nephron
Cortical nephron
Renal cortex
Renal medulla

Nephron:
Bowman's capsule
Renal cortex
Renal medulla
Collecting duct
Descending segment of loop of Henle
Ascending segment of loop of Henle

43.3a The Kidneys, Ureters, Bladder, and Urethra Constitute the Urinary System

Mammals have a pair of kidneys, located on each side of the vertebral column at the dorsal side of the abdominal cavity **(Figure 43.8)**. Internally, the mammalian kidney is divided into an outer **renal cortex** surrounding a central region, the **renal medulla**.

A **renal artery** carries blood to each kidney, where metabolic wastes and excess ions are moved into the urine by the action of the nephrons. The blood is routed away from the kidney by the **renal vein**. The urine leaving individual nephrons is processed further in **collecting ducts** and then drains into a central cavity in the kidney called the **renal pelvis**.

From the renal pelvis, the urine flows through a tube called the **ureter** to the **urinary bladder**, a storage sac located outside the kidneys. Urine leaves the bladder through another tube, the **urethra**, which opens to the outside. Two sphincter muscles control the flow of urine

from the bladder to the urethra. In human females, the opening of the urethra is just in front of the vagina; in males, the urethra opens at the tip of the penis. The two kidneys and ureters, the urinary bladder, and the urethra constitute the mammalian urinary system.

43.3b Regions of Nephrons Have Specialized Functions

Mammalian nephrons are differentiated into regions that perform successive steps in excretion. At its proximal end, a nephron forms **Bowman's capsule**, an infolded region that cups around a ball of blood capillaries called the **glomerulus (Figure 43.9)**. The capsule and glomerulus are located in the renal cortex. Filtration takes place as fluids are forced into Bowman's capsule from the capillaries of the glomerulus.

Following Bowman's capsule, the nephron forms a **proximal convoluted tubule** in the renal cortex, which descends into the medulla in a U-shaped bend called

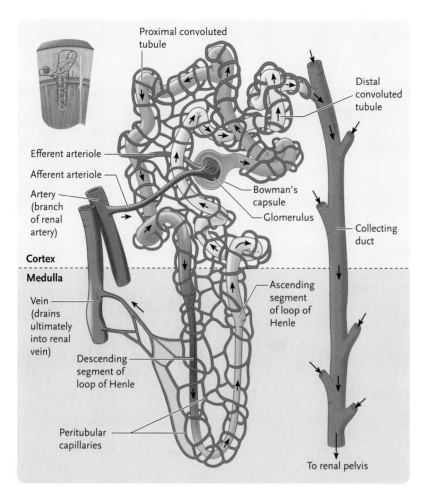

Figure 43.9
A nephron and its blood circulation.

Proximal convoluted tubule

Distal convoluted tubule

Efferent arteriole

Afferent arteriole

Artery (branch of renal artery)

Bowman's capsule

Glomerulus

Collecting duct

Cortex

Medulla

Vein (drains ultimately into renal vein)

Ascending segment of loop of Henle

Descending segment of loop of Henle

Peritubular capillaries

To renal pelvis

the **loop of Henle** and then ascends again to form a **distal convoluted tubule.** The distal tubule drains the urine into a branching system of collecting ducts that lead to the renal pelvis. As many as eight nephrons may drain into a single branch. The combined activities of the proximal convoluted tubule, the loop of Henle, the distal convoluted tubule, and the collecting duct convert the filtrate that entered the nephron at the Bowman's capsule into urine.

Unlike most capillaries in the body, the capillaries in the glomerulus do not lead directly to venules. Instead, they form another arteriole that branches into a second capillary network called the **peritubular capillaries.** These capillaries thread around the proximal and distal convoluted tubules and the loop of Henle. Some molecules and ions are reabsorbed into the peritubular capillaries, whereas others are secreted from the blood into the nephron. However, because the capillaries and the tubules are not in physical contact due to the interstitial fluid between them, this transfer is not direct. Instead, the molecules or ions leave the tubule by passing through the one-cell-thick wall, diffuse through the interstitial fluid, and then pass into the capillary through its wall.

Each human kidney has more than a million nephrons. Of these, about 20% (the *juxtamedullary*

nephrons) have long loops that descend deeply into the medulla of the kidney. The remaining 80% (the *cortical nephrons*) have shorter loops, most of which are located entirely in the cortex, and the remainder of which extend only partway into the medulla.

43.3c Nephrons and Other Kidney Structures Produce Hyperosmotic Urine

In mammals, urine is hyperosmotic to body fluids. Except for a few aquatic bird species, all other vertebrates produce urine that is hypoosmotic to body fluids or is at best isoosmotic. Production of hyperosmotic urine, a water-conserving adaptation, involves the activities of the mammalian nephron and an interaction between nephrons and the highly ordered structure of the mammalian kidney. Three features interact to conserve nutrients and water, balance salts, and concentrate wastes for excretion from the body:

- the arrangement of the loop of Henle, which descends into the medulla and returns to the cortex again;
- differences in the permeability of successive regions of the nephron, established by a specific group of membrane transport proteins in each region; and
- a gradient in the concentration of molecules and ions in the interstitial fluid of the kidney, which increases gradually from the renal cortex to the deepest levels of the renal medulla.

Researchers determined the transport activities of specific regions of nephrons by dissecting them out of an animal and experimentally manipulating them in vitro. They placed segments in different buffered solutions and passed solutions containing various components of filtrates through the segment. By labelling specific molecules or ions radioactively, the scientists followed the movements of molecules in the solution surrounding the nephron segment or in the filtrate.

43.3d Filtration in Bowman's Capsule Begins the Process of Excretion

The mechanisms of excretion (**Figure 43.10, p. 1066,** and summarized in **Table 43.1, p. 1066,**) begin in Bowman's capsule. The cells forming the walls of the capillaries surrounding Bowman's capsule, and the cells of the

Figure 43.10

The movement of ions, water, and other molecules to and from nephrons and collecting tubules in the human kidney. Nephrons in other mammals and in birds work in similar fashion. The numbers are osmolality values in mOsm/kg.

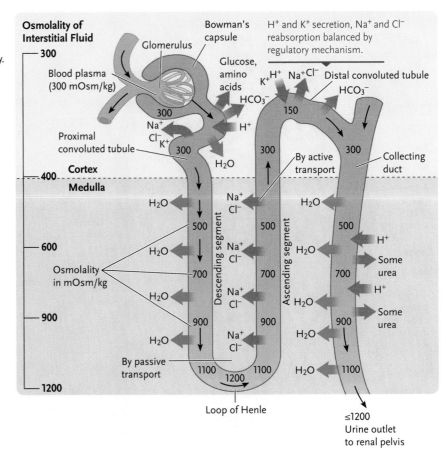

Table 43.1		Filtration, Reabsorption, and Secretion in Nephrons and Collecting Ducts		
Segment	Location	Permeability and Movement	Osmolality of Filtrate and Urine	Result of Passage
Bowman's capsule	Cortex	Water, ions, small nutrients, and nitrogenous wastes move through spaces between epithelia	300 mOsm/kg, same as surrounding interstitial fluid	Water and small substances, but not proteins, pass into nephron
Proximal convoluted tubule	Cortex	Na$^+$ and K$^+$ actively reabsorbed, Cl$^-$ follows; water leaves through aquaporins; H$^+$ actively secreted; HCO$_3^-$ reabsorbed into plasma of peritubular capillaries; glucose, amino acids, and other nutrients actively reabsorbed	300 mOsm/kg	67% of ions, 65% of water, 50% of urea, and all nutrients return to interstitial fluid; pH maintained
Descending segment of loop of Henle	Cortex into medulla	Water leaves through aquaporins; no movement of ions or urea	From 300 mOsm/kg at top to 1200 mOsm/kg at bottom of loop	Additional water returned to interstitial fluid
Ascending segment of loop of Henle	Medulla into cortex	Na$^+$ and Cl$^-$ actively transported out; no entry of water; no movement of urea	From 1200 mOsm/kg at bottom to 150 mOsm/kg at top of loop	Additional ions returned to interstitial fluid
Distal convoluted tubule	Cortex	K$^+$ and Na$^+$ secreted via active transport into urine; Na$^+$ and Cl$^-$ reabsorbed; water moves into urine through aquaporins; HCO$_3^-$ reabsorbed into plasma of peritubular capillaries	From 150 mOsm/kg at beginning to 300 mOsm/kg at junction with collecting duct	Ion balance, pH balance
Collecting ducts	Cortex through medulla, empties into renal pelvis	Water moves out via aquaporins; no movement of ions; some urea leaves at bottom of duct	From 300 mOsm/kg to 1200 mOsm/kg at junction with renal pelvis	More water and some urea returned to interstitial fluid; some H$^+$ added to urine

Aquaporins: Facilitating Osmotic Water Transport

Aquaporins are membrane proteins that form channels through which water can diffuse more rapidly than it would otherwise do. They are widely distributed, occurring in organisms from bacteria and yeast to mammals. In humans, at least 10 different aquaporins are known. The very narrow pores **(Figure 1)** permit water molecules to pass in single file, but because the molecules forming the channel are charged, other molecules of similar dimensions, such as H_3O^+, are excluded. The water moves in either direction in response to osmotic gradients: a single channel can permit as many as 3 billion molecules to cross the membrane per second.

Aquaporins are important in the functioning of mammalian kidneys. For example, one aquaporin, aquaporin-2, resides on the membranes of vesicles within the cells of the collecting ducts. If the osmotic concentration of the body fluids increases, antidiuretic hormone from the pituitary gland causes the aquaporins from the vesicles to be inserted into the membrane of the cells of the collecting ducts. The presence of more aquaporins in the membrane greatly increases the rate of osmotic reabsorption of water. The urine becomes more concentrated, and the osmotic concentration of body fluids is reduced.

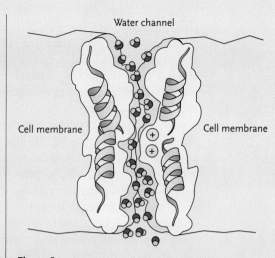

Figure 1
An aquaporin channel.

capsule itself, are separated by spaces just wide enough to admit water, ions, small nutrient molecules such as glucose and amino acids, and nitrogenous waste molecules, primarily urea. The higher pressure of the blood drives fluid containing these molecules and ions from the capillaries of the glomerulus into the capsule. A thin net of connective tissue between the capillary and Bowman's capsule epithelia contributes to the filtering process. Blood cells and plasma proteins are too large to pass and are retained inside the capillaries. The fluid entering the capsule is an ultrafiltrate of the blood.

Two factors help maintain the pressure driving fluid into Bowman's capsule. First, the diameters of the arteriole delivering blood to the glomerulus (called the **afferent arteriole**) and of the capillaries of the glomerulus itself are larger than that of arterioles and capillaries elsewhere in the body. The larger diameter maintains blood pressure by presenting less resistance to blood flow. Second, the diameter of the arteriole that receives blood from the glomerulus (called the **efferent arteriole**) is smaller than the diameter of the afferent arteriole, producing a damming effect that backs up the blood in the glomerulus and helps keep the pressure high.

In humans, Bowman's capsules collectively filter about 180 L of fluid each day, from a daily total of 1400 L of blood that pass through the kidneys. The human body contains only about 2.75 L of blood plasma, meaning that the kidneys filter a fluid volume equivalent to 65 times the volume of the blood plasma each day. On average, more than 99% of the filtrate, mostly water, is reabsorbed in the nephrons, leaving about 1.5 L to be excreted daily as urine.

43.3e Reabsorption and Secretion in the Nephron

The fluid filtered into Bowman's capsule contains water, other small molecules, and ions at the same concentrations as the blood plasma. By the time the fluid reaches the distal end of the collecting duct, reabsorption out of and secretion into the tubules and collecting duct have markedly altered the concentrations of all components of the filtrate.

The Proximal Convoluted Tubule. Reabsorption of water, ions, and nutrients back into the interstitial fluid is the main function of the proximal convoluted tubule. Na^+/K^+ pumps in the epithelium of the proximal convoluted tubule move Na^+ and K^+ from the filtrate into the interstitial fluid surrounding the tubule (see Figure 43.10). The movement of positive charges sets up a voltage gradient that causes Cl^- ions to be reabsorbed out of the tubule with the positive ions. Specific active transport proteins reabsorb essentially all the glucose, amino acids, and other nutrient molecules out of the filtrate into the interstitial fluid, making the filtrate hypoosmotic to the interstitial fluid surrounding the tubule. As a result, water moves from the tubule into the interstitial fluid by osmosis. The osmotic movement is aided by *aquaporins,* proteins that form passages for water molecules in the **transport epithelium** of the tubule cells (see *Molecule Behind Biology*). The nutrients and water that entered the interstitial fluid move into the capillaries of the peritubular network.

Some substances are secreted from the interstitial fluid into the tubule, primarily H^+ ions by active transport and the products of detoxified poisons by passive secretion (detoxification takes place in the liver).

Small amounts of ammonia are also secreted into the tubule. The secretion of H+ ions into the filtrate helps balance the acidity constantly generated in the body by metabolic reactions. H+ secretion is coupled with HCO_3^- reabsorption from the filtrate in the tubule to the plasma in the peritubular capillaries.

In all, the proximal convoluted tubule reabsorbs about 67% of the Na+, K+, and Cl- ions; 65% of the water; 50% of the urea; and essentially all the glucose, amino acids, and other nutrient molecules from the filtrate. The ions, nutrients, and water reabsorbed by the tubule are transported into the interstitial fluid and then into capillaries of the peritubular network. Although 50% of the urea is reabsorbed, the constant flow of filtrate through the tubules, and the excretion of the remaining urea in the urine, keeps the concentration of nitrogenous wastes low in body fluids.

The proximal convoluted tubule has structural specializations that fit its function. The epithelial cells that make up its walls are carpeted on their inner surface by a brush border of microvilli. These microvilli greatly increase the surface area available for reabsorption and secretion.

The Descending Segment of the Loop of Henle.

The filtrate flows from the proximal convoluted tubule into the descending segment of the loop of Henle, where water is reabsorbed. As this tubule segment descends, it passes through regions of increasingly higher solute concentrations in the interstitial fluid of the medulla (shown in Figure 43.10). (The generation of this concentration gradient is described later.) As a result, more water moves out of the tubule by osmosis as the fluid travels through the descending segment.

The descending segment has aquaporins, which allow the rapid transport of water but it has no other transport proteins. The outward movement of water concentrates the molecules and ions inside the tubule, gradually increasing the osmolality of the fluid to a peak of about 1200 mOsm/kg at the bottom of the loop. This is the same as the osmolality of the interstitial fluid at the bottom of the medulla.

The Ascending Segment of the Loop of Henle.

The fluid then moves into the ascending segment of the loop of Henle, where Na+ and Cl- are reabsorbed into the interstitial fluid. As this segment ascends, it passes through regions of gradually lessening osmolality in the interstitial fluid of the medulla. The ascending segment has membrane proteins that transport salt ions but lacks aquaporins. Because water is trapped in the ascending segment, the osmolality of the urine is reduced as salt ions, primarily Na+ and Cl-, move out of the tubule.

In the part of the ascending segment immediately following the bottom of the loop, the ion concentrations in the tubule filtrate are still high enough to move Na+ and Cl- out of the tubule by passive diffusion. Toward the top of the segment, they are moved out by active transport. Besides reducing the osmolality of the filtrate in the ascending segment, the reabsorption of salt ions from the tubule into the interstitial fluid helps establish the concentration gradient of the medulla, high near the renal pelvis and low near the renal cortex. The energy required to transport NaCl from higher levels of the ascending segment makes the kidneys one of the major ATP-consuming organs of the body.

By the time the fluid reaches the cortex at the top of the ascending loop, its osmolality has dropped to about 150 mOsm/kg. During the travel of fluid around the entire loop of Henle, water, nutrients, and ions have been conserved and returned to body fluids, and the total volume of the filtrate in the nephron has been greatly reduced. Urea and other nitrogenous wastes have been concentrated in the filtrate. Little secretion into the tubule occurs in either the descending or ascending segments of the loop of Henle.

The Distal Convoluted Tubule.

The transport epithelium of the distal convoluted tubule removes additional water from the filtrate in the tubule and works to balance the salt and bicarbonate concentrations of the filtrate against body fluids. In response to hormones triggered by changes in the body's salt concentrations, varying amounts of K+ and H+ ions are secreted into the filtrate, and varying amounts of Na+ and Cl- ions are reabsorbed. Bicarbonate ions are reabsorbed from the filtrate as in the proximal tubule.

In total, more ions move outward than inward in the distal tubule, and as a consequence, water moves out of the tubule by osmosis through aquaporins. The amounts of urea and other nitrogenous wastes remain the same. By the time the filtrate, now urine, enters the collecting ducts at the end of the nephron, its osmolality is about 300 mOsm/kg.

The Collecting Ducts.

The collecting ducts concentrate the urine. These ducts, which are permeable to water but not to salt ions, descend downward from the cortex through the medulla of the kidney. As the ducts descend, they travel through the gradient of increasing solute concentration in the medulla. This increase makes water move osmotically out of the ducts and greatly increases the concentration of the urine, which can become as high as 1200 mOsm/kg at the bottom of the medulla. Near the bottom of the medulla, the walls of the collecting ducts contain passive urea transporters that allow a portion of this nitrogenous waste to pass from the duct into the interstitial fluid. This urea adds significantly to the concentration gradient of solutes in the medulla.

In addition to these mechanisms, H+ ions are actively secreted into the fluid by the same mechanism as in the proximal and distal convoluted tubules. The balance of the H+ and bicarbonate ions established in the urine, interstitial fluid, and blood, achieved by secretion of H+ into the urine by the nephrons and collecting ducts, is important for regulating the pH of blood and body fluids. The kidneys thus provide a safety valve if the acidity of body fluids rises beyond levels that can be controlled by the blood's buffer system (see Chapter 42).

At its maximum value of 1200 mOsm/kg, reached when water conservation is at its maximum, the urine reaching the bottom of the collecting ducts is about four times more concentrated than body fluids. But it can also be as low as 50 to 70 mOsm/kg, when very dilute urine is produced in response to conditions such as excessive water intake.

The high osmolality of the interstitial fluid toward the bottom of the medulla would damage the medulla cells if they were not protected against osmotic water loss. The protection comes from high concentrations of otherwise inert organic molecules called *osmolytes* in these cells. The osmolytes, of which the most important is a sugar-alcohol called *sorbitol* (see Chapter 41, *Molecule Behind Biology*), raise the osmolality of the cells to match that of the surrounding interstitial fluid. Urine flows from the end of the collecting ducts into the renal pelvis and then through the ureters into the urinary bladder, where it is stored. From the bladder, urine exits through the urethra to the outside.

43.3f Terrestrial Mammals Have Water-Conserving Adaptations

Terrestrial mammals have other adaptations that complement the water-conserving activities of the kidneys. One is the location of the lungs deep inside the body, which reduces water loss by evaporation during breathing (see Chapter 42). Another is a body covering of keratinized skin. Skin is so impermeable that it almost eliminates water loss by evaporation, except for the controlled loss through evaporation of sweat in mammals with sweat glands.

Among mammals, water-conserving adaptations reach their greatest efficiency in desert rodents such as the kangaroo rat **(Figure 43.11)**. The proportion of nephrons with long loops extending deep into the kidney medulla of kangaroo rats is very high, allowing them to excrete urine that is 20 times more concentrated than body fluids. Further, most of the water in the feces is absorbed in the large intestine and rectum. Lacking sweat glands, they lose little water by evaporation from the body surface. Much of the moisture in their breath is condensed and recycled by specialized passages in the nasal cavities. They stay in burrows during daytime and come out to feed only at night.

About 90% of the kangaroo rat's daily water supply is generated from oxidative reactions in its cells. (Humans, in contrast, can make up only about 12% of their daily water needs from this source.) The remaining 10% of the kangaroo rat's water comes from its food. These structural and behavioural adaptations are so effective that a kangaroo rat can survive in the desert without ever drinking water.

	Kangaroo Rat	Human
Water gain (millilitres)		
From ingesting food	6.0	850
From drinking liquids	0.0	1400
By metabolism	54.0	350
	60.0	2600
Water loss (millilitres)		
In urine	13.5	1500
In feces	2.6	200
By evaporation	43.9	900
	60.0	2600

Figure 43.11
A comparison of the sources of water for a human and a kangaroo rat (*Dipodymus* species). Water conservation in the kangaroo rat is so efficient that the animal never has to drink water.

Marine mammals, including whales, seals, and manatees, eat foods that are high in salt content. They are able to survive the high salt intake because they produce urine that is more concentrated than seawater. As a result, they are easily able to excrete all the excess salt they ingest in their diet.

STUDY BREAK

1. Where does active transport of ions occur in the nephron?
2. What is the major event in the descending segment of the loop of Henle?
3. What is the major event in the collecting duct?

43.4 Kidney Function in Nonmammalian Vertebrates

Among nonmammalian vertebrates, only a few species of aquatic birds produce urine that is hyperosmotic to body fluids. The particular adaptations that maintain osmolality and water balance among these animals vary depending on whether retention of water or of salts is the major issue.

43.4a Marine Fishes Conserve Water and Excrete Salts

Marine teleosts live in seawater, which is strongly hyperosmotic to their body fluids. As a result, they continually lose water to their environment by osmosis and must replace it by continual drinking. The kidneys of marine teleosts play little role in regulating salt in their body fluids because they cannot produce hyperosmotic urine that would both remove salt and conserve water. Instead, excess Na⁺, K⁺, and Cl⁻ ions are eliminated from the body by specialized cells in the gills, called *chloride cells,* which actively transport Cl⁻ into the surrounding seawater; the Na⁺ and K⁺ ions are also actively transported to maintain electrical neutrality **(Figure 43.12a).** Divalent ions in the ingested seawater, such as Ca^{2+} and Mg^{2+}, are removed by the kidneys in an isosmotic urine. On balance, a marine teleost is able to retain most of the water it drinks and eliminate most of the salt, allowing its tissue fluids to remain hypoosmotic to the surrounding water without producing hyperosmotic urine. The kidneys play little role in the removal of nitrogenous wastes; these are released from the gills, primarily as ammonia, by simple diffusion.

Sharks and rays have a different adaptation to seawater—the osmolality of their body fluids is maintained close to that of seawater by retaining high levels of urea in body fluids, along with another nitrogenous waste, *trimethylamine oxide.* Elasmobranchs (see Chapter 27) may have concentrations of urea as high as 1300 mg per 100 mL of blood. The match in osmolality keeps sharks and rays from losing water to the surrounding sea by osmosis, and they do not have to drink seawater continually to maintain their water balance. Excess salts ingested with food are excreted in

the kidney and by specialized secretory cells in a *rectal salt gland* located near the anal opening. The importance of urea as an osmolyte is illustrated by those species of stingrays that inhabit freshwater. In such species, the concentration of urea is reduced to about 2 to 3 mg per 100 mL of blood.

43.4b Freshwater Fishes and Amphibians Excrete Water and Conserve Salts

The body fluids of freshwater fishes and aquatic amphibians (no amphibians live in seawater) are hyperosmotic to the surrounding water, which usually ranges from about 1 to 10 mOsm/kg. Water therefore moves osmotically into their tissues. Such animals rarely drink, and they excrete large volumes of dilute urine to get rid of excess water **(Figure 43.12b).** In freshwater fishes, salt ions lost with the urine are replaced by salt in foods and by active transport of Na⁺ and K⁺ into the body by the gills; Cl⁻ follows to maintain electrical neutrality. Aquatic amphibians obtain salt in the diet and by active transport across the skin from the surrounding water. Nitrogenous wastes are excreted from the gills as ammonia in both freshwater fishes and aquatic amphibians.

Terrestrial amphibians must conserve both water and salt, which is obtained primarily in foods. In these animals, the kidneys secrete salt into the urine, causing water to enter the urine by osmosis. In the bladder, the salt is reclaimed by active transport and returned to body fluids. The water remains in the bladder, making the urine very dilute; during times of drought, the water can be resorbed. Terrestrial amphibians also have behavioural adaptations that help minimize water loss, such as seeking shaded, moist environments and remaining inactive during the day. Larval amphibians, which are completely aquatic, excrete nitrogenous wastes from their gills as ammonia.

Most adult amphibians excrete nitrogenous wastes through their kidneys as urea. The leaf frog, *Phyllomedusa sauvagii* **(Figure 43.13),** however, produces uric acid as the principal nitrogenous waste. In addition, it secretes a waxy substance from glands in its skin and uses its legs to smear this over the entire surface, thereby minimizing water loss.

43.4c Reptiles and Birds Excrete Uric Acid to Conserve Water

Terrestrial reptiles and most birds conserve water by excreting nitrogenous wastes in the form of an almost water-free paste of uric acid crystals. Further water conservation occurs as the epithelial cells of the cloaca, the common exit for the digestive and excretory systems, absorb water from feces and urine before those wastes are eliminated. This arrangement is similar to the strategy used by insects, described

Figure 43.12
The mechanisms balancing the water and salt content of **(a)** marine teleosts and **(b)** freshwater teleosts.

a. Marine teleosts

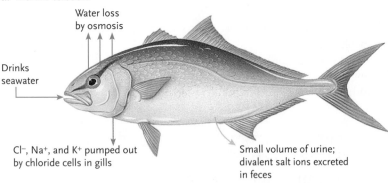

Water loss by osmosis

Drinks seawater

Cl⁻, Na⁺, and K⁺ pumped out by chloride cells in gills

Small volume of urine; divalent salt ions excreted in feces

b. Freshwater teleosts

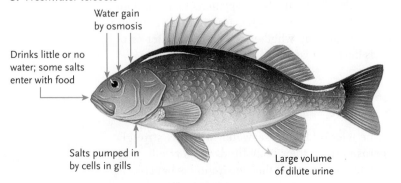

Water gain by osmosis

Drinks little or no water; some salts enter with food

Salts pumped in by cells in gills

Large volume of dilute urine

Figure 43.13

Phyllomedusa sauvagii is a tree frog that prospers in the dry woodlands of South America. Among its many adaptations to a dry environment are the production of uric acid and the secretion from skin glands of a waterproofing waxy material.

as salty tears from the eye sockets of sea turtles and crocodilians.

The adaptations described in this section allow animals to maintain the concentration of body fluids at levels that keep cells from swelling or shrinking and permit excretion of toxic wastes. An equally important challenge is maintaining an internal temperature that allows the organ systems to function with maximum efficiency. We look at these processes in the next section.

STUDY BREAK

1. What is the organ of osmoregulation in teleosts?
2. What excretory strategy is used by birds and reptiles to conserve water?
3. How do marine birds and reptiles excrete excess salts?

earlier. In reptiles, the scales covering the skin allow almost no water to escape through the body surface.

Reptiles and birds that live in or around seawater, including reptiles such as crocodilians, sea snakes, and sea turtles, and birds such as seagulls, penguins, and pelicans, take in large quantities of salt with their food and rarely or never drink freshwater. These animals typically excrete excess salt through specialized *salt glands* located in the head **(Figure 43.14)** that remove salts from the blood by active transport. The salts are secreted to the environment as a water solution in which salts are two to three times more concentrated than in body fluids. The secretion exits through the nostrils of birds and lizards, through the mouth of marine snakes, and

43.5 Introduction to Thermoregulation

Environmental temperatures vary enormously across Earth's surface. Temperatures in deserts in Australia, Africa, and the United States may reach 50°C, whereas some locations in the Antarctic experience –80°C. There are also seasonal variations. A single location in the boreal forest of Canada might experience temperatures as low as –40°C in the winter and as high as 35°C in the summer. However, animal cells can function only within a temperature range from about 0°C to 45°C. Not far below 0°C, the lipid bilayer of a biological membrane changes from a fluid to a frozen gel, which disrupts vital cell functions. Without protective measures, ice crystals will destroy the cell's organelles. At the other extreme, as temperatures approach 45°C, the kinetic motions of molecules become so great that most proteins and nucleic acids unfold from their functional form. Either condition leads quickly to cell death. Animals therefore usually maintain internal body temperatures somewhere within the 0°C to 45°C limits.

Temperature regulation (thermoregulation) is based on negative feedback pathways in which temperature receptors called *thermoreceptors* (see Chapter 34) detect changes from a temperature *set point*. Signals from the receptors trigger physiological and behavioural responses that return the temperature to the set point. The responses triggered by negative feedback mechanisms (see Chapter 32) involve adjustments in the rate of heat-generating oxidative reactions within the body, coupled with adjustments in the rate of heat gain or loss at the

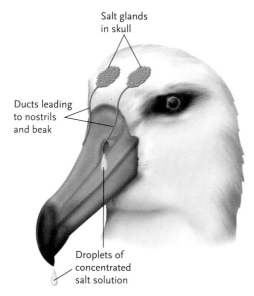

Salt glands in skull

Ducts leading to nostrils and beak

Droplets of concentrated salt solution

Figure 43.14
Salt glands in a bird living on a seacoast.

body surface. The particular adaptations that accomplish these responses vary widely among species, however. And although body temperature is closely regulated around a set point in all endotherms, the set point itself may vary over the course of a day and between seasons.

In this section, we describe the structures, mechanisms, and behavioural adaptations that enable animals to regulate their temperature.

43.5a Thermoregulation Allows Animals to Reach Optimal Physiological Performance

Within the 0°C to 45°C range of tolerable internal temperatures, an animal's *organismal performance* varies greatly. Organismal performance is a term that describes the rate and efficiency of an animal's biochemical, physiological, and whole-body processes. The speed at which the Middle Eastern lizard *Agama stellio* can sprint (one measure of organismal performance) is low when the animal's body temperature is cold, rises smoothly with body temperature until it levels to a fairly broad plateau, and then drops off dramatically with further increases in body temperature **(Figure 43.15a)**. Similar patterns of temperature dependence are observed for numerous other body functions **(Figure 43.15b)**. The range of temperatures that provides optimal organismal performance varies from one species to another.

Animals that maintain their body temperature within a fairly narrow optimal range can move quickly, digest food efficiently, and carry out necessary activities and processes rapidly and effectively (as shown in Figure 43.15b). In addition to keeping body temperatures within tolerable limits, thermoregulation allows animals to maintain an optimal level of organismal performance.

43.5b Animals Exchange Heat with Their Environments

As part of thermoregulation, animals exchange heat with their environment. Virtually all heat exchange occurs at surfaces where the body meets the external environment. As with all physical bodies, heat flows into animals if they are cooler than their surroundings and flows outward if they are warmer. This heat exchange occurs by four mechanisms: *conduction, convection, radiation,* and *evaporation* **(Figure 43.16)**.

Conduction is the flow of heat between atoms or molecules in direct contact. An animal loses heat by

a. Maximum running speed of a lizard at various body temperatures

b. Range of optimal physiological performance

Maximum speed (m/s)

Body temperature (°C)

Performance (% of maximum)

Produces excellent performance

Regulating body temperature within this range

Body temperature (°C)

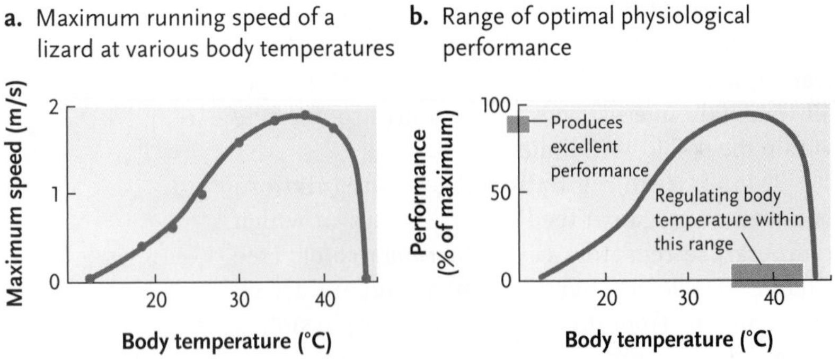

Figure 43.15
Body temperature and organismal performance. **(a)** The maximum sprint speed of a lizard (*Agama stellio*) changes dramatically with body temperature. **(b)** An animal's other behavioural and physiological processes respond to temperature changes in similar ways. The advantage of regulating body temperature within the range indicated by the bar on the horizontal axis is a high level of organismal performance, indicated by the bar on the vertical axis.

Figure 43.16
Heat flow into (in red) and out of (in blue) a marathon runner on a hot, sunny day. Unlike conduction, convection, and evaporation, which take place through the kinetic movement of molecules, electromagnetic radiation (infrared) is transmitted through space as waves of energy. (Photo: Rafael Winer/Corbis.)

conduction when it contacts a cooler object and gains heat when it contacts an object that is warmer. **Convection** is the transfer of heat from a body to a fluid (air or water) that passes over its surface. The movement maximizes heat transfer by replacing fluid that has absorbed or released heat with fluid at the original temperature. **Radiation** is the transfer of heat energy as electromagnetic radiation. Any object warmer than absolute zero (−273°C) radiates heat; as the object's temperature rises, the amount of heat it loses as radiation increases as well. Animals also gain heat through radiation, particularly by absorbing radiation from the Sun. **Evaporation** is heat transfer through the energy required to change a liquid to a gas. Evaporation of water from a surface is an efficient way to transfer heat; when the water in sweat evaporates from the body surface, the body cools down because heat is being transferred to the evaporated water in the surrounding air.

All animals gain or lose heat by a combination of these four mechanisms. A marathon runner or a bicycle racer struggling with the heat on a sunny summer day loses heat by the evaporation of sweat from the skin and from the surface of the lungs, by convection as air flows over the skin and passes out of the lungs, and by outward infrared radiation. The athlete gains heat from internal biochemical reactions (especially oxidations), by absorbing infrared and solar radiation, and by conduction as the feet contact the hot ground. To maintain a constant body temperature, the heat gained and lost through these pathways must balance.

43.5c Exothermic and Endothermic Animals

Different animals use one of two major strategies to balance heat gain and loss. Animals that obtain heat primarily from the external environment are known as **ectotherms** (*ecto* = outside); those obtaining most of their heat from internal physiological sources are called **endotherms** (*endo* = inside). All ectotherms generate at least some heat from internal reactions, however, and endotherms can obtain heat from the environment under some circumstances.

Most invertebrates, fishes, amphibians, and reptiles are ectotherms. Although these animals are popularly described as "cold-blooded," the body temperature of some, such as an active lizard, may be as high as or higher than ours on a sunny day. Ectotherms regulate body temperature by controlling the rate of heat exchange with the environment. Through behavioural and physiological mechanisms, they adjust body temperature toward a level that allows optimal physiological performance. However, most ectotherms are unable to maintain optimal body temperature when the temperature of their surroundings departs too far from that optimum, particularly when environmental temperatures fall. As a result, the body temperatures of ectotherms fluctuate with environmental temperatures, and ectotherms are typically less active when

it is cold. Nevertheless, ectotherms are highly successful, particularly in warm environments.

The endotherms—birds, mammals, some fishes, sea turtles, and some invertebrates—keep their bodies at an optimal temperature by regulating two processes: (1) the amount of heat generated by internal oxidative reactions and (2) the amount of heat exchanged with the environment. Because endotherms use internal heat sources to maintain body temperature at optimal levels, they can remain active over a broader range of environmental temperatures than ectotherms. However, endotherms require a nearly constant supply of energy to maintain their body temperatures. And because that energy is provided by food, endotherms typically consume much more food than ectotherms of equivalent size.

The difference between ectotherms and endotherms is reflected in their metabolic responses to environmental temperature **(Figure 43.17)**. The metabolic rate of a resting mouse *increases* steadily as the environmental temperature falls from 25°C to 10°C. This increase reflects the fact that to maintain a constant body temperature in a colder environment, endotherms must process progressively more food and generate more heat to compensate for their increased rate of heat loss. In this respect, an endotherm can be likened to a house in winter. To maintain a constant internal temperature, the homeowner must burn more oil or gas on a cold day than on a warm day.

By contrast, the metabolic rate of a resting lizard typically *decreases* steadily over the same temperature range.

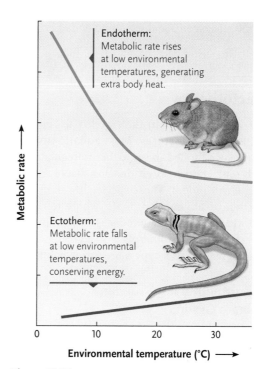

Figure 43.17

Metabolic responses of ectotherms and endotherms to cooling environmental temperatures. At any temperature, the metabolic rates of endotherms are always higher than those of endotherms of comparable size.

Because ectotherms do not maintain a constant body temperature, their biochemical and physiological functions, including oxidative reactions, slow down as environmental and body temperatures decrease. Thus, an ectotherm consumes less food and produces less energy when it is cold than when it is warm.

Ectothermy and endothermy represent different strategies for coping with the variations in environmental temperature that all animals encounter; neither strategy is inherently superior to the other. Endotherms can remain fully active over a wide temperature range. Cold weather does not prevent them from foraging, mating, or escaping from predators, but it does increase their energy and food needs—and to satisfy their need for food, they may not have the option of staying curled up safely in a warm burrow. Ectotherms do not have the capacity to be active when environmental temperatures drop too low; they move sluggishly and are unable to capture food or escape from predators. However, because their metabolic rates are lower under such circumstances, so are their food needs, and they do not have to actively look for food and expose themselves to danger to the extent that endotherms do.

Having laid the ground rules of heat transfer and weighed the relative advantages and disadvantages of ectothermy and endothermy, we now begin a more detailed examination of how individual animals actually regulate their body temperatures within these overall strategies.

STUDY BREAK

Distinguish between ectothermy and endothermy.

43.6 Ectothermy

Ectotherms vary widely in their ability to regulate internal body temperatures. Most aquatic invertebrates have such limited ability to thermoregulate that their body temperature closely matches that of the surrounding environment. These species live in or seek warm or temperate environments, where temperatures fall within a range that produces optimal physiological performance. Ectotherms with a greater ability to thermoregulate may occupy more varied habitats.

43.6a Ectotherms Are Found in All Invertebrate Groups

Most aquatic invertebrates are limited thermoregulators. Their body temperature closely follows the temperature of their surroundings. However, even among these animals, some use behavioural responses to regulate body temperature. For example, a South American intertidal mollusc, *Echinolittorina peruviana,* is longer than it is wide. Researchers in Chile have shown that this animal

orients itself as a means of thermoregulation. On sunny summer days, it faces the Sun, offering a smaller surface area for the Sun's rays. On overcast summer days, or during the winter, it orients itself with its side, which has the larger surface area, toward the Sun's rays.

Invertebrates living in terrestrial habitats regulate body temperatures more closely. Many also use behavioural responses, such as moving between shaded and sunny regions, to regulate body temperature. Some winged arthropods, including bees, moths, butterflies, and dragonflies, use a combination of behavioural and heat-generating physiological mechanisms for thermoregulation. In cool weather, these animals warm up before taking flight by rapidly vibrating the large flight muscles in the thorax, in a mechanism similar to shivering in humans. The tobacco hawkmoth (*Manduca sexta*) vibrates its flight muscles until its thoracic temperature reaches about 36°C before flying. During flight, metabolic heat generated by the flight muscles sustains the elevated thoracic temperature, so much so that a flying sphinx moth produces more heat per gram of body weight than many mammals. Honeybees (*Apis mellifera*) form masses in the hive in winter and use the heat generated by vibrating their flight muscles to maintain temperature inside the hive. Even in a Manitoba winter, with external temperatures below –20°C, the temperature in the mass of bees is normally about 30°C, and the bees may continue to raise offspring, using food stored in the hive.

43.6b Most Fishes, Amphibians, and Reptiles Are Ectotherms

Vertebrate ectotherms (most fishes, amphibians, and reptiles) also vary widely in their ability to thermoregulate. Most aquatic species have a more limited thermoregulatory capacity than that found among terrestrial species, particularly the reptiles. Some fishes, however, are highly capable thermoregulators.

Fishes. The body temperatures of most fishes remain within one or two degrees of their aquatic environment. However, many fishes use behavioural mechanisms to keep body temperatures at levels that allow good physiological performance. Freshwater species may use opportunities provided by the thermal **stratification** of lakes and ponds. They remain in deep, cool water during hot summer days, moving to the shallows to feed only during early-morning and late-evening when air and water temperatures are lower. Some fishes and sharks use endothermy: they are discussed in the next section.

Amphibians and Reptiles. The body temperature of most amphibians also closely matches the environmental temperature. The tadpoles of foothill yellow-legged frogs (*Rana boylii*) regulate their body temperature to some degree by changing their

PEOPLE BEHIND BIOLOGY

Ken Storey, Carleton University

Ken Storey, who leads a busy laboratory at Carleton University in Ottawa, explores the biochemical changes associated with hibernation and estivation in a wide range of animals, including both invertebrates and vertebrates. The wood frog, *Rana sylvatica*, is of particular interest because it spends the winter in a frozen state under the leaves on the forest floor. These frozen frogs have no heartbeat, breathing, or brain activity. Storey's research has shown that the wood frog's tolerance of freezing involves the liberation of glucose from glycogen stores in the liver and its accumulation at extremely high concentrations within the cells, resulting in a slurry of small ice crystals and glucose. This suggests a suspension of the function of insulin, which normally controls glucose concentrations. Although it may be obvious that the frogs freeze from the outside in and that heart and brain function will be the last to be suspended, it is also true that the frogs thaw from the inside out. The coordination of the events governing freezing and thawing involves a number of signal cascades, which his lab has characterized. Storey is currently exploring gene expression during the process of freezing and thawing.

location in ponds and lakes to take advantage of temperature differences between deep and shallow water or between sunny and shaded regions. Some terrestrial amphibians bask in the Sun to raise their body temperature and seek shade to lower body temperature. However, basking can be dangerous to amphibians because they lose water rapidly through their permeable skin. We have already noted that the leaf frog *Phyllomedusa sauvagii,* which often basks in sunlight, avoids this problem by coating itself with waterproofing lipids secreted by glands in its skin.

Thermoregulation is more pronounced among terrestrial reptiles. Some lizard species can maintain temperatures that are nearly as constant as those of endotherms **(Figure 43.18)**. For small lizards, the most common behavioural thermoregulatory mechanism is moving between sunny (warmer) and shady (cooler) regions. In the desert, lizards and other reptiles retreat into burrows during the hottest part of summer days. Some, such as the desert iguana (*Dipsosaurus dorsalis*),

lose excess heat by *panting*—rapidly moving air in and out of the airways. The air movement increases heat loss by convection and by evaporation of water from the respiratory tract.

Lizards also frequently adjust their posture to foster heat exchange with the environment and control the angle of their body relative to the rays of the Sun. Horned lizards (genus *Phrynosoma*) often warm up by flattening themselves against warm, sunlit rocks to maximize their rate of heat gain by conduction from the rock and radiation from the Sun. Snakes and lizards can often be found on large rocks and on roads on chilly nights, taking advantage of the heat retained by the stone or concrete. *Agama savignyi,* a lizard that lives in the Negev Desert in Israel, cools off at midday by climbing into shady bushes, moving away from the hot sand, and catching a cooling breeze.

Researchers have demonstrated experimentally that several lizard species couple physiological responses to behavioural mechanisms of thermoregulation. When a Galápagos marine iguana (*Amblyrhynchus cristatus*) is exposed to heat from infrared radiation, blood flow increases in the heated regions of the skin. The blood absorbs heat rapidly and carries it to critical organs in the core of the body. Conversely, when an area of skin is experimentally cooled, blood flow to it is restricted, thereby preventing the loss of heat to the external environment.

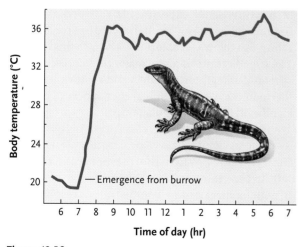

Figure 43.18

An example of excellent thermoregulation in ectotherms. The body temperature of the Australian lizard *Varanus varius* rises quickly after the animal emerges from its burrow and remains relatively stable throughout the day.

43.6c Ectotherms Can Compensate for Seasonal Variations in Environmental Temperature

Many ectotherms undergo seasonal physiological changes, called **thermal acclimatization**. These changes allow the animals to attain good physiological performance at both winter and summer temperatures.

For example, in the summer, bullhead catfish (*Ameiurus* species) can survive water temperatures as high as 36°C but cannot tolerate temperatures below 8°C.

In the winter, however, the bullhead cannot survive water temperatures above 28°C but can tolerate temperatures near 0°C. Scientists have hypothesized that the production of different versions of some enzymes (perhaps encoded by different genes or produced as a result of alternative splicing) with optimal activity at cooler or warmer temperatures underlies such acclimatization.

Another acclimatizing change involves the phospholipids of biological membranes. Membrane phospholipids have higher proportions of double bonds in carp living in colder environments than in carp living in warmer environments. The higher proportion of double bonds makes it harder for the membrane to freeze. A higher proportion of cholesterol also protects membranes from freezing.

STUDY BREAK

1. Describe two mechanisms an ectothermic animal can use to regulate its temperature.
2. What is thermal acclimatization?

43.7 Endothermy

Endotherms (mostly birds and mammals) have the most elaborate and extensive thermoregulatory adaptations of all animals. Set points vary with species and lie between about 39°C and 42°C in birds and 36°C and 39°C in mammals. We have already noted that the range of environmental temperatures that different organisms encounter is very great. A single species may encounter seasonal variations in environmental temperatures ranging over 70°C or more.

Some cold-water marine teleosts (such as tunas and mackerels) and some sharks (such as the great white) use endothermy in their aerobic swimming muscles to maintain a body core temperature as much as 10°C to 12°C warmer than their surroundings. These animals have in common the fact that they move over long distances, swimming continuously. The action of the muscles generates heat that permits the muscles and other organs to operate more efficiently.

Much of this heat would be lost at the gill–water interface. However, a *countercurrent heat exchanger* system between the swimming muscles and the gills minimizes this loss (see Chapter 42). The anatomical details of the heat exchanger vary. In principle, however, the venules containing warm blood from the muscles form a network with arterioles containing cold blood coming from the gills. The heat from the venules is transferred to the arterioles and is carried to the body tissues. Because this transfer occurs before the blood from the muscles enters the heart on its way to the gills, heat loss is minimized.

We begin by describing the basic feedback mechanisms that maintain body temperature, with primary emphasis on the human system. Later sections discuss variations in the responses of other mammals and of birds and daily and seasonal variations in the temperature set point.

43.7a The Hypothalamus Integrates Information from Thermoreceptors

Thermoreceptors are found in various locations in the human body, including the **integument** (skin), spinal cord, and hypothalamus. Two types of thermoreceptors occur in human skin (see Chapter 34). One, called a *warm receptor,* sends signals to the hypothalamus as the skin temperature rises above 30°C and reaches maximum activity when the temperature rises to 40°C. The other type, the *cold receptor,* sends signals when skin temperature falls below about 35°C and reaches maximum activity at 25°C. By contrast, the highly sensitive thermoreceptors in the hypothalamus itself produce signals when the blood temperature shifts from the set point by as little as 0.01°C.

Signals from the thermoreceptors are integrated in the hypothalamus and other regions of the brain to bring about compensating physiological and behavioural responses **(Figure 43.19)**. The responses keep body temperature close to the set point, which varies normally in humans between 35.5°C and 37.7°C for the head and trunk. The appendages may vary more widely in temperature. In very cold weather, for example, our arms, hands, legs, and feet are typically lower in temperature than the body core and the ears and nose especially so.

The hypothalamus was identified as a major thermoreceptor and response integrator in mammals by experiments on animals in which various regions of the brain were heated or cooled with a temperature probe. Within the brain, only the hypothalamus produced thermoregulatory responses such as shivering or panting. Later experiments revealed a similar response if regions of the spinal cord are cooled, indicating that thermoreceptors also occur in this location. The hypothalamus is also a major thermoreceptor and response integrator in fishes and reptiles. In birds, thermoreceptors in the spinal cord appear to be more significant in thermoregulation.

Responses When Core Temperature Falls Below the Set Point. When thermoreceptors signal a fall in core temperature below the set point, the hypothalamus triggers compensating responses by sending signals through the autonomic nervous system (see Chapter 33). Among the immediate responses is constriction of the arterioles in the skin (vasoconstriction), which reduces the flow of blood to the skin's capillary networks. The reduced flow cuts down the amount of heat delivered to the skin and therefore lost from the body surface. The reduction

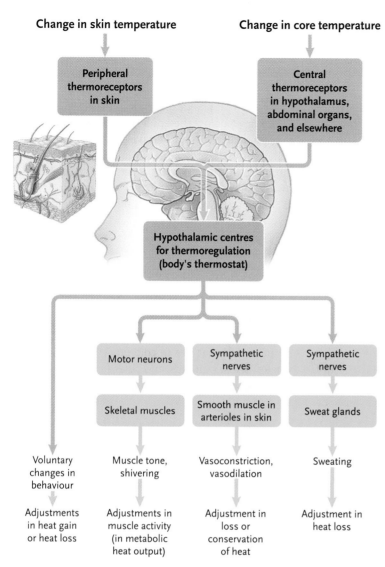

Change in skin temperature   **Change in core temperature**

Peripheral thermoreceptors in skin

Central thermoreceptors in hypothalamus, abdominal organs, and elsewhere

Hypothalamic centres for thermoregulation (body's thermostat)

Motor neurons   Sympathetic nerves   Sympathetic nerves

Skeletal muscles   Smooth muscle in arterioles in skin   Sweat glands

Voluntary changes in behaviour   Muscle tone, shivering   Vasoconstriction, vasodilation   Sweating

Adjustments in heat gain or heat loss   Adjustments in muscle activity (in metabolic heat output)   Adjustment in loss or conservation of heat   Adjustment in heat loss

Figure 43.19

The physiological and behavioural responses of humans and other mammals to changes in skin and core temperature.

in flow is most pronounced in the skin covering the extremities, where blood flow may be reduced by as much as 99% when core temperature falls.

Another immediate response is contraction of the smooth muscles that erect the hair shafts in mammals and feather shafts in birds. This traps air in pockets over the skin, reducing convective heat loss. The response is minimally effective in humans because hair is sparse on most parts of the body, but it produces the goose bumps we experience when the weather gets chilly. However, in mammals with fur coats or in birds, erection of the hair or feather shafts significantly increases the thickness of the insulating layer that covers the skin.

Immediate behavioural responses triggered by a reduction in skin temperature also help reduce heat loss from the body. Mammals may reduce heat loss by moving to a warmer locale, curling into a ball, or huddling together. We have all seen puppies huddled together to keep warm; birds such as penguins also keep warm by huddling. We humans may also put on more clothes or slip into a tub of hot water.

If these immediate responses do not return body temperature to the set point, the hypothalamus triggers further responses, most notably the rhythmic tremors of skeletal muscle we know as shivering. The heat released by the muscle contractions and the oxidative reactions powering them can raise the total heat production of the body substantially. At the same time, the hypothalamus triggers secretion of *epinephrine* (from the adrenal medulla) and *thyroid hormone* (see Chapter 35), both of which increase heat production by stimulating the oxidation of fats and other fuels. The generation of heat by oxidative mechanisms in nonmuscle tissue throughout the body is termed **nonshivering thermogenesis.**

In human newborn babies and the young of many other mammals, the most intense heat generation by nonshivering thermogenesis takes place in a specialized **brown adipose tissue** (also called brown fat) that can produce heat rapidly. Heat is generated by a mechanism that uncouples electron transport from ATP production in mitochondria (see Chapter 6); the heat is transferred throughout the body by the blood. Animals that hibernate or are active in cold regions also contain brown adipose tissue. In most mammals, brown adipose tissue is concentrated between the shoulders in the back and around the neck. In human newborns, this tissue accounts for about 5% of body weight. The tissue normally shrinks during late childhood and is absent or nearly so in most adults. However, if exposure to cold is ongoing, the tissue remains. Some Japanese and Korean divers who harvest shellfish in frigid waters and male Finlanders who work outside during the year have significant amounts of brown adipose tissue.

If none of these responses succeed in raising body temperature to the set point, the result is **hypothermia**, a condition in which the core temperature falls below normal for a prolonged period. In humans, a drop in core temperature of only a few degrees affects brain function and leads to confusion; continued hypothermia can lead to coma and death.

Responses When Core Temperature Rises Above the Set Point. When the core temperature rises above the set point, the hypothalamus sends signals through the autonomic system that trigger responses that lower body temperature. As an immediate response, the signals relax smooth muscles of arterioles in the skin (vasodilation), increasing blood flow and with it the heat lost from the body surface. In addition, in humans and other mammals with sweat glands, such as antelopes, cows, and horses, signals from the hypothalamus trigger the secretion of sweat, which absorbs heat as it evaporates from the surface of the skin.

Some endotherms, including dogs (which have sweat glands only on their feet) and many birds (which have no sweat glands), use panting as a major way to release heat. These physiological changes are reinforced by behavioural responses such as seeking shade or a cool burrow, plunging into cold water, wallowing in mud, or taking a cold drink. Elephants take up water in their trunks and spray it over their bodies to cool off in hot weather.

When the heat gain of the body is too great to be counteracted by these responses, **hyperthermia** results. An increase of only a few degrees above normal for a prolonged period is enough to disrupt vital biochemical reactions and damage brain cells. Most adult humans become unconscious if their body temperature reaches 41°C and die if it goes above 43°C for more than a few minutes.

43.7b The Skin Controls Heat Transfer with the Environment

Besides its defensive role against infection, the skin of birds and mammals is an organ of heat transfer. The arterioles delivering blood to the capillary networks of the skin constrict or dilate to control blood flow and with it the amount of heat transferred from the body core to the surface.

The outermost living tissue of human skin, the **epidermis**, consists of cells that divide and grow rapidly (Figure 43.20), becoming packed with fibres of a highly insoluble protein, *keratin*. When fully formed, the epidermal cells die and become compacted into a tough, impermeable layer that limits water loss to evaporation of the fluids secreted by the sweat glands.

The sweat glands and hair follicles are embedded in the layer below the epidermis. Called the **dermis**, it is packed with connective tissue fibres such as collagen, which resist the compression, tearing, or puncture of the skin. The dermis also contains thermoreceptors and the dense networks of arterioles, capillaries, and venules that transfer heat between the skin and the environment.

The innermost layer of the skin, the **hypodermis**, contains larger blood vessels and additional reinforcing connective tissue. The hypodermis also contains an insulating layer of fatty tissue below the dermal capillary network, which ensures that heat flows between the body core and the surface primarily through the blood. The insulating layer is thickest in mammals that live in cold environments, such as whales, seals, walruses, and polar bears, in which it is known as *blubber*.

43.7c Additional Thermoregulatory Structures and Responses

The thermoregulatory mechanisms we have described to this point are common to many birds and mammals. Many species also have specialized responses that enhance thermoregulation. In hot weather, many birds fly with their legs extended so that heat flows from their legs into the passing air. Similarly, penguins expose featherless patches of skin under their wings to cool off on days when the weather is too warm. Jackrabbits **(Figure 43.21a)** and elephants dissipate heat from their large ears, which are richly supplied with blood vessels. In times of significant heat

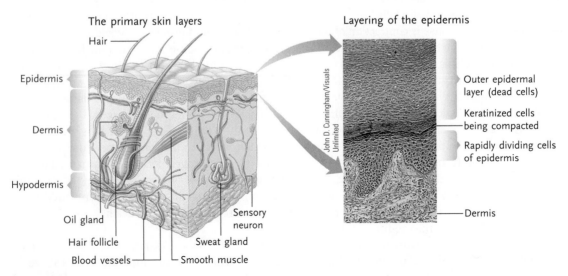

The primary skin layers

Hair

Epidermis

Dermis

Hypodermis

Oil gland

Hair follicle

Blood vessels

Sensory neuron

Sweat gland

Smooth muscle

Layering of the epidermis

Outer epidermal layer (dead cells)

Keratinized cells being compacted

Rapidly dividing cells of epidermis

Dermis

John D. Cunningham/Visuals Unlimited

Figure 43.20
The structure of human skin.

a. Dissipating heat

b. Conserving heat

Joe McDonald/Corbis

Fredrik Broman/Iconica/Getty Images, Inc.

Figure 43.21
Structural and behavioural adaptations controlling heat transfer at the body surface.
(a) A jackrabbit (*Lepus californicus*) dissipating heat from its ears on a hot summer day. Notice the dilated blood vessels in its large ears. Both the large surface area of the ears and the extensive network of blood vessels promote the dissipation of heat by convection and radiation.
(b) A husky (*Canis lupus familiaris*) conserving heat by curling up with the limbs under the body and the tail around the nose.

stress, kangaroos and rats spread saliva on their fur to increase heat loss by evaporation; some bats coat their fur with both saliva and urine.

Many mammals have an uneven distribution of fur that aids thermoregulation. In a dog, for example, the fur is thickest over the back and sides of the body and the tail and thinnest under the legs and over the belly. In cold weather, dogs curl up, pull in their limbs, wrap their tail around the body, and bury their nose in the tail so that only body surfaces insulated by thick fur are exposed to the air **(Figure 43.21b)**. When the weather is hot, dogs spread their limbs, turn on their side or back, and expose the relatively bare skin of the belly, which acts as a heat radiator. These responses are combined with seeking Sun or shade or a warm or cool surface to lie on.

The veins and arteries to the legs of mammals and birds may form a simple countercurrent heat exchange system in which cold blood returning from the foot takes heat from the arterial blood entering the foot **(Figure 43.22)**. This minimizes heat loss from the foot while maintaining a nutritive flow of blood to the extremity. This is particularly important for animals in polar regions.

In marine mammals such as whales and seals, heat loss is regulated by adjustments in the blood flow through the thick blubber layer to the skin. In cold water, blood flow is minimized by constriction of the vessels, making the skin temperature close to that of the surrounding water while the body temperature remains constant under the insulating blubber. In warmer water, blood flow to the skin above the blubber increases, allowing excess heat to be lost from the body surface.

In addition, heat loss in whales and seals is controlled by adjustments of the flow of blood to the flippers, which are not insulated by blubber and act as heat radiators. When a whale generates excessive internal heat through the muscular activity of swimming, the flow of blood from the body core to the

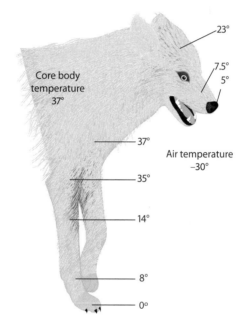

Core body temperature 37°

23°

7.5°
5°

37°

Air temperature −30°

35°

14°

8°

0°

36°   37°

7°   8°

Figure 43.22
Countercurrent circulation in the leg of an Arctic wolf. The vein and artery are parallel and close together so that heat from the warm blood in the artery is transferred to the cold blood returning from the foot, minimizing heat loss through the foot.

flippers increases. In contrast, when heat must be conserved to maintain core temperature at the set point, blood flow to the flippers is reduced.

As with ectotherms, many mammals also undergo thermal acclimatization with seasonal temperature changes. Although, in many cases, a change in day length appears to be the actual trigger, the development of a thick fur coat in winter, which is shed in summer, enables them to adapt to seasonal temperatures. Some arctic and subarctic mammals develop a thicker layer of insulating fat in winter.

43.7d Daily and Seasonal Rhythms in Birds and Mammals

The temperature set point in many birds and mammals varies in a regular cycle during the day. In some, the daily variations are relatively small. In others, larger variations are correlated with daily or seasonal temperature changes. These rhythms are a response to day length rather than temperature change.

Humans are among the endotherms for which daily variations in the temperature set point are small. Normally, human core temperature varies from a minimum of about 35.5°C in the morning to a maximum of about 37.7°C in the evening. Women also show a monthly variation keyed to the menstrual cycle, with temperatures rising about 0.5°C from the time of ovulation until menstruation begins. The physiological significance of these variations is unknown.

Camels undergo a daily variation of as much as 7°C in set point temperature. During the day, a camel's set point gradually resets upward, an adaptation that allows its body to absorb a large amount of heat. The heat absorption conserves water that would otherwise be lost by evaporation to keep the body at a lower set point. At night, when the desert is cooler, the thermostat resets again, allowing the body temperature to cool several degrees, releasing the excess heat absorbed during the day.

When the environmental temperature is cool, having a lowered temperature set point greatly reduces the energy required to maintain body temperature. In many animals, the lowered set point is accompanied by reductions in metabolic, nervous, and physical activity (including slower respiration and heartbeat), producing a sleeplike state known as **torpor.**

Entry into **daily torpor,** a period of inactivity keyed to variations in daily temperature, is typical of many small mammals and birds. These animals typically expend more energy per unit of body weight to keep warm than larger animals because the ratio of body surface area to volume increases as body size decreases. Hummingbirds feed actively during the daytime, when their set point is close to 40°C. During the cool of night, however, the set point drops to as low as 13°C. This allows the birds to conserve enough energy to survive overnight without feeding. Otherwise, they would literally starve to death. Some nocturnal animals, including bats and small rodents, such as the deer mouse, become torpid in cool locations during daylight hours when they do not actively feed. At night, their temperature set point rises and they become fully active **(Figure 43.23).**

Many animals enter a prolonged state of torpor tied to the seasons, triggered in most cases by a change in day length that signals the transition between summer and winter. The importance of day length has been demonstrated by laboratory experiments in which animals have been induced to enter seasonal torpor by changing the period of artificial light to match the winter or summer day length.

Extended torpor of small mammals during winter, called **hibernation** (*hiberna* = winter), greatly reduces metabolic expenditures when food is unobtainable. Hibernators must store large quantities of fats to serve as energy reserves. The drop in body temperature during hibernation varies with the mammal. In some, such as hedgehogs, groundhogs, and squirrels, body temperature may fall by 20°C or more. In some species of hedgehogs, body temperature falls from about 38°C in the summer to as low as 5°C to 6°C during

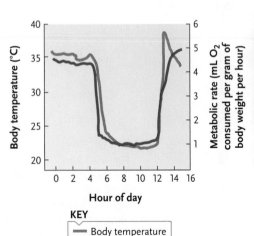

Figure 43.23
Cycle of daily torpor in a deer mouse (*Peromyscus maniculatus*).

Surviving Freezing

Many ectotherms encounter temperatures in winter that are well below the freezing point of their body fluids. The tiny second-stage caterpillar of the spruce budworm (*Choristoneura fumiferana*) spends the winter at the tips of spruce trees in Canada, where temperatures may reach –40°C. Their ability to survive depends on the production of an antifreeze protein. Antifreeze proteins also occur in marine fishes that occupy habitats where the water temperature may be below the freezing point of their body fluids. Research led by Peter Davies at Queen's University has shown that these proteins bond with forming ice crystals, limiting their growth, so although the water freezes, the crystals are small enough that cell structure is not disrupted.

The molecular structure of the antifreeze proteins is diverse, and at least four types are known. This diversity appears even within a single species: work in Garth Fletcher's laboratory at Memorial University in St. John's, Newfoundland, has shown that the protein found in plasma of several species of marine teleosts is different from that in the skin cells or gill cells. This diversity suggests that genes for the various types of antifreeze proteins in fish arose independently during the relatively recent cooling of the polar oceans.

winter hibernation. The Arctic ground squirrel's body supercools (goes to a below-freezing, unfrozen state) during hibernation, with its body temperature dropping to about –3°C. Hibernating mammals may experience brief periods of arousal during the course of the winter (see *Unanswered Questions*).

In larger mammals, such as bears and raccoons, the depth of torpor is less pronounced and is not considered by many scientists to be true hibernation. The core temperature of bears drops only a few degrees. Although sluggish, hibernating bears will waken readily if disturbed. They also waken normally from time to time, as when females wake to give birth during the winter season.

Some ectotherms, including amphibians and reptiles living in northern latitudes, also become torpid during winter. The Antarctic codfish, *Notothenia cordiceps*, spends the summer feeding on phytoplankton. In the winter, however, phytoplankton are reduced as a result of the low levels of light. The fish enter a state similar to hibernation. They remain relatively immobile in refuges, and their metabolic rate drops by about one-third. Because the temperature of the water remains constant over the seasons, the fish clearly have the capacity to control their metabolic rate independent of temperature.

Some mammals enter seasonal torpor during summer, called a **estivation** (*aestivalis* = of summer), when environmental temperatures are high and water is scarce. Some ground squirrels remain inactive in the cooler temperatures of their burrows during extreme summer heat. Many ectotherms, among them land snails, lungfishes, many toads and frogs, and some desert-living lizards, weather such climates by digging into the soil and entering a state of a estivation that lasts throughout the hot dry season.

STUDY BREAK

1. Where are thermoreceptors located in mammals?
2. What part of the brain integrates information from thermoreceptors, and what systems are activated by this region?

UNANSWERED QUESTIONS

Although active life normally occurs between relatively narrow limits of temperature and water content, there are exceptions, often involving a dormant stage. Animals can be totally frozen or almost completely desiccated, but we do not fully understand the mechanisms involved or the ultimate consequences. These phenomena of "suspended animation" have potential implications for humans. Preservation of human tissues, particularly in terms of prolonging the time that organs for transplant remain viable, is one example.

Hibernation may be of particular interest. In 1996, research by a team of Japanese scientists led by Noriaki Kondo identified a "hibernation protein complex" that appears in the blood of Asiatic chipmunks and acts on the brain as a hormone to control the onset and termination of hibernation. This raises the possibility that hibernation can be induced in nonhibernators. Can humans be induced to enter such states? Is this a potential treatment for trauma? The induction of a comalike state is already used in cases of severe brain injury. More distant, perhaps, is a potential role in exploration of space by humans. Some voyages may require months or years, and a state of suspended animation could be useful. Such questions make it important for scientists to explore more fully the states of torpor and hibernation.

Review

Go to CENGAGENOW™ at http://hed.nelson.com/ to access quizzing, animations, exercises, articles, and personalized homework help.

43.1 Introduction to Osmoregulation and Excretion

- Osmotic concentration of a solution is measured as osmolality in milliosmoles per kilogram (mOsm/kg) of solute. Both molecules and ions contribute to osmolality. Water moves osmotically from a solution of lower osmolality to one of higher osmolality. When comparing two solutions of different osmolality, the solution of higher osmolality is hyperosmotic and the solution of lower osmolality is hypoosmotic. Solutions of the same osmolality are isoosmotic.

- Osmoregulation describes the mechanisms that keep the osmolality of intracellular and extracellular fluids isotonic. Osmoregulators keep the osmolality of body fluids different from that of the environment. Osmoconformers allow the osmolality of their body fluids to match that of the environment.

- Osmoregulation is closely related to excretion because molecules and ions must be removed from the body to maintain the isotonicity of cells and extracellular fluids. Excretion also removes excess water, nitrogenous wastes, excess acid, and toxic molecules from the body.

- In most animals, tubules formed from a transport epithelium carry out the combined processes of osmoregulation and excretion. Extracellular fluids are filtered into the proximal end of the tubules; as the fluid moves through the tubules, some ions and molecules are reabsorbed from the fluid, and others are secreted into the fluid. Water is added or removed from the fluid to maintain the animal's water balance. The processed fluid is released to the exterior of the animal as urine.

- Nitrogenous wastes from the metabolism of amino acids and nucleic acids are excreted as ammonia, urea, or uric acid or as a combination of these substances.

43.2 Osmoregulation and Excretion in Invertebrates

- Most marine invertebrates are osmoconformers, whereas the invertebrates living in freshwater or terrestrial environments are osmoregulators.

- Because the body fluids of the marine osmoconformers are isoosmotic to seawater, they expend little or no energy on maintaining water balance. Body fluids in the osmoregulators living in freshwater and terrestrial environments are hyperosmotic to their surroundings. They must therefore expend energy to excrete water moving into their cells by osmosis and to obtain the salts required to maintain osmolality.

- Most invertebrates have specialized excretory tubules such as protonephridia, metanephridia, or Malpighian tubules that eliminate nitrogenous wastes and may assist in osmoregulation. A few groups eliminate wastes by diffusion.

- Protonephridia are blind tubes that produce an ultrafiltrate of the body fluids and adjust its ionic composition by secretion and reabsorption. Metanephridia are open tubes that take in coelomic fluid and adjust the ionic concentration.

- Malpighian tubules are blind tubes, and ions, uric acid, and other compounds enter by secretion.

43.3 Osmoregulation and Excretion in Mammals

- In mammals and other vertebrates, excretory tubules are concentrated in a specialized excretory organ, the kidney. Each of the pair of mammalian kidneys is divided into an outer renal cortex surrounding a central renal medulla. The mammalian excretory tubule, the nephron, has a proximal end at which filtration takes place, a middle region in which reabsorption and secretion occur, and a distal end that releases urine.

- The interstitial fluid reabsorbs ions, water, and other molecules from the nephron. A network of capillaries surrounding the nephron absorbs those materials. The urine leaving individual nephrons is processed further in collecting ducts and then pools in the renal pelvis. From there it flows through the ureter to the urinary bladder and through the urethra from the bladder to the exterior of the animal.

- At its proximal end, the mammalian nephron forms a cuplike Bowman's capsule around a cluster of capillaries, the glomerulus. A filtrate consisting of water, other small molecules, and ions is forced from the glomerulus into Bowman's capsule, from which it travels through the nephron and drains into the collecting ducts and renal pelvis.

- The proximal convoluted tubule of the nephron secretes H^+ into the filtrate and actively reabsorbs Na^+, K^+, HCO_3^-, and nutrients such as glucose and amino acids.

- In the descending segment of the loop of Henle, water is reabsorbed by osmosis.

- In the ascending segment of the loop, Na^+ and Cl^- move from the tubule by active transport.

- In the distal convoluted tubule, regulatory mechanisms operate to balance the concentrations of H^+ and salts between the urine and the interstitial fluid surrounding the nephron. K^+ and Na^+ move by active transport.

- In the collecting ducts, additional H^+ is secreted into the urine and water is reabsorbed; some urea is also reabsorbed at the bottom of the ducts.

43.4 Kidney Function in Nonmammalian Vertebrates

- Marine teleosts must continually drink seawater to replace body water lost by osmosis to their hyperosmotic environment. The Na^+, K^+, and Cl^- in the ingested seawater are excreted from the gills. Nitrogenous wastes are excreted by the gills as ammonia.

- Sharks use urea and trimethylamine oxide as osmolytes to maintain their body fluids isoosmotic with seawater. As a result, sharks and rays do not lose water by osmosis and do not drink seawater. Excess salts ingested in foods are excreted in the kidney and by a rectal salt gland.

- Body fluids of freshwater fishes and amphibians are hyperosmotic to their environment, and these animals must excrete the excess water that enters by osmosis. Body salts are obtained from food and, in fishes, by active transport through the gills. Nitrogenous wastes are excreted from the gills of fish and larval amphibians as ammonia and through the kidneys of adult amphibians as urea.

- Most reptiles and birds conserve water by secreting nitrogenous wastes as uric acid. Water is also absorbed from the urine and the feces in the cloaca.

- Marine birds and reptiles secrete excess salts through a gland in the head.

43.5 Introduction to Thermoregulation

- Animals maintain body temperature at a level that provides optimal physiological performance. Heat flows between animals and their environment by conduction, convection, radiation, and evaporation.

- Ectothermic animals obtain heat energy primarily from the environment, and endothermic animals obtain heat energy primarily from internal reactions.

43.6 Ectothermy

- Ectotherms, including all invertebrates and amphibians, reptiles, and most fishes among the vertebrates, control body temperature by regulating heat exchange with the environment. Their thermoregulatory responses may be physiological or behavioural. For most ectotherms, the ability to thermoregulate is limited, and body temperature does not differ widely from environmental temperature.

- Many animals undergo thermal acclimatization, a change in the limits of tolerable temperatures as the environment alternates between warm and cool seasons or between day and night. The acclimatization may involve structural changes and/or metabolic alterations.

43.7 Endothermy

- Endotherms, mostly birds and mammals, maintain body temperature over a narrow range by balancing internal heat production against heat loss from the body surface.

- Internal heat production is controlled by negative feedback pathways triggered by thermoreceptors in the skin, hypothalamus, and spinal cord.

- Signals from the receptors are integrated in the hypothalamus to bring about compensating responses by activating the autonomic nervous system, the endocrine system, and the motor nerves. These responses return the core temperature to a set point when deviations occur.

- When body temperature falls below a set point, responses include an increase in heat-generating metabolic reactions and a reduction of blood flow to the body surface. Behavioural responses also reduce heat loss.

- When body temperature rises above the set point, blood flow to the skin increases and sweating is induced in mammals with sweat glands. Mammals with no or few sweat glands and birds, which have no sweat glands, release heat by panting. Behavioural responses also contribute to heat loss.

- The skin of endotherms is water impermeable, reducing heat loss by direct evaporation of body fluids. Mammals with sweat glands regulate evaporative heat loss by releasing moisture to the skin surface when body temperature rises. The blood vessels of the skin regulate heat loss by constricting or dilating. A layer of insulating fatty tissue under the vessels limits losses to the heat carried by the blood. The hair of mammals and feathers of birds also insulate the skin. Erection of the hair or feathers reduces heat loss by thickening the insulating layer.

- The temperature set point in many birds and mammals varies in daily and seasonal patterns. During cooler conditions, a lowered set point is accompanied by torpor: a reduction in metabolic, nervous, and physical activity. Seasonal torpor includes winter hibernation and summer estivation.

- Some animals, such as certain cold-water marine fishes, exhibit a form of endothermy in which part, but not all, of their core is maintained at a temperature significantly higher than the surrounding environment.

Questions

Self-Test Questions

1. Which of the following statements about osmoregulation is true?
   a. In freshwater invertebrates, salts move out of the body because the body fluids are hypoosmotic to the environment.
   b. A marine teleost tends to gain water because it is isosmotic with respect to the sea.
   c. Most land animals are osmoconformers.
   d. Vertebrates are usually osmoregulators.
   e. Terrestrial animals do not expend energy to regulate their osmotic concentration.

2. Filtration and/or excretion can be carried out by
   a. ciliated metanephridia in insects.
   b. protonephridia containing flame cells in flatworms.
   c. a nephron and a bladder in insects.
   d. Malpighian tubules in the segments of earthworms.
   e. the hindgut of earthworms, which reabsorbs ions.

3. Which of the following statements are true? (Any number of the statements, from none to all five, may be correct.)
   a. The fluid that enters Bowman's capsule is an ultrafiltrate of the blood.
   b. In the ascending loop of Henle, Na$^+$ and K$^+$ enter the tubule by simple diffusion.
   c. In the descending loop of Henle, water leaves the tubule.
   d. In the proximal convoluted tubule, Na$^+$ and K$^+$ leave the tubule by active transport.
   e. In the collecting duct, water enters the tubule via aquaporins.

4. Unique to endotherms is
   a. a body temperature that does not change.
   b. torpor.
   c. thermal acclimatization.
   d. a response to seasonal temperature changes.
   e. thermoregulation by the hypothalamus.

5. Which best exemplifies ectotherms?
   a. The metabolic rate increases as the environmental temperature decreases.
   b. Body temperature remains constant when environmental temperatures change.
   c. All invertebrates are ectotherms.
   d. Food demand decreases when environmental temperatures decrease.
   e. No vertebrates are ectotherms.

Questions for Discussion

1. What is the evolutionary significance of the fact that most birds and reptiles produce uric acid as an excretory product, whereas most mammals produce urea?

2. Some insects feed on plants, such as tobacco, that contain poisons. Can you devise an experiment to test whether such poisons are eliminated by Malpighian tubules?

3. Hockey players are often advised to consume sports drinks containing salt before, during, and after a game. Why do you think this is?

4. Mammals that live in the desert are often nocturnal. What advantages are there in terms of thermoregulation?

5. The internal temperature of a reptile, such as a crocodile, varies with its developmental stage. Adult crocodiles tend to have a higher internal temperature less subject to change due to environmental variation in temperature. Why do you think this is?

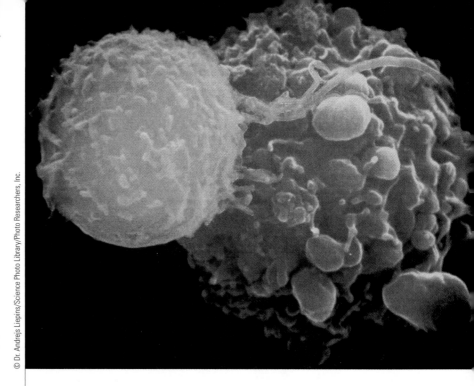

Death of a cancer cell. A cytotoxic T cell (orange) induces a cancer cell (mauve) to undergo apoptosis (programmed cell death). Cytotoxic T cells are part of the body's immune response system, programmed to seek out, attach themselves, and kill cancer cells and pathogen-infected host cells.

44 Defences against Disease

Why It Matters

Diseases have plagued all organisms including humans for billions of years. All animals, even insects, starfish, worms, and organisms too small to see with the naked eye, suffer from some sort of disease. In humans, **acquired immune deficiency syndrome (AIDS)**, first identified in the early 1980s, now affects about 40 million people worldwide and continues to spread. Malaria affects around 500 million people. Similar diseases in other organisms, such as avian influenza or avian malaria, may significantly reduce bird populations, and insect populations may be killed by outbreaks of specific viruses. Overall, bacteria, viruses, and parasites all can cause disease in their hosts. However, our immune systems are able to regulate or eliminate the majority of pathogens that cause disease. We combat the other pathogens by developing effective drug treatments that eliminate the disease-causing organism or developing vaccines that provide protection from infection.

The development of vaccines began with efforts to control smallpox, a dangerous and disfiguring viral disease that once infected millions of people worldwide, killing more than one-third of its victims. As early as the twelfth century, healthy individuals in

China sought out people who were recovering from mild smallpox infections, ground up scabs from their lesions, and inhaled the powder or pushed it into their skin. Variations on this treatment were effective in protecting many people against smallpox infection.

In 1796, an English country doctor, Edward Jenner, used a more scientific approach. He knew that milkmaids never got smallpox if they had contracted cowpox, a similar but mild disease of cows that can be transmitted to humans. Jenner decided to see if a deliberate infection with cowpox would protect humans from smallpox. He scratched material from a cowpox sore into a boy's arm. Six weeks later, after the cowpox infection had subsided, he scratched fluid from human smallpox sores into the boy's skin. (Jenner's use of the boy as an experimental subject would now be considered unethical.) Remarkably, the boy remained free from smallpox. Jenner carried out additional, carefully documented case studies with other patients with the same results. His technique became the basis for worldwide **vaccination** (*vacca* = cow) against smallpox. With improved vaccines, smallpox has now been eradicated from the human population.

Vaccination takes advantage of the **immune system** (*immunis* = exempt), the natural protection that is our main defence against infectious disease. This chapter focuses on the roles of different components of the immune system that deal with infection. We describe aspects of immune systems in different organisms, from insects to humans, to examine their similarities and differences.

44.1 Three Lines of Defence against Invasion

Every organism is constantly exposed to *pathogens*, potentially disease-causing organisms such as viruses, bacteria, protists, fungi, and parasites. Humans and other animals have three lines of defence against these threats. The first line of defence involves physical barriers that prevent the entry of pathogens. The second is the *innate immune system;* inherited mechanisms that protect the body from pathogens in a nonspecific way. The third is the *adaptive immune system,* found only in vertebrates, and involves inherited mechanisms that lead to the synthesis of molecules such as antibodies that target pathogens in a specific way. Reaction to an infection takes minutes in the case of the innate immunity system versus several days for the adaptive immune system.

44.1a Epithelium as a Barrier to Infection

An organism's first line of defence is the body surface—the skin covering the body exterior, the cuticle of an arthropod, the outer layer of a plant, and the epithelial surfaces covering internal body cavities and ducts, such as the lungs and intestinal tract. The body surface forms a barrier of tight junctions between the epithelial cells that keeps most pathogens (as well as toxic substances) from entering the body.

In the respiratory tract, ciliated cells constantly sweep the mucus with its trapped bacteria and other foreign matter into the throat, where it is coughed out or swallowed. Many of the body cavities lined by mucus membranes have environments that are hostile to pathogens. For example, the strongly acidic environment inside the stomach kills most ingested bacteria and destroys many viruses, including those trapped in swallowed mucus from the respiratory tract. Most of the pathogens that survive the stomach acid are destroyed by the digestive enzymes and bile secreted into the small intestine. Reproductive tracts may be acidic or basic, which prevents many pathogens from surviving there, and many epithelial tissues secrete enzymes such as defensins or lysozymes that are lethal to many bacteria.

44.1b Immune Systems within the Body

The body's second line of defence is a series of generalized internal chemical, physical, and cellular reactions that attack pathogens that have breached the first line. These defences include inflammation, which creates internal conditions that inhibit or kill many pathogens, and specialized cells that engulf or kill pathogens or infected body cells. These initial responses to pathogens are called **innate immunity.**

The innate immune response relies on germline-encoded receptors that recognize a set of highly conserved molecular patterns that are present on the surface of pathogens but not found on host cells. Innate immunity provides an immediate, *nonspecific* response; that is, it targets any invading pathogen and has no memory of prior exposure to that specific pathogen.

Invertebrates and plants rely solely on innate immune responses, whereas vertebrates use the innate immune system in conjunction with the adaptive response for a more powerful overall response. This most complex line of defence, found only in vertebrates, is called **adaptive** (or **acquired**) **immunity.** Adaptive immunity is *specific:* it recognizes individual pathogens and mounts an attack that directly neutralizes or eliminates them. It is stimulated and shaped by the presence of a specific pathogen or foreign molecule in the body. This mechanism, which takes several days to become protective, is triggered by specific molecules on pathogens that are recognized as being foreign to the body. The body retains a memory of the first exposure to a foreign molecule, enabling it to respond more quickly if the pathogen is encountered again in the future.

Innate immunity and adaptive immunity together constitute the immune system, and the defensive reactions of the system are termed the **immune response**. Functionally, the two components of the immune system interconnect and communicate at the chemical and molecular levels. The immune system is the product of long-term coevolutionary interactions between pathogens and their hosts. Over millions of years, the mechanisms by which pathogens attack and invade have become more efficient, but the defences of organisms against the invaders have kept pace.

44.1c How Organisms Recognize Pathogens

The basic tenet of immunity is that an organism can recognize a pathogen as being different from the host. This is often termed recognition of "nonself" and is the essential first step before any immune response can be initiated. Different organisms do this in different ways. Most organisms recognize unique *pathogen-associated molecular patterns* (PAMPs) found on many microbial organisms, using host molecules called *pattern recognition receptors* (PRRs). Common PAMPs include carbohydrates, glycoproteins, lipids, and nucleic acids. Two classic examples of PAMPs are the bacterial cell wall components lipoteichoic acid found on Gram-positive bacteria and lipopolysaccharide on Gram-negative bacteria. Major invertebrate PRRs include the Gram-negative bacteria binding protein and β1,3-glucan recognition proteins that recognize and bind to common molecules found on many groups of pathogenic organisms rather than recognizing each pathogen species individually. For example, lipopolysaccharide is found on the outer surface of all Gram-negative bacteria and serves as a general PAMP to which a PRR can bind. Once specific PRRs are activated by the presence of the PAMP, signalling cascades are initiated that activate various components of the innate or acquired immune responses.

Plants respond to pathogens in a similar manner. They have two branches of immune responses: a system that uses transmembrane PRRs that respond to the PAMPs described for invertebrates and an intracellular response that uses protein products encoded by *R* genes that respond to infection and systemic signals from nearby infected cells. Damage to the plant cells by pathogen effector molecules is believed to be the signal that activates the expression of plant R proteins.

Different PAMPs may activate the same or different signalling pathways and elicit different responses. This has been studied best in the fruitfly, *Drosophila melanogaster*. Infection of *Drosophila* with fungi and bacteria activates the Toll and immune deficiency signalling pathways, which result in nuclear factor (NF)-κB-like transcription factors being translocated to the nucleus, activating many components of the innate immune response.

44.2 Nonspecific Defences: Innate Immunity

Invertebrates only have an innate response, and the PAMP–PRR interactions activate the signalling pathways described above. These activate processes such as **phagocytosis** (the internalization and destruction of particulate matter) of small pathogens by hemocytes (blood cells) or the coagulation of the hemolymph (invertebrate blood). Larger pathogens may be encapsulated by hemocytes and covered in a melanin-like material that kills them (**melanotic encapsulation**). This may be helped by the release of reactive intermediates of nitrogen and oxygen. The third component of the innate response involves the production of small **antimicrobial peptides** that kill pathogens not eliminated by the other responses. This coordinated, multifaceted, and integrated approach eliminates potential pathogens, preventing them from harming or killing the host.

The study of invertebrate immunology began when a Russian scientist poked a rose thorn into the body of a starfish and watched how the hemocytes surrounded the thorn **(Figure 44.1)**. The innate immune responses of invertebrates, now a rapidly developing

Figure 44.1

Innate immune responses demonstrated by the role of starfish blood cells responding to a thorn that has penetrated the cuticle.

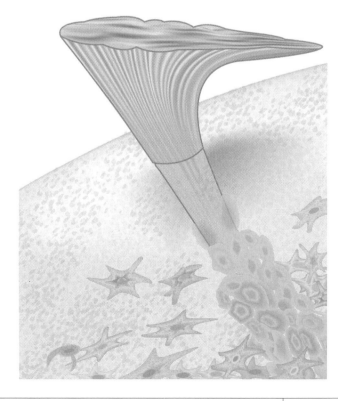

Defensins

Defensins are ubiquitous cationic molecules used in the defences of essentially all organisms. Defensins have been isolated from several orders of the higher insects, such as the Diptera (flies) and Coleoptera (beetles), and from ancient insects, such as the Odonata (dragonflies). Functional analogues have been isolated and characterized from amoebae, nematodes, scorpions, molluscs, mammals (including humans), and plants. In some organisms, these molecules are secreted into the body cavity, whereas in others, they are intracellular and are released only at a wound site.

This strong conservation suggests that this molecule is of ancient origin and has been maintained throughout evolution due to its importance in limiting the growth of microbial pathogens. Defensins are composed of a series of structures: an N-terminal loop, an α-helix, and twisted, antiparallel β-sheets **(Figure 1).** This three-dimensional shape of the molecule is stabilized by the presence of three disulphide bridges. Defensins are active against many Gram-positive bacteria, some Gram-negative bacteria, and some fungi. The lethality occurs as the one region of the peptide binds to the outer surface of the bacteria, allowing other regions to form pores in the microbial membranes, causing a permeabilization that causes a loss of cytoplasmic potassium, a depolarization of the inner membrane, reduced amounts of cytoplasmic ATP, and a reduction in respiration. This can occur in single-celled organisms, in the body cavity of an insect, in plant tissues, or in the white blood cells of a vertebrate. Defensins represent one family of molecules conserved throughout all taxa for the same function and are truly universal immune molecules.

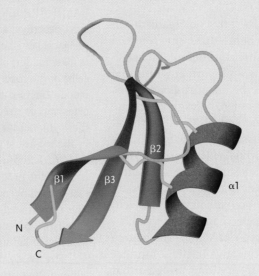

Figure 1

Computer-generated model of a defensin molecule showing the arrangement of the coiled α-helix (red) and the β-sheets (blue) that are held together by disulphide bridges (yellow).

field of research, have allowed us to understand how the innate system works in both invertebrates and vertebrates and what molecules are activated in response to different stimuli. Plants respond in a similar manner. They can wall off cells that are infected, undergo cell death to prevent pathogen development, and produce several small antimicrobial peptides that kill microbial pathogens.

Vertebrates use similar types of specific host cell-surface receptors that recognize the various PAMPs found on microbial pathogens. Some receptors activate signalling pathways that bring about the secretion of lethal antimicrobial peptides that kill the pathogen. Other receptors trigger the host cell to engulf the pathogen, as was described for phagocytosis in invertebrates, but in vertebrates, this also may initiate an inflammation response and may activate the soluble receptors of the *complement system*, described below. The innate responses that recognize and initiate immune pathways in plants, invertebrates, and vertebrates are strikingly similar.

Antimicrobial Peptides. All epithelial surfaces, namely skin; the lining of the gastrointestinal tract; the lining of the nasal passages, gills, and lungs; and the lining of the genitourinary tracts, are protected by antimicrobial peptides, some of which are called *defensins*. These epithelial cells secrete defensins upon attack by a microbial pathogen. The defensins attack the plasma membranes of the pathogens, eventually disrupting them and thereby killing the cells. In particular, defensins play a significant role in innate immunity of the intestinal tracts of vertebrates and invertebrates. Antimicrobial peptides such as the defensins are highly conserved in plants, invertebrates, vertebrates, and even single-celled organisms, indicating their important role in immunity throughout evolution. Other antimicrobial peptides are found only in specific groups, suggesting a more specialized role that has maintained their existence.

Inflammation. A tissue's rapid response to injury, including infection by most pathogens, involves **inflammation** (*inflammare* = to set on fire) the heat, pain, redness, and swelling that occur at the site of an infection.

Several interconnecting mechanisms initiate inflammation **(Figure 44.2).** Let us consider bacteria entering a tissue as a result of a wound. **Monocytes** (a type of leukocyte) enter the damaged tissue from the bloodstream through the endothelial wall of the capillary. Once in the damaged tissue, the monocytes differentiate into **macrophages** ("big eaters"), which are phagocytes that are usually the first to recognize pathogens at the cellular level. **(Table 44.1, p. 1090** lists the major types of leukocytes such as macrophages; see also **Figure 44.3, p. 1090.)** Cell-surface receptors on the macrophages recognize and bind to surface molecules on the pathogen, activating the macrophage to phago-

cytize (engulf) the pathogen (see Figure 44.2, step 1). There may not be enough macrophages present at the site of infection to eliminate all the pathogens. Activated macrophages also secrete **cytokines,** molecules that activate and recruit more immune cells to increase the system's response to the pathogen.

The death of cells caused by the pathogen at the infection site activates cells that are dispersed throughout the connective tissue, called **mast cells,** which then release histamine (see Figure 44.2, step 2). This histamine, along with the cytokines from activated macrophages, dilates local blood vessels around the infection site and increases their permeability. This increases blood flow and leakage of fluid from the vessels into body tissues (step 3). The response

Figure 44.2
The steps producing inflammation. The colourized micrograph on the left shows a macrophage engulfing a yeast cell.

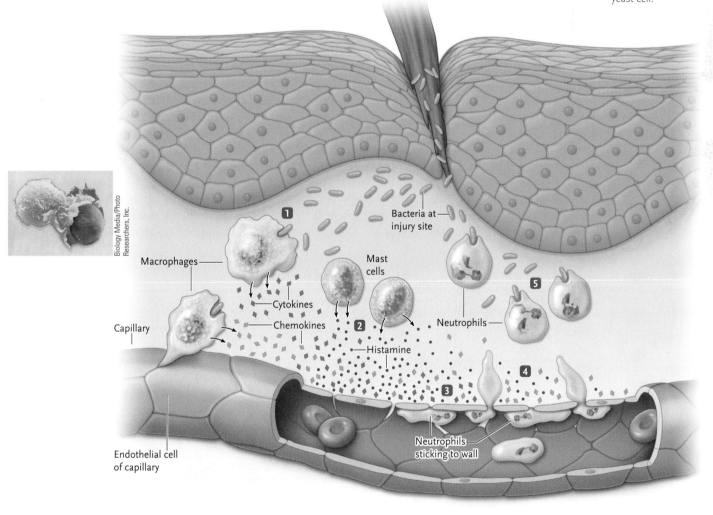

Biology Media/Photo Researchers, Inc.

1 A break in the skin introduces bacteria, which reproduce at the wound site. Activated macrophages engulf the pathogens and secrete cytokines and chemokines.

2 Activated mast cells release histamine.

3 Histamine and cytokines dilate local blood vessels and increase their permeability. The cytokines also make the blood vessel wall sticky, causing neutrophils to attach.

4 Chemokines attract neutrophils, which pass between cells of the blood vessel wall and migrate to the infection site.

5 Neutrophils engulf the pathogens and destroy them.

Table 44.1	Major Types of Leukocytes and Their Functions
Type of Leukocyte	Function
Monocyte	Differentiates into a macrophage when released from blood into damaged tissue
Macrophage	Phagocyte that engulfs infected cells, pathogens, and cellular debris in damaged tissues; helps activate lymphocytes in carrying out immune response
Neutrophil	Phagocyte that engulfs pathogens and tissue debris in damaged tissues
Eosinophil	Secretes substances that kill eukaryotic parasites such as worms
Lymphocyte	Main subtypes involved in innate and adaptive immunity are natural killer (NK) cells, B cells, plasma cells, helper T cells, and cytotoxic T cells. NK cells function as part of innate immunity to kill virus-infected cells and some cancerous cells of the host. The other cell types function as part of adaptive immunity: they produce antibodies, destroy infected and cancerous body cells, and stimulate macrophages and other leukocyte types to engulf infected cells, pathogens, and cellular debris.
Basophil	Respond to IgE antibodies in an allergy response by secreting histamine, which stimulates inflammation

initiated by cytokines directly causes the heat, redness, and swelling of inflammation.

Cytokines also make the endothelial cells of the blood vessel wall stickier, causing circulating **neutrophils** (another type of phagocytic leukocyte) to attach to them in massive numbers. From there, the neutrophils are attracted to the infection site by **chemokines**, proteins also secreted by activated macrophages (see Figure 44.2, step 4). To get to the infection site, the neutrophils pass between endothelial cells of the blood vessel wall. Neutrophils also may be attracted directly to the pathogen by molecules released from the pathogens themselves. Like macrophages, neutrophils have cell-surface receptors that enable them to recognize and engulf pathogens (step 5).

Once a macrophage or neutrophil has engulfed the pathogen, it uses a variety of mechanisms to destroy it. These mechanisms include attacks by enzymes and defensins located in lysosomes and the production of toxic compounds. The harshness of these attacks usually kills the neutrophils as well, whereas macrophages usually survive to continue their pathogen-scavenging activities. Dead and dying neutrophils, in fact, are a major component of the pus formed at infection sites. The pain of inflammation is caused by the migration of macrophages and neutrophils to the infection site and their activities there.

Some pathogens, such as parasitic worms, are too large to be engulfed by macrophages or neutrophils. In that case, macrophages, neutrophils,

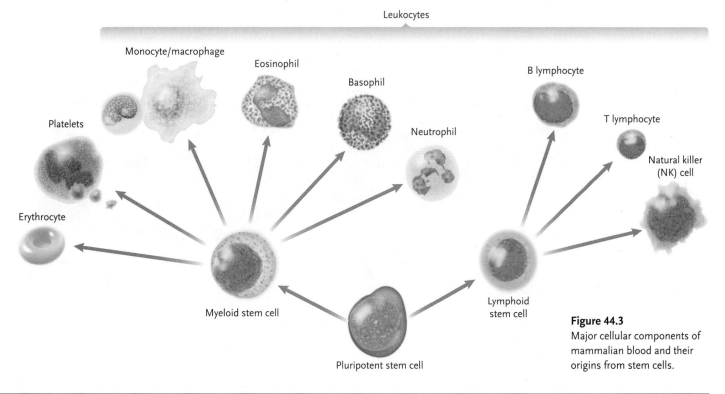

Figure 44.3
Major cellular components of mammalian blood and their origins from stem cells.

and **eosinophils** (another type of leukocyte) cluster around the pathogen and secrete lysosomal enzymes in amounts usually sufficient to kill the pathogen.

The Complement System. Another nonspecific defence mechanism activated by invading pathogens is the **complement system**, a group of more than 30 interacting soluble plasma proteins that circulate in the blood and interstitial fluid. Normally inactive, the proteins are activated when they recognize molecules on the surfaces of pathogens. Activated complement proteins participate in a cascade of reactions on pathogen surfaces, producing large numbers of different complement proteins, some of which assemble into *membrane attack complexes*. These complexes insert into the plasma membrane of many types of bacterial cells and create pores that allow ions and small molecules to pass readily through the membrane. As a result, the bacteria can no longer maintain osmotic balance, and they swell and lyse. For other types of bacterial cells, the cascade of reactions coats the pathogen with fragments of the complement proteins. Cell-surface receptors on phagocytes then recognize these fragments and engulf and destroy the pathogen.

Several activated proteins in the complement cascade also act individually to enhance the inflammatory response. For example, some of the proteins stimulate mast cells to enhance histamine release, whereas others increase the blood vessel permeability.

44.2a Combating Pathogenic Viruses

Specific molecules on pathogens such as bacteria are key to initiating innate immune responses. In contrast, the innate immunity system is often unable to distinguish between surface molecules of viral pathogens and host cells or to have access to small pathogens that live inside host cells. The host must, therefore, use other strategies to provide some immediate protection against these infections until the adaptive immunity system, which can discriminate between pathogen and host proteins, is effective. Three main strategies involve RNA interference, interferon, and natural killer cells.

RNA Interference. *RNA interference (RNAi)* is a cellular mechanism that is triggered by double-stranded (ds) RNA molecules (see Chapter 15). The dsRNA interferes with the ability of a cell to transcribe specific genes. Because dsRNA is a natural part of the life cycle of many viruses, the use of RNAi can inhibit the dsRNA found in many viruses and eliminate the infection.

Similarly, the virus can interfere with the host immune machinery, including RNAi pathways. Thus, the activities within a cell, and the success of a pathogen such as a virus, depend on the interplay between pro- and anti-infection responses. The only well-established antiviral mechanisms reported in insects such as the fruitfly, *D. melanogaster*, are RNAi and apoptosis. To combat this host response, many viruses contain genes that encode RNAi suppressors or inhibitors of apoptosis that help them survive.

Interferon. Viral dsRNA also may cause the infected host cell to produce two cytokines, interferon-α and interferon-β. **Interferons** can be produced by most cells of the body. These proteins act on both the infected cell that produces them, an autocrine effect, and neighbouring uninfected cells, a paracrine effect (see Chapter 35). They work by binding to cell-surface receptors, triggering a signal transduction pathway that changes the gene expression pattern of the cells. The key changes include the activation of a ribonuclease enzyme that degrades most cellular RNA and the inactivation of a key protein required for protein synthesis, thereby inhibiting most protein synthesis in the cell. These effects on RNA and protein synthesis inhibit replication of the viral genome while putting the cell in a weakened state from which it often can recover.

Apoptosis. Apoptosis, or programmed cell death, is a process inherent to all eukaryotic cells that has been highly conserved through evolution. It represents an intrinsic form of cell death that is tightly regulated by a variety of internal and external cellular signals to avoid killing healthy, productive cells. During development, apoptosis is required to sculpt tissues, remove old and dying cells, and eliminate embryonic cells with damaged DNA. Throughout life, apoptosis is also used as an immune response against intracellular pathogens and parasites. These organisms often trigger abnormal cellular activity that activates intrinsic apoptotic pathways, especially initiator and effector caspases, which dismantle the cell's structure. Once effector caspases have been activated, the cell is destined to die. Apoptotic cells fragment into membrane-bound apoptotic bodies that are readily phagocytosed and digested by macrophages or by neighbouring cells without generating an inflammatory response. If the pathogen is recognized as nonself and the apoptotic response is initiated in time, both the infected cell and the pathogen are eliminated, and no disease is seen. However, as discussed in Section 44.6, some pathogens have developed mechanisms to inactivate this response.

Natural Killer Cells. Cells that have been infected with a virus must be destroyed. That is the role of *natural killer (NK) cells*. NK cells are a type of *lymphocyte*, a leukocyte that carries out most of its activities in the tissues and organs of the lymphatic system (see Figure 37.17). NK cells circulate in the blood and kill target host cells—not only cells that are infected with virus but also some cells that have become cancerous.

NK cells can be activated by cell-surface receptors or by interferons secreted by virus-infected cells. NK cells are not phagocytes; instead, they secrete granules containing *perforin*, a protein that creates pores in the target cell's membrane. Unregulated diffusion of ions and molecules through the pores causes osmotic imbalance, swelling, and rupture of the infected cell. NK cells also kill target cells indirectly through the secretion of *proteases* (protein-degrading enzymes) that pass through the pores. The proteases trigger apoptosis (see Section 9.4f). That is, the proteases activate other enzymes that cause the degradation of DNA, which, in turn, induces pathways leading to the cell's death.

How does an NK cell distinguish a target cell from a normal cell? The surfaces of most vertebrate cells contain particular *major histocompatibility complex* (*MHC*) *proteins*. You will learn about the role of these proteins in adaptive immunity in the next section. NK cells monitor the level of MHC proteins and respond differently depending on their level. An appropriately high level, as occurs in normal cells, inhibits the killing activity of NK cells. Because intracellular pathogens often inhibit the synthesis of MHC proteins in the cells they infect, these cells are recognized by NK cells. Cancer cells also have low or, in some cases, no MHC proteins on their surfaces, which makes them a target for destruction by NK cells.

STUDY BREAK

1. What are the usual characteristics of the inflammatory response?
2. What processes specifically cause each characteristic of the inflammatory response?
3. What is the complement system?
4. Why does combating viral pathogens require a different response by the innate immunity system than combating bacterial pathogens? What are the four main strategies a host uses to protect against viral infections?

44.3 Specific Defences: Adaptive Immunity

Adaptive immunity is a defence mechanism that recognizes specific molecules as being foreign and clears those molecules from the body. The foreign or abnormal molecules that are recognized may be free, as in the case of toxins, or found on the surface of a virus or cell including pathogenic bacteria, cancer cells, pollen, and cells of transplanted organs. Adaptive immunity develops specifically in response to the presence of foreign molecules and therefore takes several days to become effective. This time delay to mount an adaptive response would be a significant problem in the case of invading pathogens were it not for the innate immune system, which combats the invading pathogens in its nonspecific way within minutes after they enter the body.

There are two key distinctions between innate and adaptive immunity:

- innate immunity is nonspecific, whereas adaptive immunity is specific, and
- innate immunity retains no memory of exposure to the pathogen, whereas adaptive immunity retains a memory of the foreign molecule that triggered the response, enabling a rapid, more powerful response if that pathogen is encountered again.

44.3a Antigens Cleared by B Cells or T Cells

A foreign molecule that triggers an adaptive immunity response is called an **antigen** (meaning "*anti*body *gen*erator"). Antigens are macromolecules; most are large proteins (including glycoproteins and lipoproteins) or polysaccharides (including lipopolysaccharides). Some types of nucleic acids can also act as antigens, as can various large, artificially synthesized molecules.

Antigens may be *exogenous*, meaning that they enter the body from the environment, or *endogenous*, meaning that they are generated within the body. Exogenous antigens include antigens on pathogens introduced beneath the skin, antigens in vaccinations, and inhaled and ingested macromolecules, such as toxins. Endogenous antigens include proteins encoded by viruses that have infected cells and altered proteins produced by mutated genes, such as those in cancer cells.

Antigens are recognized in the body by two types of lymphocytes, B cells and T cells. **B cells** differentiate from stem cells in the bone marrow (see Chapter 37). It is easy to remember this as "B for bone." However, the "B" actually refers to the *bursa of Fabricius*, a lymphatic organ found only in birds, where B cells were first discovered. After their differentiation, B cells are released into the blood and carried to capillary beds serving the tissues and organs of the lymphatic system. **T cells** are produced by the division of stem cells in the bone marrow. They are released into the blood and carried to the **thymus**, an organ of the lymphatic system (the "T" in "T cell" refers to the thymus).

The role of lymphocytes in adaptive immunity was demonstrated by experiments in which all of the leukocytes in mice were killed by irradiation. These mice were unable to develop adaptive immunity. Injecting lymphocytes from normal mice into the irradiated mice restored the response; other body cells extracted from normal mice and injected could not restore the response. (For more on the use of mice as an experimental organism in biology, see *Research Organisms: The Mighty Mouse and the Lowly Fruitfly*.)

Research Organisms: The Mighty Mouse and the Lowly Fruitfly

The house mouse (*Mus musculus*) **(Figure 1)** and its cells have been used as models for research on mammalian developmental genetics, immunology, and cancer and have enabled scientists to carry out experiments that would not be practical or ethical with humans. Mice are small, are easy to maintain in the laboratory, and have been used extensively as experimental animals. Gregor Mendel, the founder of genetics, kept mice as part of his studies. More recently, mouse genetic experiments have revealed more than 500 mutants that cause hereditary diseases, immunological defects, and cancer in mammals, including humans. The mouse also has been the model used to introduce and modify

a.

b.

Courtesy of Kevin Wickenheiser, University of Michigan

Edith M. Wallace

Figure 1

Two common research organisms: a mouse and a fruitfly.

genes through genetic engineering, producing giant mice by introducing a human growth hormone gene or producing "knockout" mice, in which a gene of interest is rendered nonfunctional (see Chapter 16) to determine its normal function. By knocking out mouse genes that are homologous to human genes involved in diseases such as cystic fibrosis, researchers can study human diseases in these model organisms, opening pathways to cure human genetic diseases. In 2002, the sequence of the mouse genome was published, enabling researchers to refine and expand their use of the mouse as a model organism for studies of mammalian biology and mammalian diseases.

Similarly, the fruitfly (*Drosophila melanogaster*) (Figure 1) has become a major organism to study basic genetics, aspects of gene regulation, developmental biology, and especially the role of innate immunity against pathogens. This insect can be raised quickly, cheaply, and in massive numbers. *Drosophila* has been instrumental to our understanding of dorsal–ventral patterning during development and has been one of the major organisms

used to identify pathways of immune signalling. Because insects do not have an adaptive immune system, the innate immune system can be studied by itself without the interaction with components of the adaptive system. Toll receptors, important in immune signalling, were found first in *Drosophila* and subsequently were used to identify similar molecules in vertebrates. Mutant lines lacking specific functional genes have been generated in *Drosophila* to study the roles of these genes in all organisms. Because the signalling pathways of the innate immune pathways are highly conserved, information learned on how fruitflies recognize and eliminate pathogens can be transferred to similar studies in other organisms. The genome of *Drosophila* was completed in 2000, allowing for comparisons among and between the genomes of vertebrates and invertebrates. The mighty mouse and the common fruitfly have provided researchers with amazing amounts of information to understand how our bodies work, how similar genes function in different groups of animals, and how we can apply what we learn about one animal to another.

There are two types of adaptive immune responses: **antibody-mediated immunity** (also called *humoral immunity*) and **cell-mediated immunity**. The steps involved in the adaptive immune response are similar for antibody-mediated immunity and cell-mediated immunity:

1. Lymphocyte encounter: The lymphocytes encounter, recognize, and bind to an antigen.
2. Lymphocyte activation: The lymphocytes are activated by binding to the antigen and proliferate by cell division to produce large numbers of clones.
3. Antigen clearance: The activated lymphocytes are responsible for clearing the antigen from the body.
4. Development of immunological memory: Some of the activated lymphocytes differentiate into **memory cells** that circulate in the blood and lymph, ready to initiate a rapid immune response on subsequent exposure to the same antigen.

These steps are explained in more detail in the following discussions of antibody-mediated immunity and cell-mediated immunity.

44.3b Antibody-Mediated Immunity

An adaptive immune response begins as soon as an antigen is encountered in the body and is recognized as foreign.

Antigen Encounter and Recognition by Lymphocytes. - Exogenous antigens are encountered by lymphocytes in the lymphatic system. As already mentioned, the two key lymphocytes that recognize antigens are B cells and T cells. Each B cell and each T cell is specific for a particular antigen, meaning that the cell can bind to only one particular molecular structure. The binding is so specific because the plasma membrane of each B cell and T cell is studded with thousands of identical receptors for the antigen; in B cells, they are called

a. B-cell receptor (BCR)

b. T-cell receptor (TCR)

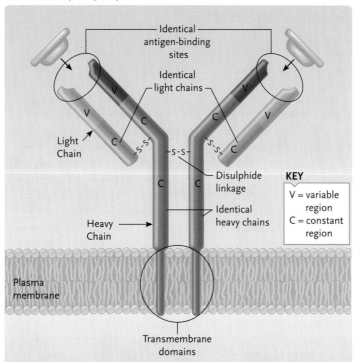
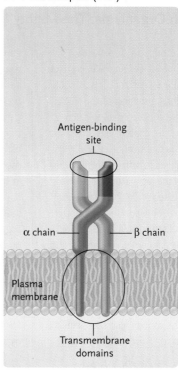

Figure 44.4

(a) Antigen-binding receptor on a B cell and the arrangement of light and heavy polypeptide chains in the antibody molecule. As shown, two sites, one at the tip of each arm of the Y, bind the same antigen. **(b)** Antigen-binding receptor on a T cell.

B-cell receptors (BCRs), and in T cells, they are called **T-cell receptors (TCRs) (Figure 44.4)**. The populations of B cells and T cells contain cells capable of recognizing any antigen, and each antigen can be recognized by multiple cells. For example, each of us has about 10 trillion B cells that collectively have about 100 million different kinds of BCRs. And all of these cells are present *before* the body has encountered the antigens.

The binding between antigen and receptor is an interaction between two molecules that fit together like an enzyme and its substrate. A given BCR or TCR typically does not bind to the whole antigen molecule but to small regions of it called **epitopes** or *antigenic determinants*. Therefore, several different B cells and T cells may bind to the population of a particular antigen encountered in the lymphatic system.

BCRs and TCRs are encoded by different genes and thus have different structures (see *The Generation of Antibody Diversity*). When the BCR on a naive B cell matches a detected antigen, it is activated and may differentiate into a plasma cell that proliferates and secretes antibodies that recognize the same antigen (see Figure 44.4a).

As you will learn in more detail, an antibody molecule is a protein consisting of four polypep-

tide chains. At one end is a region that embeds in a plasma membrane, whereas at the other end are two identical *antigen-binding sites*, regions that bind to a specific antigen. TCRs are simpler than BCRs, consisting of a protein made up of two different polypeptides (see Figure 44.4b). Like BCRs, TCRs have an antigen-binding site at one end and a membrane-embedded region at the other end.

Antibodies. Antibodies are the core molecules of antibody-mediated immunity. Antibodies are large, complex proteins that belong to a class of proteins known as *immunoglobulins* (Ig). Each antibody molecule consists of four polypeptide chains: two identical *light chains* and two identical *heavy chains* about twice or more the size of the light chain (see Figure 44.4a). The chains are held together in the complete protein by disulphide (−S−S−) linkages and fold into a Y-shaped structure. The bonds between the two arms of the Y form a hinge that allows the arms to flex independently of one another.

Each polypeptide chain of an antibody molecule has a *constant region* and a *variable region*. Each antibody type has the same amino acid sequence in the constant region of the heavy chain and likewise for the constant region of the light chain. The variable

The Generation of Antibody Diversity

The human genome has approximately 20 000 to 25 000 genes, far fewer than necessary to encode 100 million different antibodies if two genes encoded one antibody, one gene for the heavy chain and one for the light chain. The great diversity in antigen-binding capability of those receptors is generated in a different way from one gene per chain. During B-cell differentiation, the DNA segments that encode parts of the light and heavy chains undergo three rearrangements. The genes for the two different subunits of the T-cell receptor undergo similar rearrangements.

The light chain expressed by an undifferentiated B cell is encoded by three types of DNA segments, and one of each type is needed to make a complete, functional light-chain gene. In humans, about 40 different V segments encode most of the variable regions of the chain, 5 different J (joining) segments encode the rest of the variable region, only one copy of the segment makes up the constant (C) part of the chain. Thus, a complete light chain comprises one V segment, adjacent to one J segment, adjacent to the C region, which is the same for all light chains regardless of V or J segment usage (see Figure 44.4).

During B-cell differentiation, a DNA rearrangement occurs in which one random V segment and one random J segment join with the C segment to form a functional light-chain gene. During this assembly, there is a deletion of DNA between the V and J segments, and the positions at which the DNA breaks and rejoins in the V- and J-joining reaction occur randomly over a distance of several nucleotides, which adds greatly to the variability of the final gene assembly. The DNA between the J segment and the C segment becomes an intron in the final assembled gene. Transcription of this newly assembled gene produces a typical pre-mRNA molecule (see Chapter 14). The introns are removed during the production of the mRNA by RNA processing. Translation of the mRNA produces the light chain with both the variable and the constant regions.

The assembly of functional heavy-chain genes occurs similarly. However, whereas light-chain genes have one C segment, heavy-chain genes have five types of C segments, each of which encodes one of the constant regions of IgM, IgD, IgG, IgE, and IgA. The inclusion of one of the five C-segment types in the functional heavy-chain gene therefore specifies the class of antibody that will be made by the B cell.

regions of both the heavy and the light chains, by contrast, have different amino acid sequences for each antibody molecule in a population. Structurally, the variable regions are the top halves of the polypeptides in the arms of the Y-shaped molecule. The three-dimensional folding of the heavy chain and light chain variable regions of each arm creates the antigen-binding site. The antigen-binding site is identical on both arms of the same antibody molecule because both ends of the Y have the same amino acid sequences in their variable regions. However, the antigen-binding sites are different from antibody molecule to antibody molecule (produced by different B-cell clones) because of the amino acid differences in the variable regions of the two chain types.

The constant regions of the heavy chains in the tail part of the Y-shaped structure determine the *class* of the antibody, that is, its location and function. Humans have five different classes of antibodies: IgM, IgG, IgA, IgE, and IgD **(Table 44.2, p. 1096)**.

IgM antibodies are the first antibodies produced in the early stages of an antibody-mediated response after BCRs are activated and B cells differentiate into plasma cells. When they bind an antigen, IgM antibodies activate the complement system and stimulate the phagocytic activity of macrophages.

IgG antibodies circulate in the highest concentration in the blood and lymphatic system, where they also stimulate phagocytosis and activate the complement system when it binds an antigen. IgG is produced in large amounts when the body is exposed a second time to the same antigen.

IgA is found mainly in body secretions such as saliva, tears, breast milk, and the mucus coating of body cavities such as the lungs, digestive tract, and vagina. In these locations, the antibodies bind to surface groups on pathogens and block their attachment to body surfaces. Breast milk transfers IgA antibodies, and thus immunity, to a nursing infant.

IgE is secreted by plasma cells of the skin and the tissues lining the gastrointestinal tract and respiratory tract. IgE binds to basophils and mast cells, where it mediates many allergic responses, such as hay fever, asthma, and hives. When its specific antigen binds to IgE, the basophils or mast cells release histamine, which triggers an inflammatory response. IgE also contributes to mechanisms that combat infection by parasitic worms.

Table 44.2   Five Classess of Antibodies

Class	Structure	Location	Functions
IgM		Surfaces of unstimulated B cells; free in circulation	First antibodies to be secreted by B cells in primary response. When bound to antigen, promotes agglutination reaction, activates complement system, and stimulates phagocytic activity of macrophages.
IgG		Blood and lymphatic circulation	Most abundant antibody in primary and secondary responses. Crosses placenta, conferring passive immunity to fetus; stimulates phagocytosis and activates complement system.
IgA		Body secretions such as tears, breast milk, saliva, and mucus	Blocks attachment of pathogens to mucus membranes; confers passive immunity for breast-fed infants
IgE		Skin and tissues lining gastrointestinal and respiratory tracts (secreted by plasma cells)	Stimulates mast cells and basophils to release histamine; triggers allergic responses
IgD		Surface of unstimulated B cells	Membrane receptor for mature B cells; probably important in B-cell activation (clonal selection)

IgD occurs with IgM as a receptor on the surfaces of B cells; its function is not well understood but may be involved in B-cell activation.

T-Cell Activation. Let us now follow the development of an antibody-mediated immune response by linking the recognition of an antigen by lymphocytes, the activation of lymphocytes by antigen binding, and the production of antibodies. Typically, the pathway begins when a type of T cell becomes activated and follows the steps outlined in **Figure 44.5** that determine the fate of pathogenic bacteria that have been introduced under the skin. Circulating viruses in the blood follow the same pathway.

First, a type of phagocyte called a **dendritic cell** engulfs a bacterium in the infected tissue by phagocytosis (**Figure 44.6**, step 1). Dendritic cells are so named because they have many surface projections resembling the dendrites of neurons. They have the same origin as leukocytes and recognize a bacterium as foreign by the same recognition mechanism used by macrophages in the innate immune system. In essence, the dendritic cell is part of the innate immunity system, but its primary role is to stimulate the development of an adaptive immune response.

Engulfing a bacterium activates the dendritic cell; the cell now migrates to a nearby lymph node. Within the dendritic cell, the endocytic vesicle containing the bacterium fuses with a lysosome. In the

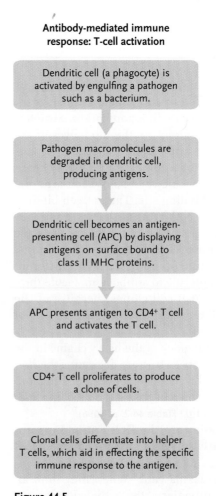

Antibody-mediated immune response: T-cell activation

Dendritic cell (a phagocyte) is activated by engulfing a pathogen such as a bacterium.

↓

Pathogen macromolecules are degraded in dendritic cell, producing antigens.

↓

Dendritic cell becomes an antigen-presenting cell (APC) by displaying antigens on surface bound to class II MHC proteins.

↓

APC presents antigen to CD4+ T cell and activates the T cell.

↓

CD4+ T cell proliferates to produce a clone of cells.

↓

Clonal cells differentiate into helper T cells, which aid in effecting the specific immune response to the antigen.

Figure 44.5
An outline of T-cell activation in antibody-mediated immunity.

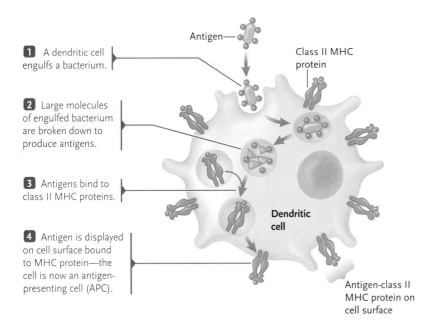

1 A dendritic cell engulfs a bacterium.

2 Large molecules of engulfed bacterium are broken down to produce antigens.

3 Antigens bind to class II MHC proteins.

4 Antigen is displayed on cell surface bound to MHC protein—the cell is now an antigen-presenting cell (APC).

Antigen—

Class II MHC protein

Dendritic cell

Antigen-class II MHC protein on cell surface

Figure 44.6
Generation of an antigen-presenting cell after a dendritic cell engulfs a bacterium.

receptor on the T cell helps link the two cells together.

When the APC binds to the CD4+ T cell, the APC secretes an *interleukin* (meaning "between leukocytes"), a type of cytokine, which activates the T cell (see Figure 44.7, step 3). The activated T cell then secretes cytokines (step 4), which act in an autocrine manner (see Chapter 35) to stimulate **clonal expansion**, the proliferation of the activated CD4+ T cell by cell division to produce a clone of cells. These clonal cells differentiate into **helper T cells** (step 5), so named because they assist with the activation of B cells. A helper T cell is an example of an **effector T cell**, meaning that it is involved in effecting—bringing about—the specific immune response to the antigen.

lysosome, the bacterium's proteins are degraded into short peptides, which are antigens (see Figure 44.6, step 2). The antigens bind to **class II major histocompatibility complex (MHC)** proteins (step 3), and the interacting molecules then migrate to the cell surface where the antigen is displayed (step 4). These steps in the lymph node, which are recapped in **Figure 44.7, p. 1098,** step 1, have converted the cell into an **antigen-presenting cell (APC)**, ready to present the antigen to T cells in the next step of antibody-mediated immunity.

MHC proteins are named for a large cluster of genes encoding them, called the **major histocompatibility complex**. The complex spans 4 million base pairs and contains 128 genes. Many of these genes play important roles in the immune system. Each individual of each vertebrate species has a unique combination of MHC proteins on almost all body cells, meaning that no two individuals of a species except identical siblings are likely to have exactly the same MHC proteins on their cells. There are two classes of MHC proteins, class I and class II, which have different functions in adaptive immunity, as we will see.

The key function of an APC is to present the antigen to a lymphocyte. In the antibody-mediated immune response, the APC presents the antigen, bound to a class II MHC protein, to a type of T cell in the lymphatic system called a **CD4+ T cell** because it has receptors named CD4 on its surface. A specific CD4+ T cell, which has a TCR with an antigen-binding site that recognizes the antigen, binds to the antigen on the APC (see Figure 44.7, step 2). The CD4

B-Cell Activation. Antibodies are produced and secreted by B cells. The activation of a B cell that makes the specific antibody against an antigen requires the B cell to present the antigen on its surface and then to link with a helper T cell that has differentiated as a result of encountering and recognizing the same antigen. The process is outlined in **Figures 44.7** and **44.8, p. 1098**.

The process of antigen presentation on a B-cell surface begins when BCRs on the B cell interact directly with soluble bacterial (in our example) antigens in the blood or lymph. Once the antigen binds to a BCR, the complex is taken into the cell and the antigen is processed in the same way as in dendritic cells, being broken down into smaller fragments, culminating with a presentation of each antigen-derived peptide fragment on the B-cell surface in a complex with class II MHC proteins (see Figure 44.7, step 6).

When one of the helper T cells produced above encounters a B cell displaying the same antigen, usually in a lymph node or in the spleen, the T and B cells become tightly linked together (see Figure 44.7, step 7). The linkage depends on the TCRs, which recognize and bind the antigen displayed by the class II MHC molecules on the surface of the B cell, and on CD4, which stabilizes the binding as it did for T-cell binding to the dendritic cell. The linkage between the cells first stimulates the helper T cell to secrete interleukins that activate the B cell and then stimulates the B cell to proliferate, producing a clone of those B cells with identical B-cell receptors (step 8). Some of the cloned cells differentiate into relatively short-lived **plasma cells**, which now secrete the same antibody that was

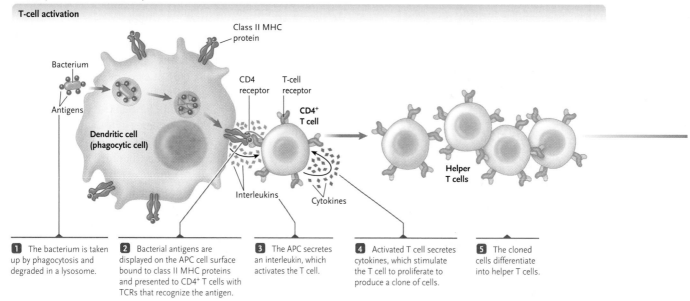

T-cell activation

Class II MHC protein

Bacterium

Antigens

Dendritic cell (phagocytic cell)

CD4 receptor

T-cell receptor

CD4⁺ T cell

Interleukins

Cytokines

Helper T cells

1 The bacterium is taken up by phagocytosis and degraded in a lysosome.

2 Bacterial antigens are displayed on the APC cell surface bound to class II MHC proteins and presented to CD4⁺ T cells with TCRs that recognize the antigen.

3 The APC secretes an interleukin, which activates the T cell.

4 Activated T cell secretes cytokines, which stimulate the T cell to proliferate to produce a clone of cells.

5 The cloned cells differentiate into helper T cells.

Figure 44.7
The antibody-mediated immune response.

displayed on the parental B cell's surface to circulate in lymph and blood. Others differentiate into **memory B cells**, which are long-lived cells that set the stage for a much more rapid response should the same antigen be encountered later in life (step 9).

Clonal selection is the process by which a lymphocyte is specifically selected for cloning when it encounters a foreign antigen from among a randomly generated, enormous diversity of lymphocytes with receptors that specifically recognize the antigen **(Figure 44.9)**. The process of clonal selection was proposed in the 1950s by several scientists, most notably F. Macfarlane Burnet, Niels Jerne, and David Talmage. Their proposals, made long before the mechanism was understood, described clonal selection as a form of natural selection operating in miniature: antigens select the cells recognizing them, which reproduce and become dominant in the B-cell population. Burnet received the Nobel Prize in 1960 for his research in immunology.

Clearing the Body of Foreign Antigens. How do the antibodies produced in an antibody-mediated immune response clear different types of foreign antigens from the body? Toxins produced by invading bacteria, such as tetanus toxin, can be *neutralized* by antibodies **(Figure 44.10a, p. 1100)**. The antibodies bind to the toxin molecules, inactivating them.

Antibodies bind to antigens on the surfaces of intact bacteria at an infection site or in the circulatory system. Because the two arms of an antibody molecule bind to different copies of the antigen molecule, an antibody molecule may bind to two bacteria with the same antigen. A population of antibodies can link many bacteria together into a lattice, causing *agglutination*, or clumping of the bacteria **(Figure 44.10b, p. 1100)**. Agglutination immobilizes the bacteria, preventing them from

Antibody-mediated immune response: B-cell activation

A BCR on a B cell recognizes an antigen on the surface of a bacterium and the bacterium is engulfed.

↓

Pathogen macromolecules are degraded in the B cell, producing antigens.

↓

B cell displays antigens on its surface bound to class II MHC proteins.

↓

Helper T cell with TCR that recognizes the same antigen links to the B cell.

↓

Helper T cell secretes interleukins that activate the B cell.

↓

B cell proliferates to produce a clone of cells.

↓

Some B-cell clones differentiate into plasma cells, which secrete antibodies specific to the antigen, and others differentiate into memory B cells.

Figure 44.8
An outline of B-cell activation in antibody-mediated immunity.

infecting cells. Antibodies can also agglutinate viruses, also preventing them from infecting cells.

More importantly, antibodies bound to antigens aid the innate immune response that was initially set off by the pathogens by stimulating the complement

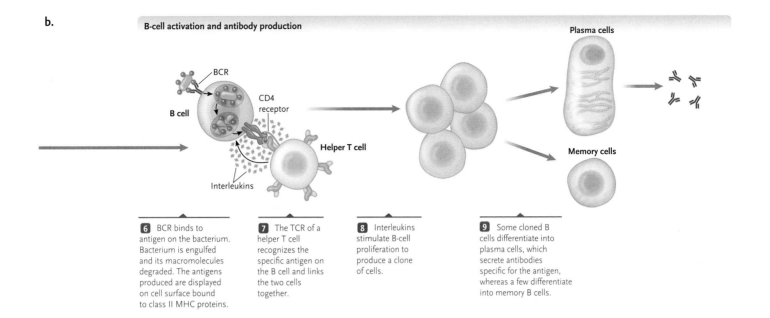

b.

B-cell activation and antibody production

Plasma cells

BCR

CD4 receptor

B cell

Helper T cell

Memory cells

Interleukins

6 BCR binds to antigen on the bacterium. Bacterium is engulfed and its macromolecules degraded. The antigens produced are displayed on cell surface bound to class II MHC proteins.

7 The TCR of a helper T cell recognizes the specific antigen on the B cell and links the two cells together.

8 Interleukins stimulate B-cell proliferation to produce a clone of cells.

9 Some cloned B cells differentiate into plasma cells, which secrete antibodies specific for the antigen, whereas a few differentiate into memory B cells.

system. Membrane attack complexes are formed and insert themselves into the plasma membranes of the bacteria, leading to their lysis and death. In the case of virus infections, membrane attack complexes can insert themselves into the membranes surrounding enveloped viruses, which disrupts the membrane and prevents the viruses from infecting cells.

Antibodies also enhance phagocytosis of bacteria and viruses. Phagocytic cells have receptors on their surfaces that recognize the heavy-chain end of antibodies (the end of the molecule opposite the antigen-binding sites). Antibodies bound to bacteria or viruses therefore bind to phagocytic cells, which then engulf the pathogens and destroy them.

For simplicity, the adaptive immune response has been described here in terms of a single antigen. Pathogens have many different types of antigens on their surfaces, which means that many different B cells are stimulated to proliferate and many different antibodies are produced. Pathogens are therefore attacked by many different antibodies, each targeted to one antigen on the pathogen's surface.

Immunological Memory. Once an immune reaction has run its course and the invading pathogen or toxic molecule has been eliminated from the body, division of the plasma cells and T-cell clones stops. Most or all of the clones die and are eliminated from the bloodstream and other body fluids. However, long-lived memory B cells and **memory helper T cells** (which differentiated from helper T cells) remain in an inactive state in the lymphatic system. Their persistence provides an **immunological memory** of the foreign antigen.

Immunological memory is illustrated in **Figure 44.11, p. 1100.** When the body is exposed to a foreign antigen for the first time, a **primary immune response** results,

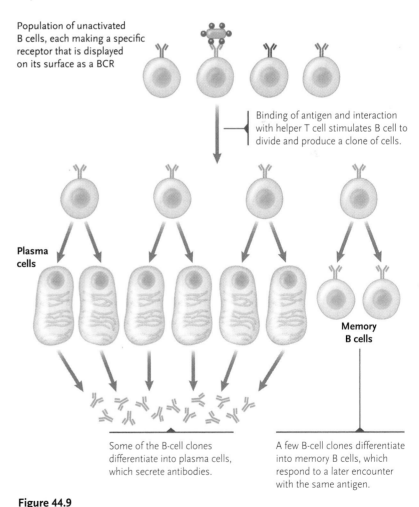

Population of unactivated B cells, each making a specific receptor that is displayed on its surface as a BCR

Binding of antigen and interaction with helper T cell stimulates B cell to divide and produce a clone of cells.

Plasma cells

Memory B cells

Some of the B-cell clones differentiate into plasma cells, which secrete antibodies.

A few B-cell clones differentiate into memory B cells, which respond to a later encounter with the same antigen.

Figure 44.9
Clonal selection. The binding of an antigen to a B cell that already displays a specific antibody to that antigen stimulates the B cell to divide and differentiate into plasma cells, which secrete the antibody, and memory cells, which remain in the circulation ready to mount a response against the antigen at a later time.

a. Neutralization

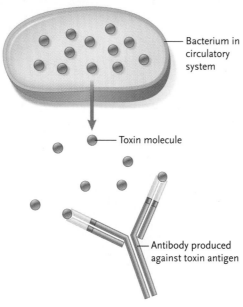

Bacterium in circulatory system

Toxin molecule

Antibody produced against toxin antigen

b. Agglutination

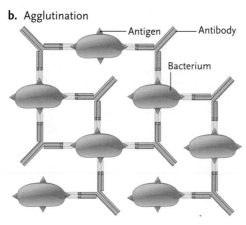

Antigen — Antibody

Bacterium

Figure 44.10
Examples of clearing antigens from the body.

following the steps already described. The first antibodies appear in the blood in 3 to 14 days, and by week 4, the primary response has essentially gone away. IgM is the first antibody type produced and secreted into the bloodstream in a primary immune response.

This primary immune response curve is followed whenever a new foreign antigen enters the body.

When a foreign antigen enters the body for a second time, a **secondary immune response** results (see Figure 44.11). The secondary response is more rapid than the primary response because it involves the memory B cells and memory T cells that have been stored. It does not have to initiate the clonal selection of a new B cell and T cell. Moreover, less antigen is needed to elicit a secondary response than a primary response, and many more antibodies are produced. The predominant antibody produced in a secondary immune response is IgG; the switch occurs at the gene level in the memory B cells.

Immunological memory forms the basis of vaccinations, in which antigens in the form of living or dead pathogens or antigenic molecules themselves are introduced into the body. After the immune response, memory B cells and memory T cells remaining in the body can mount an immediate and intense immune reaction against similar antigens. As mentioned in *Why It Matters*, Edward Jenner introduced the cowpox virus—a virus closely related to, but less virulent than, the smallpox virus—into healthy individuals who initiated a primary immune response. After the response ran its course, a bank of memory B cells and memory T cells remained in the body, able to recognize quickly the similar antigens of the smallpox virus and initiate a secondary immune response. Similarly, the polio vaccine developed by Jonas Salk uses polioviruses that have been inactivated by exposing them to formaldehyde. Although the viruses are inactive, their surface groups can still act as antigens. The antigens trigger an immune response, leaving memory B and T cells able to mount an intense immune response against active polioviruses.

Active and Passive Immunity. **Active immunity** is the production of antibodies in response to exposure to a foreign antigen, as has just been described. **Passive immunity** is the acquisition of antibodies as a result of direct transfer from another person. This form of immunity provides immediate protection against the

Figure 44.11
Immunological memory: primary and secondary responses to the same antigen.

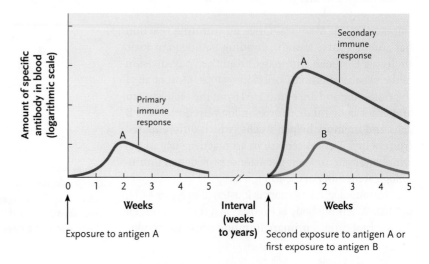

Amount of specific antibody in blood (logarithmic scale)

Secondary immune response

Primary immune response

A

A

B

Weeks

Interval (weeks to years)

Weeks

Exposure to antigen A

Second exposure to antigen A or first exposure to antigen B

antigens that the antibodies recognize without the person receiving the antibodies having developed a primary immune response. Examples of passive immunity include the transfer of IgG antibodies from the mother to the fetus through the placenta and the transfer of IgA antibodies in the first breast milk fed from the mother to the baby. Compared with active immunity, passive immunity is a short-lived phenomenon with no memory, in that the antibodies typically break down within a month. However, in that time, the protection plays an important role. For example, a breast-fed baby is protected until it is able to mount an immune response itself, an ability that is not present until about a month after birth.

Drug Effects on Antibody-Mediated Immunity. Several drugs used to reduce the rejection of transplanted organs target helper T cells. Cyclosporin A, used routinely after organ transplants, blocks the activation of helper T cells and, in turn, the activation of B cells. Although very successful, cyclosporin and other immunosuppressive drugs also leave the treated individual more susceptible to infection by pathogens.

44.3c Cell-Mediated Immunity

In cell-mediated immunity, cytotoxic T cells directly destroy host cells infected by intracellular pathogens **(Figure 44.12).** The killing process begins when some of the pathogens are broken down by cytoplasmic enzymes inside infected host cells, and the smaller protein fragments (or antigen-derived peptide fragments) themselves act as antigens. These antigens bind to class I MHC proteins, which are delivered to the cell surface by essentially the same mechanisms as in B cells (step 1). At the surface, the antigens are displayed by the class I MHC protein and the cell then functions as an APC.

The APC presents the antigen to a type of T cell in the lymphatic system called a **CD8$^+$ T cell** because it has receptors named CD8 on its surface in addition to the TCRs. The presence of a CD8 receptor distinguishes this type of T cell from that involved in antibody-mediated immunity. A specific CD8$^+$ T cell that has a TCR that recognizes the antigen binds to that antigen on the APC (see Figure 44.12, step 2). The CD8 receptor on the T cell helps the two cells link together.

The link between the APC and the CD8$^+$ T cell activates the T cell, which then proliferates to form a clone. Some of the cells differentiate to become **cytotoxic T cells** (see Figure 44.12, step 3), whereas a few differentiate into *memory cytotoxic T cells*. Cytotoxic T cells are another type of effector T cell. TCRs on the cytotoxic T cells again recognize the antigen bound to class I MHC proteins on the infected cells (the APCs) (step 4). The cytotoxic T cell then destroys the infected cell using mechanisms similar to those used by NK cells. That is, an activated cytotoxic T cell releases perforin, which creates pores in the membrane of the target cell. The leakage of ions and other molecules

Cell-mediated immune response

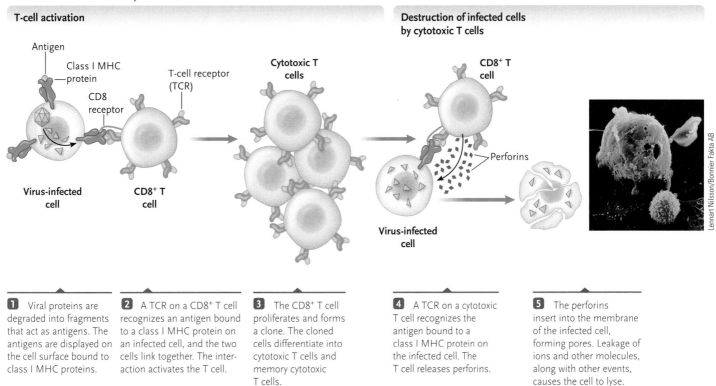

| **T-cell activation** | | | **Destruction of infected cells by cytotoxic T cells** | |

| **1** Viral proteins are degraded into fragments that act as antigens. The antigens are displayed on the cell surface bound to class I MHC proteins. | **2** A TCR on a CD8$^+$ T cell recognizes an antigen bound to a class I MHC protein on an infected cell, and the two cells link together. The interaction activates the T cell. | **3** The CD8$^+$ T cell proliferates and forms a clone. The cloned cells differentiate into cytotoxic T cells and memory cytotoxic T cells. | **4** A TCR on a cytotoxic T cell recognizes the antigen bound to a class I MHC protein on the infected cell. The T cell releases perforins. | **5** The perforins insert into the membrane of the infected cell, forming pores. Leakage of ions and other molecules, along with other events, causes the cell to lyse. |

Figure 44.12
The cell-mediated immune response.

Some Cancer Cells Kill Cytotoxic T Cells to Defeat the Immune System

Among the arsenal of weapons employed by cytotoxic T cells to eliminate abnormal, infected, or cancerous body cells is apoptosis mediated by the *Fas–FasL* system. Fas is a receptor that occurs on the surfaces of many cells; FasL is a ligand displayed on the surfaces of some cell types, including cytotoxic T cells. If a cell carrying the Fas receptor contacts a cytotoxic T cell with the FasL signal displayed on its surface, a cascade of internal reactions initiates apoptosis and kills the cell with the Fas receptor.

Surprisingly, cytotoxic T cells also carry the Fas receptor, so they can kill each other by displaying the FasL signal. This mutual killing plays an important role in reducing the level of an immune reaction after a pathogen has been eliminated. In the case of **immune privilege,** cells in specific regions such as the cornea, nervous tissue, and testes express FasL to induce the apoptosis of infiltrating cytotoxic T cells and reduce inflammation.

Some cancer cells survive elimination by the immune system by making and displaying FasL and killing any cytotoxic T cells that attack the tumour. This was found first in patients suffering from malignant melanoma, a dangerous skin cancer, who had a breakdown product associated with FasL in their bloodstream.

Proteins extracted from melanoma cells, or sections made from melanoma tissue, tested positive (using antibodies) for the presence of FasL, indicating that FasL was present in the tumour cells. Similarly, the expression of FasL mRNA was also detected in the tumour cells. However, no Fas receptor was found in these samples, suggesting that Fas synthesis was turned off in the tumour cells.

FasL in melanoma cells kills cytotoxic T cells that invade the tumour, whereas the absence of Fas receptors ensures that the tumour cells do not kill each other. The presence of FasL and absence of the Fas receptor may explain why melanomas, and many other types of cancer, are rarely destroyed by the immune system.

Melanoma cells originate from pigment cells in the skin called *melanocytes*. Normal melanocytes do not contain FasL, indicating that synthesis of the protein is turned on as part of the transformation from normal melanocytes into cancer cells.

This research could lead to an effective treatment for cancer using the Fas–FasL system. If melanoma cells could be induced to make Fas as well as FasL, for example, they might eliminate a tumour by killing each other!

---

through the pores causes the infected cell to rupture. The cytotoxic T cell also secretes proteases that enter infected cells through the newly created pores and cause it to self-destruct by apoptosis (see Figure 44.12, step 5 and photo inset). The rupture of dead infected cells releases the pathogens to the interstitial fluid, where they are open to attack by antibodies and phagocytes.

Cytotoxic T cells can also kill cancer cells if their class I MHC molecules display fragments of altered cellular proteins that do not normally occur in the body. Another mechanism used by cytotoxic T cells to kill cells, and a process used by some cancer cells to defeat the mechanism, is described in *Some Cancer Cells Kill Cytotoxic T Cells to Defeat the Immune System.*

44.3d The Use of Antibodies in Research

The ability to generate antibodies against essentially any antigen provides an invaluable research tool for scientists. Most antibodies are obtained by injecting a molecule into a test animal such as a mouse, rabbit, or goat and collecting and purifying the antibodies from the blood. Scientists can then attach a visible marker such as a dye molecule or heavy metal atom to the antibody and determine when and where specific biological molecules are found in cells or tissues. Antibodies also can be used to "grab" a molecule of interest from a mixture of molecules by attaching antibodies to that molecule to plastic beads that are packed into a glass column. When the mixture is poured through the column, the molecule remains bound to the antibody in the column. It is then released from the column in purified form by adding a reagent that breaks the antigen–antibody bonds.

Injecting a molecule of interest into a test animal typically produces a wide spectrum of antibodies that react with different parts of the antigen. Some antibodies may cross-react with other similar antigens, producing false results that can complicate the research. These problems have been solved by producing **monoclonal antibodies,** each of which reacts only against the same segment (epitope) of a single antigen. In addition to their use in scientific research, monoclonal antibodies are also widely used in medical applications such as pregnancy tests, screening for prostate cancer, and testing for AIDS and other sexually transmitted diseases.

STUDY BREAK

1. How, in general, do the antibody-mediated and cell-mediated immune responses help clear the body of antigens?
2. Describe the general structure of an antibody molecule.
3. What is clonal selection?
4. How does immunological memory work?

44.4 Malfunctions and Failures of the Immune System

The immune system is highly effective, but it is not foolproof. Some malfunctions of the immune system cause the body to react against its own proteins or cells, producing *autoimmune diseases*. In addition, some viruses and other pathogens have evolved means to avoid destruction by the immune system. A number of these pathogens, including the AIDS virus, even use parts of the immune response to promote infection. Another malfunction causes the *allergic reactions* that many of us experience from time to time.

44.4a The Immune System Normally Protects against Attack

B cells and T cells are involved in the development of **immunological tolerance**, which protects the body's own molecules from attack by the immune system. Although the process is not understood, molecules present in an individual from birth are not recognized as foreign by circulating B and T cells and do not elicit an immune response. During their initial differentiation in the bone marrow and thymus, any B and T cells that react with "self" molecules carried by MHC proteins become suppressed or are induced to kill themselves by apoptosis. The process of excluding self-reactive B and T cells goes on throughout the life of an individual.

Evidence that immunological tolerance is established early in life comes from experiments with mice. For example, if a foreign protein is injected into a mouse at birth, during the period in which tolerance is established, the mouse will not develop antibodies against the protein if it is injected later in life. Similarly, if mutant mice are produced that lack a given complement protein, so that the protein is absent during embryonic development, they will produce antibodies against that protein if it is injected during adult life. Normal mice do not produce antibodies if the protein is injected.

44.4b When Immunological Tolerance Fails

The mechanisms setting up immunological tolerance sometimes fail, leading to an **autoimmune reaction**— the production of antibodies against molecules of the body. In most cases, the effects of such antiself antibodies are not serious enough to produce recognizable disease. However, in some individuals—about 5 to 10% of the human population—antiself antibodies cause serious problems.

For example, *type 1 diabetes* (see Chapter 35) is an autoimmune reaction against the pancreatic beta cells that produce insulin. The antiself antibodies gradually eliminate the beta cells until the individual is incapable of producing insulin. *Systemic lupus erythematosus* (*lupus*) is caused by production of a wide variety of antiself antibodies against blood cells, blood platelets,

and internal cell structures and molecules such as mitochondria and proteins associated with DNA in the cell nucleus. People with lupus often become anemic and have problems with blood circulation and kidney function because the antibodies, combined with body molecules, accumulate and clog capillaries and the microscopic filtering tubules of the kidneys. Lupus patients also may develop antiself antibodies against the heart and kidneys. *Rheumatoid arthritis* is caused by a self-attack on connective tissues, particularly in the joints, causing pain and inflammation. *Multiple sclerosis* results from an autoimmune attack against a protein of the myelin sheaths that insulate the surfaces of neurons. Multiple sclerosis can seriously disrupt nervous function, producing such symptoms as muscle weakness and paralysis, impaired coordination, and pain.

The causes of most autoimmune diseases are unknown. In some cases, an autoimmune reaction can be traced to injuries that expose body cells or proteins that are normally inaccessible to the immune system, such as the lens protein of the eye, to B and T cells. In other cases, as in type 1 diabetes, an invading virus stimulates the production of antibodies that can also react with self proteins. Antibodies against the Epstein-Barr and hepatitis B viruses can react against myelin basic protein, the protein attacked in multiple sclerosis. Sometimes, environmental chemicals, drugs, or mutations alter body proteins so that they appear foreign to the immune system and come under attack.

Some viruses use parts of the immune system to get a free ride to the cell interior. For example, the AIDS virus has a surface molecule that is recognized and bound by the CD4 receptor on the surface of helper T cells. Binding to CD4 locks the virus to the cell surface and stimulates the membrane covering the virus to fuse with the plasma membrane of the helper T cell. (The protein coat of the virus is wrapped in a membrane derived from the plasma membrane of the host cell in which it was produced.) The fusion introduces the virus into the cell, initiating the infection and leading to the destruction and death of the T cell. (Further details on human immunodeficiency virus [HIV] infection and AIDS are presented in *Applied Research: HIV and AIDS*.)

44.4c Allergies: Overactivity of the Immune System

The substances responsible for allergic reactions form a distinct class of antigens called **allergens**, which induce B cells to secrete an overabundance of IgE antibodies **(Figure 44.13, p. 1104)**. IgE antibodies, in turn, bind to receptors on mast cells in connective tissue and on **basophils**, a type of leukocyte in the blood (see Table 44.1), inducing them to secrete histamine, which produces a severe inflammation. Most of the inflammation occurs in tissues directly exposed to the allergen, such as the surfaces of the eyes, the lining of the nasal passages, and the air passages of the lungs. Signal molecules

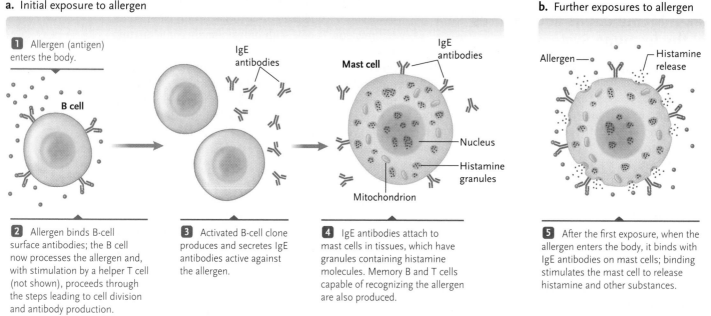

a. Initial exposure to allergen

1 Allergen (antigen) enters the body.

B cell

IgE antibodies

Mast cell

IgE antibodies

Nucleus

Histamine granules

Mitochondrion

2 Allergen binds B-cell surface antibodies; the B cell now processes the allergen and, with stimulation by a helper T cell (not shown), proceeds through the steps leading to cell division and antibody production.

3 Activated B-cell clone produces and secretes IgE antibodies active against the allergen.

4 IgE antibodies attach to mast cells in tissues, which have granules containing histamine molecules. Memory B and T cells capable of recognizing the allergen are also produced.

b. Further exposures to allergen

Allergen

Histamine release

5 After the first exposure, when the allergen enters the body, it binds with IgE antibodies on mast cells; binding stimulates the mast cell to release histamine and other substances.

Figure 44.13

The response of the body to allergens. **(a)** The steps in sensitization after initial exposure to an allergen. **(b)** Production of an allergic response by further exposures to the allergen.

released by activated mast cells also stimulate mucosal cells to secrete floods of mucus and cause smooth muscle in airways to constrict (histamine also causes airway constriction). The resulting allergic reaction can vary in severity from a mild irritation to serious and even life-threatening debilitation. *Asthma* is a severe response to allergens involving constriction of airways in the lungs. Antihistamines, medications that block histamine receptors, are usually effective in countering the effects of the histamine released by mast cells.

An individual is *sensitized* by a first exposure to an allergen, which may produce only mild allergic symptoms or no reaction at all (see Figure 44.13a). However, the sensitization produces memory B and T cells. At subsequent exposures, the system is poised to produce a greatly intensified allergic response (see Figure 44.13b).

In some persons, inflammation stimulated by an allergen is so severe that the reaction brings on a life-threatening condition called **anaphylactic shock**. Extreme swelling of air passages in the lungs interferes with breathing, and massive leakage of fluid from capillaries causes the blood pressure to drop precipitously. Death may result in minutes if the condition is not treated promptly. In persons who have become sensitized to the venom of wasps and bees, for example, a single sting may bring on anaphylactic shock within minutes. Allergies developed against drugs such as penicillin and certain foods can have the same drastic effects. Anaphylactic shock can be controlled by immediate injection of epinephrine (adrenaline), which reverses the condition by constricting blood vessels and dilating air passages in the lungs.

STUDY BREAK

1. What is immunological tolerance?
2. Explain how a failure in the immune system can result in an allergy.

44.5 Defences in Other Organisms

All organisms must be able to defend themselves, and we can compare what we know in mammals with what we know about the immune systems of other organisms. Molecular studies in sharks and rays have revealed DNA sequences that are clearly related to the sequences coding for antibodies in mammals, and sharks produce antibodies capable of recognizing and binding specific antigens. Antibody diversity is produced by the same kinds of genetic rearrangements in both sharks and mammals, although the embryonic gene segments for the two polypeptides are arranged differently in these organisms. Sharks also mount nonspecific defences, including the production of a steroid that appears to kill bacteria and neutralizes viruses nonspecifically and with high efficiency.

We have demonstrated that invertebrates lack the adaptive immune system and instead rely on innate responses that allow them to eliminate pathogens via phagocytosis, encapsulation, or the expression of antimicrobial peptides. Similarly, plants rely on innate responses only and recognize and eliminate the pathogens or infected cells, often using plant antimicrobial peptides.

Applied Research: HIV and AIDS

Acquired immune deficiency syndrome (AIDS) is a constellation of disorders that follows infection by the **human immunodeficiency virus, HIV (Figure 1).** First reported in the late 1970s, HIV now infects more than 40 million people worldwide, 64% of them in Africa. AIDS is a potentially lethal disease, although drug therapy has reduced the death rate for HIV-infected individuals.

HIV is transmitted when an infected person's body fluids, especially blood or semen, enter the blood or tissue fluids of another person. The entry may occur during vaginal, anal, or oral intercourse, or the virus may be transmitted via contaminated needles shared by intravenous drug users or from infected mothers to their infants during pregnancy, birth, and nursing. HIV is rarely transmitted through casual contact, food, or body products such as saliva, tears, urine, or feces.

The primary cellular hosts for HIV are macrophages and helper T cells, which are ultimately destroyed by the virus. Infection makes helper T cells unavailable for the stimulation and proliferation of B cells and cytotoxic T cells. The assault on lymphocytes and macrophages cripples the immune system and makes the body highly vulnerable to otherwise non–life-threatening infections.

In 1996, researchers confirmed the process by which HIV initially infects its primary target, the helper T cells. First, a glycoprotein of the viral coat, called *gp120*, attaches the virus to a helper T cell by binding to its CD4 receptor. Another viral protein triggers fusion of the viral surface membrane with the T-cell plasma membrane, releasing the virus into the cell. Once inside, a viral enzyme, *reverse transcriptase,* uses the viral RNA as a template for making a DNA copy. (When it is outside a host cell, the genetic material of HIV is RNA rather than DNA.) Another viral enzyme, *integrase,* then splices the viral DNA into the host cell's DNA **(Figure 2).** Once it is part of the host cell DNA, the viral DNA is replicated and passed on as the cell divides. As part of the host cell DNA, the virus is effectively hidden in the helper T cell and protected from attack by the immune system.

When the infected helper T cell is stimulated by an antigen, the viral

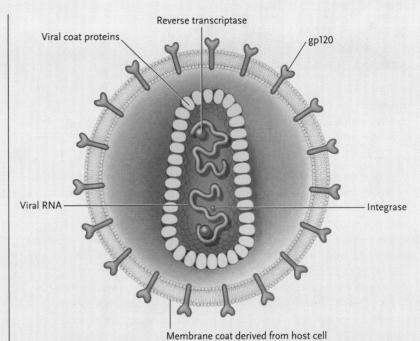

Reverse transcriptase

Viral coat proteins

gp120

Viral RNA

Integrase

Membrane coat derived from host cell

Figure 1
Structure of a free HIV viral particle.

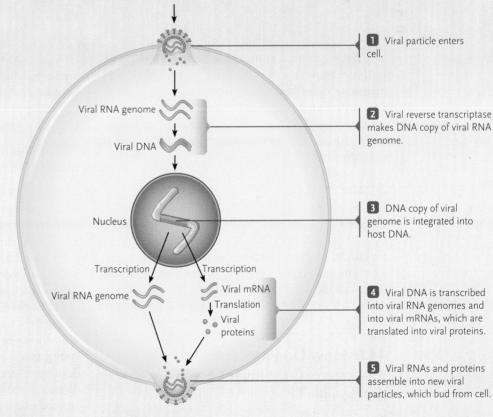

Viral RNA genome

Viral DNA

Nucleus

Transcription   Transcription

Viral RNA genome   Viral mRNA
                        Translation
                        Viral proteins

1 Viral particle enters cell.

2 Viral reverse transcriptase makes DNA copy of viral RNA genome.

3 DNA copy of viral genome is integrated into host DNA.

4 Viral DNA is transcribed into viral RNA genomes and into viral mRNAs, which are translated into viral proteins.

5 Viral RNAs and proteins assemble into new viral particles, which bud from cell.

Figure 2
The steps in HIV infection of a host cell.

(continued on page 1104)

(continued from page 1103)

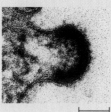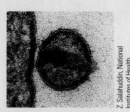

Z. Salahuddin, National
Institutes of Health

150 nm

Figure 3

An HIV particle budding from a host cell. As it passes from the host cell, it acquires a membrane coat derived from the host cell plasma membrane.

DNA is copied into new viral RNA molecules and into mRNAs that direct host cell ribosomes to make viral proteins. The viral RNAs are added to the viral proteins to make infective HIV particles, which are released from the host cell by budding **(Figure 3)**. The infection also leads uninfected helper T cells to destroy themselves in large numbers by apoptosis, through mechanisms that are still unknown.

Initially, infected people suffer a mild fever and other symptoms that may be mistaken for the flu or the common cold. The symptoms disappear as antibodies against viral proteins appear in the body, and the number of viral particles drops in the bloodstream. An infected person may remain apparently healthy for years yet can infect others. Both the transmitter and the recipient of the virus may be unaware that the virus is present, making it difficult to control the spread of HIV.

Eventually, more and more helper T cells and macrophages are destroyed, wiping out the body's immune response. The infected person becomes susceptible to secondary, opportunistic infections, such as a pneumonia caused by a fungus (*Pneumocystis carinii*); tuberculosis; persistent yeast (*Candida albicans*) infections of the mouth, throat, rectum, or vagina; and infection by many common bacteria and viruses that rarely infect healthy humans. These infections signal the appearance of "full-blown" AIDS. If untreated, this results in a steady debilitation and death, typically within 5 years.

Currently, there is no cure for HIV and no vaccine that can prevent infection. The coat proteins of HIV mutate constantly, making a vaccine developed against one form of the virus useless when the next form appears. Most mutations occur during replication of the virus during reverse transcription of viral RNA.

The development of AIDS can be greatly slowed by drugs that interfere with reverse transcription of the viral RNA. Treatment with a "cocktail" of drugs called *reverse transcriptase inhibitors* inhibits viral reproduction and destruction of helper T cells, extending the lives of people with HIV. The inhibiting cocktails are not a cure, however, because the virus is still present. If the therapy is stopped, the virus again replicates and the T-cell population drops.

The significant similarities between the innate immune responses of vertebrates, invertebrates, and plants, as described earlier in this chapter, and the similarities and differences in immune responses in "primitive" and "advanced" organisms allow us to look at the evolution of immune responses to pathogens in very diverse types of organisms.

STUDY BREAK

1. How can studies on "primitive organisms" help us understand immune responses in more advanced organisms such as humans?
2. Why is understanding the evolution of immune responses important?

44.6 How Do Parasites and Pathogens Circumvent Host Responses?

The immune systems of all organisms arose to recognize and eliminate pathogens. Some of the complexities of these systems have been detailed in this chapter. However, the fact that all organisms still suffer from diseases indicates that as strong as our immune systems are, pathogens and parasites are always looking for weak points to exploit.

44.6a Hiding, Confusing, and Manipulating the Host

Disease-causing organisms can develop in regions of the host's body that do not have a strong immune response, hide in host defence cells, confuse the immune response, or directly manipulate host responses.

Many bacteria, nematodes, trematodes, and other parasites enter via the mouth, establish in some region of the alimentary tract, and allow their eggs or offspring to exit with the feces. Although the alimentary tract of most organisms has some level of immune activity, the immune response is not as strong as it is in other regions. Thus, these organisms develop where the immune response is limited. We do not recognize broccoli as "foreign" and generate antibodies against broccoli in our guts. Organisms do not normally produce immune responses to gut contents; in fact, this might be counterproductive. Many organisms rely on symbiotic microbial organisms to

PEOPLE BEHIND BIOLOGY
Élie Metchnikoff, Pasteur Institute

In 1882, a Russian zoologist named Élie Metchnikoff, working in Italy, pierced a larva of a common starfish with a rose thorn. Later he observed cells collecting around the thorn. He recognized that these cells were trying to isolate and eliminate the thorn through what we now term phagocytosis. Although

phagocytosis was known to occur in human bacterial infections, Metchnikoff's initial studies with the starfish, and later with animal models injected with human pathogens, demonstrated that this response was used universally by animals to eliminate pathogens. These initial experiments have led to

scientific disciplines such as cellular immunology, innate immunity, and comparative immunology and have contributed to our understanding of the evolution of immune responses. For this pioneering work and keen observation, Metchnikoff shared the Nobel Prize in medicine in 1908 with Paul Ehrlich.

---

help digest food and produce vitamins. Killing these organisms could kill the host.

Some pathogens enter the cerebrospinal fluid, which often shows a reduced immune response. Although relatively protected here, the pathogens have a problem in dispersing their offspring to subsequent hosts.

Many pathogens make themselves look like the host. They invade the body cavities of vertebrates or invertebrates or the circulatory system of vertebrates and coat themselves with host factors to confuse the immune response. Trematodes, such as *Shistosoma* species, that live in our blood/lymphatic system cover themselves in host proteins so that they are hidden from the antibody response of the host and then use host factors to stimulate their development: tumour necrosis factor (TNF) stimulates egg production, and growth is stimulated by interleukin-7. Much of the pathology caused by this parasite is due to the immune response of the host rather than to any direct damage by the parasite.

Intracellular pathogens such as *Leishmania* species live and reproduce in the macrophages that normally eliminate pathogens, developing in the last place the host would look. They survive by producing antioxidant enzymes to detoxify superoxide molecules, they downregulate the signal transduction pathways that produce lethal molecules, and they produce surface glycoproteins that are refractory to host lysosomal enzymes. Therefore, the very cells the host relies on to eliminate parasites serve as incubators for these organisms. Because infected cells are considered "self," they are not eliminated by other host immune responses.

Some pathogens regularly change their surface coats to avoid recognition and destruction by the immune system. When the host develops antibodies against one version of the surface proteins, the pathogens produce different surface proteins, and the host produces new antibodies. These changes continue indefinitely, allowing the pathogens to keep one step ahead of the immune system. This strategy, called **antigenic variation**, is used by organisms such as the protozoan parasite that

causes African sleeping sickness, the bacterium that causes gonorrhea, and the viruses that cause influenza, the common cold, and AIDS.

In insect vectors of viruses such as dengue, the virus manipulates the insect's immune response. Normally, intracellular pathogens are eliminated via apoptosis (see Section 44.2). The dengue virus, however, induces the expression of *inhibitors of apoptosis* (IAPs) to prevent cell degradation until it has established and replicated, after which, it allows the cells to rupture, releasing virus particles that then enter new cells.

Many viruses also produce RNAi suppressors that prevent the host cells from preventing the viruses' multiplication using RNAi strategies, as described in Section 44.2.

44.6b Mechanisms to Evade Host Immune Responses

Several very different groups of organisms work together symbiotically to overcome the response of a host. Whereas individually they would be eliminated by the host, the combination overwhelms the ability of the host to eliminate either organism.

Nematodes + Bacteria. Soil-living nematodes that enter the body cavity of insects are killed by a combination of encapsulation and melanization described previously, and many soil bacteria cannot enter the insects. Some nematodes have teamed up with specific bacteria in a symbiotic relationship to overcome host defences. Nematodes penetrate the insect and release bacteria that are phagocytosed but are not killed. These bacteria multiply within hours and produce compounds that inactivate the insect's immune response. The nematodes feed and reproduce using the bacteria and host tissues as food. When resources for growth and reproduction become limited, millions of infective nematodes store some of these specific bacteria and leave the dead insect en masse in search of new hosts.

Wasps and Viruses. Some insect *parasitoids* such as wasps, lay their eggs in other insects. Normally, these would be eliminated via encapsulation. However, some parasitoid wasps have formed an allegiance with a polyDNA virus. The virus is injected into the host insect along with the wasp egg. This virus inactivates the host's immune response, allowing the wasp larva to develop, mature, and emerge from the host, carrying some of the virus within its reproductive tract.

In the continuing battle between host and parasite/pathogen, there are constant interactions and feedback mechanisms. Each development by the host to eliminate the pathogen is countered by pathogen factors to ensure the pathogen's survival. And some of the mechanisms used are very ingenious!

STUDY BREAK

1. Compare invertebrate and mammalian immune defences.
2. How do pathogens avoid the immune responses of their hosts?
3. What mechanisms can pathogens use to avoid detection by the host?
4. How can different organisms work together to overcome host defences?
5. How can we compare immune responses between animals, and what benefit would such comparisons provide?

Review

Go to CENGAGENOW™ at http://hed.nelson.com/ to access quizzing, animations, exercises, articles, and personalized homework help.

44.1 Three Lines of Defence against Invasion

- Humans and other vertebrates have three lines of defence against pathogens. The first, which is nonspecific, is the barrier set up by the skin and mucus membranes.

- The second line of defence, also nonspecific, is innate immunity, an innate system that defends the body against pathogens and toxins penetrating the first line. This is the only kind of immune system found in invertebrates.

- The third line of defence, adaptive immunity, is specific: it recognizes and eliminates particular pathogens and retains a memory of that exposure so as to respond rapidly if the pathogen is encountered again. The response is carried out by lymphocytes, a specialized group of leukocytes.

44.2 Nonspecific Defences: Innate Immunity

- In the innate immunity system, molecules on the surfaces of pathogens are recognized as foreign by receptors on host cells. This is common to all animals and plants. In invertebrates, this activates systems to remove the pathogen via phagocytosis, encapsulation, or the production of antimicrobial peptides. In vertebrates, the pathogen is combated by the inflammation and complement systems.

- In both vertebrates and invertebrates, epithelial surfaces and specific tissues secrete antimicrobial peptides in response to attack by microbial pathogens. These disrupt the plasma membranes of pathogens, killing them.

- Inflammation is characterized by heat, pain, redness, and swelling at the infection site. Several interconnecting mechanisms initiate inflammation, including pathogen engulfment, histamine secretion, and cytokine release, which dilates the local blood vessels, increases their permeability, and allows for leakage into body tissues.

- Large arrays of complement proteins are activated when they recognize molecules on the surfaces of pathogens. Some complement proteins form membrane attack complexes, which insert into the plasma membrane of many types of bacteria and cause their lysis. Fragments of other complement proteins coat pathogens, stimulating phagocytes to engulf them.

- Four nonspecific defences are used to combat viral pathogens: RNA interference, interferons, natural killer cells, and apoptosis.

44.3 Specific Defences: Adaptive Immunity

- Adaptive immunity, which is carried out by B and T cells, targets particular pathogens or toxin molecules.

- Antibodies consist of two light and two heavy polypeptide chains, each with variable and constant regions. The variable regions of the chains combine to form the specific antigen-binding site.

- Antibodies occur in five different classes: IgM, IgD, IgG, IgA, and IgE. Each class is determined by its constant region.

- Antibody diversity is produced by genetic rearrangements in developing B cells that combine gene segments into intact genes encoding the light and heavy chains. The rearrangements producing heavy-chain genes and T-cell receptor genes are similar. The light and heavy chain genes are transcribed into precursor mRNAs, which are processed into finished mRNAs, which are translated on ribosomes into the antibody polypeptides.

- The antibody-mediated immune response has two general phases: (1) T-cell activation and (2) B-cell activation and antibody production. T-cell activation begins when a dendritic cell engulfs a pathogen and produces antigens, making the cell an antigen-presenting cell (APC). The APC secretes interleukins, which activate the T cell. The T cell then secretes cytokines, which stimulate the T cell to proliferate, producing a clone of cells. The clonal cells differentiate into helper T cells.

- B-cell receptors (BCRs) on B cells recognize antigens on a pathogen and engulf it. The B cells then display the antigens. The TCR on a helper T cell activated by the same antigen binds to the antigen on the B cell. Interleukins from the T cell stimulate the B cell to produce a clone of cells with identical BCRs. The clonal cells differentiate into plasma cells, which secrete antibodies specific for the antigen, and memory B cells, which provide immunological memory of the antigen encounter.

- Clonal expansion is the process of selecting a lymphocyte specifically for cloning when it encounters an antigen from among a randomly generated, large population of lymphocytes with receptors that specifically recognize the antigen.

- Antibodies clear the body of antigens by neutralizing or agglutinating them or by aiding the innate immune response.

- In immunological memory, the first encounter of an antigen elicits a primary immune response. Later exposure to the same antigen elicits a more rapid secondary response with a greater production of antibodies.

- Active immunity is the production of antibodies in response to an antigen. Passive immunity is the acquisition of antibodies by direct transfer from another person.

- In cell-mediated immunity, cytotoxic T cells recognize and bind to antigens displayed on the surfaces of infected body cells or to cancer cells. They then kill the infected body cell.

- Antibodies are widely used in research to identify, locate, and determine the functions of molecules in biological systems.

44.4 Malfunctions and Failures of the Immune System

- In immunological tolerance, molecules present in an individual at birth normally do not elicit an immune response.

- In some people, the immune system malfunctions and reacts against the body's own proteins or cells, producing autoimmune disease.

- The first exposure to an allergen sensitizes an individual by leading to the production of memory B and T cells, which cause a greatly intensified response to subsequent exposures.

- Most allergies result when antigens act as allergens by stimulating B cells to produce IgE antibodies, which lead to the release of histamine. Histamine produces the symptoms characteristic of allergies (see Figure 44.13).

44.5 Defences in Other Organisms

- Antibodies, complement proteins, and other molecules with defensive functions have been identified in all vertebrates.

- Invertebrates and plants rely on nonspecific defences, including surface barriers, phagocytes, encapsulation, melanization, and antimicrobial molecules.

44.6 How Do Parasites and Pathogens Circumvent Host Responses?

- Pathogens develop in regions of the host where the immune response is limited.

- Parasites and pathogens cover themselves with host material so that they are not recognized as "nonself."

- Parasites and pathogens develop in host immune cells by inactivating the immune response.

- Some pathogens use antigenic variation to keep one step ahead of the host response.

- Some pathogens manipulate the gene expression of host molecules.

- Pathogens may work together to overcome host responses that would kill either of the participants.

Questions

Self-Test Questions

1. Viruses are controlled by
   a. CD8$^+$ T cells that bind class I MHC proteins holding viral antigens.
   b. CD4$^+$ T cells that bind free viruses in the blood.
   c. B cells secreting perforin.
   d. antibodies that bind the viruses with their constant ends.
   e. natural killer cells secreting antiviral antibodies.

2. Components of the inflammatory response include all *except*
   a. macrophages.
   b. neutrophils.
   c. B cells.
   d. mast cells.
   e. eosinophils.

3. When a person resists infection by a pathogen after being vaccinated against it, this is the result of
   a. innate immunity.
   b. immunological memory.
   c. a response with defensins.
   d. an autoimmune reaction.
   e. an allergy.

4. One characteristic of a B cell is that it
   a. has the same structure in both invertebrates and vertebrates.
   b. recognizes antigens held on class I MHC proteins.
   c. binds viral infected cells and directly kills them.
   d. makes many different BCRs on its surface.
   e. has a BCR on its surface, which is the IgM molecule.

5. Antibodies
   a. are each composed of four heavy and four light chains.
   b. display a variable end, which determines the antibody's location in the body.
   c. that belong to the IgE group are the major antibody class in the blood.
   d. that are found in large numbers in the mucous membranes belong to class IgG.
   e. function primarily to identify and bind antigens free in body fluids.

6. The generation of antibody diversity includes the
   a. joining of V to C to J segments to make a functional light-chain gene.
   b. choice from several different types of C segments to make a functional light-chain gene.
   c. deletion of the J segment to make a functional light-chain gene.
   d. joining of V to J to C segments to make a functional light-chain gene.
   e. initial generation of IgG followed later by IgM on a given cell.

7. An APC:
   a. can be a CD8$^+$ T cell.
   b. derives from a phagocytic cell and is lymphocyte stimulating.
   c. secretes antibodies.
   d. cannot be a B cell.
   e. cannot stimulate helper T cells.

8. One function of antibodies is to
   a. deactivate the complement system.
   b. neutralize natural killer cells.
   c. clump bacteria and viruses for easy phagocytosis by macrophages.
   d. eliminate the chance for a secondary response.
   e. kill viruses inside of cells.

9. Jen punctured her hand with a muddy nail. In the emergency room, she received both a vaccine and someone else's antibodies against tetanus toxin. The immunity conferred here is
   a. both active and passive.
   b. active only.
   c. passive only.
   d. first active and later passive.
   e. innate.

10. Medicine attempts to enhance the immune response when treating
    a. organ transplant recipients.
    b. anaphylactic shock.
    c. rheumatoid arthritis.
    d. HIV infection.
    e. type 1 diabetes.

Questions for Discussion

1. HIV wreaks havoc with the immune system by attacking helper T cells and macrophages. Would the impact be altered if the virus attacked only macrophages? Explain.

2. Given what you know about how foreign invaders trigger immune responses, explain why mutated forms of viruses, which have altered surface proteins, pose a monitoring problem for memory cells.

3. Cats, dogs, and humans may develop myasthenia gravis, an autoimmune disease in which antibodies develop against acetylcholine receptors in the synapses between neurons and skeletal muscle fibres. Based on what you know of the biochemistry of muscle contraction (see Chapter 36), explain why people with this disease typically experience severe fatigue with even small levels of exertion, drooping of facial muscles, and trouble keeping their eyelids open.

The large number of gulls (*Larus* species) is obvious at a landfill site near Thunder Bay, Ontario. The population of gulls reflects the local population of humans.

M.B. Fenton

45 Population Ecology

WHY IT MATTERS

Controlling rabies in wildlife involves understanding many aspects of biology, from populations to epidemiology and behaviour. Rabies, from the Latin *rabere* (to rage or rave), affects the nervous system of terrestrial mammals. Caused by a *Lyssavirus,* rabies is usually spread by bites because the virus accumulates in the saliva of infected animals. Before 1885, when Louis Pasteur in France developed a vaccine for it, rabies was common in Europe, and many people died from it every year. In 2007, the World Health Organization estimated that worldwide, more than 50 000 people die annually from rabies, usually people in the "developing world." Between 1980 and the end of 2000, 43 people in the United States and Canada died of rabies.

Animals with "furious" rabies become berserk, attacking anything and everything in their path, a behaviour that spreads the virus and helps ensure its survival. Animals with paralytic rabies ("dumb rabies") suffer from increasing paralysis that progresses forward from the hindlimbs. Animals with either manifestation of rabies can spread the disease by biting when there is virus in their saliva. Paralysis of the throat muscles means that rabid animals cannot swallow the saliva they produce, so they appear to foam at the mouth.

Rabies is almost invariably fatal once an animal or a human shows clinical symptoms of the disease, so immunization of

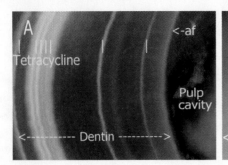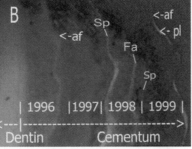

Figure 45.1

Tetracycline rings in carnivore teeth. Yellow fluorescent lines from ingestion of rabies baits with tetracycline as a biomarker. The sections are undecalcified, ultraviolet fluorescent × 100. **(a)** Coyote (*Canis latrans*) tooth with seven daily tetracycline lines from vaccine baits. **(b)** Canine tooth of a four-year-old raccoon (*Procyon lotor*) with yearly tetracycline lines in cementum. af = autofluorescent collagen; Fa = fall baits, 1998; pl = periodontal ligament; Sp = spring baits, 1998 and 1999.

someone exposed to the disease should start as soon as possible after exposure. Since 1980, human diploid vaccines have been commonly available, raising the level of protection against rabies.

From the 1960s to the 1990s, a visit to almost any rural hospital in southern Ontario (Canada) would have revealed at least one farmer receiving postexposure rabies shots. Then red foxes (*Vulpes vulpes*) were the main vector for rabies in Ontario, and cows (*Bos taurus*) exposed to rabies through fox bites in turn exposed farmers to the virus. Many farmers are accustomed to treating choking cows by reaching into the cow's gullet to clear an obstruction. A farmer dealing with a rabid cow could have been scratched and exposed to the virus, and then, after the cow died of rabies, the farmer would have received postexposure rabies shots. Rabies transmitted to cows from foxes posed a threat to human lives and was a significant drain on the economy through compensation paid to farmers whose cattle succumbed to the disease.

In 1967, 4-year-old Donna Featherstone of Richmond Hill, Ontario, died of rabies after being bitten by a stray cat. The resulting public outcry set the stage for a rabies eradication program in Ontario. Controlling fox rabies in southern Ontario was achieved by a combination of innovation and knowledge of basic biology. There were three phases: (a) developing an oral vaccine; (b) developing a means of vaccinating foxes; and (c) monitoring the impact of the program on the fox population.

First, two main baits for the oral vaccine were developed, and one, Evelyn, Rocketniki, Abelseth (ERA), was a modified live virus replicated in tissues of the mouth and throat. ERA successfully stimulated seroconversion in red foxes and vaccinated them against rabies. Second, foxes ate baits scented with chicken, and when the baits contained ERA, foxes that took the baits were vaccinated. The baits were small, the size of packets of jam one receives in restaurants, and easy to distribute widely from low-flying aircraft, allowing vaccination of foxes across large areas of southern Ontario. Third, each bait contained tetracycline, a biomarker absorbed into the system of any mammal that ate the bait. Once in the body, some tetracycline penetrated the dentine of the animals' teeth, especially in younger individuals. Biologists sectioned and stained teeth from foxes taken by trappers. In the sections, bands of tetracyline in tooth rings **(Figure 45.1)** identified foxes that had taken baits, and biologists established that over 70% of red foxes had been vaccinated by this method.

Before the bait vaccination program, on average, 211 cattle annually died of rabies in southern Ontario. The baiting program started in 1989, and by 1996, rabies in cattle dropped to an average of 11 cases a year and the levels of rabies in foxes in Ontario were dramatically reduced. The example demonstrates how problems in biology are solved by combined approaches, from population biology, behaviour, immunology, and epidemiology.

The purpose of this chapter is to introduce you to ecology in general, particularly population ecology.

45.1 The Science of Ecology

Ecology encompasses two related disciplines. In *basic ecology*, major research questions relate to the distribution and abundance of species and how they interact with each other and the physical environment. Using these data as a baseline, workers in *applied ecology* develop conservation plans and amelioration programs to limit, repair, and mitigate ecological damage caused by human activities (see also Chapter 48). Ecology has its roots in descriptive natural history dating back to the ancient Greeks. Modern ecology was born in 1870 when the German biologist Ernst Haeckel coined the term (from *oikos* = house). Contemporary researchers still gather descriptive information about ecological relationships, often as the starting points for other studies. Although ecological research is dominated by hypotheticodeductive approaches, initial inductive approaches allow biologists to generate appropriate hypotheses about how systems function. Research in ecology is often linked to work in genetics, physiology, anatomy, behaviour, paleontology, and evolution, as well as geology, geography, and environmental science. Many ecological phenomena, such as climate change, occur over huge areas and long time spans, so ecologists must devise ways to determine how environments influence organisms and how organisms change the environments in which they live.

Reprinted with the kind permission of Elsevier

Ecology can be divided into four increasingly complex and inclusive levels of organization. First, in **organismal ecology**, researchers study organisms to determine the genetic, biochemical, physiological, morphological, and behavioural adaptations to the abiotic environment (see Chapter 3). Second, in **population ecology**, researchers focus on groups of individuals of the same species that live together. Population ecologists study how the size and other characteristics of populations change in space and time. Third, in **community ecology**, biologists examine populations of different species that occur together in one area (are sympatric). Community ecologists study interactions between species, analyzing how predation, competition, and environmental disturbances influence a community's development, organization, and structure (see Chapter 47). Fourth, those studying **ecosystem ecology** explore how nutrients cycle and energy flows between the biotic components of an **ecological community** and the abiotic environment (see Chapter 47).

Ecologists can create hypotheses about ecological relationships and how they change through time or differ from place to place. Some formalize these ideas in mathematical models that express clearly defined, but hypothetical, relationships among important variables in a system. Manipulation of a model, usually with the help of a computer, can allow researchers to ask what would happen if some of the variables or their relationships change. Thus, researchers can simulate natural events and large-scale experiments before investing time, energy, and money in fieldwork and laboratory work. Bear in mind that mathematical models are no better than the ideas and assumptions they embody, and useful models are constructed only after basic observations define the relevant variables.

Ecologists use field or laboratory studies to test predictions of their hypotheses about relationships among variables in systems. In controlled experiments, researchers compare data from an experimental treatment (involving manipulation of one or more variables) with data from a control (in which nothing is changed). In some cases, "natural experiments" can be conducted because of the patterns of distribution and/or behaviour of species. This has the advantage of allowing ecologists to test predictions about how systems are operating without manipulating variables. Two species of fish, cutthroat trout (*Oncorhynchus clarki*) and Dolly Varden char (*Salvelinus malma*), live in coastal lakes of British Columbia. Some lakes have either trout or char, but others contain both species. The natural distributions of these fishes allowed researchers to measure the effect of each species on the other. In lakes in which both species live, each restricts its activities to fewer areas and eats a smaller variety of prey than it does in lakes in which it occurs alone.

45.2 Population Characteristics

Seven characteristics can be described for any population.

45.2a Geographic Range: Boundaries of Distribution

Populations have characteristics that transcend those of the individuals comprising them. Every population has a **geographic range**, the overall spatial boundaries within which it lives. Geographic ranges vary enormously. A population of snails might inhabit a small tidepool, whereas a population of marine phytoplankton might occupy an area orders of magnitude larger. Every population also occupies a **habitat**, the specific environment in which it lives, as characterized by its biotic and abiotic features. Ecologists also measure other population characteristics, such as size, distribution in space, and age structure.

45.2b Population Size and Density: Numbers of Individuals per Unit Area

Population size is the number of individuals comprising the population at a specified time (N_t). **Population density** is the number of individuals per unit area or per unit volume of habitat. Species with a large body size generally have lower population densities than those with a small body size **(Figure 45.2, p. 1114)**. Although population size and density are related measures, knowing a population's density provides more information about its relationship to the resources it uses. If a population of 200 oak trees occupies 1 hectare (10 000 m²), the population density is $200 \times 10\ 000\ \text{m}^{-2}$ or 1 tree per 50 m². But if 200 oaks are spread over 5 hectares, the density is 1 tree per 250 m². Clearly, the second population is less dense than the first, and its members will have greater access to sunlight, water, and other resources.

Ecologists measure population size and density to monitor and manage populations of endangered species, economically important species, and agricultural pests. For large-bodied species, a simple head count could provide accurate information. For example, ecologists survey the size and

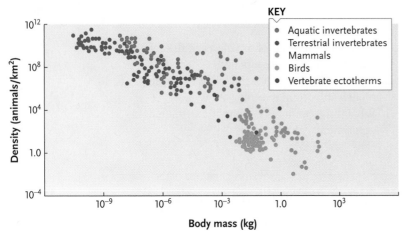

Figure 45.2

Population density and body size. Population density generally declines with increasing body size among animal species. There are similar trends for other organisms.

density of African elephant populations by flying over herds and counting individuals **(Figure 45.3)**. Researchers use a variation on that technique to *estimate* population size in tiny organisms that live at high population densities. To estimate the density of aquatic phytoplankton, for example, you might collect water samples of known volume from representative areas in a lake and count them by looking through a microscope. These data allow you to estimate population size and density based on the estimated volume of the entire lake. In other cases, researchers use the mark–release–recapture sampling technique (see *Capture–Recapture*). One ongoing challenge is measuring population size in organisms that are clones, for example, stands of poplar trees (*Populus* spp.).

45.2c Population Dispersion: The Distribution of Individuals in Space

Populations can vary in their **dispersion**, the spatial distribution of individuals within the geographic range. Ecologists define three theoretical patterns of dispersion: *clumped, uniform,* and *random* **(Figure 45.4)**.

Clumped dispersion (see Figure 45.4a) is common and occurs in three situations. First, suitable conditions are often patchily distributed. Certain pasture plants, for instance, may be clumped in small, scattered areas where cowpats had fallen for months, locally enriching the soil. Second, populations of some social animals (see Chapter 40) are clumped because mates are easy to locate within groups, and individuals may cooperate in rearing offspring, feeding, or defending themselves from predators. Third, populations can be clumped when species reproduce by asexual clones that remain attached to the parents. Aspen trees and sea anemones reproduce this way and often occur in large aggregations (see Chapter 18). Clumping may also occur in species in which seeds, eggs, or larvae lack dispersal mechanisms and offspring grow and settle near their parents.

Uniform distributions can occur when individuals repel one another because resources are in short supply. Creosote bushes are uniformly distributed in the dry scrub deserts of the American Southwest (see **Figure 45.4b**). Mature bushes deplete the surrounding soil of water and secrete toxic chemicals, making it impossible for seedlings to grow. This chemical warfare is called "allelopathy." Moreover, seed-eating ants and rodents living at the bases of mature bushes eat any seeds that fall nearby. In these situations, the distributions of species of plants and animals can

a.

b.

Figure 45.3

Counting elephants. It is easy to think that large animals such as African elephants (*Loxodonta africana*) would be easy to count from the air **(a)**. This may or may not be true, depending on vegetation. But it can be easy to overlook animals, particularly young ones **(b)** in the shade.

Capture–Recapture

Ecologists use the mark–release–recapture technique to estimate the population size of mobile animals that live within a restricted geographic range. To do this, a sample of organisms (n_1) is captured, marked, and released. Ideally, the marks (or tags) are permanent and do not harm the tagged animal. Insects and reptiles often are marked with ink or paint, birds with rings (bands) on their legs, and mammals with ear tags or collars.

Later, a second sample (n_2) of the population is captured. In the second sample, the proportion of marked (n_{2m}) to unmarked individuals is used to estimate the total population (x) of the study area by solving the equation for x.

$$n_1/x = n_{2m}/n_2.$$

Assume that you capture a sample of 120 butterflies **(Figure 1)**, mark each one, and release them. A week later, you capture a sample of 150 butterflies, 30 that you marked. Thus, you

Figure 1
This butterfly has been captured and marked before release in a capture–recapture experiment.

had marked 30 of 150, or 1 of every 5 butterflies, on your first field trip. Because you captured 120 individuals on that first excursion, you would estimate that the total population size is 120 × (150/30) = 600 butterflies.

The capture–recapture technique is based on several assumptions that are critical to its accuracy: (1) that being marked has no effect on survival; (2) that marked and unmarked animals mix randomly in the population; (3) that no migration into or out of the population takes place during the estimating period; and (4) that marked individuals are just as likely to be captured as unmarked individuals. (Sometimes animals become "trap shy" or "trap happy," a violation of the fourth assumption.)

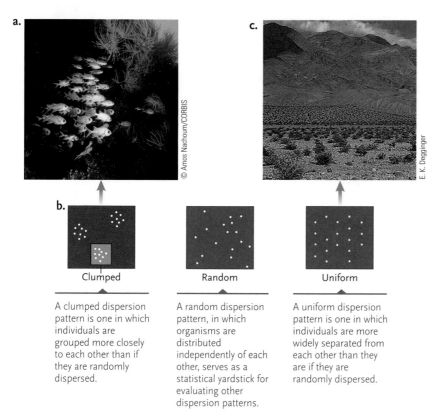

a.

c.

b.

Clumped

A clumped dispersion pattern is one in which individuals are grouped more closely to each other than if they are randomly dispersed.

Random

A random dispersion pattern, in which organisms are distributed independently of each other, serves as a statistical yardstick for evaluating other dispersion patterns.

Uniform

A uniform dispersion pattern is one in which individuals are more widely separated from each other than they are if they are randomly dispersed.

Figure 45.4
Dispersion patterns. A clumped pattern **(a)** is evident in fish that live in social groups. A random pattern **(b)** of dispersion appears to be rare in nature, where it occurs in organisms that are neither attracted to nor repelled by conspecifics. Nearly uniform patterns **(c)** are demonstrated by creosote bushes (*Larrea tridentata*) near Death Valley, California.

Black-Footed Ferret, *Mustela Nigripes*

Black-footed ferrets **(Figure 1)** are crepuscular and nocturnal hunters of the prairie. Weighing 0.6 to 1.1 kg, these weasel relatives (family Mustelidae, order Carnivora) were once abundant in western North America, from Texas in the United States to Saskatchewan and Alberta in Canada. Like other mustelids, males are larger than females. In the wild, these predators probably fed mainly on prairie dogs (*Cynomys* species) and lived around prairie dog towns. Litters range in size from one to five. Females bear a single litter a year, and males and females are sexually mature at age one year.

By 1987, *M. nigripes* was probably extinct in the wild. The last known wild population was discovered near Meeteetse, Wyoming, in 1981. Seven animals from this population were captured and brought into captivity and served as the genetic founders for a captive breeding program. Over 4800 juvenile black-footed ferrets were produced by this program, and wildlife officials began to release captive-bred animals into suitable habitat.

At Shirley Basin, Wyoming, 228 captive-born black-footed ferrets were received between 1991 and 1994. By 1996, only 25 were observed in the wild and 5 by 1997. This decline reflected the impact of diseases, specifically canine distemper and plague. In 1996, it seemed that the reintroductions would fail, and *M. nigripes* would again be extinct in the wild.

In 2003, however, 52 black-footed ferrets were observed in the field at Shirley Basin, and since then, the population has increased significantly **(Figure 2).** The increase reflects an $r = 0.47$ and can be attributed to success in the first year of life, a combination of survival and fertility.

There appears to be hope for the future of *M. nigripes*. It remains to be determined if the genetic bottleneck (see Chapter 18) that the species has endured will prove to be an important handicap to long-term survival.

Figure 1
Mustela nigripes, the black-footed ferret. This critically endangered carnivore from the North American prairie shows evidence of a comeback.

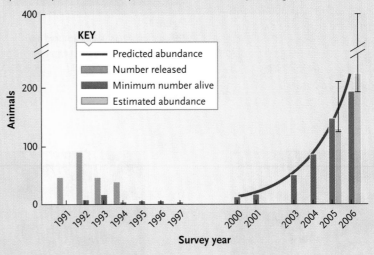

Figure 2
Population growth. Black-footed ferrets in Shirley Basin, Wyoming, have shown rapid popuation growth. The 95% confidence limits suggest a population of 192 to 401 in 2006.

---

be uniform and interrelated. Territorial behaviour, the defence of an area and its resources, also can produce **uniform dispersion** in some species of animals, such as nests in colonies of colonial birds (see Chapter 40).

Random dispersion (see Figure 45.4c) occurs when environmental conditions do not vary much within a habitat, and individuals are neither attracted to nor repelled by others of their species (conspecifics). Ecologists use formal statistical definitions of "random" to establish a theoretical baseline for assessing the pattern of distribution. In cases of random dispersion, individuals are distributed unpredictably. Some spiders, burrowing clams, and rainforest trees exhibit random dispersion.

Whether the spatial distribution of a population appears to be clumped, uniform, or random depends partly on the size of the organisms and of the study area. Oak seedlings may be randomly dispersed on a spatial scale of a few square metres, but over an entire mixed hardwood forest, they are clumped under the parent trees. Therefore, dispersion of a population depends partly on the researcher's scale of observation.

In addition, the dispersion of animal populations often varies through time in response to natural environmental rhythms. Few habitats provide a constant supply of resources throughout the year, and many animals move from one habitat to another on a seasonal cycle, reflecting the distribution of resources such as food. Tropical birds and mammals are often widely dispersed in deciduous forests during the wet season, when food is widely available. During the dry season, these species crowd into narrow "gallery forests" along watercourses, where evergreen trees provide food and shelter.

45.2d Age Structure: Numbers of Individuals of Different Ages

All populations have an **age structure**, a statistical description of the relative numbers of individuals in each age class (discussed further in Chapter 46). Individuals can be categorized roughly as prereproductive (younger than the age of sexual maturity), reproductive, or postreproductive (older than the maximum age of reproduction). A population's age structure reflects its recent growth history and predicts its future growth potential. Populations composed of many prereproductive individuals obviously grew rapidly in the recent past. These populations will continue to grow as young individuals mature and reproduce.

45.2e Generation Time: Average Time between Birth and Death

Another characteristic that influences a population's growth is its **generation time**, the average time between the birth of an organism and the birth of its offspring. Generation time is usually short in species that reach sexual maturity at a small body size **(Figure 45.5)**. Their populations often grow rapidly because of the speedy accumulation of reproductive individuals.

45.2f Sex Ratio: Females:Males

Populations of sexually reproducing organisms also vary in their **sex ratio**, the relative proportions of males and females. In general, the number of females in a population has a bigger impact on population growth than the number of males because only females actually produce offspring. Moreover, in many species, one male can mate with several females, and the number of males may have little effect on the population's reproductive output. In northern elephant seals (see Chapter 17), mature bulls fight for dominance on the beaches where the seals mate. Only a few males may ultimately inseminate a hundred or more females. Thus, the presence of other males in the group may have little effect on

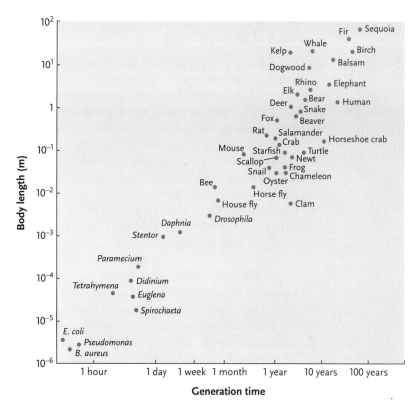

Figure 45.5

Generation time and body size. Generation time increases with body size among bacteria, protists, plants, and animals. The logarithmic scale on both axes compresses the data into a straight line.

the size of future generations. In animals that form lifelong pair bonds, such as geese and swans, the numbers of males and females influence reproduction in the population.

45.2g Proportion Reproducing: Incidence of Reproducing Individuals in a Population

Population ecologists try to determine the proportion of individuals in a population that are reproducing. This issue is particularly relevant to the conservation of any species in which individuals are rare or widely dispersed in the habitat (see Chapter 48).

STUDY BREAK

1. What is the difference between geographic range and habitat?
2. What are the three types of dispersion? What is the most common pattern found in nature? Why?
3. What is the common pattern of generation time among bacteria, protists, plants, and animals?

45.3 Demography

Populations grow larger through the birth of individuals and the **immigration** (movement into the population) of organisms from neighbouring populations. Conversely, death and **emigration** (movement out of the population) reduce population size. **Demography** is the statistical study of the processes that change a population's size and density through time.

Ecologists use demographic analysis to predict a population's future population growth. For human populations, these data help governments anticipate the need for social services such as schools, hospitals, and chronic care facilities. Demographic data allow conservation ecologists to develop plans to protect endangered species. Demographic data on northern spotted owls (*Strix occidentalis caurina*) helped convince the courts to restrict logging in the owl's primary habitat, the old growth forests of the Pacific Northwest. *Life tables* and *survivorship curves* are among the tools ecologists use to analyze demographic data.

45.3a Life Tables: Numbers of Individuals in Each Age Group

Although every species has a characteristic life span, few individuals survive to the maximum age possible. Mortality results from starvation, disease, accidents, predation, or inability to find a suitable habitat. Life insurance companies first developed techniques for measuring mortality rates (known as actuarial science), and ecologists adapted these approaches to the study of nonhuman populations.

A **life table** summarizes the demographic characteristics of a population **(Table 45.1)**. To collect life table data for short-lived organisms, demographers typically mark a **cohort**, a group of individuals of similar age, at birth and monitor their survival until all members of the cohort die. For organisms that live more than a few years, a researcher might sample the population for one or two years, recording the ages at which individuals die and then extrapolating those results over the species' life span. The approach to the timing of collection of data about reproduction and longevity will depend on the details of the species under study.

In any life table, life spans of organisms are divided into age intervals of appropriate length. For short-lived species, days, weeks, or months are useful, whereas for longer-lived species, years or groups of years will be better. Mortality can be expressed in two complementary ways. **Age-specific mortality** is the proportion of individuals alive at the start of an age interval that died during that age interval. Its more cheerful reflection, **age-specific survivorship**, is the proportion of individuals alive at the start of an age interval that survived until the start of the next age interval. Thus, for the data shown in Table 45.1, the age-specific mortality rate during the 3- to 6-month age interval is $195/722 = 0.270$, and the age-specific survivorship rate is $527/722 = 0.730$. For any age interval, the sum of age-specific mortality and age-specific survivorship must equal 1. Life tables also summarize the proportion of the cohort that survived to a particular age, a statistic identifying the probability that any randomly selected newborn will still be alive at that age. For the 3- to 6-month age interval in Table 45.1, this probability is $722/843 = 0.856$.

Table 45.1 | **Life Table for a Cohort of 843 Individuals of the Grass *Poa annua* (Annual Bluegrass)**

Age Interval (in months)	Number Alive at Start of Age Interval	Number Dying during Age Interval	Age-Specific Mortality Rate	Age-Specific Survivorship Rate	Proportion of Original Cohort Alive at Start of Age Interval	Age-Specific Fecundity (Seed Production)
0–3	843	121	0.144	0.856	1.000	0
3–6	722	195	0.270	0.730	0.856	300
6–9	527	211	0.400	0.600	0.625	620
9–12	316	172	0.544	0.456	0.375	430
12–15	144	90	0.625	0.375	0.171	210
15–18	54	39	0.722	0.278	0.064	60
18–21	15	12	0.800	0.200	0.018	30
21–24	3	3	1.000	0.000	0.004	10
24–	0	—	—	—	—	—

Source: Begon, M., and M. Mortimer. *Population Ecology.* Sunderland, MA: Sinauer Associates, 1981. Adapted from R. Law, 1975.

Life tables also include data on **age-specific fecundity**, the average number of offspring produced by surviving females during each age interval. Table 45.1 shows that plants in the 3- to 6-month age interval produced an average of 300 seeds each. In some species, including humans, fecundity is highest in individuals of intermediate age. Younger individuals have not yet reached sexual maturity, and older individuals are past their reproductive prime. However, fecundity increases steadily with age in some plants and animals.

45.3b Survivorship Curves: Timing of Deaths of Individuals in a Population

Survivorship data are depicted graphically in a **survivorship curve** that displays the rate of survival for individuals over the species' average life span. Ecologists have identified three generalized survivorship curves (blue lines in **Figure 45.6**), although most organisms exhibit survivorship patterns falling between these idealized patterns.

Type I curves reflect high survivorship until late in life (see Figure 45.6a). They are typical of large animals that produce few young and provide them with extended care, which reduces juvenile mortality. Large mammals, such as Dall mountain sheep, produce only one or two offspring at a time and nurture them through their vulnerable first year. At that time, the young are better able to fend for themselves and are at lower risk for mortality (compared with younger animals). The picture of survivorship in mammals could change if one starts with the time of conception, as opposed to birth. The change would reflect

problems of pregnancy (see Chapter 18) and health of mothers.

Type II curves reflect a relatively constant rate of mortality in all age classes, a pattern that produces steadily declining survivorship (see Figure 45.6b). Many lizards, such as the five-lined skink, as well as songbirds and small mammals, face a constant probability of mortality from predation, disease, and starvation and show a type II pattern.

Type III curves reflect high juvenile mortality, followed by a period of low mortality once offspring reach a critical age and size (see Figure 45.6c). *Cleome droserifolia*, a desert shrub from the Middle East, experiences extraordinarily high mortality in its seed and seedling stages. Researchers estimate that for every 1 million seeds produced, fewer than 1000 germinate, and only about 40 individuals survive their first year. Once a plant becomes established, however, its likelihood of future survival is higher, and the survivorship curve flattens out. Many plants, insects, marine invertebrates, and fishes exhibit type III survivorship.

STUDY BREAK

1. What is the relationship between age-specific mortality and age-specific survivorship? If age-specific mortality is 0.384, what is the age-specific survivorship?
2. What is age-specific fecundity?
3. Describe three survivorship curves. In which curve do humans fall? Songbirds? Insects?

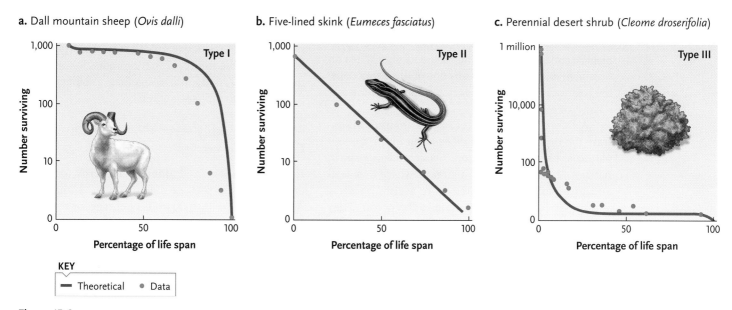

a. Dall mountain sheep (*Ovis dalli*)   **b.** Five-lined skink (*Eumeces fasciatus*)   **c.** Perennial desert shrub (*Cleome droserifolia*)

KEY
— Theoretical   • Data

Figure 45.6
Survivorship curves. The survivorship curves of many organisms (pink) roughly match one of three idealized patterns (blue).

45.4 The Evolution of Life Histories

Analysis of life tables reveals how natural selection affects an organism's **life history**, which includes the lifetime patterns of growth, maturation, and reproduction. Ecologists study life histories to understand tradeoffs in the allocation of resources to these three activities. The results of their research suggest that natural selection adjusts the allocation of resources to maximize an individual's number of surviving offspring.

Every organism is constrained by a finite **energy budget**, the total amount of energy it can accumulate and use to fuel its activities. An organism's energy budget is like a savings account. When the individual accumulates more energy than it needs, it makes deposits to this account, storing energy as starch, glycogen, or fat. When the individual expends more energy than it harvests, it makes withdrawals from its energy stores. But unlike a bank account, an organism's energy budget cannot be overdrawn, and no loans against future "earnings" are possible.

Just as humans find clever ways to finance their schemes, many organisms use different ways to mortgage their operations. Organisms that enter states of inactivity or dormancy can maximize the time over which they use stored energy. An extreme example would be animals and plants that can survive freezing, an obvious strategy for conserving energy. Hibernation and estivation in animals are other examples. Hibernating animals use periods of reduced body temperatures to survive prolonged periods of cold weather. Estivation is inactivity during prolonged periods of high temperatures. In some cases, specialized spores are resistant to heat and desiccation. Migrating birds on long flights get energy by metabolizing fat as well as other body structures, such as muscle or digestive tissue. Organisms use the energy they harvest for three broadly defined functions: maintenance (the preservation of good physiological condition), growth, and reproduction. When an organism devotes energy to any one of these functions, the balance in its energy budget is reduced, leaving less energy for other functions.

A fish, a deciduous tree, and a mammal illustrate the dramatic variations existing in life history patterns. Larval coho salmon (*Oncorhynchus kisutch*) hatch in the headwaters of a stream, where they feed and grow for about a year before assuming their adult body form and swimming to the ocean. They remain at sea for a year or two, feeding voraciously and growing rapidly. Eventually, using a Sun compass, geomagnetic, and chemical cues, salmon return to the rivers and streams where they hatched. The fishes swim upstream. Males prepare nests and try to attract females. Each female lays hundreds or thousands of relatively small eggs. After breeding, the body condition of males and females deteriorates, and they die.

Most deciduous trees in the temperate zone, such as oaks (genus *Quercus*), begin their lives as seeds (acorns) in late summer. The seeds remain metabolically inactive until the following spring or a later year. After germinating, seedling trees collect nutrients and energy and continue to grow throughout their lives. Once they achieve a critical size, they may produce thousands of acorns annually for many years. Thus, growth and reproduction occur simultaneously through much of the trees' life.

European red deer (*Cervus elaphus*) are born in spring, and the young remain with their mothers for an extended period, nursing and growing rapidly. After weaning, the young feed on their own. Female red deer begin to breed after reaching adult size in their third year, producing one or two offspring annually until they are about 16 years old, when they reach their maximum life span and die.

How can we summarize the similarities and differences in the life histories of these organisms? All three species harvest energy throughout their lives. Salmon and deciduous trees continue to grow until old age, whereas deer reach adult size fairly early in life. Salmon produce many offspring in a single reproductive episode, whereas deciduous trees and deer reproduce repeatedly. However, most trees produce thousands of seeds annually, whereas deer produce only one or two young each spring.

What factors have produced these variations in life history patterns? Life history traits, like all population characteristics, are modified by natural selection. Thus, organisms exhibit evolutionary adaptations that increase the fitness of individuals. Each species' life history is, in fact, a highly integrated "strategy" or suite of selection-driven adaptations.

In analyzing life histories, ecologists compare the number of offspring with the amount of care provided to each by the parents. They also determine the number of reproductive episodes in the organism's lifetime and the timing of first reproduction. Because these characteristics evolve together, a change in one trait is likely to influence others.

45.4a Fecundity versus Parental Care: Cutting Your Losses

If a female has a fixed amount of energy for reproduction, she can package that energy in various ways. A female duck with 1000 units of energy for reproduction might lay 10 eggs with 100 units of energy per egg. A salmon, which has higher fecundity, might lay 1000 eggs with 1 unit of energy in each. The amount of energy invested in each offspring *before* it is born is **passive parental care** provided by the female. Passive parental care is provided through yolk in an egg, endosperm in a seed, or, in mammals, nutrients that cross the placenta.

Many animals also provide **active parental care** to offspring *after* their birth. In general, species producing many offspring in a reproductive episode (e.g., the coho salmon) provide relatively little active parental care *to each offspring*. In fact, female coho salmon, each producing 2400 to 4500 eggs, die before their eggs even hatch. Conversely, species producing few offspring at a time (e.g., European red deer) provide much more care to each one. A red deer doe nurses its single fawn for up to eight months before weaning it.

45.4b How Often to Breed: Once or Repeatedly?

The number of reproductive episodes in an organism's life span is a second life history characteristic adjusted by natural selection. Some organisms, such as coho salmon, devote all of their stored energy to a single reproductive event. Any adult that survives the upstream migration is likely to leave some surviving offspring. Other species, such as deciduous trees and red deer, reproduce more than once. In contrast to salmon, individuals of these species devote only some of their energy budget to reproduction at any time, with the balance allocated to maintenance and growth. Moreover, in some plants, invertebrates, fishes, and reptiles, larger individuals produce more offspring than smaller ones. Thus, one advantage of using only part of the energy budget for reproduction is that continued growth may result in greater fecundity at a later age. However, if an organism does not survive until the next breeding season, the potential advantage of putting energy into maintenance and growth would be lost.

45.4c Age at First Reproduction: When to Start Reproducing

Individuals that first reproduce at the earliest possible age may stand a good chance of leaving some surviving offspring. But the energy they use in reproduction is not available for maintenance and growth. Thus, early reproducers may be smaller and less healthy than individuals that delay reproduction in favour of other functions. Conversely, an individual that delays reproduction may increase its chance of survival and its future fecundity by becoming larger or more experienced. But there is always some chance that it will die before the next breeding season, leaving no offspring at all. Therefore, a finite energy budget and the risk of mortality establish a tradeoff in the timing of first reproduction. Mathematical models suggest that delayed reproduction will be favoured by natural selection if a sexually mature individual has a good chance of surviving to an older age, if organisms grow larger as they age, and if larger organisms have higher fecundity. Early reproduction will be favoured if adult survival rates are low, if animals do not grow larger as they age, or if larger size does not increase fecundity. These

characteristics apply more readily to some animals and plants than they do to others. Among animals, the features discussed above apply more readily to vertebrate than to invertebrate animals. Parasitic organisms may have quite different patterns of life history.

Life history characteristics vary from one species to another, and they can vary among populations of a single species. Predation differentially influences life history characteristics in natural populations of guppies (*Poecilia reticulata*) in Trinidad (see *Life Histories of Guppies*).

STUDY BREAK

1. Organisms use energy for what three main operations?
2. Explain passive and active parental care in humans.
3. When would early reproduction be favoured?

45.5 Models of Population Growth

We now examine two mathematical models of population growth, exponential and logistic. *Exponential* models apply when populations experience unlimited growth. *Logistic* models apply when population growth is limited, often because available resources are finite. These simple models are tools that help ecologists refine their hypotheses, but neither provides entirely accurate predictions of population growth in nature. In the simplest versions of these models, ecologists define births as the production of offspring by any form of reproduction and ignore the effects of immigration and emigration.

45.5a Exponential Models: Populations Taking Off

Sometimes populations increase in size for a period of time with no apparent limits on their growth. In models of exponential growth, population size increases steadily by a constant ratio. Populations of bacteria and prokaryotes provide the most obvious examples, but multicellular organisms also sometimes exhibit exponential population growth.

Bacteria reproduce by binary fission. A parent cell divides in half, producing two daughter cells, and each can divide to produce two granddaughter cells. Generation time in a bacterial population is simply the time between successive cell divisions. If no bacteria in the population die, the population doubles in size each generation.

Bacterial populations grow quickly under ideal temperatures and with unlimited space and food. Consider a population of the human intestinal bacterium *Escherichia coli*, for which the generation time can be as short as 20 minutes. If we start with a population

Life Histories of Guppies

Figure 1
David Reznick surveys a shallow stream in the mountains of Trinidad.

Some years ago, drenched with sweat and with fishnets in hand, two ecologists were engaged in fieldwork on the Caribbean island of Trinidad. They were after guppies (*Poecilia reticulata*), small fish most of us see in pet shops. In their native habitats, guppies bear live young in shallow mountain streams **(Figure 1),** and John Endler and David Reznick were studying the environmental variables influencing the evolution of their life history patterns.

Male guppies are easy to distinguish from females. Males stop growing at sexual maturity. They are smaller, and their scales have bright colours that serve as visual signals in intricate courtship displays. Females are drably coloured and continue to grow larger throughout their lives. In the mountains of Trinidad, guppies live in different streams, even in different parts of the same stream. Two other species of fish eat guppies **(Figure 2).** In some streams, a small killifish (*Rivulus hartii*) preys on immature guppies but does not have much success with the larger adults. In other streams, a large pike–cichlid (*Crenicichla alta*) prefers mature guppies and rarely hunts small, immature ones.

Reznick and Endler found that the life history patterns of guppies vary among streams with different predators. In streams with pike–cichlids, male and female guppies mature faster and begin to reproduce at a smaller size and younger age than their counterparts in streams where killifish live. Female guppies from pike–cichlid streams reproduce more often, producing smaller and more numerous young. These dif-

Male guppy (right) that shared a stream with pike-cichlids (below)

Male guppy (right) that shared a stream with killifish (below)

Figure 2
Male guppies from streams where pike–cichlids live (top) are smaller and more streamlined and have duller colours than those from streams where killifish live (bottom). The pike–cichlid prefers to eat large guppies, and the killifish feeds on small guppies. Guppies are shown approximately life-sized; adult pike–cichlids grow to 16 cm in length, and adult killifish grow to 10 cm. (Guppy photos: David Reznick/University of California, Riverside: computer enhanced by Lisa Starr; predator photos: Hippocampus Bildarchiv.)

ferences allow guppies to avoid some predation. Those in pike–cichlid streams begin to reproduce when they are smaller than the size preferred by that predator. Those from killifish streams grow quickly to a size that is too large to be consumed by killifish **(Figure 3).**

Although these life history differences were correlated with the distributions of the two predatory fishes, they might result from some other, unknown differences between the streams. Endler and Reznick investigated this possibility with controlled laboratory experiments. They carefully shipped groups of guppies to California, where they bred guppies from each kind of stream for two generations. Both types

of experimental populations were raised under identical conditions in the absence of predators. Even in the absence of predators, the two types of experimental populations retained their life history differences. These results provided evidence of a genetic (heritable) basis for the observed life history differences.

Endler and Reznick also examined the role of predators in the *evolution* of the size differences **(Figure 4).** They raised guppies for many generations in the laboratory under three experimental conditions: some alone, some with killifish, and some with pike-cichlids. As predicted, the guppy lineage subjected to predation by killifish became larger at

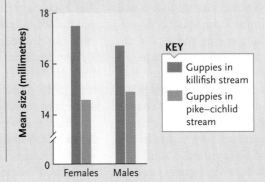

KEY

- Guppies in killifish stream
- Guppies in pike–cichlid stream

Figure 3
Guppies in streams occupied by pike–cichlids are smaller than those in streams occupied by killifish.

maturity. Individuals that were small at maturity were frequently eaten, and their reproduction was limited. The lineage raised with pike–cichlids showed a trend toward earlier maturity. Individuals that matured at a larger size faced a greater likelihood of being eaten before they had reproduced.

When they first visited Trinidad, Endler and Reznick had introduced guppies from a pike–cichlid stream to another stream that contained killifish but no pike–cichlids or guppies. There, 11 years later, guppy populations had changed. As the researchers predicted, the guppies became larger and reproduced more slowly, characteristics typical of natural guppy populations that live and die with killifish.

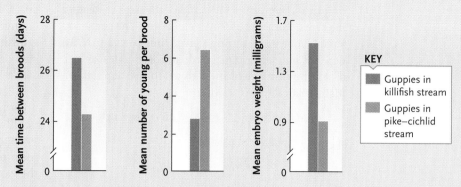

Figure 4

Female guppies from streams occupied by pike–cichlids reproduce more often (shorter time between broods) and produce more young per brood and smaller young (lower embryo weight) than females living in streams occupied by killifish.

of one bacterium, the population doubles to two cells after one generation, to four cells after two generations, and to eight cells after three generations **(Figure 45.7)**. After only 8 hours (24 generations), the population will number more than 16 million. And after a single day (72 generations), the population will number nearly 5×10^{21} cells. Although other bacteria grow more slowly than *E. coli,* it is no wonder that pathogenic bacteria, such as those causing cholera or plague, can quickly overtake the defences of an infected animal.

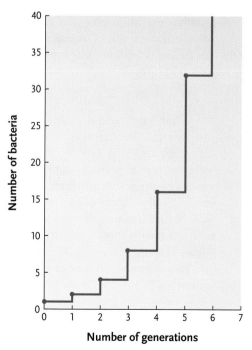

Figure 45.7

Bacterial population growth. If all members of a bacterial population divide simultaneously, a plot of population size over time forms a stair-stepped curve in which the steps get larger as the number of dividing cells increases.

When populations of multicellular organisms are large, they have the potential for exponential growth, as we shall see below for our own species. In any event, over a given time period,

change in population size = number of births − number of deaths.

We express this relationship mathematically by defining N as the population size; ΔN (pronounced "delta N") as the change in population size; Δt as the time period during which the change occurs; and B and D as the numbers of births and deaths, respectively, *during that time period*. Thus, $\Delta N/\Delta t$ symbolizes the change in population size over time, and

$$\Delta N/\Delta t = B - D.$$

The above equation applies to any population for which we know the exact numbers of births and deaths. Ecologists usually express births and deaths as *per capita* (per individual) rates, allowing them to apply the model to a population of any size. The per capita birth rate (b) is the number of births in the population during the specified time period divided by the population size: $b = (B/N)$. Similarly, the per capita death rate, d, is the number of deaths divided by the population size: $d = (D/N)$.

If in a population of 2000 field mice, 1000 mice are born and 200 mice die during 1 month, $b = 1000/2000 = 0.5$ births per individual per month, and $d = 200/2000 = 0.1$ deaths per individual per month. Of course, no mouse can give birth to half an offspring, and no individual can die one-tenth of a death. But these rates tell us the per capita birth and death rates *averaged over all mice in the population*. Per capita birth and death rates are always expressed over a specified time period. For long-lived organisms, such as humans, time is measured in years. For short-lived organisms, such as fruit flies, time is measured in days. We can calculate per capita birth and death rates from data in a life table.

Now we can revise the population growth equation to use per capita birth and death rates instead of the actual numbers of births and deaths. The change in a population's size during a given time period ($\Delta N/\Delta t$) depends on the per capita birth and death rates, as well as on the number of individuals in the population. Mathematically, we can write

$$\Delta N/\Delta t = B - D = bN - dN = (b - d)N$$

or, in the notation of calculus,

$$dN/dt = (b - d)N.$$

This equation describes the **exponential model of population growth**. (Note that in calculus, dN/dt is the notation for the population growth rate. The "d" in dN/dt is *not* the same "d" we use to symbolize the per capita death rate.)

The difference between the per capita birth rate and the per capita death rate, $b - d$, is the **per capita growth rate** of the population, symbolized by r. Like b and d, r is always expressed per individual per unit time. Using the per capita growth rate, r, in place of $(b - d)$, the exponential growth equation is written

$$dN/dt = rN.$$

If the birth rate exceeds the death rate, r has a positive value ($r > 0$), and the population is growing. In our example with field mice, r is $0.5 - 0.1 = 0.4$ mice per mouse per month. If, on the other hand, the birth rate is lower than the death rate, r has a negative value ($r < 0$), and the population is shrinking. In populations in which the birth rate equals the death rate, r is zero, and the population's size is not changing—a situation known as **zero population growth**, or ZPG. Even under ZPG, births and deaths still occur, but the numbers of births and deaths cancel each other out.

Populations will grow as long as the per capita growth rate is positive ($r > 0$). In our hypothetical population of field mice, we started with $N = 2000$ mice and calculated a per capita growth rate of 0.4 mice per individual per month. In the first month, the population grows by $0.4 \times 2000 = 800$ mice **(Figure 45.8)**. At the start of the second month, $N = 2800$ and r still $= 0.4$. Thus, in the second month, the population grows by $0.4 \times 2800 = 1120$ mice. Notice that even though r remains constant, the *increase* in population size grows each month because more individuals are reproducing. In less than two years, the mouse population will increase to more than one million! A graph of exponential population growth has a characteristic J shape, getting steeper through time. The population grows at an ever-increasing pace because the change in a population's size depends on the number of individuals in the population and its per capita growth rate.

Imagine a hypothetical population living in an ideal environment with unlimited food and shelter; no predators, parasites, or disease; and a comfortable abiotic environment. Under such circumstances (admittedly unrealistic), the per capita birth rate is very high, the per capita death rate is very low, and the per capita growth rate, r, is as high as it can be. This maximum per capita growth rate, symbolized r_{max}, is the population's **intrinsic rate of increase**. Under these ideal conditions, our exponential growth equation is

$$dN/dt = r_{max}N.$$

When populations grow at their intrinsic rate of increase, population size increases very rapidly. Across a wide variety of protists and animals, r_{max} varies inversely with generation time: species with a short generation time have higher intrinsic rates of increase than those with a long generation time **(Figure 45.9)**.

The exponential model predicts unlimited population growth. But we know from even casual observations that population sizes of most species are somehow limited. We are not knee-deep in bacteria, rosebushes, or garter snakes. What factors limit the growth of populations? As a population gets larger, it uses more vital resources,

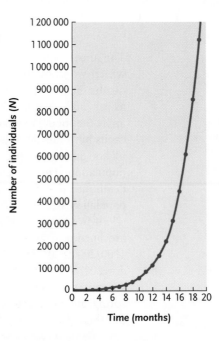

Month	Old Population Size		Net Monthly Increase		New Population Size
1	2 000	+	800	=	2 800
2	2 800	+	1 120	=	3 920
3	3 920	+	1 568	=	5 488
4	5 488	+	2 195	=	7 683
5	7 683	+	3 073	=	10 756
6	10 756	+	4 302	=	15 058
7	15 058	+	6 023	=	21 081
8	21 081	+	8 432	=	29 513
9	29 513	+	11 805	=	41 318
10	41 318	+	16 527	=	57 845
11	57 845	+	23 138	=	80 983
12	80 983	+	32 393	=	113 376
13	113 376	+	45 350	=	158 726
14	158 726	+	63 490	=	222 216
15	222 216	+	88 887	=	311 103
16	311 103	+	124 441	=	435 544
17	435 544	+	174 218	=	609 762
18	609 762	+	243 905	=	853 667
19	853 677	+	341 467	=	1 195 134

Figure 45.8

Exponential population growth. Exponential population growth produces a J-shaped curve when population size is plotted against time. Although the per capita growth rate (r) remains constant, the increase in population size gets larger every month because more individuals are reproducing.

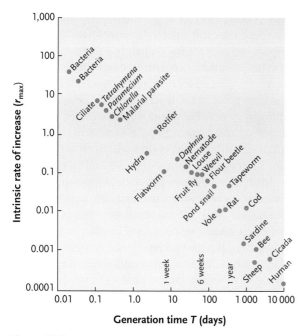

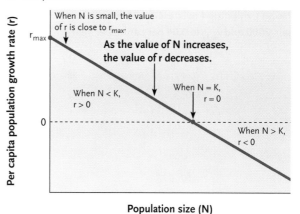

a. The predicted effect of N on r

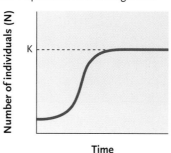

b. Population size through time

Figure 45.9

Generation time and r_{max}. The intrinsic rate of increase (r_{max}) is high for protists and animals with short generation times and low for those with long generation times.

Figure 45.10

The logistic model of population growth. The logistic model **(a)** assumes that the per capita population growth rate (r) decreases linearly as population size (N) increases. The logistic model also predicts that population size **(b)** increases quickly at first but then slowly approaches carrying capacity (K).

perhaps leading to a shortage of resources. In this situation, individuals may have less energy available for maintenance and reproduction, causing decreases in per capita birth rates and increases in per capita death rates. Energy in food is not always equally available, and when an animal spends time handling food to eat it, the ratio of cost (handling) to benefit (energy in the food) diminishes, affecting return on investment. Such rate changes can affect a population's per capita growth rate, causing population growth to slow or stop.

45.5b Logistic Models: Populations and Carrying Capacity (K)

Environments provide enough resources to sustain only a finite population of any species. The maximum number of individuals that an environment can support indefinitely is termed its **carrying capacity**, symbolized as K. K is defined for each population. It is a property of the environment that can vary from one habitat to another and in a single habitat over time. The spring and summer flush of insects in temperate habitats supports large populations of insectivorous birds. But fewer insects are available in autumn and winter, causing a seasonal decline in K for birds, and autumnal migrations occur in birds seeking more food and less inclement weather. Other cycles are annual, such as variation in water levels in wetlands from year to year.

The **logistic model of population growth** assumes that a population's per capita growth rate, r, decreases as the population gets larger **(Figure 45.10)**. In other

words, population growth slows as the population size approaches K. The mathematical expression $(K - N)$ tells us how many individuals can be added to a population before it reaches K. The expression $(K - N)/K$ indicates what percentage of the carrying capacity is still available.

To create the logistic model, we factor the impact of K into the exponential model by multiplying r_{max} by $(K - N)/K$ to reduce the per capita growth rate (r) from its maximum value (r_{max}) as N increases:

$$dN/dt = r_{max}N(K - N)/K.$$

The calculation of how r varies with population size is straightforward **(Table 45.2, p. 1126)**. In a very small population (N much smaller than K), plenty of resources are available; the value of $(K - N)/K$ is close to 1. Here the per capita growth rate (r) approaches the maximum possible (r_{max}). Under these conditions, population growth is close to exponential. If a population is large (N close to K), few additional resources are available. Now the value of $(K - N)/K$ is small, and the per capita growth rate (r) is very low. When the size of the population exactly equals K, $(K - N)/K$ becomes 0, as does the population growth rate, the situation defined as ZPG.

The logistic model of population growth predicts an S-shaped graph of population size over time, with

Table 45.2	The Effect of N on r and ΔN^* in a Hypothetical Population Exhibiting Logistic Growth in which K equals 2000 and r_{max} is 0.04 per capita per year			
N (population size)	$(K - N)/K$ (% of K available)	$r = r_{max}(K - N/K)$ (per capita growth rate)	$\Delta N = rN$ (change in N)	
50	0.990	0.0396	2	
100	0.950	0.0380	4	
250	0.875	0.0350	9	
500	0.750	0.0300	15	
750	0.625	0.0250	19	
1000	0.500	0.0200	20	
1250	0.375	0.0150	19	
1500	0.250	0.0100	15	
1750	0.125	0.0050	9	
1900	0.050	0.0020	4	
1950	0.025	0.0010	2	
2000	0.000	0.0000	0	

*ΔN rounded to the nearest whole number.

the population slowly approaching K and remaining at that level **(Figure 45.11b)**. According to this model, the population grows slowly when the population size is small because few individuals are reproducing. It also grows slowly when the population size is large because the per capita population growth rate is low. The population grows quickly (dN/dt is highest) at intermediate population sizes, when a sizable number of individuals are breeding and the per capita population growth rate (r) is still fairly high (see Table 45.2).

The logistic model assumes that vital resources become increasingly limited as a population grows.

Thus, the model is a mathematical portrait of **intraspecific** (within species) **competition**, the dependence of two or more individuals in a population on the same limiting resource. For mobile animals, limiting resources could be food, water, nesting sites, and refuges from predators. For sessile species, space can be a limiting resource. For plants, sunlight, water, inorganic nutrients, and growing space can be limiting. The pattern of uniform dispersion described earlier often reflects intraspecific competition for limited resources.

In some very dense populations, accumulation of poisonous waste products may reduce survivorship and reproduction. Most natural populations live in open systems where wastes are consumed by other organisms or flushed away. But the buildup of toxic wastes is common in laboratory cultures of microorganisms. For example, yeast cells ferment sugar and produce ethanol as a waste product. Thus, the alcohol content of wine usually does not exceed 13% by volume, the ethanol concentration that poisons yeasts that are vital to the wine-making process.

How well do species conform to the predictions of the logistic model? In simple laboratory cultures, relatively small organisms, such as *Paramecium,* some crustaceans, and flour beetles, often show an S-shaped pattern of population growth **(Figure 45.11a, b)**. Moreover, large animals introduced into new environments sometimes exhibit a pattern of population growth that matches the predictions of the logistic model **(Figure 45.11c)**.

Nevertheless, some assumptions of the logistic model are unrealistic. For example, the model predicts that survivorship and fecundity respond immediately to changes in a population's density. Many organisms exhibit a delayed response (a **time lag**) because fecundity has been determined by resource availability at some time in the past. This may reflect conditions that prevailed when individuals were adding yolk to eggs or

Figure 45.11
Examples of logistic population growth.

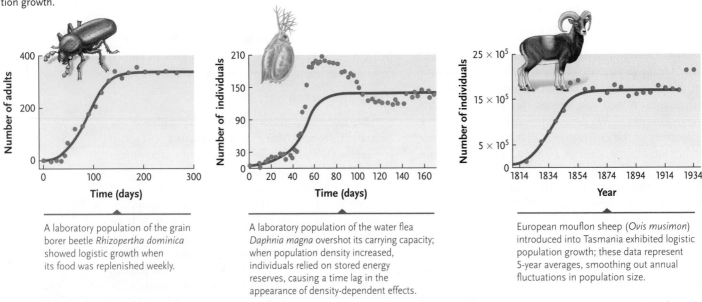

A laboratory population of the grain borer beetle *Rhizopertha dominica* showed logistic growth when its food was replenished weekly.

A laboratory population of the water flea *Daphnia magna* overshot its carrying capacity; when population density increased, individuals relied on stored energy reserves, causing a time lag in the appearance of density-dependent effects.

European mouflon sheep (*Ovis musimon*) introduced into Tasmania exhibited logistic population growth; these data represent 5-year averages, smoothing out annual fluctuations in population size.

KEY
— Theoretical • Data

endosperm to seeds. Moreover, when food resources become scarce, individuals may use stored energy reserves to survive and reproduce. This delays the impact of crowding until stored reserves are depleted and means that population size may overshoot K (see Figure 45.11b). Deaths may then outnumber births, causing the population size to drop below K, at least temporarily. Time lags often cause a population to oscillate around K.

The assumption that the addition of new individuals to a population always decreases survivorship and fecundity is unrealistic. In small populations, modest population growth may not have much impact on survivorship and fecundity. In fact, most organisms probably require a minimum population density to survive and reproduce. Some plants flourish in small clumps that buffer them from physical stresses, whereas a single individual living in the open would suffer adverse effects. In some animal populations, a minimum population density is necessary for individuals to find mates. Determining the minimum viable population for a species is an important issue in conservation biology (see Chapter 48).

STUDY BREAK

1. When do you use an exponential model rather than a logistic one?
2. Define the terms in the equation $dN/dt = (b - d)N$.
3. What does it mean when $r < 0$, $r > 0$, or $r = 0$? What is r_{max}, and how does it vary with generation time?

45.6 Population Regulation

What environmental factors influence population growth rates and control fluctuations in population size? Some factors affecting population size are **density dependent** because their influence increases or decreases with the density of the population. Intraspecific competition and predation are examples of density-dependent environmental factors. The logistic model includes the effects of density dependence in its assumption that per capita birth and death rates change with population density.

Numerous laboratory and field studies have shown that crowding (high population densities) decreases individual growth rate, adult size, and survival of plants and animals **(Figure 45.12)**. Organisms living in extremely dense populations are unable to harvest enough resources. They grow slowly and tend to be small, weak, and less likely to survive. Gardeners understand this relationship and thin their plants to achieve a density that maximizes the number of vigorous individuals that survive to be harvested.

Crowding has a negative effect on reproduction **(Figure 45.13, p. 1128)**. When resources are in short supply, each individual has less energy for reproduction after meeting its basic maintenance needs. Hence, females in crowded populations produce either fewer offspring or smaller offspring that are less likely to survive.

In some species, crowding stimulates developmental and behavioural changes that may influence the density of a population. Migratory locusts can

Figure 45.12
Effects of crowding on individual growth, size, and survival.

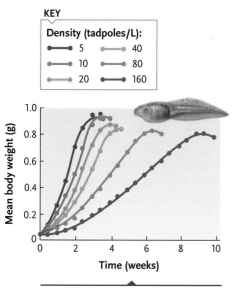

Tadpoles of the frog *Rana tigrina* grew faster and reached larger adult body size at low densities than at high densities.

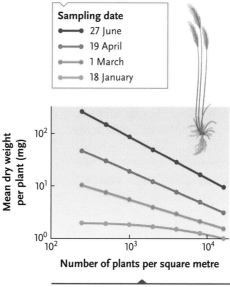

The size of the annual dune grass *Vulpia fasciculata* decreased markedly when plants were grown at high density. Density effects became more accentuated through time as the plants grew larger (indicated by the progressively steeper slopes of the lines).

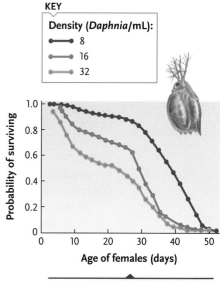

The water flea *Daphnia pulex* had higher survivorship at a density of 8/mL than at densities of 16/mL or 32/mL.

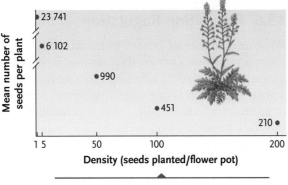

The number of seeds produced by shepherd's purse (*Capsella bursa-pastoris*) decreased dramatically with increasing density in experimental plots.

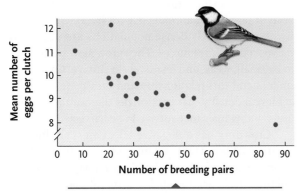

The mean number of eggs produced by the Great Tit (*Parus major*), a woodland bird, declined as the number of breeding pairs in Marley Wood increased.

Figure 45.13
Effects of crowding on fecundity.

develop into either solitary or migratory forms in the same population. Migratory individuals have longer wings and more body fat, characteristics that allow long-distance dispersal. High population density increases the frequency of the migratory form. So many locusts move away from the area of high density **(Figure 45.14)**, reducing the size and thus the density of the original population.

Although these data about locusts confirm the assumptions of the logistic equation, they do not prove that natural populations are regulated by density-dependent factors. Experimental evidence is necessary to provide a convincing demonstration that an increase in population density causes population size to decrease, whereas a decrease in density causes it to increase.

In the 1960s, Robert Eisenberg experimentally increased the numbers of aquatic snails (*Lymnaea elodes*) in some ponds, decreased them in others, and maintained natural densities in control ponds. Adult survivorship did not differ between experimental and control treatments. But there was a gradient in egg production from few eggs (snails in high-density ponds), to more (control density), to most (control density). Furthermore, survival rates of young snails declined as density increased. After four months,

Figure 45.14
A swarm of locusts. Migratory locusts (*Locusta migratoria*) moving across an African landscape can devour their own weight in plant material every day.

densities in the two experimental groups converged on those in the control, providing strong evidence of density-dependent population regulation.

At this stage, intraspecific competition appears to be the primary density-dependent factor regulating population size. Competition between populations of different species also can exert density-dependent effects on population growth (see Chapter 46).

But predation also can cause density-dependent population regulation. As a particular prey species becomes more numerous, predators may consume more of it because it is easier to find and catch. Once a prey species exceeds some threshold density, predators may consume a larger percentage of its population, a density-dependent effect. On rocky shores in California, sea stars feed mainly on the most abundant invertebrate there. When one prey species becomes common, predators feed on it disproportionately, drastically reducing its numbers. Then they switch to now more abundant alternate prey.

Sometimes several density-dependent factors influence a population at the same time. On small islands in the West Indies, spiders are rare wherever lizards (*Ameiva festiva, Anolis carolinensis,* and *Anolis sagrei*) are abundant but common where the lizards are rare or absent. To test whether the presence of lizards limits the abundance of spiders, David Spiller and Tom Schoener built fences around plots on islands where these species occur. They eliminated lizards from experimental plots but left them in control plots. After two years, spider populations in some experimental plots were five times more dense than those in control plots, suggesting a strong impact of lizard populations on spider populations **(Figure 45.15)**. In this situation, lizards had two density-dependent effects on spider populations. First, lizards ate spiders, and, second, they competed with them for food. Experimental evidence made it possible for biologists to better understand the situation.

Predation, parasitism, and disease can cause density-dependent regulation of plant and animal pop-

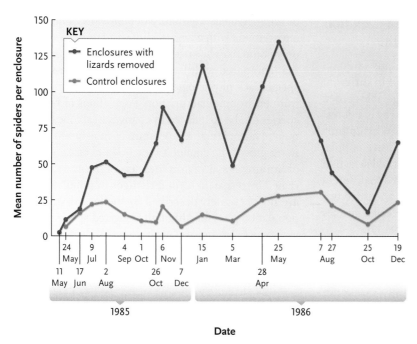

Figure 45.15

Populations of spiders (*Metepeira dantona*) on a small island in the Bahamas (Caribbean) are influenced by the presence of lizards. Note how much higher the population densities of spiders are in the absence versus the presence of lizards.

ulations. Infectious microorganisms spread quickly in a crowded population (e.g., rabies). In addition, if crowded individuals are weak or malnourished, they are more susceptible to infection and may die from diseases that healthy organisms would survive. Effects on survival can be direct or indirect.

45.6a Density-Independent Factors: Reducing Population in Spite of Density

Some populations are affected by **density-independent** factors that reduce population size regardless of its density. If an insect population is not physiologically adapted to high temperature, a sudden hot spell may kill 80% of them whether they number 100 or 100 000. Fires, earthquakes, storms, and other natural disturbances can contribute directly or indirectly to density-independent mortality. Because such factors do not cause a population to fluctuate around its *K*, these density-independent factors can reduce but do not *regulate* population size.

Density-independent factors have a particularly strong effect on populations of small-bodied species that cannot buffer themselves against environmental change. Their populations grow exponentially for a time, but shifts in climate or random events cause high mortality before populations reach a size at which density-dependent factors would regulate their numbers. When conditions improve, populations grow exponentially, at least until another density-independent factor causes them to crash again. A small Australian insect, a thrip, eats the pollen and flowers of plants in the rose family. These thrips can be abundant enough to damage blooms. Populations of thrips grow exponentially in spring,

when many flowers are available, and the weather is warm and moist **(Figure 45.16)**. But their populations crash predictably during summer because Thrips imaginis do not tolerate hot and dry conditions. After the crash, a few individuals survive in remaining flowers, and they are the stock from which the population grows exponentially the following spring.

45.6b Interactions between Density-Dependent and Density-Independent Factors: Sometimes Population Density Affects Mortality

Density-dependent factors can interact with density-independent factors and limit population growth. Food shortage caused by high population density (a density-dependent factor) may lead to malnourishment. Malnourished individuals may be more likely to succumb to the stress of extreme weather (a density-independent factor).

Populations can be affected by density-independent factors in a density-dependent manner. Some animals retreat into shelters to escape environmental stresses, such as floods or severe heat. If a population is small, most individuals can be accommodated in available refuges. But if a population is large (exceeds the capacity of shelters), only a proportion will find suitable shelter. The larger the population, the greater the percentage of individuals exposed to the stress(es). Thus, although the density-independent effects of weather limit populations of thrips, the availability of flowers in summer (a density-dependent factor) regulates the size of the starting populations of thrips the following spring. Hence, both density-dependent and density-independent factors influence the size of populations of thrips.

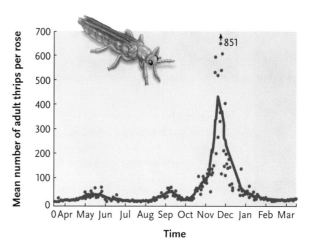

Figure 45.16

Booms and busts in a *Thrips* population. Populations of the Australian insect *thrips imaginis* grow exponentially when conditions are favourable during spring (which begins in September in the Southern Hemisphere). The populations crash in summer, however, when hot and dry conditions cause high mortality.

45.6c Life History Characteristics: Evolution of Strategies for Population Growth

Even casual observation reveals tremendous variation in how rapidly population sizes change in different species. New weeds often appear in a vegetable garden overnight, whereas the number of oak trees in a forest may remain relatively stable for years. Why do only some species have the potential for explosive population growth? The answer lies in how natural selection has moulded life history strategies adapted to different ecological conditions. Some ecologists recognize two quite different life history patterns, **r-selected** species and **K-selected** species (Table 45.3; Figure 45.17).

On the face of it, *r*-selected species are adapted to rapidly changing environments, and many have at least some of the features outlined in Table 45.3. The success

of an *r*-selected life history depends on flooding the environment with a *large quantity* of young even though only some may be successful. Small body size means that compared with larger bodied species, **r-selected species** lack physiological mechanisms to buffer them from environmental variation. Populations of *r*-selected species can be so reduced by changes in abiotic environmental factors (e.g., temperature or moisture) that they never grow large enough to reach *K* and face a shortage of limiting resources. In these cases, *K* cannot be estimated by researchers, and changes in population size are not accurately described by the logistic model of population growth. Although *r*-selected species appear to have poor tolerance of environmental change, they are said to be adapted to rapidly changing environments.

At the same time, *K*-selected species have at least some of the features outlined for them in Table 49.3. These organisms survive the early stages of life (type I or type II survivorship), and a low r_{max} means that their populations grow slowly. The success of a *K*-selected life history is linked to the production of a relatively small number of *high-quality* offspring that join an already well-established population. Generalizations about *r*-selected and *K*-selected species are misleading. We can recognize this by comparing two species of mammals.

Peromyscus maniculatus, deer mice, occur widely in North America. In southern Ontario, adults weigh 12 to 31 g, females produce average litters of 4 (range 2 to 8), and each can have 4 or 5 litters a year. Females become sexually mature at age 2 months and breed in their first year. Occasionally, deer mice mice live to age 3 years in the wild. Throughout their extensive range in North America, *Myotis lucifugus,* little brown bats, weigh 7 to 12 g; females bear a single young per litter and have one litter per year. Females may breed a year after they are born, but many may wait until they are two years old. In the wild, little brown bats can live over 30 years. Using these data, one small mammal (deer mouse) is an *r*-strategist, whereas another (little brown bat) is a *K*-strategist. To complicate matters, deer mice living in Kananaskis in the mountains near Calgary, Alberta, mature at one year and may have two litters per year, typically five young per litter. Compared to little brown bats, Kananaskis deer mice are *r*-strategists. Compared with Ontario deer mice, they are more like *K*-strategists.

Biologists may find the idea of *r*-strategists and *K*-strategists useful, but too often the idea means imposing some human view of the world on a natural system. *K*-strategists and *r*-strategists may be more like "beauty," defined by the eye of the beholder. Elephants (*Loxodonta africana, Loxodonata cyclotis, Elephas maximus*) are big and meet all *K*-strategist criteria. Many insects are small but in all other respects meet the criteria considered typical of *K*-strategists because of their patterns of reproduction. Codfish (*Gadus morhua*) are big (compared to insects or bats) but meet most of the criteria used to identify *r*-strategists, such as their patterns of reproduction.

Table 45.3	Characteristics of *r*-Selected and *K*-Selected Species	
Characteristic	**r-Selected Species**	**K-Selected Species**
Maturation time	Short	Long
Life span	Short	Long
Mortality rate	Usually high	Usually low
Reproductive episodes	Usually one	Usually several
Time of first reproduction	Early	Late
Clutch or brood size	Usually large	Usually small
Size of offspring	Small	Large
Active parental care	Little or none	Often extensive
Population size	Fluctuating	Relatively stable
Tolerance of environmental change	Generally poor	Generally good

a. An *r*-selected species

b. A *K*-selected species

Figure 45.17

Life history differences. An r-selected species, **(a)** *Chenopodium quinoa*, matures in one growing season and produces many tiny seeds. Quinoa was a traditional food staple for indigenous people of North and South America. A *K*-selected species, **(b)** *Cocos nucifera*, a coconut palm, grows slowly and produces a few large seeds repeatedly during its long life.

45.6d Population Cycles: Ups and Downs in Numbers of Individuals

Population densities of many insects, birds, and mammals in the northern hemisphere fluctuate between species-specific lows and highs in a multiyear cycle. Arctic populations of small rodents (*Lemmus lemmus*) vary in size over a 4-year cycle, whereas snowshoe hares (*Lepus americanus*), ruffed grouse (*Bonasa umbellis*), and lynx have 10-year cycles. Ecologists documented these cyclic fluctuations more than a century ago, but none of the general hypotheses proposed to date explain cycles in all species. Availability and quality of food, abundance of predators, prevalence of disease-causing microorganisms, and variations in weather can influence population growth and declines. Furthermore, food supply for a cycling population and its predators are themselves influenced by a population's size.

Theories of *intrinsic control* suggest that as an animal population grows, individuals undergo hormonal changes that increase aggressiveness, reduce reproduction, and foster dispersal. The dispersal phase of the cycle may be dramatic. When populations of Norway lemming (*Lemmus lemmus*), a rodent that lives in the Scandinavian arctic, reach their peak density, aggressive interactions drive younger and weaker individuals to disperse. The dispersal of many thousands of lemmings during periods of population growth has sometimes been incorrectly portrayed in nature films as a suicidal mass migration.

Other explanations focus on extrinsic control, such as the relationship between a cycling species and its food or predators. A dense population may exhaust its food supply, increasing mortality and decreasing reproduction. The die-off of large numbers of African elephants in Tsavo National Park in Kenya is an example of the impact of overpopulation. There elephants overgrazed vegetation in most of the park habitat. In 1970, the combination of overgrazing and a drought caused high mortality of elephants. The picture is not always clear because experimental food supplementation does not always prevent decline in mammal populations. This suggests some level of intrinsic control.

Cycles in populations of predators could be induced by time lags between populations of predators and prey and vice versa **(Figure 45.18)**. The 10-year cycles of snowshoe hares and their feline predators, Canada lynx, were often cited as a classic example of

© Ed Cesar/Photo Researchers, Inc.

Figure 45.18

The predator–prey model. Predator–prey interactions may contribute to density-dependent regulation of both populations. A mathematical model **(a)** predicts cycles in the numbers of predators and prey because of time lags in each species' responses to changes in the density of the other. (Predator population size is exaggerated in this graph.) **(b)** Canada lynx (*Lynx canadensis*) and snowshoe hare (*Lepus americanus*) were often described as a typical cyclic predator–prey interaction. The abundances of lynx (red line) and snowshoe hare (blue line) are based on counts of pelts trappers sold to the Hudson's Bay Company over a 90-year period. Recent research shows that population cycles in snowshoe hares are caused by complex interactions between the snowshoe hares, its food plants, and its predators.

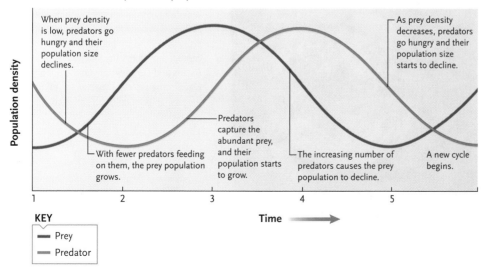

a. Predictions of a predator–prey model

When prey density is low, predators go hungry and their population size declines.

With fewer predators feeding on them, the prey population grows.

Predators capture the abundant prey, and their population starts to grow.

As prey density decreases, predators go hungry and their population size starts to decline.

The increasing number of predators causes the prey population to decline.

A new cycle begins.

Population density

Time

KEY
— Prey
— Predator

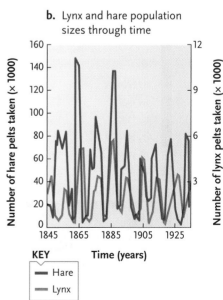

b. Lynx and hare population sizes through time

Number of hare pelts taken (× 1000)

Number of lynx pelts taken (× 1000)

Time (years)

KEY
— Hare
— Lynx

such an interaction. But snowshoe hare populations can exhibit a 10-year fluctuation even on islands where lynx are absent. Thus, lynx is not solely responsible for population cycles in snowshoe hares. To further complicate matters, the database demonstrating fluctuations was often the numbers of pelts purchased by the Hudson's Bay Company. Here, fur price influenced the trapping effort and the numbers of animals harvested. This economic reality brought into question the relationship between the numbers of pelts and actual population densities of lynx and snowshoe hares.

Charles Krebs and his colleagues studied hare and lynx interactions with a large-scale, multiyear experiment in Kluane in the southern Yukon. Using fenced experimental areas, they could add food for snowshoe hares, exclude mammalian predators, or apply both experimental treatments while monitoring unmanipulated control plots. When mammalian predators were excluded, densities of snowshoe hares approximately doubled relative to controls. Where food was added, densities of snowshoe hares tripled relative to controls. In plots where food was added *and* predators were excluded, densities of snowshoe hares increased 11-fold compared with controls. Krebs and his colleagues concluded that neither food availability nor predation is solely responsible for population cycles in snowshoe hares. They postulated that complex interactions between snowshoe hares, their food plants, and their predators generate cyclic fluctuations in populations of snowshoe hares.

STUDY BREAK

1. What are density-dependent factors? Why do dense populations tend to decrease in size?
2. Define density-independent factors and give some examples.
3. Describe two key differences between *r*-selected species and *K*-selected species.

45.7 Human Population Growth

How do human populations compare with those of other species? The worldwide human population surpassed 6 billion on October 12, 1999. Like many other species, humans live in somewhat isolated populations that vary in their demographic traits and access to resources. Although many of us live comfortably, at least a billion people are malnourished or starving, lack access to clean drinking water, and live without adequate shelter or health care.

For most of human history, our population grew slowly, reflecting the impact of a range of restraints. Over the past two centuries, the worldwide human population has grown exponentially **(Figure 45.19)**. Demographers identified three ways in which we have avoided the effects of density-dependent regulating factors.

First, humans have expanded their geographic range into virtually every terrestrial habitat, alleviating competition for space. Our early ancestors lived in tropical and subtropical grasslands, but by 40 000 years ago, they had dispersed through much of the world (see Chapter 27). Their success resulted from their ability to solve ecological problems by building fires, assembling shelters, making clothing and tools, planning community hunts, and sharing information. Vital survival skills spread from generation to generation and from one population to another because language allowed communication of complex ideas and knowledge.

Second, we have increased *K* in habitats we occupy, isolating us, as a species, from restrictions associated with access to resources. This change began to occur about 11 000 years ago, when populations in different parts of the world began to shift from hunting and gathering to agriculture (see Chapter 49). At that time, our ancestors cultivated wild grasses and other plants, diverted water to irrigate crops, and used domesticated

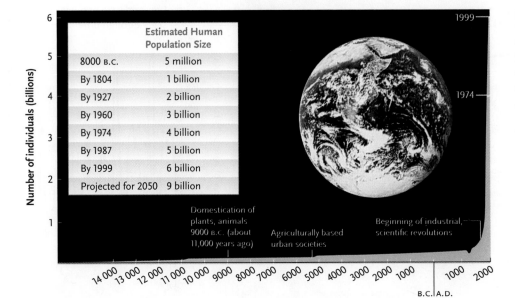

Figure 45.19

Human population growth. The worldwide human population grew slowly until 200 years ago, when it began to increase explosively. The dip in the mid-fourteenth century represents the death of 60 million Asians and Europeans from the bubonic plague. The table shows how long it took for the human population to add each billion people. (Photo: NASA.)

Estimated Human Population Size	
8000 B.C.	5 million
By 1804	1 billion
By 1927	2 billion
By 1960	3 billion
By 1974	4 billion
By 1987	5 billion
By 1999	6 billion
Projected for 2050	9 billion

animals for food and labour. Innovations such as these increased the availability of food, raising both K and rates of population growth. In the mid-eighteenth century, people harnessed the energy in fossil fuels, and industrialization began in Western Europe and North America. Food supplies and K increased again, at least in industrialized countries, largely through the use of synthetic fertilizers, pesticides, and efficient methods of transportation and food distribution.

Third, advances in public health reduced the effects of critical population-limiting factors such as malnutrition, contagious diseases, and poor hygiene. Over the past 300 years, modern plumbing and sewage treatment, removal of garbage, and improvements in food handling and processing, as well as medical discoveries, have reduced death rates sharply. Births now greatly exceed deaths, especially in less industrialized countries, resulting in rapid population growth. Note, however, that problems of hygiene and access to fresh water and food had been solved in some societies at least hundreds of years ago. Rome, for example, had a population of about 1 million people by A.D. 2, and this was supported by an excellent infrastructure for importing and distributing food, providing fresh water, and dealing with human wastes.

45.7a Age Structure and Economic Growth: Phases of Development

Where have our migrations and technological developments taken us? It took about 2.5 million years for the human population to reach 1 billion, 80 years to reach the second billion, and only 12 years to jump from 5 billion to 6 billion (see the inset table in Figure 45.19). Rapid population growth now appears to be an inevitable consequence of our demographic structure and economic development.

45.7b Population Growth and Age Structure: Not All Populations Are the Same

In A.D. 2000, the worldwide annual growth rate for the human population averaged nearly 1.26% ($r = 0.0126$ new individuals per individual per year). Population experts expect that rate to decline, but even so, the human population will probably exceed 9 billion by 2050.

In 2000, population growth rates of individual nations varied widely, ranging from much less than 1% to more than 3% **(Figure 45.20a).** Industrialized countries of Western Europe have achieved nearly ZPG, but other countries, particularly those in Africa, Latin America, and Asia, will experience huge increases over the next 20 or 25 years **(Figure 45.20b).**

For all long-lived species, differences in age structure are a major determinant of differences in population growth rates **(Figure 45.21, p. 1134).** There are three basic patterns in the graphs in Figure 45.21. In the first, in countries with ZPG, there are approximately equal numbers of people of reproductive and prereproductive ages. Second, in countries with negative growth (without immigration), postreproductives outnumber reproductives, and these populations will not experience a growth spurt when today's children reach reproductive age. Third are countries where reproductives vastly outnumber postreproductives. The ZPG

a. Mean annual population growth rates

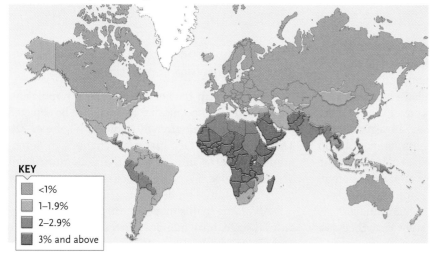

KEY
- <1%
- 1–1.9%
- 2–2.9%
- 3% and above

b. Projected population sizes for 2025

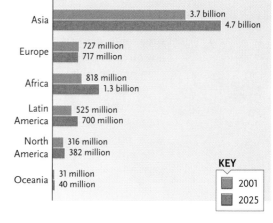

Asia — 3.7 billion / 4.7 billion
Europe — 727 million / 717 million
Africa — 818 million / 1.3 billion
Latin America — 525 million / 700 million
North America — 316 million / 382 million
Oceania — 31 million / 40 million

KEY
- 2001
- 2025

Figure 45.20

Local variation in human population growth rates. In 2001, **(a)** average annual population growth rates vary among countries and continents. In some regions **(b),** the population is projected to increase greatly by 2025 (red) compared with the population size in 2001 (orange). The population of Europe will likely decline.

a. Hypothetical age distributions for populations with different growth rates

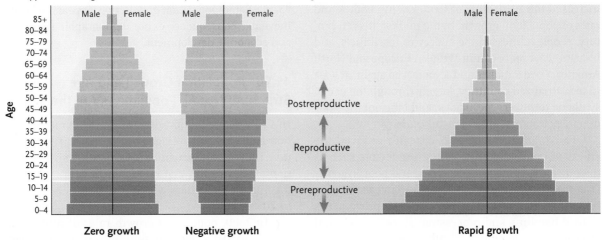

b. Age pyramids for the United States and Mexico in 2000

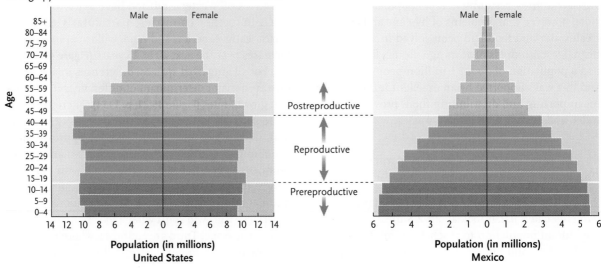

Figure 45.21

Age structure diagrams. Age structure diagrams **(a)** differ for countries with zero, negative, and rapid population growth rates. The width of each bar represents the proportion of the population in each age class. Age structure diagrams for the United States and Mexico **(b)** in 2000 (measured in millions of people) suggest that these countries will experience different growth rates.

situation is exacerbated when reproductives have very few offspring, meaning that prereproductives may not even replace themselves in the population.

Countries with rapid growth have a broad-based age structure (pattern three, above), with many youngsters born during the previous 15 years. Worldwide, more than one-third of the human population falls within this prereproductive base. This age class will soon reach sexual maturity. Even if each woman produces only two offspring, populations will continue to grow rapidly because so many individuals are reproducing. This situation can be described as a "population bomb."

The age structures of the United States and Mexico differ, which has consequences for population growth in the two jurisdictions. Remember the potential importance of immigration and emigration when considering the longer term impact of the population bomb.

45.7c Population Growth and Economic Development: Interconnections

The relationship between a country's population growth and its economic development can be depicted by the **demographic transition model (Figure 45.22).** This model describes historical changes in demographic patterns in the industrialized countries of Western Europe. Today, we do not know if it accurately predicts the future for developing nations.

According to this model, during a country's *preindustrial* stage, birth and death rates are high, and the population grows slowly. Industrialization begins a *transitional* stage, when food production rises, and health care and sanitation improve. Death rates decline, resulting in increased rates of population growth. Later, as living conditions improve, birth rates decline, causing a drop in rates of population growth. When

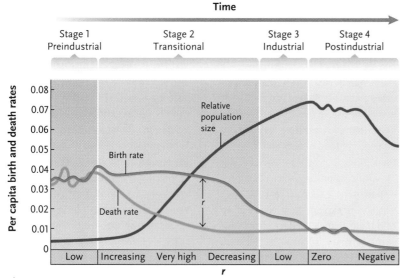

Figure 45.22

The demographic transition. The demographic transition model describes changes in the birth and death rates and relative population size as a country passes through four stages of economic development. The bottom bar describes the net population growth rate, *r*.

the *industrial* stage is in full swing, population growth slows dramatically. Now people move from countryside to cities, and urban couples often choose to accumulate material goods instead of having large families. ZPG is reached in the *postindustrial* stage. Eventually, the birth rate falls below the death rate, *r* falls below zero, and population size begins to decrease.

Today, United States, Canada, Australia, Japan, Russia, and most of Western Europe are in the industrial stage. Their growth rates are slowly decreasing. In Germany, Bulgaria, and Hungary (and other European countries), birth rates are lower than death rates, and populations are shrinking, indicating entry into the postindustrial stage. Kenya and other less industrialized countries are in the transitional stage, but they may not have enough skilled workers or enough capital to make the transition to an industrialized economy. For these reasons, many poorer nations may be stuck in the transitional stage. Third world countries experience rapid population increase because they experience declines in death rates associated with the transitional stage without decreases in birth rates typical of industrial and postindustrial stages.

45.7d Controlling Reproductive Output: Planned Reproduction

Most governments realize that increased population size is now the major factor causing resource depletion, excessive pollution, and an overall decline in quality of life. The principles of population ecology demonstrate that slowing the rate of population growth and effecting an actual decline in population size can be achieved only by decreasing the birth rate or increasing the death rate. Increasing mortality is neither a rational nor humane means of population

control. Some governments use **family planning programs** in an attempt to lower birth rates. In other countries, any form of family planning is unlawful. This topic is discussed further in Chapter 38, where we will see that education of women is a vital undertaking.

To achieve ZPG, the average *replacement rate* should be just slightly higher than two children per couple. This is necessary because some female children die before reaching reproductive age. Today's replacement rate averages about 2.5 children in less industrialized countries with higher mortality rates in prereproductive cohorts and 2.1 in more industrialized countries. However, even if each couple on Earth produced only 2 children, the human population would continue to grow for at least another 60 years (the impact of the population bomb). Continued population growth is inevitable because today's children, who outnumber adults, will soon mature and reproduce. The worldwide population will stabilize only when the age distributions of all countries resemble that for countries with ZPG.

Family planning efforts encourage women to delay their first reproduction. Doing so reduces the average family size and slows population growth by increasing generation time (see Figure 45.9). Imagine two populations in which each woman produces two offspring. In the first population, women begin reproducing at age 32 years, and in the second, they begin reproducing at age 16 years. We can begin with a cohort of newborn baby girls in each population. After 32 years, women in the first population will be giving birth to their first offspring, but women in the second population will be new grandmothers. After 64 years, women in the first population will be new grandmothers, but women in the second population will witness the birth of their first great-great grandchildren (if their daughters also bear their first children at age 16 years). Obviously, the first population will grow much more slowly than the second.

45.7e The Future: Where Are We Going?

Homo sapiens has arrived at a turning point in our cultural evolution and in our ecological relationship with Earth. Hard decisions await us, and we must make them soon. All species face limits to their population growth, and it is naive to assume that our unique abilities exempt us from the laws of population growth. We have postponed the action of most factors that limit population growth, but no amount of invention and intervention

John (Jack) S. Millar, University of Western Ontario

Jack Millar, a professor of biology at the University of Western Ontario in London, and his students study the life histories of small mammals such as mice, voles, and wood rats. They do most of this work in the field, mainly at sites in the Kananaskis Valley in south-western Alberta. The work involves trapping the small mammals and marking them so that they can recognize them later. Recaptures of known individuals allow the researchers to track the performances of individuals. Millar and his students have been following populations of deer mice for over 20 years, and their records have allowed them to ask basic questions about life history.

Most deer mice born in any year in the Kananaskis Valley are dead by the end of September of that year. Although a few females are able to bear two litters in a year, most do not. Compared with deer mice living in southwestern Ontario (see Section 45.6c), the mice in Kananaskis are barely hanging on. But Millar and his students were interested to learn what factors limit age at first reproduction

and the ability of a female to breed more than once a year. Female mice given protein-rich diets (cat food) sometimes were able to breed in their first summer, suggesting a strong influence of food quality on life history traits.

With data on deer mice covering a span of more than 20 years, Millar was able to explore the possible effects of climate change on deer mice living in the Kananaskis Valley. Specifically, between 1985 and 2003, female deer mice typically conceived their first litters on May 2 and the first births occurred on May 26. There were no statistically significant changes in the timing of first births, although the average temperatures in early May had declined by about 2°C during this period. Spring breeding of the deer mice was not related to temperatures or snowfall. The decline in temperature had no effect on the mice's reproductive success. Changes in photoperiod appear to be responsible for initiating reproductive activity in the mice. Their access to protein did affect their reproductive output.

Discoveries about diseases associated with wildlife are a side benefit of his endeavours. The work that Millar and his students have done provided a different view of the role of beavers in the spread of giardiasis, also known as "beaver fever." Giardiasis is caused by infections of a protozoan species in the genus *Giardia*. Humans can be infected if they drink water containing spores or **trophozoites** of *Giardia* species. Humans with giardiasis suffer from intestinal distress. As the name beaver fever implies, these aquatic rodents have been presumed to be the source of human infections. Using specimens, some provided by Millar and his students, P.M. Wallis and colleagues determined that 20 of 21 red-backed voles (*Clethrionomys gapperi*) were infected by *Giardia* at a much higher rate of infection than any other small mammals and also higher than beavers (2 of 50 infected).

Jack Millar's work has demonstrated how long-term experimental research involving both observations and experiments can shed light on life history strategies.

---

can expand the ultimate limits set by resource depletion and a damaged environment. We now face two options for limiting human population growth: we can make a global effort to limit our population growth, or we can wait until the environment does it for us.

Return to Figure 45.19 and observe that only the Plague (also known as the Black Death) caused any deflection from the trajectory of the curve tracking growth in the human population. The Plague appears to have been spread into Europe by the Mongols. The Plague, long established in China, arrived in the Mongol summer capital of Shangdu in 1332. By 1351, the population of China had been reduced by 50 to 66%. By 1345, the Plague had reached Feodosija in the Ukraine (then Kaffa). Between 1340 and 1400, it is estimated that the population of Africa declined from 80 million to 68 million and the world population from 450 million to between 350 million and 375 million. These data do not include the Americas because the Plague did not reach there until about 1600.

To put these percentages in context, consider human deaths associated with World War II. In this

conflict, Great Britain lost less than 1% of its population, France about 1.5%, and Germany 9.1%. In Poland and the Ukraine, where there was a postwar famine, 19% of the human populations there are said to have died.

These sobering figures remind us that we are animals, vulnerable to many of the factors that affect other species on Earth. Now, look back at the chapter opening image and note the large numbers of gulls at a landfill site. The gulls and the landfill (a polite word for "dump") illustrate a fundamental point in population biology, namely, the ability of populations to reach large numbers and have large environmental impacts whether the species are gulls or people.

STUDY BREAK

1. In what three ways have humans avoided the effects of density-dependent regulation factors?
2. What is a population bomb?
3. What does family planning encourage women to do?

Molecule Behind Biology

Progesterone

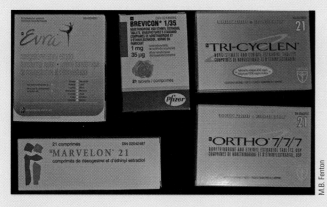

Figure 1
Birth control pills, a selection of products.

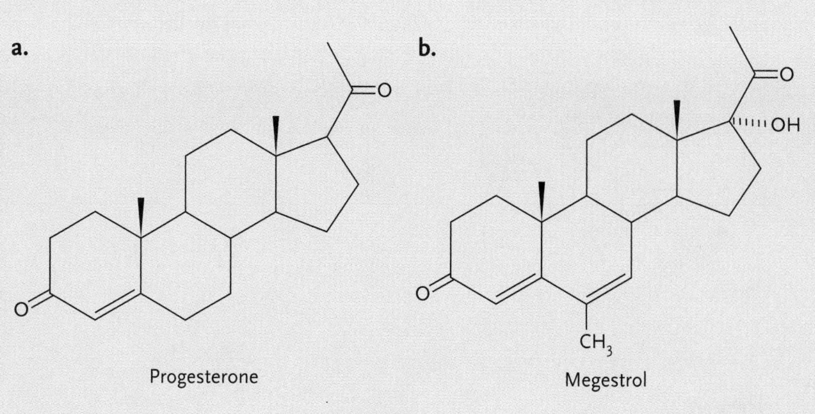

a.

Progesterone

M.B. Fenton

b.

Megestrol

Figure 2
Progesterone and the synthetic megestrol.

The advent of birth control pills **(Figure 1)** had a great impact on the behaviour of people. Women using birth control pills had more control over their fertility. Central to the development of an effective oral contraceptive was a change in the molecular structure of progesterone **(Figure 2a)**. Specifically, the addition of a CH_3 group **(Figure 2b)** meant that the new molecule, megestrol, had the same effect on a woman's reproductive system, but it was not quickly metabolized and remained in the system long enough to have the desired effect (suppressing ovulation). Similarly, slight modifications to the estradiol molecule turned it into ethinylestradiol **(Figure 3)**. Megestrol is an analogue of progesterone, and ethinylestradiol is an analogue of estradiol.

Today, biologists working in zoos use a variety of birth control methods to control the fertility of animals in their collections. For critically endangered species such as black-footed ferrets (see *Black-Footed Ferret, Mustela nigripes*) this means using information about cycles of fertility to maximize reproductive output.

For animals whose populations are growing at a rapid pace, birth control gives keepers the chance to control growth of the populations. The same principles apply to working with organisms in the wild, but getting African elephants to take their birth control pills has not proved to be easy.

Hormones and their analogues are common in untreated municipal wastewaters. In some cases, male fish exposed to these wastewaters are becoming feminized **(Figure 4)**. Specifically, some male fish produce vitellogenin mRNA and protein, substances normally associated with the maturation of oocytes in females. Males thus exposed produce early-stage eggs in their testes. This feminization occurs in the presence of estrogenic substances, including natural estrogen (17b-estradiol) and the synthetic estrogen 17a-ethinylestradiol.

Do a few feminized male fish in the population matter? Karen A. Kidd and six colleagues conducted a seven-year whole-lake experiment in northwestern Ontario (the Experimental Lakes Area). Male fathead minnows (*Pimephales promeles*) chronically exposed to low levels (5–6 $ng \cdot L^{-1}$) showed feminizing effects and the development of intersex males, whereas females had altered oogenesis. The situation led to the near-extinction of fathead minnows in the experimental lake.

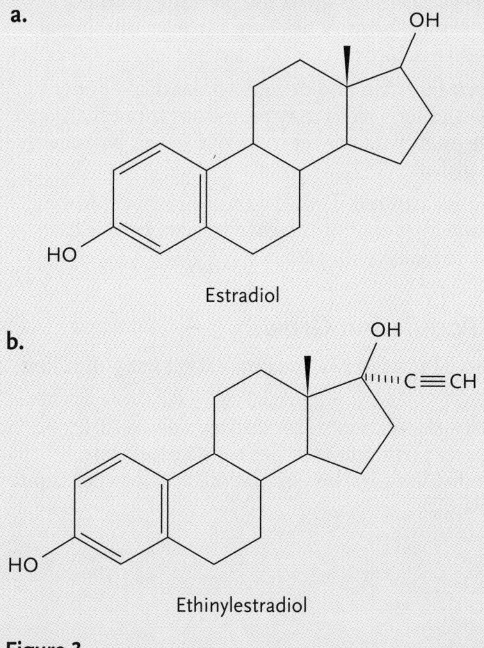

a.

Estradiol

b.

Ethinylestradiol

Figure 3
Estradiol and the synthetic ethinylestradiol.

Konrad P. Schmidt

Figure 4
Pimpheles promeles, the fathead minnow.

1. What factors establish *K* (carrying capacity) for humans living in a large (population > 1 million) city? What factors would establish *K* in a mining town? How could you define the worldwide carrying capacity for humans?

2. Choose an animal or plant species living in your neighbourhood and identify density-dependent and density-independent factors that might influence its population size. How could you demonstrate conclusively that the factors work in either a density-dependent or a density-independent fashion?

Review

Go to CENGAGENOW™ at http://hed.nelson.com/ to access quizzing, animations, exercises, articles, and personalized homework help.

45.1 The Science of Ecology

- Organismal ecology is the study of organisms to determine adaptations to the abiotic environment, including morphological, physiological, biochemical, behavioural, and genetic adaptations. Population ecologists document changes in size and other characteristics of populations of species over space and time. Community ecologists study sympatric populations, interactions among them, and how interactions affect the community's growth. Interactions may include predation and competition. Ecosystem ecologists study nutrient cycling and energy flow through the biotic and the abiotic environment.

- Mathematical models express hypotheses about ecological relationships and different variables, allowing researchers to manipulate the model and document resulting changes. In this way, researchers can simulate natural events before investing in lab work.

- Experimental and control treatments are necessary because they allow ecologists to separate cause and effect.

45.2 Population Characteristics

- Geographic range is the overall spatial boundary around a population. Individuals in the population often live in a specific habitat within the range.

- A lower population density means that individuals have a greater access to resources such as sunlight and water. The capture–mark–recapture technique assumes that (1) a mark has no effect on an individual's survival; (2) marked and unmarked individuals mix randomly; (3) there is no migration throughout the estimation period; and (4) marked and unmarked individuals are equally likely to be caught.

- Three types of dispersion are clumped, uniform, and random. Clumped is most commonly in nature because suitable conditions usually are patchily distributed and animals often live in social groups. Asexual reproduction patterns also can lead to clumped aggregations.

- Generation time increases with body size.

- The number of males in a population of mammals has little impact on population growth because females bear the costs of reproduction (pregnancy and lactation), thus limiting population growth. Sea horses are different because males get pregnant.

45.3 Demography

- Age-specific mortality and age-specific survivorship deal with age intervals. In any one interval, age-specific mortality is the proportion of individuals that died during that time. Age-specific survival is the number surviving during the interval. The two values must sum to 1. In the example, age-specific survivorship = 0.616 or 1 − 0.384.

- Age-specific fecundity is the average number of offspring produced by surviving females during each age interval.

- In a type I curve, high survivorship at a young age decreases rapidly later in life. Type I curves are common for large animals, including humans. In a type II curve, the relationship is linear because there is a constant rate of mortality across the life span. Songbirds fit in this category. A type III curve shows high mortality at a young age that stabilizes as individuals grow older and larger. Insects fall into this category.

45.4 The Evolution of Life Histories

- Maintenance, growth, and reproduction are the three main energy-consuming processes.

- Passive care occurs as nutrients cross the placenta from the mother to the developing baby. Active care involves nursing and other care provided after birth.

- Salmon have a short life span and devote a great deal of energy to reproduction. Deciduous trees may reproduce more than once and use only some energy in any reproductive event, balancing reproduction and growth.

- Early reproduction is favoured if adult survival rates are low or if, when animals age, they do not increase in size. In this case, fecundity does not increase with size.

45.5 Models of Population Growth

- An exponential model is used when a population has unlimited growth.

- dN/dt = change in a population's size during a given time period; b = per capita birth rate; d = per capita death rate; N = number of individuals in the population; $b − d$ = per capita growth rate, equals r.

- When $r > 0$, the birth rate exceeds the death rate, and the population is growing. When $r < 0$, the birth rate is less than the death rate, and the population is shrinking. When $r = 0$, the birth and death rates are equal, and the population is neither growing nor shrinking. The intrinsic rate of increase (r_{max}) is the maximum per capita growth rate. This value usually varies inversely with generation time, so a shorter generation time means a higher r_{max}.

- Intraspecific competition occurs when two or more individuals of the same species depend on the same limiting resource. For deer, this could include food, water, or refuge from predators.

- A logistic model has this pattern because when the population growth is low, the population is small. At intermediate population sizes, growth is more rapid because more individuals breed and r is high. When population growth approaches K (carrying capacity), competition increases, r decreases, and the growth of the population is reduced.

45.6 Population Regulation

- Density-dependent factors include intraspecific competition and predation. At high density, fewer resources are available for individuals, which, in turn, use more energy in maintenance needs and less in reproduction. Offspring produced at higher population densities are often smaller in number or size and less likely to survive. At high population levels, adults may be smaller and weaker.

- Density-independent factors reduce a population size regardless of density, for example, fire, earthquakes, storms, floods, and other natural disturbances.

- The r-selected species often have large numbers of small young, whereas the K-selected species usually have small numbers of larger young. Other answers may include characteristics from Table 45.3.

- Extrinsic control includes interactions between individuals in a population and their food and predators. Once a food supply is exhausted, reproduction will decrease and mortality will increase. Intrinsic control can be hormonal changes within a population that cause increased aggressiveness and faster dispersal and reduce reproduction. Aggressiveness can cause weaker individuals to be forced to disperse to reduce the population density.

45.7 Human Population Growth

- Humans have avoided the effects of density-dependent regulation factors by expanding their geographic range into virtually every habitat, increasing K through agriculture, and reducing population-limiting factors resulting from poor hygiene, malnutrition, and contagious diseases.

- A population bomb is when many offspring are born in one time period, first forming the prereproductive base. At sexual maturity, populations can grow rapidly because of the large number of individuals in this cohort.

- The preindustrial stage is characterized by slow population growth as birth and death rates are high. The transitional stage has better health care and sanitation, as well as increased food production. In the transitional stage, there is a decline in death rates, allowing population growth, but birth rates eventually decline as living conditions improve. In the industrial stage, there is slow population growth as family size decreases because couples choose to have fewer children and accumulate more material goods. In the postindustrial stage, the population size decreases as the birth rate falls below the death rate.

- Family planning encourages women to delay first reproduction, decreasing the size of the average family and, in turn, reducing the population size as generation time has increased. Decisions about reproduction should involve couples.

Questions

Self-Test Questions

1. Ecologists sometimes use mathematical models to
   a. avoid conducting laboratory studies or fieldwork altogether.
   b. simulate natural events before conducting detailed field studies.
   c. make basic observations about ecological relationships in nature.
   d. collect survivorship and fecundity data to construct life tables.
   e. determine the geographic ranges of populations.

2. The numbers of individuals per unit area or volume of habitat is called the population's
   a. geographic range.
   b. dispersion pattern.
   c. density.
   d. size.
   e. age structure.

3. One day you caught and marked 90 butterflies in a population. A week later, you returned to the population and caught 80 butterflies, including 16 that had been marked previously. What is the size of the butterfly population?
   a. 170
   b. 450
   c. 154
   d. 186
   e. 106

4. A uniform dispersion pattern implies that members of a population
   a. cooperate in rearing their offspring.
   b. work together to escape from predators.
   c. use resources that are patchily distributed.
   d. may experience intraspecific competition for vital resources.
   e. have no ecological interactions with each other.

5. The model of exponential population growth predicts that the per capita population growth rate (r)
   a. does not change as a population gets larger.
   b. gets larger as a population gets larger.
   c. gets smaller as a population gets larger.
   d. is always at its maximum level (r_{max}).
   e. fluctuates on a regular cycle.

6. A population of 1000 individuals experiences 452 births and 380 deaths in 1 year. What is the value of r for this population?
   a. 0.842/individual/year
   b. 0.452/individual/year
   c. 0.380/individual/year
   d. 0.820/individual/year
   e. 0.082/individual/year

7. According to the logistic model of population growth, the absolute number of individuals by which a population grows during a given time period
   a. gets steadily larger as the population size increases.
   b. gets steadily smaller as the population size increases.

c. remains constant as the population size increases.

d. is highest when the population is at an intermediate size.

e. fluctuates on a regular cycle.

8. Which example might reflect density-dependent regulation of population size?

a. An exterminator uses a pesticide to eliminate carpenter ants from a home.

b. Mosquitoes disappear from an area after the first frost.

c. The lawn dies after a month-long drought.

d. Northeast storms blow over and kill all willow trees along a lake.

e. A clam population declines in numbers in a bay as the number of predatory herring gulls increases.

9. A K-selected species is likely to exhibit

a. a type I survivorship curve and a short generation time.

b. a type II survivorship curve and a short generation time.

c. a type III survivorship curve and a short generation time.

d. a type I survivorship curve and a long generation time.

e. a type II survivorship curve and a long generation time.

10. One reason that human populations have sidestepped factors that usually control population growth is that

a. the carrying capacity for humans has remained constant since humans first evolved.

b. agriculture and industrialization have increased the carrying capacity for our species.

c. the population growth rate (r) for the human population has always been small.

d. the age structure of human populations has no impact on its population growth.

e. plagues have killed off large numbers of humans at certain times in the past.

Questions for Discussion

1. Do you expect to see a genetic bottleneck effect in *Mustela nigripes* populations in the wild in the future? How long will they take to appear?

2. Design an income tax policy and social services plan that would encourage people to have either larger or smaller families.

3. Many city-dwellers have noted that the density of cockroaches in apartment kitchens appears to vary with the habits of the occupants. People who wrap food carefully and clean their kitchen frequently tend to have fewer arthropod roommates than those who leave food on kitchen counters and clean less often. Interpret these observations from the viewpoint of a population ecologist.

Sphingid (hornworm) caterpillar feeding on leaves.

M.B. Fenton

46 Population Interactions and Community Ecology

WHY IT MATTERS

It is easy to believe that there are many more species (= much more species richness) in the tropics than there are in temperate zones. On a visit to tropical locations, it is easy to be impressed by the diversity of life, whether you go snorkelling at a coral reef, for a boat ride along a tropical river, or on a hike through a rain forest. Is there really such a difference between the two zones? Does it matter what organisms you look for? Consider herbivorous insects (those eating leaves), of which there are more species in the tropics than in temperate zones. Why is this the case?

Ecological theory suggests that in the tropics, greater uniformity in climate is associated with increased ecological specialization of species. For plant-eating insects in the tropics, this means that each species tends to eat a narrower range of plant species compared with their temperate counterparts. This means that each species occupies a narrower niche in the tropics, so there are more niches and hence more species.

In animals such as insects with different life stages (see Chapter 26), each life cycle stage (in some insects, egg, larva [caterpillar], pupa, and adult) fills a different niche and plays a different role in the species' life history. The eggs are development machines, the larvae are eating machines, the pupae are metamorphosis machines, and the adults are mating machines.

Biologists can obtain a picture of the feeding niche of a species of butterfly or moth by documenting the food habits of caterpillars. In this way, L.A. Dyer and 12 colleagues studied the diets of moth and butterfly (Lepidoptera) caterpillars at 8 forest sites in the New World (North, South, and Central America) between latitudes 15° S and 55° N. On average, tropical species are more specialized in their diets than temperate ones **(Figure 46.1)**. Each species of caterpillar in the tropics ate fewer species of plants than its temperate counterpart. The prediction that tropical herbivorous insects occupy narrower feeding niches than temperate ones is supported by these data for Lepidoptera. One consequence of specialization is that, there are more species of Lepidoptera in the tropics than in temperate zones.

Other tropical locations may have many species but perhaps less local variation. On the island of Papua New Guinea, many species of butterflies and moths (with herbivorous caterpillars) live in extensive geographic ranges across 75 000 km² tracts of rain forest **(Figure 46.2)**. Vojtjch Novotny and 15 colleagues reported that consistency of diversity is typical of herbivorous insects, including Lepidoptera, beetles eating wood, and fruit flies and other flies eating fruit.

Making sense of the patterns of life's diversity means exploring the details of communities of organisms. Diversity is the watchword, and it can support almost as many theories as there are ecologists. The purpose of this chapter is to explore interactions between species of organisms and place the interactions in an ecological context.

46.1 Interspecific Interactions

Interactions between species typically benefit or harm the organisms involved, although they may be neutral **(Table 46.1)**. Furthermore, where interactions with other species affect individuals' survival and reproduction, many of the relationships we witness today are the products of long-term evolutionary modification. Good examples range from predator–prey interactions to those associated with pollination or dispersal of seeds.

Interactions between species can change constantly, but remember that the interactions occur at the individual level. Some individuals of a species may be better adapted when another species exerts selection pressure on that species. This adaptation can, in turn, help those individuals exert selection pressure on the other species. This pressure can, in turn, exert selection pressure on the first species in the chain. The situation, known as **coevolution**, is defined as genetically based, reciprocal adaptation in two or more interacting species.

Some coevolutionary relationships are straightforward. Ecologists describe the coevolutionary interactions between some predators and their prey as a race in which each species evolves adaptations that temporarily allow it to outpace the other. When antelope populations suffer predation by cheetahs, natural selection fosters the evolution of faster antelopes. Faster cheetahs may be the result of this situation, and if their offspring are also fast, then antelopes will also become more fleet of foot. Other coevolved interactions provide benefits to both partners. Flower structures of different monkey-flower species have evolved characteristics that allow them to be visited by either bees or hummingbirds (see Chapter 18).

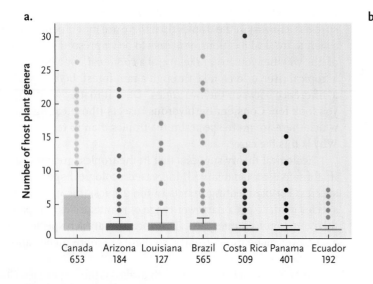

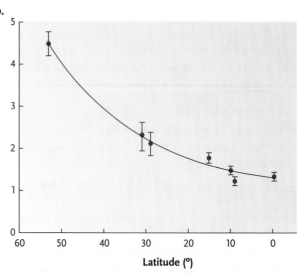

Figure 46.1
Caterpillar diet breadth. Numbers of host plant genera per caterpillar species at seven sites along a latitudinal gradient.

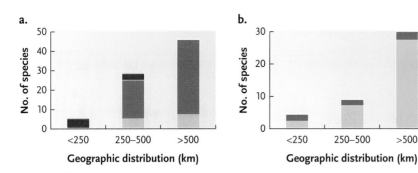
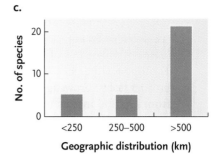

Figure 46.2

Papua New Guinea. Geographic distribution of **(a)** a caterpillar, **(b)** an ambrosia beetle, and **(c)** fruit fly species in the lowlands. Caterpillar and beetle species are classified as generalists (orange; feeding on more than one genus), clade specialists (blue; feeding on more than one species from a single genus), or monophagous (green; feeding on a single plant species). Note the differences between the caterpillars, beetles, and fruitflies.

Table 46.1	Population Interactions and Their Effects	
Interaction		Effects on Interacting Populations
Predation	+/−	Predators gain nutrients and energy; prey are killed or injured.
Herbivory	+/−	Herbivores gain nutrients and energy; plants are killed or injured.
Competition	−/−	Both competing populations lose access to some resources.
Commensalism	+/0	One population benefits; the other population is unaffected.
Mutualism	+/+	Both populations benefit.
Parasitism	+/−	Parasites gain nutrients and energy; hosts are injured or killed.

One can hypothesize a coevolutionary relationship between any two interacting species, but documenting the evolution of reciprocal adaptations is difficult. Coevolutionary interactions often involve more than two species, and most organisms experience complex interactions with numerous other species in their communities. Cheetahs take several prey species. Antelopes are prey for many species of predators, from cheetahs to lions, leopards, and hyenas, as well as some larger birds of prey. Not all predators use the same hunting strategy. Therefore, the simple portrayal of coevolution as taking place between two species rarely does justice to the complexity of these relationships.

STUDY BREAK

What is coevolution? Is it usually restricted to two species?

46.2 Getting Food

Because animals typically acquire nutrients and energy by consuming other organisms, **predation** (the interaction between predatory animals and the animal prey they consume) and **herbivory** (the interaction between herbivorous animals and the plants they eat) can be the most conspicuous relationships in ecological communities.

Both predators and herbivores have evolved characteristics allowing them to feed effectively. Carnivores use sensory systems to locate animal prey and specialized behaviours and anatomical structures to capture and consume it. Herbivores use sensory systems to identify preferred food or to avoid food that is toxic.

Rattlesnakes, such as species in the genus *Crotalus*, use heat sensors on pits in their faces (see Figure 34.32) to detect warm-blooded prey. The snakes deliver venom through fangs (hollow teeth) by open-mouthed strikes on prey. After striking, the snakes wait for the venom to take effect and then use chemical sensors also on the roofs of their mouths to follow the scent trail left by its dying prey. The venom is produced in the snakes' salivary glands. It contains neurotoxins that paralyze prey and protease enzymes that begin to digest it. Elastic ligaments connecting the bones of the snakes' jaws (mandibles) to one another and the mandibles to the skull allow snakes to open their mouths very wide to swallow prey larger than their heads (see Chapter 41).

Herbivores have comparable adaptations for locating and processing their food plants. Insects use chemical sensors on their legs and mouthparts to identify edible plants and sharp mandibles or sucking mouthparts to consume plant tissues or sap. Herbivorous mammals have specialized teeth to harvest and grind tough vegetation (see Figure 27.56). Herbivores, such as farmer ants (see Chapter 49), ruminants, or termites (see Chapter 41), may also coopt other species to gain access to nutrients locked up in plant materials.

All animals select food from a variety of potential items. Some species, described as *specialists*, feed on one or just a few types of food. Among birds, Everglades kites (*Rostrhamus sociabilis*) eat only apple snails (*Pomacea paludosa*). Koalas (see Figure 27.51) eat the leaves of only a few of the many available species of *Eucalyptus*. Other species, described as *generalists*, have broader tastes. Crows (genus *Corvus*) take food ranging from grain to insects to carrion. Bears (genus *Ursus*) and pigs (genus *Sus*) are as omnivorous as humans.

How does an animal select its food? Why pizza rather than salad? Mathematical models, collectively described as **optimal foraging theory**, predict that an animal's diet is a compromise between the costs and benefits associated with different types of food. Assuming that animals try to maximize their energy intake at any meal, their diets should be determined by the ratio of costs to benefits: the costs of obtaining the food versus the benefits of consuming it. Costs are the time and energy it takes to pursue, capture, and consume a particular kind of food. Benefits are the energy provided by that food. A cougar (*Felis concolor*) will invest more time and energy hunting a mountain goat (*Oreamnos americanus*) than a jackrabbit (*Lepus townsendii*), but the payoff for the cat is a bigger meal. One important element in food choice is the relative abundance of prey. "Encounter rate" (denoted as λ in optimal foraging theory) is usually influenced by population density and can influence a predator's diet. For the cougar, λ determines the time between jackrabbits, and when they are abundant, they can be a more economical target than larger, scarcer prey.

Food abundance affects food choice. When prey are scarce, animals often take what they can get, settling for food that has a higher cost-to-benefit ratio. When food is abundant, they may specialize, selecting types that provide the largest energetic return. Bluegill sunfishes eat *Daphnia* and other small crustaceans. When crustacean density is high, these fishes hunt mostly large *Daphnia*, which provide more energy for their effort. When prey density is low, bluegills eat *Daphnia* of all sizes **(Figure 46.3)**.

Think of yourself at a buffet. The array of food can be impressive, if not overwhelming. But your state of hunger, the foods you like, the ones you do not like, and any to which you are allergic all influence your selection. You may also be influenced by choices made by others. In your feeding behaviour, you betray your animal heritage.

STUDY BREAK

1. How do predators differ from herbivores? How are they similar?
2. Is a koala a generalist or a specialist? What is the difference?
3. What does optimal foraging theory predict? Describe the costs and benefits central to this theory.

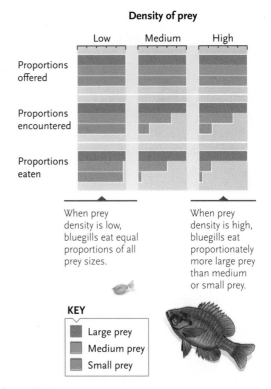

Figure 46.3

An experiment demonstrating that prey density affects predator food choice. Bluegill sunfishes (*Lepomis macrochirus*) were offered equal numbers of small, medium, and large prey (*Daphnia magna*) at three different total densities of prey. Because large prey are easy to find, the fishes encountered them more often, especially at the highest prey densities than either medium-sized or small prey. The fishes' choice of prey varied with prey density, but they always chose the largest prey available.

46.3 Defence

Predation and herbivory have a negative impact on the species being eaten, so animals and plants have evolved mechanisms to avoid being caught and eaten. Some plants use spines, thorns, and irritating hairs to protect themselves from herbivores. Plant tissues often contain poisonous chemicals that deter herbivores from feeding. When damaged, milkweed plants (family Asclepiadaceae) exude a milky, irritating sap **(Figure 46.4)** that contains poisons that affect the heart (cardiac glycosides). Even small amounts of cardiac glycosides are toxic to the heart muscles of some vertebrates. Other plants have compounds that mimic the structure of insect hormones, disrupting the development of insects that eat them. Most of these poisonous compounds are volatile, giving plants their typical aromas. Some herbivores have developed the ability to recognize these odours and avoid toxic plants. Some plants increase their production of toxic compounds in response to herbivore feeding. Potato and tomato plants damaged by herbivores have higher levels of protease-inhibiting chemicals. These compounds

Figure 46.4

Protective latex sap. Milky sap laced with cardiac glycosides oozes from a cut milkweed (*Asclepias* species) leaf. Milky sap does not always mean dangerous chemicals, for example, the sap of dandelions.

prevent herbivores from digesting the proteins they have eaten, reducing the food value of these plants.

46.3a Size: Too Big to Tackle

Size can be a defence. At one end of the spectrum, this means being too small to be considered food. At the other end, it means being so big that few, if any, predators can succeed in attacking and killing the prey. Today, elephants and some other large herbivores (megaherbivores) are species with few predators (other than humans). But 50 000 years ago, there were larger predators (see Figure 20.1), including one species of "lion" that was one-third larger than an African lion.

46.3b Eternal Vigilance: Always Be Alert

A first line of defence of many animals is avoiding detection. This often means not moving, but it also means keeping a sharp lookout for approaching predators and the danger they represent **(Figure 46.5)**.

46.3c Avoiding Detection:
Freeze—Movement Invites Detection

Many animals are cryptic, camouflaged so that a predator does not distinguish them from the background. Patterns, such as the stripes of a zebra (*Equus burchellii*), make the animal conspicuous at close range, but at a distance, patterns break up the outline, rendering the animals almost invisible. Many other animals look like something that is not edible. Some caterpillars look like bird droppings, whereas other insects look like thorns or sticks. Neither bird droppings nor thorns are usually eaten by insectivores.

46.3d Thwarting Attacks:
Take Evasive Action

Animals resort to other defensive tactics once they have been discovered and recognized. Running away is a typical next line of defence. Taking refuge in a shelter and getting out of a predator's reach is an alternative. African pancake tortoises (*Malacochersus tornieri*) are flat, as the name implies. When threatened, they retreat into rocky crevices and puff themselves up with air, becoming so tightly wedged that predators cannot extract them.

If cornered by a predator, offence becomes the next line of defence. This can involve displays intended to startle or intimidate by making the prey appear large and/or ferocious. Such a display might dissuade a predator or confuse it long enough to allow the potential victim to escape. Many animals use direct attack in these situations, engaging whatever weapons they have (biting, scratching, stinging, etc.).

46.3e Spines and Armour: Be Dangerous or Impossible to Attack

Other organisms use active defence in the form of spines or thorns **(Figure 46.6, p. 1146)**. North American porcupines (genus *Erethizon*) release hairs modified

Figure 46.5
Eternally vigilant. The sentry of a group of meerkats (*Suricatta suricatta*).

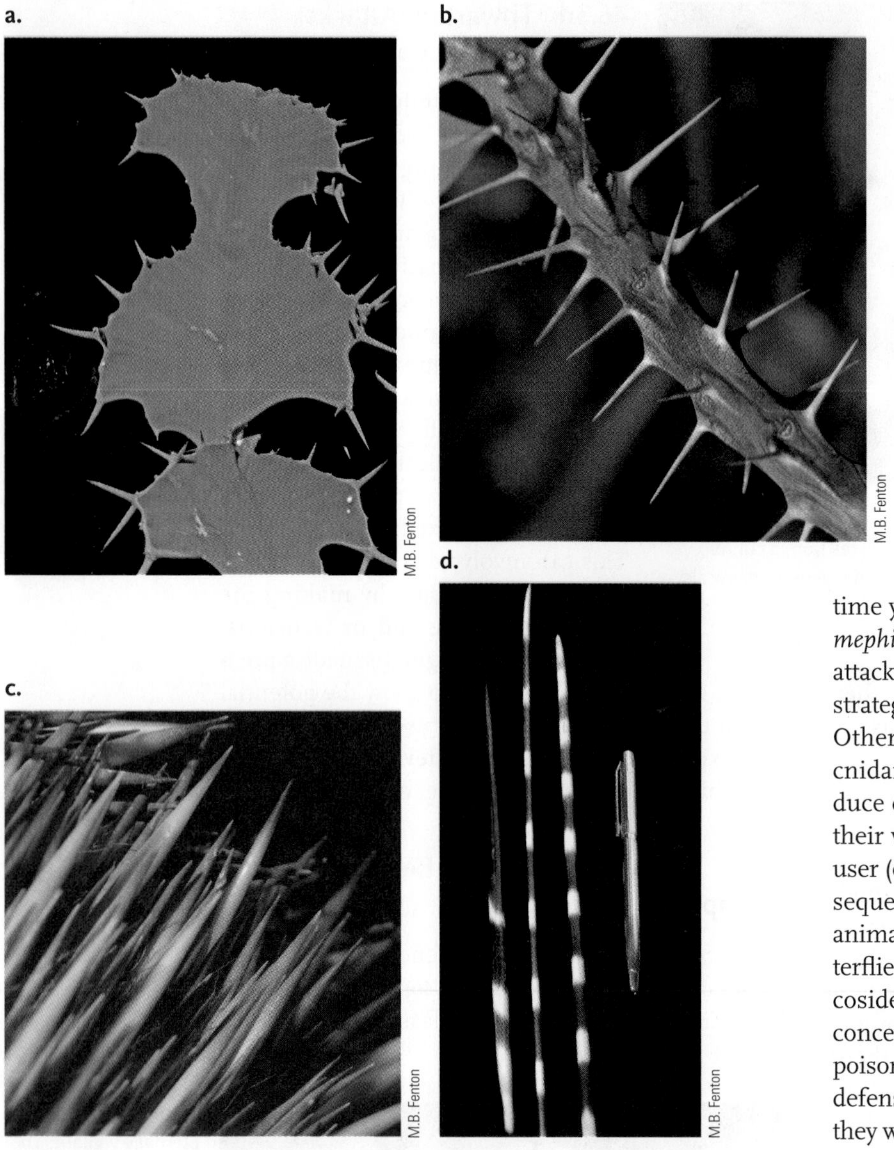

a.

b.

M.B. Fenton

M.B. Fenton

c.

d.

M.B. Fenton

M.B. Fenton

Figure 46.6

Defensive spines. Plants such as **(a)** the cowhorn euphorb (*Euphorbia gandicornis*) and **(b)** crown of thorns (*Euphorbia milli*) and animals such as **(c)** spiny anteaters (*Tachyglossus* species) and **(d)** porcupines (*Hystrix* species) both use thorns or spines in defence. Pen shown for scale with quills in **(d).**

into sharp, barbed quills that, when stuck into a predator, cause severe pain and swelling. The spines detach easily from the porcupine, and the nose, lips, and tongue of an attacker are particularly vulnerable. There are records of leopards (*Panthera pardus*) being killed by porcupine spines. In these instances, the damage to the leopards' mouths, combined with infection, was probably the immediate cause of death. Many other mammals, from monotremes (spiny anteaters) to tenrecs (insectivores from Madagascar, *Tenrec* species and *Hemicentetes* species), hedgehogs (*Erinaceus* species), and porcupines in the Old World use the same defence. So do some fishes and many plants.

Figure 46.7

Armour. Turtles and their allies (see Chapter 27) live inside shells. This leopard tortoise (*Geochelone pardalis*) is inspecting the remains of a conspecific. Armour does not guarantee survival.

Other organisms are armoured (**Figure 46.7**). Examples include bivalve and gastropod molluscs, chambered nautiluses, arthropods such as horseshoe crabs (*Limulus* species), trilobites (see Chapter 26), fishes such as catfish (*Siluriformes*), reptiles (turtles; see Figure 27.38), and mammals (armadillos, scaly anteaters; see Figure 3.3b). We know a great deal about extinct species that were armoured (see Chapter 20) because they often made good fossils.

46.3f Chemical Defence: From Bad Taste to Deadly

Like plants that produce chemicals to repel herbivores, many animals make themselves chemically unattractive. At one level, this can be as simple as smelling or tasting bad. Remember the last time your dog or cat was sprayed by a skunk (*Mephitis mephitis*)? Many animals vomit and defecate on their attackers. Skunks and bombardier beetles escalate this strategy by producing and spraying a noxious chemical. Other animals go beyond spraying. Many species of cnidarians, annelids, arthropods, and chordates produce dangerous toxins and deliver them directly into their victims. These toxins may be synthesized by the user (e.g., snake venom; see *Molecule Behind Biology*) or sequestered from other sources, often plants or other animals (see *Nematocysts*). Caterpillars of monarch butterflies (see Figure 1.29) are immune to the cardiac glycosides in the milkweed leaves they eat. They extract, concentrate, and store these chemicals, making them poisonous to potential predators. The concentrations of defensive chemicals may be higher in the animal than they were in its food. Cardiac glycosides persist through metamorphosis, making adult monarchs poisonous to vertebrate predators.

46.3g Warnings: Danger Signals

Many animals that are noxious or dangerous are **aposematic:** they advertise their unpalatability with an appropriate display (**Figure 46.8;** see Chapter 17).

M.B. Fenton

Figure 46.8

Warning colours. This arrowhead frog gets its name from toxins in its skin that were used to poison arrowheads.

Aposematic displays are designed to "teach" predators to avoid the signaller, reducing the chances of harm to would-be predators and prey. Predators that attack a brightly coloured bee or wasp and are stung learn to associate the aposematic pattern with the sting. Many predators quickly learn to avoid black-and-white skunks, yellow-banded wasps, or orange monarch butterflies because they associate the warning display with pain, illness, or severe indigestion.

But for every ploy there is a counterploy, and some predators eat mainly dangerous prey. Bee-eaters (family Meropidae) are birds that eat hymenopterans (bees and wasps). Some individual African lions specialize on porcupines, and animals such as hedgehogs (genus *Erinaceus*) seem able to eat almost anything and show no ill effects. Indeed, some hedgehogs first lick toads and then their own spines, anointing them with toad venom. Hedgehog spines treated with toad venom are more irritating (at least to people) than untreated ones, enhancing their defensive impact.

46.3h Mimicry: Advertising, True and False

If predators learn to recognize warning signals, it is no surprise that many harmless animals' defences are based on imitating (mimicking) species that are dangerous or distasteful. **Mimicry** occurs when one species evolves to resemble another **(Figure 46.9)**. **Batesian mimicry**, named for English naturalist Henry W. Bates, occurs when a palatable or harmless species (the **mimic**) resembles an unpalatable or poisonous one (the **model**). Any predator that eats the poisonous model and suffers accordingly will subsequently avoid other organisms that resemble it. However, the predator must survive the encounter. **Müllerian mimicry**, named for German zoologist Fritz Müller, involves two or more unpalatable species looking the same, presumably to reinforce lessons learned by a predator that attacks any species in the mimicry complex.

For mimicry to work, the predator must learn (see Chapter 40) to recognize and then avoid the prey. The more deadly the toxin, the less likely an individual predator is to learn by its experience. In many cases, predators learn by watching the discomfort of a conspecific that has eaten or attacked an aposematic prey.

Plants often use toxins to protect themselves against herbivores. Is this also true of toxins in mushrooms (see Chapter 18)?

46.3i No Perfect Defence: The Helmet Reality

Helmets protect soldiers, motorcyclists, and cyclists, but not completely because no defence provides perfect protection. Some predators learn to circumvent defences. Many predators learn to deal with a diversity of prey species and a variety of defensive tactics. Orb web spiders confronting a captive in a web adjust their behaviour according to the prey. They treat moths differently from beetles and bees in yet another way. When threatened by a predator, headstand beetles raise their rear ends and spray a noxious chemical from a gland at the tip of the abdomen. This behaviour deters many would-be predators. But experienced grasshopper mice from western North America circumvent this defence. An experienced mouse grabs the beetle, averts its face

a. Batesian mimicry

Drone fly (*Eristalis tenax*), the mimic

Honeybee (*Apis mellifera*), the model

b. Müllerian mimicry

Heliconius erato

Heliconius melpone

Figure 46.9

Mimicry. **(a)** Batesian mimics are harmless animals that mimic a dangerous one. The harmless drone fly (*Eristalis tenax*) is a Batesian mimic of the stinging honeybee (*Apis mellifera*). **(b)** Müllerian mimics are poisonous species that share a similar appearance. Two distantly related species of butterfly, *Heliconius erata* and *Heliconius melpone*, have nearly indistinguishable patterns on their wings.

Taipoxin: Snake Presynaptic Phospholipase A₂ Neurotoxins

Snake venoms typically are a concoction of ingredients designed to immobilize and digest prey. Like the venom of nematocysts, the effects of snake venom can include symptoms associated with neurotoxins, cardiotoxins, hemolytic actions, and digestion (necrosis) of tissues. Not all snakes (or other venomous animals) have the same venom.

Snake presynaptic phospholipase A₂ neurotoxins, or SPANS, have neurotoxic effects and work by blocking neuromuscular junctions. Phospholipase A₂ activity varies greatly among SPANS. Using mouse neuromuscular junction hemidiaphragm preparations and neurons in culture, M. Rigoni and seven colleagues explored the way in which SPANS work. They used SPANS from single-chain notexin (from *Notecis scutatis*, the eastern tiger snake), a two-subunit B-bungarotoxin (from *Bungarus multicinctus*, the many-banded krait), the three-subunit taipoxin (from *Oxyuranus scutellatus*, the taipan; **Figure 1**). and the five-unit textilotoxin (from *Pseudonaja textilis*, the eastern brown snake).

The results showed that administration of SPANS to neuromuscular junctions causes enlargement of the junctions and reduction in the contents of synaptic vessicles. SPANS also induce exocytosis of neurotransmitters. In other words, SPANS bind nerve terminals via receptors **(Figure 2),** the results indicate

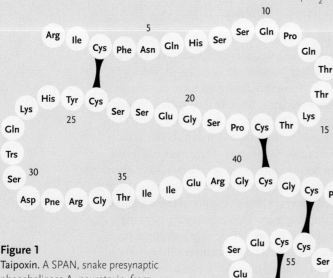

Figure 1

Taipoxin. A SPAN, snake presynaptic phospholipase A₂ neurotoxin, from *Oxyuranus scutellatus*, the taipan, a venomous elapid snake from Australia.

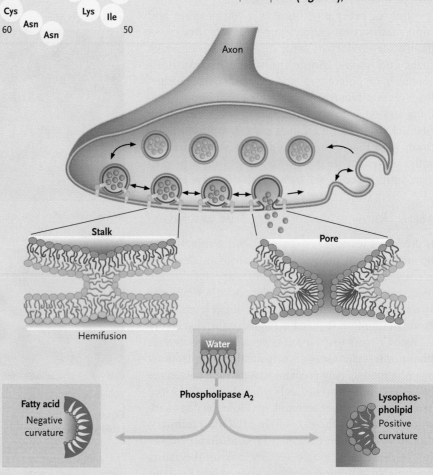

Figure 2

Under normal conditions, lipids alter membrane bending at the synapse, and this controls vesicle fusion and synaptic activity. Membrane fusion between the plasma membrane of a presynaptic neuron and a synaptic vesicle with molecules of neurotransmitter (red) leads to the formation of a pore. Mixing of lipids from inner (purple) and outer (green) leaflets is probably restricted by proteins (yellow ribbons) around the site of fusion. Phospholipase A₂ cleaves lipids, forming flat monolayers into fatty acids (negative curvature) and lysophospholipids (positive curvature). This change explains the effect of SPANS at neuromuscular junctions.

that venoms can be used to further our understanding of what happens at neuromuscular junctions.

Among extant lepidosaurian reptiles, two lineages have venom delivery systems, advanced snakes and helodermatid (gila monster) lizards. The traditional view is that the evolution of venom systems is fundamental to the radiation of snakes, although they occur in just two species of lizards. Using tools of molecular genetics, B.G. Fry and 13 colleagues explored the early evolution of venom systems in lizards and snakes. The ancestral condition (represented by venomous lizards) has lobed, noncompound, venom-secreting glands on upper and lower jaws. Advanced snakes and two lizards have more derived venom systems involving the loss of upper (maxillary) or lower (mandibular) venom glands. Analysis of venoms indicates that snakes, iguanians (monitor lizards), and anguimorphs form a single clade **(Figure 3),** suggesting that venom is an ancestral trait in this evolutionary line of reptiles.

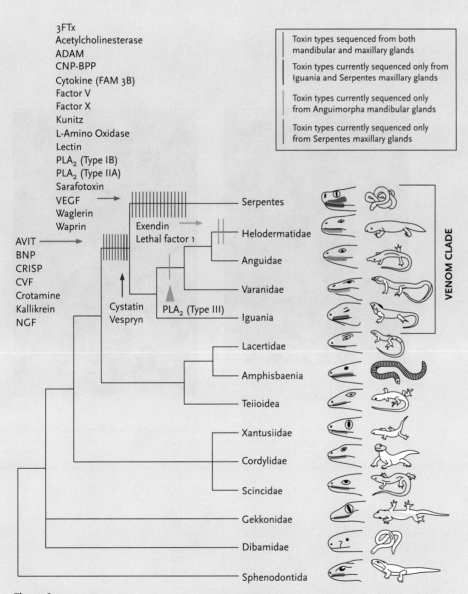

Figure 3

Snake and lizard phylogeny based on the appearance of venom. Shown here are the relative glandular development and appearance of toxin recruitment in squamate reptile phylogeny. Glands secreting mucus are blue, ancestral venom glands are red, and derived venom is orange. Elements in venom include the following: three-finger toxins (3FTx); a disintegrin and metalloproteinase (ADAM); C-type natriuretic peptide–bradykinin–potentiating peptide (CNP-BPP); cobra venom factor (CVF); nerve growth factor (NGF); and vascular endothelial growth factor (VEGF).

Nematocysts

Swimmers at ocean beaches in warmer parts of the world can be exposed to stings from the Portuguese man-of-war. In the United States, at least three human deaths have been caused by exposure to the venom of its nematocysts. First aid for someone who has been stung includes (a) using seawater to flush away any tentacles still clinging to the victim (or pick them off if necessary); (b) applying ice or cold packs to the area of the sting(s) and leaving them in place for 5 to 15 minutes; (3) using an inhaled analgesic to reduce pain; and (4) seeking additional medical aid.

Beaches where Portuguese man-of-war and other jellyfish may occur must be supervised by lifeguards who understand the danger. The inflatable bladders (sails; see **Figure 2**) make Portuguese man-of-war easy to see in the water, and swimming is not permitted when they are near.

In 1968, lifeguards and others at a beach at Port Stephens, New South Wales, in Australia were surprised and concerned when they realized that some people had been stung by Portuguese man-of-war when none of these animals had been spotted near the beach. The stinging animals turned out to be sea slugs, *Glaucus* species (see also the solar-powered sea slug, Chapter 3). Since 1903, it had been known that sea slugs use nematocysts as a defence.

Glaucus atlanticus (**Figure 3**) feed on the cnidosacs that contain the nematocysts, preferentially selecting and storing those of Portuguese man-of-war, which have two sizes of nematocysts. The sea slugs take the larger nematocysts that, when discharged, have the longest penetrants. It is likely that the same digestive processes other sea slugs use to extract chloroplasts (see Chapter 3) can be used to extract nematocysts.

This situation demonstrates the versatility of defensive systems in animals.

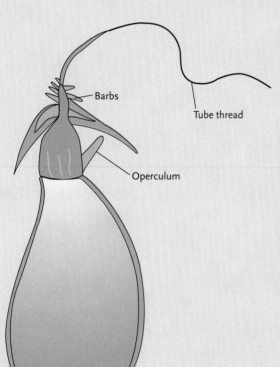

Figure 1
Nematocyst. Nematocysts (Figure 1) are stinging cells occurring in animals in the phylum Cnidaria (see Chapter 26). Nematocysts of *Physalia physalia* (Portuguese man-of-war; Figure 2) contain toxic proteins and at least six or seven enzymes that can be injurious. Unpurified nematocyst venoms have several effects, some of which can be lethal. The venoms can be neurotoxic, cardiotoxic, or myotoxic or cause lysis of red blood cells or mitochondria. Like other venoms, nematocyst venom from the Portuguese man-of-war can interfere with the transport of Na^+ and Ca^{2+} ions.

Barbs

Tube thread

Operculum

Laurence Madin, Woods Hole Oceanographic Institution

Figure 2
Physalia physalia, Portuguese man-of-war.

© Museum de Genève; photo Philippe Wagneur

Figure 3
Glaucus atlanticus, a sea slug that ingests nematocysts from *Physalia physalia*.

a. *Eleodes* bettle

b. Grasshopper mouse

Thomas Eisner, Cornell University

Thomas Eisner, Cornell University

Figure 46.10

Defence and learning. When confronted by a predator, **(a)** the headstand beetle (*Eleodes longicollis*) raises its abdomen and sprays a noxious chemical from its hind end. Experienced **(b)** grasshopper mice (*Onychomys leucogaster*) thwart the beetle's defence by grabbing it, turning it upside down, and eating it headfirst.

(to avoid the spray), turns the beetle upside down so that the gland discharges into the ground, and eats the beetle from the head down **(Figure 46.10)**.

STUDY BREAK

1. List the eight defence techniques used by animals and/or plants. Provide an example of each.
2. How do animals using chemical defences obtain the chemicals they use in this way?
3. What is the purpose of aposematic displays?

46.4 Competition

Different species using the same limiting resources experience **interspecific competition** (competition between species). Competing individuals may experience increased mortality and decreased reproduction, responses similar to the effects of intraspecific competition. Interspecific competition can reduce the size and population growth rate of one or more of the competing populations.

Community ecologists identify two main forms of interspecific competition. In **interference competition**, individuals of one species harm individuals of another species directly. Here animals may fight for access to resources, as when lions chase smaller predators such as hyenas, jackals, and vultures from their kills. Many plant species, including creosote bushes (see Figure 45.4), release toxic chemicals into the soil, which prevent other plants from growing nearby.

In **exploitative competition**, two or more populations use ("exploit") the same limiting resource, and the presence of one species reduces resource availability for others. Exploitative competition need not involve snout-to-snout or root-to-root confrontations. In the deserts of the American Southwest, many bird and ant species eat mainly seeds, and each seed-eating species may deplete the food supply available to others without necessarily encountering each other.

46.4a Competition and Niches: When Resources Are Limited

In the 1920s, the Russian mathematician Alfred J. Lotka and the Italian biologist Vito Volterra independently proposed a model of interspecific competition, modifying the logistic equation (see Chapter 45) to describe the effects of competition between two species. In their model, an increase in the size of one population reduces the population growth rate of the other.

In the 1930s, a Russian biologist, G.F. Gause, tested the model experimentally. He grew cultures of two *Paramecium* species (ciliate protozoans) under constant laboratory conditions, regularly renewing food and removing wastes. Both species feed on bacteria suspended in the culture medium. When grown alone, each species exhibited logistic growth. When grown together in the same dish, *Paramecium aurelia* persisted at high density, but *Paramecium caudatum* was almost eliminated **(Figure 46.11)**. These results

Figure 46.11
Gause's experiments on interspecific competition in *Paramecium*.

P. caudatum alone
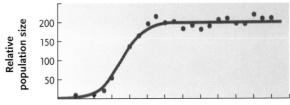

P. aurelia alone
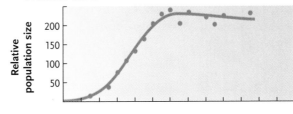

Mixed culture
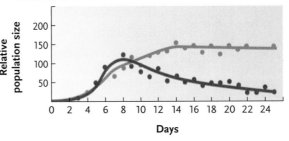

Days

Paramecium caudatum
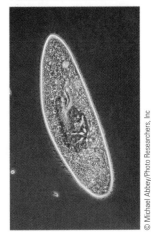

© Michael Abbey/Photo Researchers, Inc

Paramecium aurelia

© Eric V. Grave/Photo Researchers, Inc.

inspired Gause to define the **competitive exclusion principle**. Populations of two or more species cannot coexist indefinitely if they rely on the same limiting resources and exploit them in the same way. One species inevitably harvests resources more efficiently, produces more offspring than the other, and, by its actions, negatively affects the other species.

Ecologists developed the concept of the **ecological niche** to visualize resource use and the potential for interspecific competition in nature. They define a population's niche by the resources it uses and the environmental conditions it requires over its lifetime. In this context, niche includes food, shelter, and nutrients, as well as nondepletable abiotic conditions such as light intensity and temperature. In theory, an almost infinite variety of conditions and resources could contribute to a population's niche. In practice, ecologists usually identify the critical resources for which populations might compete. Sunlight, soil moisture, and inorganic nutrients are important resources for plants, so differences in leaf height and root depth, for example, can affect plants' access to these resources. Food type, food size, and nesting sites are important for animals. Often, when several species coexist, they use food and nest resources in different ways.

Ecologists distinguish the **fundamental niche** of a species, the range of conditions and resources it could tolerate and use, from its **realized niche**, the range of conditions and resources it actually uses in nature. Realized niches are smaller than fundamental niches, partly because all tolerable conditions are not always present in a habitat and partly because some resources are used by other species. We can visualize competition between two populations by plotting their fundamental and realized niches with respect to one or more resources (**Figure 46.12**). If the fundamental

niches of two populations overlap, they *might* compete in nature.

Observing that several species use the same resource does not demonstrate that competition occurs (or does not occur). All terrestrial animals consume oxygen but do not compete for oxygen because it is usually plentiful. Nevertheless, two general observations provide *indirect* evidence that interspecific competition may have important effects.

Resource partitioning occurs when several species living in the same place (sympatric) use different resources or the same resources in different ways. Although plants might compete for water and dissolved nutrients, they may avoid competition by partitioning these resources, collecting them from different depths in the soil (**Figure 46.13**). This allows coexistence of different species.

Character displacement can be evident when comparing species that are sometimes sympatric and sometimes allopatric (live in different places). Allopatric populations of some animal species are morphologically similar and use similar resources, whereas sympatric populations are morphologically different and use different resources. Differences between sympatric species allow them to coexist without competing. Allen Keast studied honey-eaters (family Meliphagidae), a group of birds from Australia, to illustrate this situation. In mainland Australia, up to six species in the genus *Melithreptus* occur in some habitats. Just off the coast of Kangaroo Island, there are two species. When two species are sympatric, each feeds in a wider range of situations than when six species live in the same area, reflecting the use of broader niches. Behavioural and

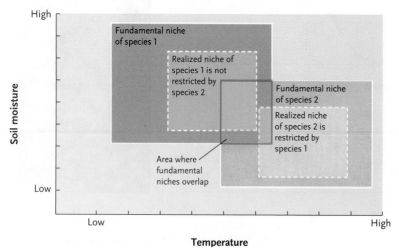

Figure 46.12

Fundamental versus realized niches. In this hypothetical example, both species 1 and species 2 can survive intermediate temperature conditions, as indicated by the shading where their fundamental niches overlap. Because species 1 actually occupies most of this overlap zone, its realized niche is not much affected by the presence of species 2. In contrast, the realized niche of species 2 is restricted by the presence of species 1, and species 2 occupies warmer and drier parts of the habitat.

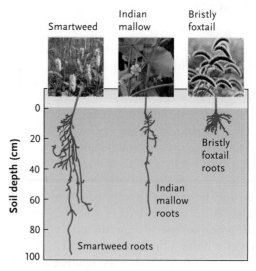

Figure 46.13

Resource partitioning. The root systems of three plant species that grow in abandoned fields partition water and nutrient resources in soil. Bristly foxtail grass (*Setaria faberii*) has a shallow root system, Indian mallow (*Abutilon theophraste*) has a moderately deep taproot, and smartweed (*Polygonum pennsylvanicum*) has a deep taproot that branches at many depths.

Photos: left, © Tony Wharton, Frank Lane Picture Agency/Corbis; middle, © Hal Horwitz/Corbis; right, © Joe McDonald/Corbis.

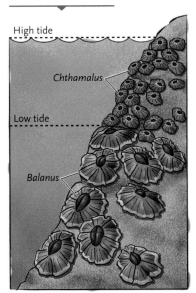

Realized niches before experimental treatments

High tide

Chthamalus

Low tide

Balanus

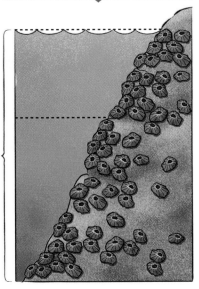

Treatment 1: Remove *Balanus*
In the absence of *Balanus*, *Chthamalus* occupies both shallow water and deep water.

High tide

Fundamental niche of *Chthamalus*

Treatment 2: Remove *Chthamalus*
In the absence of *Chthamalus*, *Balanus* still occupies only deep water.

Fundamental niche of *Balanus*

Figure 46.14
Demonstration of competition between two species of barnacles.

morphological differences are evident when species are compared between the different situations.

Data on resource partitioning and character displacement suggest, but do not prove, that interspecific competition is an important selective force in nature. To demonstrate *conclusively* that interspecific competition limits natural populations, one must show that the presence of one population reduces the population size or density of its presumed competitor. In a classic field experiment, Joseph Connell examined competition between two barnacle species **(Figure 46.14)**. Connell first observed the distributions of both species of barnacles in undisturbed habitats to establish a reference baseline. *Chthamalus stellatus* is generally found in shallow water on rocky coasts, where it is periodically exposed to air. *Balanus balanoides* typically lives in deeper water, where it is usually submerged.

In the absence of *Balanus* on rocks in deep water, larval *Chthamalus* colonized the area and produced a flourishing population of adults. *Balanus* physically displaced *Chthamalus* from these rocks. Thus, interference competition from *Balanus* prevents *Chthamalus* from occupying areas where it would otherwise live. Removal of *Chthamalus* from rocks in shallow water did not result in colonization by *Balanus*. *Balanus* apparently cannot live in habitats that are frequently exposed to air. Connell concluded that there was competition between the two species. But competition was asymmetrical because *Chthamalus* did not affect the distribution of *Balanus*, whereas *Balanus* had a substantial effect on *Chthamalus*.

46.4b Symbiosis: Close Associations

Symbiosis occurs when one species has a physically close ecological association with another (*sym* = together; *bio* = life; *sis* = process). Biologists define

three types of symbiotic interactions: *commensalism*, *mutualism*, and *parasitism* (see Table 46.1).

In **commensalism**, one species benefits and the other is unaffected by the interactions. Commensalism appears to be rare in nature because few species are unaffected by interactions with another. One possible example is the relationship between Cattle Egrets (*Bubulcus ibis*, birds in the heron family) and the large grazing mammals with which they associate **(Figure 46.15)**. Cattle Egrets eat insects and other small animals that their commensal partners flush from grass. Feeding rates of Cattle Egrets are higher when they associate with large grazers than when they do not. The birds clearly benefit from this interaction, but the presence of birds has no apparent positive or negative impact on the mammals.

In **mutualism**, both partners benefit. Mutualism appears to be common and includes coevolved relationships between flowering plants and animal pollinators. Animals that feed on a plant's nectar or

Figure 46.15
Commensalism. Cattle Egrets (*Bubulcus ibis*) feed on insects and other small animals flushed by the movements of large grazing animals such as African elephants (*Loxodonta africana*).

M.B. Fenton

a. Flowering yucca plant

Harlow H. Hadow

b. Female yucca moth

Bob and Miriam Francis/
Tom Stack & Associates

A female yucca moth uses highly modified mouthparts to gather the sticky pollen and roll it into a ball. She carries the pollen to another flower, and after piercing its ovary wall, she lays her eggs. She then places the pollen ball into the opening of the stigma.

c. Yucca moth larva

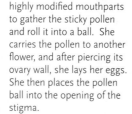

Harlow H. Hadow

When moth larvae hatch from the eggs, they eat some of the yucca seeds and gnaw their way out of the ovary to complete their life cycle. Enough seeds remain undamaged to produce a new generation of yuccas.

Figure 46.16

Mutualism between plants and animals. Several species of yucca plants (*Yucca* species) are each pollinated exclusively by one species of moth (*Tegeticula* species). The adult moth appears at the time of year when the plants are flowering. These species are so mutually interdependent that the larvae of each moth species can feed on only one species of yucca, and each yucca plant can be pollinated by only one species of moth. Most plant–animal mutualisms are less specific.

pollen carry the plants' gametes from one flower to another **(Figure 46.16)**. Similarly, animals that eat fruits disperse the seeds and "plant" them in piles of nutrient-rich feces. Mutualistic relationships between plants and animals do not require active cooperation as each species simply exploits the other for its own benefit. Some associations between bacteria and plants are mutualistic. Perhaps the most important of these associations is between *Rhizobium* and leguminous plants such as peas, beans, and clover (see Chapter 41).

Mutualistic relationships between animal species are common. Cleaner fishes, small marine species, feed on parasites attached to the mouths, gills, and bodies of larger fishes **(Figure 46.17)**. Parasitized fishes hover motionless while cleaners remove their ectoparasites. The relationship is mutualistic because cleaner fishes get a meal, and larger fishes are relieved of parasites.

The relationship between the bull's horn acacia tree (*Acacia cornigera*) of Central America and a species of small ants (*Pseudomyrmex ferruginea*) is a highly coevolved mutualism **(Figure 46.18)**. Each acacia is inhabited by an ant colony that lives in the tree's swollen thorns. Ants swarm out of the thorns to sting, and sometimes kill, herbivores that touch the tree. Ants also clip any vegetation that grows nearby. Acacia trees colonized by ants grow in a space free of herbivores and competitors, and occupied trees grow faster and produce more seeds than unoccupied trees. In return, the plants produce sugar-rich nectar consumed by adult ants and protein-rich structures that the ants feed to their larvae. Ecologists describe the coevolved mutualism between these species as *obligatory*, at least for the ants, because they cannot subsist on any other food sources.

Many animals eat honey and sometimes also the bees that produce it. In Africa, greater honey-guides (*Indicator indicator*) are birds that use a special guiding display to lead humans to beehives. In one tribe of Kenyans, the Borans, honey-gatherers use a special whistle to call *I. indicator*. Boran honey-gatherers that work with greater honey-guides are much more efficient at finding beehives than those working alone. When the honey-gatherer goes to the hive and raids it to obtain honey, greater honey-guides help themselves to bee larvae, left-over honey, and wax. Although *I. indicator* are said also to guide ratels (honey badgers, *Mellivora capensis*) to beehives, there are no firm data supporting this assumption.

In **parasitism**, one species—the parasite—uses another—the host—in a way that is harmful to the host. Parasite–host relationships are often considered to be specialized predator–prey relationships because one population of organisms feeds on another. But parasites rarely kill their hosts because a dead host is not a continuing source of nourishment.

Endoparasites, such as tapeworms, flukes, and round worms, live *within* a host (see Figure 3.18). Many endoparasites acquire their hosts passively when a host accidentally ingests the parasites' eggs or larvae. Endoparasites generally complete their life

— Cleaner wrasse

© Erik Schlogl

Figure 46.17

Mutualism between animal species. A large potato cod (*Epinephelus tukula*) from the Great Barrier Reef in Australia remains nearly motionless in the water while a striped cleaner wrasse (*Lambroides dimindiatus*) carefully removes and eats ectoparasites attached to its lip. The potato cod is a predator; the striped cleaner wrasse is a potential prey. Here the mutualistic relationship supersedes the possible predator–prey interaction.

a.

b.

c.

Marie Read Natural History Photography

M.B. Fenton

M.B. Fenton

Figure 46.18

Highly coevolved mutualisms. **(a)** Bull's horn acacia trees (*Acacia cornigera*) provide colonies for small ants (*Pseudomyrmex furruginea*) with homes in hollow enlarged thorns. The acacia also provides food for the ants (a nutrient-rich nectar) and protein-rich structures the ants feed to their larvae. In the same area in the New World tropics, **(b)** a cowhorn orchid is patrolled by ants that are also housed in **(c)** special hollows on the flower stalk.

cycle in one or two host individuals. Ectoparasites, such as leeches, aphids, and mosquitoes, feed on the exterior of the host (see Figure 3.19). Most animal ectoparasites have elaborate sensory and behavioural mechanisms allowing them to locate specific hosts, and they feed on numerous host individuals during their lifetimes. Plants such as mistletoes (genus *Phoradendron*) live as ectoparasites on the trunks and branches of trees; their roots penetrate the host's xylem and extract water and nutrients. These differ from epiphytes such as bromeliads or Spanish moss that use the host only as a base. Other plants are root parasites, for example, *Conopholis americana* (see Figure 3.17b).

Not all parasites feed directly on a host's tissues. Some bird species are brood parasites, laying their eggs in the host's nest. It is quite common for female birds such as Canvasback Ducks Brown-headed Cowbirds Kirtland's Warbler (*Aythya valisineria*) to lay their eggs in the nests of conspecifics (members of the same species). Some species of songbirds often lay some eggs in the nests of others, a variation on hedging of genetic bets and on extra pair copulations (see Chapters 38 and 40). Brood parasitism is the next level of escalation in this spectrum of parasitism. Brown-headed cowbirds (*Molothrus ater*), like other brood parasites, always lay their eggs in the nest of other species, leaving it to the host parents to raise their young. This behaviour can have drastic repercussions for host species. Brown-headed cowbirds, for instance, have played a large role in the near-extinction of Kirtland's warbler (*Dendroica kirtlandii*).

The feeding habits of insects called parasitoids fall somewhere between true parasitism and predation. A female parasitoid lays her eggs in a larva or pupa of another insect species, and her young consume the tissues of the living host. Because the hosts chosen by most parasitoids are highly specific, agricultural ecologists often try to use parasitoids to control populations of insect pests.

STUDY BREAK

1. What is interspecific competition? What two types of interspecific competition have been identified by community ecologists?
2. Describe the competitive exclusion principle.
3. What is a species' ecological niche? How does a fundamental niche differ from a realized niche?

46.5 The Nature of Ecological Communities

How do complex population interactions, such as those between predators and prey or hosts and parasites, affect the organization and functioning of ecological communities? In the 1920s, ecologists developed two hypotheses about the nature of ecological communities. Frederic Clements championed an *interactive* view, describing communities as "superorganisms," assemblages of species bound together by complex population interactions. It follows that each species in a community requires interactions with a set of ecologically different species, just as every cell in an organism requires services that other types of cells provide. Clements believed that in mature communities, component species were at *equilibrium*, meaning that the community should remain stable over time. If a fire or some other environmental factor disturbed the community, it would return to the predisturbance state.

Henry A. Gleason proposed an alternative, *individualistic* view of ecological communities. He believed that interactions between species do not always determine **species composition**; rather, a community is an assemblage of species individually adapted to similar environmental conditions. According to Gleason's hypothesis, communities do not achieve equilibrium

CHAPTER 46 POPULATION INTERACTIONS AND COMMUNITY ECOLOGY

Some Perils of Mutualism

Living organisms offer many examples of mutualistic interactions in which one species (or group of species) shows varying levels of dependence on another or others. Mutualistic situations can place species on the edge of survival. Where one species depends entirely on another, the extinction of one must lead to change or the extinction of both (e.g., Dodos, see Chapter 48, and yucca plants and their moths). There are many other examples of close relationships, including a desert melon (*Cucumis humifructus*) that depends perhaps entirely on aardvarks (*Orycteropus afer*) for dispersal of its seeds. Aardvarks can sniff out the underground melons, dig them up, and eat them to obtain water. When aardvarks bury their dung, they plant the melon's seeds and fertilize them. The survival of the melon depends on the aardvark but not vice versa.

Mutualistic interactions between species can be even more complex. In the African savannah, ants often live in mutualistic relationships with trees. In east Africa, whistling thorn acacia trees (*Acacia drepanolobium*) are host to four species of ants. One species of ant (*Crematogaster mimosae*) in particular depends on room (hollows in swollen thorns, called domatia) provided by the trees along with board (carbohydrates secreted from extrafloral glands and the bases of leaves). Another species of ant (*Cremato-gaster sjostedti*) also lives on the trees but usually nests in holes made by cerambycid beetles that burrow into and harm the trees.

The ants, particularly *C. mimosae*, attack animals that attempt to browse on the foliage or branches of *A. drepanolobium*. They deter many herbivores, from large mammals to wood-boring beetles (such as cerambycids). If large, browsing mammals are excluded from the area, *A. drepanolobium* produce fewer domatia and fewer carbohydrates for *C. mimosae*. The decline in this species of ant leads to higher damage by cerambycid beetles and increases in populations of *C. sjostedti*.

Many other plants also use ants as mercenaries (see Figure 46.18), and it is becoming clear that survival of these systems depends on the continued presence of participating species.

---

but constantly change in response to disturbance and environmental variation.

In the 1960s, Robert Whittaker suggested that ecologists could determine which hypothesis was correct by analyzing communities along environmental gradients, such as temperature or moisture **(Figure 46.19)**. Clements' interactive hypothesis predicted that species typically occupying the same communities should always occur together. Thus, their distributions along the gradient would be clustered in discrete groups with sharp boundaries between groups (see Figure 46.19a). According to Gleason's individualistic hypothesis, each species is distributed over the section of an environmental gradient to which it is adapted. Different species would have unique distributions, and species composition would change continuously along the gradient. Communities would not be separated by sharp boundaries (see Figure 46.19b).

Most gradient analyses support Gleason's individualistic view of ecological communities. Environmental conditions vary continuously in space, and most plant distributions match these patterns (see Figure 46.19c, d). But the individualistic view does not fully explain all patterns observed in nature. Ecologists recognize certain assemblages of species as distinctive communities and name them accordingly, for example, redwood forests and coral reefs.

Ecotones, the borders between communities, are sometimes wide transition zones. Ecotones are gener-ally species rich because they include plants and animals from both neighbouring communities, as well as some species that thrive only under transitional conditions. Although ecotones are usually relatively broad, places where there is a discontinuity in a critical resource or important abiotic factor may have a sharp community boundary. Chemical differences between soils derived from serpentine rock and sandstone establish sharp boundaries between communities of native California wildflowers and introduced European grasses **(Figure 46.20)**.

STUDY BREAK

1. What two hypotheses were developed by ecologists about the nature of ecological communities? How does each relate to the equilibrium state of the existing community?
2. Are ecotones generally species rich or species poor?

46.6 Community Characteristics

Growth forms (sizes and shapes) of plants vary markedly in different environments, so the appearances of plants can often be used to characterize communities. Warm, moist environments support complex vegetation

a. Interactive hypothesis

The interactive hypothesis predicts that species within communities exhibit similar distributions along environmental gradients (indicated by the close alignment of several curves over each section of the gradient) and that boundaries between communities (indicated by arrows) are sharp.

b. Individualistic hypothesis

The individualistic hypothesis predicts that species distributions along the gradient are independent (indicated by the lack of alignment of the curves) and that sharp boundaries do not separate communities.

c. Siskiyou Mountains

Most gradient analyses support the individualistic hypothesis, as illustrated by distributions of tree species along moisture gradients in Oregon's Siskiyou Mountains and Arizona's Santa Catalina Mountains.

d. Santa Catalina Mountains

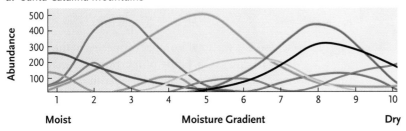

Figure 46.19
Two views of ecological communities. Each graph line indicates a different species.

with multiple vertical layers. Tropical forests include a canopy formed by the tallest trees, an understorey of shorter trees and shrubs, and a herb layer under openings in the canopy. Vinelike lianas and epiphytes grow on the trunks and branches of trees **(Figure 46.21, p. 1158)**. In contrast, physically harsh environments are occupied by low vegetation with simple structure. Trees on mountainsides buffeted by cold winds are short, and the plants below them cling to rocks and soil. Other environments support growth forms between these extremes.

Communities differ greatly in species richness, the number of species that live within them. The harsh environment on a low desert island may support just a few species of microorganisms, fungi, algae, plants, and arthropods. In contrast, tropical forests that grow under milder physical conditions include many thousands of species. Ecologists have studied global patterns of species richness (see Chapter 48) for decades. Today, as human disturbance of natural communities has reached a crisis point, conservation biologists try to understand global patterns of species richness to determine which regions of Earth are most in need of preservation.

Jasper Ridge Biological Preserve

Figure 46.20
Sharp community boundaries. Soils derived from serpentine rock have high magnesium and heavy metal content, which many plants cannot tolerate. Although native California wildflowers (bright yellow in this photograph) thrive on serpentine soil at the Jasper Ridge Preserve of Stanford University, introduced European grasses (green in this photograph) competitively exclude them from adjacent soils derived from sandstones.

Canopy

Understory

Herb layer

Epiphyte

Liana

Buttress

Figure 46.21

Layered forests. Tropical forests, such as one near the Mazaruni River in Guyana (South America), include a canopy of tall trees and an understorey of short trees and shrubs. Huge vines (lianas) climb through the trees, eventually reaching sunlight in the canopy. Epiphytic plants grow on trunks and branches, increasing the structural complexity of the habitat.

The relative abundances of species varies across communities. Some communities have one or two abundant species and a number of rare species. In others, the species are represented by more equal numbers of individuals. In a **temperate deciduous forest** in southern Quebec, red oak trees (*Quercus rubra*) and sugar maples (*Acer saccharum*) might together account for nearly 85% of the trees. A tropical forest in Costa Rica may have more than 200 tree species, each making up a small percentage of the total.

The factors underlying diversity and community structure can be expected to vary among groups

of organisms, and the interactions between very different groups of organisms can have positive effects on both. Using an experimental mycorrhizal plant system (see Chapter 24), H. Maherali and J.N. Klironomos found that after one year, the species richness of mycorrhizal fungi correlated with higher plant productivity. In turn, the diversity and species richness of mycorrhizal fungi were highest when their starting community had more distinct evolutionary lineages. This example illustrates the importance of diversity and interactions.

46.6a Measuring Species Diversity and Evenness: Calculating Indices

The number of species is the simplest measure of diversity, so a forest with four tree species has higher diversity than one with two tree species. But there can be more to measuring diversity than just counting species. Biologists use indices of diversity to facilitate comparison of data sets documenting the numbers of species and of individuals. Shannon's index of diversity (H'), one commonly used measure, is calculated using the formula

$$H' = -\Sigma_{i=1}^{S} pi \ln p_i$$

where S is the total number of species in the community (richness), p_i is the proportion of S made up by species I, and ln is the natural logarithm.

Another index, Shannon's evenness index (E_H), is calculated using the forumula

$$E_H = H'/\ln S$$

where $\ln S$ is the natural log of the number of species. Evenness is an indication of the mixture of species. Indices of diversity and evenness allow population biologists to objectively portray and compare the diversity of communities.

Use the two indices to compare the 3 forests of 50 trees each **(Figure 46.22)**. The number of species and number of individuals of each species in each forest are shown in **Table 46.2.** In Table 46.2, the values of H' and E_H provide an indication of the diversity of the three hypothetical forests and the evenness of species representations. Lower values of H' and E_H suggest com-

Figure 46.22

Species diversity. In this hypothetical example, each of three samples of forest communities (A, B, and C) contains 50 trees. Indices allow biologists to express the diversity of species and evenness of numbers (see Table 46.2).

Forest A

Forest B

Forest C

munities with few species (low H' values) or uneven distribution (low E_H values). Higher values of H' and E_H suggest a richer array of species with evenly distributed individuals.

Measures of diversity can be used to advantage. Ecologists refer to "α" diversity to represent the numbers of species sympatric in one community and "β" diversity to depict the numbers in a collection of communities. The number of herbivorous Lepidoptera species in one national park is α diversity, whereas β diversity is the number of species in the country in which the park is located. The trend to establish parks that cross international boundaries is a step toward recognizing the reality that political and biological boundaries can be quite different. Measures of diversity can be used directly in some conservation plans (see Chapter 48).

46.6b Trophic Interactions: Between Nourishment Levels

Every ecological community has trophic structure (*troph* = nourishment; see Chapter 3), comprising all plant–herbivore, predator–prey, host–parasite, and potential competitive interactions **(Figure 46.23, p. 1160).** We can visualize the trophic structure of a community as a hierarchy of trophic levels, defined by the feeding relationships among its species (see Figure 46.23a). Photosynthetic organisms are primary producers, the first trophic level. Primary producers are photoautotrophs (*auto* = self) because they capture sunlight and convert it into chemical energy that is used to make larger organic molecules that plants can use directly.

Plants are the main primary producers in terrestrial communities. Multicellular algae and plants are the major primary producers in shallow freshwater and marine environments, whereas photosynthetic protists and cyanobacteria play that role in deep, open water.

All consumers in a community (animals, fungi, and diverse microorganisms) are heterotrophs (*hetero* = other) because they acquire energy and nutrients by eating other organisms or their remains. Animals are consumers. Herbivores (primary consumers) feed directly on plants and form the second trophic level. Secondary consumers (mesopredators) eat herbivores and form the third trophic level. Animals that eat secondary consumers comprise the fourth trophic level, the tertiary consumers. At one meal, animals that are omnivores (e.g., humans, pigs, and bears) can act as primary, secondary, and **tertiary** consumers.

Detritivores (scavengers) form a separate and distinct trophic level. These organisms extract energy from organic detritus produced at other trophic levels. Detritivores include fungi, bacteria, and animals such as earthworms and vultures that ingest dead organisms, digestive wastes, and cast-off body parts, such as leaves and exoskeletons. Decomposers, a type of detritivore, are small organisms, such as bacteria and fungi, that feed on dead or dying organic material. Detritivores and decomposers serve a critical ecological function because their activity reduces organic material to small inorganic molecules that producers can assimilate (see Chapters 3 and 21).

Although omnivores obviously do not fit exclusively into one trophic level, this also can be true of other organisms. Sea slugs that use chloroplasts or carnivorous plants are examples of species that do not fit readily into trophic categories.

46.6c Food Chains and Webs: Connections in Ecosystems

Ecologists use food chains and webs to illustrate the trophic structure of a community. Each link in a food chain is represented by an arrow pointing from food to consumer (see Figure 46.23). Simple, straight-line food chains are rare in nature because most consumers feed on more than one type of food and because most organisms are eaten by more than one type of consumer. Complex relationships are portrayed as food webs, sets of interconnected food chains with multiple links.

In the food web for the waters off the coast of Antarctica (see Figure 46.23), primary producers and primary consumers are small organisms occurring in vast numbers. Microscopic diatoms (phytoplankton) are responsible for most photosynthesis, and small shrimplike krill (zooplankton) are the major primary consumers. These tiny organisms, in turn, are eaten by larger species such as fish and seabirds, as well as

Table 46.2	Shannon's Indices for Measuring Diversity and Evenness		

Numbers of Individuals Per Species

	Forest A*	Forest B*	Forest C*
Species 1	39	5	25
Species 2	2	5	25
Species 3	2	5	0
Species 4	1	5	0
Species 5	1	5	0
Species 6	1	5	0
Species 7	1	5	0
Species 8	1	5	0
Species 9	1	5	0
Species 10	1	5	0
Shannon Indices			
H' diversity	0.6	2.3	0.7
E_h evenness	0.26	1.0	1.0

*Forests from Figure 46.22.

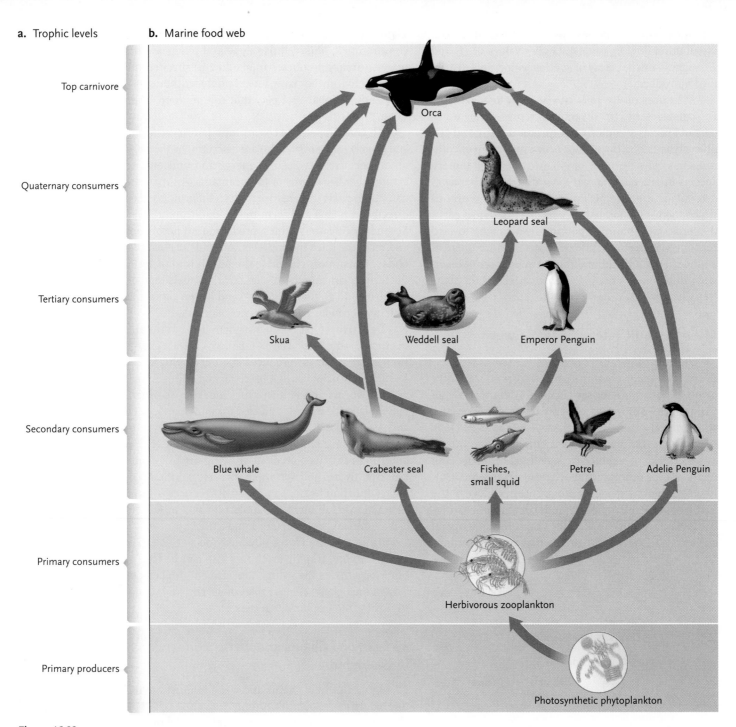

a. Trophic levels

b. Marine food web

Top carnivore

Quaternary consumers

Tertiary consumers

Secondary consumers

Primary consumers

Primary producers

Orca

Leopard seal

Skua   Weddell seal   Emperor Penguin

Blue whale   Crabeater seal   Fishes, small squid   Petrel   Adelie Penguin

Herbivorous zooplankton

Photosynthetic phytoplankton

Figure 46.23
The marine food web off the coast of Antarctica.

by suspension-feeding baleen whales. Some secondary consumers are eaten by birds and mammals at higher trophic levels. The top carnivore in this ecosystem, the orca, feeds on carnivorous birds and mammals.

Ideally, depictions of food webs would include all species in a community, from microorganisms to top consumer. But most ecologists simply cannot collect data on every species, particularly those that are rare or very small. Instead, they study links between the most important species and simplify analysis by grouping together trophically similar species. Figure 46.23 categorizes the many different species of primary pro-

ducers and primary consumers as phytoplankton and zooplankton, respectively.

Many biological "hot spots" exist, from thermal vents on the floor of some oceans to deposits of bat guano in some caves. A more recently described example is icebergs drifting north from Antarctica. The icebergs can be hot spots of enrichment because of the nutrients and other materials they shed into surrounding waters. The water around two free-drifting icebergs (0.1 km² and 30.8 km² in area) was sampled in the Weddell Sea. High concentrations of chlorophyll, krill, and seabirds extended about 3.7 km around each iceberg. These data, reported by K.L. Smith Jr. and seven colleagues, demonstrate that icebergs can have substantial effects on pelagic ecosystems.

In the late 1950s, Robert MacArthur analyzed food webs to determine how the many links between trophic levels may contribute to a community's stability. The **stability** of a community is defined as its ability to maintain species composition and relative abundances when environmental disturbances eliminate some species from the community. MacArthur hypothesized that in species-rich communities, where animals feed on many food sources, the absence of one or two species would have only minor effects on the structure and stability of the community as a whole. He proposed a connection between **species diversity**, food web complexity, and community stability.

Subsequent research has confirmed MacArthur's reasoning. The average number of links per species generally increases with increasing species richness. Comparative food web analysis reveals that the relative proportions of species at the highest, middle, and lowest trophic levels are reasonably constant across communities. In 92 communities, MacArthur found two or three prey species per predator species, regardless of species richness.

Interactions among species in most food webs can be complex, indirect, and hard to unravel. In contrast, rodents and ants living in desert communities of the American Southwest potentially compete for seeds, their main food source. Plants that produce the seeds compete for water, nutrients, and space. Rodents generally prefer to eat large seeds, whereas ants prefer small seeds. Thus, feeding by rodents reduces the potential population sizes of plants that produce large seeds. As a result, the population sizes of plants that produce small seeds may increase, ultimately providing more food for ants (see Chapter 41). Compared with the Antarctic system described above (see Figure 46.23), this community is not particularly complex.

STUDY BREAK

1. Why are indices important for population biologists? What do Shannon's indices measure?
2. Differentiate between α and β diversity.
3. Are herbivores primary or secondary consumers? Which trophic level do they form? Where do omnivores belong?

46.7 Effects of Population Interactions on Community Structure

Observations of resource partitioning and character displacement suggested that some process had fostered differences in resource use among coexisting species, and competition provided the most straightforward explanation of these patterns.

Interspecific competition can cause local extinction of species or prevent new species from becoming established in a community, reducing its species richness. During the 1960s and early 1970s, ecologists emphasized competition as the primary factor structuring communities.

46.7a Competition: More than One Species Competing for a Resource

To further explore the role of competition, ecologists undertook field experiments on competition in natural populations. The experiment on barnacles (see Figure 46.14) is typical of this approach—the impact on one species' potential competitors of adding or removing another species changed patterns of distribution or population size. The picture that emerges from the results of these experiments is not clear, even to ecologists. In the early 1980s, Joseph Connell surveyed 527 published experiments on 215 species. He found that competition was demonstrated in roughly 40% of the experiments and more than 50% of species. At the same time, Thomas W. Schoener used different criteria to evaluate 164 experiments on approximately 400 species. He found that competition affected more than 75% of species.

It is not surprising that there is no single answer to the question about how competition works in communities and influences them. Plant ecologists and vertebrate ecologists working with *K*-selected species generally believe that competition has a profound effect on species distributions and resource use. Insect ecologists and marine ecologists working with *r*-selected species argue that competition is not the major force governing community structure, pointing instead to predation or parasitism and physical disturbance. We know that even categorizing a species as "*r*-" or "*K*-" is open to discussion (see Chapter 45).

46.7b Feeding: You Are What You Eat

Predators can influence the **species richness** and structure of communities by reducing the sizes of prey populations. On the rocky coast of British Columbia, different species that fill different trophic roles compete for attachment sites on rocks, a requirement for life on a wave-swept shore. Mussels are the strongest competitors for space, eliminating other species from the community (see *Effect of a Predator on the Species Richness of Its Prey*). At some sites, predatory sea stars preferentially eat mussels, reducing their numbers and creating space for other species to grow. Because the interaction between

Effect of a Predator on the Species Richness of Its Prey

Biologists used a predatory sea star (*Pisaster ochraceus*) to assess the influence a predator can have on species richness and relative abundance of prey **(Figure 1)**. *P. ochraceus* preferentially eats mussels (*Mytilus californicus*), one of the strongest competitors for space in rocky intertidal pools. Robert Paine removed *Pisaster* from caged experimental study plots, leaving control study plots undisturbed, and then monitored the species richness of *Pisaster*'s invertebrate prey over many years.

Paine documented an increase in mussel populations in the experimental plots as well as complex changes in the feeding relationships among species in the intertidal food web **(Figure 2)**. When he removed *P. ochraceus*, the top predator in this food web, he observed a rapid decrease in the species richness of invertebrates and algae. Species richness on control plots did not change over the course of the experiment.

Predation by *P. ochraceus* prevents mussels from outcompeting other invertebrates on rocky shores.

Nancy Rotenberg/Animals, Animals—Earth Scenes

Figure 1
A predatory sea star (*Pisaster ochraceus*) feeding on a mussel (*Mytilus californicus*).

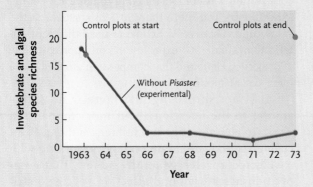

Figure 2
Changes in the species richness of invertebrates and algal species according to changes in populations of sea stars.

Pisaster and *Mytilus* affects other species as well, it qualifies as a strong interaction.

In the 1960s, Robert Paine used removal experiments to evaluate the effects of predation by *Pisaster* (see *Effect of a Predator on the Species Richness of Its Prey*). In predator-free experimental plots, mussels outcompeted barnacles, chitons, limpets, and other invertebrate herbivores, reducing species richness from 15 species to 8. In control plots containing predators, all 15 species persisted. Ecologists describe predators such as *Pisaster* as **keystone species**, defined as species with a greater effect on community structure than their numbers might suggest.

Herbivores also exert complex effects on communities. In the 1970s, Jane Lubchenco studied herbivory in a periwinkle snail, believed to be a keystone species on rocky shores in Massachusetts (see *The Complex Effects of an Herbivorous Snail on Algal Species Richness*). The features of plants and algae and the food preferences of animals that eat them together can influence community structure.

STUDY BREAK

1. How does the importance of competition vary between *K*-selected and *r*-selected species?
2. Does predation or herbivory increase or decrease species richness? Explain.
3. What is a keystone species?

The Complex Effects of an Herbivorous Snail on Algal Species Richness

Jane Lubchenco made enclosures that prevented periwinkle snails (*Littorina littorea*) from entering or leaving study plots in tidepools and on exposed rocks in rocky intertidal habitat **(Figure 1)**. She then monitored the algal species composition in the plots, comparing them to the density of the periwinkles. In this way, she examined the influence of the periwinkles on the species richness of algae in intertidal communities.

The results varied dramatically between the study plots in tidepools and on exposed rocks. In tidepools, periwinkle snails preferentially ate *Enteromorpha*, the competitively dominant alga. At intermediate densities of *Enteromorpha*, the periwinkles remove some of these algae, allowing weakly competitive species to grow. The snails' grazing increases species richness. But grazing by periwinkles when *Enteromorpha* is at low or high densities reduces the species richness of algae in tide pools. On exposed rocks, where periwinkle snails rarely eat the competitively dominant alga *Chondrus*, feeding by snails reduces algal species richness **(Figure 2)**.

Periwinkle snails (*Littorina littorea*)

Enteromorpha growing in tidepools

Chondrus growing on exposed rocks

Figure 1
The distribution of periwinkle snails and two kinds of algae.

In tidepools

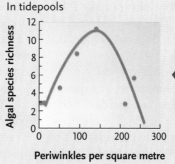

In tidepools, snails at low densities eat little algae and *Enteromorpha* competitively excludes other algal species, reducing species richness. At high snail densities, heavy feeding on all species reduces algal species richness. At intermediate snail densities, grazing eliminates some *Enteromorpha*, allowing other species to grow.

On exposed rocks

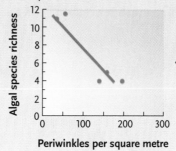

On exposed rocks, periwinkles never eat much *Chondrus*, but they consume the tender, less successful competitors. Thus, feeding by periwinkles reinforces the competitive superiority of *Chondrus*: as periwinkle density increases, algal species richness declines.

Figure 2
Density of periwinkles versus algal species richness in tidepools and on exposed rocks.

46.8 Effects of Disturbance on Community Characteristics

Recent research tends to support the individualistic view that many communities are not in equilibrium and that species composition changes frequently. Environmental disturbances such as storms, landslides, fires, floods, avalanches, and cold spells often eliminate some species and provide opportunities for others to become established. Frequent disturbances keep some ecological communities in a constant state of flux.

Physical disturbances are common in some environments. Lightning-induced fires commonly sweep through grasslands, powerful hurricanes often demolish patches of forest and coastal habitats, and waves wash over communities at the edge of the sea and sweep away organisms as well as landforms and other structures.

Joseph Connell and his colleagues conducted an ambitious long-term study of the effects of disturbance on coral reefs, shallow tropical marine habitats that are among the most species-rich communities on Earth. In some parts of the world, reefs are routinely battered by violent storms that wash corals off the substrate, creating bare patches in the reef. The scouring action of storms creates opportunities for coral larvae to settle on bare substrates and start new colonies.

From 1963 to 1992, Connell and his colleagues tracked the fate of the Heron Island Reef at the south end of Australia's Great Barrier Reef **(Figure 46.24)**. The inner flat and protected crests of the reef are sheltered from severe wave action during storms, whereas some pools and crests are routinely exposed to physical disturbance. Because corals live in colonies of variable size, the researchers monitored coral abundance by measuring the percentage of the substrate (i.e., the seafloor) that colonies covered. They revisited marked study plots at intervals, photographing and identifying individual coral colonies.

Five major cyclones crossed the reef during the 30-year study period. Coral communities in exposed areas of the reef were in a nearly continual state of flux. In exposed pools, four of the five cyclones reduced the percentage of cover, often drastically. On exposed crests, the cyclone of 1972 eliminated virtually all corals, and subsequent storms slowed the recovery of these areas for more than 20 years. In contrast, corals in sheltered areas suffered much less storm damage. Nevertheless, their coverage also declined steadily during the study as a natural consequence of the corals' growth. As colonies grew taller and closer to the ocean's surface, their increased exposure to air resulted in substantial mortality.

Connell and his colleagues also documented *recruitment*, the growth of new colonies from settling larvae, in their study plots. They discovered that the rate at which new colonies developed was almost always higher in sheltered than in exposed areas. Recruitment rates were extremely variable, depending in part on the amount of space that storms or coral growth had made available.

This long-term study of coral reefs illustrates that frequent disturbances prevent some communities from reaching an equilibrium determined by interspecific interactions. Changes in the coral reef community at Heron Island result from the effects of external disturbances that remove coral colonies from the reef as well as internal processes (growth and recruitment) that either eliminate colonies or establish new ones. In this community, growth and recruitment are slow processes, and disturbances are frequent. Thus, the community never attains equilibrium, and moderate levels of disturbance can foster high species richness.

The **intermediate disturbance hypothesis**, proposed by Connell in 1978, suggests that species richness is greatest in communities experiencing fairly frequent disturbances of moderate intensity. Moderate disturbances create openings for *r*-selected species to arrive and join the community while allowing *K*-selected species to survive. Thus, communities that

Figure 46.24

The effects of storms on corals. Five tropical cyclones (marked by grey arrows) damaged corals on the Heron Island Reef during a 30-year period. Storms reduced the percentage cover of corals in **(a)** exposed parts of the reef much more than in **(b)** sheltered parts of it.

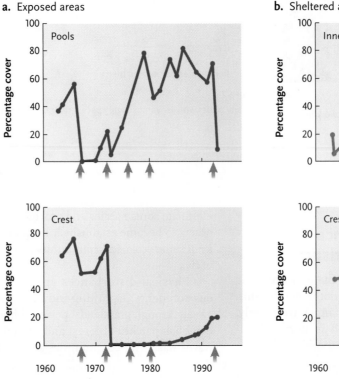

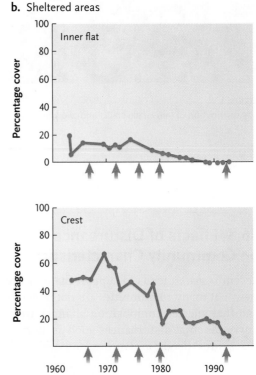

experience intermediate levels of disturbance contain a rich mixture of species. Where disturbances are severe and frequent, communities include only *r*-selected species that complete their life cycles between catastrophes. Where disturbances are mild and rare, communities are dominated by long-lived *K*-selected species that competitively exclude other species from the community.

Major hydrodynamic disturbances to coral reefs, such as tsunamis and severe storms, have important impacts on coral reefs. Using oceanographic and engineering models, it is possible to predict the degree of dislodgement of benthic reef corals and, in this way, predict how coral shape and size indicate vulnerability to major disturbances. The use of these models is particularly important during times of climate change.

Several studies in diverse habitats have confirmed the predictions of the intermediate disturbance hypothesis. Colin R. Townsend and his colleagues studied the effects of disturbance at 54 stream sites in the Taieri River system in New Zealand. Disturbance occurs in these communities when water flow from heavy rains moves rocks, soil, and sand in the streambed, disrupting animal habitats. Townsend and his colleagues measured how much the substrate moved in different streambeds to develop an index of the intensity of disturbance. Their results indicate that species richness is highest in areas that experience intermediate levels of disturbance **(Figure 46.25).**

Some ecologists have suggested that species-rich communities recover from disturbances more readily than less diverse communities. In the United States, David Tilman and his colleagues conducted large-scale experiments in midwestern grasslands. They examined relationships between species number and the ability of communities to recover from disturbance. Grassland plots with high species richness recover from drought faster than plots with fewer species.

STUDY BREAK

1. What did Connell's 30-year study of coral reefs illustrate about the ability of communities to reach a state of equilibrium?
2. What is the intermediate disturbance hypothesis? Describe one study that supports this hypothesis.
3. How does species richness affect the rate of recovery following a disturbance?

46.9 Succession

Ecosystems change over time, a process called **succession**, the change from one community type to another.

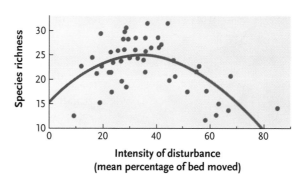

Figure 46.25
An observational study that supports the intermediate disturbance hypothesis. In the Taieri River system in New Zealand, species richness was highest in stream communities that experienced an intermediate level of disturbance.

46.9a Primary Succession: The First Steps

Primary succession begins when organisms first colonize habitats without soil, such as those created by erupting volcanoes and retreating glaciers **(Figure 46.26, p. 1166).** Lichens are often among the very first colonists (see Chapter 24), deriving nutrients from rain and bare rock. They secrete mild acids that erode rock surfaces, initiating the slow development of soil, which is enriched by the organic material lichens produce. After lichens modify a site, mosses (see Chapter 25) colonize patches of soil and grow quickly.

As soil accumulates, hardy, opportunistic plants (grasses, ferns, and broad-leaved herbs) colonize the site from surrounding areas. Their roots break up rock, and when they die, their decaying remains enrich the soil. Detritivores and decomposers facilitate these processes. As the soil becomes deeper and richer, increased moisture and nutrients support bushes and, eventually, trees. Late successional stages are often dominated by *K*-selected species with woody trunks and branches that position leaves in sunlight and large root systems that acquire water and nutrients from soil.

In the classical view of ecological succession, long-lived species, which replace themselves over time, eventually dominate a community, and new species join it only rarely. This relatively stable, late successional stage is called a **climax community** because the dominant vegetation replaces itself and persists until an environmental disturbance eliminates it and allows other species to invade. Local climate and soil conditions, the surrounding communities where colonizing species originate, and chance events determine the species composition of climax communities. We now know that even climax communities change slowly in response to environmental fluctuations.

46.9b Secondary Succession: Changes after Destruction

Secondary succession occurs after existing vegetation is destroyed or disrupted by an environmental disturbance, such as a fire, a storm, or human activity. The presence of soil makes disturbed sites ripe for colonization and may contain numerous seeds that

1 The glacier has retreated about 8 m per year since 1794.

2 This site was covered with ice less than 10 years before this photo was taken. When a glacier retreats, a constant flow of melt water leaches minerals, especially nitrogen, from the newly exposed substrate.

3 Once lichens and mosses have established themselves, mountain avens (genus *Dryas*) grows on the nutrient-poor soil. This pioneer species benefits from the activity of mutualistic nitrogen-fixing bacteria, spreading rapidly over glacial till.

4 Within 20 years, shrubby willows (genus *Salix*), cottonwoods (genus *Populus*), and alders (genus *Alnus*) take hold in drainage channels. These species are also symbiotic with nitrogen-fixing microorganisms.

5 In time, young conifers, mostly hemlocks (genus *Tsuga*) and spruce (genus *Picea*), join the community.

6 After 80 to 100 years, dense forests of Sitka spruce (*Picea sichensis*) and western hemlock (*Tsuga heterophylla*) have crowded out the other species.

Figure 46.26
Primary succession following glacial retreat. The retreat of glaciers at Glacier Bay, Alaska, has allowed ecologists to document primary succession on newly exposed rocks and soil.

germinate after disturbance. Early stages of secondary succession proceed rapidly, but later stages parallel those of primary succession.

46.9c Climax Communities: The Ultimate Ecosystems until Something Changes

Similar climax communities can arise from several different successional sequences. Hardwood forests can also develop in sites that were once ponds. During **aquatic succession**, debris from rivers and runoff accumulates in a pond, filling it to its margins. Ponds are first transformed into swamps, inhabited by plants adapted to a semisolid substrate. As larger plants get established, their high transpiration rates dry the soil, allowing other plant species to colonize. Given enough time, the site may become a meadow or forest in which an area of moist, low-lying ground is the only remnant of the original pond.

Because several characteristics of communities can change during succession, ecologists try to document how patterns change. First, because *r*-selected species are short-lived and *K*-selected species are long-lived, species composition changes rapidly in the early stages and more slowly in later stages of succession. Second, species richness increases rapidly during early stages because new species join the community faster than resident species become extinct. In later stages, species richness stabilizes or may even decline. Third, in terrestrial communities receiving sufficient rainfall, the maximum height and total mass of the vegetation increase steadily as large species replace small ones, creating the complex structure of the climax community.

Because plants influence the physical environment below them, the community itself increasingly moderates its microclimate. The shade cast by a forest canopy helps retain soil moisture and reduce temperature fluctuations. The trunks and canopy also reduce wind speed. In contrast, the short vegetation in an early successional stage does not effectively shelter the space below it.

Although ecologists usually describe succession in terms of vegetation, animals can show similar patterns. As the vegetation shifts, new resources become available, and animal species replace each other over time. Herbivorous insects, often with strict food preferences, undergo succession along with their food plants. And as herbivores change, so do their predators, parasites, and parasitoids. In old-field succession in eastern North America, different vegetation stages harbour a changing assortment of bird species **(Figure 46.27)**.

Differences in dispersal abilities (see *Dispersal*), maturation rates, and life spans among species are partly responsible for ecological succession. Early successional stages harbour many *r*-selected species because they produce numerous small seeds that colonize open habitats and grow quickly. Mature successional stages are dominated by *K*-selected species because they are long-lived. Nevertheless, coexisting populations inevitably affect one another. Although the role of population interactions in succession is generally acknowledged, ecologists debate the relative importance of processes that either facilitate or inhibit the turnover of species in a community.

46.9d Facilitation Hypothesis: One Species Makes Changes That Help Others

The **facilitation hypothesis** suggests that species modify local environment in ways that make it less suitable for themselves but more suitable for colonization by species typical of the next successional stage. When lichens first colonize bare rock, they produce a small quantity of soil that is required by mosses and grasses that grow there later. According to this hypothesis, changes in species composition are both orderly and predictable because the presence of each stage facilitates the success of the next one. Facilitation is important in primary succession, but it may not be the best model of interactions that influence secondary succession.

46.9e Inhibition Hypothesis: One Species Negatively Affects Others

The **inhibition hypothesis** suggests that new species are prevented from occupying a community by species that are already present. According to this hypothesis, succession is neither orderly nor predictable because each stage is dominated by the species that happened to have colonized the site first. Species replacements occur only when individuals of dominant species die of old age or when an environmental disturbance reduces their numbers. Eventually, long-lived species replace

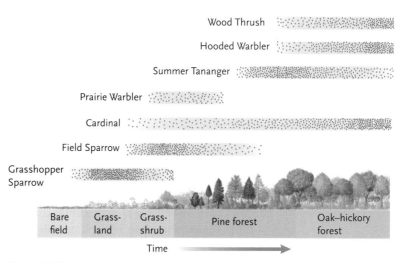

Figure 46.27

Succession in animals. Successional changes in bird species composition in an abandoned agricultural field in eastern North America parallel the changes in plant species composition. The residence times of several representative species are illustrated. The density of stippling inside each bar illustrates the density of each species through time.

short-lived species, but the precise species composition of a mature community is open to question. Inhibition appears to play a role in some secondary successions. The interactions among early successional species in an old field are highly competitive. Horseweed inhibits the growth of asters that follow them in succession by shading aster seedlings and releasing toxic substances from their roots. Experimental removal of horseweed enhances the growth of asters, confirming the inhibitory effect.

46.9f Tolerance Hypothesis: Species Tolerate One Another

The **tolerance hypothesis** asserts that succession proceeds because competitively superior species replace competitively inferior ones. According to this model, early-stage species neither facilitate nor inhibit the growth of later-stage species. Instead, as more species arrive at a site and resources become limiting, competition eliminates species that cannot harvest scarce resources successfully. In the Piedmont region of North America, young hardwood trees are more tolerant of shade than are young pine trees, and hardwoods gradually replace pines during succession. Thus, the climax community includes only strong competitors. Tolerance may explain the species composition of many transitional and mature communities.

At most sites, succession probably results from some combination of facilitation, inhibition, and tolerance, coupled with interspecific differences in dispersal, growth, and maturation rates. Moreover, within a community, the patchiness of abiotic factors strongly influences plant distributions and species composition. In deciduous forests of eastern North America, maples (*Acer* species) predominate on wet, low-lying ground, but oaks (*Quercus* species) are more abundant at higher and drier sites. Thus, a mature deciduous forest is often a mosaic of species and not a uniform stand of trees.

Disturbance and density-independent factors play important roles, in some cases speeding successional change. Moose (*Alces alces*) prefer to feed on deciduous shrubs in northern forests. This disturbance accelerates the rate at which conifers replace deciduous shrubs. On Isle Royale in Lake Superior, however, grazing by moose strongly affects balsam fir (*Abies balsamea*), their preferred food there. The net effect is a severe reduction in conifers and an increase in deciduous shrubs. Disturbance can also inhibit successional change, establishing a *disturbance climax* or **disclimax community**. In many grassland communities, periodic fires and grazing by large mammals kill seedlings of trees that would otherwise become established. Thus, disturbance prevents the succession from grassland to forest, and grassland persists as a disclimax community.

Animals such as moose can alter patterns of succession and vegetation in some communities, but the effect also extends to small mammals. Removal experiments involving kangaroo rats and plots of shrubland in the Chihuahuan Desert (southeastern Arizona) allowed J.H. Brown and E.J. Heske to demonstrate that these rodents were a "keystone guild." Kangaroo rats affect the plants in several ways. They are seed predators, and their burrowing activities disturb soils. Excluding kangaroo rats from experimental plots led to a threefold increase in the density of tall perennials and annual grasses **(Figure 46.28)**, suggesting that by predation on seeds and burrowing, these rodents affected the vegetation in the experimental areas.

On a local scale, disturbances often destroy small patches of vegetation, returning them to an earlier successional stage. A hurricane, tornado, or ava-

From James H. Brown, Edward J. Heske, "Control of a Desert-Grassland Transition by a Keystone Rodent Guild", Science, vol. 250, Dec 21, 1990, pp. 1705 - 1707. Reprinted with permission from AAAS.

Figure 46.28

Predation and succession. Kangaroo rats (*Dipodomys*) were removed from the left sides of the fence, which excluded them from the plot on the left. The top photograph was taken 5 years after the removals and the bottom one 13 years after. A large-seeded annual (after 5 years) and tall grasses are present in the *Dipodomys*-free plots.

lanche may topple trees in a forest, creating small, sunny patches of open ground. Locally occurring *r*-selected species take advantage of newly available resources and quickly colonize the openings. These local patches then undergo succession that is out of step with the immediately surrounding forest. Thus, moderate disturbance, accompanied by succession in local patches, can increase species richness in many communities.

STUDY BREAK

1. What are the two types of succession? How do they differ?
2. What is a climax community? What determines the species composition of a climax community?
3. Identify and briefly describe the three hypotheses used to explain how succession proceeds.

46.10 Variations in Species Richness among Communities

Species richness often varies among communities according to a recognizable pattern. Two large-scale patterns of species richness—latitudinal trends and island patterns—have captured the attention of ecologists for more than a century.

46.10a Latitudinal Effects: From South to North

Ever since Darwin and Wallace travelled the globe (see Chapter 20), ecologists have recognized broad latitudinal trends in species richness. For many but not all plant and animal groups, species richness follows a latitudinal gradient, with the most species in the tropics and a steady decline in numbers toward the poles **(Figure 46.29)**. Several general hypotheses may explain these striking patterns.

Some hypotheses propose historical explanations for the *origin* of high species richness in the tropics. The benign climate in tropical regions allows some tropical organisms to have more generations per year than their temperate counterparts. Small seasonal changes in temperature mean that tropical species may be less likely than temperate species to migrate from one habitat to another, reducing gene flow between geographically isolated populations (see Chapter 18). These factors may have fostered higher **speciation** rates in the tropics, accelerating the accumulation of species. Tropical communities may also have experienced severe disturbance less often than communities at higher latitudes, where periodic glaciations have caused repeated extinctions. Thus, new species may have accumulated in the tropics over longer periods of time.

Other hypotheses focus on ecological explanations for the *maintenance* of high species richness in the tropics. Some resources are more abundant, predictable, and diverse in tropical communities. Tropical regions experience more intense sunlight, warmer temperatures in most months, and higher annual rainfall than temperate and polar regions (see Chapter 3). These factors provide a long and predictable growing season for the lush tropical vegetation, which supports a rich assemblage of herbivores, and through them many carnivores and parasites. Furthermore, the abundance, predictability, and year-round availability of resources allow some tropical animals to have specialized diets. Tropical forests support many species of fruit-eating bats and birds, which could not survive in temperate forests where fruits are not available year-round.

Species richness may be a self-reinforcing phenomenon in tropical communities. Complex webs of population interactions and interdependency have coevolved in relatively stable and predictable tropical climates. Predator–prey, competitive, and symbiotic interactions may prevent individual species from dominating communities and reducing species richness.

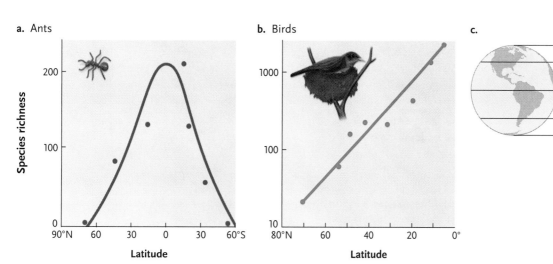

Figure 46.29

Latitudinal trends in species richness. The species richness of many animals and plants varies with latitude as illustrated here for **(a)** ants in North, Central, and South America and **(b)** birds in North and Central America. The species richness data used in **(c)** are based on records of where these birds breed.

Dispersal

Organisms often show astonishing dispersal abilities. In some cases, long-distance dispersal by plants in the Arctic is effected by the combination of strong winds and extensive expanses of ice and snow. The Svalbard Archipelago **(Figure 1)** is an interesting location for the study of plant dispersal. The islands were glaciated 20 000 years ago, and it is likely that plants did not survive this condition. The fossil record indicates that plants have been present on Svalbard for fewer than 10 000 years, although between 9500 and 4000 years ago, the climate was warmer there (by 1° to 2°C) than it is now.

Using DNA fingerprinting, I.G. Alsos and eight colleagues demonstrated that plant colonization of the Svalbard Archipelago has involved the arrival of plants from all possible adjacent regions **(Figure 2).** In eight of nine species, genetic evidence indicates

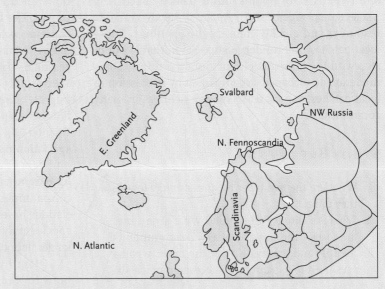

Figure 1
The location of the Svalbard Archipelago.

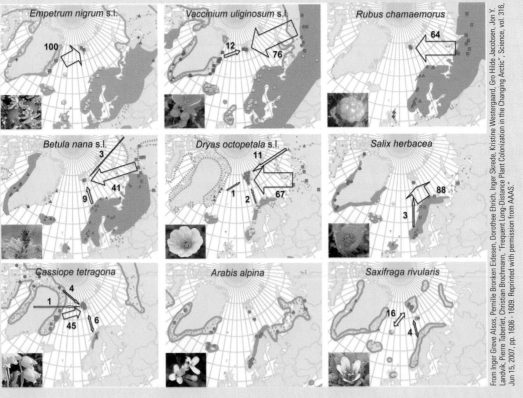

Figure 2
Source regions for Svalbard plants. Shading shows the geographic distribution of nine species of plants, and dotted lines show the distributions of related species. The main genetic groups are represented by colours, although some populations (✳) could not be assigned to a genetic group. Arrows identify source populations, and the numbers indicate the percentage allocation by source region.

From Inger Greve Alsos, Pernille Bronken Eidesen, Dorothee Ehrich, Inger Skrede, Kristine Westergaard, Gro Hilde Jacobsen, Jon Y. Landvik, Pierre Taberlet, Christian Brochmann, "Frequent Long-Distance Plant Colonization in the Changing Arctic". Science, vol. 316, Jun 15, 2007, pp. 1606 - 1609. Reprinted with permission from AAAS."

multiple colonization events. Plants can obviously disperse without assistance from animals.

In other situations, plants disperse with the assistance of animals through pollination and seeds. Using *Prunus maheleb*, the mahaleb cherry **(Figure 3),** and genetic techniques, P. Jordano and two colleagues examined the role of birds and mammals in pollination and dispersing seeds. Small passerine birds dispersed seeds short distances (most < 50 m) from the parent tree, whereas medium-sized birds (*Corvus corone* and *Turdus viscivorus*) usually dispersed seeds over longer distances (>110 m). Mammals (usually *Martes foina* and *Vulpes vulpes* but sometimes *Meles meles*) dispersed seeds ~500 m. The genetic work also provided an indi-cation of the extent of gene flow during pollination.

It is obvious that plants capable of self-fertilization or vegetative reproduction can be more effective colonists than those depending on outcrossing, especially with the help of animal pollinators.

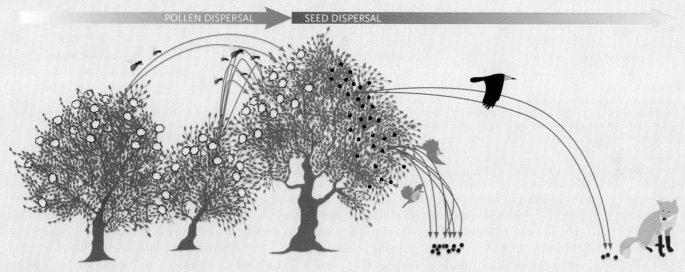

Figure 3
The movement of pollen and seeds from mahaleb cherry trees. Gene flow occurs through pollination and seed dispersal (see Chapter 18).

46.10b Equilibrium Theory of Island Biogeography

In 1883, a volcanic eruption virtually obliterated the island of Krakatoa. Within 50 years, what was left of Krakatoa had been recolonized by plants and animals, providing biologists with a clear demonstration of the dispersal powers of many living species. The colonization of islands and the establishment of biological communities there have provided many natural experiments that have advanced our knowledge of ecology and populations. Islands are attractive sites for experiments because although the species richness of communities may be stable over time, the species composition is often in flux as new species join a community and others drop out. In the 1960s, Robert MacArthur and Edward O. Wilson used islands as model systems to address the question of why communities vary in species richness. Islands provide natural laboratories for studying ecological phenomena, just as they do for evolution (see Chapter 3). Island communities can be small, with well-defined boundaries, and are isolated from surrounding communities.

MacArthur and Wilson developed the **equilibrium theory of island biogeography** to explain variations in species richness on islands of different size and different levels of isolation from other landmasses. They hypothesized that the number of species on any island was governed by give and take between two processes: the immigration of new species to an island and the extinction of species already there **(Figure 46.30, p. 1172).**

According to their model, the mainland harbours a *species pool* from which species immigrate to offshore islands. Seeds and small arthropods are carried by wind or floating debris. Animals such as birds arrive under their own power. When only a few species are on an island, the rate at which new species immigrate to the island is high. But as more species inhabit the island over time, the immigration rate declines because fewer species in the mainland pool can still arrive on the island as *new* colonizers (see Chapter 3). Once some species arrive on an island, their populations grow and persist for variable lengths of time. Other immigrants die without reproducing. As the number of species on an island increases, the rate of species extinction also rises. Extinction rates increase over time partly because more species can go extinct there. In addition, as the number of species on the island increases, competition and predator–prey interactions can reduce the population sizes of some species and drive them to extinction.

a. Immigration and extinction rates

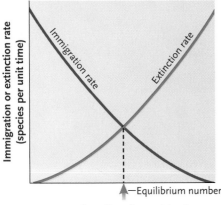

The number of species on an island at equilibrium (indicated by the arrow) is determined by the rate at which new species immigrate and the rate at which species already on the island go extinct.

b. Effect of island size

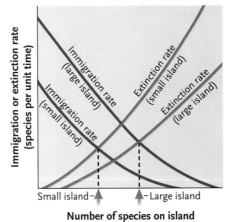

Immigration rates are higher and extinction rates lower on large islands than on small islands. Thus, at equilibrium, large islands have more species.

c. Effect of distance from mainland

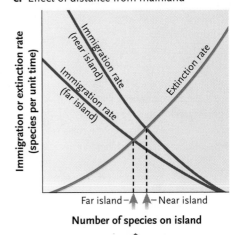

Organisms leaving the mainland locate nearby islands more easily than distant islands, causing higher immigration rates on near islands. Thus, near islands support more species than far ones.

Figure 46.30
Predictions of the theory of island biogeography. The x axes of the graphs are time.

According to MacArthur and Wilson's theory, an equilibrium between immigration and extinction determines the number of species that ultimately occupy an island. Once that equilibrium has been reached, the number of species remains relatively constant because one species already on the island becomes extinct in about the same time it takes a new one to arrive. The model does not specify which species immigrate or which ones already on the island become extinct. It simply predicts that the number of species on the island is in equilibrium, although species composition is not. The ongoing processes of immigration and extinction establish a constant turnover in the roster of species that live on any island.

The MacArthur–Wilson model also explains why some islands harbour more species than others. Large islands have higher immigration rates than small islands because they are larger targets for dispersing organisms. Moreover, large islands have lower extinction rates because they can support larger populations and provide a greater range of habitats and resources. At equilibrium, large islands have more species than small islands do (see Figure 46.30b). Islands near the mainland have higher immigration rates than distant islands because dispersing organisms are more likely to arrive at islands close to their point of departure. Distance does not affect extinction rates, so, at equilibrium, nearby islands have more species than distant islands (see Figure 46.30c).

The equilibrium theory's predictions about the effects of area and distance are generally supported by data on plants and animals **(Figure 46.31)**. Experimental work has verified some of its basic assumptions. Amy Schoener found that more than 200 species of marine organisms colonized tiny artificial "islands" (plastic kitchen scrubbers) within 30 days after she placed them in a Bahaman lagoon. Her research also confirmed that immigration rate increases with island size. Daniel Simberloff and Edward O. Wilson exterminated insects on tiny islands in the Florida Keys and monitored subsequent immigration and extinction (see *Experimenting with Islands*). Their research confirmed the equilibrium theory's predictions that an island's size and distance from the mainland influence how many species will occupy it.

The equilibrial view of species richness can also apply to mainland communities that exist as islands in a metaphorical sea of dissimilar habitat. Lakes are "islands" in a "sea" of dry land, and mountaintops are habitat "islands" in a "sea" of low terrain. Species richness in these communities is partly governed by the immigration of new species from distant sources and the extinction of species already present. As human activities disrupt environments across the globe, undisturbed sites function as islandlike refuges for threatened and endangered species. Conservation biologists apply the general lessons of MacArthur and Wilson's theory to the design of nature preserves (see Chapter 48).

The study of community ecology promises to keep biologists busy for some time to come.

STUDY BREAK

1. How does species richness change with increasing latitude?
2. In the island biogeography model proposed by MacArthur and Wilson, what processes govern the number of species on an island? What happens to the number of species once equilibrium is reached?
3. What effect do island size and distance from the mainland have on immigration and extinction of colonizing species?

PEOPLE BEHIND BIOLOGY

Bridget J. Stutchbury, York University

Bridget Stutchbury studies the behaviour and ecology of songbirds, working at sites in eastern North America (United States and Canada), as well as sites in the Neotropics. One aspect of her research is documenting the reproductive behaviour of birds. Although songbirds were thought to be monogamous over at least a breeding season, using genetic techniques, Dr. Stutchbury and others are discovering that both males and females often mate with a bird that is not their mate. This behaviour is called extrapair copulation if it is just mating or extrapair

fertilization when young result from the matings.

Using radio tracking to follow individual birds combined with DNA fingerprinting, Dr. Stutchbury and her colleagues were able to look at the movement patterns of Acadian Flycatchers (*Empidonax virescens*) and determine how far males and females travelled to meet their extrapair partners. Males travelled 50 to 1500 m from their nests to meet partners.

Work with other species such as Hooded Warblers (*Wilsonia citrina*) demonstrated that when these birds

lived in small forest fragments, their mating behaviours were disturbed compared with the behaviours of those nesting in larger tracts of forest.

Overall, her research has demonstrated that whereas some songbirds in eastern United States and Canada depend on corridors connecting habitat fragments, other species cross open habitats to use different patches of forest.

In 2007, her book *Silence of the Songbirds* reported declines in numbers of migrating songbirds and raised concerns about their future.

a. Distance effect

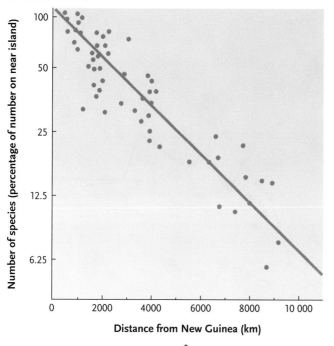

The number of lowland bird species on islands of the South Pacific declines with the islands' distance from the species source, the large island of New Guinea. Data in this graph were corrected for differences in the sizes of the islands. The number of bird species on each island is expressed as a percentage of the number of bird species on an island of equivalent size close to New Guinea.

b. Area effect

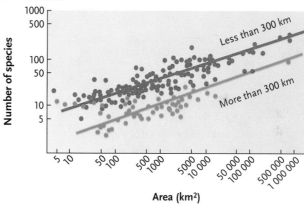

The number of bird species on tropical and subtropical islands throughout the world increases dramatically with island area. The data for islands near to a source and islands far from a mainland source are presented separately to minimize the effect of distance. Notice that the "distance effect" reduces the number of bird species on islands that are more than 300 km from a mainland source.

Figure 46.31

Factors that influence bird species richness on islands. (a) Evidence that fewer bird species colonize islands that are distant from the mainland source. **(b)** Evidence that more bird species colonize large islands than small ones.

Experimenting with Islands

Shortly after Robert MacArthur and Edward O. Wilson published the equilibrium theory of island biogeography in the 1960s, Wilson and Daniel Simberloff, one of Wilson's graduate students at Harvard University, undertook an ambitious experiment in community ecology. Simberloff reasoned that the best way to test the theory's predictions was to monitor immigration and extinction on barren islands.

Simberloff and Wilson devised a system for removing all the animals from individual red mangrove trees in the Florida Keys. The trees, with canopies that spread from 11 to 18 m in diameter, grow in shallow water and are isolated from their neighbours. Thus, each tree is an island that harbours an arthropod community. The species pool on the Florida mainland includes about 1000 species of arthropods, but each mangrove island contains no more than 40 species at one time.

After cataloguing the species on each island, Simberloff and Wilson hired an extermination company to erect large tents over each mangrove island and fumigate them to eliminate all arthropods on them **(Figure 1)**. The exterminators used methylbromide, a pesticide that does not harm trees or leave any residue. The tents were then removed.

Simberloff then monitored both the immigration of arthropods to the islands and the extinction of species that became established on them. He surveyed four islands regularly for two years and at intervals thereafter.

The results of this experiment confirm several predictions of MacArthur and Wilson's theory **(Figure 2)**. Arthropods rapidly recolonized the islands, and within eight or nine months, the number of species living on each island had reached an equilibrium that was close to the original species number. The island nearest the mainland had more species than the most distant island. However, immigration and extinction were rapid, and Simberloff and Wilson suspected that some species went extinct even before they had noted their presence. The researchers also discovered that three years after the experimental treatments, the species composition of the islands was still changing constantly and did not remotely resemble the species composition on the islands before they were defaunated.

Simberloff and Wilson's research was a landmark study in ecology because it tested the predictions of an important theory using a field experiment. Although such efforts are now almost routine in ecological studies, this project was one of the first to demonstrate that large-scale experimental manipulations of natural systems are feasible and that they often produce clear results.

Daniel Simberloff, University of Tennessee

Figure 1
After cataloguing the arthropods, Simberloff and Wilson hired an extermination company to eliminate all living arthropods.

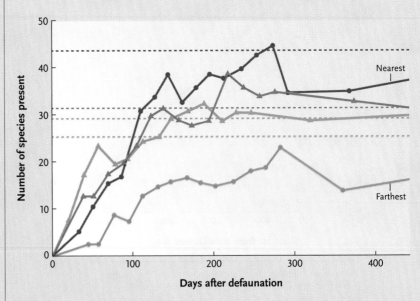

Figure 2
On three of four islands, species richness slowly returned to the predefaunation level (indicated by colour-coded dotted lines). The most distant island had not reached its predefaunation species richness after two years.

If solar slugs can take up chloroplasts and/or nematocysts and control them, could this be possible in chordates? What are the essential conditions? What are the limiting ones?

Review

Go to CENGAGENOW™ at http://hed.nelson.com/to access quizzing, animations, exercises, articles, and personalized homework help.

46.1 Interspecific Interactions

- Coevolution involves genetically based, reciprocal adaptations in two or more interacting species. Coevolution is not restricted to two species but often involves complex interactions among several species in a community.

46.2 Getting Food

- Predators eat animal prey, whereas herbivores eat plants. Predators and herbivores are animals with characteristics allowing them to feed efficiently. Predation and herbivory are the most conspicuous relationships in ecological communities.

- A koala is a specialist because it eats the leaves of only a few of the available species of *Eucalyptus*. Specialists tend to eat only a few types of food, whereas generalists take a broader diet.

- Optimal foraging theory predicts that an animal's diet is a compromise between the costs and the benefits associated with different types of food. Costs include the time and energy it takes to pursue, capture, and consume a particular kind of food. Benefits are the energy that the food provides.

46.3 Defence

- Size: the large size of elephants means that they have few natural predators.
- Eternal vigilance: meerkats are constantly on the lookout for potential predators.
- Avoiding detection: caterpillars that look like bird droppings are not recognized as edible.
- Counterattack: the sting of a bee or a scorpion, as well as the spines of porcupines and other mammals.
- Chemical defence: skunks spray a noxious chemical at potential predators.
- Warnings: black and white coloration of skunks and orange and black coloration of monarch butterflies.
- Mimicry: the harmless drone fly mimics the coloration and behaviour of the stinging bee or wasp.
- Animals using chemical defences either synthesize the chemicals themselves or sequester them from other sources. This can include plants that the organism eats.
- Aposematic displays teach would-be predators to avoid the signaller.
- Batesian mimicry occurs when an edible or harmless species mimics an inedible or a poisonous one. In Müllerian mimicry, two or more unpalatable or poisonous species have a similar appearance.

46.4 Competition

- Intraspecific competition can occur between two different species. Two types of competition are interference and exploitative. The competitive exclusion principle states that two or more species cannot coexist indefinitely if both rely on the same limiting resources and exploit them in the same way. One species will be able to harvest the available resources better and eventually outcompete the other species.

- A population's ecological niche is defined as the resources it uses and the environmental conditions it requires over its lifetime. A fundamental niche, larger than a realized niche, includes all conditions and resources a population can tolerate. A realized niche is the range of conditions and resources that a population actually encounters in nature.

- Resource partitioning occurs when sympatric species use different resources or the same resources in different ways. Plants may position their root systems at different levels, which avoids competition for water and nutrients.

- Character displacement results in sympatric species that differ in morphology and use different resources even though they would not do so in allopatric situations. An example of character displacement is the honey-eaters of Australia.

46.5 The Nature of Ecological Communities

Type of Interaction	Effect on Species Involved	Example
Commensalism	One species benefits; the other is unaffected (+/0)	Egrets and the large grazers that flush insects out of grasses during feeding
Mutualism	Both species benefit (+/+)	Bull's horn acacia tree and a species of small ants
Parasitism	One species benefits (parasite); the other is harmed (host) (+/−)	Ectoparasites such as mosquitoes and leeches and their mammalian hosts

- Two hypotheses about ecological communities have been developed by ecologists. The interactive hypothesis predicts that mature communities are at equilibrium and, if disturbed, will return to the predisturbed state. The individualistic hypothesis predicts that communities do not achieve equilibrium but rather are in a steady state of flux in response to disturbance and environmental change.

- Ecotones are generally species rich because they contain species from both communities, as well as species that occur only in transition zones.

46.6 Community Characteristics

- Indices allow population biologists to objectively compare the diversity of communities. Shannon's indices provide a measure of diversity (H') and evenness (E_H).
- Alpha (α) diversity is the number of species living in a single community. Beta (β) diversity is the number of species living in a collection of communities.
- Herbivores are primary consumers and form the second trophic level. Omnivores can be primary, secondary, and tertiary consumers (second, third, and fourth trophic levels, respectively) in a single meal.
- Generally, communities that support complex food webs are more stable. The disappearance of one or even two species does not have a major impact on the food web and thus community structure.

46.7 Effects of Population Interactions on Community Structure

- Species distribution and resource use in K-selected species are profoundly affected by competition. However, competition seems to have little effect on the community structure of r-selected species.
- Predation and herbivory can increase and/or decrease species richness, depending on the circumstances. Species richness can increase if a predator eliminates a strong competitor, allowing other organisms to exploit the available resources, for example, predatory sea stars reducing populations of mussels. Species richness can decrease when a predator eats less abundant species, further reducing their numbers.
- A keystone species has a much greater effect on the community than its numbers might suggest. Only a few individuals can have a profound impact on community structure.

46.8 Effects of Disturbance on Community Characteristics

- A community may never attain equilibrium because of disturbances such as cyclones, mortality caused by internal processes, and the recruitment of new colonies.
- The intermediate disturbance hypothesis states that species richness is greatest in communities experiencing fairly frequent disturbances of moderate intensity. Data gathered about a river system in New Zealand revealed that areas with moderate disturbance (e.g., moved rocks, soil, and sand in the streambed) had the highest species diversity.
- Generally, communities with a higher species richness recover from disturbance much more quickly than those with a low species richness.

46.9 Succession

- Primary succession begins when organisms first colonize habitats without soil, whereas secondary succession occurs after existing vegetation is destroyed or disrupted by an environmental disturbance.
- A climax community is a late successional stage that can be found in both primary and secondary succession. Climax communities are dominated by a few species that replace themselves and persist until a disturbance eliminates them. Species composition of a climax community is determined by local climate and soil conditions, surrounding vegetation, and chance events.
- The facilitation hypothesis holds that species modify the environment in a way that makes it less suitable for themselves but more suitable for those species that follow them in succession. The inhibition hypothesis contends that species currently occupying a successional stage prevent new species from occupying the same community. The tolerance hypothesis holds that early-stage species neither facilitate nor inhibit the growth of new species. Instead, succession proceeds because new species are able to outcompete and replace early-stage species.

46.10 Variance in Species Richness among Communities

- Species richness generally decreases with increasing latitude.
- The numbers of species on an island is governed by immigration of new species and extinction of species already there. Once equilibrium between immigration and extinction is reached, the number of species on an island remains relatively constant. As one species goes extinct, it is replaced by a newly arrived immigrant species.
- Large islands have higher immigration rates and lower extinction rates than small islands. Islands near the mainland have higher immigration rates than distant islands. Distance does not affect extinction rates. As a result, at equilibrium, near islands have more species than far islands.

Questions

Self-Test Questions

1. According to optimal foraging theory, predators
   a. always eat the largest prey possible.
   b. always eat the prey that are easiest to catch.
   c. choose prey based on the costs of consuming it compared to the energy it provides.
   d. eat plants when animal prey are scarce.
   e. have coevolved mechanisms to overcome prey defences.

2. Use of the same limiting resource by two species is called
   a. brood parasitism.
   b. interference competition.
   c. exploitative competition.
   d. mutualism.
   e. optimal foraging.

3. The range of resources that a population of one species can possibly use is called
   a. its fundamental niche.
   b. its realized niche.
   c. character displacement.
   d. resource partitioning.
   e. its relative abundance.

4. Differences in molar (tooth) structure of sympatric mammals may reflect
   a. predation.
   b. character displacement.
   c. mimicry.
   d. interference competition.
   e. cryptic coloration.

5. Bacteria that live in the human intestine assist human digestion and eat nutrients the human consumed. This relationship might best be described as
   a. commensalism.
   b. mutualism.
   c. endoparasitism.
   d. ectoparasitism.
   e. predation.

6. In the table below, the letters refer to five communities, and the numbers indicate how many individuals were recorded for each of five species. Which community has the highest species diversity?

	Species 1	Species 2	Species 3	Species 4	Species 5
a	90	10	0	0	0
b	80	10	10	0	0
c	25	25	25	25	0
d	2	4	6	8	80
e	20	20	20	20	20

7. A keystone species
   a. is usually a primary producer.
   b. has a critically important role in determining the species composition of its community.
   c. is always a predator.
   d. usually reduces the species diversity in a community.
   e. usually exhibits aposematic coloration.

8. Species richness can be highest in communities where disturbances are
   a. very frequent and severe.
   b. very frequent and of moderate intensity.
   c. very rare and severe.
   d. of intermediate frequency and moderate intensity.
   e. very rare frequency and mild.

9. The change in the species composition of a community from bare and lifeless rock to climax vegetation is called
   a. disturbance.
   b. competition.
   c. secondary succession.
   d. primary succession.
   e. facilitation.

10. The equilibrium theory of island biogeography predicts that the number of species found on an island
   a. increases steadily until it equals the number in the mainland species pool.
   b. is greater on large islands than on small ones.
   c. is smaller on islands near the mainland than on distant islands.
   d. can never reach an equilibrium number.
   e. is greater for islands near the equator than for islands near the poles.

Questions for Discussion

1. Many landscapes dominated by agricultural activities also have patches of forest of various sizes. What is the minimum amount of habitat required by different species? Focus on 10 species—5 animals and 5 plants. For each species, can you estimate minimum viable population?

2. Using the terms and concepts introduced in this chapter, describe the interactions that humans have with 10 other species. Try to choose at least eight species we do not eat.

3. After reading about the two potential biases in the scientific literature on competition, describe how future studies of competition might avoid such biases.

M.B. Fenton

Among the fastest growing ecosystem in the world—shopping malls and residential areas sprawl in the north part of London, Ontario.

47 Ecosystems

WHY IT MATTERS

As shown in the chapter opening photograph, the most rapidly growing habitat on the planet is urban ecosystems. They are replacing existing habitats at an astonishing rate partly because of growth in populations and in economies. The system portrayed in the photograph is low-density housing with services (water and electricity), but in many parts of the world, housing expansions are high density, with few, if any, services. Apart from humans and our domesticated plant and animal species, what components of the original flora and fauna persist? Walk around your neighbourhood and check it out.

How does the urban ecosystem differ from what was there before? In what ways does it differ? Think of runoff from rain and snow, of the heat-absorbing and reflecting properties of buildings, concrete, and asphalt. What are the effects of gardeners and landscapers, however well meaning? The urban ecosystem offers people in general, and biologists in particular, many opportunities for research and study. We must determine what changes we can effect in the construction and design of neighbourhoods to maximize their compatibility with native organisms. How can we make urban neighbourhoods more useful to migrating songbirds?

Archaeological evidence indicates that urban sprawl occurred around 6000 years before present around the site of Tell Brak in what is now northern Syria. Then the "city" that stood at Tell Brak occupied about 55 ha when other contemporary settlements rarely exceeded 3 ha and the largest of its neighbours was just 15 ha. There is evidence of spatial separation between subcommunities at Tell Brak, where neighbourhoods were divided by walls and limited points of access.

Urban sprawl may not be new, but the current scale makes it a frontier for action to achieve conservation of biodiversity. The purpose of this chapter is to explore some aspects of ecosystems and introduce them as objects of biological study.

47.1 Energy Flow and Ecosystem Energetics

Ecosystems receive a steady input of energy from an external source, usually the Sun. Energy flows through an ecosystem, but, as dictated by the laws of thermodynamics, much of that energy is lost without being used by organisms. In contrast, materials cycle between living and nonliving reservoirs, both locally and on a global scale. The flow of energy through and the cycling of materials around an ecosystem make resident organisms highly dependent on one another and on their physical surroundings.

Food webs define the pathways by which energy and nutrients move through an ecosystem's biotic components (see Chapter 3). In most ecosystems, nutrients and energy move simultaneously through a grazing food web and a detrital *food web* (**Figure 47.1**). The grazing food web includes the producer, herbivore, and secondary consumer trophic levels. The detrital food web includes detritivores and decomposers. Because detritivores and decomposers subsist on the remains and waste products of organisms at every trophic level, the two food webs are closely interconnected. Detritivores also contribute to the grazing food web when carnivores eat them.

All organisms in a particular trophic level are the same number of energy transfers from the ecosystem's ultimate energy source. Photosynthetic plants

Figure 47.1

Grazing and detrital food webs. Energy and nutrients move through two parallel food webs in most ecosystems. The grazing food web includes producers, herbivores, and carnivores. The detrital food web includes detritivores and decomposers. Each box in this diagram represents many species, and each arrow represents many arrows.

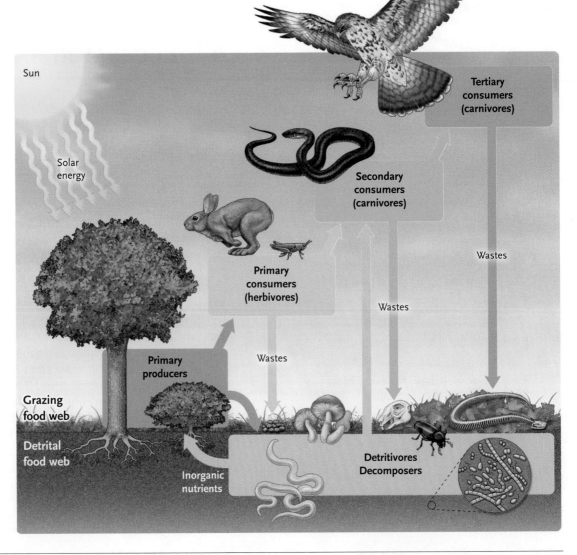

are one energy transfer removed from sunlight, herbivores (primary consumers) are two, secondary consumers are three, and tertiary consumers are four.

47.1a Primary Productivity: Fixing Carbon

Virtually all life on Earth depends on the input of solar energy. Every minute of every day, Earth's atmosphere intercepts roughly 80 kJ (kilojoules) of energy per square metre (see Chapter 1). About half of that energy is absorbed, scattered, or reflected by gases, dust, water vapour, and clouds before it reaches the planet's surface (see Chapter 3). Most energy reaching the surface falls on bodies of water or bare ground, where it is absorbed as heat or reflected back into the atmosphere. Reflected energy warms the atmosphere. Only a small percentage contacts primary producers, and most of that energy evaporates water, driving transpiration in plants (see Chapter 7).

Ultimately, photosynthesis converts less than 1% of the solar energy arriving at Earth's surface into chemical energy. But primary producers still capture enough energy to produce an average of several kilograms of dry plant material per square metre per year. On a global scale, they produce more than 150 billion tonnes of new biological material annually. Some of the solar energy that producers convert into chemical energy is transferred to consumers at higher trophic levels.

The rate at which producers convert solar energy into chemical energy is an ecosystem's **gross primary productivity.** But like other organisms, producers use energy for their own maintenance functions. After deducting energy used for these functions (see Chapter 6), whatever chemical energy remains is the ecosystem's **net primary productivity.** In most ecosystems, net primary productivity is 50 to 90% of gross primary productivity. In other words, producers use between 10 and 50% of the energy they capture for their own respiration.

Ecologists usually measure primary productivity in units of energy captured ($kJ \cdot m^{-2} \cdot yr^{-1}$) or in units of biomass created ($kg \cdot m^{-2} \cdot yr^{-1}$). **Biomass** is the dry weight of biological material per unit area or volume of habitat. (We measure biomass as the *dry* weight of organisms because their water content, which fluctuates with water uptake or loss, has no energetic or nutritional value.) Do not confuse an ecosystem's productivity with its **standing crop biomass**, the total dry weight of plants present at a given time. Net primary productivity is the *rate* at which the standing crop produces *new* biomass (see Chapter 7).

Energy captured by plants is stored in biological molecules, mostly carbohydrates, lipids, and proteins. Ecologists can convert units of biomass into units of energy or vice versa as long as they know how much carbohydrate, protein,

and lipid a sample of biological material contains. For reference, 1 g of carbohydrate and 1 g of protein each contains about 17.5 kJ of energy. Thus, net primary productivity indexes the rate at which producers accumulate energy as well as the rate at which new biomass is added to an ecosystem. Ecologists measure changes in biomass to estimate productivity because it is far easier to measure biomass than energy content. New biomass takes several forms, including

- growth of existing producers,
- creation of new producers by reproduction, and
- storage of energy as carbohydrates.

Because herbivores eat all three forms of new biomass, net primary productivity also measures how much new energy is available for primary consumers.

The potential rate of photosynthesis in any ecosystem is proportional to the intensity and duration of sunlight, which varies geographically and seasonally (see Chapters 3, 6, and 7). Sunlight is most intense and day length is least variable near the equator. In contrast, the intensity of sunlight is weakest and day length is most variable near the poles. This means that producers at the equator can photosynthesize nearly 12 hours a day, every day of the year, whereas near the poles, photosynthesis is virtually impossible during the long, dark winter. In summer, however, photosynthesis occurs virtually around the clock.

Sunlight is not the only factor influencing the rate of primary productivity. Temperature and availability of water and nutrients also affect this rate. Many of the world's deserts receive plenty of sunshine but have low rates of productivity because water is in short supply and the soil is poor in nutrients. Mean annual primary productivity varies greatly on a global scale **(Figure 47.2)**, reflecting variations in these environmental factors (see Chapter 3).

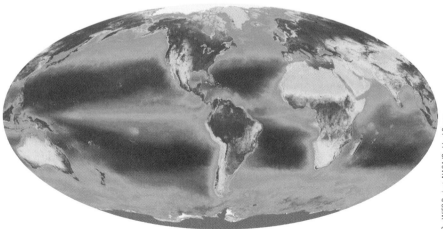

Figure 47.2

Global variation in primary productivity. Satellite data from 2002 provide a visual portrait of net primary productivity across Earth's surface. High-productivity regions on land are dark green; low-productivity regions are yellow. For aquatic environments, the highest productivity is red, down through orange, yellow, green, blue, and purple (lowest).

On a finer geographic scale, within a particular terrestrial ecosystem, mean annual net productivity often increases with the availability of water **(Figure 47.3)**. In systems with sufficient water, a shortage of mineral nutrients may be limiting. All plants need specific ratios of macronutrients and micronutrients for maintenance and photosynthesis (see Chapter 7). But plants withdraw nutrients from soil, and if nutrient concentration drops below a critical level, photosynthesis may decrease or stop altogether. In every ecosystem, one nutrient inevitably runs out before the supplies of other nutrients are exhausted. The element in shortest supply is called a **limiting nutrient** because its absence curtails productivity. Productivity in agricultural fields is subject to the same constraints as productivity in natural ecosystems. Farmers increase productivity by irrigating (adding water to) and fertilizing (adding nutrients to) their crops.

In freshwater and marine ecosystems, where water is always readily available, the depth of water

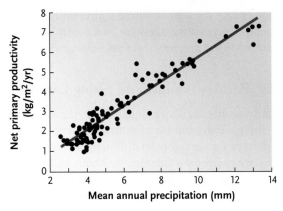

Figure 47.3

Water and net primary productivity. Mean annual precipitation among 100 sites in the Great Plains of North America. These data include only above-ground productivity.

and combined availability of sunlight and nutrients govern the rate of primary productivity. Productivity is high in near-shore ecosystems, where sunlight penetrates shallow, nutrient-rich waters. Kelp beds and coral reefs along temperate and tropical marine coastlines, respectively, are among the most productive ecosystems on Earth (**Table 47.1**; see also Figure 47.2). In contrast, productivity is low in the open waters of a large lake or ocean. There sunlight penetrates only the upper layers, and nutrients sink to the bottom; thus, the two requirements for photosynthesis—sunlight and nutrients—are available in different places.

Although ecosystems vary in their rates of primary productivity, these differences are not always proportional to variations in their standing crop biomass (see Table 47.1). For example, biomass amounts in temperate deciduous forests and **temperate grasslands** differ by a factor of 20, but the difference in their rates of net primary productivity is much smaller. Most biomass in trees is present in nonphotosynthetic tissues such as wood, so their ratio of productivity to biomass is low ($12 \text{ kg·m}^{-2}/300 \text{ kg·m}^{-2} = 0.04$). By contrast, grasslands do not accumulate much biomass because annual mortality, herbivores, and fires remove plant material as it is produced, so their productivity to biomass ratio is much higher ($6.0 \text{ kg·m}^{-2}/16 \text{ kg·m}^{-2} = 0.375$).

Some ecosystems contribute more than others to overall net primary productivity **(Figure 47.4)**. Ecosystems covering large areas make substantial total contributions, even if their productivity per unit area is low. Conversely, geographically restricted ecosystems make large contributions if their productivity is high. Open ocean and **tropical rain forests** contribute about equally to total global productivity, but for different reasons. Open oceans have low productivity, but they cover nearly two-thirds of Earth's surface. Tropical rain forests are highly productive but cover only a relatively small area.

Net primary productivity ultimately supports all consumers in grazing and detrital food webs. Consumers

Ecosystem	Mean Standing Crop Biomass (kg/m^2)	Mean Net Primary Productivity (kg/m^2/y^1)
Table 47.1 — Standing Crop Biomass and Net Primary Productivity of Different Ecosystems		
Terrestrial Ecosystems		
Tropical rain forest	450	22.0
Tropical deciduous forest	350	16.0
Temperate rain forest	350	13.0
Temperate deciduous forest	300	12.0
Savanna	40	9.0
Boreal forest (taiga)	200	8.0
Woodland and shrubland	60	7.0
Agricultural land	10	6.5
Temperate grassland	16	6.0
Tundra and alpine tundra	6.0	1.4
Desert and thornwoods	7.0	0.9
Extreme desert, rock, sand, ice	0.2	0.03
Freshwater Ecosystems		
Swamp and marsh	150	20
Lake and stream	0.2	2.5
Marine Ecosystems		
Open ocean	0.03	1.3
Upwelling zones	0.2	5.0
Continental shelf	0.1	3.6
Kelp beds and reefs	20	25
Estuaries	10	15
World Total	**36**	**3.3**

From Whittaker, R.H. 1975. *Communities and Ecosystems*. 2nd ed. Macmillan.

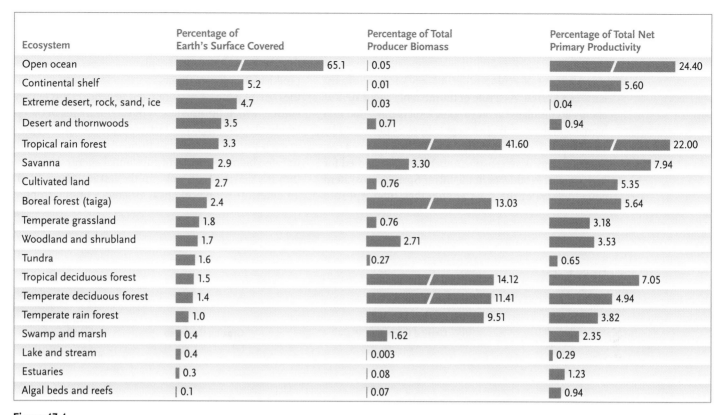

Figure 47.4

Biomass and net primary productivity. An ecosystem's percentage coverage of Earth's surface is not proportional to its contribution to total biomass of producers or its contribution to the total net primary productivity.

Ecosystem	Percentage of Earth's Surface Covered	Percentage of Total Producer Biomass	Percentage of Total Net Primary Productivity
Open ocean	65.1	0.05	24.40
Continental shelf	5.2	0.01	5.60
Extreme desert, rock, sand, ice	4.7	0.03	0.04
Desert and thornwoods	3.5	0.71	0.94
Tropical rain forest	3.3	41.60	22.00
Savanna	2.9	3.30	7.94
Cultivated land	2.7	0.76	5.35
Boreal forest (taiga)	2.4	13.03	5.64
Temperate grassland	1.8	0.76	3.18
Woodland and shrubland	1.7	2.71	3.53
Tundra	1.6	0.27	0.65
Tropical deciduous forest	1.5	14.12	7.05
Temperate deciduous forest	1.4	11.41	4.94
Temperate rain forest	1.0	9.51	3.82
Swamp and marsh	0.4	1.62	2.35
Lake and stream	0.4	0.003	0.29
Estuaries	0.3	0.08	1.23
Algal beds and reefs	0.1	0.07	0.94

in the grazing food web eat some biomass at every trophic level except the highest. Uneaten biomass eventually dies and passes into detrital food webs. Moreover, consumers assimilate only a portion of the material they ingest, and unassimilated material is passed as feces that also supports detritivores and decomposers.

47.1b Secondary Productivity: Moving Up the Trophic Scale

As energy is transferred from producers to consumers, some is stored in new consumer biomass, called **secondary productivity.** Nevertheless, two factors cause energy to be lost from the ecosystem every time it flows from one trophic level to another. First, animals use much of the energy they assimilate for maintenance and locomotion rather than for production of new biomass. Second, as dictated by the second law of thermodynamics, no biochemical reaction is 100% efficient, so some of the chemical energy liberated by cellular respiration is converted to heat, which most organisms do not use.

47.1c Ecological Efficiency: Use of Energy

Ecological efficiency is the ratio of net productivity at one trophic level to net productivity at the trophic level below. If plants in an ecosystem have a net primary productivity of $1.0 \text{ kg} \cdot \text{m}^{-2} \cdot \text{y}^{-1}$ of new tissue, and the herbivores that eat those plants produce 0.1 kg of new tissue $\text{m}^{-2} \cdot \text{y}^{-1}$, the ecological efficiency of the herbivores is 10%. The efficiencies of three processes (harvesting food, assimilating ingested energy, and producing new biomass) determine the ecological efficiencies of consumers.

Harvesting efficiency is the ratio of the energy content of food consumed compared with the energy content of food available. Predators harvest food efficiently when prey are abundant and easy to capture (see Chapter 46).

Assimilation efficiency is the ratio of the energy absorbed from consumed food to the total energy content of the food. Because animal prey is relatively easy to digest, carnivores absorb between 60 and 90% of the energy in their food. Assimilation efficiency is lower for prey with indigestible parts such as bones or exoskeletons. Herbivores assimilate only 15 to 80% of the energy they consume because cellulose is not very digestible. Herbivores lacking cellulose-digesting systems are on the low end of the scale, whereas those that can digest cellulose are at the higher end.

Production efficiency is the ratio of the energy content of new tissue produced to the energy assimilated from food. Production efficiency varies with maintenance costs. Endothermic animals often use less than 10% of their assimilated energy for growth and reproduction because they use energy to generate body heat (see Chapter 43). Ectothermic animals channel more than 50% of their assimilated energy into new biomass.

The overall ecological efficiency of most organisms is 5 to 20%. As a rule of thumb, only about 10% of energy accumulated at one trophic level is converted into biomass at the next higher trophic level, as illustrated by energy transfers at Silver Springs, Florida **(Figure 47.5).** Silver Springs is an ecosystem that has been studied for many years. Producers in the Silver Springs ecosystem convert 1.2% of the solar energy they intercept into chemical energy (represented by 86 986 kJ·m^{-2}·yr^{-1} of gross primary productivity). However, they use about two-thirds of this energy for respiration, leaving one-third to be included in new plant biomass, net primary productivity. All consumers in the grazing food web (on the right in Figure 47.5) ultimately depend on this energy source, which diminishes with each transfer between trophic levels. Energy is lost to respiration and export at each trophic level. In addition, organic wastes and uneaten biomass represent substantial energy that flows into the detrital food web (on the left in Figure 47.5). To determine the ecological efficiency of any trophic

Figure 47.5
Energy flow through the Silver Springs ecosystem.

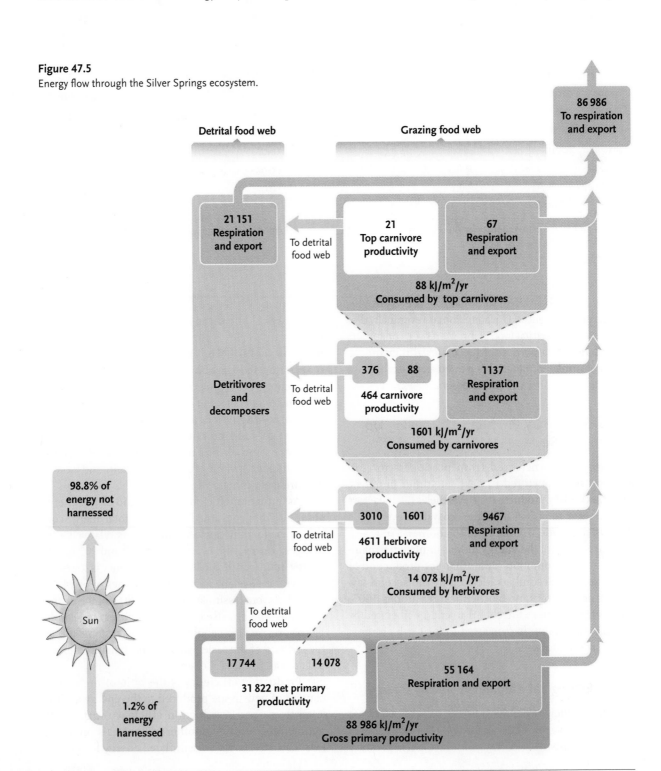

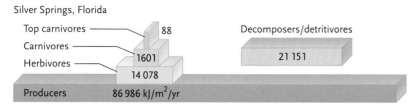

Figure 47.6

Pyramids of energy. The pyramid of energy for Silver Springs, Florida, shows that the amount of energy (kJ·m⁻²·yr⁻¹) passing through each trophic level decreases as it moves up the food web.

level, we divide its productivity by the productivity of the level below it. The ecological efficiency of midlevel carnivores at Silver Springs is 10.06%, 464 kJ·m⁻²·yr⁻¹/4611 kJ·m⁻²·yr⁻¹.

47.1d Pyramids: Energy, Biomass, and Numbers

As energy works its way up a food web, energy losses are multiplied in successive energy transfers, greatly reducing the energy available to support the highest trophic levels (see Figure 47.5). Consider a hypothetical example in which ecological efficiency is 10% for all consumers. Assume that the plants in a small field annually produce new tissues containing 100 kJ of energy. Because only 10% of that energy is transferred to new herbivore biomass, the 100 kJ in plants produces 10 kJ of new herbivorous insects, 1 kJ of new songbirds that eat insects, and only 0.1 kJ of new falcons that eat songbirds. About 0.1% of the energy from primary productivity remains after three trophic levels of transfer. If the energy available to each trophic level is depicted graphically, the result is a **pyramid of energy**, with primary producers on the bottom and higher level consumers on the top **(Figure 47.6).**

The low ecological efficiencies that characterize most energy transfers illustrate one advantage of eating "lower on the food chain." Even though humans digest and assimilate meat more efficiently than we do vegetables, we might be able to feed more people if we all ate more primary producers directly instead of first passing them through another trophic level, such as cattle or chickens, to produce meat. Production of animal protein is costly because much of the energy fed to livestock is used for their own maintenance rather than production of new biomass. But despite the economic and health-related logic of a more vegetarian diet, change in our eating habits alone will not eliminate food shortages or the frequency of malnutrition. Many regions of Africa, Australia, North America, and South America support vegetation that is suitable only for grazing by large herbivores. These areas could not produce significant quantities of edible grains and vegetables without significant additions of water and fertilizer (see Chapter 49).

Inefficiency of energy transfer from one trophic level to the next has profound effects on ecosystem structure. Ecologists illustrate these effects in diagrams called **ecological pyramids.** Trophic levels are drawn as stacked blocks, with the size of each block proportional to the energy, biomass, or numbers of organisms present. Pyramids of energy typically have wide bases and narrow tops (see Figure 47.6) because each trophic level contains only about 10% as much energy as the trophic level below it.

Progressive reduction in productivity at higher trophic levels usually establishes a **pyramid of biomass (Figure 47.7).** The biomass at each trophic level is proportional to the amount of chemical energy temporarily stored there. Thus, in terrestrial ecosystems, the total mass of producers is generally greater than the total mass of herbivores, which is, in turn, greater than the total mass of predators (see **Figure 47.7a**). Populations of top predators, from killer whales to lions and crocodiles, contain too little biomass and energy to support another trophic level; thus, they have no nonhuman predators.

Freshwater and marine ecosystems sometimes exhibit inverted pyramids of biomass (see **Figure 47.7b**). In the open waters of a lake or ocean, primary consumers (zooplankton) eat primary producers (phytoplankton) almost as soon as they are produced. As a result, the standing crop of primary consumers at any moment in time is actually larger than the standing crop of primary producers. Food webs

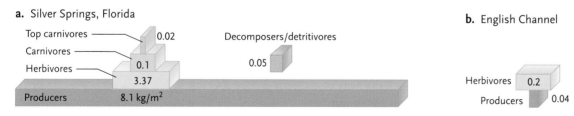

Figure 47.7

Pyramids of biomass. **(a)** The pyramid of standing crop biomass for Silver Springs is bottom heavy, as it is for most ecosystems. **(b)** Some marine ecosystems, such as that in the English Channel, have an inverted pyramid of biomass because producers are quickly eaten by primary consumers. Only the producer and herbivore trophic levels are illustrated here. The data for both pyramids are given in kg·m⁻² of dry biomass.

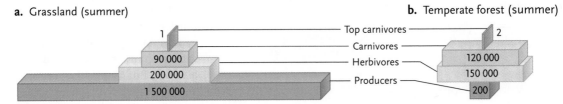

a. Grassland (summer)

b. Temperate forest (summer)

Top carnivores	1	2
Carnivores	90 000	120 000
Herbivores	200 000	150 000
Producers	1 500 000	200

Figure 47.8

Pyramids of numbers. **(a)** The pyramid of numbers (numbers of individuals per 1000 m²) for temperate grasslands is bottom heavy because individual producers are small and very numerous. **(b)** The pyramid of numbers for forests may have a narrow base because herbivorous insects usually outnumber the producers, many of which are large trees. Data for both pyramids were collected in summer. Detritivores and decomposers (soil animals and microorganisms) are not included because they are difficult to count.

in these ecosystems are stable because producers have exceptionally high **turnover rates.** In other words, producers divide and their populations grow so quickly that feeding by zooplankton does not endanger their populations or reduce their productivity. However, on an annual basis, the *cumulative total* biomass of primary producers far outweighs that of primary consumers.

The reduction of energy and biomass affects population sizes of organisms at the top of a food web. Top predators can be relatively large animals, so the limited biomass present in the highest trophic levels is concentrated in relatively few animals **(Figure 47.8).** The extremely narrow top of this **pyramid of numbers** has grave implications for conservation biology (see Chapter 48). Top predators tend to be large animals with small population sizes. And because each individual must patrol a large area to find sufficient food, members of a population are often widely dispersed within their habitats. As a result, they are subject to genetic drift (see Chapter 18) and are highly sensitive to hunting, habitat destruction, and random events that can lead to extinction. Top predators may also suffer from the accumulation of poisonous materials that move through food webs (see the next section). Even predators that feed below the top trophic level often suffer the ill effects of human activities. Consumers sometimes regulate ecosystem processes.

Numerous abiotic factors, such as the intensity and duration of sunlight, rainfall, temperature, and the availability of nutrients, have significant effects on primary productivity. Primary productivity, in turn, profoundly affects the populations of herbivores and predators that feed on them. But what effect does feeding by these consumers have on primary productivity?

Consumers sometimes influence rates of primary productivity, especially in ecosystems with low species diversity and relatively few trophic levels. Food webs in lake ecosystems depend primarily on the productivity of phytoplankton **(Figure 47.9).** Phytoplankton are, in turn, eaten by herbivorous zooplankton, themselves consumed by predatory invertebrates and fishes. The top nonhuman carnivore in these food webs is usually a predatory fish.

Herbivorous zooplankton play a central role in regulation of lake ecosystems. Small zooplankton species

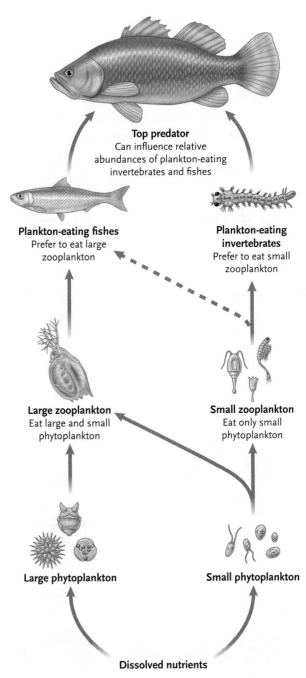

Top predator
Can influence relative abundances of plankton-eating invertebrates and fishes

Plankton-eating fishes
Prefer to eat large zooplankton

Plankton-eating invertebrates
Prefer to eat small zooplankton

Large zooplankton
Eat large and small phytoplankton

Small zooplankton
Eat only small phytoplankton

Large phytoplankton

Small phytoplankton

Dissolved nutrients

Figure 47.9

Consumer regulation of primary productivity. A simplified food web illustrates that lake ecosystems have relatively few trophic levels. The effects of feeding by top carnivores can cascade downward, exerting an indirect effect on the phytoplankton and thus on primary productivity.

Fishing Fleets at Loggerheads with Sea Turtles

Populations of loggerhead sea turtles (*Caretta caretta*) that nest on Western Pacific beaches in Australia and Japan have been in decline. Like other sea turtles, *C. caretta* hatch from eggs that females bury on sandy beaches. Immediately after hatching, the young turtles rush to the surf and the open ocean. Turtles mature at sea and return to their hatching beaches to lay eggs. Using mitochondrial DNA (mtDNA), Bruce Bowen and colleagues explored the situation sea turtles face.

The researchers took mtDNA samples from nesting populations in Australia and Japan, from populations of turtles feeding in Baja California, and from turtles drowned in fishing nets in the north Pacific. One 350-base-pair segment of mtDNA included sequence variations that are characteristic of different loggerhead populations. After samples were amplified by the polymerase chain reaction, sequencing revealed three major variants of mtDNA, which the researchers designated sequences A, B, and C. The sequences were distributed among loggerhead turtles, as shown in **Table 1.**

The mtDNA of most *C. caretta* found in Baja California and in fishing nets in the north Pacific matched that of turtles from the Japanese nesting areas. These data support the idea that loggerhead turtles hatched in Japan make the 10 000-km long migration across the North Pacific to Baja California. The data also indicate that a few turtles that hatched in Australia may follow the same migratory route.

This migration could be aided by the North Pacific Current, which moves from west to east, whereas the return trip from Baja to Japan could be made via the North Equatorial Current that runs from east to west just north of the equator. Loggerhead turtles have been found in these currents, and further tests will reveal whether they have the mtDNA sequence characteristic of the individuals nesting in Japan and feeding in Baja California.

The nesting population of *C. caretta* in Japan is 2000 to 3000 females. It is uncertain if this population can survive the loss of thousands of offspring to fishing in the North Pacific. The number of female loggerhead turtles nesting in Australia has declined by 50 to 80% in the last decade, so the loss of only a few individuals in fishing nets could also have a drastic impact on this population. To save the loggerhead turtles, wildlife managers and international agencies must establish and enforce limits on the number of migrating individuals trapped and killed in the ocean fisheries.

Like other sea turtles, the loggerheads are severely impacted by fishing. As many as 4000 loggerheads drown in nets every year, and others are caught in longline fisheries. Adoption of a new fish hook **(Figure 1)** could reduce the turtle catch in longline fishing operations.

Table 1	Sources of Turtles by Nesting Grounds		
Number of Turtles			
Location	**Sequence A**	**Sequence B**	**Sequence C**
Australian nesting areas	26	0	0
Japanese nesting areas	0	23	3
Baja California feeding grounds	2	19	5
North Pacific	1	28	5

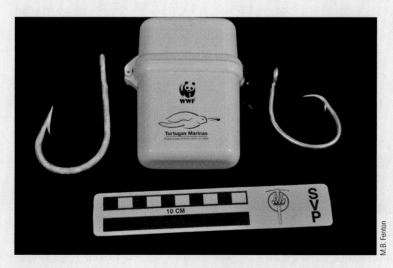

Figure 1
Sea turtles and longlining. Conventional longline hooks (left) readily catch sea turtles, whereas the hook on the right does not. The World Wildlife Fund is promoting the use of hooks that are friendly to sea turtles in an effort to reduce their losses to longline fishing.

consume only small phytoplankton. Thus, when *small* zooplankton are especially abundant, large phytoplankton escape predation and survive, and the lake's primary productivity is high. By contrast, large zooplankton are voracious, eating both small and large phytoplankton. When large zooplankton are especially abundant, they reduce the overall biomass of phytoplankton, lowering the ecosystem's primary productivity.

In this **trophic cascade**, predator–prey effects reverberate through population interactions at two or more trophic levels in an ecosystem. Feeding by plankton-eating invertebrates and fishes has a *direct* impact on herbivorous zooplankton populations and an *indirect* impact on phytoplankton populations (the ecosystem's primary producers). Invertebrate predators prefer small zooplankton. And when the invertebrates that eat small zooplankton are the dominant predators in the ecosystem, large zooplankton become more abundant; they consume many phytoplankton, causing a decrease in productivity. But plankton-eating fishes prefer to eat large zooplankton (see Figure 47.9), so when they are abundant, small zooplankton become the dominant herbivores, leading large phytoplankton to become more numerous, which raises the lake's productivity.

Large predatory fishes may add an additional level of control to the system because they feed on and regulate the population sizes of plankton-eating invertebrates and fishes. Thus, the effects of feeding by the top predator can cascade downward through the food web, affecting the densities of plankton-eating invertebrates and fishes, herbivorous zooplankton, and phytoplankton. Research

in Norway with brown trout (*Salmo trutta*), a top predator, and Arctic char (*Salvelinus alpinus*), the prey, demonstrated how culling prey can promote the recovery of top predators. In this case, Lake Takvatn was the scene of a large-scale experiment. Older, stunted prey species (*S. alpinus*) were removed. These fish had eaten small prey, so in an increase in the availability of prey and recovery of the predator resulted. In this case, *S. trutta* was the top predator, and *S. alpinus,* an introduced species, was culled to rejuvenate the system. Another process of bioremediation, the addition of piscivorous fish to a lake, also has been successful in restoring ecosystem balance in other parts of the world.

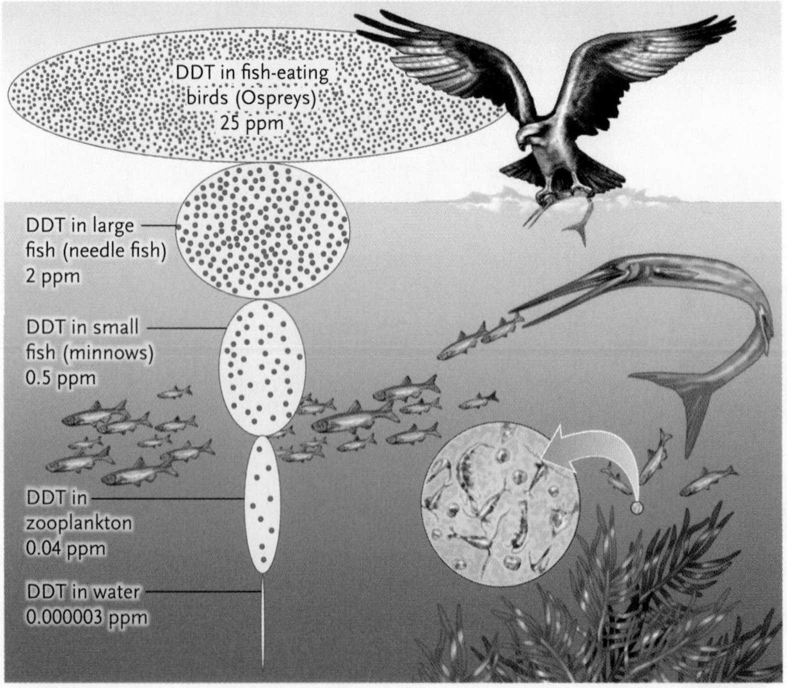

47.1e Biological Magnification: Movement of Contaminants Up the Food Chain

Ironically, DDT (a formerly popular insecticide; see *Molecule Behind Biology*) provided a clear demonstration of the interconnectedness of organisms. Consumers accumulate DDT from all the organisms they eat in their lifetimes. Primary consumers, such as herbivorous insects, may ingest relatively small amounts of DDT, but a songbird that eats many of these insects accumulates all the collected DDT consumed by its prey. A predator such as a raptor, perhaps a Sharp-shinned Hawk (*Accipiter striatus*) that eats songbirds, accumulates even more. Whether the food chain (web) is aquatic or terrestrial, the net effect on higher level consumers is the same **(Figure 47.10)**.

Natural systems have provided many examples of biological magnification. In cities where DDT was used in an effort to control the spread of Dutch elm disease, songbirds died from DDT poisoning after eating insects that had been sprayed (whether or not they were involved in spreading the disease). In forests, DDT was used in an effort to control spruce budworm moths (*Choristoneura occidentalis*), and salmon died because runoff carried DDT into their streams and rivers, where their herbivorous prey consumed it.

Despite the ban on the use of DDT in the United States in 1973, in 1990, the California State Department of Health recommended closing a fishery off the coast of California because of DDT accumulating there. DDT discharged in industrial waste 20 years earlier was still moving through the ecosystem. The half-life of DDT in an organism's body fat is eight years.

Other contaminants emulate DDT. Mercury contamination is common in many parts of the world, often as a by-product of the pulp and paper industry. Minamata, the disease humans get from mercury poisoning, is usually linked to the consumption of fish taken from contaminated watersheds. Eating fish contaminated with mercury can result in mercury concentrations in people's hair (0.9 to 94 mg $\times$ kg^{-1}) and otters (*Lontra canadensis*; 0.49 to 54.37 mg $\times$ kg^{-1}). In southern Ontario, the hair of bats that eat insects that emerge from mercury-contaminated sediments contains

Figure 47.10

Biological magnification. In this marine food web in northeastern North America, DDT concentration (measured in parts per million, ppm) was magnified nearly 10 million times between zooplankton and the Osprey (*Pandion haliaetus*).

DDT in fish-eating birds (Ospreys) 25 ppm

DDT in large fish (needle fish) 2 ppm

DDT in small fish (minnows) 0.5 ppm

DDT in zooplankton 0.04 ppm

DDT in water 0.000003 ppm

DDT: Dichloro-Diphenyl-Trichloroethane

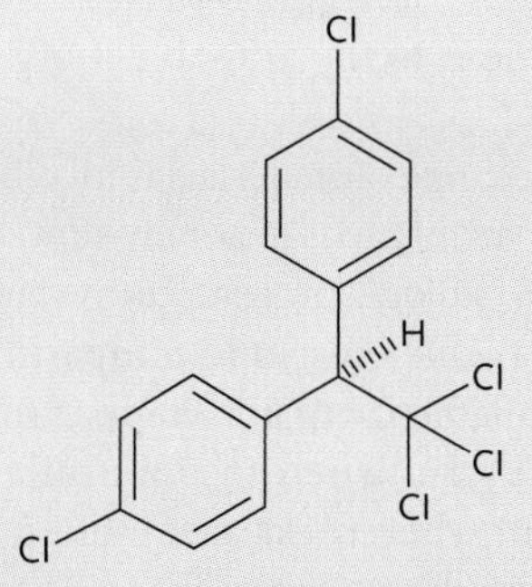

Figure 1
A molecule of DDT.

Originally formulated in 1873, DDT's potential as an insecticide was only recognized in 1939 by Paul Muller of Geigy Pharmaceutical in Switzerland. DDT, the first of the chlorinated insecticides, was used extensively in some theatres of World War II, notably in Burma (now Myanmar) in 1944, when the Japanese forces were on the brink of moving into India. There Allied forces suffered from "three m's": mud, morale, and malaria. Meanwhile, in 1943 in southern Italy, DDT was instrumental in controlling populations of lice that plagued Canadian troops there. Widespread application of DDT in Burma reduced the incidence of malaria by killing mosquitoes, the vec-

tors for the disease (see Chapter 26). After World War II, the use of DDT spread rapidly, and the World Health Organization (WHO) credited this molecule with saving 25 million human lives (mainly through control of mosquitoes that carry malaria).

At first, DDT appeared to be an ideal insecticide. In addition to being inexpensive to produce, it had low toxicity to mammals (300 to 500 $mg \cdot kg^{-1}$ is the LD_{50}, the amount required to kill half of the target population). But many insects subsequently developed immunity to DDT, reducing its effectiveness.

DDT is chemically stable and soluble in fat, so instead of being metabolized by mammals, it is stored in their fat. The biological half-life of DDT is approximately eight years (it takes about eight years for a mammal to metabolize half of the amount of DDT it has assimilated). DDT is released when fat is metabolized, so when mammals metabolize fat (for example, when humans go on a diet), they are exposed to higher concentrations of DDT in their blood. DDT also had dramatic effects on some birds, notably those higher up the food chain. Eggshell thinning was a consequence

of exposure to DDT. Populations of birds such as peregrine falcons (*Falco peregrinus*) plummeted.

Since 1985, the use of DDT has been totally banned in Canada, and it is now banned in many other countries. But DDT is still produced in counties such as the United States and still used in countries where malaria is a prominent problem because the ecological costs of DDT are considered secondary to the importance of controlling the mosquitoes. WHO estimates that every 30 seconds, a child dies of malaria. Approximately 40% of the world's population of humans is at risk of contracting malaria where they live, mainly in Africa. Malaria also remains a problem in tropical and subtropical Asia and Central and South America. People in southern Europe and the Middle East may also be at risk.

By the early 1970s, cetaceans in the waters around Antarctica had DDT in their body fat even though DDT had never been used there. The movement of DDT up the food chain and through food webs demonstrated the interconnections in biological systems. The movement of DDT also provides a graphic demonstration of the transfer of materials from one trophic level to another.

---

concentrations up to $13 \text{ mg} \times \text{kg}^{-1}$. Fish obviously are not essential to this chain of biomagnification.

STUDY BREAK

1. What is net primary productivity? How does it differ from standing crop biomass? Are the pyramids useful?
2. Many deserts have low levels of productivity, yet they receive a lot of sunlight. Why?
3. What are assimilation efficiency and production efficiency?

47.2 Nutrient Cycling in Ecosystems

The availability of nutrients is as important to ecosystem function as the input of energy. Photosynthesis requires carbon, hydrogen, and oxygen,

which producers acquire from water and air. Primary producers also need nitrogen, phosphorus, and other minerals (see Chapter 41). A deficiency in any of these minerals can reduce primary productivity.

Earth is essentially a closed system with respect to matter, even though cosmic dust enters the atmosphere. Thus, unlike energy, for which there is a constant cosmic input, virtually all the nutrients that will ever be available for biological systems are already present. Nutrient ions or molecules constantly circulate between the abiotic environment and living organisms in **biogeochemical cycles.** And unlike energy, which flows through ecosystems and is gradually lost as heat, matter is conserved in biogeochemical cycles. Although there may be local shortages of specific nutrients, Earth's overall supplies of these chemical elements are never depleted or increased.

Nutrients take various forms as they pass through biogeochemical cycles. Materials such as carbon, nitrogen, and oxygen form gases that move through global *atmospheric cycles*. Geologic processes move other materials, such as phosphorus, through local *sedimentary cycles,* carrying them between dry land and the seafloor. Rocks, soil, water, and air are the reservoirs where mineral nutrients accumulate, sometimes for many years.

Ecologists use a **generalized compartment model** to describe nutrient cycling **(Figure 47.11)**. Two criteria divide ecosystems into four compartments in which nutrients accumulate. First, nutrient molecules and ions are either *available* or *unavailable,* depending on whether they can be assimilated by organisms. Second, nutrients are present either in *organic* material, living or dead tissues of organisms, or *inorganic* material, such as rocks and soil. Minerals in dead leaves on the forest floor are in the available-organic compartment because they are in the remains of organisms that can be eaten by detritivores. Calcium ions in limestone rocks are in the unavailable-inorganic compartment because they are in a nonbiological form that producers cannot assimilate.

Nutrients move rapidly within and between the available compartments. Living organisms are in the available-organic compartment, and whenever heterotrophs consume food, they recycle nutrients within that reservoir (indicated by the circular arrow in the upper left of Figure 47.11). Producers acquire nutrients from the air, soil, and water of the available-inorganic compartment. Consumers acquire nutrients from the available-inorganic compartment when they drink water or absorb mineral ions through their integument. Several processes routinely transfer nutrients from organisms to the available-inorganic compartment. Respiration releases carbon dioxide, moving both carbon and oxygen from the available-organic compartment to the available-inorganic compartment.

By contrast, the exchange of materials into and out of the unavailable compartments is generally slow. Sedimentation, a long-term geologic process, converts ions and particles of the available-inorganic compartment into rocks of the unavailable-inorganic compartment. Materials are gradually returned to the available-inorganic compartment when rocks are uplifted and eroded or weathered. Similarly, over millions of years, the remains of organisms in the available-organic compartment were converted into the coal, oil, and peat of the unavailable-organic compartment.

Except for the input of solar energy, we have described energy flow and nutrient cycling as though ecosystems were closed systems. In reality, most ecosystems exchange energy and nutrients with neighbouring ecosystems. Rainfall carries nutrients into a forest ecosystem, and runoff carries nutrients from a forest into a lake or river. Ecologists have mapped biogeochemical cycles of important elements, often by using radioactively labelled molecules that they can follow in the environment.

47.2a Water: Staff of Life

Although it is not a mineral nutrient, water is the universal intracellular solvent for biochemical reactions. Nevertheless, only a fraction of 1% of Earth's total water is present in biological systems at any time.

The cycling of water, the **hydrogeologic cycle**, is global, with water molecules moving from oceans into the atmosphere, to land, through freshwater ecosystems, and back to the oceans **(Figure 47.12)**. Solar energy causes water to evaporate from oceans, lakes, rivers, soil, and living organisms, entering the atmosphere as a vapour and remaining aloft as a gas, as droplets in clouds, or as ice crystals. Water falls as precipitation, mostly in the form of rain and snow. When precipitation falls on land, water flows across the surface or percolates to great depths in soil, eventually reentering the ocean reservoir through the flow of streams and rivers.

The hydrogeologic cycle maintains its global balance because the total amount of water entering the atmosphere is equal to the amount that falls as precipitation. Most water that enters the atmosphere evaporates from the oceans, which are the largest reservoir of water on the planet. A much smaller fraction evaporates from terrestrial ecosystems, and most of that is through transpiration by green plants.

Constant recirculation provides fresh water to terrestrial organisms and maintains freshwater ecosystems such as lakes and rivers. Water also serves as a

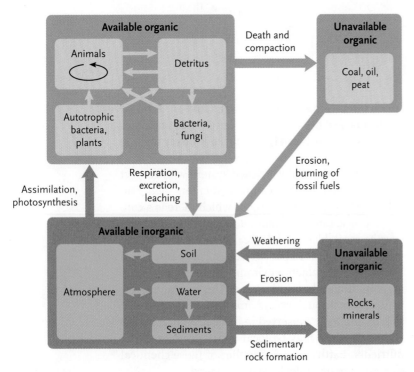

Figure 47.11
A generalized compartment model of nutrient cycling. Nutrients cycle through four major compartments within ecosystems. Processes that move nutrients from one compartment to another are indicated on the arrows. The circular arrow under "Animals" represents animal predation on other animals.

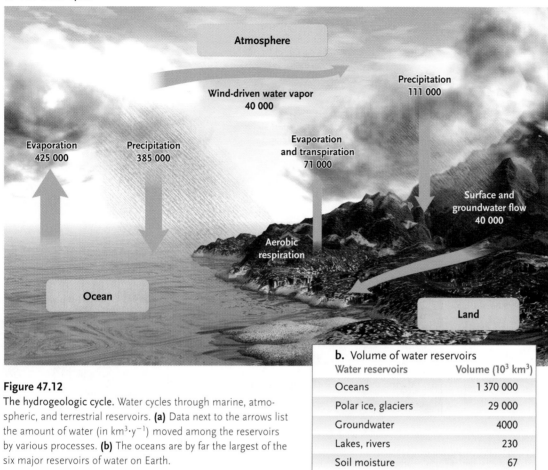

a. The water cycle

Atmosphere

Wind-driven water vapor
40 000

Precipitation
111 000

Evaporation
425 000

Precipitation
385 000

Evaporation
and transpiration
71 000

Surface and
groundwater flow
40 000

Aerobic
respiration

Ocean

Land

Figure 47.12

The hydrogeologic cycle. Water cycles through marine, atmospheric, and terrestrial reservoirs. **(a)** Data next to the arrows list the amount of water (in $km^3 \cdot y^{-1}$) moved among the reservoirs by various processes. **(b)** The oceans are by far the largest of the six major reservoirs of water on Earth.

b. Volume of water reservoirs

Water reservoirs	Volume (10^3 km³)
Oceans	1 370 000
Polar ice, glaciers	29 000
Groundwater	4000
Lakes, rivers	230
Soil moisture	67
Atmosphere (water vapor)	14

transport medium that moves nutrients within and between ecosystems, as demonstrated in a series of classic experiments in the Hubbard Brook Experimental Forest (see *Studies of the Hubbard Brook Watershed*).

47.2b Carbon: Backbone of Life

Carbon atoms provide the backbone of most biological molecules, and carbon compounds store the energy captured by photosynthesis (see Chapter 7). Carbon enters food webs when producers convert atmospheric carbon dioxide (CO_2) into carbohydrates. Heterotrophs acquire carbon by eating other organisms or detritus. Although carbon moves somewhat independently in sea and on land, a common atmospheric pool of CO_2 creates a global **carbon cycle (Figure 47.13, p. 1192).**

The largest reservoir of carbon is sedimentary rock, such as limestone. Rocks are in the unavailable-inorganic compartment, and they exchange carbon with living organisms at an exceedingly slow pace. Most *available* carbon is present as dissolved bicarbonate ions (HCO_3^-) in the ocean. Soil, atmosphere, and plant biomass are significant, but much smaller, reservoirs of available carbon. Atmospheric carbon is mostly in the form of molecular CO_2, a product of

aerobic respiration. Volcanic eruptions also release small quantities of CO_2 into the atmosphere.

Sometimes carbon atoms leave organic compartments for long periods of time. Some organisms in marine food webs build shells and other hard parts by incorporating dissolved carbon into calcium carbonate ($CaCO_3$) and other insoluble salts. When shelled organisms die, they sink to the bottom and are buried in sediments. Other animals, notably vertebrates, store calcium in bone. Insoluble carbon that accumulates as rock in deep sediments may remain buried for millions of years before tectonic uplifting brings it to the surface, where erosion and weathering dissolve sedimentary rocks and return carbon to an available form.

Carbon atoms are also transferred to the unavailable-organic compartment when soft-bodied organisms die and are buried in habitats where low oxygen concentration prevents decomposition. In the past, under suitable geologic conditions, these carbon-rich tissues were slowly converted to gas, petroleum, or coal, which we now use as fossil fuels. Human activities, especially burning fossil fuels, are transferring carbon into the atmosphere at an unnaturally high rate. The resulting change in the worldwide distribution of carbon is having profound consequences for Earth's atmosphere and

a. Amount of carbon in major reservoirs

Carbon reservoirs	Mass (10^{12} g)
Sediments and rocks	770 000 000
Ocean (dissolved forms)	397 000
Soil	15 000
Atmosphere	7500
Biomass on land	7150

b. Annual global carbon movement between reservoirs

Direction of movement	Mass (10^{12} kg)
From atmosphere to plants (carbon fixation)	1200
From atmosphere to ocean	1070
To atmosphere from ocean	1050
To atmosphere from plants	600
To atmosphere from soil	600
To atmosphere from burning fossil fuel	50
To atmosphere from burning plants	20
To ocean from runoff	4
Burial in ocean sediments	1

c. The global carbon cycle

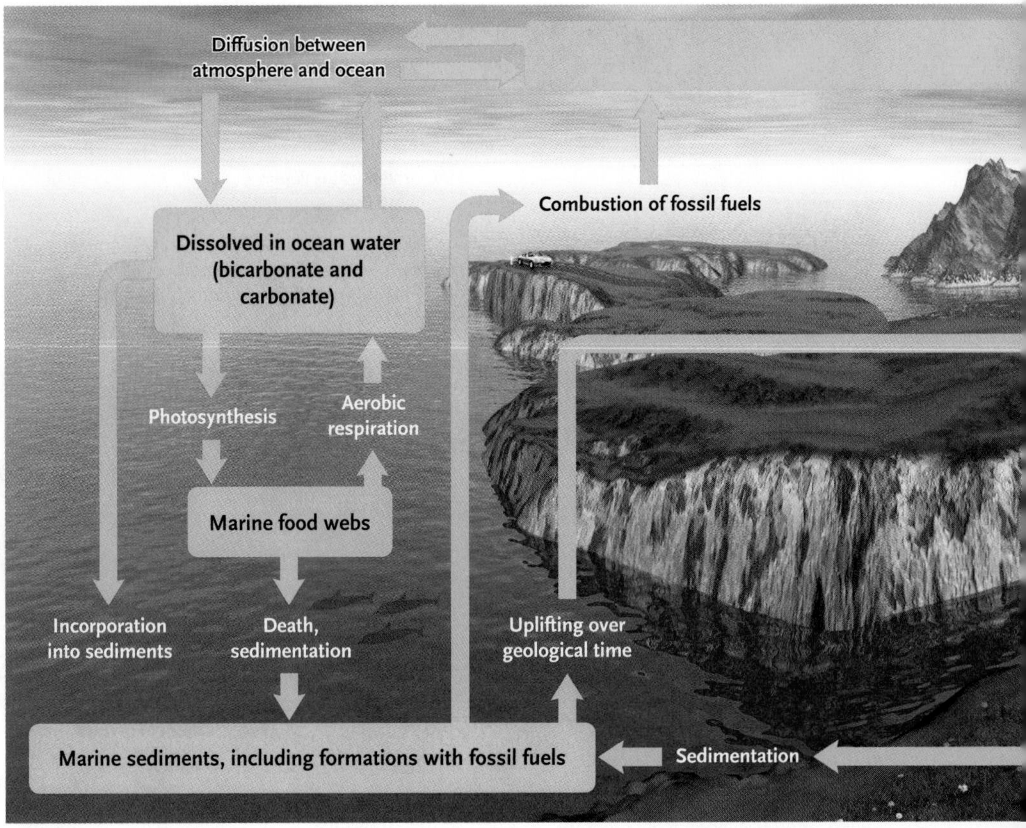

Figure 47.13

The carbon cycle. Marine and terrestrial components of the global carbon cycle are linked through an atmospheric reservoir of carbon dioxide. **(a)** By far the largest amount of Earth's carbon is found in sediments and rocks. **(b)** Earth's atmosphere mediates most of the movement of carbon. **(c)** In this illustration of the carbon cycle, boxes identify major reservoirs and labels on arrows identify the processes that cause carbon to move between reservoirs.

climate, including a general warming of the climate and a rise in sea level (see *Disruption of the Carbon Cycle*).

47.2c Nitrogen: A Limiting Element

All organisms require nitrogen to construct nucleic acids, proteins, and other biological molecules (see Chapter 41). Earth's atmosphere had a high nitrogen concentration long before life began. Today, a global **nitrogen cycle** moves this element between the huge atmospheric pool of gaseous molecular nitrogen (N_2) and several much smaller pools of nitrogen-containing compounds in soils, marine and freshwater ecosystems, and living organisms **(Figure 47.14)**.

Molecular nitrogen is abundant in the atmosphere, but triple covalent bonds bind its two atoms so tightly that most organisms cannot use it. Only certain microorganisms, volcanic action, and lightning can convert N_2 into ammonium (NH_4^+) and nitrate (NO_3^-) ions. This conversion is called **nitrogen fixation** (see Chapter 21). Once nitrogen is fixed, primary producers can incorporate it into biological molecules such as proteins and nucleic acids. Secondary consumers obtain nitrogen by consuming these molecules.

Several biochemical processes produce different nitrogen-containing compounds and thus move nitrogen through ecosystems. These processes are nitrogen fixation, ammonification, nitrification, and denitrification **(Table 47.2, p. 1195)**.

In nitrogen fixation, several kinds of microorganisms convert molecular nitrogen (N_2) to ammonium ions (NH_4^+). Certain bacteria, which collect molecular nitrogen from the air between soil particles, are the major nitrogen fixers in terrestrial ecosystems (see Table 47.2). The cyanobacteria partners in some lichens (see Chapter 24) also fix molecular nitrogen. Other cyanobacteria are important nitrogen fixers in aquatic ecosystems, whereas the water fern (genus *Azolla*) plays that role in rice paddies. Collectively, these organisms fix an astounding 200 million tonnes of nitrogen each year. Plants and other primary producers assimilate and use this nitrogen in the biosynthesis of amino acids, proteins, and nucleic acids, which then circulate through food webs.

Some plants, including legumes (such as beans and clover), alders (*Alnus* species), and some members of the rose family (Rosaceae), are mutualists with nitrogen-fixing bacteria. These plants acquire nitrogen from soils much more readily than plants

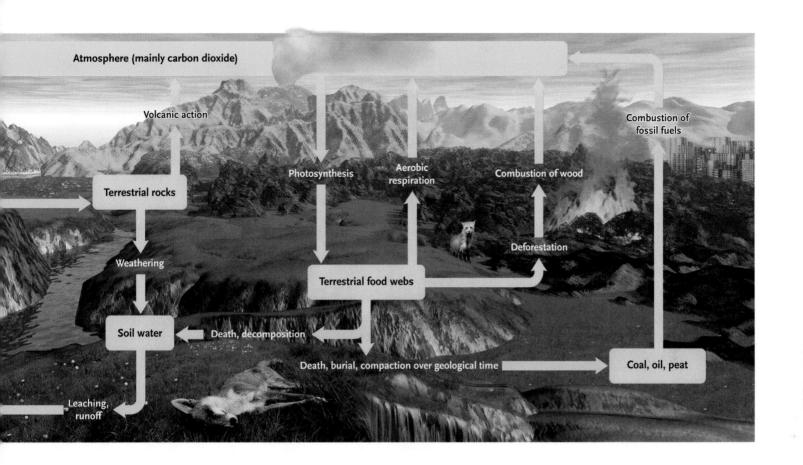

Atmosphere (mainly carbon dioxide)

Volcanic action

Combustion of fossil fuels

Terrestrial rocks

Photosynthesis

Aerobic respiration

Combustion of wood

Weathering

Deforestation

Terrestrial food webs

Soil water

Death, decomposition

Death, burial, compaction over geological time

Coal, oil, peat

Leaching, runoff

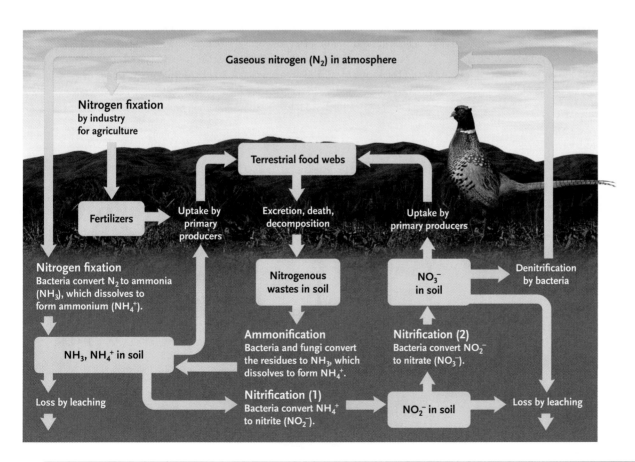

Gaseous nitrogen (N₂) in atmosphere

Nitrogen fixation
by industry
for agriculture

Terrestrial food webs

Fertilizers

Uptake by primary producers

Excretion, death, decomposition

Uptake by primary producers

Nitrogen fixation
Bacteria convert N_2 to ammonia (NH_3), which dissolves to form ammonium (NH_4^+).

Nitrogenous wastes in soil

NO_3^- in soil

Denitrification by bacteria

NH_3, NH_4^+ in soil

Ammonification
Bacteria and fungi convert the residues to NH_3, which dissolves to form NH_4^+.

Nitrification (2)
Bacteria convert NO_2^- to nitrate (NO_3^-).

Loss by leaching

Nitrification (1)
Bacteria convert NH_4^+ to nitrite (NO_2^-).

NO_2^- in soil

Loss by leaching

Figure 47.14
The nitrogen cycle in a terrestrial ecosystem. Nitrogen cycles through terrestrial ecosystems when unavailable molecular nitrogen is made available through the action of nitrogen-fixing bacteria. Other bacteria recycle nitrogen within the available-organic compartment through ammonification and two types of nitrification, converting organic wastes into ammonium ions and nitrates. Denitrification converts nitrate to molecular nitrogen, which returns to the atmosphere. Runoff carries nitrogen from terrestrial ecosystems into aquatic ecosystems, where it is recycled in freshwater and marine food webs.

Studies of the Hubbard Brook Watershed

Water flows downhill, so local topography affects the movement of dissolved nutrients in terrestrial ecosystems. A **watershed** is an area of land from which precipitation drains into a single stream or river. Each watershed represents a part of an ecosystem from which nutrients exit through a single outlet. When several streams join to form a river, the watershed drained by the river encompasses the smaller watersheds drained by the streams. The Mackenzie River watershed covers roughly 20% of Canada and includes the watersheds of the Peace and Athabasca rivers, as well as many other watersheds drained by smaller streams and rivers.

Watersheds are ideal for large-scale field experiments about nutrient flow in ecosystems because they are relatively self-contained units. Herbert Bormann and Gene Likens conducted a classic experiment on nutrients in watersheds in the 1960s. Bormann and Likens manipulated small watersheds of temperate deciduous forest in the Hubbard Brook Experimental Forest in the White Mountain National Forest of New Hampshire. They measured precipitation and nutrient input into the watersheds, the uptake of nutrients by vegetation, and the amount of nutrients leaving the watershed via streamflow. They monitored nutrients exported in streamflow by collecting water samples from V-shaped concrete weirs built into bedrock below the streams that drained the watersheds **(Figure 1).** Impermeable bedrock underlies the soil, preventing water from leaving the system by deep seepage.

Gene E. Likens from Gene E. Likens et al., Ecology Micrograph, 40(1): 23–47, 1970

Figure 1
Weir used to measure the volume and nutrient content of water leaving a watershed by streamflow.

Bormann and Likens collected several years of baseline data on six undisturbed watersheds. Then, in 1965 and 1966, they felled all of the trees in one small watershed and used herbicides to prevent regrowth. After these manipulations, they monitored the output of nutrients in streams that drained experimental and control watersheds. They attributed differences in nutrient export between undisturbed watersheds (controls) and the clear-cut watershed (experimental treatment) to the effects of deforestation.

Bormann and Likens determined that vegetation absorbed substantial water and conserved nutrients in undisturbed watersheds. Plants used about 40% of the precipitation for transpiration. The rest contributed to runoff and groundwater. Control watersheds lost only about 8 to 10 kg of calcium per hectare each year, an amount replaced by erosion of bedrock and input from rain. Moreover, control watersheds actually accumulated about 2 kg of nitrogen per hectare per year and slightly smaller amounts of potassium.

The experimentally deforested watershed experienced a 40% annual increase in runoff, a 300% increase during a 4-month period in summer. Some mineral losses were similarly large. The net loss of calcium was 10 times higher **(Figure 2)** than in the control watersheds and of potassium was 21 times higher. Phosphorus losses did not increase because this mineral was apparently retained by the soil. The loss of nitrogen, however, was very large—120 $kg \cdot ha^{-1} \cdot y^{-1}$. The washing out of nitrogen meant that the stream draining the experimental watershed became choked with algae and cyanobacteria. The Hubbard Brook experiment demonstrated that deforestation increases flooding and decreases the fertility of ecosystems.

Figure 2
Calcium losses from the deforested watershed were much greater than those from controls. The arrow indicates the time of deforestation in early winter. Mineral losses did not increase until after the ground thawed the following spring. Increased runoff also caused large water losses from the watershed.

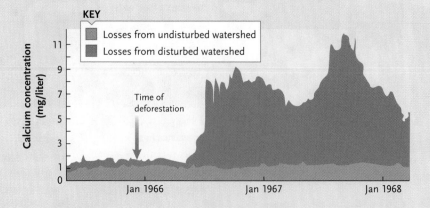

KEY
- Losses from undisturbed watershed
- Losses from disturbed watershed

Calcium concentration (mg/liter)

Time of deforestation

Jan 1966   Jan 1967   Jan 1968

Table 47.2	Biochemical Processes That Influence Nitrogen Cycling in Ecosystems			

Process	Organisms Responsible	Products	Outcome
Nitrogen fixation	Bacteria: *Rhizobium, Azotobacter, Frankia* Cyanobacteria: *Anabaena, Nostoc*	Ammonia (NH_3), ammonium ions (NH_4^+)	Assimilated by primary producers
Ammonification of organic detritus	Soil bacteria and fungi	Ammonia (NH_3), ammonium ions (NH_4^+)	Assimilated by primary producers
Nitrification			
(1) Oxidation of NH_3	Bacteria: *Nitrosomonas, Nitrococcus*	Nitrite (NO_2^-)	Used by nitrifying bacteria
(2) Oxidation of NO_2^-	Bacteria: *Nitrobacter*	Nitrate (NO_3^-)	Assimilated by primary producers
Denitrification of NO_3^-	Soil bacteria	Nitrous oxide (N_2O), molecular nitrogen (N_2)	Released to atmosphere

that lack such mutualists. Although these plants have the competitive edge in nitrogen-poor soil, nonmutualistic species often displace them in nitrogen-rich soil. In an interesting twist on the usual predator–prey relationships, several species of flowering plants living in nitrogen-poor soils capture and digest insects (see *Pitcher Plant Ecosystems*).

In addition to nitrogen fixation, other biochemical processes make large quantities of nitrogen available to producers. **Ammonification** of detritus by bacteria and fungi converts organic nitrogen into ammonia (NH_3), which dissolves in water to produce ammonium ions (NH_4^+) that plants can assimilate. Some ammonia escapes into the atmosphere as a gas. **Nitrification** by certain bacteria produces nitrites (NO_2^-), which are then converted by other bacteria to usable nitrates (NO_3^-). All of these compounds are water soluble, and water rapidly leaches them from soil into streams, lakes, and oceans.

Under conditions of low oxygen availability, **denitrification** by still other bacteria converts nitrites or nitrates into nitrous oxide (N_2O) and then into molecular nitrogen (N_2), which enters the atmosphere (see Table 47.2). This action can deplete supplies of soil nitrogen in waterlogged or otherwise poorly aerated environments, such as bogs and swamps.

In 1909, Fritz Haber developed a process for fixing nitrogen, and with the help of Carl Bosch, the process was commercialized for fertilizer production. The Haber–Bosch process has altered Earth's nitrogen cycles and is said to be responsible for the existence of 40% of the people on Earth. Before the implementation of the Haber–Bosch process, the amount of nitrogen available for life was limited by the rates at which N_2 was fixed by bacteria or generated by lightning strikes. Today, spreading fertilizers rich in nitrogen is the basis for most of the agriculture's productivity. This practice has quadrupled some yields over the past 50 years (see Chapter 49). Of all nutrients required for primary production, nitrogen is often the least abundant. Agriculture routinely depletes soil

nitrogen, which is removed from fields through the harvesting of plants that have accumulated nitrogen in their tissues. Soil erosion and leaching remove more. Traditionally, farmers rotated their crops, alternately planting legumes and other crops in the same fields. In combination with other soil conservation practices, crop rotation stabilized soils and kept them productive, sometimes for hundreds of years. Some of the most arable land in New York State was farmed by members of the Mohawk Iroquois First Nations. The evidence of this comes from the locations of palisaded villages. The people moved their villages and farming operations every 10 to 20 years, changing fields repeatedly over hundreds of years.

The production of synthetic fertilizers is expensive, using fossil fuels as both raw material and an energy source. Fertilizer becomes increasingly costly as supplies of fossil fuels dwindle. Furthermore, rain and runoff leach excess fertilizer from agricultural fields and carry it into aquatic ecosystems. Nitrogen has become a major pollutant of freshwater ecosystems, artificially enriching the waters and allowing producers to expand their populations.

47.2d Phosphorus: Another Essential Element

Phosphorus compounds lack a gaseous phase, and this element moves between terrestrial and marine ecosystems in a sedimentary cycle **(Figure 47.15, p. 1199)**. Earth's crust is the main reservoir of phosphorus, as it is for other minerals, such as calcium and potassium, that also undergo sedimentary cycles.

Phosphorus is present in terrestrial rocks in the form of phosphates (PO_4^{3-}). In the **phosphorus cycle**, weathering and erosion add phosphate ions to soil and carry them into streams and rivers, which eventually transport them to the ocean. Once there, some phosphorus enters marine food webs, but most of it precipitates out of solution and accumulates for millions of years as insoluble deposits, mainly on continental

Disruption of the Carbon Cycle

The concentrations of gases in the lower atmosphere have a profound effect on global temperature, in turn affecting global climate. Molecules of CO_2, water vapour, ozone, methane, nitrous oxide, and other compounds collectively act like a pane of glass in a greenhouse (hence the term *greenhouse gases*). They allow short wavelengths of visible light to reach Earth's surface while impeding the escape of longer, infrared wavelengths into space, trapping much of their energy as heat **(Figure 1).** Greenhouse gases foster the accumulation of heat in the lower atmosphere, a warming action known as the **greenhouse effect**. This natural process prevents Earth from being a cold and lifeless planet.

Data from air bubbles trapped in glacial ice indicate that atmospheric CO_2 concentrations have fluctuated widely over Earth's history **(Figure 2).** Since the late 1950s, scientists have measured atmospheric concentrations of CO_2 and other greenhouse gases at remote sampling sites such as the top of Mauna Loa in the Hawaiian Islands. These sites are free of local contamination and reflect average global conditions. Concentrations of greenhouse gases have increased steadily for as long as they have been monitored **(Figure 3).**

The graph for atmospheric CO_2 concentration has a regular zigzag pattern that follows the annual cycle of plant growth (see Figure 3). The concentration of CO_2 decreases during the summer because photosynthesis withdraws so much from the atmospheric available-inorganic pool. The concentration of CO_2 is higher during the winter when photosynthesis slows while aerobic respiration continues, returning carbon to the atmospheric available-inorganic pool. Whereas the zigs and zags in the data for CO_2 represent seasonal highs and lows, the midpoint of the annual peaks and troughs has increased steadily for 40 years. These data are evidence of a rapid buildup of atmospheric CO_2, representing a shift in the distribution of carbon in the major reservoirs on Earth. The best estimates suggest that CO_2 concentration has increased by 35% in the last 150 years and by more than 10% in the last 30 years.

The increase in the atmospheric concentration of CO_2 appears to result from combustion, whether we burn fossil fuels or wood. Today, humans burn more wood and fossil fuels than ever before. Vast tracts of tropical forests are being cleared and burned (see Chapter 48). To make matters worse, deforestation reduces the world's biomass of plants that assimilate CO_2 and help maintain the carbon cycle as it existed before human activities disrupted it.

The increase in the concentration of atmospheric CO_2 is alarming because plants with C_3 metabolism respond to increased CO_2 concentrations with increased growth rates. This is not true of C_4 plants (see Chapter 7). Thus, rising atmospheric levels of CO_2 will probably alter the relative abundances of many plant species, changing the composition and dynamics of their communities.

Simulation models suggest that increasing concentrations of any greenhouse gas may intensify the greenhouse effect, contributing to a trend of global warming. Should we be alarmed about the prospect of a warmer planet? Some models predict that the mean temperature of the lower atmosphere will rise by 4°C, enough to increase ocean surface temperatures. In some areas, such as the Canadian Arctic and the Antarctic, warming has occurred much more rapidly than predicted or expected. Water expands when heated, and global sea level could rise as much as 0.6 m just from this expansion. In addition, atmospheric temperature is rising fastest near the poles.

Figure 1
The greenhouse effect.

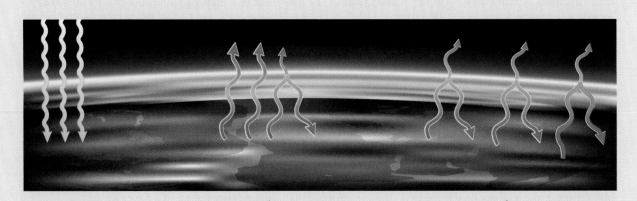

Sunlight penetrates the atmosphere and warms the Earth's surface.

The Earth's surface radiates heat (infrared wavelengths) to the atmosphere. Some heat escapes into space. Greenhouse gases and water vapor absorb some infrared energy and reradiate the rest of it back toward Earth.

When atmospheric concentrations of greenhouse gases increase, the atmosphere near the Earth's surface traps more heat. The warming causes a positive feedback cycle in which rising ocean temperatures cause increased evaporation of water, which further enhances the greenhouse effect.

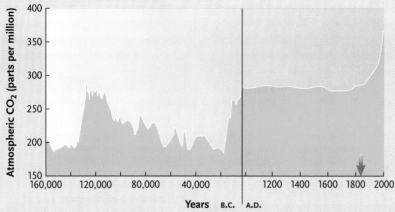

Figure 2

Carbon dioxide levels over time. The amount of atmospheric CO_2 has risen dramatically since about 1850 (arrow).

Thus, global warming may also foster melting of glaciers and the Antarctic ice sheet, which might raise sea level as much as 50 to 100 m, inundating low coastal regions. Waterfronts in Vancouver, Los Angeles, Hong Kong, Durban, Rio de Janeiro, Sydney, New York, and London would be submerged. So would agricultural lands in India, China, and Bangladesh, where much of the world's rice is grown.

Moreover, global warming could disturb regional patterns of precipitation and temperature. Areas that now produce much of the world's grains would become arid scrub or deserts, and the now-forested areas to their north would become dry grasslands.

Many scientists believe that atmospheric levels of greenhouse gases will continue to increase at least until the middle of the twenty-first century and

that global temperature may rise by several degrees. At the Earth Summit in 1992, leaders of the industrialized countries agreed to try to stabilize CO_2 emissions by the end of the twentieth century. We have already missed that target, and some countries, including the United States (then the largest producer of greenhouse gases), have now forsaken that goal as too costly. In 2008, it is likely that China and perhaps India will have surpassed the United States in production of greenhouse gases as these countries become more industrialized. Stabilizing emissions at current levels will not reverse the damage already done, nor will it stop the trend toward global warming. We should begin preparing for the consequences of global warming now. We might increase reforestation efforts because a large tract of forest can withdraw significant amounts of CO_2 from the atmosphere. We might also step up genetic engineering studies to develop heat-resistant and drought-resistant crop plants, which may provide crucial food reserves in regions of climate change.

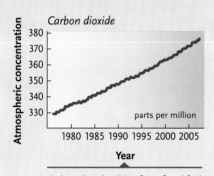

Carbon dioxide

Carbon dioxide (CO_2) from fossil fuel combustion, factory emissions, car exhaust, and deforestation contributes to an overall increase in atmospheric levels of the gas.

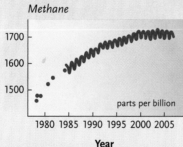

Methane

Methane (CH_4) is produced by anaerobic bacteria in swamps, landfills, and termite mounds. The microorganisms in the digestive tracts of cattle and other ruminants also produce large quantities of this gas.

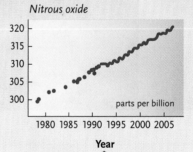

Nitrous oxide

Nitrous oxide (N_2O) is a natural gaseous by-product of denitrifying bacteria. Large quantities are also released from synthetic fertilizers and the wastes of animals in feedlots.

Figure 3

Increases in atmospheric concentrations of three greenhouse gases, mid-1970s through 2004. The data were collected at a remote monitoring station in Australia (Cape Grim, Tasmania) and compiled by scientists at the Commonwealth Scientific and Industrial Research Organization, an agency of the Australian government.

Pitcher Plant Ecosystems

Pitcher plants have modified leaves (pitchers) that act as pitfall traps for drowning and digesting insect prey. Pitchers have developed in at least five different evolutionary lines of vascular plants (see Chapter 29). Throughout much of North America, pitcher plants (the provincial flower of Newfoundland and Labrador; **Figure 1**) are common in bogs. *Sarracenia purpurea*, like other carnivorous plants, obtain much of their nitrogen from the insects they capture.

The captured arthropod prey, mainly ants and flies, is the base of a food web inside the pitchers. These are shredded and partly consumed by larvae of midges (*Metriocnemus knabi*) and sarcophagid flies (*Fletcherimyia fletcheri*; **Figure 2**). A subweb of bacteria and protozoa processes shredded prey, which are themselves prey for filter-feeding rotifers (*Habrotrocha rosa*; **Figure 3**) and mites (*Sarraceniopus gibsonii*). Mosquito larvae (*Wyeomyia smithii*) eat the bacteria, protozoa, and rotifers, whereas the larger sarcophagid fly larvae eat the rotifers and smaller mosquito larvae. Populations of bacteria, protozoa, and rotifers grow much more rapidly than populations of mosquito or midge larvae, making the system sustainable.

Pitchers are essential to the life cycles of two species of insects whose larvae live in them. A mosquito and a midge coexist in the same pitchers, and their populations are limited by the availability of insect carcasses. In any pitcher, growth in populations of the midge larvae is not affected by increases in the numbers of mosquito larvae. But populations of mosquito larvae increase as populations of midge larvae increase (see Figure 2).

The situation is an example of processing-chain commensalism because the action of one species creates opportunities for another. In this case, midge larvae feed on the hard parts of insect carcasses and break them up in the process. Mosquito larvae are filter-feeders, consuming particles derived from the decaying matter. The feeding of the midges generates additional food for the mosquito larvae. Although the populations of midge and mosquito larvae can be large in any pitcher, only a single sarcophagid fly larva occurs in any pitcher. *F. fletcheri* is a *K*-strategist (see Chapter 45) and gives birth to larvae. If you place more than one *F. fletcheri* larvae in a pitcher, a fight ensues. The larger larva either wins or leaves the pitcher to pupate in the sphagnum around it.

Figure 1
Sarracenia purpurea, a pitcher plant. The flower on a long stalk extends above the pitchers. One pitcher is shown in the photo on the left.

M.B. Fenton

Figure 2
Midge and mosquito larvae in pitchers. **(a)** The density and **(b)** total dry mass of mosquito larvae are the same whether the population of midges is low (8 midges) or high (30 midges). FH = high food availability; FL = low food availability. Error bars show standard errors of the mean.

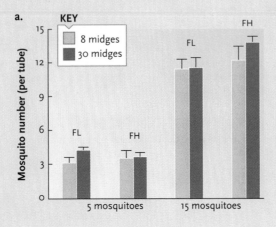

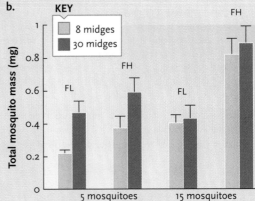

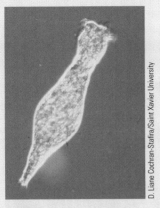

D. Liane Cochran-Stafira/Saint Xavier University

Figure 3
A bdelloid rotifer, *Habrotrocha rosa*, from a *Sarracenia purpurea* pitcher.

Jim Vargo

Jim Vargo

Figure 4
Moths whose caterpillars eat *Sarracenia purpurea*. The caterpillars of **(a)** *Exyra fax* and **(b)** *Papaipema appassionata* feed on pitcher plants, either **(a)** the lining of pitchers or **(b)** the rhizomes.

These insects do not appear to compete with their hosts, the pitcher plants. The abundance of rotifers living in the pitchers of *S. purpurea* is negatively associated with the presence of midge and mosquito larvae (which eat the rotifers). Rotifers are detritivores, and their excretory products (NO_3-N, NH_4OH, P) account for a major portion of the N acquired by the plants from their insect prey.

Two species of moths also exploit *Sarracenia purpurea* **(Figure 4)**. *Exyra fax* and *Papaipema appassionata* do not live in the pitchers. *E. fax* caterpillars eat the interior surface of the pitcher chambers, whereas *P. appassionata* caterpillars consume the rhizomes. Although predation by *E. fax* caterpillars does not kill the plants, predation by *P. appassionata* does. To what trophic level does one assign moths whose caterpillars are herbivores feeding on primary producers that eat insects?

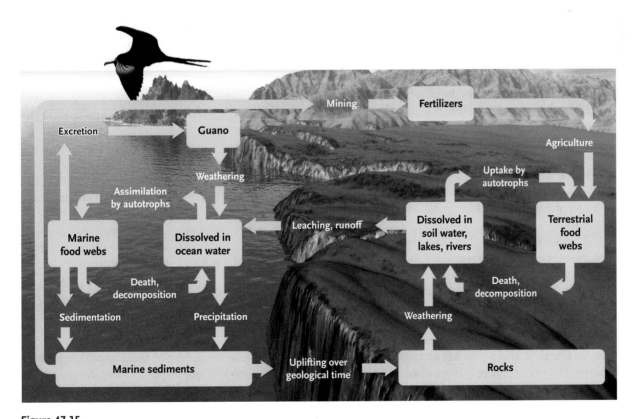

Figure 47.15
The phosphorus cycle. Phosphorus becomes available to biological systems when wind and rainfall dissolve phosphates in rocks and carry them into adjacent soil and freshwater ecosystems. Runoff carries dissolved phosphorus into marine ecosystems, where it precipitates out of solution and is incorporated into marine sediments.

Lenore Fahrig, Carleton University

The fragmentation of habitats is a ubiquitous effect of human activity on landscapes. In many parts of the world, land areas that used to be continuous forest are now large expanses of agricultural or urban landscapes dotted with small fragments of forest **(Figure 1)**. Lenore Fahrig examines the impact of landscape structure on the abundance, distribution, and persistence of organisms.

In her research, Dr. Fahrig uses a variety of organisms, from beetles to plants and birds. She considers habitats and the impacts of roads and fence lines. She and her students try to identify the habitat features associated with the persistence of species after fragmentation and the role of connectivity between fragments in the persistence of populations in the fragments.

Using a combination of theoretical and fieldwork, she has assessed the responses of species in different trophic roles to the fragmentation of habitat. Her work demonstrates that not all species respond in the same way and that some benefit from fragmentation.

The connections between theoretical work and reality emerge clearly from her research, and the implications for conservation of biodiversity (see Chapter 48) are clear.

Figure 1
An aerial view of farmland in southwestern Ontario illustrates isolated patches of forest (woodlots) and bands of woodland (riparian) along the edges of a creek. The woodlots are varied in their size and shape and in the degree of their isolation or connection to other woodlots.

---

shelves. When parts of the seafloor are uplifted and exposed, weathering releases the phosphates.

Plants absorb and assimilate dissolved phosphates directly, and phosphorus moves easily to higher trophic levels. All heterotrophs excrete some phosphorus as a waste product in urine and feces that become available after decomposition. Primary producers readily absorb the phosphate ions, so phosphorus cycles rapidly *within* terrestrial communities.

Supplies of available phosphate are generally limited, however, and plants acquire it so efficiently that they reduce soil phosphate concentration to extremely low levels. Thus, like nitrogen, phosphorus is a common ingredient in agricultural fertilizers, and excess phosphates are pollutants of freshwater ecosystems. A particularly good example is Lake Erie, one of the Great Lakes that was heavily affected by accumulations of phosphorus. The example here is more convincing because the problem has largely been resolved over the years.

For many years, phosphate for fertilizers was obtained from *guano* (the droppings of seabirds that consume phosphorus-rich food), which was mined on small islands that hosted seabird colonies, for example, in Polynesia and Micronesia. We now obtain most phosphate for fertilizer from phosphate rock mined in places such as Saskatchewan, with abundant marine deposits.

STUDY BREAK

1. How is balance maintained in the hydrogeologic cycle?
2. How do consumers obtain carbon?
3. What is the role of cyanobacteria in the nitrogen cycle? Why is their role important?

47.3 Ecosystem Modelling

Ecologists use modelling to make predictions about how an ecosystem will respond to specific changes in physical factors, energy flow, or nutrient availability. Analyses of energy flow and nutrient cycling allow us to create a *conceptual model* of how ecosystems function **(Figure 47.16)**. Energy that enters ecosystems is gradually dissipated as it flows through a food web. By contrast, nutrients are conserved and recycled among the system's living and nonliving components. This general model does not include processes that carry nutrients and energy out of one ecosystem and into another.

More importantly, the model ignores the nuts-and-bolts details of exactly how specific ecosystems function. Although it is a useful tool, a conceptual model does not really help us predict what would happen, say, if we harvested 10 million tonnes of introduced salmon from Lake Erie every year. We could simply harvest the fishes and see what happens. But ecologists prefer less intrusive approaches to study the potential effects of disturbances.

One approach to predicting "what would happen if ... " is **simulation modelling**. Using this approach, researchers gather detailed information about a specific ecosystem. They then derive a series of mathematical equations that define its most important relationships. One set of equations might describe how nutrient availability limits productivity at various trophic levels. Another might relate the population growth of zooplankton to the productivity of phytoplankton. Other equations would relate the population dynamics of primary carnivores to the availability of their food, and still others would describe how the densities of primary carnivores influence reproduction in populations at both lower and higher trophic levels. Thus, a complete simulation model is a set of interlocking equations that collectively predict how changes in one feature of an ecosystem might influence others.

Creating a simulation model is a challenge because the relationships within every ecosystem are complex. First, you must identify the important species, estimate their population sizes, and measure the average energy and nutrient content of each. Next, you would describe the food webs in which they participate, measure the quantity of food each species consumes, and estimate the productivity of each population. And, for the sake of completeness, you would determine the ecosystem's energy and nutrient gains and losses caused by erosion, weathering, precipitation, and runoff. You would repeat these measurements seasonally to identify annual variation in these factors. Finally, you might repeat the measurements over several years to determine the effects of year-to-year variation in climate and chance events.

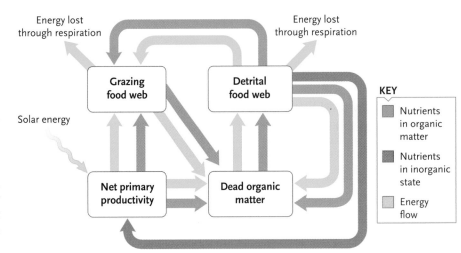

Figure 47.16
A conceptual ecosystem model. A simple conceptual model of an ecosystem illustrates how energy flows through the system and is lost from both detrital and grazing food webs. Nutrients are recycled and conserved.

After collecting these data, you would write equations that quantify the relationships in the ecosystem, including information about how temperature and other abiotic factors influence the ecology of each species. Having completed that job, you could begin to predict, possibly in great detail, the effects of adding 1000 new housing units to an area of native prairie or boreal forest. Of course, you would have to refine the model whenever new data became available.

Some ecologists devote their professional lives to studying ecosystem processes and creating simulation models. The long-term initiative at the Hubbard Brook Forest provides a good example (see *Studies of the Hubbard Brook Watershed*). As we attempt to understand larger and more complex ecosystems (and as we create larger and more complex environmental problems), modelling becomes an increasingly important tool. If a model is based on well-defined ecological relationships and good empirical data, it can allow us to make accurate predictions about ecosystem changes without the need for costly and environmentally damaging experiments. But, like all ideas in science, a model is only as good as its assumptions, and models must constantly be adjusted to incorporate new ideas and recently discovered facts.

STUDY BREAK

1. Briefly describe the process of simulation modelling.
2. Why is simulation modelling necessary?

UNANSWERED QUESTIONS

The impact of a large meteorite on Earth is said to have caused worldwide changes in climate. What biosphere and ecosystem changes would you expect to occur?

Review

Go to CENGAGENOW™ at http://hed.nelson.com/ to access quizzing, animations, exercises, articles, and personalized homework help.

47.1 Energy Flow and Ecosystem Energetics

- Net primary productivity is the chemical energy remaining in a system after energy has been used by producers to complete life processes and cellular respiration. Net primary productivity differs from standing crop biomass in that net primary productivity is a measure of energy, whereas standing crop biomass is a measure of dry weight.

- Other factors affect primary productivity, such as water and access to nutrients.

- Assimilation efficiency refers to energy absorbed from eating compared with the total energy in the food. Production efficiency is the energy content of new tissue material compared with the energy absorbed from food intake.

- Some energy is lost during transfer by consumption. The process becomes less efficient as the number of transfers increases, meaning that less energy is transferred to the final consumer.

- Biological magnification occurs when material (e.g., DDT) present in small amounts in a producer or "low-level" organism is consumed by another organism, transferring the material to the predator. DDT accumulates with each successive transfer. Top predators exhibit the highest concentrations of contaminants such as DDT.

47.2 Nutrient Cycling in Ecosystems

- The amount of water that leaves the Earth and enters the atmosphere through evaporation is equal to the amount of water reaching Earth by precipitation.

- Producers transfer atmospheric carbon (CO_2) into carbohydrates. Consumers then eat the producers and take in the carbohydrates.

- Cyanobacteria can fix nitrogen, which is crucial because although atmospheric nitrogen levels are high, this nitrogen is not accessible to plants or animals. Atmospheric nitrogen must be converted or "fixed" into a usable form such as ammonium and nitrate.

- Phosphate ions are carried to bodies of water through weathering and erosion, where most of it precipitates. Eventually, weathering releases phosphates, which are then directly absorbed by plants.

46.3 Ecosystem Modelling

- Modelling involves collecting data about an ecosystem and deriving mathematical equations about the relationships in the ecosystem. Data collected over different seasons and annual changes can be used to simulate the effects of a disruption on various levels of the ecosystem in question.

- Simulation modelling helps us understand and predict the impact of influences on certain ecosystems without actually conducting an experiment. Altering anything in an ecosystem without knowledge of its possible effects can be devastating on many or all levels.

Questions

Self-Test Questions

1. Which of the following events moves energy and material from a detrital food web into a grazing food web?
   a. A beetle eating the leaves of a living plant
   b. An earthworm eating dead leaves on the forest floor
   c. A robin catching and eating an earthworm
   d. A crow eating a dead robin
   e. A bacterium decomposing the feces of an earthworm

2. The total dry weight of plant material in a forest is a measure of the forest's
   a. gross primary productivity.
   b. net primary productivity.
   c. cellular respiration.
   d. standing crop biomass.
   e. ecological efficiency.

3. Which of the following ecosystems has the highest rate of net primary productivity?
   a. open ocean
   b. temperate deciduous forest
   c. tropical rain forest
   d. desert shrubs and thornwoods
   e. agricultural land

4. Endothermic animals exhibit a lower ecological efficiency than ectothermic animals because
   a. endotherms are less successful hunters than ectotherms.
   b. endotherms eat more plant material than ectotherms.
   c. endotherms are larger than ectotherms.
   d. endotherms produce fewer offspring than ectotherms.
   e. endotherms use more of their energy to maintain body temperature than ectotherms.

5. The amount of energy available at the highest trophic level in an ecosystem is determined by
   a. only the gross primary productivity of the ecosystem.
   b. only the net primary productivity of the ecosystem.
   c. the gross primary productivity and the standing crop biomass.
   d. the net primary productivity and the ecological efficiencies of herbivores.
   e. the net primary productivity and the ecological efficiencies at all lower trophic levels.

6. Some freshwater and marine ecosystems exhibit an inverted pyramid of
   a. biomass.
   b. energy.
   c. numbers.
   d. turnover.
   e. ecological efficiency.

7. Which process moves nutrients from the available-organic compartment to the available-inorganic compartment?
   a. respiration
   b. erosion
   c. assimilation
   d. sedimentation
   e. photosynthesis

8. Identify which of the following materials has a sedimentary cycle.
   a. water
   b. oxygen
   c. nitrogen
   d. phosphorus
   e. carbon

9. Which of the following statements is supported by the results of studies at the Hubbard Brook Experimental Forest?
   a. Most energy captured by primary producers is lost before reaching the highest trophic level in an ecosystem.
   b. Deforested watersheds experience a more significant decrease in runoff than undisturbed watersheds.
   c. Deforested watersheds lose more calcium and nitrogen in runoff than undisturbed watersheds.
   d. Nutrients generally move through biogeochemical cycles very quickly.
   e. Deforested watersheds generally receive more rainfall than undisturbed watersheds.

10. Biological magnification describes a phenomenon in which certain materials
    a. become increasingly concentrated in the tissues of animals at higher trophic levels.
    b. become most concentrated in the tissues of animals at the lowest trophic levels.
    c. accumulate only in the tissues of primary producers.
    d. accumulate only in the tissues of tertiary consumers.
    e. accumulate only in the tissues of detritivores.

Questions for Discussion

1. Identify 12 ecosystem changes associated with hydroelectric power projects. Consider upstream and downstream changes as well as those associated with transmission of generated power. How does preparing your answers draw on information presented in this chapter?

2. A lake near your home became overgrown with algae and pondweeds a few months after a new housing development was built nearby. What data would you collect to determine whether the housing development might be responsible for the changes in the lake?

3. Some politicians question whether recent increases in atmospheric temperature result from our release of greenhouse gases into the atmosphere. They argue that atmospheric temperature has fluctuated widely over Earth's history, and the changing temperature is just part of an historical trend. What information would allow you to refute or confirm their hypothesis? From another perspective, describe the pros and cons of reducing greenhouse gases as soon as possible versus taking a "wait and see" approach to this question.

A leopard photographed in the wild in South Africa.

Laura Erin Barclay

48 Conservation of Biodiversity

WHY IT MATTERS

Achieving preservation of the Earth's biodiversity is one of the most pressing challenges facing our species today. Numbers of species are a simple indicator of biodiversity, perhaps the most apparent and easy to grasp. But as we have seen, many species of organisms remain undescribed and unnamed. Without names and descriptions, how can we recognize or count them? As we shall see shortly, being unnamed means being unprotected. The Barcode of Life project (see Chapter 3) is one promising effort to better catalogue biodiversity by identifying and allowing us to name its components.

Research on ecosystems shows repeatedly how the numbers of species are associated with stability and productivity. The natural order (association between productivity and biodiversity), however, does not coincide with the productivity that our own species must achieve to feed our ever-expanding populations. Creating and maintaining agricultural monocultures is a way for us to maximize food production and efficiency of harvest. In many areas, this approach leads to the disappearance of family-operated farms. Is this progress that is justified by efforts to increase efficiency and yield? Humans also use genetically modified organisms to increase productivity and marketability, as well as other features, such as shelf life and portability. All too often, increased agricultural productivity is achieved by the use of more fertilizer, water, and energy. Does agriculture have to be the enemy of biodiversity?

Some people have connected humans' attitude to Earth and its riches with religious teachings. In 1967, Lynn White Jr., a professor of medieval history, explored the historical roots of our ecological crisis. He focused on the Christian view of creation, the importance of science, and the separation of humans from their environmental roots. A dualism between humans and nature had emerged in some Christian societies more than in others. Inherent in these societies was the prevailing idea that it is God's will that humans exploit nature for their own ends. White nominated St. Francis of Assisi **(Figure 48.1)** as the patron saint of ecologists. He said that appreciating the virtue of humility was key to understanding the teachings of St. Francis. His point was that as soon as an animal or a plant (or a meadow, lake, or grove of trees) has its own place in nature (in God's eyes), then it can become as important as we believe we are.

The onus is on us as citizens of the planet to conserve biodiversity. One of the main problems we must overcome is the attitude of many humans, as reviewed by White. Today a common reflection of this attitude is that being able to do something (afford to, have the means to) is justification enough for doing it—whether the project involves making space for a shopping mall by draining a wetland or cutting down the trees in a woodlot.

If we as a species can recognize the importance of biodiversity and accept that the world is not ours to do with as we please, what, then, is the best route to protecting and conserving biodiversity? Should we focus on species? On genetic diversity? On ecosystems? How should we blend these approaches to achieve the best support for the endeavour? How can we engage people in this important activity and perhaps move them away from a human-centric view of the world?

As we shall see, at almost every turn are examples of human activities driving other species to extinction. The motivations for human actions range from little more than greed to the daily effort to survive. The purpose of this chapter is to introduce you to a range of situations and examples associated with the reduction of biodiversity by causing extinctions and the threat of extinction. We also consider steps that can be taken to protect biodiversity, including some successes and some failures.

48.1 Extinction

Extinction is part of the process of evolution. Given that life has been on Earth for about 3 billion years, today there are more extinct than living species. Occasionally, the fossil record demonstrates a continuum in time from one species to another, sometimes blurring the boundaries between taxa (see Chapter 18). In this case, one could argue that the original species in a series lives on in its descendants. For example, the discovery that the genome of *Homo sapiens sapiens* contains some genes from *Homo sapiens neanderthalensis* leaves open the question about the distinctness of the two taxa. Although data for fossil species usually do not permit us to assess the levels of gene flow between populations, the Neanderthals provide an interesting exception. The difficulties inherent in applying the species concept to fossil material is familiar to paleontologists but less so to biologists.

Species and lineages have been going extinct since life first appeared. We should expect species to disappear at some low rate, the **background extinction rate;** as environments change, poorly adapted organisms do not survive and reproduce. In all likelihood, more than 99.9% of the species that have ever lived are now extinct. David Raup has suggested that, on average, as many as 10% of species go extinct every million years and more than 50% go extinct every 100 million years. Thus, the history of life has been characterized by an ongoing turnover of species.

The fossil record indicates that extinction rates rose well above the background rate at least five times in Earth's history. These events are referred to as **mass extinctions.** One extinction occurred at the end of the

Figure 48.1
St. Francis of Assisi.

M.B. Fenton

Ordovician and the beginning of the Devonian, the next at the end of the Devonian, then the end of the Permian, the end of the Triassic, and the end of the Cretaceous. The Permian extinction was the most severe, and more than 85% of the species alive at that time disappeared forever. This extinction was the end of the trilobites, many amphibians, and the trees of the coal swamp forests. During the last mass extinction, at the end of the Cretaceous, half of the species on Earth, including most dinosaurs, disappeared. A sixth mass extinction, potentially the largest of all, is occurring now as a result of human degradation of the environment.

Different factors were responsible for the five mass extinctions. Some were probably caused by tectonic activity and associated changes in climate. For example, the Ordovician extinction occurred after Gondwana moved toward the South Pole, triggering a glaciation that cooled the world's climate and lowered sea levels. The Permian extinction coincided with a major glaciation and a decline in sea level induced by the formation of Pangea (see Chapter 20).

Many researchers believe that an asteroid impact caused the Cretaceous mass extinction. The resulting dust cloud may have blocked the sunlight necessary for photosynthesis, setting up a chain reaction of extinctions that began with microscopic marine organisms. Geologic evidence supports this hypothesis. Rocks dating to the end of the Cretaceous period (65 million years ago) contain a highly concentrated layer of iridium, a metal that is rare on Earth but common in asteroids. The impact from an iridium-laden asteroid only 10 km in diameter could have caused an explosion equivalent to a billion tonnes of TNT that scattered iridium dust around the world. Geologists have identified the submarine Chicxulub crater, 180 km in diameter, off Mexico's Yucatán peninsula as the likely site of the impact.

Although scientists agree that an asteroid struck Earth at that time, many question its precise relationship to the mass extinction. Dinosaurs had begun their decline at least 8 million years earlier, but many persisted for at least 40 000 years after the impact. Moreover, other groups of organisms did not suddenly disappear, as one would expect after a global calamity. The Cretaceous extinction took place over tens of thousands of years. Furthermore, some organisms survived periods of extinction, such as ginkgo trees (*Ginkgo biloba*), horseshoe crabs (*Limulus polyphemus*), and coelocanths (*Latimeria chalumnae*).

Even today we cannot blame the extinctions of most species on the activities of humans. But our increasing technological capability and prowess coincide with a burgeoning population of people. This situation is exacerbated by the philosophical view that humans are disconnected from nature. Thus, we are becoming better and better at destroying the biota of the planet. Taking action requires identifying root causes and then trying to make changes that will alleviate the problems.

First, we consider extinctions not linked to humans and then review examples of situations in which our actions have either directly or indirectly led to the extinction of species. The fact that extinction is integral to the process of evolution is hardly justification or rationalization for our driving so many species there. Put another way, invoking "survival of the fittest" may not be adequate justification for eradicating other species.

48.1a Dinosaurs: The Most Notable Extinction

Why do species go extinct? There could be as many theories as there are extinct species! The disappearance of the dinosaurs is one of the best-known extinction events in the Earth's history. At the end of the Cretaceous about 65.5 million years ago, the dinosaurs disappeared. The ancestors of dinosaurs had appeared in the Triassic, and the group underwent extensive adaptive radiation reflected in body size, lifestyle, and distribution. Although people think of large and spectacular carnivorous dinosaurs such as *Tyrannosaurus rex* or the huge herbivore *Apatosaurus* (previously known as *Brontosaurus*), in reality, many species of dinosaurs were small and delicate.

Evidence from deposits in Alberta suggests that the carnivorous dinosaurs *Albertosaurus* **(Figure 48.2)** showed age-specific mortality and high juvenile survival **(Figure 48.3a, p. 1208)**. Indeed, the survivorship curves (see Chapter 45) for *Albertosaurus* resemble those for humans **(Figure 48.3b)**. The data do not

Figure 48.2

Mounted skeleton of *Albertosaurus* on display in the Royal Tyrrell Museum, Drumheller, Alberta. This late Cretaceous carnivore is abundant in fossil beds in Alberta and elsewhere.

M.B. Fenton

LIFE ON THE EDGE

Sex Determination and Global Warming

Failure to reproduce puts the survival of a species on the edge, so anything that interferes with reproduction can be threatening. Genetic recombination is a fundamental benefit of sexual reproduction, enabling it to increase genetic diversity and eliminate deleterious mutants. Effective sexual reproduction means having male and female systems, sometimes in one individual (hermaphrodites) and perhaps more often in different individuals. Males and females differ in many fundamental ways—genetically, hormonally, physiologically, and anatomically.

In humans and many other animals, gender is determined by genotype, with males having an X and a Y chromosome and females having two X chromosomes. In many reptiles, however, gender is determined environmentally. Eggs incubated at some temperatures develop into males; when incubated at other temperatures, they produce females.

In 2008, D.A. Warner and R. Shine reported the results of experiments done with jacky dragons (*Amphibolurus muricatus*), an Australian lizard in which gender is determined by temperature. Eggs incubated at 23° to 26°C or 30° to 33°C produce females; those incubated from 27° to 29°C produce males. Warner and Shine tested the hypothesis that temperature-dependent sex determination ensured production of females when they had an advantage and males when the advantage was to them. Using a combination of temperature and hormonal manipulations, Warner and Shine could produce males or females at any temperature. They used analysis of paternity to assess the reproductive output of these males and observation of eggs laid and hatching to document these females' reproductive output.

In female jacky dragons, larger body sizes occur at higher temperatures, and larger females have higher fecundity than smaller ones. Higher temperatures also correlate with larger body size in males. However, males hatched from eggs incubated between 27° and 29°C sired more offspring than those hatched from eggs incubated at lower or higher temperatures.

Change in climate, such as global warming, could put species with temperature-dependent sex determination at risk by effectively eliminating males or females from the population. Eggs incubated at the "wrong" temperatures will fail to hatch. The importance of variation in temperature during development in ectothermic organisms could explain the prevalence of genotypic-dependent sex determination in euthermic (homeothermic) viviparous animals. Viviparous or ovoviviparous ectotherms (fish, amphibians, reptiles, other animals) could also rely on temperature-dependent gender determination, provided that their developing young experience an appropriate range of temperatures.

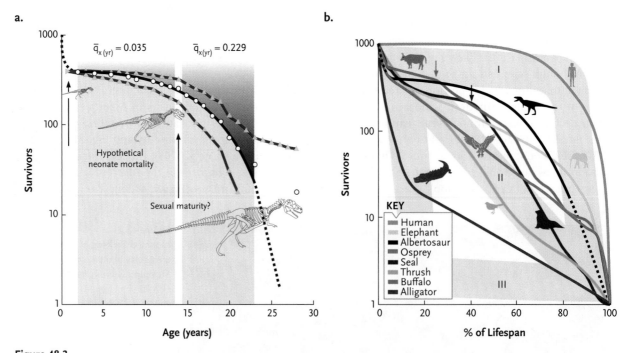

Figure 48.3

(a) Survivorship curve for a hypothetical cohort of 1000 *Albertosaurus* presumed neonatal mortality of 60%. **(b)** Survivorship of *Albertosaurus* compared with that of other animals, including humans from developed countries, short-lived birds, mammals, and lizards, as well as crocodilians and some captive mammals.

Amanda Vincent, University of British Columbia

Sea horses (see Figure 39.1) are a central focus for Professor Amanda Vincent's research. She holds a Canada Research Chair in marine conservation and is the director of Project Seahorse. Sea horses are notable for the details of their biology (see Chapter 41) and because they are big business.

Dr. Vincent has studied the behaviour of sea horses. During mating, the male's sperm fertilizes the female's eggs, but the female then transfers the fertilized eggs to the male's brood pouch; thus, the male gets "pregnant." Males and females are in regular contact during the period of pregnancy.

These contact behaviours may be key to the monogamy that appears to be typical of male–female relationships in sea horses.

The world trade in sea horses involves an estimated 20 million of them each year. In Asia, millions of dried sea horses are traded each year, mainly for use in traditional Chinese medicines. Remedies that include sea horses are said to be useful in treating symptoms from asthma and skin problems to incontinence and disorders of the thyroid. Dried sea horses may also be ingredients in aphrodisiacs.

Amanda Vincent has been very active in efforts to conserve sea horses, including working with local fishing communities to conserve them and maintain a sustainable harvest. This means establishing local protected areas and growing some sea horses in controlled conditions so that they, rather than wild stock, are harvested.

Professor Vincent is an example of a biologist who combines an academic interest with its practical applications. Her work in conservation connects the realities of harvesting animals with the demands for their conservation.

---

provide any indication of a flaw that predisposed dinosaurs to extinction. This is sobering, given the similarity between some aspects of dinosaur population biology and our own. The prevailing view today is that the disappearance of the dinosaurs is linked to the impact of an asteroid. Many other theories have been proposed to explain extinction, but the fact that birds and mammals and many other groups of organisms showed widespread extinctions at the end of the Cretaceous implies a pervasive catastrophic event.

48.1b Multituberculates: A Mammalian Example

There is more to extinction than dinosaurs. Competition (see Chapter 46) has been proposed as a mechanism that can lead to extinction. Among mammals, the adaptive radiation of rodents (order Rodentia; **Figure 48.4a**) in the early Oligocene coincides with the disappearance of multituberculates (order Multituberculata; **Figure 48.4b**). As a group, multituberculates were prominent and persisted for 100 million years (compared with 150 million years for dinosaurs), making them the most successful mammals to date.

Multituberculates ranged from small (~20 g) to medium (5 to 10 kg) in size and exhibited both terrestrial and arboreal lifestyles. We can only speculate what happened to them, and why they became extinct. The widespread success of rodents almost worldwide could lend credibility to competition as the reason for the multituberculates' demise. However, the fossil

record does not tell us what rodents and multituberculates competed for: food? nest sites?

STUDY BREAK

1. What is extinction?
2. What were multituberculates?

a.

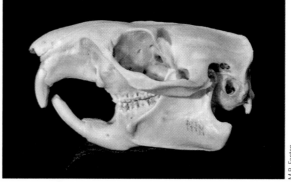

b.

Figure 48.4
(a) A skull of a groundhog (*Marmota monax*), a North American rodent, compared with **(b)** a skull of a multituberculate (*Kryptobaatar dashzevegi*) from the Mongolian late Cretaceous.

Dr. Ted Macrini, 2001, "Kryptobaatar dashzevegi" (On-line), Digital Morphology. Accessed from: http://digimorph.org/specimens/Kryptobaatar_dashzevegi/

M.B. Fenton

48.2 The Impact of Humans

When it comes to extinctions, we know most about those resulting from our activities, usually because these records are relatively recent and accessible. If you recently visited Mauritius, you might have noticed that the few remaining Mauritian calvaria trees were slowly dying of old age. Their passing will mark the extinction of this species, which has occurred even though the trees continued to bloom and produce seeds. The key to the pending extinction of *Sideroxylon majus* is the earlier extinction of Dodos. To germinate, seeds of Mauritian calvaria trees had to pass through the Dodo's digestive tract. The Dodo **(Figure 48.5)** was a medium-sized flightless bird that lived on the island of Mauritius. When European sailors first visited the island, they used Dodos as a source of fresh meat. Then, as the island was settled, the birds were exposed to introduced predators (cats, dogs, rats) and an expanding human population. Dodos vanished by 1690.

Species confined to islands often have small populations and are unaccustomed to terrestrial predators, making them vulnerable to extinction. The fossil and subfossil records show that many species of birds disappeared from islands in the South Pacific as Polynesians arrived there from the west. This occurred from Tonga to Easter Island and beyond **(Figure 48.6)**. The Galápagos, only discovered by people in 1535, was sheltered from the wave of human-induced extinctions. On Easter Island, endemic species of sea birds and other species disappeared soon after people settled there. These examples demonstrate that humans do not have to be industrial or "high tech" to effect extinctions.

Meanwhile, in the North Atlantic, people hunted *Pinguinus impennis*, the great auk, to extinction. However, land birds with large distributions and huge populations have also disappeared, such as *Ectopistes migratorius*, the passenger pigeon in eastern North America. Large-scale harvesting of these birds, combined with their low reproductive rate (clutch size: one egg), made the birds vulnerable in spite of their enormous populations. Animals that produce one young per year and suffer "normal" mortality must live at least 10 years to replace themselves in the population (see Chapter 45).

STUDY BREAK

1. When did the five mass extinctions occur? Which was the most severe? Which extinction affected the dinosaurs?
2. What caused these extinctions?
3. Why are island animals and plants particularly susceptible to extinction due to human impacts?

a.

b.

Figure 48.5
(a) A reconstruction of a *Raphus cucullatus*, the Dodo, an extinct flightless bird from **(b)** Mauritius.

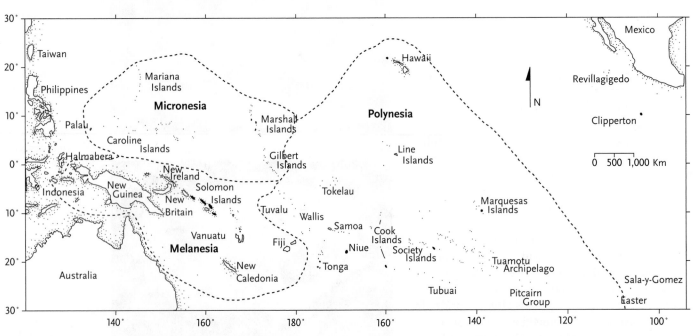

Figure 48.6
Islands in the South Pacific where the arrival of Polynesians coincided with the extinction of many island species of birds.

The 2,4-D Molecule and Resiliency

Resiliency is one of the most impressive features of life at the species and/or ecosystem levels. In one respect, this feature complicates the challenges of conserving biodiversity because introduced species can be so invasive, reflecting their adaptability.

Humans first identified 2,4-D (2,4-dichlorophenoxyacetic acid; **Figure 1**) in 1942, and from 1944, it was marketed as a herbicide more effective against broad-leaved plants than against grasses. Technically, 2,4-D is a hormone absorbed by the plant and translocated to the growing points of roots and shoots. 2,4-D kills weeds by inhibiting growth. The global market for 2,4-D is probably more than U.S.$300 million, and it is mainly used to control broad-leaved weeds in cereal crops. According to the World Health Organization (WHO), 2,4-D is a "moderately hazardous pesticide" known to affect a variety of animals (e.g., dogs but not rats). Curiously, it turns out that other animals may use 2,4-D for their own ends.

In 1971, Thomas Eisner and colleagues reported that a grasshopper (*Romalea microptera*; **Figure 2**) produced a froth of chemicals **(Figure 3)** for protection against ants. One of the main ingredients in the froth was 2,5-dichlorophenol, apparently derived from 2,4-D. This is an astonishing demonstration of adaptability that can underlie resiliency.

Resiliency and the recuperative powers of ecosystems are demonstrated by stories of "lost cities," for example, structures built by Maya in Central America, being found in a jungle. Archaeological evidence reveals that in some habitats, these buildings and pyramids were overgrown by the

rain forest in ~100 years. The Great Zimbabwe Ruins in southern Africa were overgrown by savannah woodland in a period of 100 to 200 years and only latterly "discovered" by European explorers.

2, 4-D

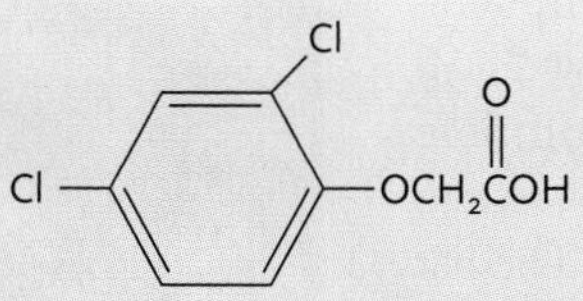

Figure 1
2,4-Dichlorophenoxyacetic acid, 2,4-D.

(2, 4-dichlorophenoxy)acetic acid

© KHALED KASSEM/Alamy

Figure 2
Romalea microptera, a grasshopper that uses an ant repellent with a 2,4-D derivative.

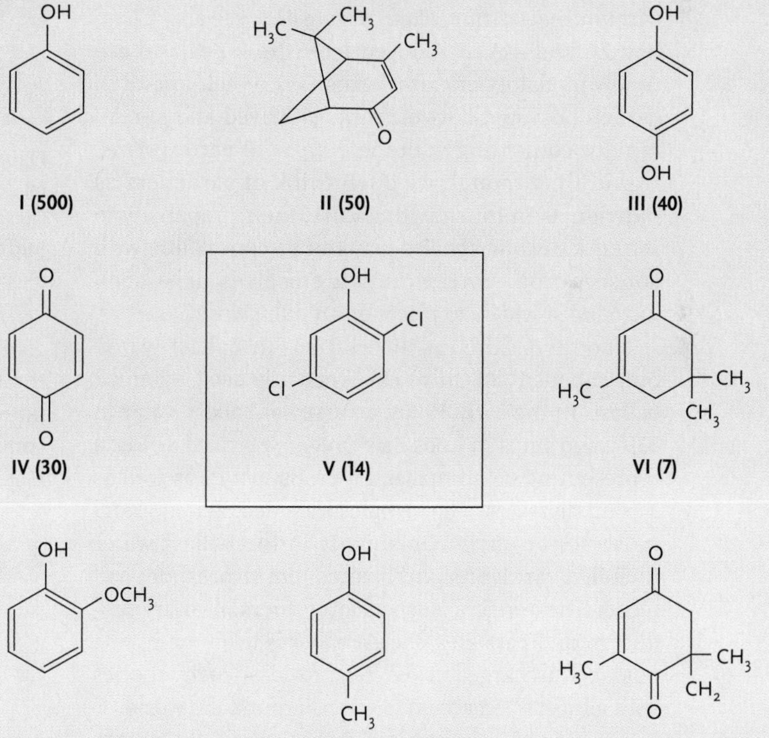

I (500) II (50) III (40)

IV (30) V (14) VI (7)

VII (4) VIII (2) IX (1)

Figure 3
Active ingredients in the defensive froth of the grasshopper, *Romalea microptera*. 2,5-Dichlorophenol (boxed) is apparently derived from 2,4-D.

48.3 Introduced and Invasive Species

Humans cause extinction through hunting and by the introduction of other species. House cats, *Felis domesticus,* are among the worst introductions people have made. Anecdotal records suggest that in 1894, one house cat (named "Tibbles") exterminated an entire population of flightless wrens **(Figure 48.7, p. 1212)** on Stephen's Island, a 2.6 km² island off the north shore of New Zealand. Fossils indicate that the wrens had occurred widely in New Zealand. This record stands for one individual, Tibbles, taking out the remaining ~10 pairs and exterminating the species.

Figure 48.7
Stephen's Island Wren, *Xenicus lyalli*. This species was exterminated by one cat.

a.

b.

Figure 48.8
(a) Zebra mussels, *Dressina polymorpha*, were introduced to the Great Lakes in North America, where they have spread rapidly. **(b)** They are directly responsible for the declines in eastern pond mussels (*Lampsilis radiata*), a local mussel species.

It should be obvious that moving species from one part of the world to another, whether done willfully or by accident, can have calamitous impacts. The invaders, once arrived and established, may outcompete resident species, laying waste to species and ecosystems. The list of introduced organisms is very long and includes many domesticated or commensal species of animals and plants. The arrival of zebra mussels **(Figure 48.8a)** in the Great Lakes is the main reason for the decline of the now endangered eastern pond mussels **(Figure 48.8b)**. The immigrant mussels outcompeted and overgrew the native ones, reducing their range and populations to levels that resulted in eastern pond mussels being recommended for listing as endangered in Canada in 2007.

Meanwhile, in parts of the British Isles, flatworms (*Arthurdendyus triangulatus*; **Figure 48.9;** see also Chapters 26 and 41) introduced from New Zealand are deadly predators of earthworms. Since their arrival in garden pots, the flatworms have thrived and spread rapidly, coinciding with the demise of earthworms. Ironically, although we often think of gardeners as individuals in touch with nature, their propensity to introduce **exotic species** may not be compatible with conservation. Earthworms themselves have been introduced widely to places around the world.

Some organisms move about in ballast water. Since about 1880, ships have regularly used water for ballast. In the early 1990s, a survey of ballast water in 159 cargo ships in Coos Bay, Oregon, revealed 367 taxa representing 16 animal and 3 protist phyla, as well as 3 plant divisions. The samples included all major and most minor phyla. Organisms in the ballast water included carnivores, herbivores, omnivores, deposit feeders, scavengers, suspension feeders, primary producers, and parasites. Ballast water is taken on in one port and discharged in another, providing many species with almost open access to waters around the world.

Meanwhile, introduced diseases (and the organisms that cause them) have decimated, if not obliterated, resident species. When Europeans arrived in the New World, *Castanea dentata*, the American chestnut tree, was widespread in forests from southern Ontario to Alabama. This large tree of the forest canopy grew to heights of 30 m. Often most abundant on prime agricultural soils, the distribution and density of the species were reduced as settlers from Europe cleared more and more land for agriculture. *Endothia parasitica*, the chestnut blight, was introduced perhaps around 1904 from Asian nursery stock. This introduced blight killed the American chestnut trees by the 1930s. By 2000, only scattered American chestnut trees remained, most of them stump sprouts.

Why are invading species so successful? Does the spread of Starlings (*Sturnus vulgaris*) or dandelions (*Taraxacum officinale*) after introduction to new continents suggest that they moved into vacant niches? Does it mean that they are better competitors? In the

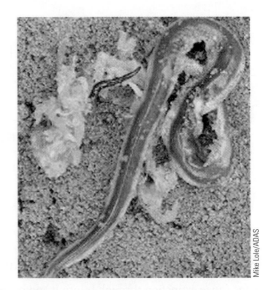

Figure 48.9
This earthworm-eating planarian (*Arthurdendyus triangulatus*) was introduced to the British Isles from New Zealand. It has had a devastating effect on local populations of earthworms.

case of starlings, 13 birds were introduced to Central Park in New York City in 1890, and they have spread far and wide. Once they are established, invading or introduced species can pose huge conservation problems because of their effects on ecosystems and diversity.

Although many invaders arrive, only a few are widely successful and become large-scale problems in their new settings. Invading plants are most often successful in nutrient-rich habitats, where they can achieve high growth rates, early reproduction, and maximal production of offspring. What happens in resource-poor settings? In the past, conventional wisdom has suggested that low-resource settings could be reservoirs for native species that could outcompete invaders.

However, an experimental examination of the responses of native and introduced species to challenging conditions revealed that invasive plant species almost always fared better **(Figure 48.10)**. *Resource use efficiency* (RUE), calculated by measuring carbon assimilation per unit of resource, provides an indicator of success. Many invasive species, such as ferns, C_3 and C_4 grasses, herbs, shrubs, and trees, were more successful in low-resource systems than native species were.

This research was conducted in Hawaii, an excellent place for studying invasive species because so many are there. Among the invaders were *Bromus tectorum* (cheatgrass), *Heracleum mantegazzianum* (cartwheel flower or giant hogweed), and *Pinus radiata* (Monterey pine). Humans have introduced these plants for gardening (cheatgrass and cartwheel flower) or commercial timber production (Monterey pine). The data demonstrate that attempting to restore ecosystems and exclude invading species by reducing resource availability does not succeed because of the efficiency with which some species use resources.

STUDY BREAK

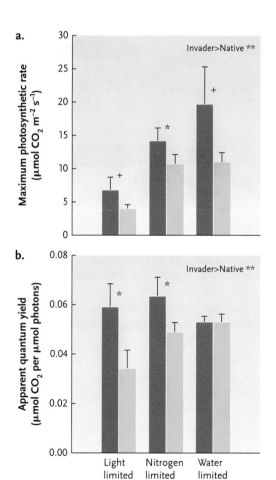

Figure 48.10

(a) Photosynthetic rates (resource use efficiency [RUE]) and **(b)** light-use efficiency of invasive plant species (blue bars) make them more competitive than native ones (yellow bars). The plants were from three different habitats in Hawaii.

48.4 How We Got/Get There

Lamentably, we know that humans can exterminate species that are populous and widespread as well as ones that have small populations and occur in a small area.

48.4a The Black Rhinoceros: Its Demise

It is estimated that 60 000 black rhinos (*Diceros bicornis*) lived in the wild in Africa in 1960 **(Figure 48.11a, p. 1214)**. This large (1.5 m at the shoulder, 1400 kg) browsing mammal was widespread in sub-Saharan Africa **(Figure 48.11b)**. Adult males and females have two distinctive "horns" **(Figure 48.12a, p. 1214;** see also Figure 48.11a), actually formed from hair. Rhinos use the horns to protect themselves and their young from predators and other rhinos. By 1981, the populations in the wild had been reduced to 10 000 to 15 000, and again reduced to about 3500 by 1987. Today only a few individuals survive in some protected areas in Africa. In less than 30 years, the species was almost exterminated in the wild.

In 1960, black rhinos were one of the "big five" on the list of big game for which hunters made safaris to Africa to shoot as trophies. Others on the list included the African lion, African elephant, Cape buffalo (*Syncerus caffer*), and leopard. Safari hunters then paid large sums of money to go to Africa and obtain licences to kill trophy specimens of each of the big five. But this hunting pressure, which has since stopped, did not lead to the extermination of black rhinos.

Figure 48.11
(a) Black rhinos (*Diceros bicornis*) were widespread and common in Africa in 1960 (orange area on the map). **(b)** Today their range (dark spots in orange areas) is much reduced, reflecting diminished poulations. Note the oxpecker (*Buphagus africana*) sitting on the rhino.

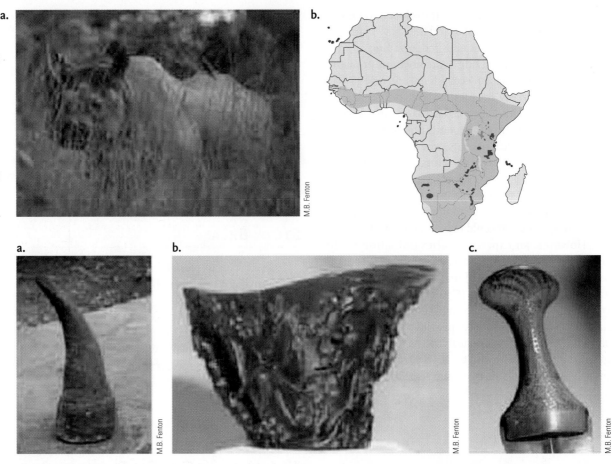

Figure 48.12
(a) A horn from a black rhino in Zimbabwe is shown with **(b)** a rhino horn bowl from China and **(c)** a jambiya with a rhino horn handle.

People have long used the horns of all species of rhino in different ways. In China, bowls made from rhino horn **(Figure 48.12b)** were believed to have magical properties in that they could remove or neutralize poisons. Travelling nobles were served wine in their own rhino horn bowls to minimize the chances of their being poisoned. In India and some other areas from India to Korea, powdered rhino horn was used as a fever suppressant. Contrary to popular belief, rhino horn does not appear to have been used as an aphrodisiac, an early version of Viagra®.

An Arabian Peninsula tradition is the carrying of a jambiya or ceremonial dagger. Jambiyas with rhino horn handles **(Figure 48.12c)** were highly prized. In 1973, when the price of oil jumped from U.S.$4 to U.S.$12 a barrel, the ensuing "energy crisis" meant a larger market for jambiyas. Increased cash flow and easy access to military weapons such as Kalashnikov assault rifles **(Figure 48.13)** provided an incentive and a means to kill rhinos. The epidemic of poaching started in northern Kenya and spread southward throughout the continent. Thus, poaching for their horns led to the catastrophic reduction in the populations of black rhinos. The large population of rhinos that had long survived in the presence of predators, including *Homo sapiens*, was not protected from extermination. In 1984,

Figure 48.13
A Kalashnikov assault rifle (an AK), a weapon widely used in the poaching of animals in many parts of the world.

going for a walk at night around the headquarters of Mana Pools National Park in Zimbabwe almost always meant meeting a black rhino. By 1987, the rhinos were very scarce, and by 1990 they did not exist in the area.

The demise of black rhinos can only be attributed to human greed.

48.4b The Barndoor Skate: Victim of Bycatch

Our harvesting of food organisms can affect more than just the target species on land and at sea. Barndoor skates (*Dipturus laevis*; **Figure 48.14**) are elasmobranchs that used to occur widely in the northwest Atlantic. With a maximum body width of ~1 m, this is one of the largest skates. Dramatic reductions

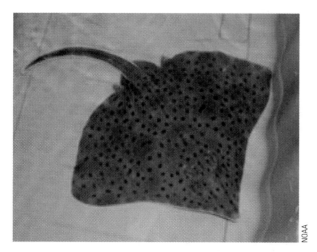

Figure 48.14

The barndoor skate (*Dipturus laevis*) was once widespread in the northwestern Atlantic Ocean. Mortality associated with bycatch has severely reduced its populations.

in the biomass of barndoor skates are obvious from locations ranging from the southern Grand Bank to southern New England **(Figure 48.15)**.

Recent captures of barndoor skates have been at depths greater than 1000 m, which may be one of the last refuges for this distinctive species. A combination of directed fishing for skates off the coasts of Newfoundland and Nova Scotia and bycatch of barndoor skates may spell the end for this species. In fishing terms, "bycatch" occurs when nontarget species are taken by fishers. Victims of bycatch include other species of fish, as well as sea turtles and marine mammals.

Removing species from ecosystems or depleting their numbers can also affect many other species in the ecosystem.

48.4c The Bay Scallop: Overfishing of Sharks

Populations of organisms we harvest for food often show marked declines. The annual harvest of bivalve molluscs has been a local fishery in

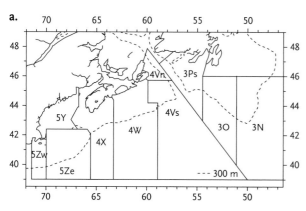

Figure 48.15

(a) Fishery subdivisions off the east coast of Canada and the U.S. providing data about the abundances of barndoor skates. **(b)** Catch records of barndoor skates from those fishing subdivisions.

Chesapeake Bay in the United States and elsewhere along the eastern seaboard for hundreds of years. In 1999, populations of bay scallops (*Agropecten irradians*; **Figures 48.16 and 48.17, p. 1216**), a main target of the fishery, were very low. The immediate reason for the low populations was the impact of predation by skates and rays that feed heavily on bivalve molluscs. Skates and rays are tertiary consumers and in turn are eaten by larger elasmobranchs, specifically various species of sharks.

Among tertiary consumers, the cownose ray **(Figure 48.18, p. 1216)** showed a marked increase in population. Evidence from surveys on the U.S. Atlantic coast estimates an order-of-magnitude increase in populations of cownose rays, and the total population of 14 species of rays and skates exceeds 40 million. So the decline in scallop (and other bivalve) populations can be explained by the increase in predation by tertiary consumers, especially skates and rays.

The picture becomes clearer when the population data for the local great sharks are added to the

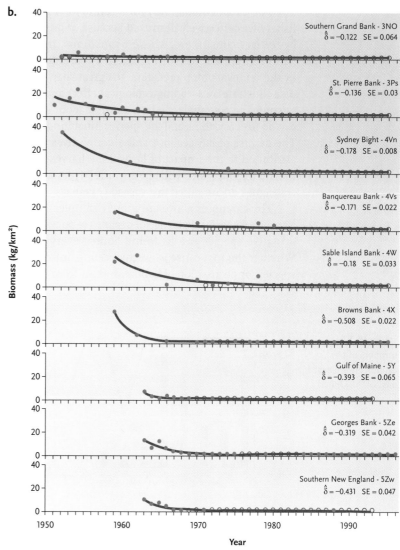

Figure 48.16
A handful of bay scallops (*Agropecten irradians*).

mix **(Figure 48.19)**. Prolonged and intensive fishing of 12 species of sharks accounts for a 35-year decline in their populations (see Figure 48.19, top row). The sharks have been taken primarily for their fins and meat. In some parts of the world, shark fins sell for ~U.S.$700 per kilogram and are used to make shark fin soup.

The data demonstrate how a century-old scallop fishery was effectively destroyed because of predation by tertiary consumers, whose populations, in turn, had been enhanced (see Figure 48.19, middle row) by the removal of top predators, the great sharks. The data illustrate a cascading ecological effect and demonstrate the potential long-term harm that our species can do to ecosystems and the species inhabiting them. The demise of bay scallops and other bivalves can be attributed to the impact of large-scale harvesting of marine resources. The late Ransome Myers and his colleagues documented this cascade of effects.

The examples above are merely samples from a long list of species. Evidence of declines of populations of native species can be found almost everywhere. Whether the root cause is overharvesting, introduced species, or destruction of habitat, species from whales

to songbirds are threatened by human activity. What can we do about it?

STUDY BREAK

1. Why was Tibbles so successful at exterminating the remaining Stephen's Island Wrens?
2. What risk does ballast water pose to native ecosystems?
3. What is the significance of RUE?

48.5 Protecting Species

The widespread recognition of trademarks such as the World Wildlife Fund (WWF) panda demonstrates how associating a cause with an icon can be very successful. It is not surprising that many conservation efforts began with a focus on one species—such as giant pandas (*Ailuropoda melanoleuca*), polar bears (*Ursus maritimus*), or redwood trees (*Sequoia sempervirens*). The lure of conservation movements that focus on charismatic species is very strong. But charismatic organisms may not need protection, whereas some species that are unattractive, dangerous, or mundane are in desperate need of our assistance. Unfortunately, mundane, ugly, and dangerous (to us) species are unlikely to serve as a call to arms (or to attract financial support). Worldwide, the WWF panda is one of the most recognized logos, whether or not pandas are in the neighbourhood.

A critical first step toward conservation is the development and adoption of objective, data-based criteria for assessing the risk posed to different species. This process has been developed on several fronts around the world. The criteria and assessment procedures perfected by the International Union for the Conservation of Nature (IUCN) are used widely. There are many records of success, but there also are many examples of species and situations in which we

Figure 48.17
Numbers of bay scallops off the east coast of the United States.

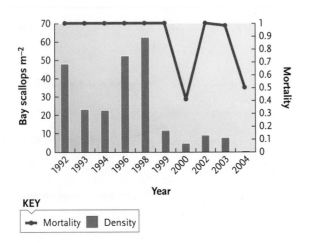

KEY
Mortality | Density

Figure 48.18
A cownose ray (*Rhinoptera bonasus*).

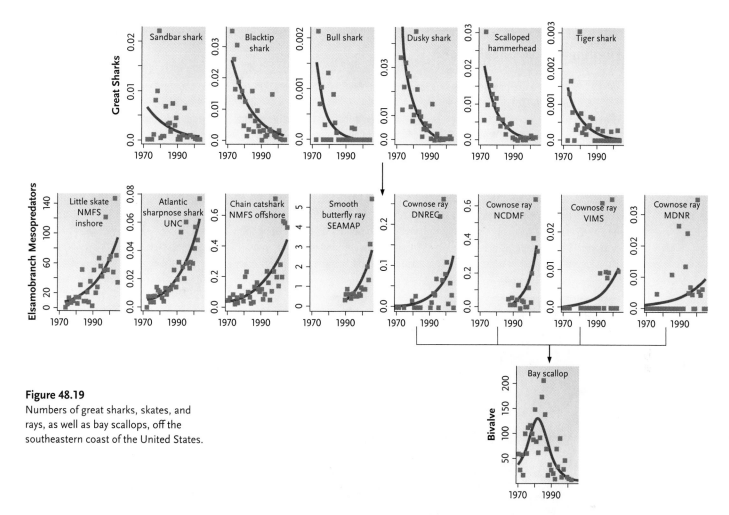

Figure 48.19

Numbers of great sharks, skates, and rays, as well as bay scallops, off the southeastern coast of the United States.

have failed. Making arguments based on data does not guarantee success. Using a data-based approach, some species emerge as being in need of protection, but others do not. Being rare or unusual, by itself, will not warrant protection. The species concept and the Linnaean system of nomenclature (see Chapter 19) are fundamental to conservation.

In Canada, recommendations about the conservation status of species involve the Committee on the Status of Endangered Wildlife in Canada (COSEWIC). The definition of wildlife includes plants and animals. Like IUCN, COSEWIC recognizes six categories for assessing species at risk:

- *Extinct:* a wildlife species that no longer exists
- *Extirpated:* a species no longer existing in one location in the wild but occurring elsewhere
- *Endangered:* a species facing imminent extirpation or extinction
- *Threatened:* a species likely to become endangered if limiting factors are not reversed
- *Special concern:* a species that may become threatened or endangered because of a combination of biological characteristics and identified threats
- *Data deficient:* a category used when available information is insufficient either to resolve a wildlife species' eligibility of assessment or to permit an assessment of its risk of extinction.

A seventh category—*not at risk*—is used to identify species not at risk of extinction under current circumstances.

COSEWIC members vote on the appropriate conservation category for each species whose status they review. The members consider the area of occupancy, an indication of the range of a species and the availability of suitable habitat. They take into consideration population information, including trends in the numbers of organisms, correcting for species that show extreme fluctuations in numbers from year to year. They consider the demographics of the species and the variability in the habitat where the species occurs. Generation time also is considered, along with specific habitat features that may be essential for the species' survival. Data on population size, particularly the numbers of reproducing adults, are important, as well as risks to the species' survival.

In a biological context, the criteria used by COSEWIC (and similar agencies elsewhere) are familiar to population biologists (see Chapter 45). The data describe the numbers of individuals in the population, fecundity, mortality, and the intrinsic rate of increase. Carrying capacity also is important, as is the area (range) over which the species occurs. These criteria are designed to promote data-based decisions about the conservation status of species.

Hunting: Threat or Salvation?

We saw earlier (see Chapter 19) how the Linnaean system of nomenclature is used to name species. Once a species has a name, however acquired, it may benefit from protection under CITES, the Convention for International Trade in Endangered Species. But will data-based decisions about what counts as endangered be consistent and predictable? The answer is "yes" and "no." The example of black rhinos showed one situation in which protection under CITES did not work. There are others.

Also in Africa, the leopard (see the chapter opening photograph) was accorded protection under CITES. The passing of the Endangered Species Act (ESA) in the United States (1972) precipitated an interesting situation: it obliged Americans to "obey" the listing of leopards on CITES Appendix 1, which banned the importation of leopard skins, including those shot on safari hunts. The rationale for the listing was the belief that leopards were endangered and their survival was threatened by hunting.

There were quick, negative responses to the ban on importing leopard skins into the United States from two different groups. First was the hunting and related associations and lobbies whose members were anxious to be able to bring home trophies. Second, leaders and governments in many African countries that benefitted from the hunts objected to the ban because safaris were (and still are) an important source of foreign exchange. In many of these countries, "safari hunting areas" were set aside to accommodate visitors, and these large tracts of land also protected populations of nongame species and appropriate habitat.

What do the data show? Leopards are 40 to 80 kg, solitary cats that hunt by stealth. They are widespread in Africa but have been little studied. The estimate is that there are more than 700 000 leopards in the wild in Africa, with resident populations in all but very small countries with high human population densities. In 2000, Zimbabwe alone had a population of more than 16 000 leopards in the wild. The 1969 safari harvest of 6100 leopards throughout Africa and the export of their skins were not a threat to the population in Zimbabwe, let alone to leopards in the whole continent.

Ecologists studied the population of leopards in the Matetsi Safari Area in Zimbabwe. Before 1974, the 4300 km² area was a cattle ranch whose operators made strong efforts to eradicate leopards to protect their livestock. After conversion to a hunting area, people on the first safaris rarely succeeded in shooting leopards. By 1984, the leopard population in the Matetsi Safari Area was 800 to 1000, and in 1988, the annual safari quota there was 3.6% (12 to 28 leopards). When leopards shot in the mid-1980s were compared with those taken in the 1970s, no change in leopard size was found. But by 1986, the average

Figure 1
Polar bear, *Ursus maritimus*.

M.B. Fenton

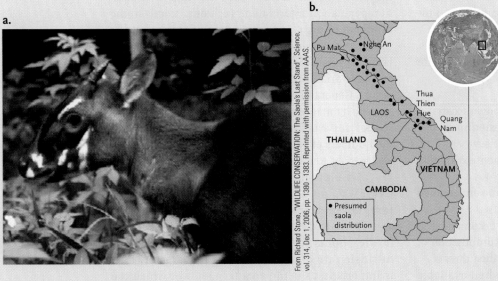

From Richard Stone, "WILDLIFE CONSERVATION: The Saola's Last Stand". *Science*, vol. 314, Dec 1, 2006, pp. 1380 - 1383. Reprinted with permission from AAAS.

Figure 2
(a) Saola (*Pseudoryx nghetinhensis*) and (b) its distribution.

age of leopards taken as trophies was 5.4 years, compared with 3.2 years from the earlier period. These data show that leopards can persist even when subjected to heavy hunting pressure. On average, leopards live longer in a safari hunting regime than when they are being hunted in the context of predator control operations. Other evidence suggests that populations of leopards persist even in urban areas—trapping evidence suggests that resident leopards live in Nairobi, the capital of Kenya.

Leopards are an interesting example of human responses to conservation. Hunting or some other form of harvesting is not necessarily a threat to the survival of some species. Indeed, some harvesting may be critical to the livelihood of some people and can advance efforts to protect some species. But decisions about harvesting made in one part of the world can influence what happens elsewhere.

Today there are quotas for the numbers of leopards that can be harvested in different countries in Africa. Safari hunters must obtain licences to take trophies, and skins exported must be accompanied by paperwork showing that the harvest was legal. The documentation allows a citizen, for example of Canada or of a European Union country, to import a leopard skin. This was not possible in the United States in the 1970s, but it is in 2009. In Africa, local farmers are permitted to kill "problem" animals that threaten their livestock or themselves and their families and may be supported in this by government officials.

Key elements in the success of harvesting include having data about the population of organisms, the rates of reproduction, and the rates of harvest. Enforcement of quotas is essential if this approach is to succeed. Legal harvest quotas do not require people who object to hunting to be hunters. Trophy

hunting is not the exclusive preserve of countries in Africa. On April 3, 2007, *The Globe and Mail* (a national newspaper published in Toronto) reported that the economy of the Canadian territory of Nunavut received Can$2.9 million from polar bear **(Figure 1)** hunting. Hunters can pay U.S.$20 000 for a polar bear hunt.

In 1992, saola **(Figure 2)** made the news as one of the first "new" species of large mammals to be discovered in recent times. These goatlike animals live in a restricted area of Vietnam, where they have been and are hunted by local people. Saolas are rare, and little is known about them. There are no quotas for the local hunters, and it is not practical to enforce a ban on their harvest. In reality, we probably lack critical information about the biology of many species of wildlife today. However, once they have names, they have a chance of being protected.

STUDY BREAK

1. What is IUCN? What role does it play in conservation?
2. What criteria would identify a species as endangered? Give an example.

48.6 Protecting What?

Before data are used to address questions of species-at-risk status, conservation biologists must decide about eligibility. The conservation jargon for this is "designatable unit." Are the organisms "real" species? Are they subspecies? Are they distinct populations? Are they really Canadian? Do they regularly occur in Canada or perhaps turn up here by accident? If the species does not breed here, is the habitat they use in Canada essential to their survival? Most species of wildlife in Canada occur close to the border with the United States, and many species widespread in the United States just make it into Canada. In some cases, a distinct population is treated as a designatable unit. Distinct populations

may be recognized by their geographic distribution and/or their genetic structure.

Questions about what units are designatable harken back to the definition of species (see Chapter 18). Off the west coast of Canada, striking differences in behaviour can be used to distinguish between two "kinds" of killer whales. The "resident" killer whales eat mainly fish and often echolocate. The "transient" killer whales eat mainly marine mammals and rarely produce echolocation signals. Furthermore, repeated sightings of recognizable individual whales indicate that different groups of these animals live in different areas along the coast **(Figure 48.20, p. 1220).**

In reviewing the conservation status of killer whales, COSEWIC recognized different designatable units based on behaviour and geography **(Figure 48.21, p. 1221).** The different units faced different threats to their survival.

Questions about what to protect often reflect different realities of biology. Migrating birds may be blown off course and end up in southern Ontario instead of their usual habitat much farther south. Marine birds or mammals may feed in Canadian waters but breed elsewhere. Many organisms commonly hitchhike, using ocean vessels, aircraft, or automobiles as vehicles of dispersal. But some hitchhikers,

An Endangered Species

Banff Springs snails, *Physella johnsoni* **(Figure 1),** live and eat algae in five hot springs on Sulphur Mountain in Banff National Park, Alberta. Not very long ago, Banff Springs snails were found in nine springs. In 1996, the total population of snails was ~5000. Water temperatures in the springs occupied by the snails range from 26° to 48°C, but temperatures less than 44°C seem best for them. Their very limited occurrence makes them vulnerable to extinction (COSEWIC, 2000).

Humans appear to be the main threat to the survival of Banff Springs snails. By discarding unsightly (to humans) accumulations of algae from pools, people have killed some snails that were in the algal mats. Changes in the patterns of water circulation may subject some snails to high temperatures that could be lethal. Well-wishers that throw copper coins into the pools may have harmed snails because of contamination arising from the interaction of copper with sulphurous water in the springs.

Other impacts of people are not clear, but one threat is entertaining to contemplate—people "skinny-dipping" in the pools are thought to threaten the snails. Some skinny-dippers have been caught and charged. Bathers in the pools, clad or unclad, may have crushed snails while getting into or out of the water. Bathers doused in sunscreen or insect repellants may have introduced chemicals into the snails' habitat and further reduced their populations.

Banff Springs snails are neither charismatic nor prominent, but the data-based approach to decision making has provided the basis for identifying them as endangered.

Figure 1
Banff Springs snail, *Physella johnsoni.*

C.M. & L. Degner/Parks Canada

a.

M.B. Fenton

b.

M.B. Fenton

c.

M.B. Fenton

Figure 48.20
Three views of a killer whale (*Orcinus orca*). **(a)** A captive animal in Vancouver, **(b)** a wild orca swimming off the Queen Charlotte Islands, and **(c)** a Haida representation.

for example, some snails, travel with birds, making the association and the dispersal more "natural."

People can be quick to try to protect species they consider to be important or distinctive. In 2003, the Ontario Ministry of Natural Resources reported four to six white-coloured moose (*Alces alces*) among the approximately 1900 moose in two wildlife management areas near Foleyet in northeastern Ontario. Should white-coloured moose be protected? There was local support for protecting the moose, animals that have cultural and spiritual significance for First Nations communities. White moose have been reported from other places in northern Ontario, Newfoundland and Labrador, and elsewhere. Although the population of white moose is small and widespread, there is no evidence that they are a designatable unit. In Canada, they have not been accorded special protection.

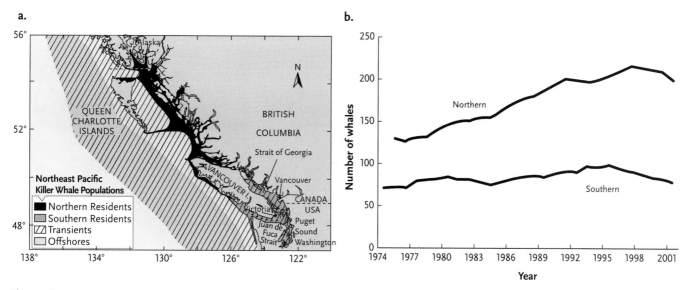

Figure 48.21

(a) The population distribution (designatable units) of killer whales off the coast of British Columbia and **(b)** estimates of population sizes of northern and southern resident killer whales (British Columbia).

STUDY BREAK

1. Give an example of a designatable unit that is a species.
2. Give an example of a designatable unit that is a population.
3. Is "designatable unit" synonymous with "species"?

48.7 The Downside of Being Rare

Whether the commodity is coins, stamps, antiques, or endangered species, as soon as something is rare enough, there is a market for it. This "get them while they last" attitude is exemplified by trade in *Leucopsar rothschildi*, Bali Starlings **(Figure 48.22a)**. This bird, another island species, faces immediate extinction, but it is in high demand as an exotic pet. In 1982, when there were fewer than 150 individuals in the wild, 35 were for sale as pets, 19 in Singapore and 16 in Bali.

Rare species also may be in demand for use of their body parts in traditional medicine. One stark example is the swim bladders of *Bahaba taipingensis*, the Chinese bahaba **(Figure 48.22b)**. At a time when fewer than 6 individuals are caught each year, more than 100 boats are trying to catch them. The swim bladders are used in traditional medicine. They are worth at least seven times their weight in gold. Shark fins are even more valuable. These are extreme examples of the earlier story about rhino horns and jambiyas.

Before criticizing and condemning the users or consumers of jambiyas or Chinese bahaba swim bladders, think about the overall impact of our lifestyle on other species of animals and plants. Of particular note is an insatiable demand for energy. Are sport utility vehicles necessary? Jet skis? Snowmobiles? All-terrain vehicles? The list goes on. Is a Canadian as justified in buying a large SUV as a North Yemenese a jambiya with a rhino horn handle? Once again, might (the ability or capacity to do something) may not be right.

a.

b.

Figure 48.22

(a) A Bali Starling (*Leucopsar rothschildi*) and **(b)** Chinese bahaba (*Bahaba taipingensis*).

Who Gets Protection?

Being recognized as rare and considered to be endangered does not necessarily translate into protection.

Because the Endangered Species Act (ESA) in the United States does not protect hybrids, this can affect the conservation of, for example, the "Florida panther" **(Figure 1)**, a subspecies of cougar. Cougars, also known as panthers, used to occur widely in North, South, and Central America. Although still widespread in some areas, the current range of cougars in most of the United States and Canada is much less than it was when Columbus arrived in the New World in 1482. Florida panthers, a small population recognized as a subspecies, occur mainly in the Florida keys. Florida panthers were protected under the ESA.

Using techniques of molecular genetics, biologists determined that Florida panthers carried the genes of cougars from South America. This situation probably arose when panthers originally caught in South America were brought to the United

Figure 1
A Florida panther (*Felis concolor coryi*).

States as zoo animals or for display in circuses or animal shows. Some of these animals escaped and interbred with local Florida panthers. Florida panthers with genes from South American cougars are technically hybrids and therefore are not protected by the ESA.

There are many other examples of situations in which genetic tools allow clearer delineation of boundaries between populations (designatable units) and species. In some cases, however, removal of protection from other "species" because of their genetic status can lead to their extinction. *Ammodramus maritimus nigrescens* or the Dusky Seaside Sparrow was previously considered to be a distinct form living in Florida. When genetic evidence showed that these darker animals were not genetically distinct, they lost their protected status and have virtually disappeared.

Other species, such as round-nosed grenadier, have suffered calamitous declines in population. These cod-like fish **(Figure 2)** were taken in large numbers after cod populations had declined **(Figure 3)**. The species was on the verge of extinction even before much was known about it. We do know that round-nosed grenadiers are late to mature, and their populations are slow to recover.

Although round-nosed grenadiers and at least four other species meet the IUCN criteria for listing as

Figure 2
Coryphaenoides rupestris, a round-nosed grenadier.

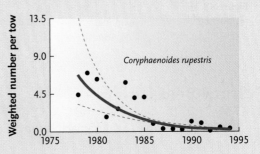

Figure 3
Captures of round-nosed grenadiers.

"endangered," these fish have not been or are not protected. Fisheries and Oceans Canada has not supported a move to protect round-nosed grenadiers. Changing fishing practices to avoid catching the few remaining round-nosed grenadiers is not economically feasible when other species are still being caught in sufficient numbers to justify a continued fishery. The situation differs only from that facing the barndoor skate in that the round-nosed grenadiers have been the targets of an active fishery.

We have seen that the hunt for polar bears can bring significant income to the economy of Nunavut, and the same is true of other jurisdictions within the bear's range. There are distinct populations of polar bears **(Table 1),** even within Canada's jurisdiction. The occurrence of bears in political jurisdictions including Canada, the United States, Russia, Iceland, Denmark (Greenland), Norway, Finland, and Sweden makes achieving their protection more difficult. The apparent vulnerability of the bears to global warming and their value as trophies may combine to hasten their demise.

Table 1 | **Status of Canadian Polar Bear Populations (January 1997).**

Population	% Females in harvest	Number	Sustainable Annual Kill	Mean Annual Kill	Environ. Concern	Status[1]	Quality of Estimate	Degree of Bias	Age of Estimate	Harvest/ Capture Data
Western Hudson Bay	31	1200	54	44	None	S[a]	Good	None	Current	Good (>15 yr)
Southern Hudson Bay	35	1000	43	45	None	S[a]	Fair	Moderate	Old	Fair (5–10 yr)
Foxe Basin	38	2300	91	118	None	S[a]	Good	None	Current	Good (>15 yr)
Lancaster Sound	25	1700	77	81	None	S[a]	Fair	None	Current	Good (>15 yr)
Baffin Bay	35	2200	94	122	None	D?[b]	Fair	None	Current	Fair (>15 yr)
Norwegian Bay	30	100	4	4	None	S[a]	Fair	None	Current	Good (>15 yr)
Kane Basin	37	200	8	6	None	S	Fair	None	Current	Fair (>15 yr)
Queen Elizabeth	–	(200?)	9?	0	Possible	S?[b]	None	–	–	–
Davis Strait	36	1400	58	57	None	S?[b]	Fair	Moderate	Outdated	Good (>15 yr)
Gulf of Boothia	42	900	32	37	None	S[a]	Poor	Moderate	Outdated	Good (>15 yr)
M'Clintock Channel	33	700	32	25	None	S[a]	Poor	Moderate	Outdated	Good (>15 yr)
Viscount Melville sound	0	230	4	0	None	I	Good	None	Current	Good (>15 yr)
Northern Beaufort Sea	43	1200	42	29	None	S	Good	None	Recent	Good (>15 yr)
Southern Beaufort Sea	36	1800	75	56	None	S	Good	Moderate	Recent	Good (>15 yr)

■ D = decreasing; I = increasing; S = stationary; ? = indicated trend uncertain
■ [a]Population is managed with a flexible quota system in which overharvesting in a given year results in a fully compensatory reduction to the following year's quota
■ [b]See text, "Population Size and Trend," for discussion.

1. What is the value of the International Union for the Conservation of Nature (IUCN)?
2. Why do species description and formal naming affect the Convention for International Trade in Endangered Species (CITES)?
3. What is the difference between an extinct species and an extirpated species? Give an example of each.

48.8 Protecting Habitat

It is obvious from many of the examples above that protecting species has not been entirely successful as a conservation strategy. As a species, we are much better at killing than we are at conserving. Whether the persecution is direct or indirect, the end result can be the same. It is also clear that destruction of habitat is an effective way to remove a species. For example, populations of mosquitoes can be limited by denying them places to lay their eggs. This is a common theme in public education programs designed to reduce the incidence of West Nile virus (or other mosquito-borne diseases).

Is protection of habitat an effective strategy? The answer can be "yes," particularly for species that are not motile. Many species of plants have specific habitat requirements. From trees to shrubs, forbs, ferns, and mosses, we know that we can protect species by protecting habitat. Furthermore, protecting large tracts of habitat can also protect large, mobile species. Rain forests, whether tropical or temperate, are examples of habitats that can be flagships for protection and conservation. They also are considered by many to be storehouses of wealth associated with biodiversity, from building materials to compounds of pharmacological value.

The case of the black rhino demonstrated how a species targeted for harvesting can be driven to the brink of extinction even when it is protected (or lives in national parks or game reserves). *Panax quinquifolius*, American ginseng, is another target species, now endangered in Canada because of harvesting. The species used to grow wild from southwestern Quebec and southern Ontario and south to Louisiana and Georgia. This 20- to 70-cm tall perennial is long-lived in rich, moist, mature, sugar maple–dominated woods. Although the species has been listed on Appendix II of COSEWIC since 1973, populations have continued to decline. In 2000, there were 22 viable populations in Ontario and Quebec, but none were secure. Black rhinos and ginseng were common ~50 years ago, but by 2008, both demonstrated the risks of being rare and expensive. They also are examples of the need for immediate on-the-ground enforcement of regulations and laws protecting species and habitats.

Protecting habitats can be most challenging in areas with larger human populations. None of the viable poplulations of American ginseng in Ontario and Quebec were far from a road, making the plants vulnerable to anyone who knew about them and wished to take advantage of the economic opportunity they presented. *Sorex bendiri*, the Pacific water shrew, is another example of a species whose future in Canada is threatened by expanding human populations and the associated value of real estate (**Figure 48.23**).

Also in British Columbia, expanding human population and the wine industry in the southern Okanagan Valley have combined to dramatically reduce a local ecosystem dominated by antelope bush (**Figure 48.24, p. 1226**). The antelope bush system, one of the most endangered ecosystems in Canada, is home to a number of species of plants and animals whose future is now threatened by the demise of the habitat they require. The boom in real estate for people looking for retirement properties, more than just the density of human populations, is a key factor in this situation. Meanwhile, in southern Ontario, the demand for real estate to accommodate the expanding housing and business market is reducing both the available natural habitats and farmland.

48.9 Effecting Conservation

Today we face many challenges when trying to protect biodiversity. Too many of the immediate threats are the direct or indirect consequences of human activities. Walt Kelly, the creator of *Pogo* (a cartoon of yesteryear), identified the problem (**Figure 48.25, p. 1226**)—us. We must protect species by acting at levels ranging from species to populations and habitats.

48.9a Human Population: A Root Problem in Conservation

One fundamental root cause of declining biodiversity is the human population and the energy and habitat consumed in trying to feed, house, and protect our flourishing species. Visit the Web site http://www.popexpo.ined.fr/english.html and use it to determine the estimated human population in the year you were born and then for the years in which your parents and grandparents were born. Even when many people are killed, the momentum of our population

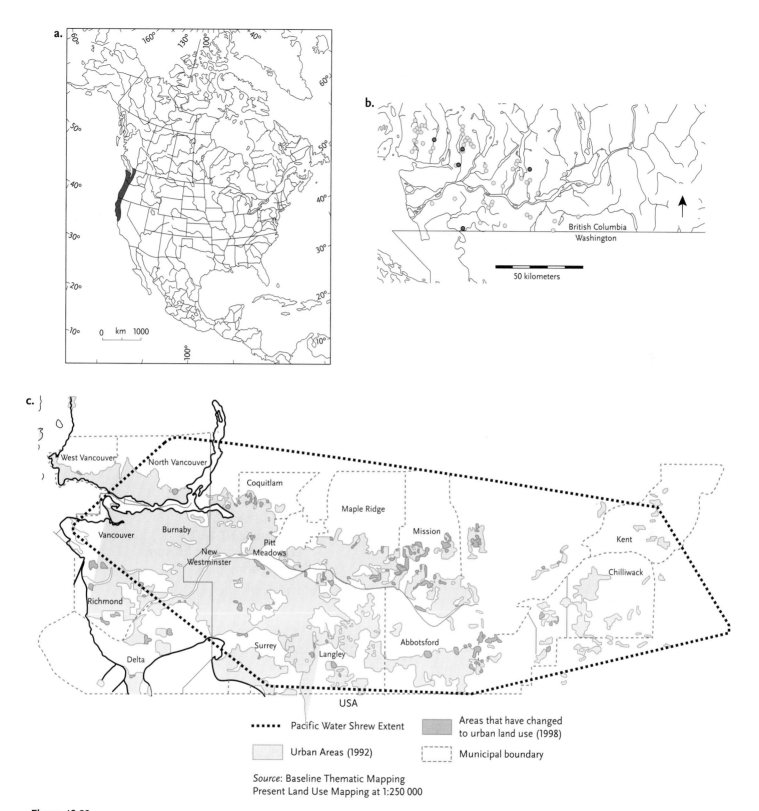

Figure 48.23

(a) The distribution of *Sorex bendiri*. **(b)** Lower Fraser Valley locations where it was found (solid circles) or not found (open circles) in recent surveys. **(c)** For comparison, the same area is shown with changes in the availability of urban lands in 1992 and 1998.

increase does not slow down. The December 2004 tsunami killed approximately 250 000 people, at a time when the world population was estimated at 6 billion. By comparison, the 1883 explosion of the island Krakatoa (and resulting tsunamis) is thought to have killed 35 000 people when the global human population was about 1.5 billion. If these estimates are correct, $4.1 \times 10^{-3}\%$ of the human population at the time was killed by the 2004 tsunami and $2.3 \times 10^{-3}\%$ by the explosion of Krakatoa. Neither calamity caused

a.

b.

M.B. Fenton

M.B. Fenton

Figure 48.24
Antelope bush, *Purshia tridentata*, showing **(a.)** the bush and **(b.)** a cross section of the stem. These woody shrubs have long life spans, and the ecosystem they typify is home to a variety of species.

the human population growth curve (see Chapter 45) to waiver.

If human population growth continues at the same rate as it grows now, it would double in 40 years. However, studies show that our population is not growing as quickly as it did during much of the twentieth century **(Figure 48.26)**. The United Nations Development Program (UNDP) has released data on human fertility (the total number of births per woman) for 162 countries **(Table 48.1)**. Compared with 1970–75, 152 countries had lower human fertility in 2000–05, 3 countries showed increases in fertility, and 7 showed no change.

Concerned about the global population and its effect on Earth, world leaders adopted the United Nations Millennium Development Goals in 2000, committing their nations to achieving the following goals by 2015:

- Ending poverty and hunger
- Universal education
- Gender equality
- Child health
- Maternal health
- Combatting HIV/AIDS
- Environmental sustainability
- Global partnerships

Figure 48.25
Walt Kelly's famous cartoon character, Pogo the Possum, in conversation with Porky (the porcupine), summed up the problem.

1971 Earth Day poster written and illustrated by Walt Kelly, featuring Pogo and Porkypine

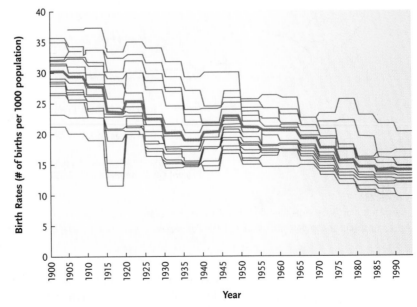

Figure 48.26
Changes in human birth rates (fertility = births per 1000 population per year) in 19 industrialized countries, plotted against increasing temperatures.

Table 48.1	Variations in Fertility Rate (Total Births per Woman): A Sample of UNDP Data for 162 Countries		
Country	Human Development Index (HDI) rank	1970–75	2000–05
Norway	2	2.2	1.8
Canada	4	2.0	1.5
U.S.A.	12	2.0	2.0
Portugal	29	2.7	1.5
Brazil	70	4.7	2.3
China	81	4.9	1.7
Indonesia	107	5.2	2.4
India	128	5.4	3.1

These goals can be achieved only if reproduction is controlled (see Chapter 38). In 1994, the United Nations held an International Conference on Population and Development (ICPD), which set a target for global investment in family planning. By 2004, the amount spent had fallen to 13% of this target. Consequently, family planning information and devices (usually for fertility control) are not readily available in many of the lowest-income countries. In 1950, Sri Lanka and Afghanistan had the same population. Sri Lanka began strong efforts to make family planning available in culturally acceptable ways. This did not happen in Afghanistan. By 2050, Afghanistan will have four times as many people as Sri Lanka. The solution centres around controlling the fertility of women, but more particularly on giving them the power to control their own fertility in culturally acceptable ways. As seen in Chapter 45, the growth potential of a population is determined by the numbers of females of reproductive age. Why females? Because females are the limiting step in reproduction, they are the ones who produce the eggs or young.

48.9b Signs of Stress: On Systems and on Species

People's demand for food, water, and energy puts thousands of other species at risk. We do not have to look far to see examples of species and ecosystems under stress (see Chapter 47). For example, we are losing birds. We know this because for years, birdwatchers and ornithologists have counted them and monitored their behaviour and activity. Locally, birds are affected by changes in habitat availability as cities and towns and their suburbs expand into adjoining land. Birds also lose habitat when agricultural operations expand to increase productivity. Birds that make annual migrations from temperate areas of the world to tropical and subtropical ones must survive the changes that accumulate across their entire circuit of habitats, each one essential to their survival.

"Bird flu" is a looming crisis for humans, one that appears to involve birds as central players. The issue here is another one involving basic biology, namely the outcome when a disease-causing organism jumps from one species (host) to another. Bird flu could have as much to do with our insatiable demand for poultry as food as it does with birds. In 2006, 12 billion chickens were farmed in China. Worldwide, poultry farms housed over 100 billion broiler chickens. Raising organisms at very high densities (see Chapter 45) provides an ideal setting for the spread of disease. Humans have responded to the threat of bird flu by wholesale slaughter of fowl, raising concerns about the roles played by migrating birds, and efforts to develop a vaccine that will protect humans from bird flu. All involve basic biology.

Drylands are arid, semiarid, and subhumid areas where precipitation is scarce and more or less unpredictable. In drylands, the combination of high temperatures, low relative humidities, and abundant solar radiation means high potential evapotranspiration. Drylands cover approximately 41% of Earth's land surface and are home to about 38% of the human population. Drylands are not just a problem of "deserts" but cover large expanses, for example, of Canada's prairie provinces. However, between 10 and 20% of the drylands are subject to some form of severe land degradation, directly affecting the lives of at least 250 million people. The complexity of the situation is clear in **Table 48.2, p. 1228.** Climate change, combined with increasing pressure on water resources for these people, their crops, and animals, compound the problems that confront them. Competition for limited resources, such as water, can generate local and international strife.

We have seen that complexity is an important and pervasive feature of ecosystems. Biodiversity is intimately associated with complexity, and disruption of this complexity often translates into reduced biodiversity and decay of ecosystems. Ironically, many social and economic systems that humans have developed are also subject to disruption by stress. This places the onus on our species to develop sustainable operations, whether in the area of agriculture, resource use and exploitation, or conservation.

STUDY BREAK

1. How is reproductive effort different between males and females in birds and in mammals?
2. List the United Nations Millennium Development Goals.
3. How are drylands at risk?

| Table 48.2 | Principles of the Drylands Development Paradigm, with a brief overview of their importance vis-à-vis the five main components of the dryland syndrome (ds-1 to ds-5, see text) and their implications for research, management, and policy, [Based on Stafford Smith and Reynolds (77)] H-E, human-environmental systems; LEK, local environmental knowledge. | | |

Principles	Why important in drylands	Links to dryland syndrome (ds-1 to ds-5)	Key implications (ki) for research management, and policy
P1: H-E systems are coupled, dynamic, and coadapting, so that their structure, function, and interrelationships change over time	The close dependency of most drylands' livelihoods on the environment imposes a greater cost if the coupling becomes dysfunctional; variability caused by biophysical factors as well as markets and policy processes, which are generally beyond local control, means that tracking the evolving changes and their functionality is relatively harder and more important in drylands.	ds-1: variability ds-4: remoteness	ki-1: Understanding dryland desertification and development issues always requires the simultaneous consideration of both human and ecological drivers and the recognition that there is no static equilibrium "to aim for."
P2: A limited suite of "slow" variables are critical determinants of H-E system dynamics	Identifying and monitoring the key slow H and E variables is particularly important in drylands because high variability in "fast" variables masks fundamental change indicated by slow variables.	ds-1: variability	ki-2: A limited suite of critical processes and variables at any scale makes a complex problem tractable.
P3: Thresholds in key slow variables define different states of H-E systems, often with different controlling processes; thresholds may change over time	Thresholds particularly matter in drylands because the capacity to invest in recovering from the impacts of crossing undesirable thresholds is usually lower per unit (area of land, person, etc.); and where outside agencies must be called upon, the transaction costs of doing so to distant policy centres are usually higher.	ds-1: variability ds-2: low productivity ds-4: remoteness ds-5: distant voice	ki-3: The costs of intervention rise nonlinearly with increasing land degradation or the degree of socioeconomic dysfunction; yet high variability means great uncertainty in detecting thresholds, implying that managers should invoke the precautionary principles.
P4: Coupled H-E systems are hierarchical, nested, and networked across multiple scales.	Drylands are often more distant from economic and policy centres, with weak linkages; additionally, regions with sparse populations may have qualitatively different hierarchical relationships between levels	ds-3: sparse population ds-4: remoteness ds-5: distant voice	ki-4: H-E systems must be managed at the appropriate scale; cross-scale linkages are important in this but are often remote and weak in drylands, requiring special institutional attention.
P5: The maintenance of a body of up-to-date LEK is key to functional coadaptation of H-E system	Support for LEK is critical in drylands because experiential learning is slower where monitoring feedback is harder to obtain (owing to more variable system, larger management units, in sparsely populated area) and, secondarily, where there is relatively less research	ds-1: variability ds-3: sparse population	ki-5: The development of appropriate hybrid scientific and LEK must be accelerated for both local management and regional policy.

48.10 Taking Action

It is easy to believe that nothing can change, that as individuals we have no power. Yet we also can think of things that have changed dramatically in a relatively short time. Two good examples are the abolition of slavery and the emancipation of women, proving humans' capacity for effecting change. On a more local level, the acceptance of the use of tobacco in public has declined remarkably in the last 20 years—in Canada and elsewhere. We also have seen the abolition of capital punishment and much more ready access to abortion in Canada.

But none of these changes are universal. Daily in the news we find stories about people living in virtual slavery, of people executed in public, of women with few or no rights in their home countries. We only need to travel short distances to learn that not everyone in Canada can eat in a smoke-free restaurant. To complicate the matter, not everyone agrees that the changes listed above are for the better.

Effecting changes in our approach to conservation means identifying the root causes for the erosion of biodiversity and the things that are impediments to conservation. This means starting by changing our own lifestyles, the food we eat, our use of energy, and our lifestyles. We must be wary of simple solutions that are often misleading and avoid blaming someone else because it is just a way of self-exoneration. Respect the rights of others. Use education and training to become informed. Learn to be objective, to examine and evaluate data or evidence. The outpouring of support for victims of the 2004 tsunami demonstrated

Figure 48.27
Eat yourself out of house and home—like this African elephant (*Loxodonta africana*) trekking across the shore to Lake Kariba.

that humans have great empathy for their fellows, and we need to extend this concern to the other species with whom we share the planet.

We have seen that action is needed at the species and the habitat level, and there is a propensity to focus more on species. But in the human view, all species are not equal. The 2006 IUCN list of threatened species shows that whereas 20% of the described species of mammals were listed as threatened, only 0.07% of the insect species received this level of attention. Other interesting numbers from this table are 12% of described species of birds listed as threatened, 4% of fish species, 3.5% of dicotyledonous plants, and 0.006% of species of mushrooms. In Canada, the same situation prevails, with mammals and birds dominating the list of threatened species, with other taxa receiving less attention. Do these data about threatened species mean that mammals are more vulnerable than insects? That we care more about mammals than about insects? Or does it mean that there are more "experts" to offer opinions and data about mammals than about insects? Are the possibilities mutually exclusive?

Biology can be at the centre of the movement to achieve conservation of biodiversity while being part of our efforts to achieve sustainable use of the resources we need as a species **(Figure 48.27)**. Conservation begins at home when we modify our lifestyles and become active on any front, from protecting local habitat and species to protecting charismatic species elsewhere. To better appreciate the situation, try to answer the questions posed in **Figure 48.28.** Elephants are an excellent example of how the objectivity that can be inherent in data is vulnerable to emotional responses.

STUDY BREAK

1. Is hunting compatible with conservation?
2. Give examples of resiliency in natural systems.
3. Are hybrids protected by the Endangered Species Act (ESA) in the United States? What are the conservation implications of this stance?

a.

b.

c.

Figure 48.28
To understand some of the dilemmas facing conservationists, use the Internet to explore the situation of African elephants. **(a)** How many species are there? What are the populations in the wild? What products from elephants do we use? **(b)** and **(c)** Are elephants endangered? How can they be protected? What are the main threats to their survival?

Should harvesting of species be permitted in protected areas such as conservation areas and provincial or national parks? Why is sport fishing permitted in some parks and protected areas, whereas hunting of birds and mammals is not?

Review

Go to CENGAGENOW™ at http://hed.nelson.com/ to access quizzing, animations, exercises, articles, and personalized homework help.

48.1 Extinction

- A species is said to be extinct when there are no living representatives known on Earth. Conservation organizations usually say that a species is extinct when it has not been seen or recorded for 50 years.

- Mass extinctions occurred at the end of the Ordovician and the beginning of the Devonian, at the end of the Devonian, at the end of the Permian, at the end of the Triassic, and at the end of the Cretaceous. The Permian extinction was the most severe, and more than 85% of the species alive at that time disappeared forever, including the trilobites, many amphibians, and the trees of the coal swamp forests. Dinosaurs did not survive the extinction that occurred at the end of the Cretaceous.

- The extinction at the end of the Cretaceous is believed to have been caused by an asteroid impact. Dust clouds resulting from the impact blocked the sunlight necessary for photosynthesis, setting up a chain reaction of extinctions that began with microscopic marine organisms and finished with dinosaurs (as well as many birds and mammals).

- Measured by time on earth, multituberculates were the most successful mammals.

48.2 The Impact of Humans

- Species (particularly flightless birds) that are confined to islands often have small populations and are unaccustomed to introduced terrestrial predators (such as cats, dogs, rats, etc.), making them vulnerable to extinction when human populations settle and expand.

- The demise of the calvaria trees (*Sideroxylum majus*) on the island of Mauritius will occur even though the trees continue to bloom and produce seeds. The extinction of the tree is linked to the earlier extinction of Dodos since *S. majus* seeds had to pass through the Dodo's digestive tract to germinate.

48.3 Introduced and Invasive Species

- Stephen's Island wrens were flightless and unaccustomed to predators. The population on the island was small, and it was easy for Tibbles to catch and kill the remaining 20 birds.

- Since about 1880, ships have regularly used ballast water. A survey of ballast water in 159 ships in Coos Bay, Oregon, revealed 367 species of organisms representing 19 animal phyla and 3 plant divisions. When ships empty their ballast, the organisms in it are introduced to the system where the ship is anchored.

- RUE, resource use efficiency, is measured as carbon assimilation per unit resource. Many invasive plant species in Hawaii are more efficient than native species. This means that conserving native biodiversity in the face of invasive and introduced organisms is a pervasive problem.

48.4 How We Got/Get There

- Horns of rhinos, particularly black rhinos (*Diceros bicornis*), have been used to make handles for ornamental daggers (jambiyas) in some parts of the Arabian peninsula. Increasing oil prices in the early 1970s increased the demand for jambiyas. The main source of rhino horn was from poaching rhinos in Africa.

- During fishing operations, nontarget species are often caught in nets. They are "bycatch." Barndoor skates once occurred across the North Atlantic. Today these skates are almost gone, although they never were the targets of a fishery.

- The demise of the bay scallops occurred because of an increase in the populations of skates and rays that are predators of bay scallops. The increase in skates and rays was attributed to the decline of populations of their predators, sharks. The sharks were extensively fished for their fins. Large-scale harvesting of the scallops for human consumption also compounded the impacts and attributed to their demise.

48.5 Protecting Species

- The International Union for the Conservation of Nature (IUCN) has established objective criteria identifying species that are at risk. Extinct means the species no longer exists; extirpated means the species is locally extinct; endangered means the species is facing imminent extirpation or extinction; threatened means the species is likely to become endangered if limiting factors are not reversed; and special concern means the species may become threatened or endangered because of biological characteristics and identified threats. The criteria take into account data on populations, their patterns of distribution, and their population status.

- The Convention for International Trade on Endangered Species (CITES) attempts to prohibit international trade in endangered species. Newly described and as yet undescribed (and therefore unnamed) species are not protected because they have no legal identity.

- Species such as passenger pigeons or Dodos that have been exterminated are extinct. Extirpated species are locally extinct. Black-footed ferrets (see Chapter 45, *Black-Footed Ferret, Mustela nigripes*) have been extirpated in Canada but still occur in the United States.

- In Canada, recommendations about the conservation status of species involve the Committee on the Status of Endangered Wildlife in Canada (COSEWIC). COSEWIC members vote on the appropriate conservation category for each species whose status they review and use IUCN criteria to assess the status of species.

48.7 The Downside of Being Rare

- An animal on the list of endangered species is more likely to become a commodity in high demand because it has become rare. The Bali starling is an example.

48.9 Effecting Conservation

- In 1994, the International Conference on Population and Development (ICPD) outlined a plan for investing in family planning.
- United Nations Millennium Development Goals of 2000 are ending poverty and hunger, universal education, gender equality, child and maternal health, combatting HIV/AIDS, environmental sustainability, and global partnerships.
- Drylands cover 41% of Earth's land surface, and 10 to 20% of drylands are subject to severe land degradation, affecting, in 2008, the lives of at least 250 million people.
- Climate change and increasing pressure on water supplies negatively affect drylands.
- The case of leopards (*Panthera pardus*) demonstrates how some species persist even in the face of considerable hunting pressure.

- Targeted hunting—selection of "trophy" or spectacular specimens—can be less threatening to a species' survival than bycatch or eradication programs (bounties on predators such as wolves). Extensive killing, even of species with large populations, can drive them to the brink of extinction. Black rhinos are a telling example.
- The overgrowth of Mayan cities or ruins in Africa demonstrates the resiliency of ecosystems. A grasshopper's use of 2,4-D to synthesize an ant repellent demonstrates the resiliency of individuals.
- Hybrids are not protected by the U.S. Endangered Species Act, putting species such as Florida panthers at risk because their populations have been genetically contaminated.

Questions

Self-Test Questions

1. Extinction is a natural part of the process of speciation. Some estimates suggest that _____ of the species that have ever lived are now extinct.
   a. >20%
   b. >30%
   c. >50%
   d. >80%
   e. >99%

2. Some researchers use evidence from a variety of sources to support the suggestion that an asteroid striking Earth in the _____ largely explains the extinction of the dinosaurs.
   a. Cambrian
   b. Ordovician
   c. Triassic
   d. Cretaceous
   e. Pleistocene

3. If our species first appeared 200 000 years before present, the multituberculates survived _____ longer than we have to date.
   a. 10
   b. 50
   c. 100
   d. 500
   e. 1000

4. Hunting by people is largely responsible for the extinction of
   a. multituberculates, Dodos, and passenger pigeons.
   b. barndoor skates, black-footed ferrets, and giant auks.
   c. Dodos, passenger pigeons, and Stephen's Island wrens.
   d. passenger pigeons, giant auks, and Dodos.
   e. black rhinos, Bali starlings, and ginseng.

5. The ballast water of ships is responsible for the spread of
   a. *Arthurdendyus triangulatus*.
   b. *Dressina polymorpha*.
   c. *Rattus norvegicus*.
   d. *Salmo salar*.
   e. *Lampsilis radiata*.

6. In Hawaii, high resource use efficiency (RUE), measured as carbon use, partly explains the success of invading
   a. rats.
   b. ferns.
   c. C_3 and C_4 grasses.
   d. flatworms.
   e. Both b and c are correct.

7. Increases in populations of tertiary consumers such as _____ appear to explain the demise of scallops off the southeastern coast of the United States.
   a. rats
   b. skates and rays
   c. sharks
   d. killer whales
   e. pelagic seabirds

8. Species such as black-footed ferrets (*Mustela nigripes*) no longer occur in Canada but still live in the United States, so they are
   a. extinct.
   b. at risk.
   c. extirpated.
   d. highly endangered.
   e. not at risk.

9. CITES is designed to stop international trade in
   a. passenger pigeons.
   b. black rhinos.
   c. Canadian beavers.
   d. Canada geese.
   e. leopards.

10. Differences in government support for family planning explain the differences in the growth of human populations in
   a. Canada and Australia.
   b. Great Britain and France.
   c. Afghanistan and Sri Lanka.
   d. Mexico and Germany.
   e. India and South Africa.

Questions for Discussion

1. Should gardeners and farmers be exempt from rules concerning the introduction of foreign species? Why? Why not?

2. In situations in which the behaviour of one endangered species threatens the survival of another (or others), how should authorities proceed?

3. What species are "rare" on your campus? What is a good working definition of rare? What steps can you take to protect rare species?

M.B. Fenton

Sunflowers. Originally from the New World, sunflowers (*Helianthus annuus*) are grown as a source of oil. In terms of harvest and area under cultivation, in 1998, sunflowers ranked twelfth in importance among domesticated plants in the world. Domesticated sunflowers often hybridize with local wild species, creating a challenge for those concerned about biodiversity.

49 Domestication

WHY IT MATTERS

In 1960, an estimated 1.8 billion people in the world (60% of the population) did not receive enough food every day to sustain themselves fully over the longer period—they were hungry. This number was reduced to 1.1 billion (17%) in 2000. Even though the world population had grown by 3 billion in the intervening period, about 700 million fewer people were hungry in 2000.

Worldwide in 2000, subsistence farmers accounted for about 66% of the hungry people. The reduction in the numbers of hungry people can be tied to changes in agriculture that have increased yields. Specifically, the combination of new genetic strains, better fertilizers, and more efficient harvesting and processing means more productivity. One indication of this change is provided by data about corn yields. In Iowa in 1935, corn yields were ~1600 kg·ha^{-1} compared with ~10 700 kg·ha^{-1} in 2000. Changes in crop yield are part of the "green revolution." Agriculture in general and the green revolution in particular have allowed humans to continue to redefine one element of carrying capacity (see Chapter 45): the amount of food available to our populations.

But agricultural improvements are not enough. Climate (see Chapter 3) also influences crop yield. In 2006 in southwestern Ontario (~42° N in Canada), the corn yield was ~10 000 $kg \cdot ha^{-1}$, whereas in Zimbabwe (~18° S), on commercial farms, it was ~5500 to 6600 $kg \cdot ha^{-1}$, compared with ~500 to 1000 $kg \cdot ha^{-1}$ on communal lands where farming was low tech. Irrigation also influences yield: in Zimbabwe, irrigated cornfields produce 8500 to 10 000 $kg \cdot ha^{-1}$, much more than nonirrigated commercial farms.

Although increases in crop yield and a reduced incidence of malnutrition and starvation sound like good news, in 2007, hunger still claimed the lives of about 20 000 children a day. Worldwide, one child in three is underweight and malnourished. Ironically, at the same time in some developed countries, obesity in children reached almost epidemic proportions.

49.1 Domesticate

The purpose of this chapter is to explore how humans have used selection to put biodiversity to work. Biologists and anthropologists believe that our ancestors originally gathered plants and hunted animals in the wild for use as food, building products, or fuel (see *Molecule Behind Biology*). From gathering, our ancestors progressed to cultivating plants, a process involving the systematic sowing of wild plant seeds. Over time, cultivation improved when people provided more care to their crops and eventually involved repetitive cycles of sowing, collecting, and sowing wild stock (**Figure 49.1**). **Domestication** is more than just taming. It occurs when people selectively bred individuals of other species (plants and animals) to increase the desirable characteristics in the progeny (e.g., in plants: yield, taste, colour, shelf life). This marked the birth of agriculture. The progression from gathering to cultivation to domestication of plants occurred independently at several locations around the world. The beginning of the Neolithic Period is often defined by the domestication of other species, and this period started at different times in different parts of the world.

But agriculture is not the exclusive domain of humans. Recall that about 50 million years ago, ants of the tribe Attini were the first to manipulate other species (fungi) to increase food availability (see Chapter 24). Today at least 200 species of ants in this tribe are obligate farmers. These early farmers have lost their own digestive enzymes and rely on fungal enzymes to digest the food for them. The ant farmers propagate their fungal crops asexually, with each colony or village working with one species. Therefore, any single species of farmer ant may propagate several different species of fungi. These ants have been involved in at least five domestication events, and there are a number of interesting parallels between these ants and people.

The list of species that humans have domesticated is long. It includes many land plants (~250 species), some yeasts, and terrestrial animals from insects to birds and mammals (~44 species). Biogeographic and genetic evidence shows that domestication of some species by humans occurred at different places and at different times. Domestication was not a one-time (or one-location) event and appears to have arisen independently in 8 to 10 environmentally and biotically diverse areas in the world.

49.1a When and Where? Tracking the History of Domestication

Data provided by the tools of molecular genetics (see Chapter 16) have made it easier to determine where and when domestication events took place. In the past, archaeologists had to try to recognize the

Figure 49.1
The way in which garlic (*Allium sativum*) is grown influences the size and development of the bulbs. From left to right, (a) one domesticated, (b) two cultivated, and (c) two wild garlic bulbs.

Salicylic Acid

Figure 1
The molecular structure of salicylic acid.

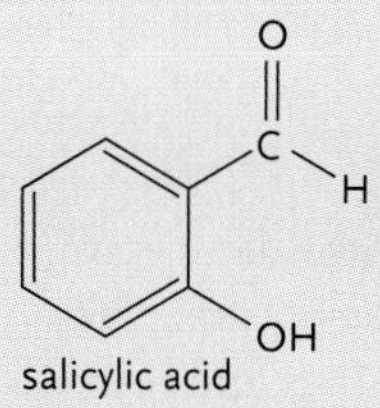

salicylic acid
2-OH-C₆C₄CO₂H

2-OH-$C_6C_4CO_2H$

The precursor of the main active ingredient in Aspirin™ is salicylic acid, which is obtained from the bark of willow trees (*Salix* species). Over 2500 years B.P., Chinese medical practitioners used an extract of willow bark to relieve pain and fever. The same kinds of extracts were used in medicine as practised in Greece and in Assyria. In Iceland 500 years ago, willow bark extracts were used to treat the symptoms of colds and headaches. Willow extract was widely used among First Nations people in North America, who commonly used it to staunch bleeding. They also used the supple willow twigs in other applications, from snares for catching mammals to nets for catching fish. This is not an example of domestication.

The spread of traditional knowledge about plants and their products among peoples is pervasive.

---

remains of domesticated species and distinguish them from wild species. This was often impossible because individual bones or pieces of plant did not always provide a clear indication of domestication. In 1973, radiocarbon dates (see the Preface and Chapter 20) suggested that the first dogs (*Canis familiaris*) were domesticated by 9500 years B.P. (Before Present), based on remains found in England and elsewhere in Europe. In 2002, mitochondrial DNA (mtDNA) evidence suggested an East Asian origin of domestication of dogs dating from 15 000 years B.P. But pictures based on genetic evidence also can change. In 2003, morphological and genetic evidence suggested a southeast Asian origin of domesticated pigs (*Sus scrofa*), whereas in 2005, new genetic data indicated multiple origins of domestication of pigs across Eurasia (**Figure 49.2**).

Worldwide, domestication of aquatic species has lagged behind that of terrestrial ones. Although there are ~180 species of domesticated freshwater animals, ~250 species of marine animals, and ~19 species of marine plants, all were domesticated in the last 1000 years, most in the last 100 years (**Figure 49.3, p. 1236**).

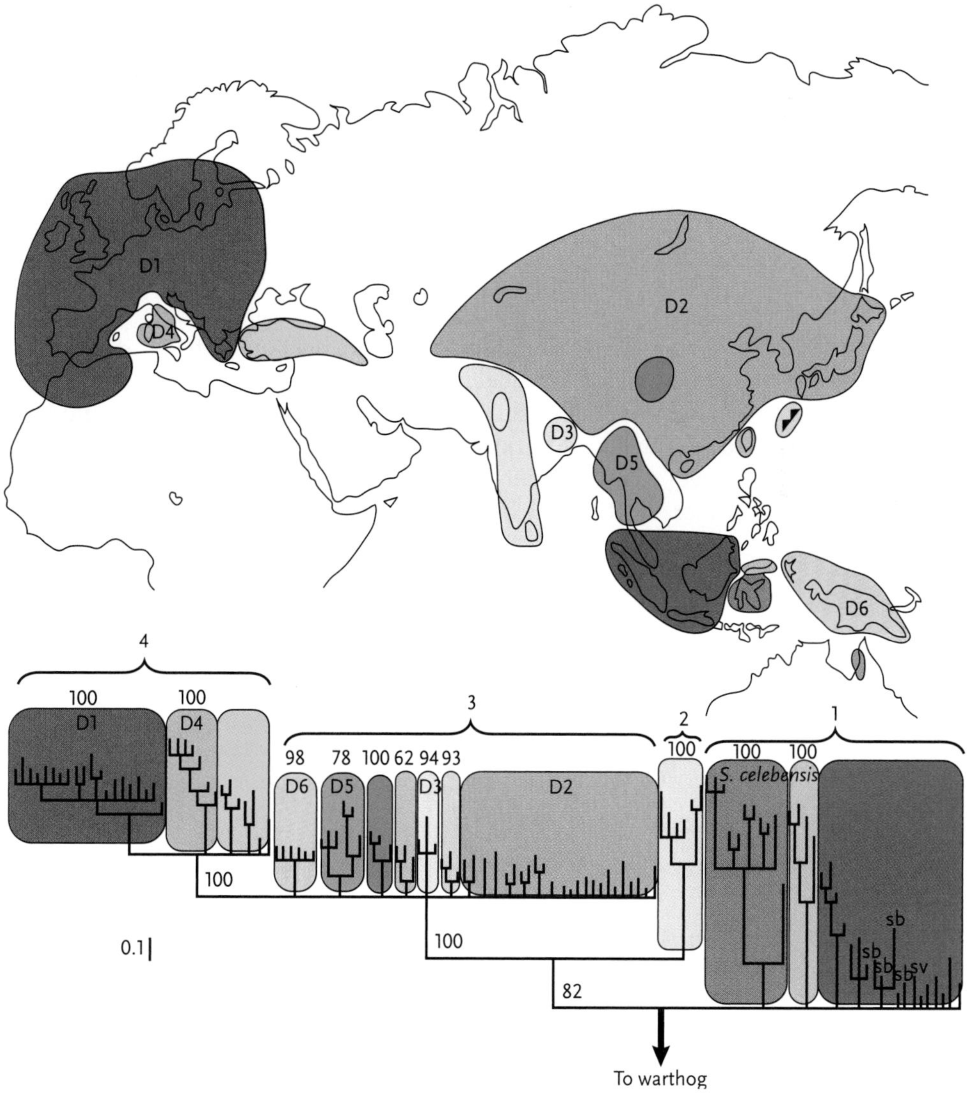

49.1b How Long Did Domestication Take? Archaeological Evidence

The time it takes to progress from harvesting tended wild crops to cultivating them βend then to domesticating them varies with species and situation. When there are clear morphological or chemical differences between cultivated and domesticated stocks, determining the place and time of domestication is

Figure 49.2

Origins of domestication of pigs. Mitochondrial DNA obtained from pigs indicates 14 clusters of related lineages, each identified by a different colour. The geographic relationships are shown with the phylogeny. Pigs were domesticated in numerous centres.

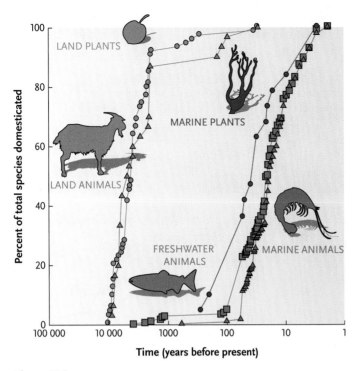

Figure 49.3
Most land species were domesticated much earlier than most aquatic ones.

possible. Wheat provides an example of such a morphological change. In wheat, as in other cereal crops, inflorescences of domesticated stocks hold on to the ripened grains (= indehiscence), whereas those of wild stocks shed them (= dehiscence) **(Figure 49.4).** Ripe indehiscent grains are easily gathered (harvested) compared with dehiscent ones that naturally scatter. Indehiscence in wheat results from a naturally occurring mutation. Material recovered from sites in northeastern Syria and Turkey has been radiocarbon

Figure 49.4
(a) Some plants readily shed ripe seeds from the inflorescence; these plants show dehiscence. **(b)** Indehiscence is the propensity to hold them. These two herbarium specimens, **(a)** bottle brush grass (*Elymus hysterix*) and **(b)** riverbank wild rye (*Elymus riparius*), illustrate dehiscence and indehiscence, respectively.

dated and shows that wild varieties of wheat were cultivated for at least 1000 years before domestication **(Figure 49.5).** When sexual reproduction is involved in the breeding process, the time to domestication is partly determined by life cycle.

When organisms reproduce asexually, domestication may occur more rapidly. Common figs (*Ficus carica* var. *domestica*) are gynodioecious and provide an example of more rapid domestication. In parthenocarpic female figs, ovaries develop without pollination and fertilization. Parthenocarpic figs can be propagated by cutting branches, sticking them in the ground, and waiting for them to grow into trees. When figs reproduce sexually, symbiotic fig wasps (*Blastophaga psenes;* **Figure 49.6**) serve as pollinators. The absence of access holes for wasps in fossil figs allows biologists to recognize parthenocarpic figs and date early incidences of fig domestication. At one site in the lower Jordan Valley (Middle East), parthenocarpic figs date to between 11 400 and 10 500 years B.P., perhaps preceding the domestication of cereal crops by about 1000 years.

49.1c In What Setting? Habitats Where Domestication Occurred

The transition from nomadic hunters and gatherers to people living more localized lives in more permanent dwellings appears to have been a prelude to cultivation and domestication. These changes meant that people would have been available to care for their "crops,"

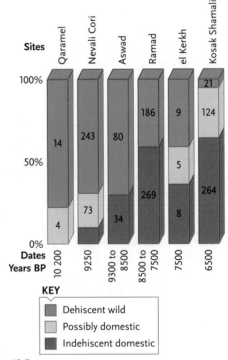

Figure 49.5
Timing of domestication of wheat. Data from archaeological digs at six locations in the Middle East demonstrate the transition from wild (dehiscent) to domesticated (indehiscent) wheat from 10 200 to 6500 years B.P.

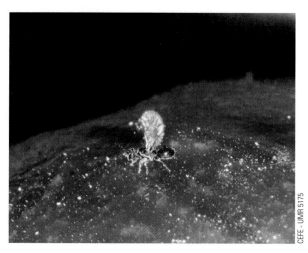

Figure 49.6
Female flowers on some fig trees (*Ficus carica*) are fertilized by symbiotic wasps, *Blastophaga psenes*.

CEFE - UMR 5175

is called "niche construction" or "ecosystem engineering," modifying the environment and setting the stage for domestication. At that lower Yangtze Neolithic site, people used fire to prepare and maintain sites in lowland swamps, where they cultivated rice **(Figure 49.7)**. This region in China was a major centre of rice domestication. The evidence suggests that rice cultivation began in coastal wetlands in an ecosystem vulnerable to coastal change. This system was very fertile and productive and used for at least 200 years before the land was inundated by seawater.

Although controlled burning has been documented at many sites in the last 10 000 years, there is evidence of it 50 000 to 55 000 years ago **(Figure 49.8, p. 1238)** at sites near Mossel Bay in South Africa. Archaeological evidence indicates that some humans were increasingly using some plant resources and used local burning to increase productivity. The increased use of plant resources and fire occurred during a period of harsh environmental conditions. These changes in human behaviour coincided with the appearance of more sophisticated tools, the use of marine organisms as food, and the first use of ochre for decoration. These modifications suggest differences in human behaviour that may have assisted the emergence of domestication.

whether grown from seeds or from parthenocarpic plants, and whether cultivated or domesticated.

Were changes in habitat associated with domestication? Evidence from pollen shows that from 7500 years B.P. in the lower Yangtze region of China, people used fire to clear alders, small woody bushes (*Alnus* species). This element in the process of domestication

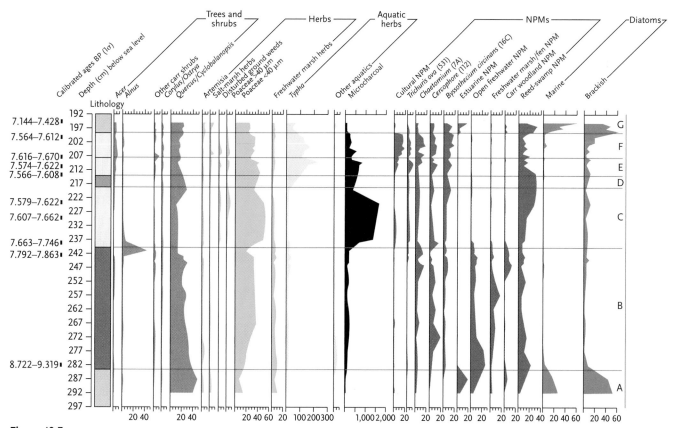

Figure 49.7
The climatic setting for rice domestication at Kuahugiao in China. Shown here are generalized stratigraphic units and their associated pollen (shown in colours) and microscopic charcoal (black) that indicate conditions of climate and habitat. The data support the use of fire to establish favourable conditions for growing rice. "NPMs" are nonpollen microfossils that provide paleoecological data. The increases in charcoal, grasses (Poaceae), and reed swamp microfossils in phases C, E, and F (right edge) indicate the use of fire to establish favourable conditions for growing rice.

Figure 49.8
In Ghana (West Africa), fire is still used to clear the underbrush from an area where forest trees have been felled.

M.B. Fenton

Meanwhile, at sites on the coast of Peru, occupied between 3800 and 3500 years B.P., people ate marine organisms as the main animal food, combined with cultivated plants (squashes, *Cucurbita* species; beans, *Phaseolus lunatus* and *P. vulgaris;* peppers, *Capsicum* species; jicama, *Pachyrrhizus tuberosus*); and wild plants (guava, *Psidium guajava;* lacuma, *Lucuma bifera;* and pacay, *Inga feuillei*). Cotton was an important crop used for making fishing tackle and clothing. The findings from these Peruvian sites and many other sites around the world suggest a progression toward domestication, including the range of foods consumed, the development of more sophisticated tools, and the use of materials from plants and animals as tools, as well as in food.

49.1d Abu Hureyra on the Euphrates: An Example of a Setting for Domestication

Figure 49.9
Abu Hureyra in Syria, the site of an early village.

This prehistoric settlement (a recent photograph is shown in **Figure 49.9**), on the south side of the Euphrates River (35° 52 N, 38° 24 E) about 130 km from Aleppo (a modern Syrian city), also illustrates progression toward domestication. The first habitations that we know of in Abu Hureyra date from about 12 000 years B.P. Its population was estimated at 100 to 200 people who lived in semisubterranean pit dwellings clustered together on a low promontory overlooking the river. By 7000 to 9400 years ago, 4000 to 6000 people lived at the same site, now in multiroomed family dwellings made of mud and brick. This settlement was built over the remains of the earlier one.

People living at Abu Hureyra about 12 000 years ago ate the fruits and seeds of over 100 species of local plants as well as local animals such as gazelles. Many of the plants and animals appear to have come from the adjoining oak-dominated park woodland. It appears to have been a time of plentiful food. The situation changed, however, and by 9400 years ago, the climate was cooler and drier, and the people relied more on cultivated plants and less on wild ones. By this time, there was little evidence of use of plants from the oak-dominated parkland, which by then was at least 14 km from the settlement. These changes were evident in pollen records and in plant and animal remains associated with the dwellings. The climate change likely triggered the start of cultivation of foods that could serve as caloric staples. Despite the changing climate and the focus on fewer food staples, the human population at the site dramatically increased.

STUDY BREAK

1. Explain the benefits of the "green revolution"? What is responsible for it?
2. How did domestication develop?
3. Why can domestication take place more rapidly in asexually reproducing organisms?

49.2 Why Some Organisms Were Domesticated

We can surmise, perhaps accurately, that securing a sustainable food supply provided an initial motivation for cultivation and domestication. It is certainly true that cultivated plants from beans to squash, corn to rice, and cereal grains all help feed many, many people worldwide. People eat different parts of plants, from flowers and fruits to seeds, leaves, stems, roots, and tubers. Plants may be a source of energy (calories), or their products may be used to enhance flavours, to control and repel pests, or as medicines. Still others, such as the bottle gourd (*Lagenaria siceraria*), are used as containers (see *People Behind Biology*). Domesticated animals provide food, but many are also used as a source of labour. The following are examples of four very different domesticated species and how people use them.

Artwork from The Tell of Abu Hureyra From The West. Reprinted with permission of Oxford University Press.

Figure 49.10

Today there are two varieties of domesticated cattle: **(a)** the humped *Bos indicus* and **(b)** the humpless Bos *taurus*.

49.2a Cattle

Cattle were among the first of the large herd mammals to be domesticated, at least 9000 years B.P. One theory proposes that the domesticators of cattle were sedentary farmers, not nomadic hunters. Some anthropologists maintain that a religious motivation was behind the domestication of cattle because the curve of their horns resembled the crescent of the moon and hence the mother-goddess. Imposing horns were particularly prominent in some male *Bos primigenius* (called "urus"), the apparent Pleistocene ancestor of domesticated cattle. Whatever the original impetus, today there are two basic stocks of cattle **(Figure 49.10),** the humped *Bos indicus* and the humpless *Bos taurus*. Cattle provide us with labour, milk, meat, hides, and blood, and in some societies, they are symbols of wealth.

49.2b Honeybees

Domestic honeybees provide us with honey and pollination services. Steps to domestication of honeybees included changes in their behaviour compatible with large population size in hives. The changes could have involved hygiene, aggression, and foraging. Although everyone recognizes honey as a product of bees, the service provided by bees is often overlooked. In 2000 in the United States, it is estimated that bees contributed about U.S.$14.5 billion through their role as pollinators. Plants such as alfalfa, apples, almonds, onions, broccoli, and sunflowers are exclusively pollinated by insects, usually more than 90% by honeybees. Many beekeepers earn significant income by moving their bees from location to location, thus providing a mobile pollinator service for farmers. Declines in populations of honeybees have serious economic implications throughout the world, but many conservationists are also concerned about the impact of populations of honeybees on native bee species.

49.2c Cotton

At least four species of cotton **(Figure 49.11)** have been domesticated: two diploid species from the Old World (*Gossypium arboreum, G. herbaceum*) and two tetraploids from the New World (*G. hirsutum* and *G. barbadense*). The domestication events appear to have been independent, and one site on the Mexican gulf coast of Tabasco shows evidence of people growing cotton by 4400 years B.P. Cotton seeds were a source of oil, whereas fibre was and is still used in applications ranging from clothing to implements.

Figure 49.11

Cotton, *Gossypium herbaceum,* showing flowers and cotton bolls.

PEOPLE BEHIND BIOLOGY

Richard Keith Downey, Research Scientist, Agriculture and Agri-Food Canada

Richard Keith Downey was born in 1927 in Saskatchewan and attended the University of Saskatchewan and Cornell University. He is best known for his work on rapeseed. In 1928, rapeseed had been brought to Shellbrook, Saskatachewan, by Fred and Olga Solovonuk when they immigrated to Canada from their native Poland. They had brought seeds of *Brassica rapa*, although *Brassica napus* from Argentina was the main traditional source of rapeseed oil. During World War II, rapeseed was grown as a source of industrial lubricants because erucic acid and glucosinolates in the oil precluded its use as food for people and livestock.

Dr. Downey was the leader of a team of researchers at Agriculture and Agrifoods Canada who developed canola (the name comes from "Canada" and "oil"), an edible, high-value crop high in proteins. The team's work resulted in a crop that went from 2600 to 4.25 million ha under cultivation in Canada. Canola is now grown around the world, and the high-protein crop is used in cooking oils and feed for livestock.

Downey and his team developed the "half-seed" method, partitioning a seed so that half could be tested for composition using gas–liquid chromatography. The researchers could then select and germinate the half-seeds with the most promising features. Using this approach, Downey and his team developed 18 varieties of canola and several of table mustard. This work

had a huge impact on the economy of Canada in general and the prairies in particular, as well as an influence on food availability throughout the world.

Dr. Downey is also known for a program that introduced children in elementary school to the scientific method. The approach involved using canola seeds that had travelled into space aboard the space shuttle *Columbia* in 1996. Children in over 2000 classrooms in Canada received space seeds along with control seeds and germinated them. Downey helped analyze the results of the experiments. Space seeds germinated faster and at higher rates than control seeds and grew more rapidly. In this way, Downey introduced elementary students to plant science and the space environment.

49.2d Yeast

Strains of the yeast *Saccharomyces cerevisiae* have been used by people in bread making beginning at least 6000 years B.P. Evidence of this is in archaeological finds in Egypt, indicating the presence of bakeries and breweries, two yeast-based operations. Analysis of 12 DNA microsatellites obtained from 651 strains of *S. cerevisiae* collected at 56 locations around the world revealed 575 distinct genotypes. Yeasts associated with bread were intermediate between wild types and those used in making beer and wine, whereas those used in the production of rice wine and saki were more similar to those used for beer. About 28% of the genetic variation in yeast genotypes was associated with geographic location. The basal group of these 12 DNA microsatellites was samples from Lebanon, suggesting a Mesopotamian origin and a spread of yeast types along the Danube River and around the Mediterranean. Different strains of yeast have different capacities for maltose fermentation. Commercial bakers' yeast strains are more effective at maltose fermentation than nonindustrial strains. Domesticated yeast makes important contributions to providing humans with food and drink and supports lucrative industries.

49.2e Rice

Rice (*Oryza sativa*) is one of the world's most important food crops, and its domestication depended on the change from dehiscence to indehiscence. Although the genetic changes involved were presumed to be minor, their exact nature has only recently become clear. Domesticated rice is derived from two wild species

(*Oryza rufipogon* and *Oryza indica*). In 2006, Changbao Li and his colleagues reported that three quantitative trait loci (QTL) in F_1 hybrids between these two species were responsible for a reduction of grain shattering (dehiscence) in rice. Specifically, *sh3*, *sh4*, and *sh8* were involved, with *sh4* explaining 64% of the phenotypic variance. In the wild species, *sh4* was dominant and caused the shattering. The genetic changes in *sh3*, *sh4*, and *sh8* affected normal development of the abscission layer, explaining the change to indehiscence **(Figure 49.12)**. We

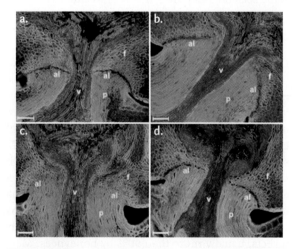

Figure 49.12

Rice dehiscence and indehiscence. Under a fluorescence microscope, a longitudinal section of the junction between the rice flower and its pedicel shows a complete abscission layer (al in a) and an incomplete one (al in b). In these figures, f = flower side; p = pedicel side; and v = vascular bundles. *Oryza nivara* is shown in **(a)**, *O. sativa* in **(b)**, *O. sativa japonica* in **(c)**, and transformed *O. sativa japonica* in **(d)**.

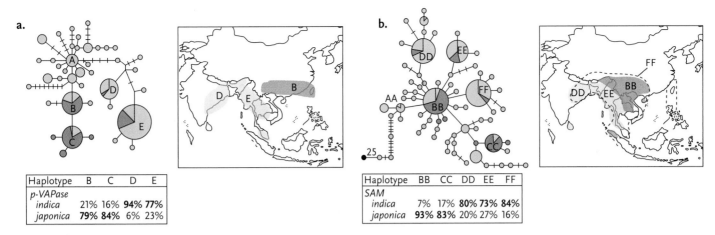

a.

Haplotype	B	C	D	E
p-VAPase				
indica	21%	16%	**94%**	**77%**
japonica	**79%**	**84%**	6%	23%

b.

Haplotype	BB	CC	DD	EE	FF
SAM					
indica	7%	17%	**80%**	**73%**	**84%**
japonica	**93%**	**83%**	20%	27%	16%

Figure 49.13

Geography of rice domestication. Genetic haplotype information allows identification of major domestication regions for rice. The haplotype network of the neutral nuclear p-VAPase region is shown in A, compared with B, the haplotype of the functional nuclear SAM region. In this diagram, purple represents *Oryza japonica*; green, *O. indica* (Aus cultivar); and pink, *O. indica*. Lines joining haplotypes represent mutational steps.

can now better understand the genetics of changes associated with domestication of rice.

These changes occur in the two major varieties of domesticated rice, *Oryza sativa japonica* and *Oryza sativa indica*. Jason Londo and his colleagues used DNA sequence variation to demonstrate multiple independent domestication events. Although *O. sativa indica* was probably domesticated in eastern India, Myanmar, and Thailand south of the Himalayan mountains, *O. sativa japonica* was domesticated in southern China **(Figure 49.13).**

49.2f Wheat

Other major domestications appear to stem from single domestication events, including wheat, barley **(Figure 49.14),** and corn. DNA fingerprinting was used by Manfred Heun and his colleagues to identify the Karacadağ mountains **(Figure 49.15, p. 1242)** in today's Turkey as *the* site of domestication of einkorn wheat. Einkorn wheat is derived from the wild *Triticum monococcum boeoticum*. Einkorn wheat is indehiscent and has a firm stalk, heavier seeds, and denser seed masses than its progenitor (see Figure 18.7), all contributing to the ease and efficiency of harvest. The genetic changes between einkorn wheat and the wild stock from which it was derived were relatively minor, and domestication probably occurred in a short period of time.

Archaeological evidence reveals that einkorn wheat spread rapidly throughout the immediate area and beyond. The same single domestication event and rapid spread of the new crop also appear to be true of barley **(Figure 49.16b, p. 1242;** *Hordeum vulgare* derived from *Hordeum spontaneum*). More recent wheat domestication is discussed in *Marquis Wheat*.

49.2g Lentils

Lentils (*Lens culinaris* from *Lens orientalis;* **Figure 49.16a, p. 1242)** were more difficult to domesticate than wheat or rice. Wild lentils are small plants, producing, on average, 10 seeds per plant. Furthermore, these seeds go through a period of programmed dormancy. The combination of low yield and dormancy means that relatively few seeds germinate. Lentils could have been cultivated only after a dormancy-free mutant had appeared. Archaeological evidence suggests that the first stages of lentil domestication (loss of dormancy) occurred in what is now southeastern Turkey and northern Syria, suggesting a single initial domestication event. The second phase was selection for strains that produced large numbers of seeds. This change may have occurred some distance (hundreds or even thousands of kilometres) from the original sites of domestication.

Figure 49.14

A barley field near London, Ontario, illustrates the consistency of a monoculture, one plant dominating an area, in this case for artificial reasons.

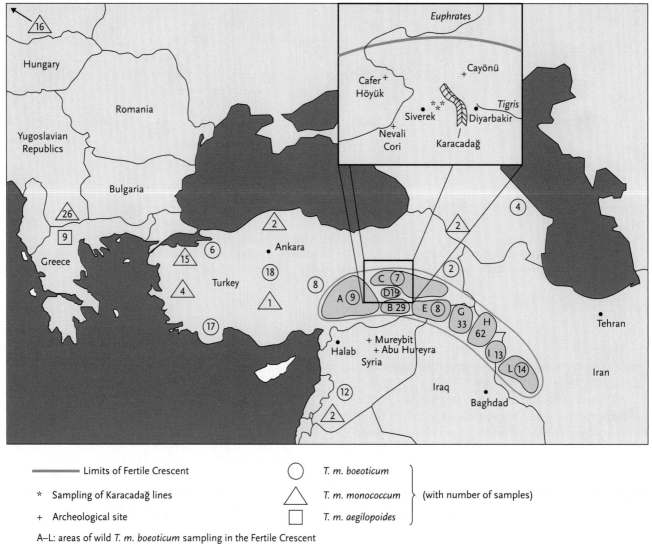

Limits of Fertile Crescent

* Sampling of Karacadağ lines

+ Archeological site

○ *T. m. boeoticum*

△ *T. m. monococcum* } (with number of samples)

□ *T. m. aegilopoides*

A–L: areas of wild *T. m. boeoticum* sampling in the Fertile Crescent

Figure 49.15
Site of wheat domestication. A phylogenetic analysis base on allelic frequency at 288 amplified fragment length polymorphism marker loci revealed the progenitor of *T. m. boeoticum* came from what is now southeastern Turkey.

49.2h Corn

Corn (*Zea mays*), like wheat and barley, appears to have been domesticated at one location. Unlike wheat or barley, domestication of corn required drastic changes from the ancestral teosinte (*Zea mays*

Figure 49.16
Two important domesticated crops are **(a)** lentils (*Lens culinaris*) and **(b)** barley (*Hordeum vulgare*). In each case, individual grains are about 6 mm long.

parviglumis; **Figure 49.17**). Corn kernels do not disassociate from the cob, presumably reflecting changes associated with domestication (and analogous to indehiscence). Analysis of some of the most ancient inflorescences of teosinte indicates that by 6250 years B.P., the kernels did not disassociate, suggesting that domestication was under way at that time. Corn appears to have been domesticated in the Central Balsas Valley of south central Mexico or in the Mexican State of Puebla at altitudes of 1000 to 1500 m. The deposits that contained the undissociated teosinte kernels also had the remains of squash (*Cucurbita pepo*). Although squash appears to have been domesticated in the same area, remains from a cave in Puebla and Oxa States in Mexico suggest that the people there were cultivating it by about 4000 years before corn. These data again demonstrate that some people were responsible for multiple domestication events.

Figure 49.17
Teosinte and domesticated corn. Ears of domesticated corn (*Zea mays*, right) are larger than those of one species of its wild relative teosinte (*Zea diploperennis*, left). Crossing domesticated corn and teosinte produces intermediate forms (centre).

49.2i Grapes

The Eurasian grape (*Vitus vinifera sylvestris*), a dioecious plant, is widespread from the Atlantic coasts of Europe to the western Himalayas, and its fruits were often eaten by Paleolithic hunter–gatherers. Domestication of *V. vinifera* involved the selection of hermaphroditic genotypes that produced larger, more colourful fruit, along with the development of techniques for vegetative propagation. Domesticated grapes have a higher sugar content than wild ones, ensuring better fermentation, greater yield, and more regular production. Domestication of grapes is associated with the production of wine, which also required storage in containers made of pottery, which appeared only about 10 500 years B.P.

Genetic analyses suggest at least two important origins of grape domestication: one in the near East and the other in the region of the western Mediterranean. Many wine-grape cultivars from Europe can be traced to western Mediterranean stock, as can over 70% of the cultivars on the Iberian Peninsula. It is possible that the original wild stock of *V. vinifera* has vanished due to genetic contamination by various cultivars.

49.2j Plants in the Family Solonaceae

Solanaceous plants include species such as deadly nightshade (*Atropa belladonna*) that produce virulent poisons, as well as food species (tomatoes, *Solanum lycopersicum*; potatoes, *Solanum tuberosum*; and eggplant, *Solanum melongena*) that are staples of many meals worldwide **(Figure 49.18)**.

Potatoes originated in western South America, specifically in areas of what is now Chile. Today, local varieties (also known as breeds, cultivars, or landraces) of potatoes are adapted to the local conditions where they grow. These landraces form the *Solanum brevicaule* complex. Genotypes from multilocus amplified fragment length polymorphisms were determined from 261 wild and 98 landrace samples. The resulting data suggest that potatoes (*S. tuberosum*) arose from one domestication event. Today, wild potatoes still occur in Chile, along with eight cultivar groups of potatoes—some diploid and some triploid, others tetraploid or pentaploid (see Chapter 18). They vary noticeably in leaf shape, floral patterns, and tuber colour. Domesticated potatoes have been selected for short stolons, large tubers, and various colours of tubers from white to yellow to black. After being introduced to Europe, potatoes became an important food staple, and millions of people were adversely affected when potato crops failed because of a blight caused by the fungus *Phytophthora infestans*. The resulting "potato famine" in Ireland had huge social repercussions: many people died, and others emigrated to Canada, the United States, and Australia.

Tomatoes (*Solanum lycopersicum*) were domesticated from plants that grew in South and Central America, but there is considerable debate about when they were domesticated. Two modern forms, wild cherry tomatoes and currant tomatoes, were recently domesticated from stock native to eastern Mexico. Eaten raw or cooked, tomatoes come in different sizes, shapes, and colours. Different strains grow well in a variety of situations, meaning that tomatoes

Figure 49.18
Solonaceae. Domesticated species include **(a)** capsicum peppers, **(b)** eggplant, **(c)** potato, and **(d)** tomato.

Marquis Wheat

Different strains of domesticated crops, such as wheat or corn, have different features that make them better suited to some areas (climates) than to others. Whereas Red Fife wheat matures in 130 days and produces 3270 kg·ha^{-1}, Hard Red Calcutta wheat matures in 110 days but yields only 1240 kg·ha^{-1}. In an effort to find a wheat variety that would grow well and produce a good yield in the Canadian Prairies, Sir Charles E. Saunders made extensive crosses and developed Marquis wheat by 1906. To do this involved selective breeding: a planned program of hybridization of different wheat varieties, rigid selection of the best available material, preliminary and final evaluations of the results from replicated trials, and extensive testing of the new varieties. In the 1880s, Dr. William Saunders (Sir Charles's father) had introduced and tested many strains of wheat, often from Russia and India.

Crosses that led to the emergence of Marquis wheat were mainly focused on Hard Red Calcutta and Red Fife. The products of the crosses were tested at stations near Agassiz, British Columbia; Indian Head, Saskatchewan; and Brandon, Manitoba. At the latter two locations, these two varieties differed by three weeks in reaching maturity.

Sir Charles is famous for the "chewing test" he used to test the products of the crosses. He observed that chewing allowed him to determine the elasticity of the gummy substance produced (gluten). Other tests included baking bread with flour ground from the different strains to ensure that the product was satisfactory.

In 1906, Marquis wheat emerged as the product of a cross between a Hard Red Calcutta female and a Red Fife male. The kernels were dark red and hard, medium in size, and short. Heads were medium in length and bearded. Marquis ripens a few days before Red Fife and produces flour that is strong and of good colour. Tested at Brandon in 1908, Marquis wheat was the earliest to ripen and yielded 4336 kg·ha^{-1}—the best among the strains compared. Note that in Ottawa, Marquis wheat was not as productive, yielding only 2522 kg·ha^{-1}, reflecting the effect of climate and conditions on yield.

---

can be grown in many different climate zones. When grown in greenhouses, tomatoes are often pollinated by resident bumblebees, although many cultivars are self-pollinating.

Eggplants include three closely related cultivated species: *Solanum melongena,* the brinjal eggplant or aubergine; *Solanum aethiopicum,* the scarlet eggplant; and *Solanum macrocarpum,* the gboma eggplant. All cultivated species are native to the Old World, with *S. macrocarpum* and *S. aethiopicum* having been domesticated in Africa. The origin of the brinjal eggplant is less certain, but it may have originated in Africa and been domesticated in India and southeast China. During the Arab conquests, it spread from there to the Mediterranean and today is cultivated around the world. Brinjal eggplants and tomatoes are autogamous diploids.

Chili peppers, *Capsicum* species, are another member of the Solonaceae that originated in the New World. Known as producers of capsaicin (see Chapter 34, *Molecule Behind Biology*), chili peppers are often used to spice food. Cultivation of *Capsicum* species was well advanced and widespread in the Americas by 6000 years B.P., and then, as now, they were used as condiments and components of complex diets.

People use other members of this family as the source of hallucinogenic compounds. Notable examples are tobacco and species in the genus *Datura.* Jimson weed or locoweed (*Datura stramonium*) contains strong poisons, including belladonna alkaloids, atropine, and scopalamine. It grows in many parts of the world and has often been used as a hallucinogenic drug because one active ingredient interferes with neurotransmitters (see Chapter 33) and can induce violent hallucinations. The name "locoweed" is a useful, important warning.

Another plant genus, *Brassica,* also includes many varieties seen on dinner tables worldwide (see *Domesticated Plants in the Genus* Brassica).

49.2k Squash

At least five species of squash (*Cucurbita;* **Figure 49.19**) were domesticated in the Americas before European settlers arrived, some of them at least 10 000 years B.P. Squash, beans, and corn were the "three sisters", staple foods farmed by many First Nations peoples in the New World. Genetic data obtained from an intron region of the mitochondrial *nad1* gene suggest that at least six independent domestication events occurred. *Cucurbita argyrosperma* appears to have been domesticated from *C. sororia,* a wild Mexican gourd that grew in the general area of Mexico as teosinte. *C. moschata* was probably domesticated somewhere in lowland South America and *C. maxima* in the humid lowlands of Bolivia from *C. andreana.* The *Cucurbita pepo* complex seems to be derived from at least two domestication events, one in eastern North America and one in northeastern Mexico.

Many people are familiar with *C. pepo* as the pumpkin or jack-o'-lantern, but the species also includes summer squashes and zucchinis. *Cucurbita maxima* is the Hubbard and other winter squashes,

Figure 49.19
Many cultivars of squash, *Cucurbita* species, are New World domesticates, some first domesticated in southwestern Mexico at least 10 000 years ago.

which also include some *C. pepo* and *C. moshata*. The diversity of these cultivars is astonishing. The intraspecific variations entailed provide another example of the difficulty of applying the species concept to the diversity of life (see Chapter 18).

49.2l Dogs

Behaviour provides clues to important aspects about domestication of dogs. Dogs are much better than chimps and wolves at understanding human communication signals. Specifically, researchers assessed the abilities of these three animals to read human signals indicating the location of food. Even young puppies with little human contact were more skillful at these social cognition skills than chimpanzees and wolves. Interspecific communication appears to have been strongly selected for during the domestication of dogs, building on the evolutionary history of social skills associated with cooperative hunting inherited from their wolf ancestors.

Dogs were among the first animals to have been domesticated, presumably to help people with hunting. Genetic, behavioural, and morphological evidence indicates that dogs were derived from wolves. The earliest morphological evidence suggests domestication of dogs by 14 500 years B.P., whereas mtDNA data suggest a date of 15 000 years B.P. Genetic data suggest an East Asian origin for dogs. Other mtDNA data from specimens in Latin American and Alaska indicate that dogs crossed into North America via the Bering Land Bridge, with people producing a group (clade) of dogs unique to the New World. The mtDNA data imply that European colonists actively prevented dogs that they brought with them from interbreeding with dogs already present in the New World.

49.2m *Salmo salar,* Atlantic Salmon

Naturally occurring widely around the North Atlantic (locations in North America, Greenland, and Europe), Atlantic salmon have been introduced to many sites around the world, from Jordan and Greece to Australia, New Zealand, Chile, Argentina, Brazil, and the Falkland Islands. There are both landlocked natural and introduced freshwater populations of Atlantic salmon. This species has been a traditional target of subsistence, sport, and commercial fishing and more recently the focus of aquaculture operations. The farmed fish have been selectively bred; thus, they are domesticated. Farmed fish are larger and more aggressive than those from wild stock, and they mature later.

Intensive fishing reduced natural stocks of Atlantic salmon, in some areas to the brink of extinction. Aquaculture operations have proven to be very lucrative, leading to a proliferation of these facilities in many areas. But there are negative impacts of some aquacultural operations. For example, escaped fish are thought to interbreed with local species (on the west coast of North America), threatening their genetic survival. This threat may be reduced by using sterile triploid Atlantic salmon for aquaculture. Sterile triploids can be mass produced, making it relatively feasible to use them in many areas. But concerns about the productivity and survival of triploid fish compared with diploid individuals have slowed the spread of their use.

Aquaculture can bring other problems. Sea lice, such as *Lepeophtheirus salmonis,* are parasitic copepod crustaceans that can cause serious problems for salmon aquacultural facilities. Recurrent sea lice infestations of aquacultured populations have spread to and decimated some wild salmon populations. Infestations by sea lice originating from cultured salmon have also caused a 99% collapse of some wild populations of pink salmon (*Oncorhynchus gorbuscha*) in coastal British Columbia.

The scale of aquaculture operations involving Atlantic salmon is astounding. In 2006, in the province of Nova Scotia (Canada) alone, 35 000 tonnes of Atlantic salmon were produced by aquaculture. Atlantic salmon also dominate farmed stock in British Columbia. The scale of production has wide social and economic implications for human nutrition, employment, and habitat.

49.2n Organisms Domesticated for More than One Use

Some animals and plants provide more than one crop. Cotton, as noted above, provides oilseed and fibre. Cattle are sources of meat, milk, blood, hides, and labour. Sheep provide wool, milk, meat, and hides. Sheep skins with the wool attached are used to make clothing, but the term "golden fleece" refers to their use to trap particles of placer gold.

Figure 49.20
Pearls for sale in the Pearl Market in Beijing, China.

M.B. Fenton

49.2o Cultivated But Not Domesticated Organisms

People cultivate many species that have not been domesticated because there is no evidence of selective breeding. Mushrooms are examples because although they have been cultivated and used as a source of protein for several thousand years, there is less information about selective breeding of specific lines (= domestication). Some species of mushrooms are also used as the source of biologically (hallucinogenic) active compounds.

Ostriches (*Strutio camelus*) are ranched (= cultivated) for their meat, hides, feathers, and eggs, but they are not domesticated. Crocodiles (*Crocodylus* species) are also ranched for their hides and meat. Oysters (*Pinctata fucata*) and other species of molluscs have been cultivated for hundreds of years mainly for pearls **(Figure 49.20)**. Other animals, such as *Python regius* (ball pythons), are bred and sold to snake fanciers. Breeders may select individuals with specific traits in their breeding programs, technically making the animals domesticated because the definition does not speak to use.

STUDY BREAK

1. What are the differences between cultivation and domestication? Give examples.
2. How has domestication affected *K*, carrying capacity?
3. Use two of the examples to compare the timing, location, and path of domestication. Be sure to explain how the domesticated organisms are used.
4. How did the domestication of rice, wheat, barley, and lentils differ from that of squash and potatoes?
5. Use an example to show how domestication of animals depends on their behaviour.

49.3 Yields

In his book *The Upside of Down: Catastrophe, Creativity and the Renewal of Civilization*, Thomas Homer-Dixon calculated the amount of energy it would have taken to build the Coliseum in Rome. He estimated that it would have taken 44 billion kilocalories of energy: 34 billion for oxen and 10 billion for human workers (assuming that both worked 220 days a year for 5 years and that the humans received $12\,500$ kJ·day^{-1}). The Roman workers would have eaten grain, mainly wheat, as well as legumes, vegetables, wine, and a little meat. The oxen would have been fed hay, mainly alfalfa, as well as legumes, millet, clover, tree foliage, and wheat chaff. Records from the time indicate a yield of wheat of about 1160 kg·ha^{-1} and alfalfa of about 2600 kg·ha^{-1}. Wheat delivers 1.0×10^7 kJ·ha^{-1} and alfalfa 1.6×10^7 kJ·ha^{-1}. Growing wheat would have required 58 days of slave labour per hectare per year. If farmers had had to pay labourers, the cost of production would have been higher.

Based on these data, Homer-Dixon calculated that building the Coliseum would have required the wheat grown on 19.8 km^2 of land and the alfalfa on 35.2 km^2. At its peak around 1 and 2 A.D., the population of the city of Rome was 1 million, and that number would have required the food produced on 8800 km^2 of land, equivalent to the area of Lebanon today. Much of the wheat that fed Rome came from North Africa, as well as Sicily, Etrusca, and Campania.

On average, one adult human needs 8300 kJ·day^{-1} to maintain a stable body mass, 3.0×10^6 kJ·yr^{-1}. This assumes a much lower level of exercise and physical exertion than the Roman worker in Homer-Dixon's calculations. If people were to meet their caloric demands from wheat alone, and if wheat delivers about 8700 kJ·kg^{-1}, at 8300 kJ·day^{-1}, each person would need to consume about 350 kg of wheat a year. In 2007 in southwestern Ontario, a wheat yield of 6600 kg·ha^{-1} meant that 1 ha (2.5 acres) of land would support 18.8 people for a year. A city of 50 000, eating only wheat, would need the wheat produced on 2660 ha **(Figures 49.21 and 49.22)**.

M.B. Fenton

Figure 49.21
The field marked with the "*" is 20 hectares (50 acres) and, if planted in wheat, should yield 13 200 kg in southwestern Ontario. Note that the area is also being used to harvest wind energy.

Table 49.1	Balancing Cost and Yield, a Farm in Southwestern Ontario (Costs as of Autumn, 2007)	
Crop/Material/Process	Yield kg·ha^{-1}	Can$·ha^{-1}
Corn	10 000	1650
Soy beans	3 500	1300
Wheat	6 600	900
Costs		
Seed		150
Fertilizer		125–225
Spray		50–100
Planting		38
Labour		40–190
Combine		100
Trucking		25
Crop insurance		20
Field rental		325

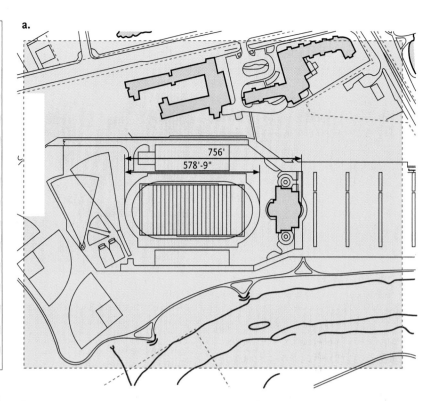

Figure 49.22

Envisioning 20 hectares (50 acres). Two views of the same area: a drawing **(a)** in which the coloured area represents 20 hectares and an aerial photograph **(b)** showing the same football stadium on the campus of the University of Western Ontario, in London, Ontario.

When put into this perspective, the variations in crop yields noted at the beginning of this chapter have huge repercussions. The data can be considered from a farmer's perspective **(Table 49.1),** and the difference in farm income would be substantial if the yield of corn were 1600 kg·ha^{-1} versus 10 000 kg·ha^{-1}. These data do not include costs of transportation (from farm to market), but they illustrate the impact of the cost of fuel. Spraying, planting, and trucking all require diesel fuel, so any change in the price of this commodity influences the costs of farm operations.

Terrain influences the level of technology that can be used in farming practices. Small terraced plots **(Figure 49.23, p. 1249)** must be worked by hand or with the help of animals. Large expanses of relatively flat land can be worked effectively and efficiently with machinery **(Figure 49.24, p. 1250),** increasing the energy consumption associated with farming but also the yield.

STUDY BREAK

1. How must crop yield be balanced against the cost of achieving it? Does this apply to a kitchen garden (as opposed to a functioning farm)?
2. What factors were fundamental in allowing the city of Rome to have a population of over 1 million people in 2 A.D.?

49.4 Complications

As anyone who has ever gardened or worked a farm well knows, there is more to growing crops than putting seeds or small plants in the ground and then harvesting the crops.

49.4a Fertilizer, Water, Yield, and Pests: Care of Crops

In the course of operating an experimental farm in the Negev Desert in southern Israel **(Figure 49.25, p. 1250),** researchers established several basic truths. By providing 20 m^3 of manure (sheep and goat) and

Domesticated Plants in the Genus *Brassica*

Some plants have provided humans with an embarrassment of riches. Imagine plants in one genus (*Brassica*) whose flowers, roots, stems, leaves, and seeds are all important food crops **(Figure 1)**. Foods from these vegetables are high in vitamin C and soluble fibre. They contain a rich mixture of nutrients, including some (diindolylmethane, sulforaphane, and selenium) thought to have anticancer effects.

A sample of an all-*Brassica* meal could include broccoli, cauliflower, Brussels sprouts, kale, and cabbage, all variations of one species, *Brassica oleracea*. You could add rutabaga (*Brassica napus*) along with turnip and rapini (*Brassica rapa*) to the menu and use canola oil (see *People Behind Biology*) and mustard (*B. rapa, B. carinata, B. elongata, B. juncea, B. nigra, B. ruprestris*) to dress the parts of the meal presented as salad.

Nuclear restriction fragment length polymorphisms (RFLPs), obtained from 10 different *Brassica rapa*, 9 cultivated

Figure 1

Brassicas in our diets. Shown here are cultivars of brassicas that commonly appear in people's diets, including **(a)** radishes, **(b)** Brussels sprouts, **(c)** cauliflower, **(d)** broccoli, **(e)** turnip, **(f)** rutabaga, **(g)** mustard, **(h)** cabbage, and **(i)** canola oil.

types of *B. oleracea*, and 6 other species in *Brassica* and related genera, suggest two basic evolutionary pathways **(Figure 2)** for diploid species: one that gave rise

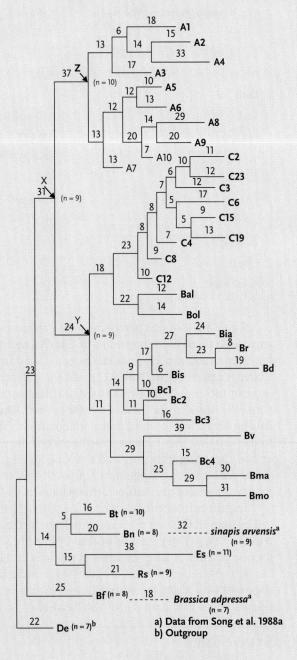

Figure 2

The shortest phylogenetic tree showing the relationships between different species of Brassica, including cultivars and wild species. A1–A5 *B. rapa* cultivars: 1 = flowering pak choi; 2 = pak choi; 3 = *B. narinosa*; 4 = Chinese cabbage; 5 = turnip; A6–A10 *B. rapa* wild: Bal = *B. alboglabra*; Bc 1–4 = *B. cretica*; Bd = *B. drepanensis*; Bia = *B. incana*; Bis = *B. isularis*; Bma = *B. macrocarpa*; Bmo = *B. montana*; Bol = *B. oleracea*; Br = *B. Rupestris*; BBv = *B. villosa*; C2–C23 *B. oleracea* cultivars: 2 = broccoli; 3 = broccoli (packman); 4 = cabbage; 8 = Portuguese tree kale; 12 = Chinese kale; 15 = kohlrabi; 19 = borecole; 23 = cauliflower; Bf = *B. fruticulosa*; Bn = *B. nigra*; Bt = *B. tournefortii*; De = *Diplotaxis erucoides*; Es = *Eruca sativa*; Rs = *Raphanus sativus*.

600 kg of ammonium sulphate per hectare, they could obtain good yields: 4800 kg·ha⁻¹ of barley (where less tended crops yielded 400 to 600 kg·ha⁻¹) and 4400 kg·ha⁻¹ (nanasit strain) or 2700 kg·ha⁻¹ (Florence strain) of wheat. They were able to produce 750 kg·ha⁻¹ of carrots and 650 kg·ha⁻¹ of onions. Achieving these yields required cultivation of the soil, irrigation, and dealing with a variety of pests. Their farm became an oasis of green that attracted hares,

gazelles, porcupines, desert partridges, and a host of insects, meaning that control of pest species had to be routine.

Irrigation was a key to good crops, and the experiment had been designed to test the prediction that by collecting runoff water and storing it in cisterns, the people there then could farm in an area with little and highly seasonal rainfall. In 2006–07, the 60-mm total rainfall in the area occurred between November 20

to *B. fruticulosa*, *B. nigra*, and *Sinapis arvensis* (*Brassica adpressa* is a close relative), and the other to *B. oleracea* and *B. rapa* (**Figure 3**). *Raphanus sativus* and *Eruca sativa* appear to be intermediate between the two lineages (see Figure 2). Europe and East Asia appear to have been centres of domestication for *Brassica* species. The related *Arabidopsis* is an important experimental tool used in understanding the genetics and selection of desirable traits in *Brassica*.

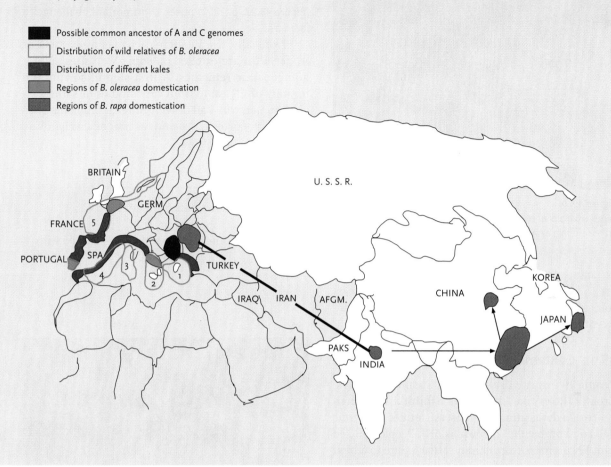

- ■ Possible common ancestor of A and C genomes
- □ Distribution of wild relatives of *B. oleracea*
- ▨ Distribution of different kales
- ▨ Regions of *B. oleracea* domestication
- ▨ Regions of *B. rapa* domestication

Figure 3
Geographic distribution and hypothetical origins and evolutionary pathways of *Brassica oleracea* and *B. rapa*, which may have shared a common ancestor in Europe.

and April 16. Water collected as runoff and stored in a 1400- m³ cistern could last the farm (people, animals, and crops) over 2 years.

49.4b Cats as Workers

The need to control rodent pests in areas where grain is stored (**Figure 49.26, p. 1250**) might be one factor explaining the domestication of cats, *Felis silvestris catus*. Carlos A. Driscoll and colleagues examined short tandem repeat (STR) and mtDNA data from 979 cats and wild progenitors to examine relationships among them. The evidence suggested at least five founder populations, including the European wildcat (*Felis silvestris silvestris*), near Eastern wildcat (*F. s. lybica*), central Asian wildcat (*F. s. ornata*), southern African wildcat (*F. s. cafra*), and Chinese desert cat (*F. s. bieti*). Each of these populations represents a distinct subspecies. Cats were thought to have been domesticated in the Near East, and their descendants were transported across the world with

Figure 49.23
A series of rock walls creates terraced areas for growing crops near Beijing in China. The terraces hold soil, but the setting precludes extensive use of mechanized farming equipment.

Figure 49.24
Large expanses of relatively flat land lend themselves to mechanized farming, allowing more uniform conditions and crops.

assistance from humans **(Figure 49.27)**. Driscoll and his colleagues proposed that domestication of cats coincided with the development of agriculture in different locations.

49.4c Contaminants in Crops

People living in rural parts of Bosnia, Bulgaria, Croatia, Romania, and Serbia exhibit a high incidence of a devastating renal disease termed "endemic Balkan nephropathy" (known as EN). People afflicted with EN progress from chronic renal failure to a high incidence of cancer of the upper urinary tract. EN and its associated cancer can be related to chronic dietary

Figure 49.25
An experimental farm plot near Avdot in the Negev Desert in southern Israel. The green areas are irrigated with water stored in an underground cistern. The experimental farm was established on an ancient farm site. Other fields previously under cultivation are shown. By collecting runoff during and after rainfall, farmers have stored water for their families and crops for hundreds of years.

poisoning by aristolochic acid. The source of the poisoning is contamination of grain crops with the plant *Aristolochia clematitis*. A clue to this situation came from horses that developed renal failure after being fed hay contaminated with *A. clematitis*. The presence of weeds or other contaminants in crops **(Figure 49.28, p. 1252)** poses an important challenge to farmers.

Ironically, the medicinal virtues of extracts of *A. clematitis* are extolled on some Web sites selling homeopathic remedies. A first step in solving the mystery of EN came from case studies of some Belgian women who had developed renal problems after taking extracts of *A. clematitis* as part of a weight loss program.

STUDY BREAK

1. What factors influence crop yield?
2. How does crop yield vary?
3. What are some of the problems associated with crops?

Figure 49.26
Harvested crops can be stored in different ways. Near Tien in China, farmers hang collections of corn cobs after harvest. This approach to crop storage suggests a dearth of local birds and rodents that might consume the corn.

Figure 49.27

The origins of domestic cats. Genetic assessment of 979 cats (*Felis silvestris catus*) based on short tandem repeats (STR) and mtDNA identifies different contributors to cat genotypes. The accompanying phenogram of 851 domesticated and wild cats illustrates the relationships between domestic cats and wild genetic contributors.

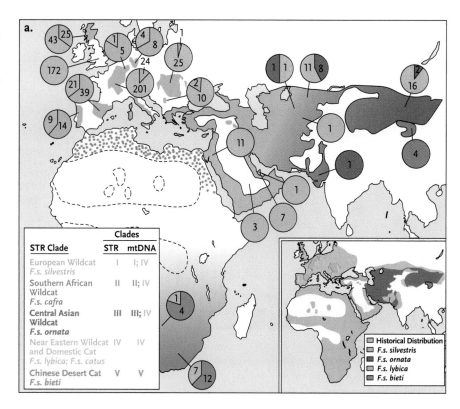

STR Clade	Clades	
	STR	mtDNA
European Wildcat *F.s. silvestris*	I	I; IV
Southern African Wildcat *F.s. cafra*	II	II; IV
Central Asian Wildcat *F.s. ornata*	III	III; IV
Near Eastern Wildcat and Domestic Cat *F.s. lybica; F.s. catus*	IV	IV
Chinese Desert Cat *F.s. bieti*	V	V

Historical Distribution
- F.s. silvestris
- F.s. ornata
- F.s. lybica
- F.s. bieti

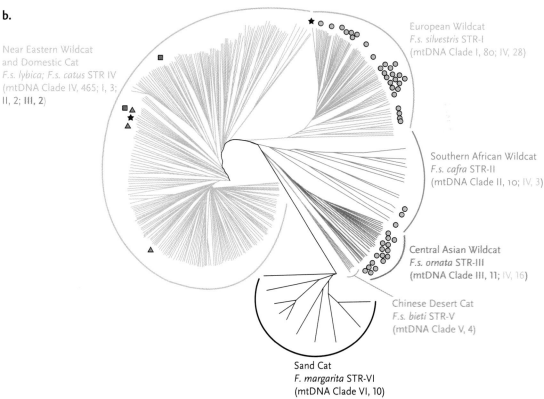

b.

Near Eastern Wildcat and Domestic Cat *F.s. lybica; F.s. catus* STR IV (mtDNA Clade IV, 465; I, 3; II, 2; **III, 2**)

European Wildcat *F.s. silvestris* STR-I (mtDNA Clade I, 80; IV, 28)

Southern African Wildcat *F.s. cafra* STR-II (mtDNA Clade II, 10; IV, 3)

Central Asian Wildcat *F.s. ornata* STR-III (mtDNA Clade III, 11; IV, 16)

Chinese Desert Cat *F.s. bieti* STR-V (mtDNA Clade V, 4)

Sand Cat *F. margarita* STR-VI (mtDNA Clade VI, 10)

49.5 Chemicals, Good and Bad

Plants are often treasured as much for the chemicals they produce as for their use as food. The chef who adds rosemary (*Rosmarinus officinalis;* **Figure 49.29,** **p. 1252**) to a dish as it is cooking knows that the flavour will enhance the final product. Other plant products are used as medicines. Phenolic compounds are responsible for the distinctive flavours of coffee, cinnamon, cloves, and nutmeg, which add flavour and

Figure 49.28
Weeds, such as the milkweeds (*Asclepias* species) growing in this barley field in south-western Ontario, can pose a problem at harvesting. If the weeds are toxic, they must be extracted from the harvested crop to ensure the safety of animals and people that consume the barley.

aroma to food. Some of these compounds, such as caffeine, can be addictive.

Ginsenosides (obtained from ginseng, *Panax quinquefolius*) appear to be used by ginseng plants as fungicides, whereas humans take them to stimulate the immune system. Other plant products are

Figure 49.29
Herbs such as rosemary (*Rosmarinus officinalis*) are used to add flavour during cooking.

Figure 49.30
Conine, the poison from poison hemlock.

toxic. Conine, produced by poison hemlock (*Conium maculatum;* not to be confused with the hemlock tree), is an active ingredient used by the plant to defend against herbivores **(Figure 49.30)**. It was the "hemlock" used to kill Socrates. Plants that look the same often produce quite different chemicals. Poison hemlock, for example, superficially resembles Queen Anne's lace (*Daucus carota*), a common weed. It also can be confused with fennel (*Foeniculum vulgare*) and parsley (*Petroselinum crispum*), two herbs often used in cooking.

The speed with which changes can occur in biological systems can be astonishing, exemplified by the assimilation of new foods into our own diets whether the product is natural or synthetic. This is not a feature unique to humans. Work on ant farmers revealed that one species (*Cyphomyrmex rimosus*) introduced to Florida quickly acquired a crop cultivated by another species of ant indigenous there (*Cyphomyrmex minutus*). Also in Florida, the grasshopper *Romalea microptera* uses a mix of chemicals to repel ants (see Chapter 48, *Molecule Behind Biology*). Soon after people began to use 2,4-D (2,4-dichlorophenoxyacetic acid) to kill weeds, the grasshopper added 2,5-dichlorophenol to its ant repellent mixture. This situation is analogous to humans cultivating and domesticating new crop species.

The pharmacological potential of plant products for treating and preventing human ailments has not been lost on three groups of people: those interested in conserving biodiversity, those concerned about alleviating human suffering, and those anxious to make money by selling biopharmaceuticals to others.

STUDY BREAK

1. Why do humans use 2,4-D? What other uses could it have?
2. Why are spices important?
3. What is poison hemlock?

49.6 Molecular Farming

Many native peoples in the New World used tobacco (*Nicotiana tabacum*) as a traditional medicine to ease the pain of childbirth, stave off hunger, and treat various ailments **(Figure 49.31)**. Tobacco was dried and smoked in ceremonies, used in poultices, brewed as a

Figure 49.31
A tobacco field in southwestern Ontario.

tea, and used as an emetic, an expectorant, and a laxative. By 1540, tobacco reached Europe, where it was said to cure illnesses ranging from epilepsy to plague. People have also used nicotine, an alkaloid in tobacco, in the same way as tobacco plants do, as an insecticide. The recreational use of tobacco has given the plant a bad name, even though other *Nicotiana* species are commonly used in gardens for their aromatic white or pink flowers.

Tobacco, however, is undergoing a rebirth as a positive contributor. *Molecular farming* involves the use of plants as producers of specific proteins useful as human medications. Plants can be genetically modified to produce large amounts of these proteins at low cost. Concerns about using plants as molecular farms reflect the possibility of biologically active products entering ecosystems and dispersing through food webs, via pollen or seeds, all of which could negatively affect existing crops. Enter tobacco, a nonfood plant that is harvested before flowering. Tobacco is not cold hardy (in Canada), but it is easy to genetically modify and is highly productive (40-day production cycle). Using appropriate DNA technology (see Chapter 16), T-DNA containing the human *IL-10* gene was inserted into the tobacco genome by connecting it to a plant promoter and an *Agrobacterium* terminator. In this way, tobacco plants are modified to produce interleukin-10 (*IL-10*), which can be used to treat irritable

bowel disease in humans. Interleukin is a cytokine involved in the regulation of inflammatory diseases, reducing the production of necrosis factor by tumours. The IL-10 was found not to enter the soil in which the tobacco was grown or the aphids or other insects that fed on the tobacco plants. The vast amounts of agricultural land that was used to grow tobacco for recreational use can now be used for molecular farming. This change to molecular farming of tobacco could produce many different useful compounds.

STUDY BREAK

1. How are plant products used in addition to their role as food?
2. What is "molecular farming"?
3. What conservation risks are associated with molecular farming? With agriculture in general?

49.7 The Future

It is tempting to believe that the diversity of life will continue to provide humans with solutions to many of their problems in the world: food for the hungry, poisons to selectively control pests, and biopharmaceuticals to cure diseases. Molecular farming potentially allows new approaches to solve old problems. Although we can continue to domesticate other species, effectively taking evolution in directions that suit us, we must remember that when new forms require higher investments in energy and fertilizer, costs can outweigh benefits.

In 2008, global increases in the price of food were partly due to our using traditional food crops (e.g., corn) to produce ethanol to fuel vehicles. In 2007, the energy return on investment (EROI) for biofuel (ethanol, or food for internal combustion engines) from corn was ~1. This means that every litre of ethanol produced from corn consumes about a litre of petroleum fuel. Other crops, for example, poplar trees, have much better EROIs. Although biofuels potentially are low-carbon-emitting energy sources, the means of production dramatically influences their "greenness." In many cases, production of biofuels generates a "biofuel carbon debt." In Brazil, Southeast Asia, and the United States, converting native habitat (rain forest, peatlands, savanna, and grasslands) to produce biofuels generates 17 to 420 times more CO_2 than the habitats normally produce. When these costs are taken into account, biofuels are not feasible environmentally friendly alternatives.

Calculating EROI and tracking it over time helps put energy use in perspective. In Rome in 1 A.D.,

Feeding People and Keeping Them Healthy

We have seen how humans' ability to harness nature has allowed us to reduce the numbers of hungry people in the world. Nevertheless, in 2008, converting food crops to biofuels in the interest of being "green" meant less food. This development coincided with climate change, which further reduced the availability of food worldwide. This impact was amplified by increases in the prices of food crops, in turn making food less accessible to many poor people.

We use plants for more than food. In 2008, over 1 million people died from malaria, caused by a blood parasite (see Chapter 26). Some 300 to 500 million people worldwide suffer from malaria. Most (over 75%) of the deaths were of children under 5 years of age and living in Africa. For years, alkaloids (quinine, quinidine, cichonidine, and cinchonine) extracted from the bark of four species of tree in the genus Cinchona have been used to treat agues and periodic fevers. Cinchona alkaloids have been particularly effective against malaria, but some strains of malaria are resistant to quinines and other treatments suggesting that they are not the solution to malaria.

Enter qinghao, also known as huang hua hao, an extract from a ubiquitous shrub, Artemisia annua. Extracts of qinghao (artemisinin) have long been used in traditional Chinese medicine. In 1971, Chinese scientists discovered its effectiveness in treating malaria and reported it to the world in 1979. In 2008, artemisinin combination treatments became first-line drugs for some forms of malaria (uncomplicated falciparum malaria), but they were not available worldwide. Using improved agriculture techniques—selection of high-yielding hybrids, microbial production, and synthetic peroxidases—could lower prices, increase the availability of artemisinin, and save many lives.

In 2007, about one-third of the annual U.S.$30 billion worldwide investment in agriculture research was aimed at solving the problems of agriculture in developing countries—home to about 80% of the global population. This investment is less than 3% of the amount that countries of the Organization for Economic Co-operation and Development (OECD) spend to subsidize their own agricultural production. Ensuring continued reduction of human hunger, OECD countries must invest more to solve the agriculture problems in the developing world.

We know that our agricultural prowess can be used to increase production of food and medicines such as artemisinin. Golden rice provides another example. In 1984, the World Health Organization estimated that 250 000 to 500 000 children a year had diets lacking in vitamin A. This deficiency damaged their retinas and corneas, so many of them went blind and half of them died.

By the early 2000s, Ingo Potrykus and Peter Byer had used genetic engineering to splice two daffodil genes and a bacterial gene into the rice genome. The genetically modified rice, "golden rice," was golden in colour and produced precursors to vitamin A in its endosperm. Golden rice offered a way to address the results of vitamin A deficiency. Growing golden rice was field tested in Louisiana in 2004 and 2005.

In 2008, it appears that even though golden rice offered the ability to prevent the suffering arising from vitamin A deficiency, it is unlikely that any will be planted before 2012. The delay reflects a combination of widespread public suspicion about genetically modified organisms (GMOs) and the high cost of obtaining approval to use GMOs. The costs associated with getting approval mean that only large companies with large budgets are likely to succeed in getting GMO products approved.

Moving more people away from "the edge" means investing more in finding solutions to agricultural operations in the developing world to ensure that discoveries such as artemisinin and golden rice are used to advantage. This requires policies and investments that support small-scale operations. Our ancestors who domesticated crops such as wheat and corn, lentils and squash, and potatoes and rice changed the world for us. We should ensure that their legacy lives on.

the EROI for wheat was 12:1, and for alfalfa, it was 27:1. In 2007, the EROI value for gasoline was ~17:1, whereas in the 1930s, it was ~100:1. The values of EROI speak to the sustainability of process, so Canadians must be concerned that the EROI on the tar sands is less than 4:1 without taking into consideration the water consumed by the process.

The development of resistance to toxins, whether of bacteria to antibiotics or of insect pests to insecticides, demonstrates that evolution works both ways. Perfecting genetic strains of crops protected by resistance to a pathogen or pest and using them exclusively can make the crops vulnerable to pathogens or pests that are resistant to the defence(s). The potato famine was exacerbated by the lack of genetic diversity of the potatoes used in Europe: they were vulnerable to blight.

Over evolutionary history, individuals able to exploit other species had an advantage over those that did not, whether within a society or between societies. The same principles apply to our own species. Our advantages of exploiting other species include increased access to food (quantity and quality), labour, materials for constructing things, and chemicals for treating disorders or controlling pests. These advantages were amplified through domestication, which meant increasing control over the other species, leading to several net effects. One effect was achieving larger populations of humans because of better access to food and/or protection from disease. Another was probably the increase in available time for the development and perfection of new tools and techniques and the emergence of groups of people in society who did not contribute directly by gathering or processing food. Such people could have contributed to society through their talents as artisans, soldiers, or even politicians.

The range of possibilities seems endless, particularly with the advent of the ability to directly modify genotypes and thus phenotypes. Perhaps we should be glad that the Attine ants have domesticated only fungi.

STUDY BREAK

What is artemisinin? What is golden rice? Why are they important?

UNANSWERED QUESTIONS

Are genetically modified organisms (GMOs) domesticated, or do they represent distinct species? What legal limits should our society impose on the use of GMOs?

Review

Go to CENGAGENOW™ at http://hed.nelson.com/ to access quizzing, animations, exercises, articles, and personalized homework help.

49.1 Domesticate

- The green revolution is credited with reducing the number of hungry people in the world by increasing agricultural productivity. Increased productivity reflects the use of improved strains of crops, increased applications of fertilizers, and more extensive irrigation. Higher productivity (crop yields) also reflects more dependence on mechanized (fossil fuel based) farming operations.

- There are three stages in the exploitation of biodiversity for human benefit. The initial stage involves "hunting and gathering," collecting organisms in the wild. The second stage is cultivation or caring for organisms under progressively controlled conditions. The third stage, domestication, involves selective breeding to enhance desired characteristics and features. The domestication process can apply to animals, plants, or other organisms, such as yeast and fungi.

- Domestication of asexually reproducing organisms does not involve selective breeding, providing more control, allowing shorter generation times, or speeding up the process of selection.

- Living in the same sites year-round allowed people to tend and protect their crops. Increased time in one place also would have facilitated the process of selective breeding.

- At least 200 species of ants in the tribe Attini are obligate farmers. Other ants also tend seed gardens.

- The first evidence of domestication appears to be figs by about 12 000 years B.P. Cultivation may have been practised for 1000 years before domestication. Many crops had been domesticated by 6000 years B.P.

49.2 Why Some Organisms Were Domesticated

- Yeast appears to have been domesticated in Egypt by 6000 years B.P., when it was used in making bread and beer. Cotton may have been domesticated in at least four sites, two in the Old World and two in the New World. Cotton is used as a source of oil seed and fibre, and some domestication had occurred by 4400 years B.P.

- The change from dehiscence to indehiscence was critical in the domestication of plants, such as grasses, whose seeds were the crop to be harvested.

- Squash and potatoes provide carbohydrates from fruits or tubers; the seeds are not the target of domestication.

- The domestication of dogs appears to have been based, in part, on their social behaviour, including their ability to communicate with people. Some experiments show that dogs are better at reading communication signals from people than other animals, such as chimps.

- Some animals are farmed for meat and hides (and for ostrich feathers and eggs), but there is no evidence of selective breeding.

49.3 Yields

- The level of farming intensity, the strains of crops, the application of fertilizers, and irrigation affect crop yields. Terrain can influence the level of mechanization.

- The data on corn and wheat yields clearly demonstrate variation over time and location.

49.4 Complications

- Rich patches of food reflect higher productivity but can attract pests.

49.5 Chemicals, Good and Bad

- Various plant chemicals are used as stimulants, medicines, hallucinogens, pesticides, and flavourings.

49.6 Molecular Farming

- Molecular farming is the use of plants to produce various useful proteins in large amounts at lower cost. Molecular farming may prove to be an inexpensive way to manufacture medicines and other substances.

49.7 The Future

- Escape of cultivated plants poses a significant risk to native species (e.g., grapes) and ecosystems. Monocultures can attract and support high levels of pests that may affect neighbouring natural systems.

Questions

Self-Test Questions

1. The first evidence of humans domesticating other organisms dates from
   a. 2000 years B.P.
   b. 6000 years B.P.
   c. 8000 years B.P.
   d. 12 000 years B.P.
   e. 20 000 years B.P.

2. The process of moving from cultivation to domestication involved dehiscence to indehiscence in
   a. rice.
   b. wheat.
   c. squash.
   d. potatoes.
   e. Both a and b are correct.

3. Parthenocarpy was important in the domestication of
   a. dogs.
   b. pigs.
   c. figs.
   d. lentils.
   e. cotton.

4. At Abu Hureyra, settlers domesticated
   a. olives.
   b. some grains.
   c. cattle.
   d. hot peppers.
   e. rice.

5. Domesticated organisms include (one or more answers may be correct)
   a. yeast, mushrooms, pigs, and oysters.
   b. yeast, pigs, rice, and corn.
   c. honeybees, yeast, rice, and ostriches.
   d. cattle, pigs, cats, and dogs.
   e. lentils, mushrooms, yeast, and crocodiles.

6. Plants in the family Solonaceae include
   a. tomatoes, potatoes, and eggplant.
   b. potatoes, hot peppers, and deadly nightshade.
   c. corn, tobacco, and hot peppers.
   d. rice, wheat, and barley.
   e. Both a and b are correct.

7. If a person needs 8300 kJ·day^{-1} and wheat yields 6,638 kg·ha^{-1}, a city of 20 000 people who ate only wheat would need the wheat produced by _____ over 1 year.
   a. 155 ha.
   b. 555 ha.
   c. 1055 ha.
   d. 2055 ha.
   e. 5055 ha.

8. Endemic Balkan nephropathy is an example of a disorder arising when people eat crops
   a. irrigated with polluted water.
   b. of grain contaminated with *Aristolochia clematitis*.
   c. of grain contaminated with fungi.
   d. of grain contaminated with *Rosmarinus officinalis*.
   e. of grain contaminated with *Asclepias exultata*.

9. *Nicotiana tabacum* is being used to produce
   a. nicotine.
   b. interleukin.
   c. acetylcholine.
   d. conine.
   e. capsaicin.

10. EROI, energy return on investment, suggests that alfalfa in Roman times was ___ times more efficient at energy production than the tar sands in Alberta in 2008.
    a. 4×
    b. 6×
    c. 10×
    d. 20×
    e. 50×

Questions for Discussion

1. When is domestication complete? At what point is a domesticated stock a separate species? Is domestication ever complete?

2. Why have so few aquatic species been domesticated? Why has there been a recent increase in the numbers of domesticated aquatic organisms?

3. How do domesticated populations threaten native species? Should this be a concern for conservation biologists? What can be done to minimize this threat to native species?

Appendix A
Answers to Self-Test Questions

Chapter 32

1. a 2. a 3. e 4. e 5. d

Chapter 33

1. d 2. a 3. c 4. b 5. c

Chapter 34

1. d 2. c 3. d 4. e 5. b 6. c 7. d 8. a 9. e 10. b

Chapter 35

1. a 2. e 3. d 4. d 5 b

Chapter 36

1. c 2. e 3. a 4. c 5. a

Chapter 37

1. e 2. b 3. a, d 4. e 5. a

Chapter 38

1. a 2. e 3. d 4. e 5. a 6. b 7. a 8. c 9. b

Chapter 39

1. a 2. e 3. c 4. c 5. e 6. e 7. d 8. e 9. b 10. c

Chapter 40

1. b 2. c 3. b 4. b 5. e 6. d 7. e 8. c 9. d

Chapter 41

1. a 2. e 3. b 4. e 5. d 6. b 7. e 8. e 9. e 10. c

Chapter 42

1. a 2. c 3. b and c 4. d 5. e

Chapter 43

1. d 2. b 3. a, c, d 4. e 5. d

Chapter 44

1. a 2. c 3. b 4. e 5. e 6. d 7. b 8. c 9. a 10. d

Chapter 45

1. b 2. c 3. b 4. d 5. a 6. e 7. d 8. e 9. d 10. b

Chapter 46

1. c 2. c 3. a 4. b 5. b 6. e 7. b 8. d 9. d 10. b

Chapter 47

1. b 2. d 3. c 4. e 5. e 6. c 7. a 8. d 9. c 10. a

Chapter 48

1. e 2. d 3. e 4. d 5. b 6. e 7. b 8. c 9. b (and e) 10. c

Chapter 49

1. d 2. e 3. c 4. b 5. d 6. e 7. c 8. b 9. b 10. b

Glossary

abdomen The region of the body that contains much of the digestive tract and sometimes part of the reproductive system; in insects, the region behind the thorax.

accommodation A process by which the lens changes to enable the eye to focus on objects at different distances.

acid precipitation Rainfall with low pH, primarily created when gaseous sulphur dioxide (SO_2) dissolves in water vapour in the atmosphere, forming sulphuric acid.

acquired immune deficiency syndrome (AIDS) A constellation of disorders that follows infection by the HIV virus.

acrosome A specialized secretory vesicle on the head of an animal sperm, which helps the sperm penetrate the egg.

acrosome reaction The process in which enzymes contained in the acrosome are released from an animal sperm and digest a path through the egg coats.

action potential The abrupt and transient change in membrane potential that occurs when a neuron conducts an electrical impulse.

active immunity The production of antibodies in the body in response to exposure to a foreign antigen.

active parental care Parents' investment of time and energy in caring for offspring after they are born or hatched.

adaptation, sensory *See* sensory adaptation.

adaptive (acquired) immunity A specific line of defence against invasion of the body in which individual pathogens are recognized and attacked to neutralize and eliminate them.

adipose tissue Connective tissue containing large, densely clustered cells called adipocytes that are specialized for fat storage.

adrenal cortex The outer region of the adrenal glands, which contains endocrine cells that secrete two major types of steroid hormones, the glucocorticoids and the mineralocorticoids.

adrenal medulla The central region of the adrenal glands, which contains neurosecretory neurons that secrete the catecholamine hormones epinephrine and norepinephrine.

adrenocorticotropic hormone (ACTH) A hormone that triggers hormone secretion by cells in the adrenal cortex.

adult stem cell Mammalian stem cells that can differentiate into a limited number of cell types associated with the tissue in which they occur.

afferent arteriole The vessel that delivers blood to the glomerulus of the kidney.

afferent neuron A neuron that transmits stimuli collected by a sensory receptor to an interneuron.

age structure A statistical description or graph of the relative numbers of individuals in each age class in a population.

age-specific fecundity The average number of offspring produced by surviving females of a particular age.

age-specific mortality The proportion of individuals alive at the start of an age interval that died during that age interval.

age-specific survivorship The proportion of individuals alive at the start of an age interval that survived until the start of the next age interval.

agonist A muscle that causes movement in a joint when it contracts.

albumin The most abundant protein in blood plasma, important for osmotic balance and pH buffering; also, the portion of an egg that serves as the main source of nutrients and water for the embryo.

aldosterone A mineralocorticoid hormone released from the adrenal cortex that increases the amount of Na reabsorbed from the urine in the kidneys and absorbed from foods in the intestine, reduces the amount of Na secreted by salivary and sweat glands, and increases the rate of K excretion by the kidneys, keeping Na and K balanced at the levels required for normal cellular function.

allantoic membrane Forms from mesoderm and endoderm that has bulged outward from the gut and encloses the allantois.

allantois In an amniote egg, an extraembryonic membrane sac that fills much of the space between the chorion and the yolk sac and stores the embryo's nitrogenous wastes.

allergen A type of antigen responsible for allergic reactions, which induces B cells to secrete an overabundance of IgE antibodies.

all-or-nothing principle The principle that an action potential is produced only if the stimulus is strong enough to cause depolarization to reach the threshold.

altruism A behavioural phenomenon in which individuals appear to sacrifice their own reproductive success to help other individuals.

alveolus (plural, alveoli) One of the millions of tiny air pockets in mammalian lungs, each surrounded by dense capillary networks.

amacrine cell A type of neuron that forms lateral connections in the retina of the eye, connecting bipolar cells and ganglion cells.

ammonification A metabolic process in which bacteria and fungi convert organic nitrogen compounds into ammonia and ammonium ions; part of the nitrogen cycle.

amnion In an amniote egg, an extraembryonic membrane that encloses the embryo, forming the amniotic cavity and secreting amniotic fluid, which provides an aquatic environment in which the embryo develops.

Amniota The monophyletic group of vertebrates that have an amnion during embryonic development.

amplification An increase in the magnitude of each step as a signal transduction pathway proceeds.

amygdala A grey-matter centre of the brain that works as a switchboard, routing information about experiences that have an emotional component through the limbic system.

anabolic steroid A steroid hormone that stimulates muscle development.

anaphylactic shock A severe inflammation stimulated by an allergen, involving extreme swelling of air passages in the lungs that interferes with breathing and massive leakage of fluid from capillaries that causes blood pressure to drop precipitously.

anatomy The study of the structures of organisms.

anchoring junction Cell junction that forms belts that run entirely around cells, "welding" adjacent cells together.

androgen One of a family of hormones that promote the development and maintenance of sex characteristics.

animal behaviour The responses of animals to specific internal and external stimuli.

animal pole The end of the egg where the egg nucleus is located, which typically gives rise to surface structures and the anterior end of the embryo.

anion A negatively charged ion.

antagonistic pair Two skeletal muscles, one of which flexes as the other extends to move joints.

anterior pituitary The glandular part of the pituitary, composed of endocrine cells that synthesize and secrete several tropic and nontropic hormones.

antibody-mediated immunity Adaptive immune response in which plasma cells secrete antibodies.

antidiuretic hormone (ADH) A hormone secreted by the posterior pituitary that increases water absorption in the kidneys, thereby increasing the volume of the blood.

antigen-presenting cell (APC) A cell that presents an antigen to T cells in antibody-mediated immunity and cell-mediated immunity.

antigenic variation The process by which an infectious organism alters its surface proteins to evade a host immune response. Parasites such as the trypanosomes that cause sleeping sickness in humans have 10% of their genes dedicated to generating new surface glycoproteins.

antimicrobial peptides Small, potent, broad-spectrum antibiotic peptides that are used by hosts collectively to eliminate bacterial and fungal pathogens. Some antimicrobial peptides also may act as immunomodulators.

aorta A large artery from the heart that branches into arteries leading to all body regions except the lungs.

aortic body One of several small clusters of chemoreceptors, baroreceptors, and supporting cells located along the aortic arch that measures changes in blood pressure and the composition of arterial blood flowing past it.

apoptosis Programmed cell death.

aposematic Refers to bright, contrasting patterns that advertise the unpalatability of poisonous or repellent species.

appendicular skeleton The bones comprising the pectoral (shoulder) and pelvic (hip) girdles and limbs of a vertebrate.

appendix A fingerlike sac that extends from the cecum of the large intestine.

aquatic succession A process in which debris from rivers and runoff accumulates in a body of fresh water, causing it to fill in at the margins.

aqueous humour A clear fluid that fills the space between the cornea and the lens of the eye.

archenteron The central endoderm-lined cavity of an embryo at the gastrula stage, which forms the primitive gut.

arteriole A branch from a small artery at the point where it reaches the organ it supplies.

artery A vessel that conducts blood away from the heart at relatively high pressure.

asexual reproduction Any mode of reproduction in which a single individual gives rise to offspring without fusion of gametes, that is, without genetic input from another individual. See also *vegetative reproduction*.

assimilation efficiency The ratio of the energy absorbed from consumed food to the total energy content of the food.

association area One of several areas surrounding the sensory and motor areas of the cerebral cortex that integrate information from the sensory areas, formulate responses, and pass them on to the primary motor area.

astrocyte A star-shaped glial cell that provides support to neurons in the vertebrate central nervous system.

atrioventricular node (AV node) A region of the heart wall that receives signals from the sinoatrial node and conducts them to the ventricle.

atrioventricular valve (AV valve) A valve composed of endocardium and connective tissue between each atrium and ventricle that prevents backflow of blood from the ventricle to the atrium during emptying of the heart.

autoimmune reaction The production of antibodies against molecules of the body.

autonomic nervous system A subdivision of the peripheral nervous system that controls largely involuntary processes, including digestion, secretion by sweat glands, circulation of the blood, many functions of the reproductive and excretory systems, and contraction of smooth muscles in all parts of the body.

axial skeleton The bones comprising the head and trunk of a vertebrate: the cranium, vertebral column, ribs, and sternum (breastbone).

axon The single elongated extension of a neuron that conducts signals away from the cell body to another neuron or an effector.

axon hillock A junction with the cell body of a neuron from which the axon arises.

axon terminal A branch at the tip of an axon that ends as a small, buttonlike swelling.

B cell A lymphocyte that recognizes antigens in the body.

background extinction rate The average rate of extinction of taxa through time.

bacteroid A rod-shaped or branched bacterium in the root nodules of nitrogen-fixing plants.

basal nucleus One of several grey-matter centres that surround the thalamus on both sides of the brain and moderate voluntary movements directed by motor centres in the cerebrum.

basal lamina A membrane secreted at the inner surface of epithelial cells.

basement membrane A membrane at the inner surface of epithelia in vertebrates. It consists of the basal lamina and a layer of connective tissue.

basilar membrane A stiff structural element within the cochlea.

basophil A type of leukocyte that is induced to secrete histamine by allergens.

Batesian mimicry The form of defence in which a palatable or harmless species resembles an unpalatable or poisonous one.

B-cell receptor (BCR) The receptor on B cells that is specific for a particular antigen.

bile A mixture of substances including bile salts, cholesterol, and bilirubin that is made in the liver, stored in the gallbladder, and used in the digestion of fats.

biogeochemical cycle Any of several global processes in which a nutrient circulates between the abiotic environment and living organisms.

biological magnification The increasing concentration of nondegradable poisons in the tissues of animals at higher trophic levels.

biomass The dry weight of biological material per unit area or volume of habitat.

bipolar cell A type of neuron in the retina of the eye that connects the rods and cones with the ganglion cells.

blastocoel A fluid-filled cavity in the blastula embryo.

blastocyst An embryonic stage in mammals; a single cell–layered hollow ball of about 120 cells with a fluid-filled blastocoel in which a dense mass of cells is localized to one side.

blastodisk A disklike layer of cells at the surface of the yolk produced by early cleavage divisions.

blastomere A small cell formed during cleavage of the embryo.

blastula The hollow ball of cells that is the result of cleavage divisions in an early embryo.

blood–brain barrier A specialized arrangement of capillaries in the brain that prevents most substances dissolved in the blood from entering the cerebrospinal fluid and thus protects the brain and spinal cord from viruses, bacteria, and toxic substances that may circulate in the blood.

bolus The food mass after chewing.

bone The densest form of connective tissue, in which living cells secrete the mineralized matrix of collagen and calcium salts that surrounds them; forms the skeleton.

Bowman's capsule An infolded region at the proximal end of a nephron that cups around the glomerulus and collects the water and solutes filtered out of the blood.

brain hormone (BH) A peptide hormone secreted by neurosecretory neurons in the brain of insects.

brain stem A stalklike structure formed by the pons and medulla, along with the midbrain, which connects the forebrain with the spinal cord.

bronchiole One of the small, branching airways in the lungs that lead into the alveoli.

bronchus (plural, bronchi) An airway that leads from the trachea to the lungs.

brown adipose tissue A specialized tissue in which the most intense heat generation by nonshivering thermogenesis takes place.

budding A mode of asexual reproduction in which a new individual grows and develops while attached to the parent.

bulbourethral gland One of two pea-sized glands on either side of the prostate gland that secrete a mucous fluid that is added to semen.

bulk feeder An animal that consumes sizable food items whole or in large chunks.

cadherin A cell surface protein responsible for selective cell adhesions that require calcium ions to set up adhesions.

calcitonin A nontropic peptide hormone that lowers the level of Ca^{2+} in the blood by inhibiting the ongoing dissolution of calcium from bone.

canines Pointed, conical teeth of a mammal, located between the incisors and the first premolars, that are specialized for biting and piercing.

capillary The smallest diameter blood vessel, with a wall that is one cell thick, which forms highly branched networks well adapted for diffusion of substances.

carbon cycle The global circulation of carbon atoms, especially via the processes of photosynthesis and respiration.

cardiac cycle The systole–diastole sequence of the heart.

cardiac muscle The contractile tissue of the heart.

carotid body A small cluster of chemoreceptors and supporting cells located near the bifurcation of the carotid artery that measures changes in the composition of arterial blood flowing through it.

carrying capacity The maximum size of a population that an environment can support indefinitely.

cartilage A tissue composed of sparsely distributed chondrocytes surrounded by networks of collagen fibres embedded in a tough but elastic matrix of the glycoprotein.

catecholamine Any of a class of compounds derived from the amino acid tyrosine that circulates in the bloodstream, including epinephrine and norepinephrine.

cation exchange Replacement of one cation with another, as on a soil particle.

CD4+ T cell A type of T cell in the lymphatic system that has CD4 receptors on its surface. This type of T cell binds to an antigen-presenting cell in antibody-mediated immunity.

CD8+ T cell A type of T cell in the lymphatic system that has CD8 receptors on its surface. This type of T cell binds to an antigen-presenting cell in cell-mediated immunity.

cecum A blind pouch formed at the junction of the large and small intestine.

cell adhesion molecule A cell surface protein responsible for selectively binding cells together.

cell body The portion of the neuron containing genetic material and cellular organelles.

cell lineage Cell derivation from the undifferentiated tissues of the embryo.

cell-mediated immunity An adaptive immune response in which a subclass of T cells—cytotoxic T cells—becomes activated and, with other cells of the immune system, attacks host cells infected by pathogens, particularly those infected by a virus.

central canal The central portion of the vertebral column in which the spinal cord is found.

cerebellum The portion of the brain that receives sensory input from receptors in muscles and joints, from balance receptors in the inner ear, and from the receptors of touch, vision, and hearing.

cerebral cortex A thin outer shell of grey matter covering a thick core of white matter within each hemisphere of the brain; the part of the forebrain responsible for information processing and learning.

cerebrospinal fluid Fluid that circulates through the central canal of the spinal cord and the ventricles of the brain, cushioning the brain and spinal cord from jarring movements and impacts, as well as nourishing the CNS and protecting it from toxic substances.

cervix The lower end of the uterus.

character displacement The phenomenon in which allopatric populations are morphologically similar and use similar resources, but sympatric populations are morphologically different and use different resources; may also apply to characters influencing mate choice.

chemical synapse A type of communicating connection between two neurons or a neuron and an effector cell in which an electrical impulse arriving at an axon terminal of the presynaptic cell triggers release of a neurotransmitter that crosses the gap and binds to a receptor on the postsynaptic cell, triggering an electrical impulse in that cell.

chemokine A protein secreted by activated macrophages that attracts other cells, such as neutrophils.

chemoreceptor A sensory receptor that detects specific molecules or chemical conditions such as acidity.

chlorosis An abnormal yellowing of plant tissues due to a lack of chlorophyll; a sign of nutrient deficiency or infection by a pathogen.

chondrocyte A cartilage-producing cell.

chorion In an amniote egg, an extraembryonic membrane that surrounds the embryo and yolk sac completely and exchanges oxygen and carbon dioxide with the environment; becomes part of the placenta in mammals.

chorionic villus (plural, villi) One of many treelike extensions from the chorion, which greatly increase the surface area of the chorion.

chylomicron A small triglyceride droplet covered by a protein coat.

chyme Digested content of the stomach released for further digestion in the small intestine.

ciliary body A fine ligament in the eye that anchors the lens to a surrounding layer of connective tissue and muscle.

circulatory system An organ system consisting of a fluid, a heart, and vessels for moving important molecules, and often cells, from one tissue to another.

circumcision Removal of the prepuce for religious, cultural, or hygienic reasons.

class II major histocompatibility complex (MHC) A collection of proteins that present antigens on the cell surface of an antigen-presenting cell in an antibody-mediated immune response.

classical conditioning A type of learning in which an animal develops a mental association between two phenomena that are usually unrelated.

cleavage Mitotic cell divisions of the zygote that produce a blastula from a fertilized ovum.

climax community A relatively stable, late successional stage in which the dominant vegetation replaces itself and persists until an environmental disturbance eliminates it, allowing other species to invade.

clitoris The structure at the junction of the labia minora in front of the vulva, homologous to the penis in the male.

clonal expansion The proliferation of the activated CD4 T cell by cell division to produce a clone of cells.

clonal selection The process by which a lymphocyte is specifically selected for cloning when it encounters a foreign antigen from among a randomly generated, enormous diversity of lymphocytes with receptors that specifically recognize the antigen.

closed circulatory system A circulatory system in which the fluid, blood, is confined in blood vessels and is distinct from the interstitial fluid.

clumped dispersion A pattern of distribution in which individuals in a population are grouped together.

cochlea A snail-shaped structure (in vertebrates) in the inner ear containing the organ of hearing.

cohort A group of individuals of similar age.

collagen Fibrous glycoprotein—very rich in carbohydrates—embedded in a network of proteoglycans.

collecting duct A location where urine leaving individual nephrons is processed further.

colon The main part of the large intestine.

commensalism A symbiotic interaction in which one species benefits and the other is unaffected.

community ecology The ecological discipline that examines groups of populations occurring together in one area.

compass orientation A wayfinding mechanism that allows animals to move in a particular direction, often over a specific distance or for a prescribed length of time.

competitive exclusion principle The ecological principle stating that populations of two or more species cannot coexist indefinitely if they rely on the same limiting resources and exploit them in the same way.

complement system A nonspecific defence mechanism activated by invading pathogens, made up of more than 30 interacting soluble plasma proteins circulating in the blood and interstitial fluid.

conduction The flow of heat between atoms or molecules in direct contact.

connective tissue Tissue with cells scattered through an extracellular matrix; forms layers in and around body structures that support other body tissues, transmit mechanical and other forces, and in some cases act as filters.

consciousness Awareness of oneself, one's identity, and one's surroundings, with understanding of the significance and likely consequences of events.

control Treatment that tells what would be seen in the absence of the experimental manipulation.

convection The transfer of heat from a body to a fluid, such as air or water, that passes over its surface.

corpus callosum A structure formed of thick axon bundles that connect the two cerebral hemispheres and coordinate their functions.

corpus luteum Cells remaining at the surface of the ovary during the luteal phase; the structure acts as an endocrine gland, secreting several hormones: estrogens, large quantities of progesterone, and inhibin.

cortical granule A secretory vesicle just under the plasma membrane of an egg cell.

cortisol The major glucocorticoid steroid hormone secreted by the adrenal cortex, which increases blood glucose by promoting breakdown of proteins and fats.

countercurrent exchange A mechanism in which the water flowing over the gills moves in a direction opposite to the flow of blood under the respiratory surface (can also apply to transfer of heat).

cranial nerve A nerve that connects the brain directly to the head, neck, and body trunk.

critical period A restricted stage of development early in life during which an animal has the capacity to respond to specific environmental stimuli.

crop Of birds, an enlargement of the digestive tube where the digestive contents are stored and mixed with lubricating mucus.

cryptic coloration Coloration that allows an organism to match its background and hence become less vulnerable to predation or recognition by prey.

cupula In certain mechanoceptors, a gelatinous structure with stereocilia extending into it that moves with pressure changes in the surrounding water; movement of the cupula bends the stereocilia, which triggers release of neurotransmitters.

cytokine A molecule secreted by one cell type that binds to receptors on other cells and, through signal transduction pathways, triggers a response. In innate immunity, cytokines are secreted by activated macrophages.

cytoplasmic determinants The mRNA and proteins stored in the egg cytoplasm that direct the first stages of animal development in the period before genes of the zygote become active.

cytoplasmic streaming Intracellular movement of cytoplasm.

cytotoxic T cell A T lymphocyte that functions in cell-mediated immunity to kill body cells infected by viruses or transformed by cancer.

daily torpor A period of inactivity and lowered metabolic rate that allows an endotherm to conserve energy when environmental temperatures are low.

demographic transition model A graphic depiction of the historical relationship between a country's economic development and its birth and death rates.

demography The statistical study of the processes that change a population's size and density through time.

dendrite The branched extension of the nerve cell body that receives signals from other nerve cells.

dendritic cell A type of phagocyte, so called because it has many surface projections that resemble dendrites of neurons, that engulfs a bacterium in infected tissue by phagocytosis.

denitrification A metabolic process in which certain bacteria convert nitrites or nitrates into nitrous oxide and then into molecular nitrogen, which enters the atmosphere.

density-dependent Description of environmental factors for which the strength of their effect on a population varies with the population's density.

density-independent Description of environmental factors for which the strength of their effect on a population does not vary with the population's density.

depolarized State of the membrane (which was polarized at rest) as the membrane potential becomes less negative.

deposit feeder An animal that consumes particles of organic matter from the solid substrate on which it lives.

dermis The skin layer below the epidermis; it is packed with connective tissue fibres such as collagen, which resist compression, tearing, or puncture of the skin.

desert A sparsely vegetated biome that forms where rainfall averages less than 25 cm per year.

determination Mechanism in which the developmental fate of a cell is set.

detritivore An organism that extracts energy from the organic detritus (refuse) produced at other trophic levels.

diabetes mellitus A disease that results from problems with insulin production or action.

diastole The period of relaxation and filling of the heart between contractions.

differentiation Follows determination and involves the establishment of a cell-specific developmental program in the cells. Differentiation results in cell types with clearly defined structures and functions.

digestive tube A tubelike digestive system with two openings that form a separate mouth and anus; the digestive contents move in one direction through specialized regions of the tube, from the mouth to the anus.

direct neurotransmitter A neurotransmitter that binds directly to a ligand-gated ion channel in the postsynaptic membrane, opening or closing the channel gate and altering the flow of a specific ion or ions in the postsynaptic cell.

disclimax community An ecological community in which regular disturbance inhibits successional change.

dispersion The spatial distribution of individuals within a population's geographic range.

dissociation The separation of water to produce hydrogen ions and hydroxide ions.

distal convoluted tubule The tubule in the human nephron that drains urine into a collecting duct that leads to the renal pelvis.

domestication Selective breeding of other species to increase desirable characteristics in progeny.

dominance hierarchy A social system in which the behaviour of each individual is constrained by that individual's status in a highly structured social ranking.

dorsal lip of the blastopore A crescent-shaped depression rotated clockwise 90° on the embryo surface that marks the region derived from the grey crescent, to which cells from the animal pole move as gastrulation begins.

duodenum A short region of the small intestine where secretions from the pancreas and liver enter a common duct.

ecdysone A steroid hormone that controls cuticle formation in insects and crustaceans and possibly nematodes.

echolocation A behaviour in which animal compares echoes of sounds it produced to the original signals. Differences between pulses and echoes allow location of obstacles and prey.

ecological community An assemblage of species living in the same place.

ecological efficiency The ratio of net productivity at one trophic level to net productivity at the trophic level below it.

ecological niche The resources a population uses and the environmental conditions it requires over its lifetime.

ecological pyramid A diagram illustrating the effects of energy transfer from one trophic level to the next.

ecological succession A somewhat predictable series of changes in the species composition of a community over time.

ecology The study of the interactions between organisms and their environments.

ecosystem ecology An ecological discipline that explores the cycling of nutrients and the flow of energy between the biotic components of an ecological community and the abiotic environment.

ecotone A wide transition zone between adjacent communities.

ectoderm The outermost of the three primary germ layers of an embryo, which develops into epidermis and nervous tissue.

ectoparasite A parasite that lives on the exterior of its host organism.

ectotherm An animal that obtains its body heat primarily from the external environment.

effector In signal transduction, a plasma membrane–associated enzyme, activated by a G protein, that generates one or more second messengers. In homeostatic feedback, the system that returns the condition to the set point if it has strayed away.

efferent arteriole The arteriole that receives blood from the glomerulus.

efferent neuron A neuron that carries the signals indicating a response away from the interneuron networks to the effectors.

eggs Nonmotile gametes.

elastin A rubbery protein in some connective tissues that adds elasticity to the extracellular matrix. It is able to return to its original shape after being stretched, bent, or compressed.

electrical signalling A means of animal communication in which a signaller emits an electric discharge that can be received by another individual.

electrical synapse A mechanical and electrically conductive link between two abutting neurons that is formed at the gap junction.

electrocardiogram (ECG) Graphic representation of the electrical activity within the heart, detected by electrodes placed on the body.

electroreceptor A specialized sensory receptor that detects electrical fields.

element A pure substance that cannot be broken down into simpler substances by ordinary chemical or physical techniques.

embryo An organism in its early stage of reproductive development, beginning in the first moments after fertilization.

embryonic stem cell Stem cells in the mammalian embryo that can differentiate into any cell type.

emigration The movement of individuals out of a population.

endangered species A species in immediate danger of extinction throughout all or a significant portion of its range.

endocrine gland Any of several ductless secretory organs that secrete hormones into the blood or extracellular fluid.

endocrine system The system of glands that release their secretions (hormones) directly into the circulatory system.

endocytic vesicle Vesicle that carries proteins and other molecules from the plasma membrane to destinations within the cell.

endoderm The innermost of the three primary germ layers of an embryo, which develops into the gastrointestinal tract and, in some animals, the respiratory organs.

endoparasite A parasite that lives in the internal organs of its host organism.

endorphin One of a group of small proteins occurring naturally in the brain and around nerve endings that bind to opiate receptors and thus can raise the pain threshold.

endoskeleton A supportive internal body structure, such as bones, that provides support.

endotherm An animal that obtains most of its body heat from internal physiological sources.

energy The capacity to do work.

energy budget The total amount of energy that an organism can accumulate and use to fuel its activities.

enzymatic hydrolysis A process in which chemical bonds are broken by the addition of H^+ and OH^-, the components of a molecule of water.

eosinophil A type of leukocyte that targets extracellular parasites too large for phagocytosis in the inflammatory response.

epiblast The top layer of the blastodisk.

epididymis A coiled storage tubule attached to the surface of each testis.

epiglottis A flaplike valve at the top of the trachea.

epinephrine A nontropic amine hormone secreted by the adrenal medulla.

epiphyte A plant that grows independently on other plants and obtains nutrients and water from the air.

epithelial tissue Tissue formed of sheetlike layers of cells that are usually joined tightly together, with little extracellular matrix material between them. They protect body surfaces from invasion by bacteria and viruses and secrete or absorb substances.

epitope The small region of an antigen molecule to which BCRs or TCRs bind.

equilibrium theory of island biogeography An hypothesis suggesting that the number of species on an island is governed by a give and take between the immigration of new species to the island and the extinction of species already there.

erythrocyte A red blood cell that contains hemoglobin, a protein that transports O_2 in blood.

erythropoietin (EPO) A hormone that stimulates stem cells in bone marrow to increase erythrocyte production.

essential element Any of a number of elements required by living organisms to ensure normal reproduction, growth, development, and maintenance.

essential fatty acid Any fatty acid that the body cannot synthesize but needs for normal metabolism.

essential mineral Any inorganic element such as calcium, iron, or magnesium that is required in the diet of an animal.

essential nutrient Any of the essential amino acids, fatty acids, vitamins, and minerals required in the diet of an animal.

estivation Seasonal torpor in an animal that occurs in summer.

estradiol A form of estrogen.

estrogen Any of the group of female sex hormones.

estuary A coastal habitat where tidal seawater mixes with fresh water from rivers, streams, and runoff.

ethology A discipline that focuses on how animals behave.

eusocial A form of social organization, observed in some insect species, in which numerous related individuals—a large percentage of them sterile female workers—live and work together in a colony for the reproductive benefit of a single queen and her mate(s).

eustachian tube A duct leading from the air-filled middle ear to the throat that protects the eardrum from damage caused by changes in environmental atmospheric pressure.

evaporation Heat transfer through the energy required to change a liquid to a gas.

excitatory postsynaptic potential (EPSP) The change in membrane potential caused when a neurotransmitter opens a ligand-gated Na^+ channel and Na^+ enters the cell, making it more likely that the postsynaptic neuron will generate an action potential.

excretion The process that helps maintain the body's water and ion balance while ridding the body of metabolic wastes.

exocrine gland A gland that is connected to the epithelium by a duct and that empties its secretion at the epithelial surface.

exoskeleton A hard external covering of an animal's body that blocks the passage of water and provides support and protection.

exotic species A non-native organism.

experimental variable The variable to which any difference in observations of experimental treatment subjects and control treatment subjects is attributed.

exploitative competition Form of competition in which two or more individuals or populations use the same limiting resources.

exponential model of population growth Model that describes unlimited population growth.

external fertilization The process in which sperm and eggs are shed into the surrounding water, occurring in most aquatic invertebrates, bony fishes, and amphibians.

external gill A gill that extends out from the body and lacks a protective covering.

extinction The death of the last individual in a species or the last species in a lineage.

extracellular fluid The fluid occupying the spaces between cells in multicellular animals.

extracellular matrix (ECM) A molecular system that supports and protects cells and provides mechanical linkages.

extraembryonic membrane A primary tissue layer extended outside the embryo that conducts nutrients from the yolk to the embryo, exchanges gases with the environment outside the egg, or stores metabolic wastes removed from the embryo.

eye The organ animals use to sense light.

facilitation hypothesis A hypothesis that explains ecological succession, suggesting that species modify the local environment in ways that make it less suitable for themselves but more suitable for colonization by species typical of the next successional stage.

family planning program A program that educates people about ways to produce an optimal family size on an economically feasible schedule.

fast block to polyspermy The barrier set up by the wave of depolarization triggered when sperm and egg fuse, making it impossible for other sperm to enter the egg.

fast muscle fibre A muscle fibre that contracts relatively quickly and powerfully.

fat Neutral lipid that is semisolid at biological temperatures.

fate map Mapping of adult or larval structures onto the region of the embryo from which each structure developed.

fat-soluble vitamin A vitamin that dissolves in liquid fat or fatty oils, in addition to water.

fatty acid One of two components of a neutral lipid, containing a single hydrocarbon chain with a carboxyl group linked at one end.

feather A sturdy, lightweight structure of birds, derived from scales in the skin of their ancestors.

feces Condensed and compacted digestive contents in the large intestine.

fertilization The fusion of the nuclei of an egg and sperm cell, which initiates development of a new individual.

fetus A developing human from the eighth week of gestation onward, at which point, the major organs and organ systems have formed.

fibre In sclerenchyma, an elongated, tapered, thick-walled cell that gives plant tissue its flexible strength.

fibrin A protein necessary for blood clotting; fibrin forms a weblike mesh that traps platelets and red blood cells and holds a clot together.

fibrinogen A plasma protein that plays a central role in the blood-clotting mechanism.

fibroblast The type of cell that secretes most of the collagen and other proteins in the loose connective tissue.

fibronectin A class of glycoproteins that aids in the attachment of cells to the extracellular matrix and helps hold the cells in position.

fibrous connective tissue Tissue in which fibroblasts are sparsely distributed among dense masses of collagen and elastin fibres that are lined up in highly ordered, parallel bundles, producing maximum tensile strength and elasticity.

filtration The nonselective movement of some water and a number of solutes—ions and small molecules, but not large molecules such as proteins—into the proximal end of the renal tubules through spaces between cells.

fixed action pattern A highly stereotyped instinctive behaviour; when triggered by a specific cue, it is performed over and over in almost exactly the same way.

flame cell The cell that forms the primary filtrate in the excretory system of many bilateria. The urine is propelled through ducts by the synchronous beating of cilia, resembling a flickering flame.

fluid feeder An animal that obtains nourishment by ingesting liquids that contain organic molecules in solution.

follicle The ovum and follicle cells.

follicle cell A cell that grows from ovarian tissue and nourishes the developing egg.

follicle-stimulating hormone (FSH) The pituitary hormone that stimulates oocytes in the ovaries to continue meiosis and become follicles. During follicle enlargement, FSH interacts with luteinizing hormone to stimulate follicular cells to secrete estrogens.

food web A set of interconnected food chains with multiple links.

forebrain The largest division of the brain, which includes the cerebral cortex and basal ganglia. It is credited with the highest intellectual functions.

foreskin A loose fold of skin that covers the glans of the penis.

formula The name of a molecule written in chemical shorthand.

fossil The remains or traces of an organism of a past geologic age embedded and preserved in Earth's crust.

fovea The small region of the retina around which cones are concentrated in mammals and birds with eyes specialized for daytime vision.

fundamental niche The range of conditions and resources that a population can possibly tolerate and use.

gallbladder The organ that stores bile between meals, when no digestion is occurring.

gametogenesis The formation of male and female gametes.

ganglion A functional concentration of nervous system tissue composed principally of nerve cell bodies, usually lying outside the central nervous system.

ganglion cell A type of neuron in the retina of the eye that receives visual information from photoreceptors via various intermediate cells such as bipolar cells, amacrine cells, and horizontal cells.

gap gene In *Drosophila* embryonic development, the first activated set of segmentation genes that progressively subdivide the embryo into regions, determining the segments of the embryo and the adult.

gap junction Junction that opens direct channels allowing ions and small molecules to pass directly from one cell to another.

gastric juice A substance secreted by the stomach that contains the digestive enzyme pepsin.

gastrula The developmental stage resulting when the cells of the blastula migrate and divide once cleavage is complete.

gastrulation The second major process of early development in most animals, which produces an embryo with three distinct primary tissue layers.

gene A unit containing the code for a protein molecule or one of its parts, or for functioning RNA molecules such as tRNA and rRNA.

generalized compartment model A model used to describe nutrient cycling in which two criteria—organic versus inorganic nutrients and available versus unavailable nutrients—define four compartments where nutrients accumulate.

generation time The average time between the birth of an organism and the birth of its offspring.

genetic engineering The use of DNA technologies to alter genes for practical purposes.

genus A Linnaean taxonomic category ranking below a family and above a species.

geographic range The overall spatial boundaries within which a population lives.

germ cell An animal cell that is set aside early in embryonic development and gives rise to the gametes.

gestation The period of mammalian development in which the embryo develops in the uterus of the mother.

gizzard The part of the digestive tube that grinds ingested material into fine particles by muscular contractions of the wall.

gland A cell or group of cells that produces and releases substances nearby, in another part of the body, or to the outside.

glans A soft, caplike structure at the end of the penis, containing most of the nerve endings producing erotic sensations.

glial cell A nonneuronal cell contained in the nervous tissue that physically supports and provides nutrients to neurons, provides electrical insulation between them, and scavenges cellular debris and foreign matter.

globulin A plasma protein that transports lipids (including cholesterol) and fat-soluble vitamins; a specialized subgroup of globulins, the immunoglobulins, constitute antibodies and other molecules contributing to the immune response.

glomerulus A ball of blood capillaries surrounded by Bowman's capsule in the human nephron.

glucagon A pancreatic hormone with effects opposite to those of insulin: it stimulates glycogen, fat, and protein degradation.

glucocorticoid A steroid hormone secreted by the adrenal cortex that helps maintain the blood concentration of glucose and other fuel molecules.

glycogen Energy-providing carbohydrates stored in animal cells.

glycolysis Stage of cellular respiration in which sugars such as glucose are partially oxidized and broken down into smaller molecules.

Golgi tendon organ A proprioceptor of tendons.

gonad A specialized gamete-producing organ in which the germ cells collect. Gonads are the primary source of sex hormones in vertebrates: ovaries in the female and testes in the male.

gonadotropin A hormone that regulates the activity of the gonads (ovaries and testes).

gonadotropin-releasing hormone (GnRH) A tropic hormone secreted by the hypothalamus that causes the pituitary to make luteinizing hormone (LH) and follicle-stimulating hormone (FSH).

graded potential A change in membrane potential that does not necessarily trigger an action potential.

grey crescent A crescent-shaped region of the underlying cytoplasm at the side opposite the point of sperm entry exposed after fertilization when the pigmented layer of cytoplasm rotates toward the site of sperm entry.

grey matter Areas of densely packed nerve cell bodies and dendrites in the brain and spinal cord.

greater vestibular gland One of two glands located slightly below and to the left and right of the opening of the vagina in women. They secrete mucus to provide lubrication, especially when the woman is sexually aroused.

greenhouse effect A phenomenon in which certain gases foster the accumulation of heat in the lower atmosphere, maintaining warm temperatures on Earth.

gross primary productivity The rate at which producers convert solar energy into chemical energy.

growth factor Any of a large group of peptide hormones that regulates the division and differentiation of many cell types in the body.

growth hormone (GH) A hormone that stimulates cell division, protein synthesis, and bone growth in children and adolescents, thereby causing body growth.

habitat The specific environment in which a population lives, as characterized by its biotic and abiotic features.

habituation The learned loss of responsiveness to stimuli.

haemolymph *See* **open circulatory system.**

haplodiploidy A pattern of sex determination in insects in which females are diploid and males are haploid.

harvesting efficiency The ratio of the energy content of food consumed compared with the energy content of food available.

haustorium (plural, haustoria) The hyphal tip of a parasitic fungus that penetrates a host plant and absorbs nutrients from it; likewise in parasitic flowering plants, a root that can penetrate a host's tissues and absorb nutrients.

head The anteriormost part of the body, containing the brain, sensory structures, and feeding apparatus.

heavy chain The heavier of the two types of polypeptide chains that are found in immunoglobulin and antibody molecules.

helper T cell A clonal cell that assists with the activation of B cells.

hepatic portal vein The blood vessel that leads to capillary networks in the liver.

herbicide A compound that, at proper concentration, kills plants.

herbivore An animal that obtains energy and nutrients primarily by eating plants.

herbivory The interaction between herbivorous animals and the plants they eat.

hermaphroditism The mechanism in which both mature egg-producing and mature sperm-producing tissue are present in the same individual.

hibernation Extended torpor during winter.

hindbrain The lower area of the brain that includes the brain stem, medulla oblongata, and pons.

hippocampus A grey-matter centre that is involved in sending information.

historical biogeography The study of the geographic distributions of plants and animals in relation to their evolutionary history.

homeodomain An encoded transcription factor of each protein that binds to a region in the promoters of the genes whose transcription it regulates.

homeostasis A steady internal condition maintained by responses that compensate for changes in the external environment.

homeostatic mechanism Any process or activity responsible for homeostasis.

homeotic gene Any of the family of genes that determines the structure of body parts during embryonic development.

horizon A noticeable layer of soil, such as topsoil, with a distinct texture and composition that varies with soil type.

horizontal cell A type of neuron that forms lateral connections among photoreceptor cells in the retina of the eye.

host A species that is fed upon by a parasite.

human chorionic gonadotropin (hCG) A hormone that keeps the corpus luteum in the ovary from breaking down.

humus The organic component of soil remaining after decomposition of plants and animals, animal droppings, and other organic matter.

hybridoma A B cell that has been induced to fuse with a cancerous lymphocyte called a myeloma cell, forming a single, composite cell.

hydrogen bond Noncovalent bond formed by unequal electron sharing between hydrogen atoms and oxygen, nitrogen, or sulphur atoms.

hydrogeologic cycle The global cycling of water between the ocean, the atmosphere, land, freshwater ecosystems, and living organisms.

hydrolysis Reaction in which the components of a water molecule are added to functional groups as molecules are broken into smaller subunits.

hydroponic culture A method of growing plants not in soil but with the roots bathed in a solution that contains water and mineral nutrients.

hydrosphere The component of the biosphere that encompasses all of the waters on Earth, including oceans, rivers, and polar ice caps.

hydrostatic skeleton A structure consisting of muscles and fluid that, by themselves, provide support for the animal or part of the animal; no rigid support, such as a bone, is involved.

hydroxyl group Group consisting of an oxygen atom linked to a hydrogen atom on one side and to a carbon chain on the other side.

hymen A thin flap of tissue that partially covers the opening of the vagina.

hyperpolarized The condition of a neuron when its membrane potential is more negative than the resting value.

hypertension Commonly called high blood pressure, a medical condition in which blood pressure is chronically elevated above normal values.

hyperthermia The condition resulting when the heat gain of the body is too great to be counteracted.

hypoblast The bottom layer of a blastodisk.

hypodermis The innermost layer of the skin that contains larger blood vessels and additional reinforcing connective tissue.

hypothalamus The portion of the brain that contains centres regulating basic homeostatic functions of the body and contributing to the release of hormones.

hypothermia A condition in which the core temperature falls below normal for a prolonged period.

hypothesis A "working explanation" of observed facts.

immigration Movement of organisms into a population.

immune privilege The situation in which certain sites in the body tolerate the presence of an antigen without mounting an inflammatory immune response. These sites include the brain, eyes, and testicles.

immune response The defensive reactions of the immune system.

immune system The combined defences, innate and acquired, a body uses to eliminate infections.

immunoglobulin A specific protein substance produced by plasma cells to aid in fighting infection.

immunological memory The capacity of the immune system to respond more rapidly and vigorously to the second contact with a specific antigen than to the primary contact.

immunological tolerance The process that protects the body's own molecules from attack by the immune system.

imprinting The process of learning the identity of a caretaker and potential future mate during a critical period.

incus The second of the three sound-conducting middle ear bones in vertebrates, located between the malleus and the stapes.

indirect neurotransmitter A neurotransmitter that acts as a first messenger, binding to a G protein–coupled receptor in the postsynaptic membrane, which activates the receptor and triggers generation of a second messenger such as cyclic AMP or other processes.

induction A mechanism in which one group of cells (the inducer cells) causes or influences another nearby group of cells (the responder cells) to follow a particular developmental pathway.

infection thread In the formation of root nodules on nitrogen-fixing plants, the tube formed by the plasma membrane of root hair cells as bacteria enter the cell.

inflammation The heat, pain, redness, and swelling that occur at the site of an infection.

ingestion The feeding methods used to take food into the digestive cavity.

inheritance The transmission of DNA (that is, genetic information) from one generation to the next.

inhibin A peptide that, in females, is an inhibitor of FSH secretion from the pituitary, thereby diminishing the signal for follicular growth. In males, inhibin inhibits FSH secretion from the pituitary, thereby decreasing spermatogenesis.

inhibiting hormone (IH) A hormone released by the hypothalamus that inhibits the secretion of a particular anterior pituitary hormone.

inhibition hypothesis A hypothesis suggesting that new species are prevented from occupying a community by whatever species are already present.

inhibitory postsynaptic potential (IPSP) A change in membrane potential caused when hyperpolarization occurs, pushing the neuron farther from threshold.

initial A plant cell that remains permanently as part of a meristem and gives rise to daughter cells that differentiate into specialized cell types.

innate immunity A nonspecific line of defence against pathogens that includes inflammation, which creates internal conditions that inhibit or kill many pathogens, and specialized cells that engulf or kill pathogens or infected body cells.

inner boundary membrane Membrane lying just inside the outer boundary membrane of a chloroplast, enclosing the stroma.

inner cell mass The dense mass of cells within the blastocyst that will become the embryo.

inner ear That part of the ear, particularly the cochlea, that converts mechanical vibrations (sound) into neural messages that are sent to the brain.

insight learning A phenomenon in which animals can solve problems without apparent trial-and-error attempts at the solution.

instinctive behaviour A genetically "programmed" response that appears in complete and functional form the first time it is used.

insulin A hormone secreted by beta cells in the islets, acting mainly on cells of nonworking skeletal muscles, liver cells, and adipose tissue (fat) to lower blood glucose, fatty acid, and amino acid levels and promote the storage of those molecules.

insulinlike growth factor (IGF) A peptide that directly stimulates growth processes.

integration The sorting and interpretation of neural messages and the determination of the appropriate response(s).

integrator In homeostatic feedback, the control centre that compares a detected environmental change with a set point.

integument Skin.

interference competition Form of competition in which individuals fight over resources or otherwise harm each other directly.

interferon A cytokine produced by infected host cells affected by viral dsRNA, which acts on both the infected cell that produces it, an autocrine effect, and neighbouring uninfected cells, a paracrine effect.

intermediate disturbance hypothesis Hypothesis proposing that species richness is greatest in communities that experience fairly frequent disturbances of moderate intensity.

internal fertilization The process in which sperm are released by the male close to or inside the entrance of the reproductive tract of the female.

internal gill A gill located within the body that has a cover providing physical protection for the gills. Water must be brought to internal gills.

interneuron A neuron that integrates information to formulate an appropriate response.

interspecific competition The competition for resources between species.

interstitial fluid The fluid occupying the spaces between cells in multicellular animals.

intertidal zone The shoreline that is alternately submerged and exposed by tides.

intestinal villus A microscopic, fingerlike extension in the lining of the small intestine.

intestine The portion of digestive system where organic matter is hydrolyzed by enzymes secreted into the digestive tube. As muscular contractions of the intestinal wall move the mixture along, cells lining the intestine absorb the molecular subunits produced by digestion.

intraspecific competition The dependence of two or more individuals in a population on the same limiting resource.

intrinsic rate of increase The maximum possible per capita population growth rate in a population living under ideal conditions.

invagination The process in which cells changing shape and pushing inward from the surface produce an indentation, such as the dorsal lip of the blastopore.

invertebrate An animal without a vertebral column.

involution The process by which cells migrate into the blastopore.

ion A positively or negatively charged atom.

iris Of the eye, the coloured muscular membrane that lies behind the cornea and in front of the lens, which by opening or closing determines the size of the pupil and hence the amount of light entering the eye.

islets of Langerhans Endocrine cells that secrete the peptide hormones insulin and glucagon into the bloodstream.

isotonic Equal concentration of water inside and outside cells.

juvenile hormones A family of fatty acid hormones that govern metam\orphosis and reproduction in insects and crustaceans.

juxtaglomerular apparatus A group of receptors that monitor the pressure and flow of fluid through the distal tubule of the kidney.

keystone species A species that has a greater effect on community structure than its numbers might suggest.

kilocalorie (kcal) The scientific unit equivalent to a calorie and equal to 1000 small calories.

kin selection Altruistic behaviour to close relatives, allowing them to produce proportionately more surviving copies of the altruist's genes than the altruist might otherwise have produced on its own.

kinesis A change in the rate of movement or the frequency of turning movements in response to environmental stimuli.

K-selected species Long-lived, slow reproducing species that thrive in more stable environments.

labia majora A pair of fleshy, fat-padded folds that partially cover the labia minora.

labia minora Two folds of tissue that run from front to rear on either side of the opening to the vagina.

landscape ecology The field that examines how large-scale ecological factors—such as the distribution of plants, topography, and human activity—influence local populations and communities.

larynx The voice box.

lateral geniculate nuclei Clusters of neurons located in the thalamus that receive visual information from the optic nerves and send it on to the visual cortex.

lateral inhibition Visual processing in which lateral movement of signals from a rod or cone proceeds to a horizontal cell and continues to bipolar cells with which the horizontal cell makes inhibitory connections, serving both to sharpen the edges of objects and enhance contrast in an image.

lateralization A phenomenon in which some brain functions are more localized in one of the two hemispheres.

lateral-line system The complex of organs and sensory receptors along the sides of many fishes and amphibians that detects vibrations in water.

leaching The process by which soluble materials in soil are washed into a lower layer of soil or are dissolved and carried away by water.

learned behaviour A response of an animal that is dependent on having a particular kind of experience during development.

learning A process in which experiences stored in memory change the behavioural responses of an animal.

leghemoglobin An iron-containing, red-pigmented protein produced in root nodules during the symbiotic association between *Bradyrhizobium* or *Rhizobium* and legumes.

lek A display ground where males each possess a small territory from which they court attentive females.

lens The transparent, biconvex intraocular tissue that helps bring rays of light to a focus on the retina.

leukocyte A white blood cell, which eliminates dead and dying cells from the body, removes cellular debris, and participates in defending the body against invading organisms.

Leydig cell A cell that produces the male sex hormones.

life history The lifetime pattern of growth, maturation, and reproduction that is characteristic of a population or species.

life table A chart that summarizes the demographic characteristics of a population.

ligament A fibrous connective tissue that connects bones to each other at a joint.

ligand-gated ion channel A channel that opens or closes when a specific chemical, such as a neurotransmitter, binds to the channel.

light The portion of the electromagnetic spectrum that humans can detect with their eyes.

light chain The lighter of the two types of polypeptide chains found in immunoglobulin and antibody molecules.

limbic system A functional network formed by parts of the thalamus, hypothalamus, and basal nuclei, along with other nearby grey-matter centres—the amygdala, hippocampus, and olfactory bulbs—sometimes called the "emotional brain."

limiting nutrient An element in short supply within an ecosystem, the shortage of which limits productivity.

limnetic zone The sunlit, open water in a lake, beyond the zone where plants rooted in the bottom can grow.

lithosphere The component of the biosphere that includes the rocks, sediments, and soils of the crust.

liver A large organ whose many functions include aiding in digestion, removing toxins from the body, and regulating the chemicals in the blood.

loam Any well-aerated soil composed of a mixture of sand, clay, silt, and organic matter.

logistic model of population growth Model of population growth that assumes that a population's per capita growth rate decreases as the population gets larger.

long-term memory Memory that stores information from days to years or even for life.

long-term potentiation A long-lasting increase in the strength of synaptic connections in activated neural pathways following brief periods of repeated stimulation.

loop of Henle In mammals, a U-shaped bend of the proximal convoluted tubule.

loose connective tissue A tissue formed of sparsely distributed cells surrounded by a more or less open network of collagen and other glycoprotein fibres.

lumen The inside of the digestive tube.

lung One of a pair of invaginated respiratory surfaces, buried in the body interior where they are less susceptible to drying out; the organs of respiration in mammals, birds, reptiles, and most amphibians.

luteinizing hormone (LH) A hormone secreted by the pituitary that stimulates the growth and maturation of eggs in females and the secretion of testosterone in males.

lymph The interstitial fluid picked up by the lymphatic system.

lymph node One of many small, bean-shaped organs spaced along the lymph vessels that contain macrophages and other leukocytes that attack invading disease organisms.

lymphatic system An accessory system of vessels and organs that helps balance the fluid content of the blood and surrounding tissues and participates in the body's defences against invading disease organisms.

lymphocyte A leukocyte that carries out most of its activities in the tissues and organs of the lymphatic system. Lymphocytes play major roles in immune responses.

lysosome Membrane-bound vesicle containing hydrolytic enzymes for the digestion of many complex molecules.

macromolecule A very large molecule assembled by the covalent linkage of smaller subunit molecules.

macronutrient In humans, a mineral required in amounts ranging from 50 mg to more than 1 g per day. In plants, a nutrient needed in large amounts for the normal growth and development.

macrophage A phagocyte that takes part in nonspecific defences and adaptive immunity.

magnetoreceptor A receptor found in some animals that navigate long distances that allows them to detect and use Earth's magnetic field as a source of directional information.

magnification The ratio of an object as viewed to its real size.

major histocompatibility complex A large cluster of genes encoding the MHC proteins.

malleus The outermost of the sound-conducting bones of the middle ear in vertebrates.

malnutrition A condition resulting from a diet that lacks one or more essential nutrients.

marsupium An external pouch on the abdomen of many female marsupials, containing the mammary glands, and within which the young continue to develop after birth.

mass extinctions The disappearance of a large number of species in a relatively short period of geologic time.

mast cell A type of cell dispersed through connective tissue that releases histamine when activated by the death of cells, caused by a pathogen at an infection site.

maternal-effect gene One of a class of genes that regulate the expression of other genes expressed by the mother during oogenesis and that control the polarity of the egg and, therefore, of the embryo.

mating The pairing of a male and a female for the purpose of sexual reproduction.

mating systems The social systems describing how males and females pair up.

mechanoreceptor A sensory receptor that detects mechanical energy, such as changes in pressure, body position, or acceleration. The auditory receptors in the ears are examples of mechanoreceptors.

melanocyte-stimulating hormone (MSH) A hormone secreted by the anterior pituitary that controls the degree of pigmentation in melanocytes.

melatonin A peptide hormone secreted by the pineal gland that helps maintain daily biorhythms.

melanotic encapsulation The mechanism by which hemocytes move toward and form a capsule around pathogens that are too big to phagocytose. The capsule may then be melanized by the deposition of phenolic compounds that further isolate the pathogen.

membrane attack complexes An abnormal activation of the complement (protein) portion of the blood, forming a cascade reaction that brings blood proteins together, binds them to the cell wall, and then inserts them through the cell membrane.

membrane potential An electrical voltage that measures the potential inside a cell membrane relative to the fluid just outside; it is negative under resting conditions and becomes positive during an action potential.

memory The storage and retrieval of a sensory or motor experience or a thought.

memory B cell In antibody-mediated immunity, a long-lived cell expressing an antibody on its surface that can bind to a specific antigen. A memory B cell is activated the next time the antigen is encountered, producing a rapid secondary immune response.

memory cell An activated lymphocyte that circulates in the blood and lymph, ready to initiate a rapid immune response on subsequent exposure to the same antigen.

memory helper T cell In cell-mediated immunity, a long-lived cell differentiated from a helper T cell, which remains in an inactive state in the lymphatic system after an immune reaction has run its course and ready to be activated on subsequent exposure to the same antigen.

meninges Three layers of connective tissue that surround and protect the spinal cord and brain.

menstrual cycle A cycle of approximately 1 month in the human female during which an egg is released from an ovary and the uterus is prepared to receive the fertilized egg; if fertilization does not occur, the endometrium breaks down, which releases blood and tissue breakdown products from the uterus to the outside through the vagina.

mesoderm The middle layer of the three primary germ layers of an animal embryo, from which the muscular, skeletal, vascular, and connective tissues develop.

metabolism The biochemical reactions that allow a cell or organism to extract energy from its surroundings and use that energy to maintain itself, grow, and reproduce.

metamorphosis A reorganization of the form of certain animals during postembryonic development.

metanephridium (plural, metanephridia) The excretory tubule of most annelids and molluscs.

metapopulation A group of neighbouring populations that exchange individuals.

micelle A sphere composed of a single layer of lipid molecules.

micronutrient Any mineral required by an organism only in trace amounts.

microscope Instrument of microscopy with different magnifications and resolutions of specimens.

microscopy Technique for producing visible images of objects that are too small to be seen by the human eye.

microvilli Fingerlike projections forming a brush border in epithelial cells that cover the villi.

midbrain The uppermost of the three segments of the brain stem, serving primarily as an intermediary between the rest of the brain and the spinal cord.

middle ear The air-filled cavity containing three small, interconnected bones: the malleus, incus, and stapes.

migration The predictable seasonal movement of animals from the area where they are born to a distant and initially unfamiliar destination, returning to their birth site later.

mimic The species in Batesian mimicry that resembles the model.

mimicry A form of defence in which one species evolves an appearance resembling that of another.

mineralocorticoid A steroid hormone secreted by the adrenal cortex that regulates the levels of Na and K in the blood and extracellular fluid.

minimum viable population size The smallest population size that is likely to survive both predictable and unpredictable environmental variation.

mitochondrion Membrane-bound organelle responsible for synthesis of most of the ATP in eukaryotic cells.

mitosis Nuclear division that produces daughter nuclei that are exact genetic copies of the parental nucleus.

model The species in Batesian mimicry that is resembled by the mimic.

molecule A unit composed of atoms combined chemically in fixed numbers and ratios.

moult-inhibiting hormone (MIH) A peptide neurohormone secreted by a gland in the eye stalks of crustaceans that inhibits ecdysone secretion.

monoclonal antibody An antibody that reacts only against the same segment (epitope) of a single antigen.

monocyte A type of leukocyte that enters damaged tissue from the bloodstream through the endothelial wall of the blood vessel.

monogamy A mating system in which one male and one female form a long-term association.

monosaccharides The smallest carbohydrates, containing three to seven carbon atoms.

monotreme A lineage of mammals that lay eggs instead of bearing live young.

morula The first stage of animal development, a solid ball or layer of blastomeres.

motif A highly specialized region in a protein produced by the three-dimensional arrangement of amino acid chains within and between domains.

motor neuron An efferent neuron that carries signals to skeletal muscle.

motor unit A block of muscle fibres that is controlled by branches of the axon of a single efferent neuron.

moult-inhibiting hormone (MIH) A peptide neurohormone secreted by cells in the eyestalks (extensions of the brain leading to the eyes).

mucosa The lining of the gut that contains epithelial and glandular cells.

Müllerian duct The bipotential primitive duct associated with the gonads that leads to a cloaca.

Müllerian mimicry A form of defence in which two or more unpalatable species share a similar appearance.

multicellular organism Individual consisting of interdependent cells.

muscle fibre A bundle of elongated, cylindrical cells that make up skeletal muscle.

muscle spindle A stretch receptor in muscle; a bundle of small, specialized muscle cells wrapped with the dendrites of afferent neurons and enclosed in connective tissue.

muscle tissue Cells that have the ability to contract (shorten) forcibly.

muscle twitch A single, weak contraction of a muscle fibre.

muscularis The muscular coat of a hollow organ or tubular structure.

mutualism A symbiotic interaction between species in which both partners benefit.

mycorrhiza A mutualistic symbiosis in which fungal hyphae associate intimately with plant roots.

myoblast An undifferentiated muscle cell.

myofibril A cylindrical contractile element about 1 m in diameter that runs lengthwise inside the muscle fibre cell.

myogenic heart A heart that maintains its contraction rhythm with no requirement for signals from the nervous system.

myoglobin An oxygen-storing protein closely related to hemoglobin.

natural killer (NK) cell A type of lymphocyte that destroys pathogen infected cells.

navigation A wayfinding mechanism in which an animal moves toward a specific destination, using both a compass and a "mental map" of where it is in relation to the destination.

negative feedback The primary mechanism of homeostasis, in which a stimulus—a change in the external or internal environment—triggers a response that compensates for the environmental change.

negative pressure breathing Muscular contractions that expand the lungs, lowering the pressure of the air in the lungs and causing air to be pulled inward.

nephron A specialized excretory tubule that contributes to osmoregulation and carries out excretion, found in all vertebrates.

nerve A bundle of axons enclosed in connective tissue and all following the same pathway.

nervous tissue Tissue that contains neurons, which serve as lines of communication and control between body parts.

net primary productivity The chemical energy remaining in an ecosystem after a producer's cellular respiration is deducted.

neural crest A band of cells that arises early in the embryonic development of vertebrates near the region where the neural tube pinches off from the ectoderm; later, the cells migrate and develop into unique structures.

neural plate Ectoderm thickened and flattened into a longitudinal band, induced by notochord cells.

neural signalling The process by which an animal responds appropriately to a stimulus.

neural tube A hollow tube in vertebrate embryos that develops into the brain, spinal cord, spinal nerves, and spinal column.

neurogenic heart A heart that beats under the control of signals from the nervous system.

neuromuscular junction The junction between a nerve fibre and the muscle it supplies.

neuron An electrically active cell of the nervous system responsible for controlling behaviour and body functions.

neuronal circuit The connection between axon terminals of one neuron and the dendrites or cell body of a second neuron.

neuropile The region of a ganglion in which branching axons and dendrites make interconnections.

neuroscience The integrated study of the structure, function, and development of the nervous system.

neurosecretory neuron A neuron that releases a neurohormone into the circulatory system when appropriately stimulated.

neurotransmitter A chemical released by an axon terminal at a chemical synapse.

neurulation The process in vertebrates by which organogenesis begins with development of the nervous system from ectoderm.

neutrophil A type of phagocytic leukocyte that attaches to blood vessel walls in massive numbers when attracted to the infection site by chemokines.

nitrification A metabolic process in which certain soil bacteria convert ammonia or ammonium ions into nitrites that are then converted by other bacteria to nitrates, a form usable by plants.

nitrogen cycle A biogeochemical cycle that moves nitrogen between the huge atmospheric pool of gaseous molecular nitrogen and several much smaller pools of nitrogen-containing compounds in soils, marine and freshwater ecosystems, and living organisms.

nitrogen fixation A metabolic process in which certain bacteria and cyanobacteria convert molecular nitrogen into ammonia and ammonium ions, forms usable by plants.

nitrogenous base A nitrogen-containing molecule with the properties of a base.

nociceptor A sensory receptor that detects tissue damage or noxious chemicals; their activity registers as pain.

node of Ranvier The gap between two Schwann cells, which exposes the axon membrane directly to extracellular fluids.

nonshivering thermogenesis The generation of heat by oxidative mechanisms in nonmuscle tissue throughout the body.

norepinephrine A nontropic amine hormone secreted by the adrenal medulla.

nuclear envelope In eukaryotes, membranes separating the nucleus from the cytoplasm.

nucleoplasm The liquid or semiliquid substance within the nucleus.

nucleotide The monomer of nucleic acids consisting of a five-carbon sugar, a nitrogenous base, and a phosphate.

nucleus The central region of eukaryotic cells, separated by membranes from the surrounding cytoplasm, where DNA replication and messenger RNA transcription occur.

nutrition The processes by which an organism takes in, digests, absorbs, and converts food into organic compounds.

ocellus (plural, ocelli) The simplest eye, which detects light but does not form an image.

olfactory bulb A grey-matter centre that relays inputs from odour receptors to both the cerebral cortex and the limbic system.

oligodendrocyte A type of glial cell that populates the CNS and is responsible for producing myelin.

ommatidium (plural, ommatidia) A faceted visual unit of a compound eye.

oocyte A developing gamete that becomes an ootid at the end of meiosis.

oogenesis The process of producing eggs.

oogonium A cell that enters meiosis and gives rise to gametes, produced by mitotic divisions of the germ cells in females.

open circulatory system An arrangement of internal transport in some invertebrates in which the vascular fluid, hemolymph, is released into sinuses, bathing organs directly, and is not always retained within vessels.

operant conditioning A form of associative learning in which animals learn to link a voluntary activity, an operant, with its favourable consequences, the reinforcement.

optic chiasm Location just behind the eyes where the optic nerves converge before entering the base of the brain, a portion of each optic nerve crossing over to the opposite side.

optimal foraging theory A set of mathematical models that predict the diet choices of animals as they encounter a range of potential food items.

organ Two or more different tissues integrated into a structure that carries out a specific function.

organ of Corti An organ within the cochlear duct that contains the sensory hair cells detecting sound vibrations transmitted to the inner ear.

organ system The coordinated activities of two or more organs to carry out a major body function such as movement, digestion, or reproduction.

organic molecule Molecule based on carbon.

organismal ecology An ecological discipline in which researchers study the genetic, biochemical, physiological, morphological, and behavioural adaptations of organisms to their abiotic environments.

organogenesis The development of the major organ systems, giving rise to a free-living individual with the body organization characteristic of its species.

oscula One or more openings in a sponge through which water is expelled.

osmoconformer An animal in which the osmolarity of the cellular and extracellular solutions matches the osmolarity of the environment.

osmolality A measure of the osmotic concentration of a solution. It is measured in osmoles (the number of solute molecules and ions) per kilogram of solvent.

osmoreceptor A chemoreceptor in the hypothalamus that responds to changes in the osmolarity of the fluid surrounding it, which reflects the osmolarity generally of the body fluids.

osmoregulation The regulation of water and ion balance.

osmoregulator An animal that uses control mechanisms to keep the osmolarity of cellular and extracellular fluids the same but at levels that may differ from the osmolarity of the surroundings.

osmotic pressure A state of dynamic equilibrium in which the pressure of the solution on one side of a selectively permeable membrane exactly balances the tendency of water molecules to diffuse passively from the other side of the membrane due to a concentration gradient.

osteoblast A cell that produces the collagen and mineral of bone.

osteoclast A cell that removes bone minerals and recycles them through the bloodstream.

osteocyte A mature bone cell.

osteon The structural unit of bone, consisting of a minute central canal surrounded by osteocytes embedded in concentric layers of mineral matter.

otolith One of many small crystals of calcium carbonate embedded in the otolithic membrane of the hair cells.

outer ear The external structure of the ear, consisting of the pinna and meatus.

oval window An opening in the bony wall that separates the middle ear from the inner ear.

ovarian cycle The cyclic events in the ovary leading to ovulation.

ovary In animals, the female gonad, which produces female gametes and reproductive hormones. In flowering plants, the enlarged base of a carpel in which one or more ovules develop into seeds.

overexploitation The excessive harvesting of an animal or plant species, potentially leading to its extinction.

overnutrition The condition caused by excessive intake of specific nutrients.

oviduct The tube through which the egg moves from the ovary to the outside of the body.

oviparous Referring to animals that lay eggs containing the nutrients needed for development of the embryo outside the mother's body.

ovoviviparous Referring to animals in which fertilized eggs are retained within the body and the embryo develops using nutrients provided by the egg; eggs hatch inside the mother.

ovulation The process in which oocytes are released into the oviducts as immature eggs.

ovum A female sex cell, or egg.

oxytocin A hormone that stimulates the ejection of milk from the mammary glands of a nursing mother.

pacemaker cell A specialized cardiac muscle cell in the upper wall of the right atrium that sets the rate of contraction in the heart.

pair-rule genes In *Drosophila* embryonic development, the set of segmentation regulatory genes activated by gap genes that divide the embryo into units of two segments each.

paleobiology The study of ancient organisms.

pancreas A mixed gland composed of an exocrine portion that secretes digestive enzymes into the small intestine and an endocrine portion, the islets of Langerhans, that secretes insulin and glucagon.

parasitism A symbiotic interaction in which one species, the parasite, uses another, the host, in a way that is harmful to the host.

parasitoid An insect species in which a female lays eggs in the larva or pupa of another insect species, and her young consume the tissues of the living host.

parasympathetic division The division of the autonomic nervous system that predominates during quiet, low-stress situations, such as while relaxing.

parathyroid gland One of a pair of glands that produce parathyroid hormone (PTH) (found only in tetrapod vertebrates).

parathyroid hormone (PTH) The hormone secreted by the parathyroid glands in response to a fall in blood Ca^{2+} levels.

parental investment The time and energy devoted to the production and rearing of offspring.

parthenogenesis A mode of asexual reproduction in which animals produce offspring by the growth and development of an egg without fertilization.

partial pressure The individual pressure exerted by each gas within a mixture of gases.

parturition The process of giving birth.

passive immunity The acquisition of antibodies as a result of direct transfer from another person.

passive parental care The amount of energy invested in offspring—in the form of the energy stored in eggs or seeds or energy transferred to developing young through a placenta—before they are born.

pattern formation The arrangement of organs and body structures in their proper three-dimensional relationships.

pelagic province The water in a marine biome.

pepsin An enzyme made in the stomach that breaks down proteins.

per capita growth rate The difference between the per capita birth rate and the per capita death rate of a population.

perfusion The flow of blood or other body fluids on the internal side of the respiratory surface.

pericycle A tissue of plant roots, located between the endodermis and the phloem, which gives rise to lateral roots.

peripheral nervous system (PNS) All nerve roots and nerves (motor and sensory) that supply the muscles of the body and transmit information about sensation (including pain) to the central nervous system.

peristalsis The rippling motion of muscles in the intestine or other tubular organs characterized by the alternate contraction and relaxation of the muscles that propel the contents onward.

peritubular capillary A capillary of the network surrounding the glomerulus.

peroxisome Microbody that produces hydrogen peroxide as a by-product.

phagocytosis Process in which some types of cells engulf bacteria or other cellular debris to break them down.

pheromone A distinctive volatile chemical released in minute amounts to influence the behaviour of members of the same species.

phosphate group Group consisting of a central phosphorus atom held in four linkages: two that bind —OH groups to the central phosphorus atom, a third that binds an oxygen atom to the central phosphorus atom, and a fourth that links the phosphate group to an oxygen atom.

phosphorus cycle A biogeochemcial cycle in which weathering and erosion carry phosphate ions from rocks to soil and into streams and rivers, which eventually transport them to the ocean, where they are slowly incorporated into rocks.

phosphorylation The addition of a phosphate group to a molecule.

photoautotroph A photosynthetic organism that uses light as its energy source and carbon dioxide as its carbon source.

photophosphorylation The synthesis of ATP coupled to the transfer of electrons energized by photons of light.

photopsin One of three photopigments in which retinal is combined with different opsins.

photoreceptor A sensory receptor that detects the energy of light.

photosynthesis The conversion of light energy to chemical energy in the form of sugar and other organic molecules.

photosystem A large complex into which the light-absorbing pigments for photosynthesis are organized with proteins and other molecules.

photosystem II In photosynthesis, a protein complex in the thylakoid membrane that uses energy absorbed from sunlight to synthesize ATP.

physiological respiration The process by which animals exchange gases with their surroundings—how they take in oxygen from the outside environment and deliver it to body cells and remove carbon dioxide from body cells and deliver it to the environment.

physiology The study of the functions of organisms—the physicochemical processes of organisms.

pigment A molecule that can absorb photons of light.

piloting A wayfinding mechanism in which animals use familiar landmarks to guide their journey.

pinacoderm In sponges, an unstratified outer layer of cells.

pineal gland A light-sensitive, melatonin-secreting gland that regulates some biological rhythms.

pinna The external structure of the outer ear, which concentrates and focuses sound waves.

pituitary A gland consisting mostly of two fused lobes suspended just below the hypothalamus by a slender stalk of tissue that contains both neurons and blood vessels; it interacts with the hypothalamus to control many physiological functions, including the activity of some other glands.

plasma The clear, yellowish fluid portion of the blood in which cells are suspended. Plasma consists of water, glucose and other sugars, amino acids, plasma proteins, dissolved gases, ions, lipids, vitamins, hormones and other signal molecules, and metabolic wastes.

plasma cell A large antibody-producing cell that develops from B cells.

platelet An oval or rounded cell fragment enclosed in its own plasma membrane, which is found in the blood; they are produced in red bone marrow by the division of stem cells and contain enzymes and other factors that take part in blood clotting.

pleura The double layer of epithelial tissue covering the lungs.

polar body A nonfunctional cell produced in oogenesis.

polarity The unequal distribution of yolk and other components in a mature egg.

pollutant Materials or energy in a form or quantity that organisms do not usually encounter.

polyandry A polygamous mating system in which one female mates with multiple males.

polygamy A mating system in which either males or females may have many mating partners.

polygyny A polygamous mating system in which one male mates with many females.

polypeptide The chain of amino acids formed by sequential peptide bonds.

polysaccharide Chain with more than 10 linked monosaccharide subunits.

population density The number of individuals per unit area or per unit volume of habitat.

population ecology The ecological discipline that focuses on how a population's size and other characteristics change in space and time.

population size The number of individuals in a population at a specified time.

positive feedback A mechanism that intensifies or adds to a change in internal or external environmental condition.

positive pressure breathing A gulping or swallowing motion that forces air into the lungs.

posterior pituitary The neural portion of the pituitary, which stores and releases two hormones made by the hypothalamus, antidiuretic hormone and oxytocin.

postsynaptic cell The neuron or the surface of an effector after a synapse that receives the signal from the presynaptic cell.

postsynaptic membrane The plasma membrane of the postsynaptic cell.

predation The interaction between predatory animals and the animal prey they consume.

prediction A statement about what the researcher expects to happen to one variable if another variable changes.

pregnancy The period of mammalian development in which the embryo develops in the uterus of the mother.

prepuce Foreskin; a loose fold of skin that covers the glans of the penis.

presynaptic cell The neuron with an axon terminal on one side of the synapse that transmits the signal across the synapse to the dendrite or cell body of the postsynaptic cell.

presynaptic membrane The plasma membrane of the axon terminal of a presynaptic cell, which releases neurotransmitter molecules into the synapse in response to the arrival of an action potential.

primary cell layers The ectoderm, mesoderm, and endoderm layers that form the embryonic tissues.

primary consumer A herbivore, a member of the second trophic level.

primary immune response The response of the immune system to the first challenge by an antigen.

primary motor area The area of the cerebral cortex that runs in a band just in front of the primary somatosensory area and is responsible for voluntary movement.

primary producer An autotroph, usually a photosynthetic organism, a member of the first trophic level.

primary somatosensory area The area of the cerebral cortex that runs in a band across the parietal lobes of the brain and registers information on touch, pain, temperature, and pressure.

primary structure The sequence of amino acids in a protein.

primary succession Predictable change in species composition of an ecological community that develops on bare ground.

primitive groove In the development of birds, the sunken midline of the primitive streak that acts as a conduit for migrating cells to move into the blastocoel.

primitive streak In the development of birds, the thickened region of the embryo produced by cells of the epiblast streaming toward the midline of the blastodisk.

production efficiency The ratio of the energy content of new tissue produced to the energy assimilated from food.

progesterone A female sex hormone that stimulates growth of the uterine lining and inhibits contractions of the uterus.

progestin A class of sex hormones synthesized by the gonads of vertebrates and active predominantly in females.

prokaryote Organism in which the DNA is suspended in the cell interior without separation from other cellular components by a discrete membrane.

prolactin (PRL) A peptide hormone secreted by the anterior pituitary that stimulates breast development and milk secretion in mammals.

promiscuity A mating system in which individuals do not form close pair bonds, and both males and females mate with multiple partners.

propagation In animal nervous systems, the concept that the action potential does not need further trigger events to keep going.

proprioceptor A mechanoreceptor that detects stimuli used in the CNS to maintain body balance and equilibrium and to monitor the position of the head and limbs.

prostaglandin One of a group of local regulators derived from fatty acids that are involved in paracrine and autocrine regulation.

prostate gland An accessory sex gland in males that adds a thin, milky fluid to the semen and adjusts the pH of the semen to the level of acidity best tolerated by sperm.

protein Molecules that carry out most of the activities of life, including the synthesis of all other biological molecules. A protein consists of one or more polypeptides depending on the protein.

protonephridium The simplest form of invertebrate excretory tubule.

proximal convoluted tubule The tubule between the Bowman's capsule and the loop of Henle in the nephron of the kidney, which carries and processes the filtrate.

pulmocutaneous circuit In amphibians, the branch of a double blood circuit that receives deoxygenated blood and moves it to the skin and lungs or gills.

pulmonary circuit The circuit of the cardiovascular system that supplies the lungs.

pupa The nonfeeding stage between the larva and adult in the complete metamorphosis of some insects, during which the larval tissues are completely reorganized within a protective cocoon or hardened case.

pupil The dark centre in the middle of the iris through which light passes to the back of the eye.

pyramid of biomass A diagram that illustrates differences in standing crop biomass in a series of trophic levels.

pyramid of energy A diagram that illustrates the amount of energy that flows through a series of trophic levels.

pyramid of numbers A diagram that illustrates the number of individual organisms present in a series of trophic levels.

radiation The transfer of heat energy as electromagnetic radiation.

random dispersion A pattern of distribution in which the individuals in a population are distributed unpredictably in their habitat.

rapid eye movement (REM) sleep The period during deep sleep when the delta wave pattern is replaced by rapid, irregular beta waves characteristic of the waking state. The person's heartbeat and breathing rate increase, the limbs twitch, and the eyes move rapidly behind the closed eyelids.

reabsorption The process in which some molecules (for example, glucose and amino acids) and ions are transported by the transport epithelium back into the body fluid (animals with open circulatory systems) or into the blood in capillaries surrounding the tubules (animals with closed circulatory systems) as the filtered solution moves through the excretory tubule.

realized niche The range of conditions and resources that a population actually uses in nature.

receptor protein Protein that recognizes and binds molecules from other cells that act as chemical signals.

reciprocal altruism Form of altruistic behaviour in which individuals help nonrelatives if they are likely to return the favour in the future.

recognition protein Protein in the plasma membrane that identifies a cell as part of the same individual or as foreign.

recombination The physical exchange of segments between the chromatids of homologous chromosomes or between the chromosomes of prokaryotic cells or viruses.

rectum The final segment of the large intestine.

reflex A programmed movement that takes place without conscious effort, such as the sudden withdrawal of a hand from a hot surface.

refractory period A period that begins at the peak of an action potential and lasts a few milliseconds, during which the threshold required for generation of an action potential is much higher than normal.

relative abundance The relative commonness of populations within a community.

release The process in which urine is released into the environment from the distal end of the excretory tubule.

releasing hormone (RH) A peptide neurohormone that controls the secretion of hormones from the anterior pituitary.

renal artery An artery that carries bodily fluids into the kidney.

renal cortex The outer region of the mammalian kidney that surrounds the renal medulla.

renal medulla The inner region of the mammalian kidney.

renal pelvis The central cavity in the kidney where urine drains from collecting ducts.

renal vein The vein that routes filtered blood away from the kidney.

residual volume The air that remains in lungs after exhalation.

resource partitioning The use of different resources or the use of resources in different ways by species living in the same place.

respiratory medium The environmental source of O_2 and the "sink" for released CO_2. For aquatic animals, the respiratory medium is water; for terrestrial animals, it is air.

respiratory surface A layer of epithelial cells that provides the interface between the body and the respiratory medium.

respiratory system All parts of the body involved in exchanging air between the external environment and the blood.

resting potential A steady negative membrane potential exhibited by the membrane of a neuron that is not stimulated—that is, not conducting an impulse.

retina A light-sensitive membrane lining the posterior part of the inside of the eye.

reversible The term indicating that a reaction may go from left to right or from right to left, depending on conditions.

rhodopsin The retinal–opsin photopigment.

root nodule A localized swelling on a root in which symbiotic nitrogen-fixing bacteria reside.

round window A thin membrane that faces the middle ear.

r-selected species A short-lived species adapted to function well in a rapidly changing environment.

ruminant An animal that has a complex, four-chambered stomach.

saccule A fluid-filled chamber in the vestibular apparatus that provides information about the position of the head with respect to gravity (up versus down), as well as changes in the rate of linear movement of the body.

salivary amylase A substance that hydrolyzes starches to the disaccharide maltose.

salivary gland A gland that secretes saliva through a duct on the inside of the cheek or under the tongue; the saliva lubricates food and begins digestion.

salt marsh A tidal wetland dominated by emergent grasses and reeds.

saltatory conduction A mechanism that allows small-diameter axons to conduct impulses rapidly.

sarcomere The basic unit of contraction in a myofibril.

sarcoplasmic reticulum In vertebrate muscle fibres, a complex system of vesicles modified from the smooth endoplasmic reticulum that encircles the sarcomeres. The sarcoplasmic reticulum is part of the pathway for the stimulation of muscle contraction by neural signals.

savanna A biome comprising grasslands with few trees, which grows in areas adjacent to tropical deciduous forests.

Schwann cell A type of glial cell in the PNS that wraps nerve fibres with myelin and also secretes regulatory factors.

scrotum The baglike sac in which the testes are suspended in many mammals.

secondary consumer A carnivore that feeds on herbivores, a member of the third trophic level.

secondary immune response The rapid immune response that occurs during the second (and subsequent) encounters of the immune system of a mammal with a specific antigen.

secondary productivity Energy stored in new consumer biomass as energy is transferred from producers to consumers.

secondary succession Predictable changes in species composition in an ecological community that develops after existing vegetation is destroyed or disrupted by an environmental disturbance.

secretion A selective process in which specific small molecules and ions are transported from the body fluids (in animals with open circulatory systems) or blood (in animals with closed circulatory systems) into the excretory tubules.

segment polarity genes In *Drosophila* embryonic development, the set of segmentation regulatory genes activated by pair-rule genes that set the boundaries and anterior–posterior axis of each segment in the embryo.

segmentation genes Genes that work sequentially, progressively subdividing the embryo into regions, determining the segments of the embryo and the adult.

segregation The separation of the pairs of alleles that control a character as gametes are formed.

selective cell adhesion A mechanism in which cells make and break specific connections to other cells or to the extracellular matrix.

selectively permeable Membranes that selectively allow, impede, or block the passage of atoms and molecules.

semen The secretions of several accessory glands in which sperm are mixed prior to ejaculation.

semicircular canal A part of the vestibular apparatus that detects rotational (spinning) motions.

semilunar valve (SL valve) A flap of endocardium and connective tissue reinforced by fibres that prevent the valve from turning inside out.

seminal fluid Fluid secreted by the seminal vesicles that contains prostaglandins, which, when ejaculated into the female, trigger contractions of the female reproductive tract that help move the sperm into and through the uterus.

seminal vesicle A vesicle that secretes seminal fluid.

seminiferous tubule One of the tiny tubes in the testes where sperm cells are produced, grow, and mature.

sensitization Increased responsiveness to mild stimuli after experiencing a strong stimulus; one of the simplest forms of memory.

sensor A tissue or organ that detects a change in an external or internal factor such as pH, temperature, or the concentration of a molecule such as glucose.

sensory adaptation A condition in which the effect of a stimulus is reduced if it continues at a constant level.

sensory hair cell A hair cell that sends impulses along the auditory nerve to the brain when alternating changes of pressure agitate the basilar membrane on which the organ of Corti rests, moving the hair cells.

sensory neuron A neuron that transmits stimuli collected by their sensory receptors to interneurons.

sensory receptor (transducer) A receptor formed by the dendrites of afferent neurons or by specialized receptor cells making synapses with afferent neurons that pick up information about the external and internal environments of the animal.

sensory transduction The conversion of a stimulus into a change in membrane potential.

sequential hermaphroditism The form of hermaphroditism in which individuals change from one sex to the other.

serosa The serous membrane: a thin membrane lining the closed cavities of the body; has two layers with a space between that is filled with serous fluid.

Sertoli cell One of the supportive cells that completely surrounds developing spermatocytes in the seminiferous tubules. Follicle-stimulating hormone stimulates Sertoli cells to secrete a protein and other molecules that are required for spermatogenesis.

set point The level at which the condition controlled by a homeostatic pathway is to be maintained.

sex chromosomes Chromosomes that are different in male and female individuals of the same species.

sex ratio The relative proportions of males and females in a population.

sexual reproduction The mode of reproduction in which male and female parents produce offspring through the union of egg and sperm generated by meiosis.

short-term memory Memory that stores information for seconds.

sign stimulus A simple cue that triggers a fixed action pattern.

simple diffusion Mechanism by which certain small substances diffuse through the lipid part of a biological membrane.

simulation modelling An analytical method in which researchers gather detailed information about a system and then create a series of mathematical equations that predict how the components of the system interact and respond to change.

simultaneous hermaphroditism A form of hermaphroditism in which individuals develop functional ovaries and testes at the same time.

single-lens eye An eye type that works by changing the amount of light allowed to enter into the eye and by focusing this incoming light with a lens.

sinoatrial node (SA node) The region of the heart that controls the rate and timing of cardiac muscle cell contraction.

sinus A body space that surrounds an organ.

skeletal muscle A muscle that connects to bones of the skeleton, typically made up of long and cylindrical cells that contain many nuclei.

slow block to polyspermy The process in which enzymes released from cortical granules alter the egg coats within minutes after fertilization so that no other sperm can attach and penetrate to the egg.

slow muscle fibre A muscle fibre that contracts relatively slowly and with low intensity.

smooth muscle A relatively small and spindle-shaped muscle cell in which actin and myosin molecules are arranged in a loose network rather than in bundles.

social behaviour The interactions that animals have with other members of their species.

soil solution A combination of water and dissolved substances that coats soil particles and partially fills pore spaces.

solute The molecules of a substance dissolved in water.

solution Substance formed when molecules and ions separate and are suspended individually, surrounded by water molecules.

solvent The water in a solution in which the hydration layer prevents polar molecules or ions from reassociating.

somatic cell Any of the cells of an organism's body other than reproductive cells.

somatic nervous system A subdivision of the peripheral nervous system controlling body movements that are primarily conscious and voluntary.

somites Paired blocks of mesoderm cells along the vertebrate body axis that form during early vertebrate development and differentiate into dermal skin, bone, and muscle.

spatial summation The summation of EPSPs produced by firing of different presynaptic neurons.

speciation The process of species formation.

species A group of populations in which the individuals are so closely related in structure, biochemistry, and behaviour that they can successfully interbreed.

species composition The particular combination of species that occupy a site.

species diversity A community characteristic defined by species richness and the relative abundance of species.

species richness The number of species that live within an ecological community.

sperm Motile gamete.

spermatocyte A developing gamete that becomes a spermatid at the end of meiosis.

spermatogenesis The process of producing sperm.

spermatogonium (plural, spermatogonia) A cell that enters meiosis and gives rise to gametes, produced by mitotic divisions of the germ cells in males.

spermatozoon Also called sperm; a haploid cell that develops into a mature sperm cell when meiosis is complete.

sphincter A powerful ring of smooth muscle that forms a valve between major regions of the digestive tract.

spinal cord A column of nervous tissue located within the vertebral column and directly connected to the brain.

spinal nerve A nerve that carries signals between the spinal cord and the body trunk and limbs.

spiracle An opening in the chitinous exoskeleton of an insect through which air enters and leaves the tracheal system.

stability The ability of a community to maintain its species composition and relative abundances when environmental disturbances eliminate some species from the community.

standing crop biomass The total dry weight of plants present in an ecosystem at a given time.

stapes The smallest of three sound-conducting bones in the middle ear of tetrapod vertebrates.

starch Energy-providing carbohydrates stored in plant cells.

statolith A movable starch- or carbonate-containing stonelike body involved in sensing gravitational pull.

stele The central core of vascular tissue in roots and shoots of vascular plants; it consists of the xylem and phloem together with supporting tissues.

stem cell Undifferentiated cells in most multicellular organisms that can divide without differentiating and also can divide and differentiate into specialized cell types.

stereocilia Microvilli covering the surface of hair cells clustered in the base of neuromasts.

steroid A type of lipid derived from cholesterol.

stomach The portion of the digestive system in which food is stored and digestion begins.

stratification Horizontal layering of sedimentary rocks beneath the soil surface.

stretch receptor A proprioceptor in the muscles and tendons of vertebrates that detects the position and movement of the limbs.

submucosa A thick layer of elastic connective tissue that contains neuron networks and blood and lymph vessels.

subsoil The region of soil beneath topsoil, which contains relatively little organic matter.

substrate The particular reacting molecule or molecular group that an enzyme catalyzes.

succession The change from one community type to another.

surface tension The force that places surface water molecules under tension, making them more resistant to separation than the underlying water molecules.

survivorship curve Graphic display of the rate of survival of individuals over a species' life span.

suspension feeder An animal that ingests small food items suspended in water.

symbiosis An interspecific interaction in which the ecological relations of two or more species are intimately tied together.

symmetry (adj., symmetrical) Exact correspondence of form and constituent configuration on opposite sides of a dividing line or plane.

sympathetic division Division of the autonomic nervous system that predominates in situations involving stress, danger, excitement, or strenuous physical activity.

synapse A site where a neuron makes a communicating connection with another neuron or an effector such as a muscle fibre or gland.

synaptic cleft A narrow gap that separates the plasma membranes of the presynaptic and postsynaptic cells.

synaptic vesicle A secretory vesicle in the cytoplasm of an axon terminal of a neuron, in which neurotransmitters are stored.

systemic circuit In amphibians, the branch of a double blood circuit that receives oxygenated blood and provides the blood supply for most of the tissues and cells of a body.

systole The period of contraction and emptying of the heart.

T cell A lymphocyte produced by the division of stem cells in the bone marrow and then released into the blood and carried to the thymus. T cells participate in adaptive immunity.

T (transverse) tubule The tubule that passes in a transverse manner from the sarcolemma across a myofibril of striated muscle.

tactile signal A means of animal communication in which the signaller uses touch to convey a message to the signal receiver.

taxis A behavioural response that is directed either toward or away from a specific stimulus.

T-cell receptor (TCR) A receptor that covers the plasma membrane of a T cell, specific for a particular antigen.

temperate deciduous forest A forested biome found at low to middle altitudes at temperate latitudes, with warm summers, cold winters, and annual precipitation between 75 and 250 cm.

temperate grassland A nonforested biome that stretches across the interiors of most continents, where winters are cold and snowy and summers are warm and fairly dry.

temperate rain forest A coniferous forest biome supported by heavy rain and fog, which grows where winters are mild and wet and the summers are cool.

template A nucleotide chain used in DNA replication for the assembly of a complementary chain.

temporal summation The summation of several EPSPs produced by successive firing of a single presynaptic neuron over a short period of time.

tendon A type of fibrous connective tissue that attaches muscles to bones.

territory A plot of habitat, defended by an individual male or a breeding pair of animals, within which the territory holders have exclusive access to food and other necessary resources.

tertiary consumer A carnivore that feeds on other carnivores, a member of the fourth trophic level.

testis (plural, testes) The male gonad. In male vertebrates, they secrete androgens and steroid hormones that stimulate and control the development and maintenance of male reproductive systems.

testosterone A hormone produced by the testes, responsible for the development of male secondary sex characteristics and the functioning of the male reproductive organs.

tetanus A situation in which a muscle fibre cannot relax between stimuli, and twitch summation produces a peak level of continuous contraction.

thalamus A major switchboard of the brain that receives sensory information and relays it to the regions of the cerebral cortex concerned with motor responses to sensory information of that type.

thermal acclimatization A set of physiological changes in ectotherms in response to seasonal shifts in environmental temperature, allowing the animals to attain good physiological performance at both winter and summer temperatures.

thermoreceptor A sensory receptor that detects the flow of heat energy.

thermoregulation The control of body temperature.

thick filament A type of filament in striated muscle composed of myosin molecules; they interact with thin filaments to shorten muscle fibres during contraction.

thin filament A type of filament in striated muscle composed of actin, tropomyosin, and troponin molecules; they interact with thick filaments to shorten muscle fibres during contraction.

thorax The central part of an animal's body, between the head and the abdomen.

threshold potential In signal conduction by neurons, the membrane potential at which the action potential fires.

thymus An organ of the lymphatic system that plays a role in filtering viruses, bacteria, damaged cells, and cellular debris from the lymph and bloodstream and in defending the body against infection and cancer.

thyroid gland A gland located beneath the voice box (larynx) that secretes hormones regulating growth and metabolism.

thyroid-stimulating hormone (TSH) A hormone that stimulates the thyroid gland to grow in size and secrete thyroid hormones.

thyroxine (T_4) The main hormone of the thyroid gland, responsible for controlling the rate of metabolism in the body.

tidal volume The volume of air entering and leaving the lungs during inhalation and exhalation.

tight junction Region of tight connection between membranes of adjacent cells.

time lag The delayed response of organisms to changes in environmental conditions.

tolerance hypothesis Hypothesis asserting that ecological succession proceeds because competitively superior species replace competitively inferior ones.

topsoil The rich upper layer of soil where most plant roots are located; it generally consists of sand, clay particles, and humus.

torpor A sleeplike state produced when a lowered set point greatly reduces the energy required to maintain body temperature, accompanied by reductions in metabolic, nervous, and physical activity.

trace element An element that occurs in organisms in very small quantities (0.01%); in nutrition, a mineral required by organisms only in small amounts.

trait A particular variation in a genetic or phenotypic character.

transmission In neural signalling, the sending of a message along a neuron and then to another neuron or to a muscle or gland.

transport epithelium A layer of cells with specialized transport proteins in their plasma membranes.

triglyceride A nonpolar compound produced when a fatty acid binds by a dehydration synthesis reaction at each of glycerol's three —OH-bearing sites.

triiodothyronine (T_3) A hormone secreted by the thyroid gland that regulates metabolism.

trimester A division of human gestation, three months in length.

trophic cascade The effects of predator–prey interactions that reverberate through other population interactions at two or more trophic levels in an ecosystem.

trophic level A position in a food chain or web that defines the feeding habits of organisms.

trophoblast The outer single layer of cells of the blastocyst.

trophozoite Motile, feeding stage of *Giardia* and other single-celled protists.

tropic hormone A hormone that regulates hormone secretion by another endocrine gland.

tropical deciduous forest A tropical forest biome that occurs where winter drought reduces photosynthesis and most trees drop their leaves seasonally.

tropical forest Any forest that grows between the Tropics of Capricorn and Cancer, a region characterized by high temperature and rainfall and thin, nutrient-poor topsoil.

tropical rain forest A dense tropical forest biome that grows where some rain falls every month, mean annual rainfall exceeds 250 cm, mean annual temperature is at least 25°C, and humidity is above 80%.

turnover rate The rate at which one generation of producers in an ecosystem is replaced by the next.

tympanum A thin membrane in the auditory canal that vibrates back and forth when struck by sound waves.

umbilical cord A long tissue with blood vessels linking the embryo and the placenta.

umbilicus Navel; the scar left when the short length of umbilical cord still attached to the infant after birth dries and shrivels within a few days.

undernutrition A condition in animals in which intake of organic fuels is inadequate or whose assimilation of such fuels is abnormal.

unicellular organism Individual consisting of a single cell.

uniform dispersion A pattern of distribution in which the individuals in a population are evenly spaced in their habitat.

ureter The tube through which urine flows from the renal pelvis to the urinary bladder.

urethra The tube through which urine leaves the bladder. In most animals, the urethra opens to the outside.

urinary bladder A storage sac located outside the kidneys.

uterine cycle The menstrual cycle.

uterus A specialized saclike organ, in which the embryo develops in viviparous animals.

utricle A fluid-filled chamber of the vestibular apparatus that provides information about the position of the head with respect to gravity (up versus down), as well as changes in the rate of linear movement of the body.

vaccination The process of administering a weakened form of a pathogen to patients as a means of giving them immunity to subsequent infection, and disease, caused by that pathogen.

vagina The muscular canal that leads from the cervix to the exterior.

variable An environmental factor that may differ among places or an organismal characteristic that may differ among individuals.

vas deferens The tube through which sperm travel from the epididymis to the urethra in the male reproductive system.

vegetal pole The end of the egg opposite the animal pole, which typically gives rise to internal structures such as the gut and the posterior end of the embryo.

vein In a plant, a vascular bundle that forms part of the branching network of conducting and supporting tissues in a leaf or other expanded plant organ. In an animal, a vessel that carries the blood back to the heart.

ventilation The flow of the respiratory medium (air or water, depending on the animal) over the respiratory surface.

ventricle In the brain, an irregularly shaped cavity containing cerebrospinal fluid. In the heart, a chamber that pumps blood out of the heart.

venule A capillary that merges into the small veins leaving an organ.

vertebrae The series of bones that form the vertebral column of vertebrate animals.

vertebral column The series of vertebrae that surrounds and protects the dorsal nerve cord and forms the supporting axis of the body.

vertebrate A member of the monophyletic group of tetrapod animals that possess a vertebral column.

vesicle A small, membrane-bound compartment that transfers substances between parts of the endomembrane system.

vestibular apparatus The specialized sensory structure of the inner ear of most terrestrial vertebrates that is responsible for perceiving the position and motion of the head and, therefore, for maintaining equilibrium and for coordinating head and body movements.

vibrio Any of various short, motile, S-shaped or comma-shaped bacteria of the genus *Vibrio*.

virus An infectious agent that contains either DNA or RNA surrounded by a protein coat.

visual signal A means of communication in which animals use facial expressions or body language to send messages to other individuals.

vital capacity The maximum tidal volume of air that an individual can inhale and exhale.

vitamin An organic molecule required in small quantities that the animal cannot synthesize for itself.

vitamin D A steroidlike molecule that increases the absorption of Ca^{2+} and phosphates from ingested food by promoting the synthesis of a calcium-binding protein in the intestine; it also increases the release of Ca^{2+} from bone in response to PTH.

vitelline coat A gel-like matrix of proteins, glycoproteins, or polysaccharides immediately outside the plasma membrane of an egg cell.

vitreous humour The jellylike substance that fills the main chamber of the eye, between the lens and the retina.

voltage-gated ion channel A membrane-embedded protein that opens and closes as the membrane potential changes.

vulva The external female sex organs.

watershed An area of land from which precipitation drains into a single stream or river.

water-soluble vitamin A vitamin with a high proportion of oxygen and nitrogen able to form hydrogen bonds with water.

wavelength The distance between two successive peaks of electromagnetic radiation.

wax A substance insoluble in water that is formed when fatty acids combine with long-chain alcohols or hydrocarbon structures.

wetland A highly productive ecotone often at the border between a freshwater biome and a terrestrial biome.

white matter The myelinated axons that surround the grey matter of the central nervous system.

Wolffian duct A bipotential primitive duct associated with the gonads that leads to a cloaca.

X chromosome Sex chromosome that occurs paired in female cells and single in male cells.

Y chromosome Sex chromosome that is paired with an X chromosome in male cells.

yolk The portion of an egg that serves as the main energy source for the embryo.

yolk sac In an amniote egg, an extraembryonic membrane that encloses the yolk.

zero population growth A circumstance in which the birth rate of a population equals the death rate.

zona pellucida A gel-like matrix of proteins, glycoproteins, or polysaccharides immediately outside the plasma membrane of the egg cell.

zygote A fertilized egg.

Credits

CHAPTER 32 **765:** Simon Fraser/SPL/Photo Researchers, Inc. **766:** David Macdonald **769:** (top left) Yuri Arcurs/Shutterstock **769:** (left) Ray Simmons/Photo Researchers, Inc. **769:** (centre) Ed Rescheke/Peter Arnold, Inc. **769:** (right) Don Fawcett **770:** (left) Gregory Dimijian/Photo Researchers, Inc. **771:** Photo: Darren Wong, Dr David Merrit. **772:** (a) Ed Reschke **772:** (b) Ed Reschke **772:** (c) Fred Hossier/Visual Unlimited **772:** (d) Ed Reschke **772:** (e) Ed Reschke **772:** (f) Ed Reschke **774:** (left) Ed Reschke **774:** (centre) Ed Reschke **774:** (right) BioPhoto Associates/Photo Researchers, Inc. **775:** (bottom) Lennart Nilsson from Behold Man, ©1974 Albert Bonniers Forlag and Little, Brown and Company, Boston **778:** Fred Bruemmer

CHAPTER 33 **783:** © C. J. Guerin, Ph. D., MRC Toxicology Unit/SPL/Photo Researchers, Inc. **784:** Photo by Drees **786:** (top) Triarch/Visuals Unlimited **786:** (bottom) Society for Neuroscience **787:** C. Raines/Visuals Unlimited **795:** Dennis Kunkel/Visuals Unlimited **798:** E.R. Lewis, T.E. Everhart, Y. Y. Zevi/Visuals Unlimited **800:** (top left) Reproduced with the permission of the Minister of Public Works and Government Services Canada, 2009. **800:** (bottom left) From RUPPERT. Invertebrate Zoology. © 1994 Nelson Education Ltd. Reproduced by permission. www.cengage.com/permissions **801:** (top) Reproduced with the permission of the Minister of Public Works and Government Services Canada, 2009. **802:** Used with permission by Cornell University Press **803:** From Pough, Heiser, McFarland. Vertebrate Life, 4E. Published by Prentice-Hall. Reprinted by permission of Pearson Education, Ltd. **806:** Courtesy of Dr. Marcus Raichle, courtesy of Washington University School of Medicine, St. Louis **810:** (top left) Yuri Arcurs/Shutterstock

CHAPTER 34 **819:** Photo by M.B. Fenton **820:** (top left) Photo by Gord Temple **820:** (bottom left) Data by M.B. Fenton **823:** (centre right) BIOSCIENCE by Narins. Copyright 1990 by American Institute of Biological Sciences (AIBS). Reproduced with permission of American Institute of Biological Sciences (AIBS) in the format Textbook, CD-ROM and DVD via Copyright Clearance Center. **823:** (bottom) With kind permission from Springer Science+Business Media: Journal of Comparative Physiology A, "Detection of vibrations in sand by tarsal sense organs of the nocturnal scorpion, Paruroctonus mesaensis", volume 131, Mar 1, 1979, pp. 23–30, Philip Brownell. **824:** Reprinted by permission of Blackwell Publishing **825:** From Kurt Schwenk, "Why Snakes Have Forked Tongues", Science, Mar 18, 1994, vol. 263, pp. 1573–1577. Reprinted with permission from AAAS. **826:** Herve Chaumeton/Agence Nature **827:** bottom, From Sanjay P. Sane, Alexandre Dieudonne, Mark A. Willis, Thomas L. Daniel, "Antennal Mechanosensors Mediate Flight Control in Moths", Science, Feb 9, 2007, vol. 315, pp. 863–866. Reprinted with permission from AAAS. **827:** (top left and right) Photo by A Percival-Smith **828:** (left) Photo by M.B. Fenton **828:** (top right) Photo: William Pflieger **828:** (bottom right) © William R. Elliot **830:** Reprinted by permission from Macmillan Publishers Ltd: Nature, F. Spoor, S. Bajpai, S. T. Hussain, K. Kumar and J. G. M. Thewissen, "Vestibular evidence for the evolution

of aquatic behaviour in early cetaceans", Vol. 417, pp. 163–166, copyright (2002). **831:** Andrew Syred/Photo Researchers, Inc. **833:** (all photos) M.B. Fenton **834:** (all photos) M.B. Fenton **835:** Reprinted by permission from Macmillan Publishers Ltd: Nature, Gebhard F. X. Schertler, "Signal transductionThe rhodopsin story continued", Vol. 453, pp. 292–293, copyright (2008). **839:** (top left) Yuri Arcurs/Shutterstock **840:** © A & J Visage / Alamy **841:** (top) Dr. M.V. Parthasarathy/Cornell Integrated Microscopy Center **841:** (bottom left) © A. Shay/OSF/Animals Animals—Earth Scenes **841:** (bottom right) Louisa Howard, Dartmouth College EM Facility **843:** Reprinted by permission from Macmillan Publishers Ltd: Nature, Gordon M. Shepherd, "Smell images and the flavour system in the human brain", Vol. 444, pp. 316–321, copyright (2006). **844:** (left) Reprinted by permission from Macmillan Publishers Ltd: Nature, Kenneth C. Catania, "OlfactionUnderwater 'sniffing' by semi-aquatic mammals", Vol. 444, pp. 1024–1025, copyright (2006). **844:** (top right) David Hosking/Frank Lane Picture Agency **844:** (bottom right) Photo by G.G. Carter **846:** Kenneth Lohmann/University of North Carolina **847:** (top and bottom) M.B. Fenton

CHAPTER 35 **851:** Dr. Richard B. Dominick **861:** Mirror Syndication International Ltd., 1986 **862:** © John Shaw/Tom Stack & Associates, Inc. **867:** (top left) Yuri Arcurs/Shutterstock **868:** (top left) © DLILLC/Corbis

CHAPTER 36 **873:** G. Delpho/Peter Arnold, Inc. **874:** Kiisa Nishikawa/Northern Arizona University **875:** (inset photos) Don Fawcett/Visuals Unlimited **880:** (bottom) From RUPPERT. Invertebrate Zoology. © 1994 Nelson Education Ltd. Reproduced by permission. www.cengage.com/permissions **881:** Linda Pitkin/Planet Earth Pictures **884:** (all photos) M.B. Fenton **886:** (top left) Yuri Arcurs/Shutterstock **887:** (left) Jupiter Images **887:** (right) Petr Kratochvil/Public Domain Pictures

CHAPTER 37 **891:** (top) © The Print Collector / Alamy **891:** (bottom) © The Print Collector / Alamy **892:** From A. D. Waller, Physiology, The Servant of Medicine, Hitchcock Lectures, University of London Press, 1910 **896:** Reprinted by permission of McGraw-Hill Companies **897:** National Cancer Institute/Photo Researchers, Inc. **899:** Professor P. Motta/Department of Anatomy/University La Sapienca, Rome/SPL/Photo Researchers, Inc. **901:** (top left) Yuri Arcurs/Shutterstock **905:** Lennart Nilsson from Behold Man © 1974 published by Albert Bonniers Forlag and Loitt, Brown and Company **907:** (top left) © DLILLC/Corbis **907:** (bottom) © Rolf Hicker

CHAPTER 38 **911:** ZEOvit.com **912:** Photo: Robert Poulin **913:** Dr. Stanley Flegler/Visuals Unlimited **914:** Nature's Images, Inc. / Photo Researchers, Inc. **916:** (left) Dr. David M. Phillips / Visuals Unlimited **916:** (right) Ryuzo Yanagimachi **918:** (top) Hans Pfletschinger **918:** (bottom) David M. Phillips/Visuals Unlimited **919:** © Copyright (2001) National Academy of Sciences, U.S.A. **920:** From David N. Reznick, Mariana Mateos, Mark S. Springer, "Independent Origins and Rapid Evolution of the Placenta in the Fish Genus Poeciliopsis", Science, Nov 1, 2002, vol. 298, pp. 1018–1020. Reprinted with permission from AAAS. **921:** From David N. Reznick, Mariana Mateos, Mark S. Springer, "Independent Origins and Rapid Evolution of the Placenta in the Fish Genus Poeciliopsis", Science, Nov 1, 2002, vol. 298, pp. 1018–1020. Reprinted with permission from AAAS. **922:** (left) John Cancalosi/Peter Arnold, Inc. **922:** (top right) Robin Chittenden; Frank Lane Picture Agency/Corbis **922:** (bottom right) Andrew J.

Martinez / Photo Researchers, Inc. **924:** (left) Photo by M.B. Fenton **924:** (top right) Photo by M.B. Fenton **924:** (bottom right) Copyright © 1987 Society for Reproduction and Fertility **925:** Lennart Nilsson from A Child is Born, © 1966, 1977 Dell Publishing Company, Inc. **928:** (top) Michael C. Webb/Visuals Unlimited **928:** (bottom) Photo by M.B. Fenton **929:** (top left) Yuri Arcurs/Shutterstock **931:** Lennart Nilsson From A Child Is Born, © 1966, 1977 Dell Publishing Company, Inc.

CHAPTER 39 **935:** Photo by M.B. Fenton **936:** © Rudie Kuiter/OceanwideImages.com **937:** (top left) Joseph G. Kunkel **937:** (bottom left) Photo by M.B. Fenton **937:** (top right) Reprinted by permission from Macmillan Publishers Ltd: Nature, Alexander Kupfer, Hendrik Muller, Marta M. Antoniazzi, Carlos Jared, Hartmut Greven et al., "Parental investment by skin feeding in a caecilian amphibian", Vol. 440, pp. 926–929, copyright (2006). **937:** (bottom right) Reprinted by permission from Macmillan Publishers Ltd: Nature, Alexander Kupfer, Hendrik Muller, Marta M. Antoniazzi, Carlos Jared, Hartmut Greven et al., "Parental investment by skin feeding in a caecilian amphibian", Vol. 440, pp. 926–929, copyright (2006). **938:** Reprinted by permission from Macmillan Publishers Ltd: Nature, Alexander Kupfer, Hendrik Muller, Marta M. Antoniazzi, Carlos Jared, Hartmut Greven et al., "Parental investment by skin feeding in a caecilian amphibian", Vol. 440, pp. 926–929, copyright (2006). **939:** (all bottom photos) Carolina Biological Supply Company **943:** (all top photos) Carolina Biological Supply Company **944:** Peter B. Armstrong, University of California, Davis. **945:** Carolina Biological Supply Company **946:** Carolina Biological Supply Company **947:** (top left) Yuri Arcurs/Shutterstock **950:** (all photos) Lennart Nilsson, A Child is Born, © 1966, 1977, Dell Publishing Company, Inc. **952:** (top left) Photo by M.B. Fenton **952:** (bottom left) Copyright (2001) National Academy of Sciences, U.S.A. **961:** (top) Oliver Meckes / Nicole Ottawa / Photo Researchers, Inc. **961:** (bottom) UCSF Computer Graphics Laboratory, National Institutes, NCRR Grant 01081 **962:** (top left) Carolina Biological Supply Company **962:** (top right) Lennart Nilsson from A Child is Born, © 1966, 1977 Dell Publishing Company, Inc.

CHAPTER 40 **967:** Photo by M.B. Fenton **968:** (left) A. & J. Binns/ Vireo **968:** (centre) © J. Schumacher/ VIREO **968:** (right) © G. McElroy/ VIREO **968:** (bottom) Dr. Stephen Yezerinac, Bishop's University, Lennoxville, Quebec, Canada **969:** Marie Read Natural History Photography **970:** (top left) Evan Cerasoli **970:** (bottom left) © Stephen Dalton/Photo Researchers, Inc. **970:** (top right) Eugene Kozloff **970:** (centre right) Stevan Arnold **970:** (bottom right) Stevan Arnold **972:** © Nina Leen/Time and Life Pictures/Getty Images **973:** © imagebroker / Alamy **975:** (bottom left) Russell Fernald, Stanford University **975:** (all top photos) Russell Fernald, Stanford University **976:** (bottom right) American Scientist 79:316–329 article by M. May 1991 **977:** © Jeff Foott/Dcom/ DRK Photo **978:** (left) Rod Planck/ Photo Researchers, Inc. **978:** (centre) Kenneth Catania/ Department of Behavioral Sciences/Vanderbilt University **978:** (right) Kenneth Catania/Department of Behavioral Sciences/Vanderbilt University **979:** (left) Photo by M.B. Fenton **979:** (right) © E. Mickleburgh/Ardea, London **980:** © Kenneth W. Fink/ Photo Researchers, Inc. **981:** (bottom) Tom and Pat Leeson **983:** (bottom left) © Alan Williams / Alamy **983:** (centre left) Mark Hamblin/ Oxford Scientific/ Index Stock **983:** (bottom right) Picture courtesy of: www.amazilia.net **984:** (bottom right) San Diego State University Biology Department **985:** Howard Hall/ Oxford Scientific/

and b) NOAA **1212:** (bottom right) Photo:Mike Lole/ADAS **1214:** (top left) Photo by M.B. Fenton **1214:** (centre a–c) Photo by M.B. Fenton **1214:** (bottom right) Cpl. D.A. Haynes **1215:** (top) Photo: NOAA **1215:** (bottom left) From Jill M. Casey, Ransom A. Myers, "Near Extinction of a Large, Widely Distributed Fish", Science, vol. 281, Jul 31, 1998, pp. 690–692. Reprinted with permission from AAAS. **1215:** (bottom right) From Jill M. Casey, Ransom A. Myers, "Near Extinction of a Large, Widely Distributed Fish", Science, vol. 281, Jul 31, 1998, pp. 690–692. Reprinted with permission from AAAS. **1216:** (top left) Photo: Warren Gaines **1216:** (bottom right) NOAA **1216:** (top right) Photographer : Degner, M. & L, used with permission of Parks Canada **1217:** (bottom right) From Ransom A. Myers, Julia K. Baum, Travis D. Shepherd, Sean P. Powers, Charles H. Peterson, "Cascading Effects of the Loss of Apex Predatory Sharks from a Coastal Ocean", Science, vol. 315, Mar 30, 2007, pp. 1846–1850. Reprinted with permission from AAAS. **1218:** (centre) Photo by M.B. Fenton **1218:** (bottom) From Richard Stone, "WILDLIFE CONSERVATION: The Saola's Last Stand", Science, vol. 314, Dec 1, 2006, pp. 1380–1383. Reprinted with permission from AAAS. **1220:** (bottom a–c) Photo by M.B. Fenton **1221:** (top) COSEWIC **1221:** (centre right) © Genevieve Vallee/Alamy **1221:** (bottom right) Photo: Chinese Academy of Fishery Sciences **1222:** (bottom left) © Arco Images GmbH/Alamy **1222:** (top right) PROMA/Marine Girard **1223:** COSEWIC **1225:** COSEWIC **1226:** (top) Photo by M.B. Fenton **1226:** (bottom left) 1971 Earth Day poster written and illustrated by Walt Kelly, featuring Pogo and Porkypine **1229:** (left) Photo by M.B. Fenton **1229:** (right a–c) Photo by M. B. Fenton

CHAPTER 49 **1233:** Photo by M.B. Fenton **1234:** Photo by M.B. Fenton **1235:** (bottom) From Greger Larson, Keith Dobney, Umberto Albarella, Meiying Fang, Elizabeth Matisoo-Smith, Judith Robins, Stewart Lowden, Heather Finlayson, Tina Brand, Eske Willerslev, Peter Rowley-Conwy, Leif Andersson, Alan Cooper, "Worldwide Phylogeography of Wild Boar Reveals Multiple Centers of Pig Domestication", Science, vol. 307, Mar 11, 2005, pp. 1618–1621. Reprinted with permission from AAAS. **1236:** (top left) From Carlos M. Duarte, Nuria Marba, Marianne Holmer, "ECOLOGY: Rapid Domestication of Marine Species", Science, vol. 316, Apr 20, 2007, pp. 382–383. Reprinted with permission from AAAS. **1236:** (bottom left) Photo by M.B. Fenton **1236:** (bottom right) From Ken-ichi Tanno, George Willcox, "How Fast Was Wild Wheat Domesticated?", Science, vol. 311, Mar 31, 2006, pp. 1886. Reprinted with permission from AAAS. **1237:** (top left) CEFE–UMR 5175 **1237:** (bottom) Reprinted by permission from Macmillan Publishers Ltd: Nature, Vol. 449: pp.

459–462, Fire and flood management of coastal swamp enabled first rice paddy cultivation in east China by Y. Zong, Z. Chen, J. B. Innes, C. Chen, Z. Wang et al., copyright (2007). **1238:** (top left) Photo by M.B. Fenton **1238:** (bottom left) Photo by M.B. Fenton **1239:** (top left) Tom McHugh / Photo Researchers, Inc. **1239:** (top right) Photo by M.B. Fenton **1239:** (bottom right) Photo by M.B. Fenton **1240:** (top) Yuri Arcurs/Shutterstock **1240:** (bottom) From Changbao Li, Ailing Zhou, Tao Sang, "Rice Domestication by Reducing Shattering", Science, vol. 311, Mar 31, 2006, pp. 1936–1939. Reprinted with permission from AAAS. **1241:** (top) Copyright (2001) National Academy of Sciences, U.S.A. **1241:** (bottom left) Photo by M.B. Fenton **1241:** (bottom right) Photo by M.B. Fenton **1242:** (top) From Manfred Heun, Ralf Schafer-Pregl, Dieter Klawan, Renato Castagna, Monica Accerbi, Basilio Borghi, Francesco Salamini, "Site of Einkorn Wheat Domestication Identified by DNA Fingerprinting", Science, vol. 278, Nov 14, 1997, pp. 1312–1314. Reprinted with permission from AAAS. **1242:** (bottom) Photo by M.B. Fenton **1243:** (top) John Doebley **1243:** (bottom) Photo by M.B. Fenton **1245:** Photo by M.B. Fenton **1246:** (top) Photo by M.B. Fenton **1246:** (bottom) Photo by M.B. Fenton **1247:** (top) Drawing courtesy of University of Western, Ontario **1247:** (bottom) Photo by M.B. Fenton **1248:** (top left) Photo by M.B. Fenton **1248:** (top right) With kind permission from Springer Science+Business Media: Theoretical and Applied Genetics, "Brassica: taxonomy based on nuclear restriction fragment length polymorphisms (RFLPs)", volume 79, Apr 1, 1990, pp. 497–506, K. Song. **1249:** (top) With kind permission from Springer Science+Business Media: Theoretical and Applied Genetics, "Brassica: taxonomy based on nuclear restriction fragment length polymorphisms (RFLPs)", volume 79, Apr 1, 1990, pp. 497–506, K. Song. **1249:** (bottom) Photo by M.B. Fenton **1250:** (top) Photo by M.B. Fenton **1250:** (bottom left) Photo by M.B. Fenton **1250:** (bottom right) Photo by M.B. Fenton **1251:** From Carlos A. Driscoll, Marilyn Menotti-Raymond, Alfred L. Roca, Karsten Hupe, Warren E. Johnson, Eli Geffen, Eric H. Harley, Miguel Delibes, Dominique Pontier, Andrew C. Kitchener, Nobuyuki Yamaguchi, Stephen J. O'Brien, David W. Macdonald, "The Near Eastern Origin of Cat Domestication", Science, vol. 317, Jul 27, 2007, pp. 519–523. Reprinted with permission from AAAS. **1252:** Mark Bernards **1252:** (top) Photo by M.B. Fenton **1252:** (bottom) Photo by M.B. Fenton **1253:** Photo by M. B. Fenton **1254:** © DLILLC/Corbis

ICONS
People Behind Biology: Yuri Arcurs/Shutterstock
Life on the Edge: © DLILLC/Corbis

Index

Dodos, extinction of, 1210
dogs
    body axes in, 939i
    body temperature regulation of, 778, 778i
    digestion in, 1020i
    domestication of, 1245
    experiments with, 972
    odour communication in, 843
    threat display in, 981i
dolphins, echolocation used by, 977
domestic animals, 1238, 1239, 1239i
domestication, history of, 1234–1246
dorsal lip of blastosphere, 942
Downey, Richard Keith, 1240
Driscoll, Carlos, 1249, 1250
drought, surviving, 1063
Drylands Development Paradigm, 1228t
dual olfactory system, 843i, 843t
Dvl genes, 971

ear, structures of, 831–833, 832i
earthworms *(Lumbricus)*
    closed circulatory system of, 893i
    digestion in, 1018, 1018i
    hermaphroditism in, 922, 922i
    movement of, 881i
    predators of, 977
    threats to, 1212
    vibrations, response to, 831
eastern red bat, 819–820, 819i
ecdysone, 854, 869, 869i
echinoderms, 883
echolocation, 819–820, 820i, 839, 976, **977**
ecological communities, **1155–1156**
    boundaries of, 1157i
    characteristics of, 1156–1161, 1157i, 1163–1165
    disturbance effect on, 1163–1165
ecological efficiency, **1183–1185**
ecological niche, **1152**
ecological pyramids, **1188**
ecology, **1112–1113**
economic growth and development, 1133, 1134–1135
ecosystem, 1179–1201
    connections in, 1159–1161
    ecology of, **1113**
    energetics of, 1180–1186, 1188–1189
    modelling of, **1200–1201**
    nutrient cycling in, 1189–1192, 1195, 1200
ecotones, 1156
ectoderm, **939**, 940i
ectotherms, 1073–1074, 1073i, 1075i, 1081
ectothermy, **1074–1076**
effectors, 777, **785**
effector T cells, **1097**
efferent arteriole, 1067
efferent neurons, **785**
egg-eating snakes, 1034–1035, 1036i
eggplants, 1244
eggs, **914**, 916, 917i
    components of, 938
    human, 930
Ehrlich, Paul, 1107
Eimer's organs, 977–978
Eisenberg, Robert, 1128
elastin, **770**, 771
electrical fields, receptors detecting, 821
electrical signalling, 780
electrical synapses, **787**, 788i
electric fishes, 847, 847i
electric light bulb. *See* light pollution
electrocardiogram (ECG), 892, 892i, **903**
electroencephalogram, **813**
electrophysiology, 891
electroreceptors, **847**
elephants

thermoregulatory structures and responses in, 1078
trunk, movement of, 881–882
elimination (as step in digestion), 1018, 1027
Elvin, Christopher, 771
embryonic development
    in animal kingdom, 947, 948–949, 952
    housing for, 936
    human, 947–952, 950i
    mechanisms of, 938–940
embryonic stem cells, 779
embryos, teeth of, 921i
emigration, **1118**
emulsifer secretion (as step in digestion), 1018
endangered species, 1220, 1221–1223
endemic Balkan nephropathy (EN), 1250
Endler, John, 1122, 1123
endocrine cells, 857i
endocrine glands, **769–770**, 770i, 852, 857i, 858t–859t
endocrine hormones, 858t–859t
endocrine system, **852**
    blood flow and pressure control, role in, 906
    cell signalling involvement of, 852–853, 852i
    information flow, role in, 784
endocrinology, 1028
endoderm, **939**, 940i
endolymph, **829**
endoparasites, 1154–11555
endorphins, 797, **861**
endoskeleton, **883**
endotherms, 1073, 1073i, 1074
endothermy, **1076–1081**
energy. *See also* free energy
    in sound, 831
energy return on investment (EROI), 1253–1254
environment
    heat exchange with, 1072–1073, 1072i
enzymatic hydrolysis (as step in digestion), 1018
enzymes
    secretion of, 1024
enzyme secretion (as step in digestion), 1018
epiblast, **943**
epidermis, **1078**
epididymis, **927**
epiglottis, 1022
epinephrine
    secretion of, 864
epiphytes, 1030i, 1031
epithelial cells, 768–769
epithelial tissues, **767–770**, 775
epithelium
    as infection barrier, 1086
    glands formed by, 769–770
epitopes, **1094**
equilibrium theory of island -biogeography, **1171–1172**
erythrocytes, **898–899**, 899i, 905i
erythropoietin, **899**
escape behaviour, 976–977
Escherichia coli
    population growth in, 1121, 1123
    reselin in, 771
esophagus, 1022, 1022i
essential elements, **1003**, 1004t–1005t
estivation, **1081**
estradiol, 853, **865**, 1137, 1137i
estrogens, **865**, 974
ethanol, 1253
ethinylestradiol, 1137, 1137i
eusocial animals, **993–994**
eustachian tube, **833**

evaporation, 1072, **1073**
evasive action, 1145
Ewer, Denis William (Jakes), 997
Ewer, R.F. (Griff), 997
excitatory postsynaptic potential (EPSP), **798**, 799i, 813
excretion, 1060–1061
excretory tubules, 1061–1063
exocrine glands, **769**, 770i, 852
exocytosis
    neurotransmitter release by, 794–795
exoskeleton, **882**
exploitative competition, **1151**
exponential population growth, **1121, 1123–1125**, 1124i
extensor muscles, 885
external fertilization, **916**
external gills, **1043**, 1043i
extinction, **1206–1207, 1209**
    human impact and, 1210, 1213–1216
    mutualism and, 1156
    preventing, 931–932
extracellular digestion, **1017**
extracellular fluid, **766**
extracellular matrix (ECM), **767**, 768
extraembryonic membranes, **943–944**, 944i
extreme environments, survival in, 1052, 1053
eye
    development of, 946–947, 946i
    image-forming, 834–836
    structure of, 835i

facial expressions
    communication role of, 979
    human infant response to, 969–970, 970i
facilitation hypothesis, **1166**
factor IX, 908
factor VIII, 908
Fahrig, Lenore, **1200**
family planning, 931, 1135
fast block, **918**
fast muscle fibres, **879**, 879t, 887
fathead minnows, 1137, 1137i
fats
    enzymatic digestion of, 1026
fat-soluble (hydrophobic) products, absorption of, 1027i
fat-soluble (hydrophobic) vitamins, **1009–1010**
fatty acids, **853**
    compounds formed with, 1025
    origin and evolution of, 870
feces, 1027
fecundity
    age-specific, 1119
    crowding effects on, 1128i
    versus parental care, 1120–1121
feedback pathways, hormones regulated by, 853–854
feeding, 1161–1162
female circumcision, 928
females, reproduction role of, 923–926
female sexual arousal, 929–930
fermentation, **1021**
fertilization
    human, 930
    in animals, **916–918**, 917i
fertilizers, 1006–1007, 1012
fetus, **948**, 951–952
fibrin, 899, 899i
fibrinogen, **897**
fibroblasts, 771
fibronectin, **770**
fibrous connective tissue, **771**

Herrmann, Esther, 995
hibernation, **1080–1081**
hibernation protein complex, 1081
high altitudes, survival at, 1052, 1053
hindbrain, **803**
hippocampus, **807**
hoatzin *(Opisthocomus hoatzin)*, 1021, 1021*i*
Hochachka, Peter, **1052**
homeostasis, **765–766, 776–779**
  requirements for, 1002
  of terrestrial organisms, 1058
homeostatic mechanisms, **766**, 778*i*
homeotic genes, **961**
home range and territory, 983–984
Homer-Dixon, Thomas, 1246
homing pigeons, magnetic sense in, 847
honeybees
  age and task specialization of, 974, 974*i*
  body temperature regulation of, 778
  communication in, 780*i*
  domestic, 1239
  hormones of, 974
  magnetic sense in, 847
hormones
  action mechanisms of, 854–857
  and behaviour, 973–976
  behavioural and developmental events, role in, 851–852
  cardiac output and arteriole diameter regulated by, 906
  inactive forms of, 854–855
  and local regulators, 853
  origin and evolution of, 870
  receptors, binding to, 855–856, 855*i*
  secretion of, 852–854
hovering flight, **1029**
Hox genes, 961, 962*i*
Hubbard Brook Watershed, 1194
human behaviour
  annoyance, signs of, 997
  evolutionary analysis of, 996
  kin selection impact on, 995–996
human childbirth
  oxytocin role in, 862
  positive feedback mechanisms in, 779
human chorionic gonadotropin (hCG), **930**
human digestive system, 1019*i*, 1020*i*, 1025*i*, 1026*i*
human immunodeficiency virus 1 (HIV-1)
  prevention of, 928
  stages of, 1105–1106, 1105*i*
human impact, 1210, 1213–1216
human infants, response to facial expressions, 969–970, 970*i*
human–nature dualism, 1206
human nutrition, essential elements, 1008–1010, 1008*i*
human population
  animal populations affected by, 1111*i*, 1136
  conservation and, 1224–1227
  future projections of, 1133*i*, 1135–1136
  growth of, 1132–1136, 1132*i*, 1133*i*
human respiratory system, 1046–1050, 1048*i*, 1049*i*
humans
  at high altitudes, 1053
  instinctive responses in, 969–970, 970*i*
human social behaviour, **994–997**
hummingbirds
  fuel required by, 1029
humus, **1010**, 1012
hunting, 1216–1217
husky, body temperature regulation of, 778*i*
hydras

digestion in, 1017
endocrine systems of, 867
nervous system of, 800*i*
hydrogeologic cycle, **1190**
hydrophobic hormones, **855–856**
hydroponics, **1002–1003**, 1003*i*
hydrostatic movement, 881–882
hydrostatic pressure, **903**
hydrostatic skeleton, **881–882**, 1063
hydroxyapatite, **772**
hymen, **923**
hyperosmotic urine, 1065
hyperpolarization, **789**
hypoblast, **943**
hypodermis, **1078**
hypothalamus, 808, 859–862, 860*i*, 864
  digestion control centres in, 1027, 1029–1030, 1032
  oxidative metabolism controlled through, 1029–1030
  physiological role of nuclei in, 1028
  thermoreceptor information integrated by, 1076–1078
hypothalamus grey-matter centres, **807**, 807*i*
H zone, **874**

I bands, **874**, 875
IgA, 1095
IgD, 1095, 1096
IgE, 1095
IGF II (peptide), 865
IGF I (peptide), 865
IgG, 1095
IgM, 1095
immigration, **1118**
immune response, **1087**
immune system, 1085, **1086**
  versus cancer, 1102
  malfunctions and failures of, 1102, 1103–1104
  T cell role in, 1085*i*
immunoglobulins, **1094**
immunological memory, **1099–1100**, 1100*i*
imperfect defence, 1147, 1151
incus (anvil), **832**
indigo buntings, 986*i*
indirect neurotransmitters, **794**
induction, **939–940**, 955–957
industrial stage, 1134–1135
infanticide, in African lions, 990
infection thread, **1014**, 1015*i*
inferior vena cava, **900**
inflammation, **1088–1091**, 1089*i*
inhibiting hormones (IHs), **859**
inhibition hypothesis, **1166–1167**
inhibitory postsynaptic potential (IPSP), **798**, 799*i*, 813
inland garter snakes, 970, 971
innate immunity, **1086**, 1087–1092, 1087*i*
inner cell mass, **948**
inner ear, **832–833**
insects
  body temperature regulation of, 778
  digestion in, 1018, 1018*i*
  exoskeletons of, 882, 882*i*
  gas exchange in, 1044–1045
  growth and development, hormones regulating, 868–869
  hearts of, 902
  life cycle of, 1141–1142
  movement of, 771
  mucins in, 1049
  muscles of, 774
  nervous system of, 802*i*
  osmoregulation and excretion in, 1061
  reproduction of, 917*i*

tracheal system in, 1044–1045, 1045*i*
walking in, 886
insight learning, **972**
instinct, **969–971**
instinctive behaviour, **968**, 969, 970, 973
insulin, 865, **866**, 866*i*
insulinlike growth factor (IGF), **860**
insulinlike hormones in animal kingdom, 868
insulinlike peptides (ILPs), 865, 870
integration, **785, 799**
integrator, **777**
integument, **1076**
intercalated disk, **773**
intercostal muscles, contractions of, 1047–1048
interference competition, **1151**
interferon, **1091**
internal fertilization, **916**, 919
internal fluid environment, **766**
internal gills, **1043–1044**, 1043*i*
interneurons, **785**
interspecific competition, **1151**
interspecific interactions, **1142–1143**
interstitial fluids, **766**, 905, 907, 908
intestinal cells, 958*i*
intestinal villi, 1024, 1024*i*, 1025
intracellular digestion, **1017**
intraspecific competition, **1126**, 1128
intrinsic control (of population), **1131**
intrinsic rate of increase, 1124
invasive species, 1211–1213
invertebrates
  circulatory systems of, 894
  as ectotherms, 1074
  endocrine systems in, 867–869
  fibrous connective tissue in, 771
  ganglia in, 800–801
  hearing in, 831
  muscles of, **774**
  organ systems in, 775
  osmoregulation and excretion in, 1061–1063
  smell in, 841
  striated muscles in, 880
Irish potato famines, 1254
iron deficiency in plants, 1006
irrigation, 1248–1249
island biogeography, **1171–1172**
islands
  experimenting with, 1174
islets of Langerhans, **865–867**

jellyfish (Scyphozoa), 800
Jensen, Keith, 995
Jimmie (bulldog), 891, 892, 892*i*, 903
joints, **883**, 885–886
juvenile hormone (of insects), **868**, 869*i*, 974

kangaroos, thermoregulatory structures and responses in, 1079
kardia (Greek term), 893
katydids, mating of, 824
keratin, 770
kidney, 1057*i*, 1063–1064, 1064, 1064*i*, 1066*i*
killer whales *(Orcinus orca)*, 977, 1220*i*
killifish, guppies with, 1122–1123, 1122*i*, 1123*i*
kin selection, **992**, 995–996
Kirshenbaum, Lorrie, **901**
knee joint ligaments, 885*i*
knockout mice, 971, F-40
Kondo, Noriaki, 1081

labia minora, **923**, 929
labour, domestic animals as source of, 1238